Potential Plant Protection Strategies

Editors

D. PRASAD

Division of Nematology
Indian Agricultural Research Institute (IARI)
New Delhi

RAJVIR SHARMA

Division of Agronomy
Indian Agricultural Research Institute (IARI)
New Delhi

I.K. International Publishing House Pvt. Ltd.

NEW DELHI • BANGALORE

Published by

I.K. International Publishing House Pvt. Ltd.
S-25, Green Park Extension,
Uphaar Cinema Market,
New Delhi–110 016 (India)
E-mail: info@ikinternational.com
Website: www.ikbooks.com

ISBN 978-93-80578-29-3

10 9 8 7 6 5 4 3 2 1

Published by Krishan Makhijani for I.K. International Publishing House Pvt. Ltd., S-25, Green Park Extension, Uphaar Cinema Market, New Delhi–110 016 and Printed by Rekha Printers Pvt. Ltd., Okhla Industrial Area, Phase II, New Delhi–110 020.

कृषि वैज्ञानिक चयन मंडल
भारतीय कृषि अनुसंधान परिषद्
कृषि अनुसंधान भवन-I, पूसा, नई दिल्ली-110012
AGRICULTURAL SCIENTISTS' RECRUITMENT BOARD
INDIAN COUNCIL OF AGRICULTURAL RESEARCH
KRISHI ANUSANDHAN BHAVAN-I, PUSA, NEW DELHI-110012
Telephone: (O) 25843295, 25846540, Fax : 91-11-25846540
Telegram : AGRECBOARD E-mail : cdmayee@icar.org.in

डॉ. चारू दत्त मायी
अध्यक्ष
DR. C. D. MAYEE
CHAIRMAN

Foreword

The use of chemical pesticides has been the most potent method for controlling crop pests but the excessive and indiscriminate use of hazardous chemicals has led to various problems of health of not only human beings but also of livestocks, fish and birds due to environmental pollution, loss of biodiversity, etc. Besides the development of resistance in pests, decimation of beneficial insects such as parasitoids, predators and honeybees often cause colossal loss to the productivity. It has been amply realized that it is always wise to make judicious and need based use of pesticides in combination with other safer methods of pest management.

The editors of the book **Potential Plant Protection Strategies, Dr. D. Prasad**, Division of Nematology and **Rajvir Sharma**, Division of Agronomy, IARI, New Delhi have made concerted efforts to compile the work done at various institutions under the National Agricultural Research System (NARS) into various thought provoking chapters. The authors have put forth new impetus on the plant protection subject as a whole. This book not only serves as a reference source but also as a guide for shaping future research in the area of plant protection to achieve the most cherished goal of ecologically and economically sustainable agriculture in general and plant protection in particular.

(C.D. Mayee)
13.07.2009

Preface

Plant protection has become one of the important inputs in crop production. In the context of the changing cropping pattern and the environment degradation, strategies are needed to minimize use of organic inputs hence, eco-friendly means and options are presently required and have special significance. During the past few decades, the science of plant protection has developed rapidly resulting in many diverse techniques to manage a variety of pests, pathogens and nematodes. The prospects of use of pesticides do not appear good because of accidental and incidental trauma encountered in the field of chemical pesticides. Therefore, vigorous research efforts for the control of various pests become imperative.

A number of methods like cultural, biological and physical means of them are modification of the agro-practices being followed by the farmers since ages, were also developed either for economic or environmental reasons, but very soon thought and realized that none of these techniques individually could serve as a panacea against target pests and pathogens including nematodes. After invoking them resulting in resurgence of the target species or suppression of useful enemies. Therefore, the approach should be multifaceted to combat the disease complex.

The book "**Potential Plant Protection Strategies**" is an attempt to analysis the problems and purpose, suitable measures for the effective management of the insect pests, plant diseases and plant parasitic nematodes. Therefore, an effort was made to compile possible data base information in this book which contains 39 exhaustive and thought provoking chapters written by specialists in their field of plant protection.

There are 15 chapters on Entomological problems as a whole covering chemo sterility, Transgenic technology, Insect pests of Mung and Urdbean, IPM, modified atmosphere for control of insects in storage. Eco-friendly management of potato pests, use of biotechnological tools in pest management and tissuc culture and transgenic crops in pest management.

There are 18 chapters deals with Plant Diseases like Wilt of chick pea, Citrus canker, Bacterial diseases of fruit crops. Biocontrol of chick pea with and soil pathogens, Antifungal activity of essential oils, Management of rice diseases through induced resistance crop loss by viral diseases, Viral diseases of orchids, *Trichoderma* spp. for disease management, Pathotype, Management of red rot in sugarcane, Biocontrol agents and Mechanisms of action. There are 4 articles on Nematology that highlights eco-froendly management of nematode in vegetables, Nematophagous fungi of plant parasitic nematodes, Biocontrol agents and Bacterial antagonists of plant parasitic nematodes.

There are few chapters of common interest such as paradigm shift in Lac application, Seed production in sugar cane, Genetic engineering approaches for detoxification and decontamination of pesticides and Attraction behaviour of *Discolaimus* major to prey nematodes.

Besides these, 3 chapters on Importance of weeds, Integrated weed management and Role of composting in organic farming.

The book "**Potential Plant Protection Strategies**" not provides references but also serve as guide and inspiration for future research into the realm of biological, chemical, physical and quarantine aspects and onslaughts of modern agriculture on it.

The scientist, teachers, students, scholars, administrators and policy makers dealing with pest and disease management strategies in particular and plant protection in general expected to be, this book very useful and informative.

I am grateful to all the contributors and their valuable articles and kind cooperation, I am also thankful to I.K. International Publishing House Pvt. Ltd., New Delhi for the technical guidance and help in bringing on this book.

D. Prasad
Rajvir Sharma

Contents

I

Fusarium Wilt of Chickpea and its Management

S.C. Dubey, Birendra Singh and R. Bhavani

Division of Plant Pathology,
Indian Agricultural Research Institute, New Delhi–110 012

Pulses are an important source of protein. After achieving food security in the country, the nutritional security continues to be a cause of our concern. The pulses have the promise with 20-25% protein, which is 2.5-3 times more than the cereals. While cereals are relatively deficient in lysine and rich in the sulphur containing amino acids, the reverse is true of many pulses. Chickpea contains 6.46% lysine as compared to 2.5% in cereals. Pulses also have immense importance in sustainable agriculture by way of fixing atmospheric nitrogen and improving soil health. The use of inputs like irrigation, fertilizers and pesticides has been increased tremendously during last three decades to increase crop production and mechanization in agriculture. Unfortunately, these efforts are focused on staple food crops. Chickpea cultivation gradually moved into more marginal lands and yields unlike those of many other crops could not increase.

Chickpea is a crop of both tropical and temperate regions. *Kabuli* type is grown in temperate regions while the *Desi* type chickpea is grown in the semi-arid tropics. Low yield of chickpea is attributed to its susceptibility to several fungal, bacterial and viral diseases. In general, estimates of yield losses by individual insects and diseases range from 5 to 10% in temperate regions and 50 to 100% in tropical regions (Van Emden *et al.,* 1988). Among the diseases affecting chickpea, wilt caused by *Fusarium oxysporum* f. sp. *ciceris* (Padwick) Matuo and K. Sato is considered one of the limiting factors for its low productivity (Haware and Nene, 1982; Dubey *et al*., 2007). In India, wilt alone causes on an average 10% loss annually and it is prevalent in all chickpea growing states (Singh and Dahiya, 1973). However, it was observed that early wilting causes 77-94% losses while late wilting causes 24-65% loss (Haware and Nene, 1980). Recently, we made an extensive survey in the states of Delhi, Punjab, Haryana, Rajasthan and Jharkhand and observed the wilt incidence from 14.1-32.0% (Dubey *et al.,* 2008).

SYMPTOMS

The disease appears at any stage of crop growth. Early wilt symptoms can be observed on susceptible cultivars within 4 weeks of sowing. The infected seedlings show sudden drooping of leaves and petioles. These seedlings retain their dull green colour. They usually show uneven shrinking of the stem above and below the collar region. Whole seedling collapse and lie on the ground. These plants do not show any external rotting of roots. Wilt pathogen has also been reported to produce yellowing symptoms resulting in a progressive foliar yellowing followed by necrosis, 30-40 days after inoculation from southern Spain (Trapero-Cases and Jimenez-Diaz, 1985). Late wilt also occurred during reproduction growth, but affected plants did not show external root-rot symptoms and exhibited internal discoloration of the xylem and pith, if the collar region is cut transversely or diseased stem split vertically. Partial wilting also occurs occasionally.

PATHOGEN

Chickpea wilt was first reported from India (Butler, 1918). However, Rhind (1926) expressed the doubt about fungal nature of the disease, and in 1928 Robertson also failed to isolate a specific organism from wilted chickpea plant. It was McRae (1929) who confirmed the association of *Fusarium* species with wilted chickpea plants. Earlier it was believed that two different causes were associated with chickpea wilt, one was the species of *Rhizoctonia* resembling *R. bataticola,* while other was of physiological disorder (Dastur, 1935). Finally, the causal agent was confirmed as *Fusarium oxysporum* f. sp. *ciceri* (Chattopadhyay and Sen Gupta, 1967).

A new fungus *Fusarium solani* was also found associated with wilted chickpea plants (Grewal *et al.,* 1974a and 1974b). They collected large number of wilted chickpea plants of different cultivars from IARI (New Delhi), Hisar (Haryana) and Ludhiana (Punjab) and *Fusarium oxysporum* f. sp. *ciceri* (15.5%), *Rhizoctonia solani* (8.5%), *R. bataticola* (8.3%), *Operculella padwickii* (6.2%) and *F. moniliforme* (1.6%) were found to be associated with these samples. *Fusarium solani, Fusarium oxysporum* f. sp. *ciceri* and *Rhizoctonia solani* proved potential pathogens and caused 82-100% mortality.

Considering the complex nature of the wilt disease, a symposium was organized by Dr. H.K. Jain, the then Director of IARI, New Delhi in 1973 on *Problems of wilt and breeding for wilt resistance in Bengal gram.* During the symposium it was concluded that gram wilt comprises two components, one being principally affected by environmental factors, like soil moisture, soil and atmospheric temperature, soil aeration, nutrient status of soil, plant density, date of sowing, etc. and second, the most important, being contributed by the pathogen wherein *Fusarium oxysporum* f. sp. *ciceri* and *Fusarium solani* were involved. In nature, however, there is no demarcation between these two components and the different causal factors interact freely and hence the role of each factor remains to be determined (Jain and Behl, 1974). Subsequently, Fusarial component was recognized as one of the most important part of chickpea wilt (Pal, 1998). Later on the diseases involved in wilt complex were characterized and clearly distinguished on the basis of symptoms caused by them and were designated as black root rot (*Fusarium solani*), dry root rot (*Rhizoctonia bataticola*), wet root rot (*Rhizoctonia solani*), collar rot (*Sclerotium rolfsii*), wilt (*Fusarium oxysporum* f. sp. *ciceri*) and foot rot (*Operculella padwickii*) (Nene *et al.,* 1978).

The mycelium of *F. oxysporum* f. sp. *ciceris* is delicate, white, cottony, becoming felted and wrinkled with age. Hyphae are septate and profusely branched. Microconidia are oval to cylindrical, straight to curved, 2.5 – 3.5 × 5-11 μm on simple short conidiophores. Macroconidia develop on the same conidiophores; thin walled, 3-5 septate, fusioid, pointed at both ends and measure 3.5-4.5 × 25-65 μm in size. Chlamydospores are smooth or rough walled, terminal or intercalary and may develop singly, in pairs or in a chain. Recently, Dubey *et al.* (2008) observed high level of variability in morphological characters of 112 isolates of *F.oxysporum* f. sp. *ciceris.*

PHYSIOLOGICAL SPECIALIZATION

Wilt disease, one of the most important factors limiting chickpea production worldwide, has been the target of breeding for resistance (Kumar *et al.,* 1985). Cultivation of resistant cultivars is one of the important means for the management of *Fusarium* wilt of chickpea, but the deployment of resistant cultivars has not been extensive because of undesirable agronomic characteristics. However, the high pathogenic variability in *F. oxysporum* f. sp. *ciceris* may limit the effectiveness of resistance (Haware and Nene, 1982; Dubey *et al.,* 2008).

The isolates of the pathogen were variable in their pathogenic behaviour and 300 chickpea isolates of *Fusarium* were categorized into three groups as non-pathogenic, pathogenic and those causing seed-rot (Padwick, 1939). Internationally, at present 8 physiological races of the pathogen (0, 1A, 1B/C, 2, 3, 4, 5, and 6) have been identified by reaction on a set of differential chickpea cultivars (Haware and Nene, 1982; Jimenez-Diaz *et al.,* 1993). The distribution pattern of these races in different parts of the world indicates regional specificity for their occurrence. Races 2, 3 and 4 have been reported only from India (Haware and Nene, 1982), whereas, 0, 1 B/C, 5 and 6 are found mainly in the Mediterranean region and United States (Jimenez-Gasco *et al.,* 2001). Unlike the other races, race 1A is more widespread and has been reported in India, California (USA) and the Mediterranean region (Haware and Nene, 1982; Jimenez-Gasco *et al.,* 2001). Pathotypes also exhibit differences in disease symptoms. Races 0 and 1 B/C cause yellowing, whereas 1A, 2, 3, 4, 5 and 6 showed wilting syndrome (Trapero-cases and Jimenez-Diaz, 1985). The races causing yellowing symptoms are not prevalent in India. The yellowing pathotypes are less aggressive than the wilting ones.

Honnareddy and Dubey (2006) established 3 new races of the pathogen in India based on the reactions on set of known differential cultivars. The reactions noticed on the cultivars were not similar to the earlier report even with the isolates of reported races taken from culture collections (Haware and Nene, 1982). This indicated that the present set of differential cultivars should be re-standardized with a new set of cultivars to get clear cut virulence differentiation of present pathogenic population. Recently, study of Dubey and Singh (2008) and Dubey *et al.* (2008) on the virulence analysis of 64 isolates collected from major chickpea growing states of India on 14 varieties including 10 international differentials revealed that more than one race is prevalent in one state. Majority of the isolates did not match with the race specific reactions, therefore, it was suggested that the cultivar GPF-2, DCP 92-3 and KWR 108 should be included in the set of differentials to get clear cut differential reactions.

The identification of pathogenic races is generally based on the differential reactions to selected host genotypes. Several biotechnological tools like Randomly Amplified Polymorphic DNA (RAPD), Amplified Fragment Length Polymorphism (AFLP) and Simple Sequence Repeat (SSR) have been increasingly used to study variability in pathogenic population. Barve *et al.,* (2001) distinguished four races prevalent in India based on oligonucleotide probes. Sivaramakrishanan *et al.* (2002) studied genetic variability among 43 Indian isolates of chickpea wilt pathogen by using RAPD and AFLP, and reported that molecular marker gave three different clusters. Of these, two clusters represented race-1 and race-2, and third cluster consisted of race-3 and race-4. They also observed that the high levels of DNA polymorphism with the molecular markers suggest the rapid evolution of new recombinants of the pathogen in the chickpea growing fields. A phylogenetic relationship among the eight known pathogenic races of chickpea wilt pathogen showed that the inferred intraspecific phylogeny observed in each of those races forms a monophyletic lineage. Moreover, virulence of races to resistant chickpea cultivars has been acquired in a simple stepwise pattern, with few parallel gains or losses (Jimenez-Gasco *et al.,* 2004). Sharma and Muehlbauer (2007) identified Sequence Tagged Microsatellite Site (STMS) markers linked to six genes governing resistance to six races (0, 1A, 2, 3, 4 and 5) of the pathogen.

Randomly Amplified Polymorphic DNA (RAPD), Inter-Simple Sequence Repeat (ISSR) and Simple Sequence Repeat (SSR) markers were also used to assess the genetic diversity of 64 isolates of the pathogen collected from 5 major chickpea growing states of the country (Dubey and Singh, 2008). The cluster generated by RAPD grouped all isolates into 2 major groups at 30 per cent genetic similarity, whereas, ISSR and SSR analysis also grouped all isolates into 2 major clusters. Majority of Punjab state isolates and few from Rajasthan were distinguished separately by the molecular marker evaluated, whereas, second group contains isolates from all other states suggest the existence of diverse genetic population of the pathogen at same location. The RAPD (OPM6, OPI 9, OPN 4, OPF 1, P 17, P 21 and SC 1), ISSR (ISSR 7, ISSR 11 and ISSR 12) and SSR (MB 17) markers clearly distinguished area specific isolates.

Differences in the stages of plant-pathogen interaction account for the difference in aggressiveness between *F. oxysporum* f. sp. *ciceris* races (Jimenez-Diaz *et al.,* 1989). The faster and more extensive colonization of xylem vessels of chickpea cultivar P-2245 in roots and stem was observed by race 5 than by race 0. Genetics of resistance in chickpea cultivar WR-315 against races 1A, 2, 3, 4, and 5 of the pathogen concluded that the resistance was governed by single genes to each of the 5 races (Sharma *et al.*, 2005).

MANAGEMENT

The *Fusarium* wilt is soil and internally seed-borne disease. However, the chlamydospore like structure was detected in the helium region of the seed (Haware *et al.,* 1978). The disease development is faster at 24-27°C soil and air temperature. Below 17°C, infection remains restricted in the root without any wilt symptoms. The importance of the soil temperature has also been substantiated by the observation that late sowing of the crop reduces the incidence of the disease. A PCR-based method was developed for the detection of pathogen in natural and artificial soils (Pedrajas *et al.,* 1999).

Deep ploughing during summer and removal of host debris from the field reduces inoculum levels. The soil inoculum can be reduced by addition of 15-20 tonnes of farmyard manure with *Trichoderma* sp. @ 4-5 kg/ha before sowing (Singh and Dubey, 2007). Seed dressing with benlate T (Benomyl + thiram) eradicated seed-borne inoculum (Haware *et al.,* 1978). The disease can be managed by seed treatment with various seed dressing fungicides. Two years field data clearly indicates that seed treatment with Bavistin + thiram (1:1) @ 2.5 g/kg seed before sowing decreased seedling mortality by 7% increased seed germination and grain yield by 11.2% and 2.8 Q/ha, respectively.

The identification and use of host plant resistance has the great potential in the long-term management of wilt. The genotypes H99-9, Pusa 212, JG 315, JG 322, PCS 1(Sel.ICCV-11), PCS 2 (Sel.KPG 142-1), PCS 5 (Sel.BGD-112), and PCS 6 (Sel.Pusa-1073) showed resistant (<10% wilt incidence) reaction during 3 years (2001-2004) and were deposited to Gene Bank, NBPGR, New Delhi with ACC No. 405202-405209 (Dubey and Singh, 2004). Recently, two genotypes, namely, H01-36 and PCS 8 (Sel.H82-2) showing ≤ 10 per cent wilt were graded as resistant during three consecutive years, i.e., 2004-07 crop seasons of testing under wilt sick field and deposited to Gene Bank, NBPGR, New Delhi with ACC No. 553468 and 553469 (Dubey and Singh, 2008). In addition to these, a large number of wilt resistant cultivars, namely, Avrodhi, Haryana Channa 1, BGD 72, BGM 547, GNG 469 (Smrat), GNG 663 (Vardan), RSG 693 (Aadhar), KPG-59 (Uday), K 3256 (Pragati), Phule G-87207 (Vishal), Phule G 9425-9, JG 322, GPF 2, PBG 1, Pusa 372 and Pusa 1053 (Chamtkar) were identified. The resistance to wilt for race 2 in WR 315 is controlled by a single recessive gene (Sharma *et al.,* 2005). Tullu *et al.* (1998) reported single recessive gene for resistance to race 4 and identified a RAPD marker linked with resistance.

In recent times, there has been a worldwide swing in the use of eco-friendly methods for protecting the crops from pests and diseases. The use of potential harmful chemical sprays is viewed with dissatisfaction in many countries. As such in the present context, biological control of wilt with bioagents offers a great promise. A biological control agent colonizes the rhizosphere, the site requiring protection and leaves no toxic residues as opposed to chemicals. The first requirement of biological control is the identification and deployment of highly effective strains. The filamentous fungi, *Trichoderma* have attracted the attention because of their multiprong action against various plant pathogens (Harman *et al*., 2004). The species of *Trichoderma* have been evaluated against the wilt pathogen and have exhibited greater potential in managing chickpea wilt under glasshouse and field conditions (Kaur and Mukhopadhayay, 1992). Earlier, Padwick (1941) also observed that a species of *Trichoderma* was highly antagonistic to wilt pathogen of chickpea. Mane (1995) isolated and evaluated 88 different fungal and bacterial agents against *F. oxysporum* f. sp. *ciceris* and four isolates of *Trichoderma* spp., three isolates of fluorescent Pseudomonas, one isolate of *Acrophilophora* sp. and three isolates of *Gliocladium* spp. showed antagonistic activity *in vitro*. Under field conditions, maximum wilt reduction (28.3%) in cultivar Pusa 256 was observed when *T. viride* was applied as seed coating with talc as carrier with gum.

Recently, 10 isolates belonging to three species of *Trichoderma* i.e., *T. viride, T. harzianum* and *T. virens* were evaluated against four isolates of the pathogen representing 4 different races commonly prevalent in India. *T. viride* (IARI P-1) followed by *T. harzianum* (IARI P-4) and *T. viride* (IARI P-19) inhibited maximum mycelial growth of the pathogen (Fig. 1.1). These bioagents also enhanced seed germination, root and shoot length and decreased wilt incidence under green-house condition. The isolates proved potential in *in vitro* tests and were evaluated individually and in combination with carboxin under field condition. The efficacy of *Trichoderma* species was enhanced in combination with carboxin. The integration of *T. harzianum* (10^6 spores/ml/10g seed) and carboxin (2.0 g/kg seed) enhanced seed germination by 12.0-14.0 per cent and grain yield by 42.6-72.9 per cent and reduced wilt incidence (44.1-60.3 per cent)

Table 1.1 Effect of various seed treatments on seed germination, wilt incidence and grain yield of chickpea under sick field conditions during cropping seasons of 2002/03, 2003/04, 2004/05.

Treatment	Seed germination (%)				Wilt incidence (%)				Grain yield (Kg ha^{-1})			
	2002-03	2003-04	2004-05	Mean	2002-03	2003-04	2004-05	Mean	2002-03	2003-04	2004-05	Mean
Trichoderma harzianum (IARI P-4)	91.2 (72.8)	90.3 (71.9)	90.4 (71.9)	90.6 (72.2)	44.3 (41.7)	59.1 (50.2)	50.6 (45.4)	51.3 (45.8)	1388.9	430.6	1386.7	1068.7
T. viride (IARI P-1)	91.1 (73.0)	86.4 (68.3)	90.6 (72.2)	89.4 (71.2)	55.9 (48.4)	61.4 (51.6)	57.5 (49.3)	58.3 (49.8)	1305.6	416.7	1450.0	1057.4
Bacillus subtilis	88.7 (71.0)	81.4 (64.4)	83.2 (65.8)	84.4 (67.1)	79.9 (63.5)	89.2 (70.8)	69.0 (78.5)	79.4 (70.9)	611.1	236.1	1000.0	615.7
Kalisena™	89.0 (70.8)	82.1 (65.0)	84.9 (67.1)	85.3 (67.6)	76.9 (61.3)	85.0 (67.2)	70.5 (57.1)	77.5 (61.9)	763.9	277.8	1050.0	697.2
Carboxin	90.7 (72.4)	84.6 (66.9)	85.9 (68.1)	87.1 (69.1)	67.9 (55.5)	79.0 (62.8)	67.2 (55.1)	71.4 (57.8)	972.2	333.3	1353.3	886.3
T. harzianum + carboxin	93.2 (75.1)	93.1 (75.3)	96.1 (79.0)	94.1 (76.5)	32.3 (34.6)	47.5 (43.6)	39.7 (39.1)	39.8 (39.1)	1847.2	625.0	1736.7	1403.0
T. viride + carboxin	92.9 (74.7)	91.9 (73.6)	95.5 (77.9)	93.4 (75.4)	33.4 (35.3)	50.1 (45.1)	42.5 (40.6)	42.0 (40.3)	1750.0	513.9	1690.0	1318.0
B. subtilis + carboxin	90.0 (71.7)	82.5 (65.3)	85.0 (67.2)	86.2 (68.1)	69.6 (56.5)	81.0 (64.2)	66.2 (54.5)	72.3 (58.4)	902.8	319.4	1130.0	784.1
Kalisena™ + carboxin	90.9 (72.6)	85.9 (68.0)	87.3 (69.2)	88.0 (69.9)	65.5 (54.0)	75.0 (60.0)	65.0 (53.8)	68.5 (55.9)	1250.0	361.1	1126.7	912.6
Carbendazim + TMTD	91.4 (73.0)	90.8 (72.5)	93.3 (75.1)	91.8 (73.5)	41.3 (40.0)	54.2 (47.4)	41.0 (39.8)	45.5 (42.4)	1527.8	472.2	1526.7	1175.6
Control (No seed treatment)	80.2 (64.4)	80.1 (63.6)	84.5 (66.9)	81.6 (64.9)	81.3 (64.4)	91.0 (72.6)	71.0 (57.5)	81.1 (64.8)	500.0	208.3	996.7	568.3
SEm ±	(1.4)	(1.3)	(1.2)	(1.3)	(1.1)	(0.5)	(1.1)	(0.9)	63.9	13.9	48.4	42.1
LSD ($p = 0.05$)	(4.1)	(3.8)	(3.5)	(3.8)	(3.2)	(1.5)	(3.4)	(2.7)	188.5	27.8	145.1	120.5

SEm ±: Standard Error of Mean
The figures in parentheses are transformed angular values.
Kalisena™ = Commercial formulation of *Aspergilus niger,* carboxin (vitavax ™), carbendazim (Bavistin™), tetramethyl thiram disulphide (Thiram™)
(Source: Dubey *et al.* (2007). Biological control 40: 118-127)

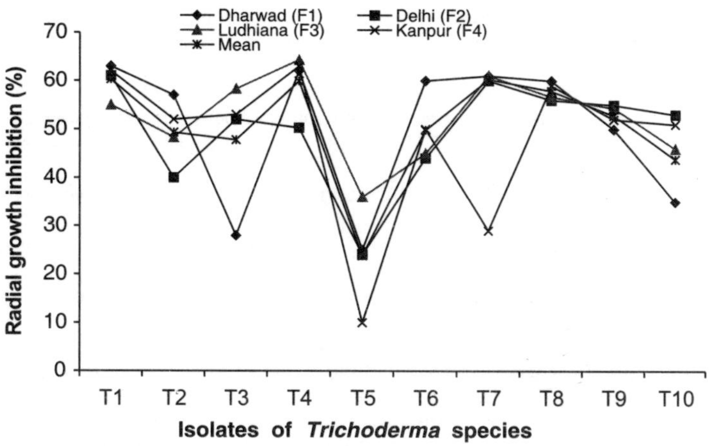

Fig. 1.1. Radial growth inhibition of isolates of *F. oxysporum* f. sp. *ciceris* by isolates of *Trichoderma* T_1= *T. viride* (IARIP- 01), T_2= *T. viride* (IARIP- 02), T_3= *T. viride* (IARIP- 20), T_4= *T. viride* (IARIP-19), T_5= *T. viride* (IARIP-5), T_6= *T. viride* (IARIP-6), T_7= *T. harzianum* (IARIP-4), T_8= *T. harzianum* (IARIP-18), T_9= *T. virens* (IARIP-3), T_{10}= *T. virens* (IARIP-17) at 6 days of incubation in dual culture.
(Source: Dubey *et al.* (2007). Biological control 40: 118-127).

during 3 years (2002-05) of field experimentations (Table 1.1) under wilt sick field (Dubey *et al.,* 2007). The isolates of *Trichoderma* could be distinguished among the species by using oligonucleotide primers. Primer OPA 13 efficiently differentiated all the isolates of *Trichoderma* into their species except T_2 of *T. viride*. Some of the primer as primer OPA 10 also showed species specific banding pattern (Dubey and Suresh, 2006).

REFERENCES

Barve, M.P., Haware, M.P., Sainani, M.N., Ranjekar, P.K. and Gupta, V.S. (2001). Potential of microsatellites to distinguish four races of *F. oxysporum.* f. sp. *ciceri* prevalent in India. *Theoretl. Appl. Genet.* **102:** 138-147.

Butler, E.J. (1918). *Fungi and Plant Disease*. Bishen Singh Mahendra Pal Singh and Periodical Experts, Dehradun and Delhi. pp. 547.

Chattopadhyay, S.B. and Sen Gupta, P.K. (1967). Studies on wilt diseases of pulses, I. variation and taxonomy of *Fusarium* spp. associated with wilt diseases of pulses. *Ind. J. Mycol. Res.* **5:** 45.

Dastur, J.F. (1935). Gram wilt in the central provinces–*Agric.liv-stk*, India. **5:** 615-627.

Dubey, S.C. and Birendra Singh (2004). Reaction of chickpea genotypes against *Fusarium oxysporum* f. sp. *ciceris* causing vascular wilt. *Indian Phytopath.* **57**: 233.

Dubey, S.C. and Birendra Singh (2008). Evaluation of chickpea genotypes against *Fusarium oxysporum* f. sp. *ciceris* causing wilt. *Indian Phytopath.* **61:** 280-281.

Dubey, S.C., Birendra Singh and Bahadur, P. (2007). Diseases of pulse crops and their ecofriendly management In: *Ecofriendly Management of Plant Diseases.* (Eds. S. Ahamad and U. Narain) Daya Publishing House, Tri Nagar, Delhi. pp. 16-44.

Dubey, S.C. and Singh, S.R. (2008). Virulence analysis and oligonucleotide fingerprinting to detect diversity among Indian isolates of *Fusarium oxysporum* f. sp. *ciceris* causing chickpea wilt. *Mycopathologia* **165:** 389-406.

Dubey, S.C., Singh, S.R. and Birendra Singh (2008). Morphological and pathogenic variability of Indian isolates of *Fusarium oxysporum* f. sp. *ciceris* causing chickpea wilt. *Archives of Phytopathology and Plant Protection* (Accepted for publication).

Dubey, S.C. and Suresh, M. (2006). Randomly amplified polymorphic DNA markers for *Trichoderma* species and antagonism against *Fusarium oxysporum* f. sp. *ciceris* causing chickpea wilt. *J. Phytopathol.* **154:** 663-669.

Dubey, S.C., Suresh, M. and Birendra Singh, (2007). Evaluation of *Trichoderma* species against *Fusarium oxysporum* f. sp. *ciceris* for integrated management of chickpea wilt. *Biological Control* **40**: 118-127.

Grewal, J.S., Pal, M. and Kulshrestha, D.D. (1974a). A new record of wilt of gram caused by *Fusarium solani. Curr. Sci.* **43**: 767.

Grewal, J.S., Pal, M. and Kulshrestha, D.D. (1974b). Fungi associated with gram wilt. *Ind. J. Genet. Pl. Breed.* **34**: 242-248.

Harman, G.E., Howell, C.R., Viterbo, A., Chet, I. and Lorito, M. (2004). *Trichoderma* species–opportunistic, avirulent plant symbionts. *Nature Reviews* **2**: 43-56.

Haware, M.P. and Nene, Y.L. (1980). Influence of wilt at different growth stages on yield loss of chickpea. *Trop. Grain Legume Bull.* **19**: 38-40.

Haware, M.P. and Nene, Y.L. (1982). Races of *Fusarium oxysporum* f. sp. *ciceri. Pl. Dis.* **66**: 809-810.

Haware, M.P., Nene, Y.L. and Rajeshwari, R. (1978). Eradication of *Fusarium oxysporum* f. sp. *ciceri* transmitted in chickpea seeds. *Phytopathology* **68**: 1364-1367.

Honnareddy, N. and Dubey, S.C. (2006). Pathogenic and Molecular characterization of Indian isolates of *Fusarium oxysporum* f. sp. *ciceris* causing chickpea wilt. *Curr. Sci.* **91**: 661-666.

Jain, H.K. and Bahl, P.N. (1974). Recommendations of symposium on gram wilt. *Ind. J. Genet. Pl. Breed.* **34**: 236-238.

Jimenez-Diaz, R.M., Trapero-Casas, A. and Cabrea de la Colina, J. (1989). Races of *Fusarium oxysporum* f. sp. *ciceri* infecting chickpeas in southern Spain. In: *Vascular wilt diseases of plants. NATO ASI Ser.* (Tjamos, E.C. and Beckman, C.H. Eds.) Springer-Verlag, Berlin H. **28:** 515-520.

Jimenez-Diaz, R.M., Alcala-Jimenez, A.R., Hervas, A. and Trapero-Casas, J.L. (1993). Pathogenic variability and host resistance in the *Fusarium oxysporum* f. sp. *ciceris* / *Cicer arietinum* pathosystem. In: Proc. 3rd Eur. Semin. Fusarium Mycotoxins, Taxonomy Pathogenicity and Host Resistence. Plant breeding and Acclamatization Institute, Poland. pp. 87-94

Jimenez-Gasco, M.M., Perez-Artes, E. and Jimenez-Diaz, R.M. (2001). Identification of pathogenic races 0, 1B/C, 5 and 6 of *Fusarium oxysporum* f. sp. *ciceri* with Random Amplified Polymorphic DNA (RAPD), *Euro. J. Pl. Pathol.* **107**: 237-248.

Jimenez-Gasco, M.M., Navas-Cortes, J.A. and Jimenez-Diaz, R.M. (2004). The *Fusarium oxysporum f. sp. ciceris/Cicer arietinum* pathosystem: a case study of the evolution of plant-pathogenic fungi into races and pathotypes. *Int. Microbiol.* **7**: 95-104.

Kaur, N.P. and Mukhopadhayay, A.N. (1992). Integrated control of chickpea wilt complex by *Trichoderma* spp. and chemical methods in India. *Trop. Pest Management.* **38**: 372-375.

Kumar, J., Haware, M.P. and Smithson, J.B. (1985). Registration of four short duration *Fusarium* wilt resistant kabuli chickpea germplasm. *Crop Sci.* **25**: 576-577.

Mane, S.S. (1995). *Studies on Fusarium oxysporum f. sp. ciceri causing chickpea wilt with special reference to its management by bio-agents.* Ph.D. thesis, IARI, New Delhi. p.72.

Mc Rae, W. (1929). India; new disease reported during the year 1928. *Int. bull. Pl. Prot.* III: 21-22.

Nene, Y.L., Haware, M.P. and Reddy, M.V. (1978). *Diagnosis of some wilt like disorders of chicpea,* ICRISAT Information Bulletin-**3**, p. 44.

Padwick, G.W. (1939). *Report of the Imperial Mycologist*. Scient. Rept. Agric. Res. Inst., New Delhi, 1937-38; 105-112.

Padwick, G.W. (1941). *Report of the Imperial Mycologist*. Scient. Rept. Agric. Res. Inst., New Delhi, 1937-38; 94-101.

Pal, M. (1998). Diseases of pulse crops, their relative importance and management. *J. Mycol. Pl. Pathol.* **28**: 114-122.

Pal, M. and Birendra Singh (1993). Channe koo Uktha Rog Se Bachayen. *Kheti* **47**(6): 24-25.

Pedrajas, M.D.G., Bainbridge, B.W., Heale, J.B., Artes, E.P. and Diaz, R.M.J. (1999). A simple PCR-based method for the detection of the chickpea wilt pathogen *F. oxysporum* f. sp. *ciceris* in artificial and natural soils. *Euro. J. Pl. Pathol.* **105**: 251-259.

Rhind, D. (1926). *Annual report of the Mycologist of Burma, for the year ended the 30th June 1925.* Rangoon. Supdt. Govt. Printing and Stationary, Burma. p. 5.

Robertson, H.F. (1928). *Mycology-Report on the operation of the Dept. of Burma for the year ended 30th June, 1927.* pp. 11-12.

Singh, K.B. and Dahiya, B.S. (1973). Breeding for wilt resistance in Chickpea. *Symposium on problem and breeding for wilt resistance in Bengal gram.* Sept. 1973 at IARI, New Delhi, pp. 13-14.

Singh, Birendra and Dubey, S.C. (2007). Channe Kaa Mallyni Rog: Bachaw Ke Uppaya. *Kheti* **60**(8): 18-20.

Sharma, K.D., Chen, W.D. and Muehlbauer, F.J. (2005). Genetics of chickpea resistance to five races of Fusarium wilt and a concise set of race differentials for *Fusarium oxysporum* f. sp. *ciceris*. *Pl. Dis.* **89:** 385-390.

Sharma, K.D. and Muehlbauer, F.J. (2007). Fusarium wilt of chickpea: physiological specialization, genetics of resistance and resistance gene tagging. *Euphytica* **157:** 1-14.

Sivaramakrishnan, S., Kannan, S. and Singh, S.D. (2002). Genetic variability of Fusarium wilt pathogen isolates of chickpea (*Cicer arietinum* L.) assessed by molecular markers. *Mycopathologia* **155:** 171-178.

Trapero-Casas, A. and Jimenez-Diaz, R.M. (1985). Fungal wilt and root rot diseases of chickpea in southern Spain. *Phytopathology,* **75:** 1146-1151.

Tullu, A., Muehlbauer, F.J., Simon, C.J., Mayer, M.S., Kumar, J., Kaiser, W.J. and Kraft, J.M. (1998). Inheritance and linkage of a gene for resistance to race 4 of Fusarium wilt and RAPD markers in chickpea. *Euphytica* **102:** 227-232.

Van Emden, H.F., Ball, S.L. and Rao, R. (1988). Pest diseases and weed problems in pea lentil and faba bean and chickpea. In: (Ed. Summerfield, R.J.) *World Crops: Cool season Food Legumes*. ISBN 90-247-3641-2. Kluwer Academic Publishers. Dordrecht, The Netherlands. pp. 519-534.

2
Paradigm Shift in Applications of Lac

S. Srivastava, Govind Pal and D. Prasad[1]

Indian Institute of National Rasins and Gums, Ranchi, Jharkhand
[1]Division of Nematology, I.A.R.I., New Delhi–110 012

Lac is a product of commerce secreted by an insect known as *Kerria lacca* (Kerr.) thriving on the tender twigs of specific host trees *viz. palas* (*Butea monosperma*), *ber* (*Zizyphus mauritiana*), *kusum* (*Schleichera oleosa*), *Ficus sp*., etc. Three natural, renewable, non-toxic and eco-friendly products, i.e., resin, dye and wax are derived from the lac. *Rangeeni* and *Kusmi* are the two strains of lac insect which are based on preference of the insect for specific host plants. Lac is mainly cultivated in India, Myanmar, Thailand, parts of China, Taiwan, Indonesia, Philippines, Vietnam, Cambodia, etc. Lac cultivation is an important source of income for the forest and sub-forest dwellers. Besides it has high potential for generating employment for both men and women. It is a very remunerative crop, paying high economic returns to the farmers and also foreign exchange to the country through its export.

The purified, commercial form of lac resin is commonly called shellac; however, the term lac and shellac are often used synonymously. Lac resin is unique in the sense that even with its low molecular weight it has the property of forming films on a wide variety of surfaces. These films possess outstanding adhesion, gloss, scratch hardness and resistance to petrochemical solvents. The resin can also be hydrolysed to its component acids—some of these acids serve as an important starting materials in the synthesis of wide variety of compounds for industry. Its thermal plasticity and ready fusibility, coupled with its ability to absorb large amount of filler make it valuable for moulding industry. Its high dielectric strength, moderate dielectric constant and freedom from tracking increases its usefulness in electrical and electronic industry.

Similarly, lac dye, which is a polyhydroxy anthraquinone derivative, shows brilliance and fastness, especially when used for dying silk and wool. Presently, lac dye is also used as food colour in soft drinks, wines, sauces and bean jams by countries like China, Japan and Thailand. Lac wax is an important component in many formulations for floor polishing, automobile, cosmetics, fruit and confectionery coating, additive for ink ribbon and it is also employed as an ingredient in dental impression compositions.

LAC PRODUCTION IN INDIA

India is the largest producer of lac in the world and the national production of lac was approximately 23,229 tons during 2006-07. Chhattisgarh state ranks 1st followed by Jharkhand, Madhya Pradesh, Maharashtra and West Bengal. The above five states contributed around 95 per cent of the national lac production during 2006-07. Lac culture is practiced mostly in the Indian states of Chhattisgarh, Jharkhand, Madhya Pradesh, West Bengal, Maharashtra, Orissa and parts of Uttar Pradesh, Andhra Pradesh, Gujarat and NEH region and forms an earning support for the forest and sub-forest dwellers of these regions (Table 2.1).

Table 2.1. State-wise production of sticklac in India (in tons).

Sl. No.	States	Annual Production of sticklac (tons)		
		2004-05	2005-06	2006-07
1.	Andhra Pradesh	40	15	60
2.	Assam and Meghalaya	85	165	145
3.	Bihar	15	15	15
4.	Chhattisgarh	4,270	6,015	8,875
5.	Gujarat	75	50	50
6.	Jharkhand	10,780	6,414	7,490
7.	Madhya Pradesh	2,870	2,515	3,164
8.	Maharashtra	875	1,160	1,300
9.	Orissa	405	415	440
10.	Uttar Pradesh	490	310	450
11.	West Bengal	1,395	926	1,240
	Total	**21,300**	**18,000**	**23,229**

EXPORT OF LAC

Lac is an export oriented commodity and around 70 per cent of lac produced in India is exported to about more than 75 different countries in refined and semi-refined form. The top ten importing countries of Indian lac are U.S.A., Egypt A RP, Germany, Indonesia, Spain, Pakistan, Italy, Bangladesh, Japan and U.K. that accounted for around 80 per cent of the total exports by quantity and value. Shellac, seedlac, dewaxed shellac, aleuritic acid, bleached lac, gasket shellac, lac dye, molamma lac and shellac wax are the forms of lac exported from India. The export earning from lac and lac products during the year 2005-06 was 159.62 crores (Table 2.2). Analysis of last five year export data show much increase in export of value added products of lac, namely, bleached lac, dewaxed shellac and aleuritic acid in comparison to traditional products of lac (seedlac/shellac/button lac).

Lac, as such, has utilization in various areas, and modification of its functional groups with some chemicals and synthetic resins has resulted in its innumerable utilization in the area of surface coating, adhesive, electrical and electronic, food, leather, pharmaceutical and cosmetic industries. The paradigm shift in the applications of lac from conventional to future application is summarized here.

Table 2.2. Export of lac and its value-added products.

Sl.	Name of product	Export in 2005-06		% Change in last 5 years	
		Quantity (tonnes)	Value (Rs. lakh)	Quantity (tonnes)	Value (Rs. lakh)
1.	Shellac	5,120.97	8,854.48	+ 11.14	+ 62.10
2.	Seedlac	1,325.29	2,043.66	+ 143.26	+ 323.95
3.	Bleached Lac	418.76	1,126.01	+ 403.68	+ 725.16
4.	Dewaxed Shellac	244.17	731.29	+ 164.05	+ 298.68
5.	Aleuritic acid	162.35	2,488.84	+ 166.15*	+ 268.21*
6.	Gasket Shellac	4.54	6.50	– 66.17	– 51.78
7.	Shellac wax	4.50	11.30	+ 78.57*	+ 174.94*
8.	100% E.O.Us	-	700.00	-	-
9.	Grand Total	7,280.58	15,962.08	+ 30.24	+ 144.75

* % Change in last 4 years

CONVENTIONAL APPLICATIONS

1. Surface Coating Industry

(i) Heat and waterproof French polish (melfolac)

A French polish composition based on shellac and synthetic resin, popularly known as Melfolac, has been developed, which may be applied either by brush or spray. The varnish is not only heat and waterproof, but also gives a very glossy and attractive finish on wooden furnitures, radio and TV cabinets, musical instruments, etc.

When pigmented, melfolac produces quick-drying, waterproof paints suitable for application on shop windows, display panels, kitchen furniture, etc.

(ii) Picture varnish

Pictures and paintings need protection against dust, abrasion and humid atmosphere for their long durability. Lac-based varnish composition, which can be sprayed, has been developed for protection of pictures and paintings.

(iii) Book varnish

An insecticidal varnish based on lac has been developed especially to be used for bound books and journals to save them from the damage due to insect attack.

(iv) Water-soluble lac

Lac is not water-soluble as such, but it can be dissolved in water in presence of some chemicals. The aqueous lac solution is used for coating earthen pots, imparting glaze to surface and providing protection against humidity. It may also be used by small-scale industry entrepreneurs in leather finishes and in photoengraving, replacing the imported glue. This is also suitable for making emulsion paints and inks. It also finds use in metal lacquering ornamental brassware for protection against tarnishing.

(v) Shellac etch primer

An anti-corrosive primer for the exposed surface of metals like aluminium, zinc, etc. has been developed, which possesses outstanding adhesion property and forms smooth adherent film with hard matt appearance.

Lac-based antifouling paints for shipping industries are also in use.

2. Adhesive Industry

(i) Gasket shellac compound

Gasket shellac compound is a viscous solution of shellac and other additives in organic solvents, mainly denatured spirit. It is used to fix in position all kinds of gaskets in between two metallic parts of internal combustion engine in the automobile industry. It provides sealing of joints against leakage of oil. It is also used for coating rivets to strengthen the joints, prevents corrosion, as a jointing compound in ordinary cargo and passenger vessels for plate seams and water tight bulk heads.

(ii) Sealing wax

Shellac-based sealing wax formulations have been developed in which lac resin provides excellent adhesion to paper, contributed to the luster, clarity of impression and mechanical strength of the seal.

(iii) Making of bangles

A number of small-scale industries in India are engaged in making attractive lac bangles and wooden turnery articles.

3. Electrical Industry

(i) Insulating varnish

Shellac possesses good electrical properties like dielectric strength, anti-surface carbonization, arc resistance apart from its ability to adhere to a variety of surfaces. Both air-drying and baking-type insulating varnishes for coating of coil/armatures of electric motors, transformers have been developed.

(ii) Micanites

The adhesive, moulding and insulating properties of shellac are amply utilized in making sheets of micanite and also moulded micanite products.

(iii) Coating of isolators, PCBs, spark plugs, etc.

Use of lac in printed circuit boards (PCBs), coating of silicon cells, spark plugs, etc. is also known.

4. Pharmaceutical Industry

Lac-based coating compositions are available for enteric coating of medicinal tablets and sustained release capsules. In such compositions, lac is used either as a solution or in the form of an emulsion. Lac functions as a moisture barrier for protecting the core ingredient; it also control disintegration and serves as a finishing coat over wax, prior to printing of the trade mark on the medicinal tablet. Modern use of lac in medicine is for controlled release and micro encapsulation of drugs. It is also used in coating of vitamin C and garlic tablets.

5. Confectionery Industry

The main uses of lac in the confectionery industry are for coating chocolates, such as extruded chocolates, chocolate covered nuts and raisins and similar products. The lac coating enhances and protects the gloss, serves as a barrier to moisture either entering or leaving the confections, prevents blocking together and extends shelf-life.

6. Cosmetic Industry

Lac is used in mascara and lipsticks as a binder, additive in nail lacquers and adhesive in bangles and wooden turnery. It is also used in hair spray formulations.

7. Miscellaneous

Moulding and adhesive properties of lac extends its application in manufacturing jewellery and ornaments from coloured stones. It is also used for polishing and sharpening stones (diamond), making of impact and water resistant coal block, coating of fertilizer for slow release and proofing and stiffening of hat.

The use of lac in pyrotechnics and ammunitions continues. It finds extensive use as a water proofing varnish for cartridges, shell cases, etc., as a retarder in gunpowder and as a general adhesive and binding material in fire crackers.

Future Applications

A few fields in which lac and lac-based products possess potential for future applications are described below:

(i) Fruit and vegetable coating

India produces annually about 150 million tons of fruits and vegetables. The installed capacity of fruits and vegetables processing industry has increased from 11.08 lakh tons in January 1993 to 21.18 lakh tons in January 2006. However, processing level of fruits and vegetables is very low, estimated to be around only 2.20 per cent of the total production. Total export for processed fruits and vegetables in 2004-05 was of Rs 1462.72 crores. The food processing industry is one of the largest industries in India. It is ranked fifth in terms of production, consumption, export and expected growth. It is projected that the market for Asian Food Processing sector, currently valued at US $400 million, is likely to proliferate rapidly at a rate of 12-15% per annum for the next 5 years. India is projected to remain as a net exporter of fruits and vegetables. In Europe and USA, shellac wax is used for coating fruits (citrus, apples, etc.) and vegetables; unfortunately, such a huge maiden field of application is yet to take a desired entrepreneurship in the country. Under the circumstances, lac-based coating formulation is likely to have a potentially viable market as a coating material for fruits and vegetables for extending post-harvest life. A few coating compositions have been developed at ILRI for extending post-harvest life of certain fruits and vegetables; rigorous evaluation is under process. Effort is to be continued to formulate suitable high performance coating compositions for use in this highly potential application field, as a few coating compositions based on natural material are already in use.

(ii) Soft drinks

Our country produced 6230 million bottles of soft drinks during 1998-99. Its production might have increased further during this period. Our country exported Rs. 114 crore worth alcoholic and non-alcoholic beverages in 2004-05. There is a report that China is already using lac dye in soft drink formulations. The Food Additive Hygiene Standards of China approves the use of lac dye (code no. 08.104) as a colouring agent in fruit/vegetable drinks, carbonated drinks, blended wines, candies, fruit jams, condiments and sauces, with a maximum permissible dosage of 0.5 g/kg body weight. This use is yet to start in our country, an appreciable scope of use is awaited for food grade lac dye in this sector. Soft drink bottling is expected to be one of the most promising sub-sectors in the coming years.

(iii) Chocolate and candy coating

Shellac is used for coating of chocolates and candies for preservation and protection from moisture. It is reported that income potential of manufacture of chocolate and candy is literally unlimited. It is projected that chocolate retail is a $19 billion market and candy is a $9 billion market worldwide. Demand for chocolate and candy is increasing by 3.5% annually. As people are getting more aware about antioxidant health benefits of chocolate, it is expected that customers asking for this kind of more organic products will further enhance the utilization of lac in this sector.

(iv) Pharmaceutical glaze

Shellac is being used as pharmaceutical glaze for tablets and pills since long, although some alternative coating materials are also used presently, but its natural origin might give an edge over animal by-product gelatin. Uninterrupted supply of lac will sustain its share in the export market.

(v) Electronic components

Printed circuit board assembly and original design manufacturing are set to grow to $2.5 billion by 2010. This is because of the increasing demand for telecom equipment like mobile handsets. Global electronics manufacturing service providers like Flextronics International, Jabil Circuits, Elcoteq SE and Soleqtron are expected to establish manufacturing bases in India. Hon Hai Precision Industry, the largest manufacturer of connectors for use in PCs in Taiwan and San Jose, California headquartered, Sanmina-SCI leading electronic manufacturing service provider Sanmina-SCI are in the process of setting up their units. Jabil Circuits has plans to invest up

Applications of lac in fruit and candy coating.

Applications of lac in cosmetics.

Applications of lac in fruit coating.

Applications of lac in textiles.

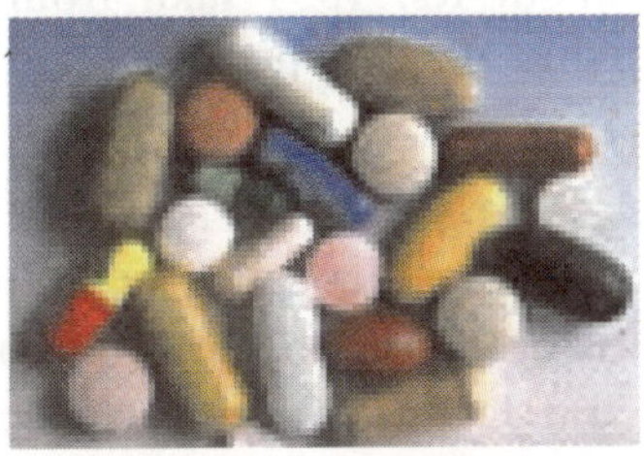

Applications of lac in medicines.

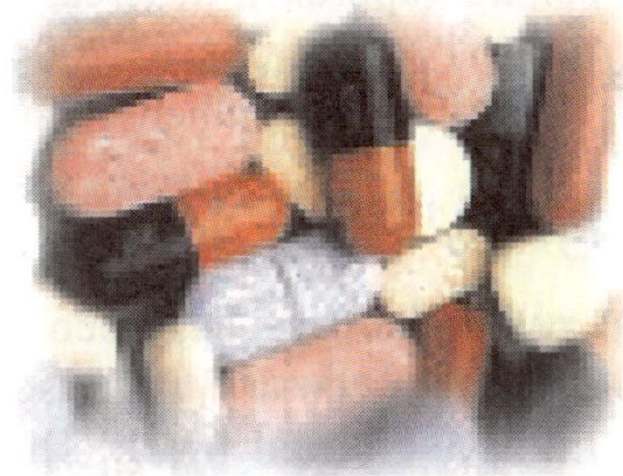

Applications of lac in medicines.

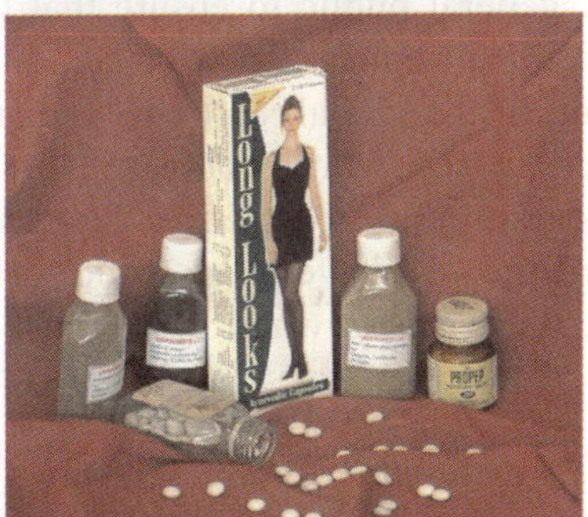

Applications of lac in pharmaceuticals.

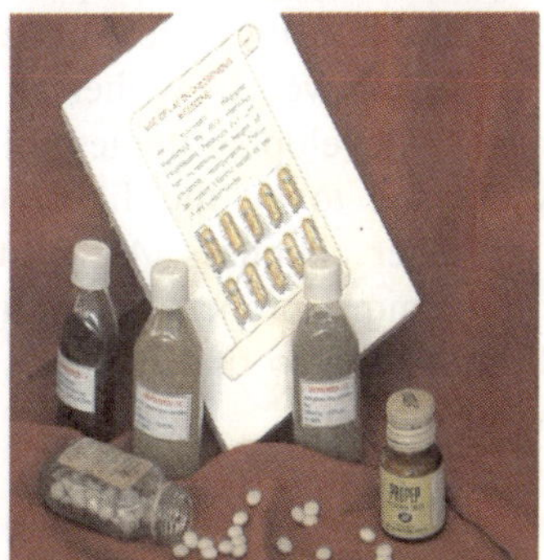

Applications of lac in pharmaceuticals.

Fig. 2.1.

to $100 million in India over the next few years. Domestic companies like TVS Electronics, Chennai, capital city of the south Indian state of Tamil Nadu-based computer peripherals manufacturer are also expanding their operations.

AT&S, Europe and India's largest producer of printed circuit boards and one of the world's most successful suppliers to the mobile phone industry, posted sales of EUR 375m in 2005/06, a 13% increase. [Operating profit (EBIT) for the year was up 24%, to about EUR 26m. There is an ample scope to promote shellac based insulating coating formulations in this sector.]

(vi) Blend of shellac with different polymers

Researches already done have revealed that shellac is compatible with a number of synthetic resins/polymers. Some more researches are needed to be done on the possibility of utilization of shellac polymer blends, so that shellac can find application in this fast developing polymer sector in which even a small percentage of use of shellac will ultimately result in a new avenue for substantial consumption of shellac.

(vii) Synthesis of fine chemicals (aleuritic acid, perfumery chemicals, etc.)

Aleuritic acid is a major constituent acid of lac resin and starting material for the synthesis of various products of industrial importance such as isoambrettolide, ambrettolide, civetone and exaltone. These compounds find application in perfumery industry as fixative. Aleuritic acid has been used for synthesis of compounds viz., insect sex pheromone components, plant growth regulators, insect repellents, etc. Aleuritic acid is a crystalline solid and its cost of production is Rs.1,000/- per kg (approx.), whereas market price varies from Rs. 1800-2000/- per kg. It is an export material (92.5 tons in 2004-05); country earned about Rs. 1551 lakh from export (2004-05). Thus, there is immense potential for entrepreneurship development.

(viii) Lac dye for textile industry

Natural red colour lac dye has been known for its use in textile industry for dying silk and wool for more than a century. It is used in China and Japan as colouring material for food and beverages. Lac dye isolated by ILRI is identical to Chinese and Japanese dyes (evidence from analysis results). Very good prospect is envisaged in the export market.

To conclude, it is up to researchers, industry and the general public to ensure that this gift of nature is not allowed to wither and die. A consortium, of all concerned with lac, is needed to be formed to develop strategies for achieving leadership in the global lac trade. Indian lac is inherently blessed to be the finest quality lac produced in the world compared to the lac produced by other countries. A worldwide awareness programme is required to be undertaken regarding this unique property of Indian lac. Concerted efforts will have to be made for further value-addition of lac so that it may survive in the future as it has survived in the past against continuous challenges.

3

Citrus Canker Research: Global Scenario

S.G. Borkar

Department of Plant Pathology, Post Graduate Institute
Mahatma Phule Agricultural University, Rahuri–413 722
Dist. Ahmednagar, Maharashtra

The citrus fruits have a prominent place among popular fruits and are extensively grown in tropical and subtropical regions. They possess greater adaptability to different climatic conditions and so are grown with equal success even in some favourable parts of temperate regions of the world. It is a crop adaptable to wide range of soils, terrain, planting and cultural arrangements, and over 100 nations are reported to grow citrus (Reitz, 1984).

Many taxonomists have a general agreement that Himalayan region and South China are the places of origin for most citrus fruits (Rajput and Haribabu, 1985). The oranges are believed to have been originated in China, the lime in India, the citron in Persia and India, the lemon in Arabia and the pummelo in Malayan Archipelago (De Condolle, 1886).

The world citrus production in early 21st century was 93.75 million tonnes wherein the contribution of sweet orange was 71%, tangerine 18%, lemon and limes 10% and grape fruit 6%.

Citrus fruits are one of the best source of vitamin C and also a good source of vitamin A. It also contains citric acid, riboflavin, thiamin and niacin. The citrus industry is flourishing on account of demand for fruit juice and peel oil. The industrial peel oil, pectins, citric acid and essences are used in flavouring extracts and to some extent in perfume and soap industry.

The citrus crop suffers from number of diseases caused by several fungi, bacteria and viral pathogens. The important fungal diseases are gummosis *(Phytophthora palmivora* Butl.), anthracnose *(Colletotrichum gloeosporioides* periz.), citrus scab *(Elsinoe fawcetti* Bitancourt and Jenkins), sooty mould *(Capnodium citri* Berk), bacterial disease citrus canker *(Xanthomonas axonopodis* pv. *citri}*, viral disease tristeza and phytoplasma disease citrus greening as well as disease complex like citrus dieback. However, citrus canker *(Xanthomonas axonopodis* pv. *citri)* is one of the most prevalent and destructive disease of citrus, particularly lime in the world.

It is evident from the fact that the Florida Department of Agriculture and Consumer Services and the United State Department of Agriculture, Animal and Plant Health and Inspection Service

(USDA, APH15) was engaged in what might be the largest single regulatory agriculture programme to eradicate a plant disease like citrus canker ever undertaken in the history of the United States, known as the Citrus Canker Eradication Programme (CCEP) and was an attempt to mitigate the serious consequences this disease can have on the 8.5 billion dollar Florida commercial citrus industry (Schubert *et al.* 2001).

The state witnessed several outbreaks of this disease; the first in 1912, second in 1986, third in 1995, fourth in 1997 and fifth in 1999 and the cost of Citrus Canker Eradication programme exceeds: $200 million (Mavrodieva *et al.* 2004). This indicates the seriousness and destructive nature of the disease. The disease might have originated in India because herbarium specimens sent from Dehradun to Kew (prior to 1831) show the disease canker symptoms. The disease is caused by the bacterial agent, which has been recently reclassified as *Xanthomonas; axonopodis* pv. *citri* (Synonym = *X. campestris* pv. *citri* (Vauterin *et al.*) 1995). The disease is widely distributed and endemic in many tropical and subtropical citrus growing areas (Goto, 1992; Stall and Civerolo, 1997). Asiatic Citrus Canker (ACC) is the most aggressive strain of the bacterium (Vaniere *et al.* 2003).

OUTBREAK AND EPIDEMIC

Citrus canker caused by *Xanthomonas axonopodis* pv. *citri* (Hasse) dowson is a widespread disease among various species and varieties of citrus. In many important commercial varieties, heavy losses are reported in the quality and quantity of the fruit yield. These losses occur due to defoliation and death of attacked young twigs. When the fruits are severely infected, they either shed prematurely or develop corky blemishes, remain small and fetch lower price. The affected tree becomes a perennial source for the spread of infection.

Stevens (1915) reported first outbreak of this disease in United States in Florida in 1912, which threatened its $9 billion industry. Fawcett and Jenkins (1933) found herbarium specimens identical with citrus canker, which had been collected from North-West India (probably between 1827 and 1831) and Java, Indonesia (between 1842 and 1844), indicating that the disease had been probably originated in the Indo-Malay region of Asia and spread to other Asian countries during the last century. Similarly, herbarium specimens available at the Royal Botanical Gardens in Kew, England and which are from India (collected in 1830) and Java (collected in 1840) confirm the centre of origin of the disease as South-East Asia (Civerolo, 1984), probably from some places in India or Java.

By the twentieth century, citrus canker was present in Asia/Japan, South and Central Africa, the middle East, Australia, New Zealand, the Pacific Island, South America and Florida (Koizumi, 19852). Luthra and Sattar (1940) reported the first appearance of citrus canker in Punjab (India) in 1940. Nirvan (1961) reported the dissemination of citrus canker bacterium through transfer of diseased planting material. Rangaswami (1962) reported that rain plays an important role in dissemination of the bacterium from tree to tree.

Gottawald and Graham (1992) reported second outbreak of citrus canker in Florida around 1986.

Goto (1992) reported that free moisture and strong wind favours the spread of the bacteria. The first flush in spring is infected by the bacterium splashed by rains from the canker lesions to over winter shoots. The disease cycle on the spring shoot may be limited to a rather short period of time unless the leaves are injured by storms.

Goto (1992) further reported that the disease once reported to be eradicated from Australia, New Zealand, South Africa and USA during 1980s was again reported in Australia as well as in Mexico in 1981 and in Florida in 1984.

Schubert *et al.* (1996) and Gottawald *et al.* (1997) analysed epidemiological aspects from the spread of citrus canker in Urban Miami Florida and synergetic interaction with the Asian citrus leaf miner. They also reported third outbreak in Urban Miami in 1995. Stall and Civerolo (1997) reported fourth outbreak of citrus canker on the west coast of Florida, which was the result of incomplete eradication of the 1986 outbreak. Davis *et al.* (2000) identified *Xanthomonas axonopodis* pv. *citri* in small and well-separated infection foci in citrus by polymerase chain reaction in Indonesia and adjacent regions of Papua New Guinea. Sun *et al.* (2000) reported an unique isolate of citrus canker bacterium on key lime in Wellington and Lake Worth, Florida, which caused fifth outbreak. This new strain of citrus canker has been termed as the "Wellington" strain (AW), which was found in Florida in 1999. Schubert (2001) suggested the origin of the disease in the tropical areas of Asia such as South China, Indonesia and India where citrus species are presumed to have originated. The disease is presently prevalent in Africa, Asia, Australia, Oceania and South America. Pruvost *et al.* (2002) evaluated the dynamics of *Xanthomonas axonopodis* pv. *citri* populations in leaf lesions exposed to mild winter temperatures prevailing in a tropical environment.

SYMPTOMATOLOGY AND FAVOURABLE WEATHER

The bacterium affects the leaves, twigs, thorns and fruits. The typical symptom of canker is a crateriform lesion surrounded by a water soaked margin. This first appears as small yellowish spots, which enlarge to a diameter of 3 to 4 mm and become raised, rough or spongy and tan or brown in colour. These cankerous spots are surrounded by characteristic yellow areas or halos. In severe cases, lesions become irregular and may coalesce. As these lesions persist on branches, it is very likely that these may be invaded by other organisms and by their synergetic action may be aggravating the symptom development. These situations might lead to death of twigs, branches and cracking of fruit causing considerable losses. Citrus canker causes reduction in yield as well as deteriorates quality of fruits, which are important from commercial point of view, i.e., infected fruits fetch less prices in the market.

On the leaves the disease first appears as small, translucent spots, which are yellow in colour. With the advancement of the disease, the spots become greyish, spongy with rough corky surface. The spots may coalesce to form elongated lesions, which are surrounded by a yellowish brown raised margin and a yellow halo. In severe cases, lesions are also formed on the petioles and midribs of the leaves causing defoliation. The spots on the young twigs are like those on the leaves. On the old twigs they are, however, more prominent, regular in shape and corky in appearance. Lesions on fruits are similar to those on leaves except that the yellow halo is absent and the centre is prominently raised (Chari Ram *et al.,* 1972).

Koizumi (1972) reported that the typical symptoms of Asian citrus canker is crateriform lesion surrounded by a water soaked margin and can be observed on leaves, fruits and stems. When the bacterium is inoculated on tolerant hosts or mature fruits, the symptoms can differ slightly and are relatively erumpent. Goto and Yaguchi (1979) described that as the lesions aged, they first turn grey, then tan to brown and a water soaked margin appears, surrounded by a chlorotic halo. The centre of lesion becomes raised and corky. Eventually, the centres of leaf lesions become crater-like, may fall out creating shot hole effect. Stall *et al.* (1982), observed that *Xanthomonas axonopodis* pv. *citri* is capable of naturally infecting green citrus tissues most readily while they are in the last half of expansion phases of growth. Once leaves, twigs and fruit reach mature size and begin to harden off physiologically, they become more resistant to infection.

The earliest symptoms on leaves appear as tiny, slightly raised blister-like lesions around 7 days after inoculation under optimum conditions. Optimum temperature for infection falls between 20 and 30°C (Krizumi, 1985). Aiyappa (1958) postulated that heavy rainfall, high humidity and low temperature may enhance the disease. Pruvost *et al*. (1999) studied the effect of irrigation practices on the spatio-temporal increase of Asiatic citurs canker in stimulated nursery plot in reunion island.

Peltier (1920) reported a temperature of 22-30°C as favourable for the citrus canker development. Nirvan (1963) reported that the infection takes place either through natural opening or through wounds caused by various agencies. Sothisorubini *et al.* (1986) stated that ideal conditions for infection of citrus canker bacterium were 30°C and 100% R.H., although the disease can be severe during both dry and wet conditions. Bacteria get entry predominantly through wounds and occasionally through natural openings. Infection takes 7 days to develop from time of inoculation under laboratory conditions and about 18 days in the field. The severity of infection is reported to be more during rainy season as there are more chances of injuries made to the leaf and twig surface due to rubbing of plant parts to each other by wind blown dust/sand particles and thunderstorms, etc. as compared to non-rainy season (Anonymous, 2003). Rangaswami (1962) also reported that rain plays an important role in the development of symptoms. Goto (1992) also reported that the disease cycle on spring shoot may be limited to a rather short time unless the leaves are injured by the storms. These observations were also confirmed by Vanterzet and Keil (1972) and Crosse *et al.* (1972). Injury made by leaf miner on the citrus leaves also enhances and spreads the citrus leaf canker disease (Rodrigues *et al.* 1998). Koizumi and Grieson (1979), Takashi and Doke (1987) and Palazzo *et al.* (1984) reported that the citrus canker bacteria enter host plant through wound. Khan *et al.* (2002) used stepwise regression to evaluate the influence of maximum and minimum air temperature, rainfall, relative humidity, wind speed and clouds on citrus canker disease. He found that out of these six variables, minimum temperature and wind speed significantly influenced the citrus canker disease development.

Gottawald and Graham (1992) stated that all above ground tissues of citrus are susceptible to *X. axonopodis* pv. *citri* when they are young and have maximum susceptibility during the last half of the expansion phase of growth. Schubert *et al.* (2001) reported that on twigs and fruits, citrus canker symptoms are similar, i.e., raised corky lesions surrounded by an oily or water soaked margin. No chlorosis surrounds twig lesions but may be present on fruit lesions. Twig dieback, fruit blemishes and early fruit drop are major economic impacts of the disease in advanced stages. If twigs are not killed back by girdling infections, the lesions can persist for many years, causing raised corky patches in the smooth bark. Braithwaite *et al.* (2002) observed yellow and brown raised, corky lesions, which darkened with age and had central depression on leaves, twigs and fruits on cultivated citrus. These water soaked lesions, 2-3 mm in diameter, developed at inoculation sites after 10 days and bacteria were consistently reisolated from the affected tissues. Further testing and observation revealed that the causal organism was *X. axonopodis* pv. *citri.* This was the first report of the citrus canker caused by *X. axonopodis* pv. *citri* on *Citrus sinensis* in Bolivia.

Vernier *et al.* (2003) reported infection on immature fruits and leaves. Water soaked lesions surrounded by yellow halo are observed on leaves, and cracking and malformation of the immature fruit are also seen. Heavy infection on fruits and leaves causes premature fruit drop and defoliation.

Any above ground part of a tree may be affected due to this disease, but the symptoms are commonly seen on leaves, twigs and fruits. On the leaves the spots are first white, changing to grey or tan and finally the spongy lesions ruptures to expose a typical crater-like appearance. The lesions have a oily margin or yellow halo. The size of the lesions varies with the citrus species attacked, but may be as large as 12 mm in diameter on grape fruit. On fruit the lesions are spongy, raised, have oily margins and after eruption show a typical crater-like appearance (Knorr *et al.* 1957; Klotz, 1973; Fawcett, 1936).

BACTERIUM

Hasse (1915) was the first to establish the bacterial nature of the disease and describe the organism as *Pseudomonas citri.* The organism was subsequently studied by Breed *et al.* (1998) who included it under *Xanthomonas citri* (Hasse) Dowson. In India citrus canker bacterium has been reported to be quite serious. (Rangaswami and Rajagopalan, 1973).

Hamlin (1967) noticed clear cut differences amongst the four isolates of *Xanthomonas citri* in their ability to infect various species and varieties of citrus and indicated the possibility of classifying *Xanthomonas citri* into pathotypes.

Three distinct forms (A, B and C) of the bacterium have been reported to exist on the basis of geographical distribution and host range of the pathogen (Stall and Seymour, 1983). The canker bacterium known as Asiatic canker or true canker is widely distributed and has a broad host range (Goto, 1992). Brdbent *et al.* (1992) identified *Xanthomonas axonopodis* pv. *citri* (Asiatic group or group A) using fatty acids profiles and DNA fingerprints besides using pathogenicity in a series of hosts. Vauterin *et al.* (1995) reclassified the bacterial agent causing citrus canker disease as *Xanthomonas axonopodis* pv. *citri* (Synonym = *Xanthomonas campestris* pv. *citn).*

Jayasekara and Suctsugu (1998) found that bacterial isolate obtained from diseased lemon fruits showing brownish to blackish cankers mostly 3-8 mm in diameter on their surface was linked to pathogenicity, bacteriological and serological characteristics of *X. axonopodis* pv. *citri.* In slide agglutination test and agar gel diffusion test, the pathogen consistently showed specific reaction with this antiserum and homologous antiserum of *X. axonopodis* pv. *citri* obtained from Japan indicating that the bacterium is similar to the Japanese isolate. Tung-ShuYun and Kuo-TsongTeh (1999) reported induction of mutant in *Xanthomonas axonopodis* pv. *citri* by insertion of the transposon Tn5tacl. This mutant did not grow or illicit canker disease in citrus leaves but was still able to induce hypersensitive response in a non-host plant (the common bean). They further demonstrated that phosphoglucose isomerase (PGI) gene is required for pathogenicity of *Xanthomonas axonopodis* pv. *citri.* Kalita *et al.* (1999) collected twenty isolates of *Xanthomonoas axonopodis* pv. *citri* from different citrus growing areas of Assam, India and pathogenicity was tested on 8 citrus cultivars specifically Assam lemon (*Citrus, limon*), sweet orange (*C. sinensis*), rangpur lime (*C. limonia*), rough lemon (*C. jambhiri*), Lummelo (*C.

grandis), grape fruit (*C. paradise*), Volkamer lemon (*C. volkameriana*) and Cleopatra orange (*C. reticulata*). Khodakaramian *et al.* (1999) studied phenotypic characteristics, host range and distribution of the strain of *Xanthomonas axonopodis* inducing citrus canker in Southern Iran. Khodakaramian and Swings (2002) compared 73 isolates of *Xanthomonas axonopodis pv. citri* (57 isolates from affected citrus trees in Southern Iran and remaining 16 were representative of all five currently described pathogenicity groups of pathogen causing citrus canker and spot disease) by amplified fragment length polymorphism fingerprinting.

Das (2002) determined the pathogenic variability in the bacteria. The bacterial isolates Xac 2 and Xac 6 from Maharashtra were most pathogenic to acid lime causing upto 75% canker severity. Xac 11 from Tamil Nadu was least pathogenic, i.e., canker severity was upto 10% only.

Mavrodieva *et al.* (2004) developed a fast, sensitive and reliable real time polymerase chain reaction (PCR) using a portable, field hardened RAPID machine and primers designed to detect all canker causing strains.

CONTROL METHODS

Various kinds of control measures have been studied for the pathogen (Fang and Zang, 1992; Khodakaramian *et al.* 2002, Graham *et al*., 2004, Pan Xiabiang, 2004), but these have not proved much effective in the management of the disease under field condition.

Effect of several chemicals and agrochemicals have been studied on the bacterium (Ram Kishun and Chhnd, 1987, Timmer, 1989). Thrirumalachar *et al.* (1956)tested five commercially available antibiotics, i.e., Aureomycin, Terramycin, sodium penicillin, *G.* dihydrostreptomycin and chloromycetin *in vitro* against thirty-two *Xanthomonas* species occurring in India and concluded that Aureomycin was the best antibiotic. Rangaswami (1962) evaluated the efficacy of streptomycin sulphate *in vitro* and found that it was effective against *Xanthomonas citri.* Desai *et al.* (1967) reported that plant disease control was based on external chemical protectants. The antibiotics were effective within the plant. Considering the cost of antibiotics and their availability and stability under field conditions, it may be stated that they have proved superior to conventional bactericidal chemicals in case of bacterial diseases. Mathur *et al.* (1973) found that all the five treatment, *viz.,* Bordeaux mixture (1%), copper oxychloride (0.25%), streptocycline (0.0025%), Agrimycin (200 ppm) and neem cake significantly reduced the disease on acid lime (*Citrus aurantifolia*). Streptocycline alone was distinctly superior to all the treatments. Pal *et al.* (1981) tested the efficacy of some antibiotics, fungicides and different combinations of these including four per cent kerosene oil against five phytopathogenic bacteria including *Xanthomonas citri.*

Four per cent kerosene oil emulsion was found to be superior to plantomycin and paushamycin. Krishna and Nema (1983) achieved the best control of *X. axanopodis* pv. *citri* with streptocycline (500 ppm) with 4 spray schedule followed by Paushamycin, Plantomycin, Bordeaux mixture and Agrimycin. Sawant *et al.* (1985) reported that Agrimycin 100 at 500 ppm was the best in checking the spread of the disease. This was followed by Aureofungin at 500 ppm. Neem cake at 2500 ppm was helpful to a limited extent while at 5000 ppm it showed phytotoxic effect. Urrutia and Stapleton (1986) sprayed copper oxychloride and copper sulphate both at 2.5 g a.i./lit after shoot emergence. In winter 2, 3 and 4 sprays of copper oxychloride reduced leaf infection by 60, 60-80 and 90 per cent, respectively as compared to control. In spring the disease intensity was much higher than in winter and no treatment reduced leaf infection by more than 50 per cent of the control. Spraying every 5 days, 2 to 4 times beginning 5 to 10 days after shoot emergence was reported to reduce infection. Vegetative shoots that had emerged in winter were susceptible during the entire growth. Sothisorubini *et al.* (1986) indicated that tetracycline inhibit symptom development of citrus canker at 100 ppm in detached leaves, while chloramphenicol, streptomycin, copper sandoz and cupravit (copper oxychloride) were effective at 1000 ppm. Leite *et al.* (1987) reported that copper based bactericides were most effective on moderately resistant cultivars, incidence being reduced upto 90 per cent as compared with unsprayed plants. Mc Guire (1988) found that copper containing products were generally superior in reducing epiphytotic population of *Xanthomonas citri* in nursery. Kale *et al.* (1994) investigated the effect of foliar spray of 100 ppm. Streptocycline + 0.1 per cent copper oxychloride on *Xanthomonas citri* infection of 6 years old kagzilime seedlings in Maharashtra. Sprays were applied at an interval of 7, 15 and 21 days and the most cost-effective chemical control was achieved by spraying at an interval of 7 or 15 days. Pabitra *et al.* (1996) have observed *in vitro* inhibition of *X. campestris* pv. citri by *Bacillus subtilis*, *B. polymyxa*, *P. fluorescens*, *Serratia marcescens*, *Aspergillus terreus*, *Trichoderma viridae* and *T. harzianum* isolated from phylloplane of lemon.

Ota (1993) studied interaction between *Xanthomonas* and antagonist pseudomonas *in vivo* and *in vitro*.

Bora *et al.* (2001) used plant extract from 30 medicinal plants at 20% strength for suppressing growth of *Xanthomonas axonopodis* pv. citri and found that the extracts of *L. indica* was the most effective in suppressing the disease and demonstrated 78.46 disease control compared to control.

Zhong Dezhi and Ling Nan Chang (2002) showed that spraying spring shoots with 77% kocide (copper hydroxide) wettable powder at 20-30 days after bud burst and spraying summer autumn shoots at 10-15 days after bud burst provides good citrus canker control.

An application of BM (4 : 4 : 50) or copper compounds (Cu content 50%) with $CaCO_3$ in late March has been recommended by Goto (1992) against the disease. In addition two prunings along with four sprays of copper oxychloride (0.5%) or BM (1%) have been reported to be effective against the disease under Indian condition (Kishun and Chand, 1987). Streptocycline sulphate is specifically recommended against this disease and six sprays at 1000 ppm alongwith two prunings reduced the canker in acid lime (Balaraman and Purshothman, 1981). Foliar sprays of streptocycline (100 ppm) plus copper oxychloride (0.1%) at 7 or 15 days interval have also been found effective in reducing the disease (Kale *et al.* 1994). Gottawald and Timmer (1995) suggested use of windbreaks along with the application of copper bactericides as effective control measures for citrus canker.

REFERENCES

Aiyappa, K.M. (1958). Citrus canker *Xanthomonas citri* (Hasse) Dowson. *Mysore Agric. J.* **33:** 164-167.

Anonymous, (2003). Review of current citrus phytosanitary problems and control polices employed. Committee on Commodity problems Intergovernmental group on citrus fruit. Thirteen Session Havana Cub, 2003.

Balaraman, K. and Purshotham, R. (1981). Control of citrus canker on acid lime. *South Indian Hortic.* **29**(4): 175-177.

Basim, E. and Basim, H. (2003). Antibacterial activity of Rosa damascene essential oil. *Fitoterapia.* **74**(4): 394-396.

Bora L.C., Minku Das and Samuel, J. (2001). Medicinal plant extracts for management of citrus canker and bacterial leaf spot 0 beteline. *J. Agril. Sci. Soc. of North East India.* **14**(2): 159-164.

Braithwaite, M., Leite, R.P., Smith, J.J., Boa, E. and Saddler, G.S. (2002). First report of citrus canker caused by *Xanthomonas campestris* pv. *citri* on citrus sinensis in Bolivia. *Plant Pathology.* **51**(3): 383.

Breed, R.S., Murray, E.G.D. and Smith, N.R. (1998). Bergeys Manual of Determinative Bacteriology Willams and Wilkins Co., Baltimore.

Broadbent, P., Pahy, P.C., Gilling, M.R., Brandley, J.K. and Barnes, D. (1992). Asiatic citrus canker detected in Pummelo orchard in Northern Australia. *Plant Disease.* **76**(8): 824-829.

Chari Ram, Nirvan, R.S. and Saxena, M.L. (1972). Citrus canker and its control with fungicides. *Punjab Hort. J.* **12:** 240-243.

Civerolo, E.L. (1984). Bacterial canker disease of citrus. *J. Rio Grande Valley Hortic. Assoc.* **37:** 127-146.

Das, A.K. (2002). Pathogenic variability in *Xanthomonas axonopodis* pv. *citri,* causal agent of citrus canker. *J. Mycol. and Pl. Pathol.* **32**(2): 247-248.

Davis, R.H., Hailstones, D.L., Jacobson, S.C., Eichner, R.H., Gunua, T.G., Rahamma, S. and Broadbent, P. (2000). Surveillance for citrus canker disease in New Guinea and Northern Australia. *Australian Plant Pathology.* **29**(3): 222.

De Condolle, A. (1886). Origin of cultivated plants. Kagan Paul, London.

Desai, S.G., Patel, M.L. and Desai, M.V. (1967). *In vitro* activity of Streptocycline against bacterial plant pathogen. *Indian Phytopath.* **20:** 296-300.

Fang, Y.M. and Zhang, Y.X. (1992). Study on integrated management of citrus canker. *Acta Phytophylactica Sinica.* **19**(2): 101-105.

Fawcett, H.S. (1936). Citrus diseases and their control. McGraw Hill, New York. pp. 656.

Faweett, H.S. and Jehkins, A.E. (1933). Records of citrus canker from herbarium specimens of genus citrus in England and the United States. *Phytopathology.* **23:** 820-824.

Goto, M. (1992). Citrus canker. *In:* Plant Disease of International Importance. Disease of fruit crops. Vol. III. Edited by J. Kumar, H.S. Chaube, U.S. Singh and A.N. Mukhopadhyay. Prentice Hall, Englewood Cliffs, New Jersey. pp. 170-208.

Gota, M. and Yaguchi, Y. (1979). Relationship between defoliation and disease severity in citrus, canker (*Xanthomonas citri*). *Ann. Phytopathol. Soc. Jpn.* **45:** 689-694.

Gottawald, T.R. and Graham, J.H. (1992). A device for precise and non-disruptive stomatal inoculation of leaf tissue with bacterial pathogens. *Phytopathology.* **82:** 930-935.

Gottawald, T.R. and Timmer, L.W. (1995). The efficacy of windbreaks in reducing the spread of citrus canker caused by *Xanthomonas campestris* pv. *citri. Trap. Agric.* **72:** 194-201.

Gottawald, T.R., Graham, J.H. and Schubert, T.S. (1997). An epidemiological analysis of the spread of citrus canker in urban Miami, Florida and synergistic interaction with the Asian citrus leaf miner. *Fruits.* **52:** 371-378.

Gottawald, T.R., Sun, X., Riley, T., Grahman, J.H., Ferrandino, F. and Taylor, E.L. (2002). Geo-referenced spatiotemporal analysis of the urban citrus cancker epidemic in Florida. *Phytopathology.* **92**(4): 361-377.

Grahman, J.H., Gottwald, T.R., Cubero, J., Achor, D.S. (2004). *Xanthomonas axonopodis* pv. *citri* : factors affecting successful eradication of citrus canker. *Molecular Plant Pathology.* **5**(1): 1-15.

Hamlin, S.A. (1967). Studies on occurrence of pathotypes in *X. citri* (Hasse) Dowson. *Punjab Horticultural J.* **7:** 90-93.

Hasse, C.H. (1915). *P. citri,* the cause of citrus canker. *J. Agril. Research,* **4:** 97-100.

Jayasekara, J..S. and Suctsugu, T. (1998). Studies on the diagnosis of *Xanthomonas campestris* pv. *citri* (Hasse) Dye 1978 : 1. the serological diagnosis of citrus canker. *JICA/NGQS*-Occasional paper. **2:** 18.

Kale, K.B., Kolte, S.O. and Peshney, N.L. (1994). Economics of chemical control of citrus canker caused by *Xanthomonas axonopodis* pv. *citri* under field conditions. *Indian Phytopathlogy.* **47**(3): 253-255.

Kalita, P. Bora, L.C. and Bhagabati, K.N. (1999). Variability of *Xanthomonas axonopodis* pv. *citri* (Hasse) Dye, the incitant of citrus canker in Assam. *J. Agril. Sci. Soc. of North-East India.* **12**(1): 136-138.

Khan, M.A., Rashid, A. and Ikram, F. (2002). Weekly minimum temperature, one variable model to predict citrus canker disease. *Pakistan J. Phytopathol.* **14**(1): 7-10.

Khodakaramian, G. and Ghasemi, A. (2002). Effects of two antibactericidal chemicals and pruning on control of citrus bacterial canker disease in garden in South of Iran. *Seed and Plant.* **18**(3): 316-327.

Khodakaramian, G. and Swings, J. (2002). AFLP fingerprinting of the strains of *Xanthomonas axonopodis* inducing citrus canker disease in Southern Iran. *J. Phytopathol.* **150**(4/5): 227-231.

Khodakaramian, G., Rahimian, H., Mohamadi, M. and Allameh, A. (1999). Phenotypic characteristics, host range and distribution of the strains of *Xanthomonas axonopodis* including citrus canker in southern Iran. *Iranian J. Pl. Pathol.* **35**(1/4): 111.

Kishun, R. and Chand, R. (1987). Studies on germplasm resistance and chemical control of citrus canker. *Indian J. Hortic.* **44:** 126-132.

Klotz, L.J. (1973). Colour Handbook of Citrus Diseases. University California, Riverside. pp. 122.

Knorr, L.C., Suit, R.F. and Ducharme, E.P. (1957). Handbook of citrus disease. *Florida Bull. Univ. Fla. Agri. Exp. Sta.* pp. 587.

Koizumi, M. (1972). Studies on the symptoms of citrus canker on mandarin fruit and existence of causal bacteria in the tissues. *Bull. Hortic. Res. Stn. Ser.* **12:** 229-243.

Koizumi, M. (1985). Citrus canker : The world situation. Pages 2-7 in : citrus canker: An International Perspective. L.W. Timmer, ed. University of Florida, Lake Alfred.

Koizumi, M., and Grieson, W. (1979). Relation of temperature to the development of citrus canker lesion in the spring. Proc. Int. Soc. *Citri* fruit. **3:** 924-928.

Krishna, A. and Nema, A.G. (1983). Evaluation of chemicals for control of citrus canker. *Ind. Phytopath.* **36**(2): 348-350.

Leite Junior, R.P., Mohan, S.K., Pereira, A.L.G. and Campacci, C.A. (1987). Integrated control of citrus canker effect of genetic resistance and application of bactericides. *Fitopatologia Brasileira.* **12**(3): 257-263.

Loucks, K.W. (1934). Citrus canker and its evaluation in Florida. *Archives of the Florida Departement of Agriculture.* Division of Plant Industry, Grainesville.

Luthra, J.C. and Sattar, A. (1940). A note on citrus canker *(Phytomonas Citri)* in the Punjab. Punjab Proc. 27th Indian Sci. Congr. Part III : 228-229 (Abst.).

Mathur, A.S., Iralappan, I., Krishnamurthy, C.S., Rajappan, P.V. and Gowder, R.B. (1973). Efficacy of different fungicides and antibiotics on the control of citrus canker caused by *X. citri* (Hasse) Dawson. *Madras Agric. J.* **60:** 626.

Mavrodieva, V., Levy, L. and Gabriel, D.W. (2004). Improved sampling methods for real time polymerase chain reaction diagnosis of citrus canker from field samples. *Phytopathology.* **94**(1): 61-68.

Mc Guire, R.G. (1988). Evaluation of bacterial chemicals for control of *Xanthomonas citri* on citrus. *Plant Disease.* **72**(12): 1016-1020.

Nirvan, R.S. (1961). Citrus canker and its control. *Hortic. Advance.* **5:** 171-175.

Nirvan, R.S. (1963). Citrus canker and its control. *Gardening,* **4:** 52-58.

Ota, (1983). Interaction *in vitro* and *in vivo* between *Xanthomonas axonopodis* pv. *citri* and antagonistic pseudomonas sp. *Ann. Phytopathol. Soc. Japan.* **49:** 308.

Pabitra, K., Bora, L.C., Bhagabati, K.N. and Kalita, P. (1996). Phylloplane microflora of citrus and their role in management of citrus canker. *Indian Phytopathol.* **49:** 234-237.

Pal, V., Neelam Rani and Chand, J.N. (1981). Sensitivity of five phytopathogenic bacteria to some antibiotics and fungicides. *Pestology.* **5:** 28-30.

Palazzo, D.A., Malavalta, V.A. Jr. and Nogueira, E.M. (1984). Effect of some climatic factors on the index of infection of citrus canker caused by *Xanthomonas campenstris* pv. *citri* on Valencia orange (*Citrus sinensis),* in Bataguassu, M.S. *Fitopatol. Bras.* **9:** 283-290.

Pan-Xiao Qiang, (2004). Experiment of control of citrus canker disease by copper hydroxide, fungicide. China-Fruits. **5:** 57.

Peltier, G.L. (1920). Influence of temperature and humidity on the growth of *Pseudomonas citri* and its host plants and on infection and development of the disease. *J. Agric. Res.* **20:** 447-506.

Pruvost, O., Boher, B., Brocherieux, C., Nicole, M. and Chiroleu, F. (2002). Survival of *Xanthomonas axonopodis* pv. *citri* in leaf lesions under tropical environment conditions and simulated splash dispersal of inoculum.

Pruvost, O., Gottawald, T.R. and Brocherieux, C. (1999). The effect of irrigation practices on the spatio-temporal increase of Asiatic citrus canker in simulated nursery plots in Reunion Island. *European J. Plant Pathology.* **105**(1): 23-37.

Rajput, C.B.S. and Haribabu, R.S. (1985). Citriculture. Kalyani Publishers. p. 368.

Ram Kishun and Ramesh Chand, (1987). Studies on germplasm resistance and chemical control of citrus canker. *Indian J. Horticulture,* **44**(1-2): 126-132.

Rangaswami, G. (1962). Bacterial Plant Diseases in India. Asia Publishing House, Bombay.

Rangaswami, G. and Rajagopalan, S. (1973). Bacterial plant pathology. Paublished by Tamil Nadu. Agric. Univ. Coimbatore. pp. 171-180.

Reitz, H.J. (1984). Outlook of Agriculture, **13:** 140-146.

Rodrigues, J.C.V., Rossetti, V., Machado, M.A. and Sobrinho, J.T. (1998). Citrus leaf miner: A factor for increase of pests and citrus canker, *Laranja.* **19**(1): 49-60.

Sawant, D.M., Ghawate, A.G., Jadhav, J.V. and Choudhari, K.G. (1985). Control of citrus canker in acidlime. *Maharashtra J. Hort.* **2:** 55-58.

Schubert, T.S., Gottawald, T.R., Graham, J.H., Rizvi, S.A., Sun, X. and Dixon, W.N. (2001). Meeting the challenges of Eradicating citrus canker in Florida-Again. *Plant Disease.* **85**(4): 340-356.

Schubert, T.S., Miller, J.W. and Gabriel, D.W. (1996). Another outbreak of bacterial canker on citrus in Florida. *Plant Dis.* **80:** 1208.

Schubert, T.S., Rizvi, S.A., Sun, XiaoAn, Gottawald, T.R., Graham, J.H. and Dixon, W.N. (2001). Meeting the challenges of Eradicating citrus canker in Florida-Again. *Plant Disease.* **85**(4): 340-356.

Singh, S. (2000). Citrus in India. Pages 278-303 *In:* Hi-Tech Citrus Management. Proc. of Int. Sym on *Citri* culture, November 23-27, (1999). National Research Centre for Citrus, Nagpur, p. 275.

Singh, S. (2001). Citrus industry of India. Pages 3-41 In citrus (Eds. Shyam Singh and Naqvi, S.A.M.H.). International Book Pub. Co., Lucknow. p. 558.

Sothisarubini Nalliah, Sundaresan, R.S.P. and Sivapalan, A. (1986). Studies on *X. citri* (Hasse) Dowson causing canker disease on citrus. *Vingnanam J. Sci.* **1**(1): 19-25.

Stall, R.E. and Civerolo, E.L (1997). Research relating to the recent outbreak of citrus canker in Flarida. *Ann. Rev. Phytopathol.* **29:** 399-420.

Stall, R.E. and Seymour, C.P. (1983). Canker: a threat to citrus in the Gulf coast states. *Plant Dis.* **67:** 581.

Stall, R.E., Marco, G.M. and Echenique, B.I. (1982). Importance of mesophyll in mature leaf resistance to cancrosis of citrus. *Phytopathology.* **72:** 1097-1100.

Stevens, H.E. (1915). Studies of citrus canker. *Fla. Agric. Exp. Stn.* **124**(3): 1-53.

Sun, X., Stall, R.E., Cubero, J., Gattawald, T.R., Graham, J.H., Dixan, W.D., Schubert, T.S., Peacack, M.E., Dickstein, E.R. and Chaloux, P.H. (2000). Detection of a unique isolate of citrus canker bacterium from key lime in Wellington and Lake Warth, Flarida (Abstr.). *Prac. Int. Citrus Canker Res. Workshop*, June 20-22, 2000. Florida Department of Agriculture and Consumer Services.

Takahishi, T. and Doke, N. (1984). A role of extracellular polysaccharide of *X. campestris* pv. *citri* in bacterial adhesion to citrus leaf tissues in preinfections stage. *Ann. Phytopathol. Soc. Jpn.* **50:** 565-573.

Thirumalachar, M.I., Patel, M.K., Kulkarni, N.B. and Dhande, D.W. (1956). Effects *in vitro* of some antibiotics on thirty two *Xanthomonas* species occurring in India. *Indian Phytopath.* **46:** 486-488.

Timmer, L.W. (1989). Evaluation of bactericides for control of citrus canker in Argentina. *Florida State Horticultural Society.* **101:** 6-9.

Tung-ShuYun and Kuo-TsongTeh, (1999). Requirement for phosphoglucose isomerase of *Xanthomonas campestris* is pathogenesis of citrus canker. *Applied and Environmental Microbiology.* **65**(12): 5564-5570.

Urrutia, M.V.M. and Stapleton, J.J. (1986). Effect of the number of copper sprays on leaf bacteriosis of Mexican lime. *Revista Mexicana defitopatogia*, **4**(1): 13-18.

Veuterin, L., Hoste, B., Kersters, K. and Swings, T. (1995). Reclassification of *Xanthomonas. Int. J. Syst. Bacteriol.* **45:** 472-489.

Vernier, C.J., Gottawald, T.R. and Pruvost, O. (2003). Disease development and symptom expression of (*Xanthomonas axonopodis* pv. *citri*) in various citrus plant tissues. *Phytopathology.* **93:** 832-843.

Zhong Dezhi and Ling Nan Chang, (2002). Observation on the occurrence regulation of citrus canker disease and its control. *South China Fruits.* **31**(2): 20-21.

4

Significance of Transgenic Crops in Pest Management

S.K. Mandal and R.K. Mandal

Subject Matter Specialist (Plant Protection), Krishi Vigyan Kendra, Saraiya–843 126, Muzaffarpur, Bihar

SMS (Plant Breeding) Krishi Vigyan Kendra, Saraiya–843 126, Muzaffarpur, Bihar

The major challenges facing our agriculture today are the production of sufficient food to feed the growing population and the prevention of environmental degradation. Damage by crop pests, which include insects, disease-causing organisms, nematodes, rodents, etc. is a major constraint to increased food production. The reduction of losses caused by pests is the obvious strategy for increased food supply. In addition, quality aspects such as pest-free and residue-free agricultural products are becoming increasingly important in light of market liberalization and increasing importance of agricultural exports. Pest control has traditionally meant the use of chemicals pesticides. There is no doubt that chemicals have played a significant role in pest control for the protection of crops. However, their excessive and inappropriate use in our agro-ecosystem in the last two decades or so has resulted in degradation of our environment, while our pest problems seem greater than ever. There are more and more reports of resistance of pests to pesticides, the number of pest outbreaks have increased and many innocuous species have attained the status of serious pests. The extent of pesticide residue in the environment is also a matter of great concern. Surveys have indicated that food commodities are highly contaminated with persistent pesticide residues. Therefore, there is an urgent need to evolve strategies and technologies that will not only meet the increasing demands for food and fibre, but will also provide us pesticide residues free agricultural commodities.

The various facets of pest management in agriculture should, therefore, be suitably engineered to maintain the productivity in a suitable manner. Looking into the growing importance attached by all nations to the sustainable use of renewable natural resources, it is difficult to see that anything other than IPM will provide an acceptable or affordable basis for pest control in the future. IPM based on recent advances in molecular biology have combined these two approaches into a third approach by exploiting biotechnology and genetic engineering in host plant resistance against insect pests. Genetic engineering science includes transgenic crops that contribute substantially in this regard. Several successes have been achieved in developing transgenic crops for their application in integrated pest management programme (Table 4.1).

Table 4.1. Application of transgenic plants in IPM.

Technique	Application	Examples
Embryo rescue and protoplast fusion	Overcoming incompatibility to produce hybrids (cultivated x wild sp.)	Transfer of resistance in rice against BPH and fruit borer resistant in brinjal.
Genetic transformation	Transfer of novel genes for HPR from diverse plants and bacterial systems.	Bt gene in tomato, tobacco, potato, cotton, populas, maize, CpTi gene in tobacco.
RFLP	Tagging genes govering resistance to pests with molecular markers to facilitate selection.	WBPH-1 linked with RG 146 in rice. Resistance to BPH linked with RG 457 in rice.
		Resistance to bruchid beetle linked with *pA* 882 in mungbean.
Dot blot hybridization	Precise diagnosis of viruses in insect population.	Tomato yellow leaf curl virus.

Source: Singh (1997).

GLOBAL ADOPTION OF TRANSGENIC CROPS

The people's Republic of China was the first country to commercialize transgenic crops in the early 1990s with the introduction of a virus resistant tobacco, followed by virus resistant tomato. In 1994, the USA followed when the Calgene Company got the first approval to commercialize a genetically modified tomato 'Flavr Saver TM', the delayed ripening tomato. From then onwards, the development and use of transgenic crops gained momentum. Recently, James (2005) has reviewed status of transgenic crops during the first decade of commercialization (1996-2005). The global area under transgenic crops increased more than fifty-fold, from 1.7 million ha in 1996 to 90.0 million ha in 2005 (Table 4.2). The transgenic crops were grown in 21 countries by 8.5 million farmers in 2005; 90 per cent of the beneficiary farmers were resource poor farmers from developing countries. The largest area under transgenic crops is in USA, followed by Argentina, Brazil, Canada and China—the principal adopters of transgenic crops globally. In 2005, the 21 countries growing transgenic crops included 11 developing countries and 10 industrial countries. More than one-third (33.9 million ha, 38%) of the global transgenic area was grown in developing countries. The total number of EU countries commercializing modest areas of Bt maize are five, viz. Spain, Germany, Portugal, France and the Czech Republic. In 2005, Bt rice was grown commercially for the first time on approximately 4,000 ha in Iran by several hundred farmers. In the first decade, the accumulated biotech crop area was 475 million ha, equivalent to almost half of the total land area of the USA or China (James, 2005).

Of the total global area planted to soybean, cotton, canola and maize, the area under transgenic crops in 2005 was 60, 28, 18 and 14 per cent, respectively (Table 4.3).

Table 4.2. Global area (million ha) of transgenic crops, 1996-2005.

Country	1996	1997	1998	1999	2000	2001	2002	2003	2004	2005
USA*	1.5	8.1	20.5	28.7	30.3	35.7	39.0	42.8	47.6	49.8
Argentina*	0.1	1.4	4.3	6.7	10.0	11.8	13.5	13.9	16.2	17.1
Brazil*	0.0	0.0	0.0	0.0	0.0	0.0	0.0	3.0	5.0	9.4
Canada*	0.1	1.3	2.8	4.0	3.0	3.2	3.5	4.4	5.4	5.8
China*	<0.1	<0.1	<0.1	0.3	0.5	1.5	2.1	2.8	3.7	3.3
Paraguay*	0.0	0.0	0.0	0.0	0.0	0.0	0.0	0.0	1.2	1.8
India*	0.0	0.0	0.0	0.0	0.0	0.0	<0.1	0.1	0.5	1.3
South Africa*	0.0	0.0	<0.1	0.1	0.2	0.2	0.3	0.4	0.5	0.5
Uruguay*	0.0	0.0	0.0	0.0	<0.1	<0.1	<0.1	0.1	0.3	0.3
Australia*	<0.1	0.1	0.1	0.1	0.2	0.2	0.1	0.1	0.2	0.3
Mexico*	<0.1	<0.1	<0.1	<0.1	<0.1	<0.1	<0.1	<0.1	0.1	0.1
Romania*	0.0	0.0	0.0	<0.1	<0.1	<0.1	<0.1	<0.1	0.1	0.1
Philippines*	0.0	0.0	0.0	0.0	0.0	0.0	0.0	<0.1	<0.1	0.1
Spain*	0.0	0.0	<0.1	<0.1	<0.1	<0.1	<0.1	<0.1	<0.1	0.1
Colombia	0.0	0.0	0.0	0.0	0.0	0.0	<0.1	<0.1	<0.1	<0.1
Iran	0.0	0.0	0.0	0.0	0.0	0.0	0.0	0.0	0.0	<0.1
Honduras	0.0	0.0	0.0	0.0	0.0	0.0	<0.1	<0.1	<0.1	<0.1
Portugal	0.0	0.0	0.0	0.0	0.0	0.0	0.0	0.0	0.0	<0.1
Germany	0.0	0.0	0.0	0.0	<0.1	<0.1	<0.1	<0.1	<0.1	<0.1
France	0.0	0.0	<0.1	<0.1	<0.1	0.0	0.0	0.0	0.0	<0.1
Czech Republic	0.0	0.0	0.0	0.0	0.0	0.0	0.0	0.0	0.0	<0.1
Total	**1.7**	**11.0**	**27.8**	**39.9**	**44.2**	**52.6**	**58.7**	**67.7**	**81.0**	**90.0**

*Biotech mega-countries, which grow 50,000 ha or more of transgenic crops in 2005.
Source: James (2005)

Table 4.3. Global adoption rates of major transgenic crops in 2005.

Sl. No.	Crop	Adoption rates of transgenic crops (%)	Area of transgenic crops (m ha)	Area of non-transgenic crops (m ha)	Total area of transgenic + non-transgenic crops (m ha)
1.	Soybean	60	54.4	36.6	91.0
2.	Cotton	28	9.8	25.2	35.0
3.	Canola	18	4.6	21.4	26.0
4.	Maize	14	21.2	125.8	147.0

Source : James (2005).

Transgenic soybean continued to be the principal crop in 2005, occupying 54.4 million ha (60% of global transgenic area), followed by maize (21.2 million ha – 24%), cotton (9.8 million ha – 11%) and canola (4.6 million ha – 5%). Herbicide tolerance, deployed in soybean, maize, canola and cotton was the most dominant trait occupying 63.7 million ha (71%) followed by Bt insect resistance at 6.2 million ha (18%) and 10.1 million ha (11%) to the stacked genes. Thus, the cultivation of insect protective transgenic crops has increased from about 1.7 million ha in 1996 to 23.6 million ha (inclusive of transgenic with herbicide tolerance) in 2005 (James, 2005). Currently, more than 20 transgenic insect protective events for potato, maize and cotton are registered with the Environmental Protection Agency of USA (Table 4.4).

Table 4.4. Insect protective transgenic crop events registered by US Environmental Protection Agency (February 17, 2006).

Transgenic crop event	Registrant	Date of registration	Date of expiry
Bt potato Cry3A	Monsanto 524-474	May 1995	No expiration date
Bt corn event 176 Cry1Ab	Mycogen 68467-l	August 1995	April 2001
Bt corn event 176 Cry1Ab (2 products–field corn, popcorn)	Syngenta 66736-l	August 1995 March 1998	June 30, 2001
Bt cotton Cry1Ac	Monsanto 524-478	October 1995	September 30, 2006
Bt corn event MON 801 Cry1Ab	Monsanto 524-492	May 1995	Voluntarily cancelled May 8, 1998
Bt corn 11 CryAb (Field and sweet corn no refugia for sweet corn)	Syngenta field corn 67979-l, sweet corn 65269-1	August 1996; February 1998	October 15, 2008
Bt corn Cry1Ab (Mon 801)	Monsanto 524-489	December 1996	October 15, 2008
Bt corn Cry9C (domestic field corn for feed and non-food uses)	Aventis 264-669	May 1998	Voluntarily cancelled
Replicase for potato leaf roll	Monsanto 524-474	November 1998	No expiration date
Bt corn POCry1F	Dow/Mycogen 68467-2	May 2001	October 15, 2008
Bt corn POCry1F	Pioneer/Dupont 29964-3	May 2001	October 15, 2008
Bt cotton Cry2Ab2 in comb with Cry1Ac	Monsanto 524-522	December 2002	September 30, 2006
Bt corn Cry3bl	Monsanto 524-528	February 2003	July 31, 2006
Bt corn stack Cry3bl+Cry1ab	Monsanto 524-545	October 31, 2003	July 31, 2006
Bt cotton Cry1Ac+Cry1F (Wide strike)	Dow Agro Sciences 68467-3	September 30, 2004	September 30, 2009
Bt corn MOCry1F Event DAS-06275-8	Dow Agro Sciences 68467-4	May 27, 2005	October 15, 2008
Bt corn Cry34Abl + Cry35Abl	Dow Agro Sciences 68467-5	August 31, 2005	September 30, 2010
Bt corn Cry34Abl + Cry35Abl	Pioneer/Dupont 29964-4	August 31, 2005	September 30, 2010
Bt corn Cry34Abl + Cry35Abl with POCry1F	Dow Agro Sciences 68467-6	October 27, 2005	October 15, 2008
Bt corn Cry34Abl + Cry35Abl with POCry1F	Pioneer/Dupont 29964-5	October 27, 2005	October 15, 2008
Bt corn Cry3bl MON88017	Monsanto 524-551	December 13, 2005	September 30, 2010
MON 88017 + MON810 AKA Cry3Bbl + Cry1A	Monsanto 524-552	December 13, 2005	October 15, 2008

Source : http://www.epa.gov/pesticides/biopesticides/pips/pip_list.htm

Monsanto markets Bt maize as MON810 and Bt cotton as Event 531 under different trade names in several countries (Table 4.5). Syngenta seeds (Formerly Novartis Seeds) has two maize events in the regulatory area as Event 176 and Bt 11. The trade name for Event 176 is Knockout®, whereas Bt 11 is sold as NK-brand, Bt corn with YieldGard® (a trademark of Monsanto Company), both of which contain *Cry1Ab* gene. Aventis marketed StarLink® corn, which contains *Cry9C* protein and is approved only for animal feed and ethanol production. Its registration was voluntarily cancelled in USA in October 2000 because of its inadvertent introduction into human food supplies. There were three Bt cotton events in China in 1999, one from Monsanto and two developed in China. Event GK has a modified *Cry1A* gene and SGK has two insecticidal genes, *Cry1A* and *CpTi*. A Monsanto-affiliated company, NatureMark, marketed *Cry3A* potatoes in the USA, Romania and Canada under variations of the trade name, NewLeaf®. However, the company stopped marketing of Bt potatoes in 2001 (Shelton *et al.*, 2002; Shelton and Bellinder, 2006). So far nearly 105 events of 20 transgenic crops have been released for commercial cultivation in 21 countries (Khadi, 2006).

Table 4.5. Transgenic crops registered in different countries.

Crop/Company	Event	Country	Year of approval
Yield Gard® Corn/Monsanto	MON810	Argentina	1998
	MON810	Bulgaria	2000
	MON810	Canada	1997
	MON810	EU	1998
	MON810	South Africa	1999
	MON810	USA	1996
NK Brand Bt Corn with	Bt11	Canada	1996
YieldGard®/Syngenta	Bt11	USA	1996
Knockout® Corn/Syngenta	Event176	Argentina	1998
	Event176	Canada	1996
	Event176	EU	2001
	Event176	USA	1995
Bollgard® Cotton/Monsanto	531	Argentina	1198
	531	Australia	1996
	531	China	1997
	531	India	2002
	531	Indonesia	2001
	531	Mexico	1996
	531	South Africa	1997
	531	USA	1995
Bt Cotton/CAAS*	GK	China	1997
Bt Cotton/CAAS*	sGK	China	1999
New Leaf Potato/Monsanto	Russet Burbank, Atlantic, Superior	USA	1995
	Russet Burbank	Canada	1995
	Russet Burbank, Superior	Romania	1999
New Leaf Plus/Monsanto	Russet Burbank 350, 129	USA	1998
New Leaf Y/Monsanto	Russet Burbank, Shepody	USA	1999
	Russet Burbank, Shepody	Canada	1999

*CAAS = Chinese Academy of Agricultural Sciences

Source : Shelton *et al.* (2002).

The global market value of transgenic crops in 2005 was $5.25 billion, representing 15 per cent of the $34.02 billion global crop protection market and 18 per cent of approximately $30 billion global commercial seed market. The accumulated global value for the ten-year period, since transgenic crops were first commercialized in 1996, is estimated at $29.3 billion. The global number and proportion of small farmers from developing countries growing transgenic crops is expected to increase dramatically in the next decade (2006-2015). A similar trend is likely to occur in the less effluent and more agriculturally-based countries of Eastern Europe, such as the Czech Republic, which has recently joined the European Union (EU) and those expected to join later. There were also signs of progress in EU in 2005, as five countries commercialized Bt maize and European Commission approved 17 maize varieties with insect resistance conferred by MON810, making it the first transgenic crop to be approved for planting in all 25 EU countries. Taking all these developments in view, the global area under transgenic crops is projected to increase up to 150 million ha in 2010, with at least 15 million farmers growing transgenic crops in up to 30 countries or even more (James, 2005).

ADOPTION OF TRANSGENIC CROPS IN INDIA

The Government of India's Genetic Engineering Approval Committee (GEAC) approved the commercial cultivation of Bt transgenic cotton in the country in 2002. Two Bt cotton varieties, viz. MECH-162 and MECH-184 were recommended for the Central Zone and MECH-12 for the South Zone. Six Bt cotton varieties, viz. RCH-134, RCH-317, Ankur-651, Ankur-2534, MRC-6301 and MRC-6304 were approved in March 2005 for commercial cultivation under North Indian conditions (Dhaliwal *et al.*, 2006). Bt cotton was planted in an area of 50,000 ha in 2002, which went upto about 1,00,000 ha in 2003, 0.5 million ha in 2004 and 1.3 million ha in 2005. India had by far the largest year-on-year proportional increase, with almost three-fold increase from 2004 and is now one of the 14 biotech mega-countries, i.e., growing 50,000 ha or more of transgenic crops. A total of 20 Bt cotton hybrids were approved for cultivation in 2005. The largest hectarage under Bt cotton was in Maharashtra (0.607 million ha), followed by Andhra Pradesh (0.283 million ha), Gujarat (0.164 million ha), Madhya Pradesh (0.148 million ha), Northern states (0.064 million ha), Karnataka (0.040 million ha) and Tamil Nadu (0.029 million ha). The number of farmers growing Bt cotton hybrids in India has increased from 300,000 small farmers in 2004 to 1 million in 2005. According to a survey conducted by Indian Market, Research Bureau in 2004, cultivation of Bt cotton resulted in five-fold decrease in insecticides, an average of 58 per cent increase in yield and an average of 163 per cent increase in profit for Bt cotton farmers. India is estimated to have enhanced farm income from Bt cotton by $124 million between 2002-2004 (James, 2005).

In a recently concluded meeting, GEAC approved 19 Bt cotton hybrids for commercial cultivation in India. Of these, 13 contain the BG-1 event belonging to nine local companies; 2 contain the BG-II event of MAHYCO; 2 contain the *Cry1Ac* gene of JK seeds; and one contains a fusion of *Cry1Ab* and *CryAc*, a GFM event of Nath seeds. While these 18 varieties are approved for cultivation in Southern cotton growing zone, the remaining one containing the MAHYCO BG-II event is approved for cultivation in India's central cotton growing zone. This brings India's total to 63 new hybrids to be sold in the 2006 season giving the country's farmers more choices on which varieties to cultivate (Anonymous, 2006).

REFERENCES

Anonymous (2006). India approves GM soyabean oil, more GM cotton. *Crop Biotech Update* (June 2, 2006): 5.

Dhaliwal, G.S., Singh, R. and Chhillar, B.S. (2006). *Essential of Agricultural Entomology*, Kalyani Publisher, New Delhi.

James, C. (2005). Global status of commercialized Biotech/GM crops : 2005. ISAAA Briefs No. 34. International Service for the Acquisition of Agri-biotech. Applications. Ithaca, New York.

Khadi, M.B. (2006). Transgenic and pest management, In : *Brainstorming on Integrated Pest Management and Biopesticides* : Summary of Presentations. April, 26, 2006. National Centre for Integrated Pest Management, New Delhi, pp. 16-20.

Shelton, A.M., Zhao, J.Z. and Rough, R.T. (2002). Economic, ecological, food safety, and social consequences of the development of Bt transgenic plants. *Ann. Rev. Entomol.* **47:** 845-881.

Shelton, A.M. and Bellinder, R.R. (2006). Role of biotechnological advances in shaping the future of integrated pest management. In: O. Koul and G.W. Cuperus (Eds.). *Ecological Based Integrated Pest Management.* CAB International, Wallingford, UK, pp. 269-288.

Singh, R. (1997). Biotechnology in Insect-Pest Management In: Harvir Singh and H.R. Rohilla (Ed). Advanced Training Course on Insect Pest Management. March, 1997, ICAR Centre of Advanced Studies, CCSHAU, Hissar, Harayana, India, pp. 75-80.

5

Chemosterility: Approach for Insect Pest Management

S.K. Mandal

Subject Matter Specialist (Plant Protection), Krishi Vigyan Kendra, Saraiya–843 126, Muzaffarpur, Bihar

In the modern context of pest management, control of pests has been aimed to be brought about by newer concepts without disruption of ecosystem. These concepts constitute mainly the integrated pest management and the various methods that bring about reproductive suppression rather than mortality of an already existing population. These advanced methods include the ionizing radiations, chemosterilants, antibiotics, hormones, genetic manipulations, use of attractants, repellents, antifeedants, utilizing the knowledge of radio frequencies, infra radiation, visible ultraviolet radiation and animal behaviour in controlling pest species have also received considerable interest in the recent years.

CHEMOSTERILITY FOR PEST MANAGEMENT

Application of synthetic chemical pesticides is an effective and fastest method of pest control. In this pest control method, most of the natural enemies of insects and useful insects are also killed along with the pests. Synthetic pesticides, being non-biodegradable, accumulate in the environment, causing lots of pollutions and hence disturbing the ecological balance. So, the need of the hour is to adopt eco-friendly techniques to control the insect pests. Chemosterility is one such technique, which is effective and gaining value these days.

Chemosterility is the method in which the chemicals are administered to the insects for depriving them of their ability to reproduce. Any chemical that can inhibit the growth of gonads or interfere with the reproductive capacity of an insect in any of the way such as prevent copulation and production of viable eggs, induce dominant lethal mutations or inhibit development of progeny at any state—egg, larval or pupal—is called a chemosterilant. This idea of controlling insects by sterilizing techniques was first conceived by E.F. Knipling in 1955 on the screw worm fly, *Cochliomyia hominivorax*. This was a sexual sterilization aids in their own species destruction. This technique involves artificial rearing and releasing of sterilized males, which are otherwise physically healthy, to compete with the natural males in the field. After copulating with such sterilized males, the females are deprived of the opportunity to produce fertilized and viable eggs. Consequently, a certain proportion of the expected pest population does not appear and the pest is suppressed in numbers and eventually may even be eradicated.

CLASSIFICATION OF CHEMOSTERILANTS

On the basis of possessing sterilizing activities, these chemicals are categorized as alkylating agents, antimetabolites and miscellaneous compounds.

Alkylating Agents

These compounds were discovered by LaBrecque in 1961 in the USA. They contain several groups of Aziridimyl (aziridines or ethylenimines) and can be classified into three categories: Apholate, Aphomide and Aphoxide, whose examples are described below:

(i) **Apholate:** Chemically this compound is 2,2,4,4,6,6-hexakins (1-aziridinyl) 2,2,4,4,6,6-hexahydro-1,3,5,2,4,6-triazatriphosphorine. It is a white odourless and crystalline substance, partially soluble in water and organic solvents (alcohol, chloroform, acetone), stable in cool dry condition and affected by humidity and high temperature.

(ii) **Aphomide:** Chemically this compound is N, N'-ethylenebis, [P-P-bis(1-aziridinyl)N"-methylphosphinic amide].

(iii) **Aphoxide:** Chemically this compound is tris (1-aziridinyl) phosphine oxide, also called TEPA. It is a crystalline, colourless and odourless substances, extremely soluble in water, and highly soluble in alcohol, ether and acetone, and hygroscopic and unstable. Its other derivatives are METEPA, phosphine oxides, tris (2-methyl, 1-aziridinyl) and thio-TEPA (Phosphine sulphide, tris(1-aziridinyl). The former is a straw coloured liquid completely soluble in water and solvents, and the latter, a white crystalline solid relatively less soluble.

Biological properties: The most important biological property of all alkylating agents is that they are highly reactive and readily combine with a variety of chemicals and biochemicals, (nucleic acids and proteins) replacing a hydrogen atom by an alkyl or a substituted alkyl group such as aminoalkyl, hydroxyalkyl and thioalkyl. The process of replacement of the 'H' atom is called alkylation, which is actually responsible for the sterility in insects. The agents produce permanent sterility in both sexes and are readily absorbed through the skin.

Mechanism of action: These alkylating agents acts on insects in the following ways:

1. **By inhibiting nucleic acid (DNA, RNA) synthesis:** Insects treated with alkylating agents like apholates had their eggs deficient in DNA and RNA leading to complete sterility. It is, however, not known if this effect is direct at the molecular level or indirect through ovarian damage, hormonal imbalance, or any other metabolic abnormality.
2. **By inhibiting gonadial development:** Many alkylating agents like TEPA, chloroethylamine or nitrogen mustard when given in high dose, prevent the development of ovaries in houseflies.

3. **By producing mutagenic effect:** Many alkylating agents produce effects similar to those produced by ionic radiation such as mutation, chromosomal aberration, cancer and sterility. Some chemicals like thio-TEPA induce mutagenic change only in oocytes, and others like tretamine, both in oocytes and sperms. The effects are called "Dominant lethal mutations" which express themselves during the development of zygote or any time thereafter.

Antimetabolites

These are the chemicals that are structurally related to biologically active substances. Due to their structural similarity, the biological systems fail to distinguish them from their own natural substances and commit the mistake of utilizing them in place of the latter. For example, if bacteria are grown in a medium containing 5-fluorouracil, the latter will replace a large percentage of the normal metabolite, uracil, in the RNA of the former. Three groups of antimetabolites are more important viz., purine compounds, pyrimidine compounds and folic acid compounds. They may compete with the normal nucleotides for enzymes and thus prevent orderly incorporation of the nucleotides into the nucleic acids, and they may exert a feedback control on the synthesis of the nucleotides themselves. Use the antimetabolic as well as the normal metabolite is "Amethoprotein". It is a folic acid analogue, which interferes with the formation of vitamin-BC (folic acid) in higher organisms. Though some antimetabolites are known to induce sterility in insects, and they are not as potent as alkylating agents, some produce temporary sterile females are most susceptible when compared with males.

Mechanism of action: They may be doing so in the following three ways:

1. They may compete with the normal nucleotides for enzymes and thus prevent the orderly incorporation of the nucleotides into the nucleic acids.
2. They may exert a feedback control on the synthesis of the nucleotides themselves.
3. They may get directly incorporated into the nucleic acids.

The foregoing effects in some way, not yet clearly understood, sterile the insects. If 5-fluorouracil is fed to insects, it replaces the normal uracil from the RNA of the eggs and this may be the cause of the inviability of the eggs. This effect may be temporary, for when the chemical is depleted in the body, the eggs become viable again.

Miscellaneous Groups of Chemosterilants

Under this group structurally unrelated chemosterilants are kept, which are categorized as follows:

1. **Non-alkylating analogues of aziridinyl compounds:** These chemicals were discovered in an attempt to find safer chemicals. They are HEMPA (hexamethylphosphoramide), a dimethylamine of TEPA and HEMEL (hexamethylemelamine), a dimethylamine analogue of treatamine, both being male housefly sterilants.
2. **Organotin compounds:** Numerous triphenyltin derivatives such as triphenyltin hydroxides, triphenyltin chloride (chlorotriphenyltin), triphenyltin acetate (acetoxy triphenyltin), alkyltriphenyltin, etc. have been found to be reproduction inhibitors of houseflies. They sterile both males and females but the females are more sensitive because they are affected at a much lower concentration than the males.

3. **Insect hormones:** Juvenile hormones produced by the corpora allata of insects produced sterility in female insects. For example, Altosid (T.N-Methoprene) used against mosquitoes, hornfly, stored tobacco pests and *Bombyx mori*.
4. **Plant products:** Due to pollution and resistance problems posed by chemicals, Entomologists of late have been turning their attention to plants as source both for chemosterilants and insecticides. Saxena and Koul (1978) from the Regional Research Laboratory, Jammu-Tawi, have reported on ovicidal effects of several essential oils extracted from plants such as eucalyptus, cassia, anise, clove, coriander, lemon-grass, an aroid, *Acorus*, etc. Alkaloids like monocrotaline and colchicines are already known for their antimitotic effects and therefore, considered as potent chemosterilants.
5. **Plant Growth Regulator:** PGRs are also act same like which also induce the suppression of reproductivity rate of agricultural pests. For example, GA_3 effects to bean mite (*Tetranychus telarius*) fecundity. GA_3 also effects to *Spodoptera litura* reproduction rate.

WAY TO STERILIZATION

- **As sprays and dusts:** Like conventional insecticides, chemosterilants can also be applied as sprays and dusts. This technique will be particularly useful for treating ponds, lakes, reservoirs and open ditches to sterile aquatic insects. Mosquito larvae reared in water containing 10 ppm TEPA were found to develop into sterile adults.
- **As insecticide chemosterilant mixture:** This method will serve dual purpose : first the insect population will be reduced to give quick protection to crops, and second, those few pests left would be sterilized thereby providing a further reduction in insect numbers. Such an approach will be more acceptable to farmers who are more concerned with immediate results.
- **Direct contact with residue:** Instead of applying chemosterilants to the insects, they are applied to the place where the insects like to rest, congregate or breed such as decaying materials, manure piles, etc. Also pretreated screens of glass plates, mesonite sheets, boards etc. could be fixed at places where insects might be expected to come in contact with them. Obviously, for such purpose, quick sterilizing chemicals will be needed and there are chemicals available now which can sterilize within 4-48 hours. The method would be more profitable if the insects could be lured to selected and treated areas by the use of baits and sex attractants. An interesting method was devised for sterilizing houseflies. A small piece of soft leather (Chamics) pad impregnated with METEPA was attached to the abdomen of the female insect, so that during copulation, the male contacted it and got sterilized.
- **Oral ingestion:** Oral administering of chemosterilants is preferred. Chemosterilants may be sprayed on the natural food of insect or mixed with baits and kept in fields. The insects will ingest the chemical along with their food and get sterilized. In case of the houseflies, chemosterilants are mixed in the refuge on which they breed.

REQUIREMENTS OF CHEMOSTERILIZATION TECHNIQUE

This technique is practicable in controlling only those insects which satisfy certain specific requirements. It is successful in islands or among otherwise isolated populations. The species against which it is applicable should be amenable to artificial rearing on a mass scale. A thorough knowledge of the population dynamics and sexual behaviour of the insect is another prerequisite. It is necessary that sterilized insects are not affected in their vigour, longevity, behaviour and mating competitiveness.

PRACTICAL APPLICATION OF CHEMOSTERILANTS

Agricultural Pests

Not much efforts has been made to control agricultural pests by chemosterilants. Most of the work that has been carried out relates to cage experiments performed in various laboratories. Such experiments have been tried with lepidopteran insects like cabbage looper (*Trichoplusia ni*), pink cotton bollworm (*Pectinophora gossypiella*), etc. However, some actual field trials have been attempted in California on Mexican fruitflies (with TEPA) and in Alabama on the cotton boll weevil (with apholate). In spite of the success in the case of the former, chemosterilants were abandoned in favour of irradiation for the sake of safety. The latter could not meet with complete success due to insufficient isolation of the treated areas.

Public Health Pests

The houseflies were the first to be tried with TEPA mixed in food bait and used in one and a half acre refuse dump in Florida in 1962. Nine weekly treatments brought the insect population down. Measurements were done by counting the number of flies landing per minute on a grid $18^2 \times 18^2$ kept at ten scattered localities. However, the population soon increased because the area was not isolated. Attempts to control houseflies have also been made in Italy (1964), Sicily (1967) and Japan (1968). Testse flies were chemosterilized in Zimbabwe and 98 per cent control was achieved complete eradication could not be possible due to reinvasion from the adjoining areas. Mosquitoes (*Aedes, Cules*) were also put to chemosterilant treatments. The first field release of chemosterilized mosquitoes was made in seahorse key islands off the coast of Florida in 1968, another in Kenya (1971) and then in India at villages Bamnauli (1971) and Dhulsiras (1972) near Delhi. Controls were only partial again due to the lack of isolation of these places.

ADVANTAGES AND DISADVANTAGES OF CHEMOSTERILANTS

Advantages: (i) chemosterilants offer the possibility of 99 per cent pest control as against only 90 per cent by synthetic insecticides; (ii) they can be applied in a number of different ways—orally, topically, by injection, as dusts, sprays, fumigants and even for treating insects habitats and resting places; (iii) they offer means of sterilizing wild population in the field itself, which is not possible with irradiation; (iv) it is cheaper to sterilize with chemosterilants than by irradiation; and (v) the problem of resistance in the case of chemosterilants may not be as acute as with insecticides.

Disadvantages: Sterilizing and mutagenic effects of alkylating agents extend to higher animals including man, and some are believed to be carcinogenic and phytotoxic.

CONCLUSION

Although chemosterility technique has been one of the potent ways for controlling insects, till now it has not emerged as a method of extensive usage. This is probably because of high cost and laborious work. Other important factor responsible is the possible hazard of mutagenic, carcinogenic, teratogenic types with these chemicals to higher animals including man. Nowadays there are more than a thousand of chemosterilants, but a very limited of these are on the verge of practical application. Invention of less hazardous chemosterilants offers a great possibility in future.

REFERENCES

Knipling, E.F. (1955). Possibility of insect control or eradication through the use of sexually sterile males. *J. Econ. Entomol.* **48**: 459-462.

Saxena, B.P. and O. Koul (1978). Utilization of essential oils for insect control. *Indian Perfumer.* **22**(3): 139-149.

6

Transgenic Technology: A Modern Weapon in Pest Management

S.K. Mandal[1] and V.B. Jha[2]

Subject Matter Specialist (Plant Protection), Krishi Vigyan Kendra, Saraiya–848 126
S.M.S. (Plant Breeding), Krishi Vigyan Kendra, Sipaya–841 501, Gopalganj, Bihar

The advent of Green Revolution during the 1960s in India has lead to quantum jump in agricultural production through the adoption of modern agricultural technology comprising of high-yielding varieties, use of chemical fertilizers, assured irrigation and improved agronomic practices. However, the dramatically increasing human population in the 21st century needs a 2nd Green Revolution that could only be achieved by the development of resistant varieties through transgenic technologies.

Transgenic crops represent a promising technology that can make a vital contribution of global food, feed and fibre security and also play a vital role in the alleviation of poverty, particularly in the third world. Despite inherent opposition from the environmentalist from around the world, the biotech crop cultivation is growing in double digit figure. According to the International Service for Acquisition of Agri-biotech Applications (ISAAA), biotech plant cultivation has increased by 15 per cent to 67.7 million hectares. About 7 million farmers in 18 countries are growing biotech crop compared to 6 million farmers in 16 countries in 2002. The acreage of biotech crop in developing countries also increased to one-third from one quarter in 2002. According to the ISAAA production in next five years, 10 million farmers in 25 countries will plant biotech crop in 100 million hectares.

Insects have always been a major cause of yield loss in agriculture. They account for 20-30 per cent of crop loss. Several approaches like physical, chemical and biological have been used with varying degree of success to reduce such losses. To increase the net productivity by preventing such losses in an eco-friendly manner is the need of the hour. Natural resistance that provides reasonable protection against insect pests is available in some crops. It means that crop or variety has a specific gene, which confers the plant resistance through production of certain protein (genes are the unit of inheritance). Through breeding the resistance can be transferred to different varieties of the same crop. But such type of transfer of resistance from one specie to another is not possible through breeding programme. With the advancement of technologies in the recent past, it is now possible to identify and isolate a particular gene of one organism, and then to introduce it into another. Such techniques are referred to as genetic engineering. Such gene transfer can be done across the species boundaries. The organism that receives the gene is called transgenic. So it is now possible to introduce the insect resistance gene of one species into a desired plant, which never had such type of resistance gene in its gene pool. Once introduced, the new trait will be inherited generation after generation and can also be transferred to other varieties of that crop by hybridization.

This new technology is seen as an additional tool for the control of insect pests and could offer certain advantages over conventional use of insecticides, such as more effective targetting on specific insects, greater resilience to weather condition, no question of seasonal application, operator exposure and most importantly eco-friendly system.

TRANSGENIC TECHNOLOGY

A transgenic plant is simply a normal plant with one or more additional genes from diverse source. Scientists have bred crops for resistance against insect pests for a long time. The application of transgenic plants through genetic engineering is the latest concept in insect pest management. These transgenic plants produce insecticidal or antifeedant proteins continuously in the plants under field conditions (Meeusen and Warren, 1989). The particular advantages genetic engineering offers are that derived genes can be transferred without co-transfer of undesirable characteristics, and it enables the transfer of genes across species barriers. In this approach, identification of useful genes to be transferred is a limiting factor. Because of this reason, the genes conferring insect resistance transferred by this approach are mostly limited to Bt endotoxin gene and cowpea protease inhibitor (*CpTi*) gene (Table 6.1).

Table 6.1. Transgenic plants carrying genes for insect resistance.

Crop	Foreign gene	Gene transfer method	Target insect pest(s)
Cotton	*Bt*	*Agrobacterium*-mediated	*Helicoverpa zea* (Boddie), *Trichoplusia ni* (Hubner), *Spodoptera exigua* (Hubner)
Maize	*Bt*	Microprojectile bombardment of immature embryos	*Ostrinia nubilalis* (Hubner)
Pea	*α Al-Pv*	*Agrobacterium*-mediated	*Callosobruchus chinensis* (Linnaeus), *Callosobruchus maculatus* (Fabricius)
Populus	*Bt*	Microprojectile bombardment	*Lymantria dispar* (Linnaeus), *Malacosoma disstria* (Hubner)
Potato	*Bt*	*Agrobacterium*-mediated	*Leptinotarsa decemlineata* (Say), *Phthorimaea operculella* (Zeller)
Tobacco	*Bt*	*Agrobacterium*-mediated	*Manduca sexta* (Johannsen)
	CpTi	*Agrobacterium*-mediated	*Heliothis virescens* (Fabricius), *M. sexta*
	CryIA(c)	Plasmid transformation	*H. virescens, H. zea, S. exigua*
	P-lec *α*-ai	*Agrobacterium*-mediated	*H. virescens, Tenebrio molitor* Linnaeus
Tomato	*Bt*	*Agrobacterium*-mediated	*H. zea, Keiferia lycopersicella* (Walsingham), *M. Sexta*
White spruce	*Bt*	Particle acceleration	*Choristoneura fumiferana* (Clemens)

Source : Aronson (1994); Brar and Khush (1994); Panda and Khush (1995)

The efficiency of tissue culture and transformation protocols is one of the most important components for successful generation of transgenic crops (Sharma and Ortiz, 2000).

The major components for the development of transgenic plants are:

1. Development of reliable tissue culture and regeneration systems.
2. Proportion of gene constructs and transformation with suitable vector.

3. Efficient techniques of transformation for introduction of genes into the crop plants.
4. Recovery and multiplication of transgenic plants.
5. Molecular and genetic characterization of transgenic plants for stable and efficient gene expression.
6. Transfer of genes into elite cultivars by conventional breeding methods.
7. Evaluation of transgenic plants for their effectiveness in alleviating the biotic and abiotic stresses without being an environmental hazard.

Before taking up any attempt to produce transgenic plants to counter insect attack, the following requirements and priorities need to be identified:

1. The factors for resistance should be controlled by single genes.
2. Standardization of methods for transfer of such genes can be easily accomplished.
3. Expression of transformed gene should occur in the desired tissues at the appropriate time.
4. The transgenic plant should be safe for consumption.
5. Inheritance of the gene in the successive generations should be very stable.
6. There should be no penalty for yield in terms of other quantitative characters.

The outline of production of a typical insect-resistant transgenic plant is given in Table 6.2.

Table 6.2. Steps involved in production of transgenic plants against insect attack.

Step	Objective	Approach
1.	Field resistance of plant to insect attack	• Purify compounds
2.	Identification of insecticidal compound	• Bioassay vs test species • Bioassay vs pests • Toxicity trials vs mammals
3.	Prove utility and viability of insecticidal compound	• Purify appropriate protein if necessary, sequence and/or raise antibodies • Make cDNA genome library probe with oligonucleotide or via expression system
4.	Obtain encoding gene (or gene encoding biosynthetic enzyme)	• Engineer into suitable vector • Transform test species
5.	Make suitable construct	• Assay plants for expression
6.	Produce transgenic test plants expressing foreign compound	• Bioassay with suitable pests
7.	Demonstrate insect resistance in transgenic plants	• File patent(s) – Apply for approval
8.	Obtain regulatory body's approval for delicate release	• Field trials – Choose crop plant species
9.	Produce transgenic crop plants expressing foreign compound	• Bioassay vs major pests • Plant breeding to transfer foreign gene to commercial lines
10.	Test in national coordinated trials, bulk up seed	• Obtain regulatory body's approval for commercial use

Source : Gatehouse *et al.* (1991)

Peculiar achievements have been made in the production, characterization and field evaluation of transgenic plants. Both *Agrobacterium*-mediated gene transfer and direct DNA transfer methods have been used to produce transgenic plants with new genetic properties. Genes conferring resistance to insects have been inserted into crop plants such as maize, rice, wheat, sorghum, sugarcane, cotton, potato, tobacco, broccoli, cabbage, chickpea, pigeonpea, cowpea, groundnut, tomato, brinjal and soya bean.

Development of insect-resistant transgenic crop cultivars has focused on two distinct approaches:

1. Integration of microbial (bacterial) genes encoding for production of toxic proteins, especially from *Bacillus thuringiensis* (Bt).
2. Integration of plant genes encoding for production of enzyme inhibitors and sugar-binding lectins.

Both the approaches were pioneered in the mid-1980s and thus have developed in parallel. However, the first approach, based in particular on integration of δ-endotoxin genes derived from various subspecies of Bt, has undoubtedly received more attention and thus enjoyed greater progress. To date all commercially available insect-resistant transgenic cultivars express semi-active Bt toxins, whereas cultivars expressing insecticidal plant proteins are not currently available outside research institutions (Brar and Khush, 2001, Koul and Dhaliwal, 2004a, Bernal *et al.*, 2004, Sharma *et al.*, 2004).

MICROBIAL-DERIVED GENES

Bt δ-endotoxins

Bacillus thuringiensis Berliner (Bt) is a Gram-positive entomocidal spore-forming bacterium. Sporulating cells of Bt produce an inclusion body called parasporal crystal. The crystal proteins (cry) are grouped into different classes based on sequence homology and insecticidal specificity (Crickmore *et al.*, 1998). Bt toxins are thus highly specific; some strains are specific to Lepidoptera, while others are specific to Diptera and Coleoptera.

Crystal proteins and their genes have been classified based on their structure and activities spectrum into CryI (Lepidoptera species), CryII (Lepidoptera and Diptera), CryIII (Coleoptera) and CryIV (Diptera). Each of these are functionally divided into several toxin types. This heterogeneity in toxin production is responsible for some of the diversity in activity spectrum among strains. There is also a high degree of heterogeneity among binding sites in some species, suggesting that some sites may bind a single toxin whereas others may bind two or more toxins (McGaughey and Whalon, 1992).

Bt toxins are stomach poisons. They act by disrupting the membrane of the cell in the insect gut, disturbing the ionic balance and thereby paralyzing the gut. *B. thuringiensis* was recognized as a pathogen of certain caterpillars as early as 1901. Of all the biological agents that have been evaluated as pest control products, *B. thuringiensis* has been the most successful insect pathogen.

Bacillus products are of two types The most important are those produced from a strain *Kurstaki* called Btk effective against Lepidopteran pests of forest trees, vegetables, cotton and ornamental plants. These products have been sold since the early 1960s and account for bulk of Bt applications. In the late 1970s, another strain called Bti (i for *israelensis*) was identified. The products of this strain are effective against Dipteran insects. It is estimated that more than 2000 tonnes of these preparations have been applied worldwide with no undesirable effects. However, these formulations are expensive to produce, are of very poor persistence under field conditions and there is a need of carefully timed repeated treatments. In order to overcome these limitations, it was recognized that expression of *B. thuringiensis* toxin genes in transgenic plants might increase their utility in insect control.

Table 6.3. Transgenic crops carrying Bt genes for insect resistance.

Transgenic crop	Transgene(s)	Target insect pest(s)
Alfalfa	*Cry1C* (Synthetic gene)	*Spodoptera littoralis* (Biosduval)
Broccoli	*Cry1C*	*Plutella xylostella* (Linnaeus), *Trichoplusia ni* (Hubner), *Pieris rapae* (Linnaeus)
Canola	*Cry1A(c)*	*Helicoverpa zea* (Boddie), *Spodoptera exigua* (Hubner)
Chickpea	*Cry1A(c)*	*Helicoverpa armigera* (Hubner)
Cotton	*Cry1A(b), Cry1A(c)*	*H. armigera, H. zea, Heliothis virescens* (Fabricius), *Pectinophora gossypiella* (Saunders), *S. exigua, T. ni*
Groundnut	*Cry1A(c)*	*Elasmopalpus lignosellus* (Zeller)
Egg plant	*Cry(III)b*	*Leptinotarsa decemlineata* (Say)
Maize	*Cry1A(b),Cry1A(c), Cry9C*	*Ostrinia nubilalis* (Hubner), *Chilo partellus* (Swinhoe), *Busseola fusca* Fuller, *H. zea*
Poplar	*Cry1A(a)*	*Lymantria dispar* (Linnaeus)
Potato	*Cry1A(b),Cry1A(b)6, CryIIIA, CryIIIB, Cry1Ac9, CryV-Bt*	*Phthorimaea operculella* (Zeller), *L. decemlineata* *P. operculella*
Rice	*Cry1A(b), Cry1A(c), CryII(a)*	*Chilo suppressalis* (Walker), *Cnaphalcrosis medinalis* (Guenee), *Scirpophaga incertulas* (Walker)
Sorghum	*Cry1A(c)*	*C. partellus*
Soybean	*Cry1A(c)*	*H. virescens, H. zea*
Sugarcane	*Cry1A(b)*	*Diatraea saccharalis* (Fabricius)
Tobacco	*Cry1A(b), Cry1A(c)*	*H. virescens, M. sexta*
	CryIIa5	*H. armigera*
	Cry1Aa2	*H. zea*
Tomato	*Cry1A(c)*	*M. sexta*
	Bt(k)	*H. zea, M. sexta, Keifera lycopersicella* (Walsingham)

Source : Hilder and Boulter (1999), Brar and Khush (2001), Sharma *et al.* (2004)

Bt endotoxins are attractive candidates for insect-resistant crop development using transgenic technology because (i) they have an established safety record, (ii) they act rapidly and are completely biodegradable and proven safe to human beings and non-target organisms and the environment, and (iii) the endotoxins are the products of single genes and are highly effective against the larvae of Lepidoptera, a major group of destructive insect pests. Therefore, the genes encoding the δ-endotoxins were cloned in the early 1980s, and expression of modified toxin genes in transgenic tobacco and tomato provided the first examples of genetically engineered insect resistance in plants (Barton *et al.*, 1987, Fischhoff *et al.*, 1987, Vaeck *et al.*, 1987). Since then, Bt gene have been introduced into and expressed in a wide range of crop species (Table 6.3).

Although, the early transgenic plants containing Bt genes demonstrated some enhanced resistance to target pests, it was soon apparent that expression levels from the native bacterial genes were too low to provide adequate protection from pest species in the field. Hence, *Cry* genes were modified for plant preferred codon usage, i.e., removal of polyadenylation signals, a higher GC content and removal of secondary structure and mRNA destabilization motifs while keeping the amino acid sequence unaltered. For the expression of *Cry* genes in plant, suitable plant expression promoters such as the constitutive promoters, which allow expression in all plant parts at all times, e.g., cauliflower mosaic virus (CaMV), and ubiquitin promoter have been used. On the contrary, for a more controlled expression of the transgene in only certain specified plant parts, a tissue specific promoter, e.g., phosphoenol pyruvate carboxylase (PEPC) promoter, which expresses only in green tissue and maize, pollen-specific promoter have been used. For maternal transmission of the transgene, chloroplast transformation vectors have been developed such that the expression of the transgene is confined and does not spread through pollen. As a result, majority Bt genes now in use for plant genetic engineering have been substantially modified and essentially rebuilt, with expression levels in transgenics much higher (100x) than was obtainable using native Bt genes (Hilder and Boulter 1999, Kaur and Gujar, 2004).

(i) Cotton

Considerable progress has been made in developing cotton cultivars with Bt genes for resistance. The first generation Bt cotton varieties were developed by Monsanto and their seed partners to express the *Cry1Ac* protein. Worldwide, several different *Cry1Ac* events are grown commercially. In the USA and some other countries, Bollgard® event MON531 was used by seed company breeding programmes to develop the first commercial Bt cotton varieties. In Australia and other countries, Ingard® event MON757 was used. The tobacco budworm, *Heliothis virescens* (Fabricius) and pink bollworm, *Pectinophora gossypiella* (Saunders) are very well controlled by Bollgard® cotton, whereas the cotton bollworm, *Helicoverpa zea* (Boddie) is controlled satisfactorily except during the bloom stage when feeding on reproductive parts of flowers, which tend to have a lower concentration of *Cry* protein than other plant parts. The introduction of Bt cotton in 1996 and 1997 was very timely since *P. gossypiella* and *H. virescens* in the USA and *H. armigera* in China and Australia had become resistant to many of the conventional insecticides and Bt cotton offered a reasonable solution to the problem (Benedict and Ring, 2004).

Cotton cultivar Coker 312, transformed with the *Cry1A(c)* gene (having 0.1% toxin protein), has shown high levels of resistance to cabbage looper, *Trichoplusia ni* (Hubner), tobacco caterpillar, *Spodoptera exigua* (Hubner) and cotton bollworms, *H. zea* and *H. virescens.* In transgenic cotton, bollworm damage was reduced to 2.3 per cent in flowers and 1.1 per cent in bolls compared to 23 per cent damage in flowers and 12 per cent damage in the commercial cultivar, Coker 312. The cottonseed yield was 1950 kg ha^{-1} in Coker 312 as compared to 1460 kg in Bt cotton (Benedict *et al.*, 1996).

Bt cotton, along with event MON531 or MON757 expresses *Cry1A(c)* proteins in leaves at about 12.6 µg/g of fresh leaf tissue and 9.9 µg/g of fresh seed tissue (http://agrobios.com/). Very little *Cry* protein can be detected in the pollen, and none in the nectar. Studies of Bt protein expression show that the concentration varies as much as 3-to 5-fold during the season in the plants' growth, before flower buds develop, so that by the time the plants are blooming, and fruits have began to mature in mid to late season, *Cry1Ac* concentration in the leaves starts to decline (Greenplate, 1999). It has been shown that the level of control of *H. armigera*, *H. zea* and *S. frugiperda* is positively correlated with the concentration of *Cry1Ac* (Greenplate *et al.*, 2003). Recently, both laboratory and field trials have demonstrated that NC33B expressing *Cry1Ac* protein, and GK-2 expressing *Cry1A* protein, are highly resistant to the Asian corn borer, *Ostrinia furanacalis* (Guenee), an important component of Lepidopteran pest complex of cotton in China. Both the varieties have provided season-long protection against this pest in the field, and in addition, higher larval survival was observed on GK-2 than NC33B in assays with late season tissues (He *et al.*, 2006).

The second-generation Bt crops, developed with stacked *cry* genes, have been shown to improve the level of control and broaden the spectrum of caterpillar pests controlled. China has developed the new events, GK (expressing fused Cry1Ab and Cry1Ac proteins) and SGK (expressing the fused Cry1Ac-Cry1Ab and stacked CpTi proteins). Monsanto has created Bollgard® II by inserting a synthetic gene from *B. thuringiensis*, Cry2Ab, expressing Cry2Ab protein, into the Bollgard® cotton variety, DP50B, already expressing Cy1Ac through event MON531, creating the new stacked gene event 15985 (Benedict and Ring, 2004). Bollgard® II is much more effective than Bollgard® at controlling all the caterpillar pests of cotton worldwide, especially *H. armigera, S. exigua, S. frugiperda* and *H. zea* (Greenplate *et al.*, 2003). Laboratory and field studies have shown that Bollgard® II increased mortality of *H. zea* from 84.2 to 92.2 per cent, *S. fugiperda* from 16.1 to 100 per cent, *S. exigua* from 50.1 to 94.9 per cent and *T. ni* from 1.2 to 97.4 per cent. This high level of pest control is due, in part, to the very high concentration of Cry2Ab protein in Bollgard® II (Benedict and Ring, 2004). In October 2004, Dow has received full registration by US Environmental Protection Agency (EPA) for Wide Strike®, a cotton product that expresses both Cry1F and Cry1C (Shelton and Bellinder, 2006).

(ii) Maize

The first generation Bt proteins engineered into maize were Cry1Ab (from *B. thuringiensis*) and Cry9c (from *B. thuringiensis* subsp. *Tolworthi*) and these were highly effective against European corn borer, *Ostrinia nubilalis* (Hubner). The dominant Cry1Ab events are Monsanto's YieldGard® maize, event MON810; Syngenta's BT11 sold under the NK-brand as YieldGard®, and Syngenta's event 176, known as Knockout®. The level of expression of Bt proteins tends to be lower in fresh leaves of 176 than MON810 and BT11 (Benedict and Ring, 2004). Recently, Cry1F has been introduced in maize (Dow's Herculex 1®), and maize with a Cry3Bb gene or a binary toxin genetic system for control of the maize rootworm (*Diabrotica*) complex is being developed (Shelton and Bellinder, 2006).

Maize plants transformed with Bt genes have also been found to be effective against the spotted stem borer, *Chilo partellus* (Swinhoe) and the maize stalk borer, *Busseola fusca* Fuller in Southern Africa. Maize plants with *Cry1Ab* gene are also resistant to the sugarcane borers, *Diatraea grandiosella* Dyar and *D. saccharalis* (Fabricius). Transgenic tropical maize inbred lines with *Cry1Ab* or *Cry1Ac* genes with resistance to corn earworm, *H. zea;* fall armyworm, *S. frugiperda*; Southwestern corn borer, *Diabrotica undecimpunctata howardi* Barber and Western maize rootworm, *D. virgifera* LeConte have also been developed (Sharma *et al.*, 2004).

The level of target insect suppression provided by current Bt maize events varies according to the pest species, transformation event, hybrid background, and environmental factors that influence concentration of Bt protein in plant tissues that need protection (Williams *et al.*, 1998). The Cry1Ab protein levels in events BT11 and MON810 in fresh leaf tissue range from 7 to 30 µg/g and in grain from 0.2 to 5 µg/g, with the event MON810 at the lower end and event BT11 in the upper end of the range (http://www.agbios.com/). The level of *Cry1Ab* protein expression in leaves of event 176 is much lower, ranging from 0.6 to 3.02 µg/g, with no detectable Cry protein in grain. In leaf tissue, *Cry1Ab* provides good to excellent control of *O. nubilalis, O. furnacalis* (Guenee) and *D. grandiosella,* but is less effective for *H. zea* and *S. frugiperda,* and has little or no effect on *Agrotis ipsilon* (Hufnagel) (Shelton *et al.*, 2002). As suggested by the protein concentration data above, the level of target pest suppression is less for kernels in ears than for green leaf tissue. Therefore, increasing the expression of Cry1Ab in leaves and kernels above 30 μg/g fresh tissue could significantly improve control of the target lepidopteran pests (Benedict and Ring, 2004).

(iii) Rice

Various Chinese Universities and research institutes, in cooperation with the International Rice Research Institute (IRRI) and universities in many countries have transformed rice by inserting the *Cry1Ab* gene alone and a fused *Cry1Ab/Cry1Ac* into conventional rice varieties and hybrids. Initial field and laboratory tests of these Bt lines against lepidopteran pests have shown good season long control (Brar and Khush, 2001; Benedict and Ring, 2004). Rice plants having 0.05 per cent toxin of the total soluble leaf protein have shown high levels of resistance to the striped stem borer, *Chilo suppressalis* Walker and rice leaf folder, *Cnaphalocrocis medinalis* Guenee (Fujimoto *et al*., 1993). Tu *et al*. (2000) produced transgenic indica rice CMS restorer line Mingui 63 (T51-1) expressing a Bt fusion gene derived from *Cry1Ab* and *Cry1Ac* under the control or rice actin 1 promoter. The level of Bt fusion protein detected was 20 μg/mg soluble protein. Field testing of transgenic rice lines showed high protection against *Scirpophaga incertulas* (Walker) and *C. medinalis*. The percentage of plants with white heads was significantly lower on the Bt Shanyou 63 (11%) as compared to control Shanyou 63 (44%) plants. Similarly, transgenic plants showed no damage by *C. medinalis* as compared to 47.9 per cent damaged plants in non-transgenic rice.

Scented varieties of rice (Basmati 370 and M7) have been transformed with *CryII(a)* and are resistant to *S. incertulas* and *C. medinalis* (Mqbool *et al*., 1998). Khanna and Raina (2002) developed Bt-transgenics of elite indica rice breeding lines (IR-64, Pusa Basmati-1 and Karnal Local) with synthetic *Cry1Ac* gene. Selected Bt-lines of IR-64 and Pusa Basmati-1, having Bt-titres of 0.1 per cent (of total soluble protein) showed 100 per cent mortality of *S. incertulas* larvae within 4 days of infection in cut-stems as well as at the vegetative stage in whole plant assays. Hussain *et al*. (2002) expressed *Cry1Ab* in Basmati rice under the control of three promoters (PEPC, ubiquitin, and pollen-specific promoter derived from *Bp10* gene of *Brassica napus* in *pGEM 4Z*). Toxin protein expression was 0.05 per cent of the total protein under the control of PEPC promoter alone or in combination with the pollen-specific promoter. However, the expression was nearly 0.15 per cent of the total protein under the control of ubiquitin promoter, suggesting that a specific promoter can be used to limit the expression of *Cry1Ab* gene in desired plant parts.

(iv) Potato

A modified *Cry3A* gene has been expressed in potato plants with resistance to Colorado potato beetle, *Leptinotarsa decemlineata* (Say) (Perlak *et al*., 1993). Transgenic potato plants containing the *Cry1A(b)* gene (Bt 884) and a truncated gene *Cry1A(b)6* resulted in less damage to the leaves by the potato tuber moth, *Phthorimaea operculella* Zeller (Jansens *et al*., 1995). However, the size of the leaf tunnels increased over time in plants containing only the Bt 884 gene, while there was no increase in tunnel length in those containing *Cry1A(b)6*. The latter also resulted in 100 per cent mortality of the insects in tubers stored up to six months. Several other Bt potato lines showing complete mortality of larvae of *P. operculella* have also been developed (Sharma *et al.*, 2004). However, the marketing of Bt potatoes was stopped in 2001 due to opposition from several food producers not to use Bt potatoes in their products (Shelton *et al.*, 2002; Benedict and Ring, 2004; Shelton and Bellinder, 2006).

2. Cholesterol Oxidase

Screening of filtrates from microbial fermentation often reveals presence of certain proteins that attack some insects. The protein present in two streptomyces culture filtrates, which show acute toxicity to boll weevil larvae, has turned out to be a cholesterol oxidase (CO). Cholesterol is required for integrity and normal function of virtually all cellular membranes, therefore any interference in its availability leads to toxicity. Histological studies on boll weevil larvae fed on diet containing CO show cellular attenuation accompanied by local cytolysis of midgut epithelial cells. CO catalyses the oxidation of cholesterol already incorporated into the membrane to produce ketosteroids and hydrogen peroxide. Midgut epithelium is the primary target for CO and thus it interferes with the process of digestion and absorption (Purcell *et al.*, 1993). Although newly emerged adult boll weevils exposed to CO do not exhibit acute toxicity. They display severe oostatic effects (i.e., reduction in fecundity and oviposition) due to poor development of ovaries in adult females. LC_{50} of first and second instar larvae of boll weevil is comparable to that of δ-endotoxin. Enzymatic activity of CO has been detected in protoplast of transgenic tobacco with native CO gene (Estruch *et al.*, 1997).

3. Isopentenyl Transferase

The gene for isopentenyl transferase (ipt), a key enzyme in the cytokinin biosynthesis, was isolated from *Agrobacterium tumefaciens*. Expression of *ipt* in transgenic tobacco and tomato plants by a wound inducible promoter has resulted in a decrease in leaf consumption by tobacco hornworm (*M. sexta*) and reduced survival of the peach potato aphid, *Myzus persicae* (Schuler *et al.*, 1998).

4. Vegetative Insecticidal Proteins (VIPs) from Bacillus sp.

In search of novel insecticides proteins, it is discovered that culture supernatant fluid collected during vegetative (i.e., long phase) growth of *Bacillus* sp. is a rich source of insecticidal substances. Supernatant fluid of *Bacillus cereus* isolates possesses acute bioactivity against Northern and Western corn rootworm. Here insecticidal properties are associated with a binary system, whose two proteins are termed vip1 and vip2. Similarly, supernatant fluid of certain *B. thuringiensis* cultures has potent insecticidal properties against black cutworm and fall armyworm. Potency-wise it has been shown that VIPs are comparable to δ-endotoxin. The cloning and characterization of the gene encoding this activity has yielded a novel insecticidal protein vip3A, which is active against lepidopteran insects, particularly black cutworm, fall armyworm and beet armyworm. This protein results in complete lysis of gut epithelial cells, particularly columnar cells (Estruch *et al.*, 1997). Transgenic maize with *vip3A* gene is found to be resistant to underground pests like western and northern corn rootworm (Estruch *et al*, 1997 and Koziel *et al.*, 1998).

PLANT-DERIVED GENES

The search for insect control proteins alternative to Bt has concentrated, with a few exceptions, on those derived from plants. In the course of their co-evolution over a period of about 300 million years, plant have evolved some very effective measures to counter the predatory insects. Indeed, plant genes expressing chemical products responsible for plant resistance to insects are probably available against all pest species because they presumably evolved in response to herbivory by insects. Many different classes of plant proteins have been shown to have some toxic or anti-metabolic effect on insects in artificial diet or *in vitro* investigations and have been proposed as targets for crop genetic engineering. On major advantage of using plant genes as a source of crop resistance is their broad spectrum of activity across several orders, including sap-sucking insect pests, such as Hemiptera, in addition to Lepidoptera, Coleoptera and Orthoptera. Thus far, a number of economically important crop plants, such as oilseeds, potato, rice, sugarcane, tobacco, among others have been genetically transformed to express various genes of plant origin (Table 6.4). In many of these cases, significant effects of the plant genes were evident on target pest mortality rates and/or developmental and reproductive parameters (Gatehouse and Gatehouse, 1998; Hilder and Boulter, 1999; Legaspiltal, 2004).

Largely, plant transformation involving plant genes has focused on: (i) protease inhibitors, which disrupt amino acid metabolism; (ii) α-amylase inhibitors, which target carbohydrate metabolism; (iii) lectins, which cause agglutination and cell aggregation; and (iv) enzymes such as chitinase, which target insect exoskeleton. The varied modes of action and levels of specificity of these gene products increase the potential target-pest range of transgenic cultivars and allow the possibility of combining (pyramiding) genes that are active at various target sites within an insect pest or against various pests. However, levels of protection provided by genes of plant origin are typically lower than those provided by genes expressing Bt toxins. Frequently, effects on targets insects are sublethal, including reduction in feeding, weight gain, developmental rates and fecundity. Even then, transgenic cultivars with sublethal or chronic effects on target pests may be more attractive components of IPM strategies than cultivars with acute toxic effects because they are more likely to be compatible or act synergistically with other biopesticides strategies (Gatehouse and Gatehouse, 1998, 2000; Bernal *et al*., 2004; Legaspi *et al*., 2004).

Table 6.4. Transgenic crops expressing insecticidal plant genes.

Transgenic crop	Transgene(s)	Origin of transgene	Target insect pest(s)
Alfalfa	*Manduca E-1*	*Manduca sexta* (Johanssen)	Thrips
Apple	*CpTi*	Cowpea	*Cydia pomonella* (Linnaeus)
Adzuki bean	*α-AI*	Bean α-amylase I	*Callosobruchus chinensis* (Linnaeus), *C. maculatus* (Fabricius)
Cabbage	Modified *CpTi (sck)*	Cowpea	*Pieris rapae* (Linnaeus), *Helicoverpa armigera* (Hubner)
Cotton	*CpTi Manduca serpin*	Cowpea	*H. armigera, M. sexta*
Lettuce	*Pot PI-II*	Potato	*Teleogryllus commodus* (Walker)
Maize	*Avidin*	Chicken avidin	Storage insect pests
	WGA	Wheat	*Ostrinia nubilalis* (Hubner), *Diabrotica* spp.
Mustard	*WGA*	Wheat	*Lipaphis erysimi* (Kaltenbach)
	OC-I	Rice	Coleoptera
	C-II	Soybean	Lepidoptera, Diptera

Table 6.4. Contd.

Transgenic crop	Transgene(s)	Origin of transgene	Target insect pest(s)
Pea	*α-AI*	Bean α-amylase I	*Bruchus pisorum* (Linnaeus) *Zabrotes subfasciatus* (Boheman)
Poplar	*OC-I*	Rice	*Chrysomela tremulae*
	C-II	Soybean	Lepidoptera
	Pot PI-II	Potato	*Plagiodera versicolora* (Laicharting)
Potato	*CpTi*	Cowpea	*Lacanobia oleracea* (Linnaeus)
	GNA	Snowdrop	*L. oleracea, Myzus persicae* (Sulzer), *Aulacorthum solani* (Kaltenbach)
	BCH	Bean chitinase	M. persicae, A. solani
	OC-I	Rice	*Leptinotarsa decemlineata* (Say)
	Kti3, C-II, PI-IV	Soybean	*Spodoptera littoralis* (Boisduval)
Rice	*CpTi*	Cowpea	*Chilo suppressalis* (Walker) *Sesamia inferens* (Walker)
	Pot PI-II	Potato	*C. suppressalis, S. inferens*
	Pin 2	Potato	*C. suppressalis*
	mwti 1b	Synthetic	*C. suppressalis*
	GNA	Snowdrop	*Nilaparvata lugens* (Stal), *Nephotettix virescens* (Distant)
	SBTI	Soybean	*N. lugens*
	CC	Corn	*Sitophilus zeamais* Motschulsky
Strawberry	*CpTi*	Cowpea	*Otiorhynchus sulcatus* (Fabricius)
Sugarcane	*GNA*	Snowdrop	*Eoruma loftini* (Dyar), *Diatraea saccharalis* (Fabricius), *Antitrogus consanguineus*
	Pot PI-II	Potato	*A. consanguineus*
Sweet potato	*CpTi*	Cowpea	*Euscepes postfasciatus* (Fairmaire),
Tobacco	*CpTi*	Cowpea	*Heliothis virescens* (Fabricius) *H. armigera, H. Zea, M. sexta*
	Pot PI-II	Potato	*M. sexta, Chrysodeixis eriosoma* (Doubleday)
	Tom PI-II	Tomato	*M. sexta*
	α-AI	*Bean α-amylase*	*Agrotis ipsilon* (Hufnagel), *Tenebrio molitor* Linnaeus
	GNA	Snowdrop	*H. virescens, M. persicae*
	p-Lec	Pea	*H. virescens*
	GTPI	Giant taro	*H. armigera*
	NaPI	*Nicotiana alata*	*Helicoverpa punctigera* (Wallengren)
	Kti3, C-II, PI-IV	Soybean	*S. littoralis*
	SBTI	Soybean	*H. armigera, H. Zea, H. virescens,* *S. littoralis, M. sexta*
	Manduca serpin	*M. sexta*	*Bemisia tabaci* (Gennadius)
	SpTi	Sweet potato	*S. litura*
	Insect chitinase gene	Insect chitinase	*H. virescens*
	AaIT	Scorpion (*Androctonus australis*)	*H. armigera*
Tomato	*CpTi*	Cowpea	*L. oleracea*
	GNA	Snowdrop	*L. oleracea*
	Pot PI-I	Potato	*H. armigera, T. commodus*
	Pot PI-II	Potato	*H. armigera, T. commodus*
Wheat	*CMe*	Barley	*Sitotroga cerealella* (Olivier)
	GNA	Snowdrop	*Sitobion avenae* (Fabricius)

α-AI = α-Amylase inhibitor, *BCH* = Bean chitinase, *C-II* = Double headed serine protease inhibitor from soybean, *CC* = Corn cystatin, *CpTi* = Cowpea trypsin inhibitor, *GNA* = Snowdrop lectin, *GTPI* = Giant taro (*Alocasia macrorrhiza*) proteinase inhibitor, *NaPI* = *Nicotiana alata* protease inhibitor, *OC-I* = Oryzacystatin I, *p-lec* = Pea lectin, *Pot PI-I* = Potato proteinase inhibitor I, *Pot PI-II* = Potato proteinase inhibitor II, *SBTI* = Soybean Kunitz-type trypsin inhibitor, *SpTi* = Sweet potato trypsin, *Tom PI-II* = Tomato proteinase inhibitor II, *WGA* = Wheat germ agglutinin.
Source: Gatehouse and Gatehouse (1998), Hilder and Boulter (1999), Brar and Khush (2001), Sharma *et al.* (2004).

1. Protease Inhibitors

The presence of antimetabolic proteins, which interfere with the processes of digestion in insects, is a strategy for defence that plants have used extensively. Proteins can occur constitutively in tissues that are particularly vulnerable to attack, such as seeds, or mechanical wounding in tissues attacked by chewing insects can induce them. Analysis of the effects of dietary protease inhibitors has shown that these are detrimental to the growth and development of insects from a variety of genera including *Helicoverpa, Spodoptera, Diabrotica* and *Tribolium.* Plants can now be transformed with protease inhibitor gene with strong promoters to express the inhibitor proteins in relatively high levels at specific times (Brar and Khush, 2001).

Several classes of protease inhibitors corresponding to different types of insect gut proteases have been characterized. Many insects, particularly members of Lepidoptera, depend on serine proteases (trypsin, chymotrypsin and elastase like endoproteases) as their primary protein digestive enzymes, and genes encoding members of various serine protease inhibitor families have been cloned and introduced into transgenic plants. Insects also produce their own serine protease inhibitors for the regulation of their digestive protease. It has been suggested that these could be turned against the insects by expressing them in transgenic plants. Other pests rely on cystein proteases (thiol proteases) as their primary digestive proteases, which act on papain and cathepsin. These have been targeted with cysteine protease inhibitors, which have been shown to exert chronic effects on important pests such as corn root worm, *Diabrotica* spp. against which there are no effective Bts. Some unusual protease inhibitors, e.g., metallo and aspartate inhibitors, inhibiting carboxypeptidase and cathepsin also occur (Hilder and Boulter, 1999, Kaur and Gujaar, 2004).

(i) Serine protease inhibitors

The first gene of plant origin to be used in transgenic crop protection was that isolated from cowpea encoding a double-headed trypsin inhibitor (*CpTi*) and transferred in tobacco (Hilder *et al.*, 1987). A simple construct was prepared in which a full length coding sequence derived from a cDNA alone was placed under the control of the constitutively expressed cauliflower mosaic virus (CaMV) 35S promoter. Transformants were screened for *CpTi* expression, which showed that many of the resulting plants expressed *CpTi* at levels greater than 0.1 per cent of total soluble protein. Subsequent experience has shown that this is generally the case for expression of genes of plant origin encoding defensive proteins in transgenic plants, in contrast to the very low level of expression observed for unmodified toxin genes of bacterial origin. *CpTi* is a small polypeptide of about 80 amino acids, homologous sequences are encoded by a moderately repetitive gene family in the cowpea genome. This protein is considered to be particularly suitable candidate for transfer to other species through genetic engineering because: (i) it is an effective antimetabolite against a wide range of field and storage pests belonging to Lepidoptera, Coleoptera and Orthoptera; (ii) has no deleterious effects on mammals; and (iii) also has a small polypeptide of about 80 amino acids (Brar and Khush, 2001).

Biossays against *H. virescens* caterpillars showed that transgenics expressing *CpTi* at the highest levels (about 1% of total soluble protein) caused increased mortality, reduced growth and reduced plant damage. The antimetabolic effects of *CpTi* expressed in transgenic tobacco have also been observed with other lepidopteran pests including *H. zea, S. littoralis* and *M. sexta.*

Subsequent trials carried out in California showed that expression of *CpTi* in tobacco afforded significant protection against *H. zea* in the field (Hilder and Boulter, 1999). Following on from the study using tobacco as a model system, the gene encoding *CpTi* has been expressed in a range of different crops (Table 6.3). For example, constitutive expression of *CpTi* in rice conferred significantly enhanced levels of resistance towards two species of rice stem borer, viz. *Chillo suppressalis* (Walker) and *Sesamia inferens* (Walker) in the field. Furthermore, the trials with *CpTi* transgenic strawberry plants suggested that these plants were highly resistant to the vine weevil, *Otiorhynchus sulcatus* Fabricius (Hilder and Boulter, 1999; Gatehouse and Gatehouse, 2000).

Other serine protease inhibitor encoding genes have also been used as protective agents for crops. The tomato inhibitor II gene, PI-II (which encodes a trypsin inhibitor with some chymotrypsin inhibitor activity), when expressed in tobacco was also shown to confer insect resistance. The bioassays showed that the decrease in larval weight in insects reared on transgenic plants was roughly proportional to the level of PI-II being expressed. Several of the transgenic plants were shown to contain inhibitor levels over 200 μg/g tissue, these levels are within the range that is routinely induced by wounding leaves of either tomato or potato plants. However, tobacco plants expressing tomato inhibitor I at similar levels had no deleterious effect upon larval development, showing the specificity of interactions between inhibitors and insect species (Gatehouse and Gatehouse, 1998, 2000). Three soyabean protease inhibitor genes (KTi3, C-II and PI-IV), when transformed into tobacco and potato, showed variable expression among different plants. The level of resistance to *S. littoralis* was particularly high in tobacco, where many plants caused complete mortality of the larvae, while in potatoes the larval mortality was much less achieved, but resulted in a reduction of larval weight gain by 50 per cent (Marcherti *et al.*, 2000).

Trypsin inhibitor was isolated from cowpea, purified and tested in artificial diets against a wide range of both field and storage insect pests including members of Lepidoptera (*Helicoverpa* sp., *Spodoptera* sp.), Coleoptera (*Diabrotica* sp., *Anthonomus* sp.) and Orthoptera (*Locusta* sp.). Purified *CpTi*, when added to artificial diets, had detrimental effects on larval development, only when applied at concentrations found in resistant seeds. In all the cases, it was found to be an effective insecticide and also possessing no mammalian toxicity. *CpTi* becomes a ideal candidate for genetic transformation. It has been transferred into *Nicotiana tabaccum* and has been found to impact resistance against *Heliothis virescens* Fabricius, *H. zea*, *Spodoptera littoralis* Boisduval and *Manduca sexta* Johannsen (Table 6.5).

Table 6.5. Effect of CpTi-expressing transgenic tobacco on different insect species. (Figures are % of controls)

Insect	Survival	Insect biomass	Plant damage
Helicoverpa zea Boddie in laboratory	60	64	34
H. zea in field	41	50	48
Heliothis virescens Fabricius	65	27	40
Spodoptera littoralis Boisduval	69	63	49
Manduca sexta Johannsen	55	18	44

Source: Hilder and Boulter (1999)

(ii) Cysteine protease inhibitors

Cysteine proteases are used by plants for protein mobilization and by animals for intracellular lysozomal protein digestion and protein inhibitors of cysteine proteases (cystatins) are widely distributed throughout all living organisms to regulate these endogenous proteases, even if they are usually present in small amount. The genes encoding cysteine protease inhibitors have been suggested for use in transgenic plants for the control of coleopteran pests. Although, there have been several studies carried out demonstrating *in vitro* inhibition of insect digestive proteases by cysteine protease inhibitors, with a few example of their deleterious effects against insects when incorporated into artificial diets (Gatehouse and Gatehouse, 1998, 2000), as yet there are few published reports describing their insecticidal effects *in planta*.

A gene encoding a rice cystein proteinase inhibitor oryzacystatin has been expressed constitutively in transgenic poplar trees, conferring resistance to the coleopteran pest, *Chrysomela tremulae* (Lepte *et al*., 1995). Irice *et al*. (1996) introduced corn cystatin (CC) into protoplasts of rice and assayed the cystatin activity of the transgenic rice plants against a crude midgut proteinase fraction from *Sitophilus zeamais* Motschulsky. The result showed that 50 per cent of the midgut protease activity in *S. zeamais* was inhibited by 2 μg and completely inhibited by 5 μg of transgenic seed protein fraction, whereas untransformed rice seeds had no significant effect. Cysteine protease inhibitor from potato (multicystatin, PMC) has been expressed in chrysanthemum at a level of 0.13 per cent of total protein. However, no correlation between reduction in oviposition rate by the western flower thrips, *Frankliniella occidentalis* (Pergande) and PMC expression could be established, which may be due to the relatively low expression level of PMC in chrysanthemum (Seetharam *et al*., 2002).

2. Alpha-amylase Inhibitors

Plants produce inhibitors of insect gut alpha-amylases, which are required for the digestion of starch, a major energy source, particularly for the weevils. The alpha-amylase inhibitors produced by plants have different types of structure and different mode of action and target specificity, and hence can be used for insect control in transgenic plants (Franco *et al*., 2002). Transgenic tobacco plants expressing amylase inhibitors from wheat (Wheat α-amylase inhibitor WAAI) increase the mortality of lepidopteran larvae by 30-40 per cent. Similarly, transgenic pea seeds expressing alpha-amylase inhibitor derived from common bean (BAAI) were found to exhibit increased resistance against bruchid beetles, *Callosobruchus* spp. and pea weevil, *Bruchus pisorum* Linnaeus (Sharma *et al*., 2004).

Enhanced levels of resistance to the bruchids have been observed in transgenic adzuki beans expressing the alpha-amylase inhibitor of common bean. While even low levels of the amylase inhibitor were sufficient to provide resistance to the adzuki bean weevil, higher levels of the protein make the seeds resistant to the cowpea weevil, *C. maculatus* (Fabricus) and pea weevil, *C. chinensis* (Linnaeus) as well (Ishimoto *et al*., 1999). Two alpha-amylase inhibitors, α-AI-1 and α-AI-2, from the common bean inserted into pea, are effective in protecting peas against the pea weevil, *B. pisorum* under field conditions. α-AI-1 provided complete protection from *B. pisorum* by inhibiting alpha-amylase by 80 per cent, while α-AI-2 inhibits the enzyme by 40 per cent. α-AI-1 result in larval mortality, whereas α-AI-2 delays the maturation of the larvae (Morton *et al*., 2000).

3. Lectins

Lectins are plant-derived proteins that bind to oligo-and polysaccharides, and cause agglutination and cell aggregation. Lectins have been isolated and characterized from a wide variety of plants such as pea, rice, wheat, snowdrop, castor, soybean, mungbean, garlic, sweet potato, tobacco, chickpea and groundnut (Sharma *et al.*, 2000). A number of plant lectins exhibiting insecticidal characteristics are being evaluated as alternatives to Bt δ-endotoxins. The precise mode of action of insecticidal lectins is unknown. However, binding to specific carbohydrates and agglutination (fusion by adhesive substances) in the insect midgut has been clearly demonstrated. Specifically, lectins may interfere with development and structural integrity of the midgut peritrophic membrane, bind to glycosated targets in the insect midgut, thereby inhibiting nutrient absorption or cell disruption in the midgut, or bind or block the peritrophic membrane protecting the insect midgut surface (Legaspi *et al.*, 2004).

Lectins from wheat (wheat germ agglutinin, WGA) and the snowdrop plant (*Galanthus nivalis* agglutinin, GNA) are inhibitory to the sap-sucking homopteran pests such as aphids, leaf hoppers and plant hoppers, which feed on the phloem exudates and against which there are no known *Cry* proteins (Table 6.6). In addition, lectins have also exhibited inhibitory activity against several lepidopteran and coleopteran pests. A gene encoding the pea lectin (*p-lec*) has been expressed in transgenic tobacco, and the plants expressing pea lectin upto 1 per cent of total protein reduced the larval biomass of *Heliothis virescens* (Fabricius) and leaf damage (Boulter *et al.*, 1990). Transgenic tobacco plants containing both *CpTi* and P-Lec were obtained

Table 6.6. Transgenic plants developed with insect resistance genes.

Gene product	Target insect	Transformed plants
Transgenic plants with crystal protein genes of B. thuringiensis		
Cry1Ab	Lepidoptera	Apple, cotton, maize, rice, potato, tomato, tobacco, white clover
CryIH	Lepidoptera	Maize
Cry3A	Coleoptera	Tobacco, potato, egg plant
Cry9C	Lepidoptera	Maize
Transgenic plants with protease inhibitor gene		
C-11(Soybean serine protease inhibitor)	Coleoptera, Lepidoptera	Potato, tobacco, poplar, oilseed, rape
CpTi (Cowpea trypsin inhibitor)	Coleoptera, Lepidoptera	Apple, lettuce, potato, rice, sunflower, sweet potato, tobacco, tomato
OC-I (Rice-cysteine protease inhibitor)	Coleoptera, Homoptera	Poplar, tobacco, oilseed, rape
Pot PI-II (Potato protease inhibitor II)	Lepidoptera, Orthoptera	Birch, tobacco, lettuce, rice
Transgenic plant with lectin gene		
Snowdrop lectin (GNA)	Homoptera, Lepidoptera	Tobacco, potato, rice, oilseed, rape, sweet potato, sugarcane, sunflower, tomato
Pea lectin (*p-lec*)	Homoptera, Lepidoptera	Potato, tobacco
Wheat germ agglutinin (WGA)	Lepidoptera, Coleoptera	Maize
Rice lectin	Lepidoptera, Coleoptera	Maize
Transgenic plants with α-amylase inhibitors		
α-amylase inhibitor of common bean (αAI-Pv)	Coleoptera	Pea, tobacco
α-amylase inhibitor from cereals (WMAI-I)	Lepidoptera	Tobacco

Source : Das (2005)

through hybridization of two primary transformed lines. The plants expressing two insecticidal proteins reduced the insect biomass by 90 per cent as compared to 50 per cent reduction in plants expressing either *CpTi* or P-Lec. This study demonstrated that not only the products of lectin genes could enhance resistance to insect attack in transgenic plants, but also showed that additive protective effects could be obtained from different plant-derived insect-resistant genes (Gatehouse and Gatehouse, 2000).

Survival of the brown plant hopper, *Nilaparvata lugens* (Stal) decreased to 60 per cent in transgenic rice expressing GNA protein (2% of total protein), under the control of phloem tissue-specific sucrose synthase promoter which direct the expression of the gene in the phloem of leaves, stems, petioles and roots for protection from the phloem feeding pests (Rao *et al.*, 1998). Tingjuangjuin *et al.* (2000) developed GNA-transgenic lines of two elite rice varieties, 'Khao Dawk Mali 105' (KDLM105) and 'Supanburi 60' (SP60) known to resist transformation. Over 30 transgenic lines were produced with highest level of expression of 0.25 per cent of the total soluble protein. Constitutive and phloem-specific expression of GNA reduced nymphal survival of *N. lugens* by 37 and 42 per cent, respectively. In plants where GNA expression is tissue-specific (phloem and epidermal layer) or constitutive, survival of green leaf hopper, *Nephotettix virescens* (Distant) has been reduced by 23 and 53 per cent, respectively (Foissac *et al.*, 2000).

Fitches *et al.* (2004) reported that a recombinant SF11/GNA fusion protein, comprising of snowdrop lectin (*Galanthus nivalis* agglutinin, GNA) fused to an insecticidal spider venom neurotoxin (*Segestria florentina* 1, SF11), killed 100 per cent first instar tomato moth, *Lacanobia oleracea* (Linnaeus), larvae after 6 days of feeding, exhibiting a dose-dependent toxicity, and resulted in reduction in growth when fed to larvae in the later stadia. More recently, significant effects on the mortality of *N. lugens* were observed, with 100 per cent of the insects fed on the SF11/GNA fusion protein dead after 7 day. The survival of the aphid, *Myzus persicae* (Sulzer) was also reduced when fed on the SF11/GNA fusion protein; after 14 days, only 49 per cent of the aphids that were fed on the fusion protein were still alive compared with about 90 per cent of the aphids fed on the control diet or on diet containing GNA only. The SF11/GNA fusion protein also showed the development of *M. persicae*, and the reproductive capacity of the aphids fed on the fusion protein was severely reduced (Down *et al.*, 2006).

Wheat germ agglutinin (WGA) is antimetabolic, antifeedant and insecticidal to the mustard aphid, *Lipaphis erysimi* (Kaltenbach). Bioassays using leaf discs showed that feeding on transgenics induced high mortality and significantly reduced fecundity of aphids (Kanrar *et al.*, 2002). Thus, plant lectins have shown biological activity against a wide range of insects. However, consideration should be given with regard to their deployment in transgenic plants because of their known toxicity to mammals and humans (Sharma *et al.*, 2004).

4. Enzymes

Transgenic expression of various enzymes has been proposed as a crop protection strategy (Gatehouse and Gatehouse, 2000). The most obvious candidate is chitinase, which is an important structural component of insects. Expression of an insect chitinase in transgenic tobacco enhances resistance to some lepidopterans (Ding *et al.*, 1998). Similar marginal protective effects have been observed from expression of bean chitinase (BCH) in transgenic tobacco. Transgenic potato plants expressing a gene encoding BCH were found to reduce fecundity of the glasshouse potato aphid, *Aulacorthum solani* (Kaltenbach), though this reduction was not statistically significant. However, nymphs produced on these BCH expressing plants were significantly smaller compared to those on control, non-transformed plants (Gatehouse and Gatehouse, 1998).

A bacterial endochitinase from *Serratia marceasens* has been shown to act synergistically with Bt toxin against *Spodoptera littoralis* (Boisduval) larvae, but not yet in transgenic plants. Cholesterol oxidase secreted from *Streptomyces* has been shown to be acutely toxic to larvae of *Anthonomus grandis* Boheman. *A. streptomyces* cholesterol oxidase gene has been expressed in protoplasts, but insecticidal activity in transgenics has yet to be reported. Lipoxygenase has been shown to have some insecticidal effects. Although introduced lipoxygenases have been expressed in transgenic plants, enhanced resistance of such plants to insects has not been reported (Hilder and Boulter, 1999; Sharma *et al.*, 2004).

5. Other Insecticidal Protein from Plants

Tryptophan decarboxylase (TDC) from periwinkle expressed in tobacco is known to reduce the reproduction of white fly (*Bemisia tabaci*) by upto 97 per cent. TDC converts tryptophan to indole alkaloid tryptamine. Tryptamine and tryptophan derived alkaloids in plants may act as antioviposition and antifeedant agents or as inhibitors of larval and pupal development (Schuler *et al.*, 1998).

OTHER NOVEL GENES

The genes derived from other sources such as chicken, scorpion and spider are also being screened for their insecticidal potential. Avidin, a glycoprotein found in chicken egg white, sequesters the vitamin biotin. Kramer *et al.* (2000) produced transgenic maize containing *avidin gene*. The *avidin* at 100 ppm is toxic to and prevents development of insects that damage grains during storage. Insect-specific neurotoxin AaIT from the venom of the scorpion, *Androctonus australis* (Linnaeus), in tobacco has shown insecticidal activity against *H. armigera* larvae (up to 100% mortality after 6 days). Transgenic plants of tobacco have been obtained containing an insecticidal spider peptide gene, and some of these plants have exhibited resistance to *H. armigera*. The role of neurotoxins from insects and spiders need to be studied in greater detail before they are deployed in other organisms and plants because of their possible toxicity to mammals (Sharma *et al.*, 2004).

PYRAMIDING GENES

The effectiveness and durability of resistance in transgenic crops is likely to be greater if they are engineered with multi-gene, multi-mechanistic resistance. In order to increase the protective efficacy, spectrum of activity and durability of resistance, it is envisaged that 'packages' of different genes will be introduced into crops. The components of such packages should each act on a different target within the insect, thus mimicking the multi-mechanistic resistance which occurs in nature. Protease inhibitors should be particularly valuable in this respect since, apart from their inherent insecticidal effects, they would protect other introduced gene products from premature digestion in the insect gut.

The first report of such an approach has been the introduction of both *CpTi* and pea lectin (P-lec) into tobacco. These plants were obtained by cross-breeding plants derived from the two primary transformed lines. Although, the insecticidal effects of the two genes were not synergistic, they were additive, with insect biomass on the double expressions being reduced by nearly 90 per cent compared to those from control plants and 50 per cent to those from plants expressing either CpTi or P-lec alone. Leaf damage was also least double expressing plants (Boulter *et al.*, 1990).

It has been found that low levels of serine protease inhibitors enhanced the insecticidal activity of *Cry1A*, Cry3 and Cry4 against their respective target insects. A synthetic sequence incoding the 29 amino acid squash (*Cucurbita maxima*) trypsin inhibitor (CMTI) was fused to a truncated Cry1Ac, Bt gene and introduced into tobacco. Larval growth reduction bioassays against *H. virescens* should about 6-fold increase in Bt activity in leaf extracts from transgenics expressing the CMTI-Bt fusion protein compared with transgenics expressing the truncated Bt alone.

In order for pyramiding to be successful, it is essential that the mechanisms of resistance are compatible. Introduction of an serine protease inhibitor and the snowdrop lectin into the same transgenic tobacco does not result in additivity of protective effect. Since one of the effects of GNA is believed to be antifeedant, it presumably reduces the intake level of serine proteinase inhibitor to below the threshold for the latter to have any effect (Hilder and Boulter, 1999). A combination of *Cry1Aa* and *Cry1Ac* has a synergestic effect, while a combination of *Cry1Aa* and *Cry1Ab* produces an antagonistic effect against the gypsy moth, *Lymantria dispar* (Linnaeus). However, in tobacco plants, a combination of *Cry1Ab* and *Cry1Ac* genes has shown to be effective for controlling the lepidopteran insects (Sharma *et al.*, 2004).

Several attempts have been made to pyramid insect-resistance genes with differing modes of action. Zhao *et al.* (1998) generated transgenic tobacco plants expressing the Bt gene *Cry1A* and cowpea trypsin inhibitor gene *CpTi*. The insecticidal activity of transgenic plants expressing both genes was significantly higher against *H. armigera* than that of the plants expressing the Bt gene alone. Moreover, even after 11 generations of selection of *H. armigera* for resistance to *Cry1A* and *CpTi*, there was significantly less resistance to the insecticidal proteins in the larvae selected on double transgenic than those selected on Bt plants. Transgenic rice plants expressing GNA showed enhanced resistance to *N. lugens*, while the plants expressing *Cry1Ac* were protected against *C. suppressalis*. Expression of both transgenes gave protection against both the pests but did not increase protection against either pest significantly over the levels observed in plants containing a single insecticidal transgene (Loc *et al.*, 2002). Similarly, the activity of Bt in transgenic plants against insect pests can be enhanced by the addition of genes expressing serine protease inhibitors and tannic acid (Koul and Dhaliwal, 2004b, Rajyashri and Mohan, 2004).

There have also been successful attempts of pyramiding different Bt toxin genes in order to increase the durability and range of resistance. Transgenic potato plants containing *Cry1Ab*, gene Bt 884 and *Cry1Ab6* resulted in increased resistance to potato tuber moth, *Phthorimaea operculella* (Zeller). While the effectiveness of Bt 884 gene alone reduced with time, there was no decrease in resistance in transgenics carrying both genes (Jansens *et al.*, 1995). Rice plants of nine different varieties have been transformed with plant codon optimized versions of two synthetic *Cry1Ab* and *Cry1Ac* coding sequences from Bt (Cheng *et al.*, 1998). Accumulations of high levels (upto 3% of soluble protein) of *Cry1Ab* and *Cry1Ac* proteins have been detected in transgenic plants. The transgenic plants are highly toxic to *C. suppressalis* and *S. incertulas*, with mortalities of 97-100 per cent within 5 days of infestation. Similarly, it has been suggested that *Cry1Ac* and *Cry1F* can be expressed together in transgenic cotton plants, decreasing the EC_{50} to *H. armigera* by 26 times (Chakrabarti *et al.*, 1998).

Multiple genes conferring resistance to insects and viruses can also be engineered into crops very effectively. Transgenic potato carrying the Bt gene *CryV* (toxic to Lepidoptera and coleopteran) and a potato potyvirus y coat protein gene PVy, were most resistant to the potato tuber moth, *Phthorimaea operculella* (Zeller) and potato virus y infection rather than non-transgenic lines. The double transgenics were as lethal to *P. operculella* as a *CryV* transgenic line (Li *et al.*, 1999).

Gene pyramiding can be used to counteract the biotypes of insect pests. Recently, attempts have been made to pyramid two insect resistance genes, GM2 and GM4, which provide resistance to rice gall midge, *Orseolia oryzae* (Wood Mason), biotypes 1, 2 and 4, and 1, 2, 3 and 4, respectively. The resultant rice plants would be effective against the most prevalent biotypes of gall midge in India (Rajyashri and Mohan, 2004).

RESISTANCE IN INSECT PEST TO TRANSGENIC CROPS

The insects are exposed to the toxin proteins in transgenic crops throughout the feeding cycle and season, and as a result, the insect populations are under continuous selection pressure (Gould, 1998). As the toxins are expressed in all parts of the plant, there is a serious threat of rapid development of resistance in insect pests. Toxin production may also decrease over the crop growing season. Low doses of the toxins eliminate the most sensitive individuals of a populations, leaving a population in which resistance can develop much faster (Sharma *et al.*, 2004).

Mechanisms of Resistance

Two main strategies have been employed to genetically engineer plants with resistance to insect herbivores. By far, the most important is the insertion of insecticidal protein genes from *B. thuringiensis* and the second is the use of plant-derived compounds found in high levels in many plants and believed to be for defuse against attack. These compounds generally induce starvation or reduce nutrient uptake and growth in herbivores by inhibiting the active sites of digestive enzymes or binding to midgut membranes (Chilcutt and Johnson, 2004).

1. Bt endotoxins

It is already known that insects can develop resistance to Bt endotoxin and hence there is no reason not to expect this to happen in case of Bt transgenic crops. Numerous Bt-resistant populations were developed in the laboratory, however, it is only the diamond back moth, *Plutella xylostella* (Linnaeus), which has developed resistance under natural conditions following sprays of Bt products. High level of resistance to *Cry1A* toxins have been found in populations from the Philippines, Hawaii, Florida and Asia, with cross-resistance to *Cry1F* at least in Hawaii (Rousch, 1997, Kumar, 2002). Many possible mechanisms can be envisaged to explain the development of resistance in insects to Bt transgenic plants (Frutos *et al.*, 1999, Chilcutt and Johnson, 2004).

(i) Lower level of toxin activation

The most important mechanism is the slower activation of the protoxins by gut proteinases. The protoelytic processing of protoxins into toxins is essential for the insecticidal activity of Bt formulations. It is also required by some transgenic plants expressing combinations of Bt toxins as fusion proteins or toxins extending beyond the protease activation site. This mechanism of resistance was first described in the Indian meal moth, *Plodia interpunctella* (Hubner). The Bt resistant strains of *P. interpunctella* displayed slower processing and activation of *Cry1* protoxins. The slower protoxin activation of the resistant strain led to a reduced quantity of toxin resulting in a survival advantage. Some of the resistant strains lacked a major trypsin like gut enzyme, and absence of the gut protease and resistance to the toxin were genetically linked (Oppert *et al.*, 1997). Similarly, a Bt-resistant strain of *Heliothis virescens* (Fabricius) exhibited a slower activation of the protoxin as well as a faster rate of degradation of the toxin by midgut extracts (Forcada *et al.*, 1996).

(ii) Reduced binding to midgut membrane

The second most frequently observed mechanism of resistance among insect pests is the modification of the receptor site. This mechanism has been studied extensively on a large range of insect species and toxins, and most of these studies pointed out a change in the level of affinity of the receptor for the toxin or to a decreased number of receptor site. Several studies have provided convincing evidence that in *P. interpunctella* and *P. xylostella*, resistance is partly due to changes in the binding affinity of receptors on the brush border membrane of the insect midgut. The first study of mechanism of resistance in a field-evolved Bt resistant strain was made by Ferre *et al.* (1991) using a colony of *P. xylostella* from the Phillippines. In the resistant strain, a loss of specific binding to *Cry1Ab* suggested that the resistance was due to a change in the *Cry1Ab* binding site. A field collected strain of *P. xylostella* from Florida selected for resistance to HD-1 was highly resistant to *Cry1Aa*, *Cry1Ab* and *Cry1Ac*, but not to *Cry1B*, *Cry1C* and *Cry1D*. Analysis of binding characteristics of biotinylated toxins on brush border membrane vesicles showed a loss of binding to *Cry1Ab* in the resistant strain, whereas binding properties of *Cry1B* and *Cry1C* remained unchanged (Tang *et al.*, 1996). Recent studies implicate the binding site receptor to be aminopeptidase-N in *P. xylostella* and *Lymantria dispar* (Linnaeus), and N-acetylgalactosamine in *Manduca sexta* Johannsen (Chilcut and Johanson 2004).

(iii) Other mechanism

This mechanism corresponds to a still unknown or perhaps to several unknown phenomena and was reported for the first time by Gould *et al.* (1992) in *H. virescens*. Although *H. virescens* population was selected for resistance to *Cry1Ac*, it also showed resistance to *Cry1Ab* and *Cry2Aa*. The resistance strain also grew faster than the susceptible one in the presence of *Cry1Aa*, *Cry1B* and *Cry1C*. An insect species selected for resistance to only one toxin can become resistant to other toxins sharing few sequence homologies for this broad resistance mechanism. Liu *et al.* (2000) found that there was no difference in binding of *Cry1C* between susceptible and resistant *P. xylostella*. They also found that *Cry1C* protoxin was significantly less toxic to resistant larvae than *Cry1C* toxin, although there was no difference in mortality of susceptibles between the protoxin and toxin. These results suggest that reduced conversion of protoxin to toxin is a minor mechanism of resistance to *Cry1C*, and neither reduced conversion nor reduced binding are major resistance mechanisms. It can, therefore, only be assumed that at least part of these reported cases of resistance correspond to altered post-binding events when resistant cannot be related to reduced binding or to lower level of toxin activitation (Kumar, 2002).

2. Plant-derived Genes

The plants contain a variety of compounds that negatively impact many species of insects by interfering with their growth and development, even causing death. For example, lectins, protease inhibitors and alpha-amylase inhibitors generally starve insect herbivores by inhibiting the active sites of digestion, and hence these inhibitors are part of the natural defense mechanisms of plants that apparently evolved against herbivores (Koul and Dhaliwal, 2004a).

(i) Protease inhibitors

Increasing evidence suggests that insects can adapt to the ingestion of protease inhibitors. This should have been expected because insect species that feed on plants naturally exhibiting a specific inhibitor are generally not susceptible to that particular inhibitor. Jongsma *et al.* (1995) found that if lepidopteran larvae fed on transgenic plants expressing the potato inhibitor II protease, only 18 per cent of the protease activity of their gut abstracts was sensitive to inhibition by inhibitor II from the transgenic plants. In contrast, 78 per cent of gut abstracts was sensitive in larvae fed on control plants. They showed that increased activity of a protease insensitive to inhibitor II compensated for the decline in inhibitor II sensitivity. The insects become resistant to protease inhibitors by induction of protease inhibitor-insensitive protease activity that allows the insects to digest proteins independent of the level of protease inhibitors.

The lepidopteran and coleopteran species can overexpress the existing gut proteases or induce the production of new types that are insensitive to the introduced protease inhibitors (Jouanin *et al.*, 1998). The ability of some insect species to compensate for protease inhibition by switching to an alternative proteolytic activity or overproducing the existing protease, limits the application of their use in pest control. This also leads to the possibility of susceptible pests evolving resistance to protease inhibitors either by activating genes to produce alternate proteases or by activating genes to increase the levels of existing protease and overcome protease inhibitors by saturation (Jouanin *et al.*, 1998). The other possibilities for evolution of resistance include the increased use of minor proteases already present in the insect, improving the affinity of target proteases or by inactivating the protease inhibitors before they can reach their target enzymes (Chilcutt and Johnson, 2004).

(ii) Alpha-amylase inhibitors

Two alpha-amylase inhibitors, α-AI-1 and α-AI-2 from the common bean were inserted into pea, which lacks alpha-amylase activity. Morton *et al.* (2000) found that α-AI-1 was effective against the pea weevil, *Bruchus pisorum* (Linnaeus) whose larvae mature in the developing peas. It was also found that α-AI-2 only weakly inhibited *B. pisorum* alpha-amylase at pH 5.5-6.5, but substantially inhibited at pH 4.0-4.5, it is likely that *B. pisorum* possesses multiple alpha-amylase enzymes and not all are inhibited by α-AI-2. This suggests several mechanisms by which pea weevils could evolve resistance to alpha-amylase inhibitors. One way could be to activate genes that produce alternative alpha-amylase enzymes, and other could be to increase levels of currently expressed alpha-amylases so as to overcome the inhibitors. Many insects have a gut pH level that inhibits the optimal activity of amylase inhibitors. Additionally, insects produce a variety of amylases, any of which may not be affected by different amylase inhibitors.

(iii) Lectins

In insects, lectins are known to bind to glycoproteins, which constitute a major part of the insect digestive tract membranes. It is possible that the insect gut contains specific ligand-bonding molecules that are targets of plant defensive lectins. Lectins probably affect insects by binding to the chitin in the peritrophic membrane or binding to the midgut epithelium (Zhu-Salsman *et al.* 1998). Probably because of this binding, nutrient uptake is reduced, causing reduced growth and/or starvation. Insect resistance to lectins could evolve through changes in the target receptor sites, such as a change in their structure so that lectins could not bind. Another possibility is that the changes in the peritrophic membrane could prevent lectins from reaching the midgut epithelium. Lastly, resistance to lectins could possibly evolve through increase in their number of target receptor sites as well as increase or activation of enzymes that could inactive the lectins so that they would not be able to bind (Chilcutt and Johnson, 2004).

Resistant Management Strategies

Insects have demonstrated a high capacity to develop resistance to a wide array of chemical insecticides. Resistance against microbial spray of δ-endotoxin has already been observed. There has been some concern that use of δ-endotoxin genes for insect control in transgenic plants may lead to rapid development of resistance (McGaughey and Whalon, 1992 and Alstad and Andow, 1995). Such concern is, however, based largely on computer models.

To minimize the development of resistance to transgenic monitoring of pest population should be combined with a variety of strategies:

(i) ***High dose strategy*:** The strategy to produce transgenic plants synthesizing very high level of insecticidal proteins has been found to be highly effective in preventing or delaying development of resistant variety of insects. This strategy may help eliminate heterozygotes (i.e., carrying only one resistant allele). Although there is a chance of decline in titre level of the protein in successive generations or with ageing, there is no report of decline of titre level below the important desired level in Bt-cotton and maize in the trials before commercial release (Alstad and Andow, 1996). Again it is apprehended that high dose of toxin in transgenic plant product may be harmful for human health. But Bt gene has been found to be species specific and is thus harmless to other non target insects, vertebrates including users and the environment.

(ii) ***Refuse strategy*:** It is adopted to provide a source of susceptible insects for mating with the resistant ones, of any. This will prevent the fixation of resistance gene. Refuges can naturally occur when a certain percentage of an agricultural acreage is planted with non-transgenic plants. Refuges may also be established by using mixture of transgenic and non-transgenic seeds, or by growing border rows of susceptible variety, or by growing susceptible variety a little away from the crop field for highly mobile insect species (Shelton *et al.*, 2000). However, the point of concern regarding this strategy is the size of the refuge. In India, for Bt cotton cultivation, experts have suggested a separate refuge of 30 per cent of the land cultivation.

(iii) ***Multigene strategy***: Transgenic plants producing additional insecticidal proteins with either different modes of action, different targets or both can solve the problem. Recent discovery of a number of new insecticidal strains of *B. thuringiensis* and new insecticidal proteins from such strains may provide more active components to be used for the purpose. New insecticidal components from sources other than *B. thuringiensis* may also be helpful for this. Several options should be available that include introduction of multiple genes with unique mode of action into the same plant, or alternatively, independent plants each with a unique insecticidal protein could be used in rotation.

(iv) ***Integrated tactics***: The most important approach to manage resistance to transgenic plants is to diversify pest management techniques as well as resistance management techniques. This is particularly so in case of multiple crops and multiple pests prevalent in tropical and sub-tropical countries like India. A toxin may be highly active against a given pest, but moderately active against another pest also present in the same crop or in the same area, and for which the dose delivered by the transgenic plant does not correspond to a high dose. In that case, the second pest will be exposed to a moderate dose and may develop resistance. As there is no single answer or strategy to delay resistance, only a logical combination of various tactics will provide sustainability (Kumar, 2002, Benedict and Ring, 2004, Chilcutt and Johnson, 2004).

Hokkanen and Wearing (1995) suggested multiple tactics for Bt resistance management in oilseed crops, which could be applied to any transgenic crop: (i) provide refuge for susceptible, (ii) do not use pesticides that kill susceptibles, (iii) do not use pesticides that kill natural enemies, (iv) enhance natural control through crop rotation or tillage, and rotate between susceptible and resistant crop genotypes over large areas. The high-dose refuge strategy, and especially the association of this approach with combinations of toxins is advocated by scientists, seed companies and administrators as the best approach of controlling evolution of resistance. This composite strategy allows flexibility in terms of spatial organization of refuge areas and choice of toxins that will help adaptation and implementation in various agronomic systems (Gould, 1998, Kumar, 2002, Matten *et al.*, 2004).

When Bt cotton, Bt corn and Bt potato were first registered by Environmental Protection Agency (EPA), Monsanto Company (developer of these products) was required to file insecticide-resistance management (IRM) plans in order to delay the development of target pest resistance to the specific *Cry* proteins present in their Bt crops (Table 6.7). Since then, the EPA has made IRM plans a requirement for registration of all transgenic crops. These plans act as useful proactive tools to avoid/delay the development of resistance in transgenic crops. As may be seen (Table 6.7), IRM plans for Bt cotton require a gradual phasing in all of the known tactics to avoid/delay the development of pests that are resistant to Bt crops. The stacking of multiple *Cry* proteins and increasing *Cry* protein dosage in Bt crops is a very positive step to improving the IRM. Pyramided traits are expected to be most effective when combined with refugees and a fully-implemented integrated pest management programme that uses all available pest suppression methods in a truly holistic crop management approach (Benedict and Ring, 2004).

Table 6.7. Current Monsanto deployment strategy to minimize the development of pests with insecticide resistance to Cry1Ac in Bt cotton.

Short term (at commercialization 1996-97)
- High expression of crystal proteins to control insects heterozygous for resistance alleles.
- Refuges of non-Bt cotton to produce susceptible insects (4 unsprayed non-Bt acres per 100 Bt acres or 25 sprayed non-Bt 100 Bt acres).
- Agronomic practices that minimize insect exposure to crystal protein in Bt cotton.
- Integrated pest management strategies to increase beneficial insect effectiveness, and reduce conventional insecticide use.
- Monitoring target insect populations for susceptibility to crystal protein.
- Report on Bt cotton performance, especially any 'failures' investigate the cause.

Medium term (2-5 years after commercialization)
- Continue above plus.
- Combine 2 insecticidal genes within the plant with different target sites/modes of action for the target lepidopterans.

Long term (> 5 years after commercialization)
- Continue all short and medium term strategies plus.
- Incorporate natural plant resistance to insect traits into Bt cotton.
- Incorporate other novel insecticidal genes into cotton that control the insects.

Source: Benedict and Ring (2004).

INTEGRATION WITH OTHER APPROACHES

Transgenic plants are compatible with all other approaches for management of insect pests. In a recent study, the effects of snowdrop lectin (*Galanthus nivalis* agglutinin: GNA) on the biology of larval development, pupation and adult emergence of 2-spot ladybird beetle, *Adalia bipunctata* L., fed on aphids, *Myzus persicae* (Sulzer) loaded with GNA, were studied in detail (Down *et al.*, 2000). There was no evidence for deleterious effects of the transgene-encoded resistance factor on the predator, indicating thereby that transgenic plants can be integrated with biocontrol agents.

Generally, transgenic plants do not need any insecticide application against target pests. An initial assessment of the benefits from transgenic crops has been reviewed by James (1997). Virus resistant tobacco in China increased leaf yield by 5 to 7 per cent and resulted in savings of 2 to 3 insecticide applications. Insect resistant Bt cotton in USA in 1996 resulted in insecticide savings as high as US $ 140-280 per ha, with 70 per cent of Bt cotton planted in 1996 requiring no insecticides to control the target pests. This resulted in an average yield increase of 7 per cent with a net benefit of US $ 60 million for the 0.73 million ha of Bt cotton. Borer resistant Bt corn in USA produced an average yield increase of 9 per cent both in 1996 and 1997. Benefits from the use of Bt maize in the USA were estimated at US $ 19 million on 0.28 million ha in 1996, and US $ 190 million on 2.8 million ha in 1997. Similarly, herbicide tolerant soybean in USA in 1996 resulted in 10 to 40 per cent less herbicide requirements.

ADVANTAGES AND DISADVANTAGES OF TRANSGENIC CROPS

Advantages of Transgenic Crops

Transgenic crop protection has provided new avenues for management of insect pests and it holds great advantages to be included in IPM system.

- The low toxicity of proteinase inhibitors and Bt δ-endotoxin as compared to conventional insecticides would reduce the selection pressure and may slow down the development of resistance.
- Since all plant parts including growing points would remain covered with toxins, dependence on weather for efficacy of the sprays would be eliminated.
- Since toxins will always be there, there will be no need of continuous monitoring of pests.
- Transgenic plants would also provide protection to those plant parts, which are difficult to be treated with pesticides. Thus, transgenics may prove useful for controlling bollworms and borers that are difficult to control by means of insecticides.
- The cost of application in the form of equipment and labour will be nil or negative.
- The development cost is only a fraction of the cost of development of a conventional pesticides.
- There would be no problem of contamination in the form of drift and groundwater contamination.
- Insecticidal activity would be restricted to those insects, which actually attack the plants. Transgenic plants would be safe to non-target species and human beings.
- Transgenic plants will have inbuilt resistance to various insects replacing some of the current pesticide usage with protection, which is intrinsically biodegradable, thus reducing the use of chemical insecticides and minimizing the problem of environmental pollution.

Disadvantages of Transgenic Crops

- Reducing the biodiversity: The transgenic technique reduces the biodiversity of plants. This mostly happens in the cross-pollinated crops. Pollen grains of cross-pollinated transgenic crops reach the stigma of neighbouring fields, wild relatives. After fertilization, the genotype of new seed is same as the transgenic plants.
- Development of insect resistance to transgenic plant: Resistance to Bt toxins has developed in fields population of the various insects and moth in response to repeated use of Bt techniques, for example, Diamond Back Moth.
- Transgenic plants affect non-target or beneficial insects: Bt toxin may also kill the beneficial and non-target insects like honeybees, natural pollinates and disturb our ecosystem.
- The resistance against δ-endotoxin is not ruled out, though it may be slow as compared to insecticides because of continuous exposure of the insect to the toxin.
- Some weeds may introgress an insect resistance gene from related transgenic plants and make them less susceptible to their usual herbivores, thus exhibiting greater reproductive success and may create a weed problem.
- If lepidopteran herbivores were removed from plant species, other insects might experience competitive release and become more common.
- The use of transgenic plants possessing resistance factor in one locality may affect the insect population dynamics in other areas.
- Although Lepidoptera are herbivores as larvae, many are important pollinators of wild species as adults. Therefore, it larval mortality is high, plant communities might also be disrupted by decreasing availability of pollinators. Further, disruption of pollinators and plant communities could occur if the toxin is expressed in plant nectar.

The stability of engineered genes such as the Bt gene for insect resistance has often been questioned. Constitutive expression of the transgenic may exert great selection pressure on the insect population. However, the selection pressure may be minimized by expressing the insecticidal genes only in certain plant parts (tissue specific), or only in response to insect feeding (wound specific) or only at particular growth stages (temporal specific). It has been shown that *pin-2* promoter when attached to the insecticidal gene directs the synthesis of insecticidal proteins in the tissues of transgenic tobacco preferentially consumed by tobacco hornworm larvae. The *pin-2* promoter dose not express the insecticidal gene in the leaves, stems, or roots of potato plants until the plants are mechanically wounded. Similarly, if Bt endotoxin is made to express in the boll tissues of cotton, there would be no selection for resistance in early generation of *Helicoverpa armigera* (Hubner). Transgenic tobacco plants expressing snowdrop lectin coupled to the RSsl promoter demonstrate phloem-specific expression of this transgene. Similarly, the phloem-specific promoters for use in transgenic rice to control brown plant hopper and other phloem feeders have been isolated (Panda and Khush, 1995).

Resistance of insects can be overcome by plants expressing low levels of toxin. The tissue specific expression of insecticidal genes supplemented by occasional insecticide use would lead to durable resistance. Choice of most appropriate enzyme inhibitors, viz. trypsin inhibitors and α-amylase inhibitors to control specific insect pests is important. Recent strategies employed are to produce gene packages, whose products are targeted to different biochemical and physiological processes within the insect.

Plants can now be genetically engineered with two or more insecticidal genes with different modes of action on the target insects. Pyramiding of two Bt genes through genetic engineering has been proposed to prolong the useful life of transgenic resistance. Combining genes encoding a toxin and a gene encoding a repellent may offer longer lasting resistance than either approach alone. For example, it is known that insects that are resistant to *Bt kurstaki* type insecticides, with *Cry1A* protein, are resistant to other Bt isolates producing *Cry1A* toxins, but are not cross-resistant to *Cry1B* or *Cry1C* types. These other Bt types might, therefore, be useful as a second tier in a resistance pyramid if combined into the same plant variety already carrying the *Cry1A* toxin gene. Resistance pyramiding can also be achieved by engineering plants that already possess a certain level of natural insect resistance with selected toxin genes.

In contrast to the highly efficient, but narrow range Bt toxin, proteinase inhibitors (PIs) are effective to a lesser extent but have the advantage of a broader toxicity spectrum. An interesting question is how double Bt/PI transgenic crops would perform. Indeed cotransforming a pea lectin gene (LecA) and *CpTi* generated transgenic plants with increased resistance compared to plants, which harboured only either of both genes, is likely to provide stable protection to transgenic crop plants. In this way, it is expected to provide a multimechanistic form of resistance, which can be tailored to different crops and prevailing insect pests.

CONCLUSION

The take-off stage in transgenic plant technology in insect pest management has been initiated, but increased efforts have to be made towards a better understanding of the biological regulation of plant-gene action in order to produce ecologically accepted crops. Emphasis should be on the genetics of plant-herbivore interaction in natural and agricultural systems and on the genetics of plant secondary metabolism (Gould, 1988). Genetically engineered, pest resistant cultivars, would tend to limit the use of synthetic chemicals, making pest resistance genes as an important natural resource. While the challenge of moving toxin-coding microbial genes or plant genes sequences into several crops has been met, there is a need to more these toxin-coding groups into more crops enabling these genes to be expressed at higher levels. However, the production of crops that express resistance coding microbial genes or plant genes sequences should be undertaken only when needed. Considering the rapid pace at which transgenic technologies are emerging, the development of plants that produce toxins in specific tissues during the growing season, to control insects, deserves high consideration.

Resistance management should involve tactics which are coordinated against the pest, in particular transgenic plants, where a combination with Bt or other plant derived genes mortality factors could be most useful. The increasing attention given to transgenic technology may enable more effective production of predators and parasites and genetic manipulation may produce more competitive organisms. The incorporation of genes conferring pesticide resistance into predators and parasites would facilitate their integration with the use of pesticides.

While genetically modified plants can express a transgenic stably without affecting other agronomic characters, cases of instability and pleiotropic effects exist, resulting in a transgenic becombing silent for several generations and then becoming activated (Rogers and Parkes, 1995). The likelihood of transgenes coding for toxins through time, becoming incorporated within the microbial genes or plant genes of wild relatives and increasing the opportunities for pest species exposure to the toxin, has to be kept in mind. It has been suggested that by intercropping transgenic plants with unmodified plants, it should be possible to create an environment unfavourable to pest colonization (Thacker, 1993, 1994).

REFERENCES

Abhijit, Das (2005). Transgenic plants : An emerging approach for pest management. *Everyman's Science.* **4**(1): 15-22.

Alstad, D.N. and Andow, D.A. (1995). Managing the evolution of insect resistance in transgenic plants. *Science.* **268**: 1894-1896. In: Abhijit Das (Eds.) Transgenic plants: An emerging approach for pest management, *Everyman's Science*, April-May, 2005, **40**(1): 15-22.

Alstad, D.N. and Andow, D.A. (1996). *Ag. Biotech. New Inform.* **8**: 177N. In: Abhijit Das (Eds) Transgenic plants: An emerging approach for pest management, *Everyman's Science*, April-May, 2005, **40**(1): 15-22.

Aronson, A.I. (1994). *Bacillus thuringiensis* and its use as a biological insecticides. *Pl. Breed. Rev.* **12:** 19-45.

Barton, K., Whitely, H. and Yang, N.S. (1987). *Bacillus thuringiensis* δ-endotoxin in transgenic. *Nicotiana tabacum* provides resistance to lepidopteran insects. *Pl. Physiol.* **85**: 1103-1109.

Benedict, J.H., Sachs, E.S., Altman, D.W., Diaton, D.R., Kohel, R.J., Ring, D.R. and Berberich, B.A. (1996). Field performance of cotton expressing cry a insecticidal crystal protein for resistance to *Heliothis Virescens* and *Helico urpa zia* (Lipidoptera : Noctuidae). *J. Econ. entomol.* **89**: 230-238.

Benedict, J.H. and Ring, D.R. (2004). Transgenic crops expressing Bt proteins : Current status, challenges and outlook. *In:* O. Koul and G.S. Dhaliwal (Eds.). *Transgenic Crop Protection. Concepts and Strategies.* Science Publisher, Inc, Enfield (NH), USA, pp. 15-84.

Bernal, J.S., Prasifka, J., Setamou, M. and Heinz, K.M. (2004). Transgenic insecticidal cultivars in IPM : Challenges and opportunities. *In:* O. Koul, G.S. Dhaliwal and G.W. Cuperus (Eds.). Integrated Pest Management : Potential, Constraints and Challenges. CAB International, Wallignford, U.K. pp. 123-145.

Boulter, D., Edwards, G.A., Gatehouse, A.M.R., Gatehouse, J.A. and Hilder, V.A. (1990). Additive protective effects of incorporating two different higher plant derived insect resistance genes in transgenic plants. *Crop Prot.* **9**: 351-354.

Brar, D.S. and Khush, G.S. (1994). Cell and tissue culture for plant improvement. *In:* A.S. Basra (Ed.). *Mechanisms of Plant Growth and Improved Productivity.* Marcel Dekker, New York, USA, pp. 229-278.

Brar, D.S. and Khush, G.S. (2001). Transgenics in insect pest management. *In:* S.P. Singh, B.S. Bhumannavar, J. Poorani and D. Singh (Eds.). Biological Control Lead Paper, Symp. On Biocontrol Based Pest Management for Quality Crop Protection in the Current Millennium, July 18-19, 2001, Ludhiana, pp. 113-131.

Chakraborti, S.K., Mandaokar, A.D., Ananda Kumar, P. and Sharma, R.P. (1998). Synergistic effect of Cry1A and Cry1F δ-endotoxin of *Bacillus thuringiensis* on cotton bollworm, *Helicoverpa armigera. Curr. Sci.* **75**: 663-664.

Cheng, X.G., Sardana, R., Kalpan, H. and Altosaar, I. (1998). *Agrobacterium* – transformed rice plants expressing synthetic Cry1A(b) and Cry1A(c) genes are highly toxic to striped stem borer and yellow stem borer. *Proc. Natl. Acad. Sci. USA.* **95**: 2767-2772.

Chilcutt, C.F. and Johnson, M.W. (2004). Development of resistance in pests to transgenic plants : Mechanisms and management strategies. *In:* O. Koul and G.S. Dhaliwal (Eds.). *Transgenic Crop Protection : Concepts and Strategies.* Science Publishers Inc., Enfield, New Hampshire, USA, pp. 177-217.

Crickmore, N., Zeigler, D.R., Feitelson, J., Schnepf, E., Van Rie, J., Lereclus, D., Baum, J. and Dean, D.H. (1998). Revision of the nomenclature for the *Bacillus thuringiensis* pesticidal crystal proteins. *Microbial. Mol. Biol. Rev.* **62**: 807-813.

Ding, X., Gopalakrishnan, B., Johnson, L.B., White, F.E., Wang, X., Morgan, T.D., Kramer, K.J. and Muthukrishnan, S. (1998). Insect resistance of transgenic tobacco expressing an insect chitinase gene. *Transgenic Res.* **7**: 77-84.

Down, R.E., Fitches, E.C., Wiles, D.P., Corti, P., Bell, H.A., Gatehouse, J.A. and Edwards, J.P. (2006). Insecticidal spider venom toxin fused to snowdrop lectin is toxic to the peach-potato aphid, *Myzus persicae* (Hemiptera:Aphidae) and the rice brown plant hopper, *Nilaparvata lugens* (Hemiptera:Delphacidae). *Pest Manage. Sci.* **62**: 77-85.

Down, R.E., Ford, L., Woodhouse, D., Raemaekers, J.J.M., Leitch, B., Gatehouse, J.A. and Gatehouse, A.M.R. (2000). Snowdrop lectin (GNA) has no acute toxic effects on a beneficial insect predator, the α-spotted ladybird (*Adalia bipunctata* L.). *J. Insect. Physiol.* **46**: 379-391.

Estruch, J.J., Carozzi, N.B., Desai, N., Duck, N.B., Warren, G.W. and Koziel, M.G. (1997). *Nat. Biotechnol.* **15**: 137. In: Abhijit Das (Eds) Transgenic plants: An emerging approach for pest management., *Everyman's Science* April-May 2005, **40**(1): 15-22.

Ferre, J., Real, M.D., Van Rie, J., Jansens, S. and Peferoen, M. (1991). Resistance to *Bacillus thuringiensis* bioinsecticides in a field population of *Plutella xylostella* is due to a change in a midgut membrane receptor. *Proc. Natl. Acad. Sci. USA.* **88**: 5119-5123.

Ferutos, R., Rang, C. and Royer, M. (1999). Managing insect resistance to plants producing *Bacillus thuringiensis* toxins. *Crit. Rev. Biotech.* **19:** 227-276.

Fischhoff, D.A., Bowdish, K.S., Perlak, F.J., Marrone, P.G., McCormick, S.M., Niedermeyer, J.G., Dean, D.A., Kusano-Kretzmer, K., Mayer, F.J., Rochester, D.E., Rogers, S.G. and Fraley, R.T. (1987). Insect tolerant transgenic tomato plants. *Bio/Technology.* **5**: 807-812.

Fitches, E., Edwards, M.G., Mee, E., Grishin, C., Gatehouse, A.M.R., Edwards, J.P. and Gatehouse, J.A. (2004). Fusion proteins containing insect-specific toxins as pest control agents. Snowdrop lectin delivers fused insecticidal spider venom toxin to insect haemolymph following oral ingestion. *J. Insect Physiol.* **50:** 61-71.

Foissac, X., Nguyen, T.L., Christen, P., Gatehouse, M.R. and Gatehouse, J.A. (2000). Resistance to green leaf hopper (*Nephottix virescens*) and brown plant hopper (*Nilaparvata lugens*) in transgenic rice expressing snowdrops lectin (*Galandhus nivelis* agglutinin, GNA). *J. Insect Physiol.* **46**: 573-583.

Forcada, C., Alcacer, E., Garcera, D. and Martiniz, R. (1996). Differences in the midgut proteolytic activity of two *Heliothis virescens* strains, one susceptible and one resistant to *Bacillus thuringiensis* toxins. *Arch. Insect Biochem. Physiol.* **31**: 257-272.

Franco, O.L., Rigden, D.J., Melo, M.R. and Grossi De-Sa, M.F. (2002). Plant alpha-amylase inhibitors and their interaction with insect alpha-amylase. *Eur. J. Biochem.* **269**: 397-412.

Fusimoto, H., Itoh, K., Yamamoto, M., Kayozuka, J. and Shimamoto, K. (1933). Insect resistant rice generated by introduction of a modified δ-endotoxin gene of *Bacillus thuringiensis. Bio/Technology.* **11**: 1151-1155.

Gatehouse, A.M.R. and Gatehouse, J.A. (1998). Identifying proteins with insecticidal activity : Use of encoding genes to produce insect resistant transgenic crops. *Pestic. Sci.* **52**: 165-175.

Gatehouse, A.M.R., Hilder, V.A. and Gatehouse, J.A. (1991). Genetic engineering of plants for insect resistance. *In:* D. Grierson (Ed.). *Plant Biotechnology Series.* Vol. 1 *Plant Genetic Engineering.* Blackie Publications, London, pp. 105-135.

Gatehouse, J.A. and Gatehouse, A.M.R. (2000). Genetic engineering of plants for insect resistance. *In:* J.E. Rechcigl and N.A. Rechcigl (Eds.). *Biological and Biotechnological Control of Insect Pests.* Lewis Publishers, Boca Raton, Florida, USA, pp. 211-241.

Gosh, P.K. (1997). Transgenic plants and biosafety concerns in India. *Curr. Sci.* **72**: 172-179.

Gould, F. (1988). Evolution of biology and genetically engineered crops. *Bio. Science.* **38**: 26-33.

Gould, F. (1998). Sustainability of transgenic insecticidal cultivars : Integrating pest genetics and ecology. *Annu. Rev. Entomol.* **43**: 701-726.

Gould, F., Martinez-Ramirez, A., Andreson, A., Ferre, J., Silva, F.J. and Moar, W.J. (1992). Broad spectrum resistance to *Bacillus thuringiensis* toxins in *Heliothis virescens*. *Proc. Natl. Acad. Sci. USA*. **89**: 701-726.

Greenplate, J.T. (1999). Qualification of *Bacillus thuringiensis* insect control protein cry Ac over time Ballgard cotton fruit and terminals. J. Econ. Entomol. **22**: 1377-1383.

Greenplate, J.T., Mullins, J.W., Penn, S.R., Dahm, A., Rich, b.J., Osborn, J.A., Rahn, P.R., Ruscheke, R.R. and Shapple, R.W. (2003). Partial characterization of cotton plants expressing two toxin proteins from *Bacillus thuringiensis*: Relative toxin contribution, toxin interaction, and resistance management. *J. Apple Entomol.* **127**: 340-347.

He, K., Wang, Z., Bai, S., Zheng, L., Wang, Y. and Cui, H. (2006). Efficacy of transgenic Bt cotton for resistance to the Asian corn borur (Lipidoptra: Crambidae). *Crop. Prot.* **25**: 167-173.

Hilder, V.A. and Boulter, D. (1999). Genetic engineering of crop plants for insect resistance. A critical review. *Crop Prot.* **18**: 177-191.

Hilder, V.A., Gatehouse, A.M.R., Sherman, S.E., Barker, R.F. and Boulter, D.A. (1987). A novel mechanism for insect resistance engineered into tobacco. *Nature* (London). **330**: 160-163.

Hokkanen, H.M.T. and Wearing, C.H. (1995). Assessing the risk of pest resistance evolution to *Bacillus thuringiensis* engineered into crop plant. A case study of oilseed rape. *Field Crops Res.* **45**: 171-179.

Husmaom. T., Asad, J., Mqbool, S.B., Datta, S.K. and Raizuddin, S. (2002). Variability in expression of Cry/Ab gene in indica basmati rice. *Euphytica* **128**: 121-128.

Irie, K., Hosoyama, H., Takeuchi, T., Iwabuchi, K., Watanabe, H., Abe, M. and Arai, S. (1996). Transgenic rice established to express corn cystatin exhibits strong inhibitory activity against insect gut proteinases. *Pl. Molec. Biol.* **30**: 149-157.

Ishimoto, M., Yamada, T. and Kaga, K. (1999). Insecticidal activity of an α-amylase inhibitor like protein resembing a pupative precursor of α-amylase inhibitor in the commons bean, *Phaseolus vulgaris*. I. *Biochem. Biophys. Acta*. **1432**: 104-112.

James, C. (1997). *Global Status of Transgenic Crops in 1997.* ISAAA Briefs No. 5, International Service for the Acquisition of Agri-biotech Applications (ISAAA), Ithaca, New York, USA.

Jansens, S., Cornelissen, M., de Clercq, R., Reynaerts, A. and Peferoen, H. (1995). *Phthorimaea operculella* (Lepidoptera:Gelechiidae) resistance in potato by expression of *Bacillus thuringiensis* Cry1A(b) insecticidal crystal protein. *J. Econ. Entomol.* **88**: 1469-1476.

Jongsma, M.A., Bakker, P.L., Peters, J., Bosch, D. and Stiekema, W.J. (1995). Adaption of *Spodoptera exigua* to plant proteinase inhibitors by induction of proteinase activity insensitive to inhibition. *Proc. Natl. Acad. Sci. USA*. **92**: 8041-8045.

Jouanin, L., Bonade-Bottino, M., Girad, C., Morrot, G. and Giband, M. (1998). Teransgenic plants for insect resistance. *Plant. Sci.* **131**: 1-11.

Kanrar, S., Venkateswari, J., Kirti, P.B. and Chopra, V.L. (2002). Transgenic Indian mustard (*Brassica juncea*) with resistance to the mustard aphid (*Lipaphis erysimi* Kalt). *Plant Cell Rep.* **20**: 976-981.

Kaur, S. and Gujar, G.T. (2004). Contemporary approaches for genetically engineered insect-resistant transgenic crop. *In:* G.S. Dhaliwal and R. Singh (Eds.). *Host Plant Resistance to Insects : Concepts and Applications.* Panima Publishing Corporation, New Delhi. pp. 492-516.

Khanna, H.K. and Raina, S.K. (2002). Elite indica transgenic rice plants expressing modified cry/Ac endotoxin of *Bacillus thuringiensis* show enhanced resistance to yellow stem boner (*Scirpophaga incertulas*) *Transgenic Res.* **11**: 411-423.

Koul, O. and Dhaliwal, G.S. (2004b). Transgenic crop protection : An introduction. *In:* O. Koul and G.S. Dhaliwal (Eds.). *Transgenic Crop Protection : Concepts and Strategies.* Science Publishers, Enfield, New Hampshire, USA, pp. 1-14.

Koul, O. and Dhaliwal, G.S. (Eds.) (2004a). *Transgenic Crop Protection : Concepts and Strategies.* Science Publishers, Enfield (NH), USA.

Koziel, M.G., Carozzi, N.B. and Warren, G.W. (1998). *In:* Agricultural Biotechnology. A. Altem (Ed.), 283 Macel Dekker Inc, New York.

Kramer, K.J., Morgan, T.D., Thorne, J.E., Dowell, F.E., Bailey, M. and Howard, J.A. (2000). Transgenic avidin maize is resistant to storage insect pests. *Nature Biotechnol.* **18**: 670-674.

Kumar, P.A. (2002). Insect pest resistant transgenic crops. *In:* R.K. Upadhyay (Eds.). *Advances in Microbial Control of Insect Pests.* Kluwer Academic Publishers, New York, pp. 71-82.

Legaspi, J.C., Legaspi, B.C. Jr. and Setamou, M. (2004). Insect-resistant transgenic crops expressing plant lectins. *In:* O. Koul and G.S. Dhaliwal (Eds.). *Transgenic Crop Protection : Concepts and Strategies Science.* Publishers, Inc., Enfield (NH), USA, pp. 85-116.

Lepte, J.C., Bonade-Bottino, M., Augustin, S., Pilate, G., LeTan, V.D., Delplanque, A., Cornu, D. and Jounanin, Z. (1995). Toxicity of *Chrysomela tremulae* (Coleoptera:Chrysomelidae) of transgenic poplars, expressing a cysteine proteinase inhibitor. *Mol. Breed.* **1**: 319-328.

Li, W.B., Zarka, K.A., Douches, D.S., Coombs, J.J., Pett, W.L., Grafius, E.J. and Li, W.B. (1999). Co-expression of potato PVY coat protein and CryV-Bt genes in potato. *J. Am. Soc. Hort. Sci.* **124**: 218-223.

Liu, J., Poiner, G.O. and Berry, R.E. (2000). Control of insect pests with entomopathogenic nematodes : The impact of molecular biology and phylogenetic reconstruction. *Annu. Rev. Entomol.* **45**: 287-306.

Loc, N.T., Tinjuangjun, P., Gatehouse, A.M.R., Christou, P. and Gatehouse, J.A. (2002). Linear transgene constructs lacking vector backbone sequences generate transgenic rice plants which accumulate higher levels of proteins conferring insect resistance. *Mol. Breed.* **9**: 231-244.

Marchetti, S., Delledone, M., Fogher, C., Chiaba, C., Chiesa, F., Savazzini, F. and Giordaneo, A. (2000). Soybean kunitz, C-II and PI-IV inhibitor genes confer different levels of insect resistance to tobacco and potato transgenic plants. *Theor. Appl. Genet.* **101**: 519-526.

Matten, S.M., Hellmich, R.L. and Reynolds, A. (2004). Current resistance management strategies for Bt corn in the United States. *In:* O. Koul and G.S. Dhaliwal (Eds.). *Transgenic Crop Protection : Concepts and Strategies.* Science Publishers, Inc., Enfield, New Hampshire, USA, pp. 261-288.

McGaughey, W.H. and Whalon, M.E. (1992). Managing insect resistance to *Bacillus thuringiensis*. *Toxins Sci.* **258**: 1451-1455. *In:* Abhijit Das (Eds.) Transgenic plants: An emerging approach for pest management, *Everyman's Science*, April-May, 2005, **40**(1): 15-22.

Meeusen, R.L. and Warren, G. (1989). Insect control with genetically engineered crops. *A. Rev. Ent.* **34**: 373-381.

Morton, R.L., Shroeder, H.E., Bateman, K.S., Chrispeels, M.J., Armstrong, E. and Higgins, T.J.V. (2000). Bean alpha-amylase inhibitor 1 in transgenic pea (*Pisum sativum*) provides complete protection from Pea weevil (*Bruchus pisorum*) under field conditions. *Proc. Natl. Acad. Sci.* USA **97**: 3820-3825.

Mqbool, S.B., Husnain, T., Raizuddin, S. and Christon, P. (1998). Effective control of yellow rice boner and rice leaf folder in transgenic rice indica varieties Basmati 370 and M7 using noual δ-endotoxin cry 2 A *Bacillus thuringiensis* gene *Mol. Breed* **4**: 501-507.

Oppert, B., Kramer, K., Beeman, R.W., Johnson, D. and McGaughey, W.H. (1997). Proteinase-mediated insect resistance to *Bacillus thuringiensis* toxins. *J. Biol. Chem.* **272 :** 23473-23476.

Panda, N. and Khush, G.S. (1995). *Host plant resistance to insects.* CAB International, Wallingford, UK.

Perlak, F.J., Stone, T.B., Muskopf, Y.N., Peterson, L.J., Perker, G.B., Mcpherson, S.A., Wyman, J., Loue, S., Reed, G., Biever, D. and Fischoff, D.A. (1993). Genetically improved potatoes: protection from damage by colorado potato beeths. *Plant Mol. Biol.* **22**: 313-321.

Purcell, J.P., Greenplate, J.T., Jennings, M.G., Pyerse, J.S., Pershing, J.C., Sims, S.R., Prinsen, M.J., Corbin, D.R., Tran, M., Sammons, R.D. and Stonard, R.J. (1993). *Biochem. Biophys. Res. Commun.* **196**: 1406. In Abhijit Das (Eds.) Transgenic plants: An emerging approach for pest management, *Everyman's Science*, April–May, 2005, **40**(1): 15-22.

Rajyashri, K.R. and Mohan, M. (2004). Gene pyramiding : A transgenic approach to enhancing resistance durability in plants. *In:* O. Koul and G.S. Dhaliwal (Eds.). *Transgenic Crop Protection: Concepts and Strategies.* Science Publishers, Inc., Enfield, New Hampshire, USA, pp. 119-260.

Rao, K.V., Rathore, K.S., Hodges, T.K., Fu X Stoger, E., Sudhakar, D., Williams, S., Christou, P., Bharathi, M. and Brown, D.P. (1998). Expression of snowdrop lectin (GNA) in transgenic rice plants confers resistance to rice brown plant hopper. *Plant J.* **15**: 469-477.

Rogers, H.J. and Parkes, H.C. (1995). Transgenic plants and the environment. *J. Expt. Biol.* **46**: 467-488.

Roush, R. (1997). Managing resistance to transgenic crops. *In:* N. Carozzi and M. Koziel (Eds.). *Advances in Insect Control : The Role of Transgenic Crops.* Taylor and Francis, London, pp. 271-294.

Schuler, T.H., Poppy, G.M., Kerry, B.R. and Denholm, I. (1998). *Trends Biotechnol.* **16**: 168. *In:* Abhijit Das (Eds.) Transgenic plants: An emerging approach for pest management, *Everyman's Science*, April-May, 2005, **40**(1): 15-22.

Seetharam, A., Kuiper, G., Visser, P.B., Kogel de, W.J., Udaya Kumar, M. and Jongsma, M.A. (2002). Expression of potato multicystatin in florets of chrysanthemum and assessment of resistance to western flower thrips, *Frankliniella occidentalis. Acta. Hort.* **572**: 121-129.

Shade, R.E., Schroeder, H.E., Pueyo, J.J., Tabe, L.M., Murdock, L.L., Higgins, T.J.V. and Chrispeels, M.J. (1994). *Biotechnology.* **12**: 793.

Sharma, H.C. and Ortiz, R. (2000). Transgenics, pest management, and the environment. *Curr. Sci.* **79**: 421-437.

Sharma, O.P., Sehgal, M., Jeswani, M.D., Trivedi, T.P., Garg, D.K., Srivastava, C.P., Benagi, V.I., Nigam, G.L. and Lavekar, R.C. (2004). Pigeonpea. *In:* A. Singh, H.R. Sardana and N. Sabir (Eds.). Validated IPM Technologies for Selected Crops. National Centre for Integrated Pest Management, New Delhi, pp. 77-89.

Shelton, A.M., Tang, J.D., Roush, R.T., Metz, T.D. and Earle, E.D. (2000). *Nat. Biotechno.* **18**: 339. *In:* Abhijit Das (Eds.) Transgenic plants: An emerging approach for pest management, *Everyman's Science*, April-May, 2005, **40**(1): 15-22.

Shelton, A.M., Rhao, J.R. and Roush, R.T. (2002). Economic, ecological, food safety, and social consequences of the development of Bt transgenic plants. *Annu. Rev. Entomol.* **47**: 845-881.

Shelton, A.M. and Bellinder, R.R. (2006). role of biotechnoligical advances in shaping the future of integrated pest management. *In:* O. Koul and G.W. Cuperus (Eds.) *Ecologically Based Integrated Pest Management.* CAB International, Wallingford, UK, pp. 269-288.

Tang, J.D., Shelton, A.M., Van Rie, J., de Rosck, S., Moar, W.J., Roush, R.T. and Peferoen, M. (1996). Toxicity of *Bacillus thuringiensis* spore and crystal protein to resistant diamond back moth (*Plutella xylostella*). *Appl. Environ. Microbiol.* **62**: 564-569.

Thacker, J.R.M. (1993). Transgenic crop plants and pest control. *Sci. Prog.* **776**: 207-219.

Thacker, J.R.M. (1994). Transgenic plants. *Trends Ecol. Evol.* **9**: 486.

Tingjuanjun, P., Loc, N.T., Gatehouse, A.M.R., Gatehouse, J.A. and Christou, P. (2000). Enhanced insect resistance in Thai rice varieties generated by particle bombardment. *Mol. Breed.* **6**: 391-399.

Tu, J.G., Rhang, K., Datta, C., Xu, Y., He, Q., Rhang, Khush, G.S. and Datta, S.K. (2000). Field performance of transgenic elite commercial hybrid rice expressing *facillus thuringiensis* δ–endotoxin. Nature Biotechnol. **18**: 1101-1104.

Vaeck, M., Reynaerts, A., Hofte, H., Jansens, S., DeBruckleer, M., Dean, C., Zabeau, M., Van Montagu, M. and Leemans, J. (1987). Transgenic plants protected from insect attack. *Nature.* **327**: 33-37.

Williams, W.P., Buckly, P.M., Sagens, J.B. and Hanten, J.H. (1998). Evaluation of transgenic corn for resistance to corn earworm (Lipidopfera: Noctuidae), Fall armyworm (Lipidoptera: Noctidae) and southestern corn borer (Lipidoptera : Crambidae) in a laboratory bioassay. *J. Agric. Entomol.* **15**: 105-112.

Zhao, J.Z., Zhao, K.J., LU, M.J., Fan, X.L. and Guo, S.D. (1998). Interactions between *Helicoverpa armigera* and transgenic Bt cotton in North China. *Scient. Agric. Sinica.* **31**: 1-6.

Zhu-Salsman, K., Shade, R.E., Koiwa, H., Salzman, R.A., Narasimhan, M., Bressan, R.A., Hasegawa, P.M. and Murdoch, L.L. (1998). Carbohydrate binding and resistance to proteolysis control insecticidal activity of *Griffonia simplicifolia* lectin II. *Proc. Natl. Acad. Sci.* USA **95**: 15123-15128.

7

Bacterial Diseases of Fruit Crops and their Management

Dinesh Singh[1] and R.R. Sharma[2]

[1]Division of Plant Pathology, IARI, New Delhi–110 012
[2]Division of Post Harvest Technology, IARI, New Delhi–110 012

India is the second largest producer of fruits (55.4 MT) in the world (NHB, 2008). Among different fruits, India is the largest producer of mango, banana, sapota and acid lime. Our country produces about 39% of world's mangoes and 23% bananas. The fruit production in India has increased dramatically in the recent years, but due to increasing population, the per capita availability of fruits is very low (93 g/day) while ICMR recommended 120 g/day/person. To fulfil the requirement of fruits, either we have to increase the production or reduce the pre- and post-harvest losses caused by diseases. Fungi, bacteria, phytoplasma, viruses, viroids, nematodes and angiosperm plants are responsible for causing various diseases in fruits. However, more attention has been paid in the past to manage fungal and virus diseases and bacterial diseases have usually been ignored. It is well known that bacterial diseases like citrus canker, moko disease in banana and bacterial black spot/canker in mango has national importance. Bacterial diseases in some fruits are quite common and may cause excessive losses to them under favourable conditions (Shekhawat and Patel, 1975; Kishun, 1981) and can reduce market value of the fruits. In this chapter, bacterial diseases of important temperate fruits such as apple, pear and stone fruits (peach, plum, cherry, etc.), and tropical and subtropical fruits like mango, banana, citrus, grape, pomegranate and karonda have been described with importance of diseases, symptoms, causal organisms, disease cycle and control measures.

A. TEMPERATE FRUITS

Many bacterial diseases occur on temperate fruits, but only a few are of commercial importance. Among these, fire blight, crown gall, canker, gummosis and some leaf spots cause major losses. A brief description about the disease, causal agent and control measures has been given here under.

1. Apple and Pear

Apple and pear are considered as most important temperate fruits throughout the world. Both are delicious nutritive as they contain antioxidants. Several bacterial diseases infect them, however, fire blight, bacterial blossom blight or blast of pear, crown gall and hairy root diseases are common in apple and pear fruits.

(i) Fire blight

The name 'fire blight' was first used in 1817 to describe the sudden browning of leaves as if they have passed through a hot flame and causing a morbid matter to exude from the pores of the bark. It is considered as the most destructive bacterial disease of temperate fruits. It was the first plant disease reported, which was caused by bacteria. It is known to occur in 19 countries and 4 continents of the world. Fire blight (*Erwinia amylovora* (Burril) Winslow) was reported on apple in Kumaon Hills by Singh (1943), but Byrde and Hutcheon (1986) had shown doubt about it.

Symptoms

Infection occurs on flowers, twigs, fruits, branches and the trunk. Usually, the first symptom appears on flowers. The most obvious symptom is the wilting and browning or blackening of the infected blossoms or leaves. The affected parts become water soaked from which white to clear droplets ooze out. The flowers shrivel and fall or hang on the trees. The bacterial ooze also dries out. The infection progresses on to leaves and then to twigs. The leaves curl, shrivel, and usually hang down like flowers. Current season's growth also gets infected and wilts from tip downwards. The infection slowly reaches the branches and trunk, and kills most of the parts slowly and steadily. These infections form the canker on branches and stem, which girdles them. If the branches are not girdled completely, the cankers remain dormant as shunken cracked margins. Bacteria invade the fruits through pedicels or lenticels. Developing fruits become water soaked, turn brown, shrivel, get mummified and hang on trees for months together. Finally, they turn dark brown to black (as in pear), reddish (as in apple) and fall down the tree.

Causal organisms

The causal organism of fire blight is *Erwinia amylovora* pv. *amylovora* (Burrill) Winslow. It is Gram negative, rod-shaped bacteria having cell size of 0.8 x 1-3 µm, occurring singly, in pairs or short chains. Bacterium is motile by 1-8 peritrichous flagella. Bacteria are facultative anaerobes.

Disease cycle

Erwinia amylovora pv. amylovora, overwinters at the edges of cankers. Bacteria may also move into the orchard via insects or wind from neighbouring infection sites, including ornamental plants and wild hosts. Open blossoms are the most susceptible parts on both apple and pear trees. Further spread of the pathogen occurs when pollinating insects (mainly honey bees) carry it. Secondary infections can occur throughout the growing season and arise from the spread of *E. amylovora* as a result of primary infections in blossoms and shoots, as well as from oozing cankers. Large numbers of bacteria are spread by rain, wind, and piercing-sucking insects. Pruning with contaminated equipment can also hasten the spread of fire blight. The spread of the pathogens is favoured by warm humid weather. The development of the disease requires a temperature of about 18.5°C during flowering season with high humidity.

Control measures

Fire blight is a threatening disease, which cannot be easily controlled by a few simple measures. Hence, the following measures should be followed in an integrated manner to achieve the desirable results:

- Strict quarantine measures should be adopted during the import of fruits, seed, bud wood and other plant parts.
- Grow resistant varieties, whenever possible. For example, Delicious group of apple is inherently resistant to fire blight.
- Ensure better drainage in the orchards.
- Sprinkler system of irrigation should be avoided in vicinities having problem of fire blight.
- Excessive use of nitrogenous fertilizes should be avoided.
- Pruning should not be heavy, because it favours vigorous growth, which is susceptible to fire blight attack.
- Spray streptomycin (100 ppm) or oxytetracycline (100 ppm) during flowering. It should be repeated after 4-5 days.
- Cell suspension sprays of *Pantoea agglomerans*, (Syn. *Erwinia herbicola), Pseudomonas fluorescens* antagonistic bacteria are quite effective (Vanneste, 1996).

(ii) Hairy root

In certain areas, hairy root is also a serious problem in apple and pear. It is caused by *Agrobacterium rhizogenes.*

Symptoms

The bacterium invades the roots through wounds and cause excessive growth of the fibrous roots, which originate from one place. It appears like a broom. The affected plants show stunted appearance. It mainly affects the plants in the nursery.

Control measures

The following measures are useful for the control of this disease effectively:

- Follow quarantine measures strictly to avoid the entry of the bacterium into new locations.
- Mechanical injury to the plants should be avoided at every cost.
- Root chewing insects should be controlled in time.
- Infested fields should be planted with maize or other grain crops for several years.
- The crops like tomato and marigold should not be planted near the vicinity of orchards, having infestation of crown gall.
- Infected nursery stock should be destroyed.
- Biocontrol of hairy root with the use of *Agrobacterium radiobacter* has been very successful in many countries of world, which should be tried in our country as well.

2. Stone Fruits

In our country, many stone fruits like peach, nectarine, plum cherry, apricot, etc. are grown commercially. Out of these fruits, cherry and apricots are exclusively grown in temperate climate, whereas peach and plum grow profitably well even under subtropical climate and Indian plains. Under all types of climatic conditions, different bacterial diseases such as bacterial spot, bacterial canker hyperplastic bacterial canker of almond and crown gall infect stone fruits, which are briefly described below:

(i) Bacterial canker or gummosis

Bacterial canker or gummosis is the most serious bacterial disease of stone fruits, though it appears on apple and pear also. The pathogen causes disease in stone fruits such as sweet cherry, almond, apricot, peach, and plum. When cankers girdle branches or stems, substantial damage to trees may occur. *Pseudomonas syringae* pv. *morsprunorum* produces coronatine toxin, but not syringomycin (Liang *et al.,* 1994). The disease caused by *P. syringae.* pv. *morsprunorum* is called bacterial canker of stone fruits; while *P. syringae* pv. *syringae* infection on stone fruits is known by various names *viz.*, bacterial canker, gummosis, shoot blight, spur blight, sour sap, whither tip, die back and blast. *P. syringae* pv. *syringae* has wide host range causing diseases in fruit trees, dicot and monocot herbs, but *P. syringae* pv. *morsprunorum* affects only the stone fruits. In India, it was first reported that bacterial canker is caused by *Pseudomonas syringae* (Durgapal, 1971a; Agarwal *et al.,* 1970), but later it was referred to as *P. morsprunorum* (Durgapal, 1971b).

Symptoms

The disease affects all the aerial parts of the tree. The characteristic symptom of the disease is the formation of cankers, accomplished by gum exudation. These cankers usually develop at the base of an infected spur, bud union, and in the pruning wounds. The bark of the spurs first shows narrow brown streaks, which develop into canker. From these cankers, gum breaks through the bark during spring and runs down the surface of the limbs. After sometimes the branch or the entire tree above the canker dies. Thus, dieback of the terminal shoot or twigs is also a common symptom of gummosis. Sometimes, it appears on flowers, which when seriously affected, show bighted appearance. Leaf infection results in angular or circular, water soaked spots about 0.5–3.0 mm in diameter. They later become brown, dry and brittle and ultimately fall out leaving shot holes. The spots are frequently surrounded by chlorotic haloes. The old twigs, infected leaves may become pale yellow and shed prematurely. On fruits, flat, round, superficial, dark brown, scabby lesions may develop. The spots may be sunken in some cases. The bud and flower infection is commonly caused by *P. syringae* pv. *syringae.* The dormant flower and leaf buds become blighted. Small cankers develop at the base of the bud. Infection of flowers occurs only under very favourable conditions. The flowers become water soaked, turn brown, wilt and hang on the twig. Infection from flowers may spread into the twig and cause twig blight.

Causal organism

The disease is caused by *Pseudomonas syringae* pv. *syringae* and the more specialized *P. syringae* pv. *morsprunorum* (Wormald, 1931; Young, Dye and Wilkie 1978), which is restricted predominantly to cherry and plum and differ from pv. *syringae* not only in a narrow host range, but also in being able to infect through leaf scars in autumn without prior wounding (Crosse, 1966). Most strains of *P. syringae* pv. *syringae* produce the phytotoxins syringomycins, which appear to play a role in the virulence of the pathogen. Many strains of *P. syringae* are ice nucleation active, i.e., they serve as nuclei for ice formation, and therefore cause frost injury to plants at relatively high freezing temperatures. Some bacteria also produce bacteriocins toxic against non-ice nucleation-active strains, thus assuring a comparative advantage for them.

Disease cycle

Bacteria overwinter in active cankers, infected buds and leaves, systematically in the xylem of some hosts, epiphytically on buds and limbs of infected or healthy trees, and possibly on weeds and non-susceptible hosts. The pathogen infects mainly through leaf scars and wounds. It is spread from cankers and other diseased tissues and as an epiphyte by splashing rain. Bacteria can survive in cankers or dormant buds. During the growing season, epiphytic populations are present on all parts. Frost enhances the infection, as in the case of infections by *P. syringae pv. syringae*. Hail and storms enhance formation of stomatal infection and leaf spots. Races with some host specificity have been determined by using bacteriophages (Crosse, 1966; Sundin *et al*., 1988; Hatting *et al.,* 1989).

Control measures

The following practices have been reported to be useful for controlling bacterial canker or gummosis in the fruits.

- Ensure proper drainage in the orchards.
- Use healthy bud-wood for raising new plants.
- Use resistant rootstocks for propagation.
- Grow resistant varieties. For example, World's Earliest and Fair Howen (peach); Flemish, Napolean, Rainbow, Lambert (cherry), Bluce, Katha (almond), Black Champa, Burbank, Mariposa (plum) varieties show resistance against gummosis.
- Spray the trees with copper oxychloride or Bordeaux mixture once before the winter starts and again in spring, but before flowering.
- Spray streptomycin (100-200 ppm) in spring.

(ii) Leaf spot, shot-hole, black spot

The disease was first described by E. F. Smith, 1903, from the USA. It is present in areas where stone fruits are grown and may cause serious in losses by reducing the marketability of the fruits and by weakening trees through leaf spots, defoliation, and lesions on twigs. In some stone fruits, leaf spot has been recorded.

Symptoms

The bacterium attacks the leaves and form angular dark brown spots, which are surrounded by light yellowish green leaf tissue. The necrotic tissues or spots cut off and shed, resulting in the appearance of shot holes on the leaves. Usually, several spots coalesce together. Severely affected leaves turn yellow and drop off. The spots are most frequent near the leaf tip, due to accumulation of bacteria in drops of rain or dew. Twig lesions are also formed on the badly defoliated plants. On fruits, small circular, slightly sunken, water soaked brown spots can be observed, sometimes surrounded by a light green halo. Gum can exude from the spots and symptoms may look very similar to insect's damage. In severe cases, when fruits are infected in early phase, they become malformed and show cracks.

Causal organism

Disease is caused by *Xanthomonas arboricola* pv. *pruni* (Smith, 1903) Vauterin, Hoste, Kersters, Swings, 1995.

Disease cycle

The bacterium overwinters during summer in the cankers and twig tip from where bacteria may spread by rain, wind and insects to other parts of the tree and infect through leaf scars, lenticels, stomata and insect wounds (Moffett, 1973). The disease may be confused with symptoms caused by *P. syringae* pv. *morsprunorum* or *P. s.* pv. *syringae*. Temperature of 15–28°C, heavy rain and wind in spring favour epidemic infection. There seems to be no strain differentiation between hosts (Scortichini *et al.,* 1996).

Control measures

- Select budwood from disease free trees.
- Use of resistant cultivars is best way to control the disease (Young, 1987; Ritcie, 1995)
- Remove the diseased plant parts and burn them.
- Spray Bordeaux mixture during January and again in spring.

(iii) Crown gall

The disease was known in USA in 1870 and the bacterium was isolated from stem gall of grapes in 1897 from Cavara (Italy). Smith and Townsend correctly identified the bacterium in 1907. In India, the disease was reported by Singh in 1943 and organism was isolated and pathogenicity was established by Durgapal in 1971a at Shimla (H.P.). The disease can affect many woody and herbaceous plants belonging to 140 genera of more than 61 families (Stapp, 1961). Out of them apple, peach, apricot, almond, grape, rose, pear, cherry and plum are important hosts of this disease. However, in India, it is so far observed on cherry, plum and peach (Durgapal, 1971a). First report of this disease on almond was made by Jindal and Sharma (1988). During 1997, as many as 8,58,567 grafted plants of stone fruits were rejected on account of crown galls causing losses to the tune of Rs. 6.00 millions (Anonymous, 1997).

Symptoms

The disease first appears as small overgrowths on the stem and roots particularly near the soil line. Young galls are more or less spherical, white or flesh coloured and soft. As the growth enlarges, either surface become convoluted. In advanced stages, the outer tissues turn dark brown or black due, which results in death and decay of peripheral cells. The galls usually vary in size from 0.25–4 inches in diameter but in extreme cases, they may be a foot or more in diameter. They are generally knobby and knotty and become more kleft as they grow older. In fruit trees, the galls may have a corky exterior but the inner tissues become woody. Occasionally complex tumours are formed from which roots and buds may arise giving rise to shoots carrying leaves. This type of tumour is called 'teratomas'. Tumours may also appear on branches of trees, petioles and leaf veins. In some hosts, secondary tumors may develop at points distant from the primary tumours. They are free of bacteria and their tissues can be grown on chemically defined media. The tissues when transplanted into healthy plants produce sterile tumours. The affected plants may become stunted, produce small chlorotic leaves and in general are more susceptible to adverse environmental conditions, especially winter. Aerial galls are formed in grapevine (*Vitis vinifera*) and *Rubus* spp. There is some evidence that these host the bacteria can move through vascular system and that gall formation is associated with frost to the stem or canes (Fahy and Persley, 1983).

Causal organism

The disease is caused by *Agrobacterium tumefaciens* (Smith and Townsend, 1907) Conn 1942. Bacteria are rod-shaped, 0.8 × 1.5–3.0 µm, motile by 1-4 peritrichous flagella; when growing on carbohydrate containing media bacteria produce abundant polysaccharide slime. Colonies are not pigmented and usually smooth; bacteria are rhizosphere and soil inhabitant (Schroth *et al.*, 1965).

Disease cycle

The bacterium is a soil inhabitant, remaining viable for several years (Patel, 1929, Banfield, 1934). It infects the host only through fresh wounds created by cultural practices, grafting, insects, etc. (Stapp, 1961). Deeper wounds provide better channels of infection and result in formation of larger galls. Under suitable condition, even a single bacterium can cause the infection. The bacterium multiplies in watery intercellular spaces in the wounds. The presence of living bacteria is necessary for only about three days. During these periods, part of the Ti plasmid, known as T-DNA, is transferred from bacterium to plant cell. The T-DNA becomes integrated into the plant nuclear DNA and directs the plant cell to (a) grow and divide out of the control and (b) synthesize crown gall specific chemicals called "opines", which can be catabolised only by bacteria harbouring a Ti plasmid. It is a unique parasitic system and has been called "genetic colonization" (Schell *et al.,* 1979). The parasite colonizes the host with a piece of DNA, which redirects the metabolic activities of the host to benefit the parasite.

Control measures

The following control measures are considered important for controlling crown gall effectively.

- Follow quarantine measures strictly to avoid the entry of the bacterium into new locations.
- Mechanical injury to the plants should be avoided at every cost.
- Root chewing insects should be controlled in time.
- Infested fields should be planted with maize or other grain crops for several years.
- The crops like tomato and marigold should not be planted near the vicinity of orchards having infestation of crown gall.
- Infected nursery stock should be destroyed.
- Biocontrol of crown gall with the use of *Agrobacterium radiobacter strain* K84 or K1026 has been very successful in many countries of world where sensitive (biovar 2) strains *A. tumefaciens* occur (Jones and Kerr, 1989; Vicedo *et al.,* 1993).
- *A. radiobacter* is a good root colonizer and produces a bacteriocin, Agrocin 84, which is toxic for the crown gall bacterium *A. tumefaciens*.

TROPICAL AND SUBTROPICAL FRUITS

1. Mango

Mango is called 'king of fruits' in our country. India is the largest producer of mangoes in the world. However, its productivity is only about 7 tonne/ha, which is quite low. Many bacterial diseases such as bacterial black spot/ canker crown gall and fruit rot are common during field and storage equally responsible.

(i) Bacterial black spot (bacterial canker)

The bacterial canker was first reported as bacterial leaf spot from Pune by Patel *et al.,* (1948). The disease is prevalent in Maharashtra, Tamil Nadu, Delhi, Haryana and Uttar Pradesh. It probably existed since 1881 in Bihar (Patel *et al.,* 1948, Shekhawat and Patel, 1972, Kishun and Sohi, 1983). In nature, the disease has been observed to cause 20-50% fruit infection in Langra and 10-85% in seedling mangoes. The diseased fruits are highly vulnerable to insect and fungal invasion through the cankers. The market value of infected fruits is considerably reduced.

Symptoms

The disease attacks all aerial parts of the plants like leaves, petioles, twigs, stems and fruits. On leaves, minute, water-soaked irregular lesions appear in groups or any parts of the leaf blade, which enlarge, turn–brown or dark brown, become angular due to vein limitation and finally cankerous and raised. The disease is mostly surrounded by chlorotic haloes. Several lesions may coalesce to form irregular cankerous patches. In old lesions, the upper epidermis becomes loose, greyish white, and sometimes drop-off. Severe infection may result in yellowing of the leaf and defoliation.

On fruits, the initial water soaked lesions gradually develop into cankers. These cankers are much more raised and dark than those on the leaves and often burst open releasing a highly contagious gummy substance containing bacterial cells. Infected fruits may develop cracks in the peel and those badly affected drop prematurely. Under natural conditions, the pathogen causes water-soaked spots, which turn into cankers on fruits, leaves, stalks and branches, the disease is called bacterial canker (Sekhawat and Patel, 1975; Kishun and Sohi, 1983). Black stem cankers are filled with gummy exudates (Moffett *et al.*, 1979).

Causal organism

The disease is caused by *Xanthomonas campestris pv. mangiferae-indicae* (Patel, Moniz, and Kulkarni, 1948) Robbs, Ribeiro and Kimura 1974. It is Gram negative, rod-shaped bacterium, motile by monotrichous flagellum and size 0.36-0.54 × 0.45-1.44 μm.

Disease cycle

The disease is non-systemic and pathogen survives one fruiting season to next in localized cankers on twigs and leaves. Survival of *X. c.* pv. *mangiferae-indicae* is very efficient in leaf lesions. Bacterial populations decrease relatively slowly in leaf lesions on trees maintained under controlled conditions. Three-month-old leaf lesions on a susceptible cultivar may contain up to 10^7 CFU, and up to 10^5 CFU were detected in 18-month-old lesions. Incidence of diseased fruit is highly correlated with leaf spot severity. Twig cankers also serve as a source of inoculum, but their relative importance compared with bacterial inoculum in leaves is difficult to assess. Bacterial populations on mango surfaces can average as high as 10^6 CFU per leaf or fruit. Surface populations vary depending on climatic conditions, but are correlated to the number of leaf lesions. The presence of free water allows the release and redistribution of bacteria from the ruptured epidermis. Bacterium is also found pathogenic on *Spondias mangifera*, *Anacardium occidentale* (Patel *et al.*, 1948b), *Acanthosperma hispidum, Lantana camara, Ficus glomerata, Solanum tarbum, Psophocarpus fellagonolobus* and *Caesalpinia minosoldes* (Kishun, 1981). Role of mango stones in survival of pathogen has been established by Kishun and Chand (1988). Pathogen has also been found to survive on weed hosts (Kishun and Chand, 1988). Disease spread is rapid during rains. In new areas, disease spreads through infected planting materials. *Myllocerus discolor* var. *variegate, Orthaga evadrusalis* and *Canthecondita furcellate* (Kishun, 1986) are found associated in mechanical transmission of the disease. The pathogen enters intact leaves through stomata, in fruit through stomata and lenticels and in twig through lenticels (Sekhawat and Patel, 1975). It is an exclusively intercellular parenchymatous pathogen and causes chloronemea, separation, and collapse of the cells to form bacterial pockets.

Control measures

Following measures have been considered useful for the control of this disease—

- Any potential source of inoculum of pathogen should be destroyed.
- Orchard sanitation is recommended to avoid further spread of the disease.
- Planting material should be taken from certified nurseries.
- Dipping the fruits in Agrimycin-100 @ 250 ppm proved effective (Roa *et al.,* 1977). Two spray of Streptocycline (250-300 ppm) at 20 days interval reduce fruit infection (Bose and Singh, 1980). The disease can also be reduced with monthly spraying of 1000 ppm Bavistin (Kishun and Sohi, 1984).
- Varieties like Bombay Green, Fazli and Suvarnrekha (Shekhawat and Patel, 1975; Kishun and Sohi, 1983) are reported to be resistant.
- *Bacillus coagulans*, a bacterium antagonist, provides effective biocontrol against canker.

(ii) Bacterial fruit rot

Soft rot of mango fruits was first reported by Patel and Padhya (1948) in Bombay. The pathogens also affect carrot, cucumber, okra, French bean, tomato, radish, banana, sweet potato, peas, beetroot, potato, cauliflower, brinjal, cowpea and sweet orange. The bacterium occurs worldwide, but is most prevalent in warm and humid climates.

Symptoms

Bacterium enters the fruits through wounds or lenticels and dissolves the middle lamella by producing the enzyme pectinase. As a result, some brownish lesions appear on the fruits. Later, these lesions increase in size, coalesce to form big patches and causes complete rotting of the fruits.

Causal organism

Soft rot of mango is caused by *Pectobacterium carotovorum subsp. carotovorum* (Jones 1901) Hauben *et al.* 1999, *Erwinia carotovora pv. carotovora* (Jones 1901) Bergey *et al.* 1923, *Erwinia herbicola* (Löhnis 1911) Dye 1964. The soft rot bacteria can grow and are active over range of temperatures from 5-35°C. They are killed with extended exposure at about 50°C. The colonies of *E. carotovora pv. carotovora* on most media are greyish-white to creamy white, smooth, round, glistening and slightly raised and visible on isolation plates after about 24 h (Bradbury, 1977 a & b).

Disease cycle

Soft rot bacteria survive in infected fleshy organs during storage. When soft rot bacteria enter wounds, they feed and multiply at first on the liquids released by the broken cells on the wound surface. There they produce increasing amounts of pectolytic enzymes that break down the pectic substances of the middle lamella and bring about maceration of the tissues. Because of the high osmotic pressure of the macerated tissue, water from the cells diffuses into the intercellular spaces. As a result, the cells plasmolyze, collapse, and die. Bacteria continue to move and multiply into the intercellular spaces, while their enzymes advance ahead of them and prepare the tissues for invasions. The invaded tissues become soft and are transformed into a slimy mass consisting of innumerable bacteria swimming about in the liquefied substances. The epidermis of the most tissues are not attacked by the bacteria; however, cracks are usually present, and the slimy mass extrudes through them into the soil or in storage, where it comes into contact with other fleshy organs, which are subsequently infected.

Control measures

Following measures are considered necessary to control soft rot of mango:

1. Avoid mechanical injury to the fruits.
2. Dip the fruits in streptocycline sulphate solution (0.03%) before storage.
3. Avoid cultivation of vegetables like tomato, chillies, peas, beet, cauliflower, brinjal etc., near orchards.
4. Spray streptocycline (100 ppm) during fruit development stage in May.

2. Banana

Banana is the most important tropical fruit of world. In India, there has been a phenomenal increase in its production during the last decade. It has happened because the farmers are now using tissue culture plants for commercial purposes. In addition, adoption of high-density planting techniques and drip irrigation has further increased the production. Some bacterial diseases of banana such as bacterial wilt, soft rot of rhizome, bacterial leaf spot, blood disease and bugtok are affected in different countries. However, in India, bacterial wilt is common in banana growing areas but situation is not very bad.

(i) Bacterial wilt or Moko disease

Bacterial wilt or Moko disease was first reported from Trinidad, West Indies during 1890. It is a serious disease of banana in many countries of world wherever banana is cultivated, but in India, it is not so serious.

Symptoms

It is a wilt disease, more often confused with Panama wilt of banana. The disease, in general, is recognized by rapid wilting and collapse of leaves, pre-mature ripening of fruits, discoloration of vascular strands, and wilting and blackening of the suckers. The symptoms start on rapidly growing young plants. The youngest three leaves turn pale green or yellow and collapse near the junction of lamina and petiole. Most leaves collapse within 3-7 days. The characteristic symptoms appear on young suckers, which are blackened, stunted and twisted. In severe case, plants collapse before the emergence of bunches. The presence of yellow fingers in otherwise green stem often indicates the occurrence of Moko disease. Initially, a firm dry rot is found within the fruits of infected plants. Fruit stalk also show discolored strands.

Causal organism

It is caused by a bacterium, *Ralstonia solanacearum*, which was earlier known as *Pseudomonas solanacearum.* Colony on TTC medium for 48 hr at 30°C is fluidal, > 5mm, white or pink, EPS producing colonies, motile with flagellum, grow at 37°C, it does no grow at 41°C, NaCl tolerance up to < 2.0%, change the nitrite to nitrate, gas form nitrate, hypersensitive reaction shows on tobacco plants. Usually soil borne.

Disease cycle

The pathogen is soil borne. The banana pathotype of *R. solanacearum* survive in a wholesome manner in the presence of a susceptible host (Govindarajan and Gnanamanickam, 1980). Bacteria survive in host plants and, for several months at least, in soil. Bacteria enter roots through wounds, reach the xylem vessels and multiply, and move through them. Bacteria are transmitted through infected banana rhizomes and contaminated tools and equipment.

Control measures

The control of Moko disease is very difficult. However, some remedial measures are advised for the control of disease:

(i) Grow resistant varieties like, Poovan and Montha etc., in the infected fields.
(ii) Eradicate the infected plants and destroy them.
(iii) Use disease-free planting material for raising new plantations.
(iv) Follow 3 year's crop rotation of banana, sugarcane and rice.
(v) Disinfect the planting holes with chloropicrin.
(vi) Microbial antagonist, *Pseudomonas fluorescence*, is quite effective against the causal agent of Moko disease.
(vii) Bacteria that have been used with reasonable success in biocontrol of *R. solanacearum* in banana and some other crops are *Bacillus polymoxa, B. subtilis, Bacillus* spp., *Burkholderia glumae, Erwinia* spp., *Pseudomonas aeruginosa, P. fluorescens, R. solanacearum* (spontaneous avirulent mutants- EPS negative), *Stenotrophomonas maltophilia, Streptomyces mutabilis* (Trigalet *et al.,* 1994).

(ii) Soft rot of rhizome

Soft rot of rhizome can cause severe losses in particular localities. It is caused by *Erwinia chrysanthemi* var. *paradisiaca.*

Symptoms

A massive soft odorous rot of the centre or a portion of a rhizome characterizes the disease. It progresses upward the pseudostem and destroys the growing point, causing internal decay and discolouration of the vascular tissues. Leaves show yellowing, which wilt very soon. Sometimes, it is confused with fusarium wilt.

Control measures

(i) Uproot and destroy the affected plants.
(ii) Soil and plant drenching with bleaching powder at an interval of 10-15 days is quite effective in controlling the disease.

(iii) Bacterial leaf spot

The disease was first reported on *Musa paradisiaca* L. from Tamil Nadu by Rangaswami and Rangarajan (1965). Most of the cultivars of *M. paradisiaca* were found susceptible. The disease was widespread on cultivars Monthan, Poovan, Rasthali and Peyan. Bacterial leaf spot is a serious bacterial disease in some varieties of banana in certain areas.

Symptoms

The disease affects only the leaf blades in which it appears as chlorotic linear streaks along the veins. Later, these steaks coalesce to form large, irregular chlorotic patches. In severe infestations, the leaves may roll and dry up in the affected portions, presenting a scorched appearance. The affected plants become stunted, which affects the size of bunch and fingers, reducing the yield and quality invariably.

Causal organism

The disease is caused by *Xanthomonas musicola.*

Control measures

Try the following measures to control this disease effectively:

- Cut and burn the affected leaves.
- New plantings should be established from disease free planting material.
- Grow resistant varieties and avoid Poovan, Rasthali, Peyan and Monthan cultivars in affected areas.

3. Citrus

Citrus is a leading group of fruits in about 100 countries of the world. This group constitutes a number of fruits like mandarin, oranges, lemons, limes, grapefruits, pummelo, etc. Different *Citrus* spp. are affected by different bacterial diseases to a variable extent. However, canker, bacterial blast and black pit (*Pseudomonas syringae* pv. *syringae*), citrus variegated chlorosis (*Xylella fastidiosa*) and citrus greening diseases are of worldwide occurrence.

(i) Citrus canker

The disease was first reported by Abe in 1899 from Japan and was first described by Stevens in 1914 and the bacterium by Hasse in 1915 from the USA. Canker is a widespread disease in all the citrus growing areas of the world. In many countries of the world, severe attack of the disease has resulted in the mass eradication of millions of trees and nursery stock. In many advanced countries like Australia, New Zealand, South Africa and the USA, it was completely eradicated, but it again reappeared in early eighties. In India, citrus canker was first reported from Punjab (Luthra and Sattar, 1940). Heavy losses have been reported in epidemics due to leaf drop and fruit loss due to premature fruit drop and fruits with spots that cannot be marketed or start to rot and have to be destroyed. Furthermore quarantine measures such as burning of trees and destruction of fruits may add to these losses (Goto, 1992).

Symptoms

The bacterium infects all the aerial parts of plants and hence the symptoms appear on leaves, twigs, thorns, older branches and fruits as necrotic brown spots with rough surface. Lesions first appear as small, round watery translucent spots on lower surface of the leaf and then on upper surface. As the disease advances, spots become white or greyish and give a rough, corky appearance. The lesions are surrounded by watery yellowish halo. It may cause premature defoliation of the leaves. Cankers on the twigs are elongated and may cause them to break. On larger branches, the cankers are irregular, rough and more prominent. Cankers on fruits are similar to those on leaves, except that the yellow halo is absent and crater-like depression in the centre is more pronounced and injury to the fruit is only in the peel and not on juice sacs. The lesions on fruit remain confined to the rind, but sometimes, it causes cracks and fissures on fruit peel.

Causal organism

The disease is caused by *Xanthomonas axonopodis.* It is a rod-shaped bacterium and Gram-negative, motile with monotrichous polar flagella. The bacterium has a genome length approximately 5 Mbp. There are various types of citrus canker disease caused by different pathovars and variants of the bacterium. The Asiatic type of canker (Canker A), *X. axonopodis* pv. *citri*, caused by a group of strains originally found in Asia, is the most widespread and severe form of the disease. Cancrosis B, caused by a group of *X. axonopodis* pv. *aurantifolii* strains originally found in South America is a disease of lemons, key lime, bitter orange, and pummelo. Cancrosis C, also caused by strains within *X. axonopodis pv. aurantifolii*, only infects key lime and bitter orange. "A" strains, discovered in Oman, Saudi Arabia, Iran, and India, only infects key lime. Straight rods, 0.4–1.0 × 1.5–4.0, motile with polar flagellum, growth on agar medium is usually yellow, and most are slow growing. All species are plant pathogens found only in association with plants and plant materials. Khan and Hingorani (1970) identified 6 strains of the bacterium, but their number was reduced to 3 when they were tested on *Murraya exotica*. Prasad *et al.* (1978) also reported varability in *X. campestris* pv. *citri.*

Disease cycle

The bacterium propagates in lesions formed on leaves, stems, and fruit. When there is free moisture on the lesions, bacteria ooze out and can be dispersed to new growth and other plants. Rainwater collected from foliage with lesions contains between 10^5 and 10^8 cfu/ml. Wind-driven rain is the main natural dispersal agent, and wind speeds 18 mph (8 m/s) aid in the penetration of bacteria through the stomatal pores or wounds made by thorns, insects such as the Asian leafminer. Water congestion of leaf tissues can be seen following rainstorms with wind. Citrus foliage can hold 7 microlitres/cm^2 of leaf area. Studies of inoculum associated with water congestion have demonstrated that as few as 1 to 2 bacterial cells forced through stomatal openings can lead to infection and lesion formation. Pruning causes severe wounding and can be a site for infection. The long distance dissemination takes place through diseased planting material (Reddy and Murthy, 1996).

Control measures

Following measures should be tried for the effective control of citrus canker:

(i) Uproot the affected plants or plant parts and destroy them.

(ii) Use disease free nursery stock for the establishment of new orchards.

(iii) Spray Bordeaux mixture (1%) before monsoon.

(iv) Proper care should be taken for the control the leaf winner. For this, systemic insecticide like rogor or metasystox (0.05%) is quite effective.

(v) Streptomycin sulphate (500 ppm) or phytomycin (2,500 ppm) sprayed at fortnightly interval can effectively check the disease.

(vi) Some bioagents like *Pseudomonas fluorescens, Bacillus subtilis* and *Aspergillus terreus* have shown strong antagonistic effect on *X. campestris* pv. *citri*. Though, these agents have not yet been used commercially for the control of citrus canker (Kalita *et al.*, 1996).

(vii) *Pseudomonas syringae* prevented enlargement of canker lesions as well as subsequent defoliation of infected leaves. The bacterium has ability to stimulate production of phytoalexin "citrusnins" in the infected tissues (Ota, 1983).

(ii) Citrus greening

Citrus greening is a common disease in many citrus growing countries of the world. In India, it is particularly serious in northern states, where it is considered more dangerous than tristeza virus.

Symptoms

Leaf chlorosis is the major symptom of greening, which resembles zinc deficiency symptoms. Characteristically, green dots or islands appear against yellow background of leaves. A characteristic feature of greening is that the yellow areas are surrounded on one side by the mid-rib and on the other side by lateral veins. The yellowing expands towards the margins and size of the leaves is reduced drastically. The internodes of the branches are shortened, giving a bushy appearance to the branch. The diseased plants look stunted, they flower earlier and produce smaller fruits of poor quality.

Causal organism

Citrus greening was earlier considered as a viral disease for a long time. Later, some pathologists claimed that MLOs are responsible for this disease, but in 1996, it was finally accepted that it is a bacterial disease and *Candidatus Libreobacter asiatium* was identified as causal agent for Asian citrus greening and *Candidatus Libreobacter african* for African citrus greening. These are phloem-limited bacteria. The African strain does not require high temperature for optimum expression as the Asiatic strain. Neither of them can be cultured on artificial media.

Disease cycle

Vegetative propagation and two psyllid insects spread the pathogen. The Asian strain is spread primarily by *Diaphorina citri*, whereas the primary vector for the African strain is *Trioza erytreae*, but both insect vectors can spread either strain of the bacterium.

Control measures

Following control measures can be applied to control greening disease effectively:
(i) Uproot and destroy the infected plants.
(ii) Raise new plants from disease free indexed stock or from nucellar seedlings.
(iii) *Citrus psylla* should be kept under control by spraying systemic insecticides like rogor or metasystox (0.05%).
(iv) Spray bavistin + ledermycin (500 ppm each) six times at 10 days interval.

4. Grapevine

Grape is basically a crop of temperate region. But in India, it is mainly grown in the tropical climate. At present, the productivity of Indian grape is highest (30.0 t / ha) in the world. Different bacterial diseases affect grape, but they do not cause much loss to the growers. However, some common bacterial diseases like bacterial canker/blight and Pierce's disease have been reported to occur on grape in different countries of the world.

(i) Bacterial canker

The disease was first observed on Anab-e-Shahi variety during 1969, but its seriousness was noticed only in 1984, when it appeared in epidemic form on almost all the grape growing areas.

Symptoms

The bacterium infects all the aerial parts of the plant. Initially, small water-soaked spots appear on the lower side of the leaves, which are surrounded by yellow halo. These spots enlarge in size and become dark brown, angular and cankerous. Sometimes these spots coalesce and form larger patches. Leaves also show vein infection. Infected leaves after drying usually remain attached to the stem and give a blighted appearance. In advance stages of infection, stunting, cracking and abnormal growth of canes is also noticed. Lesions on berries are brown to black, cankerous. Severely affected berries become small and shrivelled. It may cause premature drop of berries. Further, the affected berries are not fit for marketing.

Causal organism

Initially *Pseudomonas viticola* was identified as its causal organism, but later, *Xanthomonas campestris* pv. *viticola* was reported to be its casual organism. However, not it is established that the disease is caused by *Xylophilus ampelinus* (Panagopoulos, 1969) Willems, Gillis, Kersters, Van den Broecke and De Ley 1987. The pathogen is a Gram negative, rod-shaped bacterium with rounded ends, motile by single polar flagellum and measure 0.4–1.2 × 2.3 µm. The bacteria do not grow at 41°C. Yellow pigment is insoluble on nutrient agar medium. Carbon source like L-arabitol and D-glucose is not utilized by bacteria.

Disease cycle

X. ampelinus is transmitted by pruning tools and water (irrigation). Infection occurs in wet and warm conditions. The bacterium can occur as an epiphyte and survive in wood of producing plants or planting material and can spread with cuttings (Rao, 1988; Grall and Manceau, 2003). The pathogen also infects neem, *Phyllanthus maderaspatensis* (Nayudu, 1972) and mango, which may serve as alternate host. A temperature range of 25-30°C seems to be more favourable for the spread of this disease. The disease frequency increases under the conditions of maximum rainy days (Chand and Kishun, 1988).

Control measures

Following control measures have been suggested to control bacterial canker of grapevine:

- Destroy the affected vines.
- Pruning should be done in late October.

- Alternate hosts like neem, *Phyllanthus* spp., etc., in the vicinity of vineyards should not have any infection.
- Use disease free cuttings for propagation.
- Avoid excessive use of irrigation.
- Give prophylactic sprays of copper oxychloride or Bordeaux mixture before monsoon.
- Usually antibiotics have poor efficiency, but streptomycin sulphate (500 ppm) may be sprayed at 5-7 days interval in areas receiving low rainfall.
- Excessive use of nitrogenous fertilizers should be avoided.

(ii) Pierce's disease

Pierce's disease has been named after N.B. Pierce, who first reported it. It is a devastating disease of grape in the USA, killing plants, especially in California wine production areas. The disease was described for the first time by Wells *et al.* (1987) from the USA. Grapevine, peach and orange are main hosts of the disease. *Medicago sativa* (alfalfa: dwarf), *Prunus dulcis* (almond: leaf scorch), *P. armeniaca* (apricot), *P. salicina, P. domestica* (plum: leaf scald), *P. angustifolia, Acer rubrum, Morus rubra, Platanus occidentalis, Quercus rubra, Ulmus americana* and *Vinca minor* have also been reported as natural hosts (Janse, 2006). Many crops and wild plants carry Pierce's disease strains without symptoms (*e.g.* wild grasses, sedges and trees), systemically or non-systemically infected (Hill and Purcell, 1995). Strain from different hosts may differ slightly in characteristics and host range (Almeida and Purcell, 2003).

Symptoms

The first evidence of Pierce's disease infection usually is a drying or "scorching" of leaves. The leaves become slightly chlorotic along the margins before drying, or the outer leaf may dry suddenly while still green. Typically, the leaf dries progressively over a period of days to weeks, leaving a series of concentric zones of discoloured and dead tissue. "Scorched" leaves detach from the distal end of the petiole (leaf stem) rather than from the base of the petiole, leaving the bare petioles attached to canes, often well after normal leaf fall. The bark on affected canes often matures unevenly, leaving islands of mature (brown) bark surrounded by immature (green) bark or the reverse.

Causal organism

The disease is caused by *Xylella fastidiosa* Wells, Raju, Hung, Weisburg, Mandelco-Paul and Brenner, 1987. The pathogen is a typical fastidious xylem–inhabiting bacterium. It can be cultured on special nutrient media. It is single, straight rod, 0.3 × 1–4 μm, producing long filamentous strands under some cultural conditions. Colonies are small, smooth or finally undulated margins. Habitat is xylem of plant tissue. It is a Gram-negative bacterium, non-motile, aflagellate, strictly aerobic and non-pigmented.

Disease cycle

The pathogen is transmitted by grafting and many species of leafhoppers, such as the blue-green sharpshooter (*Graphocephala atropunctata*) and the glassy-winged sharpshooter. Leafhopper vector required the pathogen after feeding on infected hosts for about two hours and may continue to transmit it to healthy hosts for the rest of their lives.

Control measures

There is no practical control of Pierce's disease of grape in the field. However, the precautionary measures like uprooting and burning of infected vine, production and use of healthy planting material and certification schemes, insect and wild host control, etc. can be taken up. Normal dormant season pruning of grapevines can eliminate many incipient infections of *X. fastidiosa.* However, when the infection is widespread, pruning is no longer effective (Janse, 2006).

5. Pomegranate

Pomegranate (*Punica granatum*) has been exploited commercially for juice. Its products are gaining popularity day-by-day. In the global functional food industry, pomegranate is often mentioned among a novel category of exotic plant sources called "superfruits". Pomegranate is infected by bacterial leaf spot, which is now very common in Maharashtra and Karnataka states of India.

(i) Bacterial leaf spot

Bacterial leaf spot is a serious bacterial disease of pomegranate in certain areas, which was first described by Hingorani and Mehta in 1952 but they did not identify the pathogen. The disease was also observed in Himachal Pradesh (Sohi *et al.,* 1964) and Bangalore (Chand and Kishun, 1987, Kishun, 1993). Ramesh *et al.,* (1991, 1993) reported different isolates of pathogenic bacteria from various parts of pomegranate cultivars and their variability in growth rate under exposure of different antibiotics. Bacterial leaf spot is a serious bacterial disease of pomegranate in certain areas.

Symptoms

The disease appears as few to numerous small deep red spots of 2-5 mm diameter with indefinite margins on leaf blade. The affected leaves may be distorted or malformed, which may shed in severe infestation. On fruits, there is development of raised, dark brown lesions of indefinite margins. Later, these spots turn dark-red. Symptoms may also appear on branches and stem. Cracking of the affected plant parts has also been observed in severe infestation. Rani *et al.*, (2001) reported that *X. axonopodis* pv. *punicae* produced typical black spot symptoms on different parts of pomegranate. Water-soaked turns brown, which are raised and oily in appearance (Rani *et al.,* 2002).

Causal organism

Xanthomonas axonopodis pv. *punicae* (Hingorani and Singh, 1959) Dye is a Gram-negative, rod-shaped bacterium, motile with single polar flagella and size $0.5 \times 1.0–2.5$ mm size.

Disease cycle

Bacterium survives on the tree as well as in diseased fallen leaves. Probably high temperature and low humidity favours the disease. The disease spreads to healthy plants through winds, splashed rains and in new areas through infected cuttings.

Control measures

Adopt following measures to control bacterial blight of pomegranate:

- Destroy the affected plants or their parts.
- Avoid injury to the plants.
- Apply Bordeaux paste to the cuts after pruning.
- Spraying of copper fungicides provides some control.

6. Karonda

Although, karonda (*Carissa carandus* L.) is not grown commercially in our country, but stray plantings can be seen. Its fruits are mainly used for pickle making. Only one bacterial disease (bacterial canker) has been reported from India, which affect karonda.

(i) Bacterial canker

This disease was noticed for the first time during monsoon season of 1962 at Pune (Moniz *et al,* 1964). Later, the disease was also recorded from Rajasthan (Kumar and Patel, 1968; Rathore, 1970).

Symptoms

The disease first appears on lower surface of leaves as small, round, water soaked, translucent dots, gradually expanding, turn dark brown and surrounded with halo. Spots on under surface of the leaf enlarge and become raised. The upper surface may be slightly raised. The central tissue of the spots finally collapses and forms a depression. Twigs show similar symptoms but flowers and fruits do not show any symptoms (Patel and Shekhawat, 1972).

Causal organism

Pathogen of this disease is *Xanthomonas campestris pv. Carissae* (Moniz *et al.*, 1964). The bacterium is Gram negative; rod-shaped, 0.66 × 2.5 mm size, motile with monotrichous polar flagellum.

Disease cycle

The pathogen survives in disease plant debris. *Cestrum nocturnum* L., *Thevetia nerifolia* Juss. are also infected by the bacterium under artificial inoculums (Kumar and Patel, 1968). The bacterium enters the host mainly through stomata but can also enter through wounds. A mean temperature of 24°C and relative humidity of 56 per cent favour the disease.

Control measures

The disease can be controlled by removal of all diseased leaves and pruning the shrub at a height of 45 cm from the ground level. Spraying with 200 pppm phytomycin also reduced the disease (Prasada *et al.,* 1971).

REFERENCES

Agarwal, R.K. and Sharma, V.C. (1970). Natural varietal infectibility of stone fruits with bacterial canker. *Indian J. Agirc. Sci.* **40**: 559-561.

Agnihotri, H., Sarbboy, A.K. and Singh, D.V. (1997). Management of threatening plant diseases of national importance. Malhotra Publishing Houses, New Delhi, India.

Almeida, R.P.P. and Purcell, A.H. (2003). Biological traits of *Xylella fastidiosa* strains from grapes and almonds. *Appl. and Environ. Microbiol.* **69**: 7447-7452.

Anonymous (1997). Area and production under fruit plants. Directorate of Horticulture, Navbahar, Shimla, H.P. India.

Banfield, W.M. (1934). Life history of the crown gall organism in relation to its pathogensis on the red raspberry. *J. Agri. Res.* **48**: 761-787.

Bose, P.C. and Singh, C. (1980). Chemical control of bacterial canker of mango. *Pesticide*, **14**: 30-31.

Bradbury, J.F. (1977a). *Xanthomonas axonopodis.* CMI descriptions of pathogenic fungi and bacteria No. 558. Kew: Commonwealth Mycological Institute.

Bradbury, J.F. (1977b). *Xanthomonas fraariae.* CMI descriptions of pathogenic fungi and bacteria No. 557. Kew: Commonwealth Mycological Institute.

Byrde, R.J.W. and Hutcheon, J.A. (1986). Apple diseases in North India. *Rev. Trop. Plant Path.*, **2**: 107-127.

Chand, R. and Kishun, R. (1987). Studies on bacterial diseases of fruit crops. *Annul Report, IIHR, Bangalore,* pp. 58.

Chand, R. and Kishun, R. (1988). April pruning and bacterial blight in grapevine. *Drakas. Suvenir*, pp 49-51.

Crosse, J.E. (1966). Epidemiological relations of the Pseudomonads pathogens of deciduous fruit trees. *Ann. Rev. of Phytopath.* **4:** 291-310.

Durgapal, J.C. (1971a). A preliminary note on bacterial diseases of temperate plants in India. *Indian Phytopathol.* **24:** 379-382.

Durgapal, J.C. (1971b). Half yearly progress report (Jan.–June, 1971). Division of Plant Pathology, IARI, New Delhi.

Fahy, P.C. and Persley, G.J. (1983). *Plant Bacterial Diseases. A Diagnostic Guide.* New York, USA.

Givindarajan, G and Gnanamanickam, S.S. (1980). Survival of moko wilt pathogen *Pseudomonas fluorescence* strain in the soil of Madras, Tamil Nadu, *India. J. Microbiol.* **20:** 234-235.

Goto, M. (1992). Citrus canker. *In:* Kumar J., Chaube, H.S., Singh, U.S. Mukhopadhyay, A.N., eds. *Plant Disease of International importance*, Vol. III, *Disease of Fruit Crops.* Englewood Cliffs, USA; Prentice Hall, Inc.

Grall, S. and Manceau, C. (2003). Colonization of *Vitis venifera* by a green fluorescence protein-labeled, *gfp* marked strain of *Xylophilus ampelinus*, the causal agent of bacterial necrosis of grapevine. *Appl. and Environ. Microbiol.* **69:** 1904-1912.

Hancock, J.F. (1999). Strawberries. CABI Publishing, New York, USA.

Hasse, C.H. (1915). *Pseudomonas citri*, the cause of citrus canker. A preliminary report. *J. Agril. Res.* **4:** 97-100.

Hatting, M.J., Roos, I.M.M. and Mansvelt, E.L. (1989). Infection and systemic invasion of deciduous fruit trees by *Pseudomonas syringae* in South Africa. *Plant Dis.* **73**: 784-789.

Hill, B.L. and Purcell, A.H. (1995). Acquisition and retention of *Xylella fastidiosa* by an efficient vector, *Graphocephala atropunctata. Phytopathology,* **85:** 209-212.

Hingorani, M.K. and Mehta, P.P. (1952). Bacterial leaf spot of pomegranate. *Indian Phytopath.* **5:** 55-56.

Hingorani, M.K. and Singh, N.J. (1959). *Xanthomonas punicae* sp. nov. on *Punica granatum. Indian J. Agric. Sci.* **29:** 45-48.

Janse, J.D. (2006). *Phytobacteriology: Principal and Practice.* CABI Publishing, CAB International Wallingford, Oxfordshire OX10 8DE, UK.

Jindal, K.K. and Sharma, R.C. (1988). Occurrence of biovar-1 of *Agrobacterium tumefacience* in almonds in India. *Indian Phytopath.* **41:** 614-615.

Jones, D.A. and Kerr, A. (1989). *Agrobacterium radiobacter* strain K1026, genetically engineered derivative of strain K 84, for biological control of crown gall. *Plant Dis.* **73:** 15-18.

Kalita, Pabitra, Bora, L.C. and Bhagabati, K.N. (1996). Phylloplane microflora of citrus and their role in management of citrus canker. *Indian Phytopath.* **49**(3): 234-237.

Kanwar, Z.S. (1976). A note on diseases of pomegranate (*Punica granatum*) in Haryana. *Hary. J. Hort. Sci.* **5:** 171-180.

Kishun, R. (1981). Bacterial canker of mango. *Annul Report, IIHR, Bangalore*, pp. 70.

Kishun, R. (1986). Role of insects in transmission and survival of *Xanthomonas campestris* pv. *mangiferaeindicae.* In *Advances in Research on Plant Pathogenic Bacteria* (S.S. Gnanamanickam and Mahadevan, A. eds), Today and tommorow Printers and Publishers, New Delhi, pp. 33-35.

Kishun, R. (1993). Bacterial diseases of fruits. **In:** Advances in Horticulture, Vol. 3- Fruit crops (Eds., Chadha, K.L and Pareek, O.P.). Malhotra Publishing House, New Delhi, India, pp. 1384-1406.

Kishun, R. and Chand, R. (1988). Survival of *Xanthomonas campestris* pv. *mangiferaeindicae* on weed hosts. *Indian Phytopath.* **41:** 269.

Kishun, R. and Sohi, H. (1983). Bacterial canker in mangoes. *Indian Farmer's Digest*, **14:** 21-23.

Kishun, R. and Sohi, H. (1984). Control of bacterial canker of mango by chemicals. *Pesticides,* **8:** 32-33.

Kumar, V. and Patel, P.N. (1968). Symptoms and host range in a bacterial diseases on *Carissa carandas* L. *Indian Phytopath.* **21:** 315-317.

Liang, L.Z., Sobiczewski, P., Paterson, J.M., Jones, A.L. (1994). Variation in virulence, plasmid content and genes for coronatine synthesis between *Pseudomonas syringae* pv. *morsprunorum* and *P. syringae pv. syringae* from *Prunus. Plant Dis.* **78:** 389-392.

Luthra, J.C. and Sattar, A. (1940). A note on citrus canker (*Phytomonas citri*) in the Punjab. *Proc. 27th Int. Sci. Congr. Part III,* pp. 228-229.

Moffett, M.L. (1973). Bacterial spot of stone fruit in Queesland. *Aust. J. Biological Sci.* **26:** 171-179.

Moniz, L., Sabley, J.E. and More, W.D. (1964). A new bacterial canker of *Carissa congesta* in Maharashtra. *Indian Phytopath.* **17:** 256.

Nayudu, M.V. (1972). *Pseudomonas viticola* sp. nov. incitant of a new bacterial diseases of grapevine. *Phytopath. Z.* **73:** 183-186.

Nambiar, K.K.N. (1993). Diseases and disorders of coconut. ***In***: Advances in Horticulture, Vol. 10 – Plantation crops (Eds., Chadha, K.L. and Rethinam, P.). Malhotra Publishing House, New Delhi, India, pp. 857-882.

Ota, T. (1983). Interaction *in vitro* and *in vivo* between *Xanthomonas campestris pv. citri* and antagonistic *Pseudomonas* sp. *Ann. Phytopath.. Soc. Japan,* **49:** 308.

Patel, M.K. (1929). Viability of certain plant pathogens in soil. *Phytopathology,* **19:** 295-300.

Patel, M.K. and Padhya, A.C. (1948). Bacterial soft rot of mango in Bombay. *Indian Phytopath.* **1:** 127-128.

Patel, P.N. and Shekhawat, G.S. (1972). Bacterial canker of karonda edited In: Bacterial diseases of plant in India, Plant Bacteriology, Division of Plant Pathology, IARI, New Delhi, Vol. I. 199-200.

Prasad, M.V.R., Moses, G.J. and Reddy, G.S. (1978). Variability in *Xanthomonas citri,* the incitant of citrus canker. *Indian Phytopath.* **31:** 227-229.

Prasad, R., Pathak, V.N., Goyal J.P. and Sharma, H.C. (1971). Control of bacterial canker of Karonda (*Carissa carandas). Proc. 2nd Int. Symp. Plant Pathol.,* New Delhi, pp. 145-146.

Ramesh, C., Ram, R.K. , Chand, R. and Kishun, R. (1991). Studies on bacterial blight (*Xanthomonas campestris pv. punicae* (Hingorani and Singh) Dye from leaf to node in pomegranate. *Indian Phytopath.* **44**(3): 370-372.

Ramesh, C., Ram, R.K., Chand, R. and Kishun, R. (1993). Systemic movement of *Xanthomonas campestris pv. punicae* of pomegranate. *Inter. J. Trop. plant Dis. Res.* **11**(1): 85-90.

Rangaswami, G. and Rangarajan, M. (1965). A bacterial leaf spot disease of banana. *Phytopathology,* **55:** 1035-1036.

Rani, U., Verma, K.S., Sharma, K.K. and Rani, U. (2001). Pathogenic potential of *Xanthomonas axonopodis* pv. *punicae* and field response of different pomegranate cultivars. *Plant Dis. Res.* **16**(2): 198-202.

Rani, U., Verma, K.S. and Rani, U. (2002). Perpetuation of and spread of *Xanthomonas axonopodis pv. punicae* causing black spot of pomegranate cultivars. *Plant Dis. Res.* **12**(1): 46.

Rao, A.P., Rao, V.V.R. and Pandith, S.V. (1977). A note on mango canker in Andhra Pradesh. *Indian J. Mycol. Plant Pathol.* **7:** 71.

Rao, M.N. (1988). *Xanthomonas. Drakshavrita,* **12:** 11-12.

Rathore, R.S. (1970). Further studies of bacterial leaf spot of Carissa caranda L. M. Sc. Thesis, University of Udaipur, Jobner Campus.

Reddy, G.S. and Murthi, V.D. (1996). Citrus diseases and their control. ICAR, New Delhi, India.

Ritchie, D.E. (1995). Bacterial spot. *In:* Ogawa, J.M., Zehr, E.I., Bird, G.W., Ritchie, D.F., Uriu, K., Uyemoto, J.K. eds. *Compendium of Stone fruit diseases.* St Paul, USA: *Amer. Phytopath. Soc.* 50-52.

Shekhawat, G.S. and Patel, P.N. (1972). Studies on bacterial canker of mango. edited In: Bacterial diseases of plant in India, Plant Bacteriology, Division of Plant Pathology, IARI, New Delhi, Vol. I. 199-200.

Shekhawat, G.S. and Patel, P.N. (1975). Bacterial diseases of mango. Studies on bacterial canker of mango. *Z. Ptlkrankh Pitschutz,* **82:** 129-138.

Schroth, M.N., Thompson, J.P. and Hildebrand, D.C. (1965). Isolation of *Agrobacterium tumefaciens–A. radiobacter* group from soil. *Phytopathology,* **55:** 645-647.

Scortichini, M., Janse, J.D., Rossi, M.P. and Derks, J.H.J. (1996). Characterization of *Xanthomonas campestris* pv. *pruni* strains from different hosts by pathogenecity tests and analysis of whole-cell fatty acid and whole cell proteins. *J. Phytopath.* **144:** 69-74.

Singh, U.B. (1943). Some diseases of fruits and fruit trees in Kumaon-1. *ICAR Misc. Bull.* **51:** 5 & 11.

Smith, E.F and Townsend, C.O. (1907). A plant tumor of bacterial origin. Science, N.S. **26:** 671-673.

Sohi, H.S., Jain, S.S., Sharma, S.L. and Verma, B.R. (1964). New records of plant diseases from Himachal Pradesh. *Indian Phytopath.* **17:** 35-41.

Stapp, C. (1961). Bacterial plant pathogens. Oxford University Press. U.K.

Stevens, H.E. (1914). Citrus canker. A preliminary bulletin. *Florida Agril. Expt. StationBulletin,* **122:** 113-118.

Sundin, G.W., Jones, A.L. and Olson, B.D. (1988). Overwintering and population dynamics of *Pseudomonas syringae* pv. *syringae* and *P. s.* pv. *morsprunorum* on sweet and sour cherry trees. *Canadian J. Plant. Path.* **10:** 281-288.

Trigalet, A., Frey, P. and Trigalet-Demerey, D. (1994). Biological control of bacterial wilt caused by *Pseudomonas solanacearum*: state of the art and understanding. *In:* Hayward A.C., Hartman, G.L. Eds. Bacterial Wilt: the disease and its causative agent. Wallingford, UK: CAB International.

Vanneste, J.L. (1996). Honey bees and epiphytic bacteria to contro fire blight, a bacterial diseases of apple and pears. *Biocont. News and inform.* **17:** 67N-78N.

Vicedo, B., Penalver, R., Asins, M.J. and Lopez, M.M. (1993). Biological control of *Agrobacterium tumefaciens,* colonization, and pAgK84 transfer with *Agrobacterium radiobacter* K84 and the Tra-mutant strain K1026. *Appl. and Environ. Microbiol.* **59:** 309-315.

Wells, J.M., Raju, B.C., Hung, H.Y., Weisburg, W.G., Mandelco-Paul, L. and Brenner, D.J. (1987). *Xylella fastidiosa* gen. nov.,sp. nov.: Gram negative, xylem–limited, fastidious plant bacteria related to *Xanthomonas* spp. *Int. J. Syst. Bacteriol.* **37:** 136-143.

Young, J.M. (1987). Orchard management and bacterial diseases of stone fruit. *New Zealand J. Exptl. Agril.* **15:** 257-266.

8

Insect Pests of Mungbean and Urdbean and their Integrated Management: A Review

S.K. Singh and D.K. Yadav[1]

[1]Division of Crop Protection, Indian Institute of Pulses Research, Kanpur–208 024, U.P., India
[1]Project Coordinating Unit (Linseed), C.S.A.U.A.& Technology, Kanpur–208 002

Pulses are an important component of the predominantly vegetarian diet of the Indian society. Besides being rich in protein, they are also important for sustainable agricultural production system. Mungbean, *Vigna radiata* (L.) Wilczek and urdbean *V. mungo* (L.) Hepper are widely grown food legumes in the entire Indian subcontinent during the spring/summer season and the rainy seasons. In India, mungbean and urdbean are grown in about 60 lac ha land (Singh *et al.* 2006). About 271 species of insect pests are known to damage these crops during the entire growing season and also as store grain pests (Bhatnagar and Dahia, 2005). The number of insect species found infesting mungbean is more than that in urdbean, which varies with the stages of crop growth (Patnaik *et al.*, 2000). Severe losses are caused by a complex of stem flies during the seedling stage, a large defoliator fauna consisting of grasshopper, flea beetles, hairy caterpillars, semi-loopers and other lepidopteran caterpillars during the vegetative stage and by a varied flower and pod feeding fauna consisting of sucking bugs, thrips, flower and pod webber and lepidopteran pod borers. Singh and Yadav (2002-03) reported bruchid-infesting mungbean with 1-4 eggs/grain and egg hatchability of 85.6% on it. This large insect pest fauna is normally regulated by an equally large and varied parasitoid and predator fauna and also by disease pathogens, which if conserved, provide an effective natural control of crop losses. Insect pest problem is relatively much less in summer crops due to hot and dry weather conditions as compared to the rainy season crop. In rainy season crops, profuse vegetative growth is attained during September, which coincides with the peak activities of a variety of insects. The article encompasses a review of pest array on mungbean and urdbean and their management.

Mungbean is spread all over the country in about 3.4 million hectares. India is a leading country in the production of mungbean with about 65% of the world hectarage and about 54% of the world production (Pohelman, 1991). Urdbean is a highly priced pulse and very rich in phosphoric acid. The important states producing these pulse crops are Madhya Pradesh, Uttar Pradesh, Maharashtra, Punjab, Andhra Pradesh, Karnataka and Tamil Nadu. Under total pulses, both mungbean and urdbean occupy more than 30% of the area and about 69% of total dry bean grown in the country. As regard the productivity levels, the former crop has 0.38 t ha^{-1} while the later has 0.42 t ha^{-1} (Anonymous, 1993) are extremely low in view of the realistic on farm yield potential of >1 to 2 t ha^{-1} for most of the cultivated varieties of these crops. Low crop yield is largely due to crop losses by major insect pests. These crops can be grown on a wide range of soil types in semi-arid tropical environment and several soil moisture regimes, but a warm, deep well drained loam or sandy soil is most suitable for the higher yield. Successful cultivation of mungbean and urdbean crops must, therefore, include their efficient management, and this alone has the potential to more than double the present levels of on farm productivity.

PEST COMPLEX OF MUNGBEAN AND URDBEAN

A thorough scan of literature revealed that among various insect pests of mungbean and urdbean, two and half dozen (Table 8.1) are assumed as major pests of economic importance (Arora and Pajni, 1957; Babu *et al.* 1978, Sinha *et al.* 1975; Srivastava *et al.* 1977, Singh and Singh 1977; Saxena, 1978; Chhabra and Kooner, 1980, 1985a, Chhabra *et al.* 1981 a & b, 1993a; Dhuri and Singh, 1983; Hameed and Dubey, 1983; Sehgal and Ujagir, 1988; Sontakke and Muduli, 1990; Thapa and Timsina, 1999; Natrajan and Subbarao, 1992; Hussain and Saharia, 1994; Borah 1995; Ahmad *et al.* 1998; Lion-Xuming *et al.,* 1998; Ujagir *et al.,* 1999; Reddy *et al.,* 1999; Chandrakar and Srivastava, 2000; Lambrides and Imrie, 2000; Sethuram *et al.,* 2001; Dhakshinamoorthy and Selvanarayanan, 2002 and Jagruti *et al.,* 2002).

INSECT PEST PROBLEM OF MUNGBEAN AND URDBEAN CROPS VIS-À-VIS CROP PHENOLOGY

By and large mungbean and urdbean are grown during rainy season, however, in India these are also sown during spring/summer seasons, mid-February to mid-April in irrigated areas and

Table 8.1. Insect-pest complex of mungbean and urdbean.

Common Name	Scientific Name	Family	Reference
	ORDER ORTHOPTERA		
Deccan Grasshopper	*Colemania sphenariodies* Bol	Acridiidas	Nair, 1975
Grasshopper	*Oxya velox* Fb (=O chinensis Thunb)	Acridiidas	Singh and Singh, 1977
Acridaid	*Chortogonus trachypterus* Blanch	Acridiidas	Lal, 1983
	ORDER HEMIPTERA		
Mealy bug	*Ceococcuscoffeaie* Green	Coccidae	Subbao Roa *et al.* 1976
Cow bug	*Tricentrus bicolor* Dist.	Mambracidae	Mayar *et al.*, 1976
Lygaeid bug	*Chauliops fallax* Scott.	Lygaeidae	Rawat and Sahu, 1968
Bean aphid	*Aphis craccivora* Koch	Aphididate	Nayar *et al.*, 1976
Aleyrodid	*Zaphanera publica* (Singh)	Aleyrodidate	Nayar *et al.*, 1976
Cotton whitelly	*Bemisia tabaci* Genn.	Aleyrodidate	Nena, 1972
Coreid bug	*Riptortus pedestris* Fb	Coreidae	Nayar *et al.*, 1976
	R Linearis Fb	Coreidae	Nayar *et al.*, 1976
	R Fuscus Fb	Coreidae	Nayar *et al.*, 1976
Brown pod bug	*Anoplocnemis phasiana* F	Coreidae	Nayar *et al.*, 1976

Table 8.1 contd.

Table 8.1. Contd.

Common Name	Scientific Name	Family	Reference
Lab bug	*Coptosoma cribaria* Fb.	Pentatomidae	Nayar *et al.*, 1976
Green bug	*Nezara viridula* L.	Pentatomidae	Nayar *et al.*, 1976
Leaf hopper	*Empoasca kerri* Pruthi	Jassidae	Pruthi *et al.*, 1981
	E Moti Pruthi	Jassidae	Chhabra *et al.*, 1981
	E Terminalis Distant	Jassidae	Chhabra *et al.*, 1981
	ORDER: THYSANOPTERA		
Thrip	*Caliothrips indicus* (Bagnall)	Thripidae	Lal *et al.*, 1981
	Megalurothrips distalis (Karny)	Thripidae	Lal *et al.*, 1981
	ORDER: LEPIDOPTERA		
Shoot borer	*Anarsia ephippias* Meyr.	Gelechiidae	Fletcher, 1914
Leaf webber	*Eucosma melanaula* Meyr.	Eucosmidae	Nayar *et al.*, 1976
	Lamprosema indicate F.,	Pyralidae	Nair, 1975
		Pyralidae	Nair, 1975
Spotted caterpillar	*Maruca (testulalis) vitrata*	Pyralidae	Nair, 1975
Pirk borer	*Cydia ptychora* Meyr.	Pyralidae	Lal *et al., 1980*
Pyralid	*Nicoleia vulgalis* Gn.	Pyralidae	Nair, 1975
Blue butterfly	*Euchrysops cnejus* F.	Lycaenidae	Nayar *et al.*, 1976
	Lampides boeticus L.	Lycaenidae	Nayar *et al.*, 1976
Sphingid caterpillar	*Herse convolvuli* L.	Sphingidae	Nayar *et al.*, 1976
Bihar hairy caterpillar	*Diacrisia obliqua* Wlk.	Arctiidae	Nair, 1975
Red hairy caterpillar	*Amsacta moorei* Butl.	Arctiidae	Bindra and Kittur, 1961
	A. albistriga Wlk.	Arctiidae	Anon, 1970
	A. lactinea (Cam) (=Estigmene lactinea)	Arctiidae	Subba Rao *et al.,* 1976
Tussock caterpillar	*Euproctis scintillans* (Wlk.)	Lymantridae	Subba Rao *et al.*, 1976
Leaf miner	*Gracillaria soyella V.D.*	Gracillariidae	Nair, 1975
	Acrocercops phaseospora Myer.	Gracillariidae	Yyas, 1978
	Acrocercops sp.	Gracillariidae	Gupta and Singh, 1980
Green leaf caterpillar	*Antricarsia irrotata B.*	Noctuidae	Nayar *et al.*, 1976
Leaf caterpillar	*Eublema hemirhoda* Wlk.	Noctuidae	Nayar *et al.*, 1976
Pod borer	*Heliothis armigera* Hb.	Noctuidae	Nayar *et al.*, 1976
Semilooper	*Diachrysia orichalcea* F.	Noctuidae	Babu *et al.,* 1978
	Plusia peponis F.	Noctuidae	Nair, 1975
	P. chalcytis F.	Noctuidae	Nair, 1975
	P.daubei F.	Noctuidae	Nair, 1975
	Autographa nigrisigna Wlk.	Noctuidae	Nair, 1975
Tobacco caterpillar	*Spodoptera litura* F.	Noctuidae	Nair, 1975 Babu *et al.*, 1978
	*S. exigua (*Hb.)	Noctuidae	Singh and Singh, 1978
	ORDER: DIPTERA		
Stem fly	*Ophiomyia phaseoli* Coq.	Agromyzidae	Saxena, 1973b
Pea leaf miner	*Phytomyza horticola* Gour	Agromyzidae	Singh and Singh, 1978
	ORDER: HYMENOPTERA		
Ant	*Ceratina binghami* Ckll.	Apidae	Nayar *et al.*, 1976
	Ceratina binghami Ckll.		
	ORDER: COLEOPTERA		
Blister beetle	*Mylabris pustulata* Th.	Meloidae	Nayar *et al.*, 1976
Weevil	*Apion ampulum* Fst.	Apionidae	Subramaniam, 1959
Galerucid beetle	*Luperodes* sp.	Chrysomelidae	Lal *et al.*, 1980
	Madurasia obscurella Jacoby	Chrysomelidae	Menon and Saxena, 1970
Girdle beetle	*Oberea brevis* S.	Cerambycidae	Nair, 1975; Odak and Thakur, 1978; Singh *et al.*, 1978

Table 8.1 contd.

Table 8.1. Contd.

Common Name	Scientific Name	Family	Reference
	ORDER: ORTHPOTERA		
Deccan Grasshopper	*Colemania sphenariodies* Bol	Acridiidas	Nair, 1975
Grasshopper	*Oxya velox* Fb(=*O. chinensis* Thunb)	Acridiidas	Singh and Singh, 1977
Acridaid	*Chortogonus trachypterus* Blanch	Acridiidas	Lal, 1983
	ORDER: HEMIPTERA		
Mealy bug	*Ceococcuscoffeaie* Green	Coccidae	Subbao Rao *et al.* 1976
Cow bug	*Tricentrus bicolor* Dist.	Mambracidae	Mayar *et al.*, 1976
Lygaeid bug	*Chauliops fallax* Scott.	Lygaeidae	Rawat and Sahu, 1968
Bean aphid	*Aphis craccivora* Koch	Aphididate	Nayar *et al.*, 1976
Aleyrodid	*Zaphanera publica* (Singh)	Aleyrodidate	Nayar *et al.*, 1976
Cotton whitelly	*Bemisia tabaci* Genn.	Aleyrodidate	Nena, 1972
Coreid bug	*Riptortus pedestris* Fb	Coreidae	Nayar *et al.*, 1976
	R. Linearis Fb	Coreidae	Nayar *et al.*, 1976
	R. Fuscus Fb	Coreidae	Nayar *et al.*, 1976
Brown pod bug	*Anoplocnemis phasiana* F	Coreidae	Nayar *et al.*, 1976
Lab bug	*Coptosoma cribaria* Fb.	Pentatomidae	Nayar *et al.*, 1976
Green bug	*Nezara viridula* L.	Pentatomidae	Nayar *et al.*, 1976
Leaf hopper	*Empoasca kerri* Pruthi	Jassidae	Pruthi *et al.*, 1981
	E. Moti Pruthi	Jassidae	Chhabra *et al.*, 1981
	E. Terminalis Distant	Jassidae	Chhabra *et al.*, 1981
	ORDER: THYSANOPTERA		
Thrip	*Caliothrips indicus* (Bagnall)	Thripidae	Lal *et al.*, 1981
	Megalurothrips distalis (Karny)	Thripidae	Lal *et al.*, 1981
	ORDER: LEPIDOPTERA		
Shoot borer	*Anarsia ephippias* Meyr.	Gelechiidae	Fletcher, 1914
Leaf webber	*Eucosma melanaula* Meyr.	Eucosmidae	Nayar *et al.*, 1976
	Lamprosema indicate F.,	Pyralidae	Nair, 1975
		Pyralidae	Nair, 1975
Spotted caterpillar	*Maruca (testulalis) vitrata*	Pyralidae	Nair, 1975
Pirk borer	*Cydia ptychora* meyr.	Pyralidae	Lal *et al.,* 1980
Pyralid	*Nacoleia vulgalis* Gn.	Pyralidae	Nair, 1975
Blue butterfly	*Euchrysops cnejus* F.	Lycaenidae	Nayar *et al.*, 1976
	Lampides boeticus L.	Lycaenidae	Nayar *et al.*, 1976
Sphingid caterpillar	*Herse convolvuli* L.	Sphingidae	Nayar *et al.*, 1976
Bihar hairy caterpillar	*Diacrisia obliqua* Wlk.	Arctiidae	Nair, 1975
Red hairy caterpillar	*Amsacta moorei* Butl.	Arctiidae	Bindra and Kittur, 1961
	A. albistriga Wlk.	Arctiidae	Anon, 1970
	A. lactinea (Cam) (=*Estigmene lactinea)*	Arctiidae	Subba Rao *et al.,* 1976
Tussock caterpillar	Euproctis scintillans (Wlk.)	Lymantridae	Subba Rao *et al.*, 1976
Leaf miner	*Gracillaria soyella V.D.*	Gracillariidae	Nair, 1975
	Acrocercops phaseospora Myer.	Gracillariidae	Yyas, 1978
	Acrocercops sp.	Gracillariidae	Gupta and Singh, 1980
Green leaf caterpillar	*Antricarsia irrotata* B.	Noctuidae	Nayar *et al.*, 1976
Leaf caterpillar	*Eublema hemirhoda* Wlk.	Noctuidae	Nayar *et al.*, 1976
Pod borer	*Heliothis armigera* Hb.	Noctuidae	Nayar *et al.*, 1976
Semilooper	*Diachrysia orichalcea* (F.)	Noctuidae	Babu *et al.*, 1978
	Plusia peponis F.	Noctuidae	Nair, 1975

Table 8.1 contd.

Table 8.1. Contd.

Common Name	Scientific Name	Family	Reference
	P. chalcytis F.	Noctuidae	Nair, 1975
	P.daubei F.	Noctuidae	Nair, 1975
	Autographa nigrisigna Wlk.	Noctuidae	Nair, 1975
Tobacco caterpillar	*Spodoptera litura* (F.)	Noctuidae	Nair, 1975 Babu *et al.*, 1978
	S. exigua (Hb.)	Noctuidae	Singh and Singh, 1978
	ORDER: DIPTERA		
Stem fly	*Ophiomyia phaseoli* Coq.	Agromyzidae	Saxena, 1973b
Pea leaf miner	*Phytomyza horticola* Gour	Agromyzidae	Singh and Singh, 1978
	ORDER: HYMENOPTERA		
Ant	*Ceratina binghami* Ckll. *Ceratina binghami* Ckll.	Apidae	Nayar *et al.*, 1976
	ORDER: COLEOPTERA		
Blister beetle	*Mylabris pustulata* Th.	Meloidae	Nayar *et al.*, 1976
Weevil	*Apion ampulum* Fst.	Apionidae	Subramaniam, 1959
Galerucid beetle	*Luperodes* sp.	Chrysomelidae	Lal *et al.*, 1980
	Madurasia obscurella Jacoby	Chrysomelidae	Menon and Saxena, 1970
Girdle beetle	*Oberea brevis* S.	Cerambycidae	Nair, 1975; Odak and Thakur, 1978; Singh *et al.*, 1978

harvested by mid-June before the onset of monsoon. During this period, the insect pest activity is restricted to merely those, which can withstand hot and dry weather. Stem flies cause seedling mortality at the initial stage of crop growth. Further, defoliators like flea bettle, hairy caterpillar, semiloopers, and hawk moth and til caterpillar are spectacular at vegetative stage of the crops. The thrips initiates at the bud stage and poses a serious threat to the peak flowering stage, eventually resulting in flower drop. Subsequently, pod borer damage initiate at flowering and become alarming at the green pod stage. Flea beetle damage commenced from 2nd std. week and continued till 7th std. week with maximum incidence (12.7%) in 5th std. week at Vamban where lepiderpteran borer initiated from 5th std. week having maximum population during 8th std. week and Blister beetle damage initiated from 6th std. week which persisted upto 9th week when senescence set in (Anonymous, 2002). Singh and Rai (2005) reported the *C. chinensis* as serious pest and found its reproductive potential success as high as 77.87% in young parents matings. Bruchid ravages the standing crops in the field as primary infestation and subsequent damage in the storage.

After the initiation of monsoon showers, sowing of rainy season crop starts. So, major area with these legumes are sown during July and harvested by the end of October. Since in rainy season profuse vegetative growth is attained during September, hence it coincides with the insect pests viz. defoliators, flower and pod feeders activity during the period.

ESTIMATION OF LOSSES CAUSED BY INSECT PEST COMPLEX

The assessment of crop losses is very essential to infer and determine the relative importance of insect pests so as to supervise the resource allocation for research and pest management of any crop. Since the reliable and accurate estimation of crop losses due to pest is very difficult as the pest attacks vary not merely across the space but in time as well, hence the pest inflicted losses differ from one agro-ecological zone to another and even in a particular zone between the localities in different years.

In India, serious attempts have not so far been made towards comprehensive assessment of crop losses in pulse except in chickpea and pigeonpea. However, individual attempts have been made to assess crop losses, which are enumerated below.

General Assessment

Sinha (1981) reported from Dholi (Bihar) that urd (variety Navin) yielded 1056 kg/ha under protected condition as compared to only 689 kg/ha in check. The avoidable loss was calculated to be 35%. At Raipur, grain yield of mungbean resulted in 577 kg/ha under full protection, 433 kg/ha need based protection and 267 kg/ha in untreated control (Anon., 2002). Assessment of yield losses in mungbean due to insect pest reflected 465 kg/ha by seed treatment with carbofuran @ 8 ml/kg of seed + 2 spray of monocrotophos at 30 and 50 DAS and control plot registered merely 315 kg/ha (Anon., 2002). Saxena (1983) found that urdbean variety S1 yielded 874 kg/ ha when protected against pest complex in comparison to only 259 kg/ha without protection. Increase in yield was manifold and the avoidable loss was calculated as 71 per cent. Avoidable losses in Haryana have been estimated to be 48% in mungbean, and 27% in urdbean (Yadav and Yadav, 1979). In Bihar, avoidable losses were found to vary from 7 to 35% in urdbean against pest complex (Hameed and Durbey, 1983). In Karnataka, nearly 15% crop was lost to pest complex in mungbean and urdbean (Thondadarya and Hiremath, 1983). Patel and Patel (1983) reported that in Maharashtra one crore rupees worth pulses (including mungbean and urdbean) are lost annually due to pest complex. Acephate application (0.07%) registered 625 kg/ha yields against 416 kg yield in control in urdbean (Anon., 2002) by suppressing pod borers at Berhampore (Orissa).

Pest-wise Assessment

Stemfly, Ophiomyia centrosemalis *de Mejre*

Stemfly infestation begins as soon as the plants appear above ground. The attacked plant becomes stunted and yellow, and finally dies. The plants that survive an attack produce few pods, many of which are empty or contain small seeds. Stemfly damage is more severe during the rainy season than in summer season crop (Sehgal and Ujagir, 1988). Talekar (1990) reported 20% yield loss in mungbean in Taiwan. The pest under population pressure can cause withing of plants to the extent of 99.3% (Brar *et al.*, 1993b). Stemfly in mungbean was to the extent of 39% infestation and in urdbean was as high as 47.7% during *Kharif* 2002 at Coimbatore (Anonymous, 2002). At Pantnagar, infestation due to this pest was as high as 80% (Anonymous, 2001). Stemfly damage was 26.6% during Rabi at Vamban (Anonymous, 2002).

Leaf Hopper, Empoasca kerri *Pruthi*

The infested plant becomes stunted and a definite reduction in yield occurs. Gupta and Saxena (1971) reported that every unit increase in number of Jassids per plant reduces the yield by 443 kg/ha in mungbean.

Whitefly, Bemisia tabaci Gennadius

The pest sucks the sap of leaves and secrets honeydew that causes shooty moulds and hampers the photosynthetic activity of leaves. The whitefly is a potent vector of yellow mosaic virus (YMV) both in mungbean and urdbean, as most of the crop loss is due to YMV. The YMV can cause losses to the tune of 30-70% in mungbean (Marimuthu *et al.* 1981). In case of severe infection of YMV, a very few pods are produced. The pods are reduced in size and grains are shriveled and reduced in size as well. About 57.7% MYMV damage was reported during *Kharif* when the pest intensity was 41.3 whiteflies/cage, in BM 4 (AVT) genotype of mungbean (Anonymous, 2001) in Hisar, while 36.5% infestation was recorded in the same season in following year (Anonymous, 2002). The maximum whitefly population/cage was recorded 24.7 in Pusa 105 (Anonymous, 2002). Basheer *et al.* 2002 found virus disease incidence ranging from 0 to 66.6%.

Black Aphid, Aphis craccivora Koch

The feeding by aphids results in depletion of plant sap by sucking tender leaves, and yield may be upto 30 to 70 per cent. Besides, aphid can transmit banding cucumber mosaic virus in mungbean. Patel (1976) reported 40 to 80% damage in pea by *A. craccivora.* The greenish black aphid, *Aphis craccivora* Koch was reported from Kanpur in Uttar Pradesh (Srivastava and Singh, 1976), Punjab (Chhabra *et al.* 1986a) and Orissa (Raju and Panda, 1983) in India, causing damage to mungbean and blackgam. While recording reaction of mungbean entries against major insect pest during Kharif 2002 at Badnapur, aphid infestation was found with an average of 2.63 aphid (Anonymous, 2002). The damage varied from 4.26 to 34.78 per cent at Shillongani (Anonymous, 2002).

Galerucid Beetle, Madhurasia obscurella Jacoby

This is one of the major pest of mungbean and urdbean found in Uttar Pradesh, Haryana, Orissa and Delhi in India and causes 20 to 60% damage to the crop from germination till harvest (Srivastava and Singh, 1976). The larvae also damage root nodules of the crop at rates between 25 to 60% (Sahoo and Patnaik, 1994). While screening of mungbean entries (AVT_1 and AVT_2) for resistance to *Madurasia obscurella* during Kharif 2002 at Lam, the maximum mean infestation was recorded 66.02% against Pusa 0172 (Anonymous, 2002) and the number of holes/leaf by the beetle was 3.76 at Badnapur in TARM-18 entry of mungbean. At Khargone% damaged by this pest was found varying significantly at all the three intervals of 15, 30 and 45 DAS. The infestation varied from 30.83 and 39.27 per cent in JV 2 and IU 31-13, respectively (Anonymous, 2002) and from 37.96 to 70.86% infestation by the pest was reported from Lam (Anonymous, 2002).

Spotted Borer, Maruca (testulalis) vitrata Fab.

The legume pod borer or spotted pod borer *M. (testulalis) vitrata* is a serious pest of grain legumes in tropics and subtropics. This pest is able to establish on the crop early. Its caterpillar feeds on buds, flower, and bore the pods, make webs and causes damage. According to an estimate, the extent of damage to pods alone varies widely from 5.7 to 89.4% in Bihar

(Sinha *et al.*, 1979). Mungbean cultivars registered 7.52 to 8.93% damage by the pest in Vamban (Anonymous, 2002).

Gram Pod Borer, Helicoverpa armegera Hubner

The young larva feeds on tender leaves, buds, flowers and subsequently it bores into the pods and feeds on the seeds with its head and part of the body only thrust inside, the nest remains outside. A single larva may destroy 30-40 pods before its reaches maturity, and 44.96% damage was reported from Ranchi (Anonymous, 2002) against genotype MH 96-1.

Thrips, Scirtothrips dorsalis and Megalurothrips distalis Karney

This is one of the major insect pests of mungbean and urdbean in India, especially in northern states of the country. Both adult and nymph suck the buds and the flowers shed before opening. Infested inflorescence becomes abnormal showing symptoms of flower drop. Excessive dry and drought weather resulted in severe infestation. Losses in summer mungbean in the Punjab state, India were dominated by thrips, *Megalurothrips distalis* (Karny), and have been reported to be 54.3% losses by this pest (Chhabra and Kooner, 1985a). Chhabra (1992) reported avoidable losses in mungban from 20 to 64.7% and in urdbean from 18.1 to 61.5% from reports of different research stations of the All India Coordinated Pulses Improvement Project (AICPIP), India.

Bihar Hairy Caterpillar Spilaractia obliqua Walker

Caterpillars cause damage. When young, they feed gregariously but on few plants in scattered manner and eat the green matter of leaves, leaving behind only the midrib. The grown up caterpillars feed voraciously on leaves, soft stems, branches, and may totally denude the crop within few days.

Bruchid Callosobruchus spp.

Among the insect pests that attack pulses, bruchids are the most destructive. They inflict damage not only on the standing crops in the field, but also cause substantial loss in storage. Grubs damage the pods in the field and seeds under storage condition. Pre-harvest infestation, usually of low level, is very serious as it multiplies in severe proportion in storage. Sparse information is available to quantify field infestation in different crops. Prevett (1961) reported that in cowpea bruchid complex caused 3.1 and 11.0% infestation of pods collected as they dried or as single harvest, respectively. The extent of pod damage by *C. maculatus* varied from 0.3% to 10% (Lal *et al.* 1980; Latif and Reed 1983; Bindra and Jakhmola (1967). Reported damage due to pulse beetles varied from 1.2 to 10.2 per cent. Dias (1987) reported on the basis of samples collected from 8 different ecological zones that Bruchids infected seeds to the extent of 19.3%. Damage due to stored grains pest, *Callosobruchus chinensis* in mungbean starts from the field. The secondary infestation is more damaging (Talekar, 1990). The infestation has been reported upto 100% of the stored seeds of mungbean and cowpea within 3 to 5 months and under ordinary storage conditions. (Caswell and Akibus, 1980). The losses are almost 100% if pulses are stored beyond rainy season (Anonymous, 2004). Gujar and Yadav (1978) reported a weight loss in mungbean of 55.6 to 73.5% in individual seed damaged by single *C. maculatus* and 30 to 70% by *C. chinensis* in one generation. The losses in seed by bruchid due to improper storage have been reported to be 50 per cent after only three months of storage (Hussain and Abdel, 1982). Rathore and Sharma (2002) reported an estimated loss of nearly 0.21 million tones in pulses caused by bruchids, costing 315 million rupees. *Vigna* pulses are ravaged by bruchid genera of *Callosobruchus, Bruchidius, Speularius* and *Zabrotes.*

Damaging Stages

Insect Pests of Root and Seedling Stage

Stemflies (Agromyzid) cause serious damage at seedling stage to the crop in India, Malaysia and Taiwan (Ho 1967; Sehgal and Ujagir 1988; Talekar, 1990). Stemfly damage in mungbean is mainly due to *Ophiomyia phaseoli* (Tryon). (Srivastava and Singh, 1976; Saxena 1978; Islam *et al.* 1984, Gupta and Singh 1984b; Lal 1985; Rajapakse and Jayasena 1989). There are reports on five species of agromizid stemflies, *Melanagromyza sojae* (Zehntner), *M. dolichostigma* de Meijere and *M. obtusa* Malloch *Ophiomyia phaseoli* (Tryon) and *O. spencerelia* and *O. centrosemalis* (Sehgal and Ujagir, 1988; Talekar, 1990). This pest under population pressure can cause wilting of plants to the extent of 99.3% (Brar *et al.* 1993b). The flies are active in summer and mate 2-6 days after emergence (Panwar, 1995). The pest is also recorded to remain active during winter as larvae or pupae in peas (Ram Ujagir, 1997). Singh (2003) reported summer crop in northern irrigated plains to be more vulnerable to stemfly and it is one of the major pest in India.

As regards the grubs of galerucid beetle, *Madurasia obscurella,* it feeds on the root hair nodules of the plants. (Saxena 1973; Gupta and Singh 1984a). Manjunath *et al.* (1988) reported that a wireworm, *Agriotes* sp., feeds on the roots of mungbean plants. Nath and Singh *et al.* (1998) noticed its peak during the flowering stage. Singh (2003) reported the nature of damage of adult beetles to feed root hairs and root nodules and also on the foliage and make small circular holes in the leaves at seedling stage. The caterpillars of cutworms cut stems and branches of seedling at night (Anonymous, 1976). *Amsacta moorei* also cause extensive damage.

Insect Pests of Vegetative Stages

(i) Lepidoptera, Coleoptera, Orthoptera, Foliage Feeders

Insect pest species of the order Lepidoptera, Coleoptera and Orthoptera cause maximum damage at the vegetative stage. Insect pests' defoliating these crops feed primarily on leaves, flower and even pods as well. It is difficult to estimate damage inflicted by the defoliating insects as it depends upon the extent and duration of damage to the production of photosynthate. Normally, it will be rather little to that of the proportion, etc. loss in foliage as new leaves, flowers and pods may develop, compensating partially for the destroyed parts. Removal of 25 to 50% of the area of each leaf at the flowering stage of the mungbean plant to stimulate loss for insect feeding did not reduce the total dry weight or seed yield, but seed yields were reduced with 75% reduction in leaf area (Robertson, 1981).

As regards the foliage damage, numerous lepidopterans have been reported damaging mungbean and urdbean. Ooi (1973) reported from Malaysia about the damage of leaf rollers and leaf folders. Besides, *Hedylepta* (*Lamprosema*) *indicala* (F) also damage mungbean crop in India (Sehgal and Ujagir, 1988) and Philippines (Catipon, 1986). *Homona coffearia* Nietner, *Sylepta sabinersalis*, Wlk, *Agrius convolvuli* Linn) and *Helicoverpa armigera* (Hub.) are the other foliage feeders in India and Philippines (Chhabra and Kooner 1980; Nair 1986; Litsinger *et al.* 1988; Sehgal and Ujagir 1988). Pupation of *A. convoluli* takes place in soil, sometimes 8-10 cm down and pupal period ranges from 17-26 days to 4-6 months depending upon the climate (Hill, 1993). Damage by *Cydia ptychora* (Meyrick) has been reported from Uttar Pradesh (Lal *et al.*, 1980) and Southern India (Nair, 1986). Subba Rao *et al.* (1974) reported caterpillars of *Amsacta. lactinea* (Cramier) as a major defoliator in South India. A severe outbreak of hairy caterpillars *A. moorei* and *Spilosoma obliqua* (Wlk) has been reported from Rajasthan (Parihar, 1979) and northern Bihar (Sinha *et al.*, 1985). The seriousness of *Spilosoma obliqua* has been reported from Bihar, M.P., U.P. and Punjab (Atwal and Dhaliwal, 1986) and (Raman Goud, 1993), Delhi (Singh *et al.* 1990), from Myanmar, China, Japan and Bangladesh (Hazarika, 1952). Singh and Pandey (1995) studied the biology of *A. moorei* on different varieties of groundnut in Delhi and common incidence was recorded from AP, TN, Karnataka and Maharashtra.

Adult galarucide beetle, *Madurasia obscurella* feed on the leaves (Yein and Singh, 1981; Gupta and Singh, 1984a). Galerucid beetle in green gram infestation was reported during Rabi 2001-02 at Manipale (Anonymous, 2002). Various species of beetles are leaf feeders or they may serve as virus vectors. Lal *et al.* (1980) reported damage by *Epilachna* sp. from Kanpur, Uttar Pradesh. Urdbean mottled virus reported from India and Thailand was transmitted by bean leaf beetle, *Cerotoma trifurcata* (Forsti) and Mexican bean beetle, *Epilachna varivestis* (Mulsi). *Monolepta signatata* Oliver was found to transmit these disease, while *Aphis craccivora* Koch did not transmit this disease (Honda *et al.*, 1982).

Lal *et al.* (1980) reported that the weevil *Apion sp.* feeds on foliage, stem, buds, flowers and pod of mungbean and urdbean (Dhuri and Singh, 1983) in India. *A. ampulum* Fst and *A. collaris* Mshill have been reported by Sakamoto *et al.* (1983) feeding on buds, flowers and pods of mungbean in Japan. Three chrysomelid viz. *Altica coerulea* Oliver, *Luperodes* sp. and *Madurasia obscurella* have been reported to feed on foliage of mungbean and urdbean (Lal *et al.* 1980; Dhuri and Singh, 1983).

In the crops, grasshopper damage is mostly confirmed to the rainy season in India. So, *Chrotogonus trachypterus* Blanch, and *Oxya velox* (Feb.) usually cause damage as reported by Singh and Singh 1977; Dhuri and Singh 1983; Sehgal and Ujagir (1988). Patnaik, Mohanti and Chhabra (2000) reported *Gastrimagus marmoratus* (Thunb); *Colemania sphenariodes* Boliv; *Chrotogonus trachypterus* Blanch as minor pests of mungbean and urdbean. Foliage feeding Coccinelids like *Epilachna dodecastigma* Wied; *E. Varivestis* Muls; *Leis dimidiata* Fabr; are also of minor importance prevailing on these legumes.

(ii) Hermipteran, Thysanopteran and Acarina Floage, Flower and Pod Feeders

Insect pests of order Hamiptera, Thysanoptera and mite species (order: Acarina) damage the tender foliage, flowers and pod by sucking the cell sap. Their damage, by and large, leads to flower drop and poor seed quality. Attack of sucking pests on summer/spring season crops in India usually occur with the emergence of reproductive stage of the plant. This stage coincides with the dry periods from May to mid-June and later with rainy season and it causes major yield loss in mungbean and urdbean. However, rainy season crop can withstand the feeding stress caused by the sucking pests due to increased vigour and relatively greater compensating ability than the summer season crops.

Thrips species attack urdbean crop in Delhi (Dhuri and Singh, 1983). Two species of thrips, *Caliothrips indicus* (Bognall) and *M. distalis* have been reported to attack tender buds and flowers of mungbean in Punjab, Uttar Pradesh and Maharashtra in India (Kooner *et al.*, 1983; Lal 1985; Dahaya *et al.* 1986; Sehgal and Ujagir 1988). Chhabra and Kooner (1985b) reported extensive thrips damage to summer mungbean crop in Punjab and observed 19 thrips/25 flower, resulting in 82.4% flower drop. Three species of thrips viz., *Caliothrips* sp., *Scirtothrips* sp. and *Thrips palami* Karney have been found to cause damage to mungbean crop in Philippines (Litsinger *et al.* 1988). *M. usitatus* Zehntrer has been reported by Sepswasdi *et al.* (1990) causing leaves and pod damage to mungbean in Thailand. As regards biology and bionomics of *M. distalis,* it has not been studied in detail, and its one generation is completed within three weeks (Reed *et al.* 1989). Sehgal *et al.* (2000) reported that thrips *Caliothrips indicus* and *Megalurothrips distalis* are the dominant. SML 112 and SML 103 were rated resistant and had the least inflorescence deformity due to *M. distalis* (Chhabra and Kooner, 1994).

Whitefly, *Bemisia* tabaci Gennadius, a polyphogous pest, has been reported to cause direct and indirect damage by transmitting mungbean yellow mosaic virus (MYMV) (Srivastava and Singh, 1976; Chhabra and Kooner 1980; Singh and Sehgal 1983). The host range, biology and field incidence has been studied widely in India (Nene, 1972; Murugesan and Chelliah 1978). Kooner and Chhabra (1980) observed 12 overlapping generations per year and it was most active during May to September in Punjab. Chhabra *et al.* (1984 and 1988) reported that the mechanism of resistance in the crops against whitefly, *B. tabaci;* Jassids, *Empoasca moti* and *E. terminalis* and MYMV is due to antibiosis. Incidence of MYMV caused by *B. tabaci* was less in summer than in the rainy season crop due to higher temperature (40 $\pm$ 5°C) (Sharma and Verma, 1982). Chhabra and Kooner (1993) tested thirty nine accession of mungbean *Vigna radiata,* for their reaction to *Bemisia tabaci* and found ML 370, ML 428, ML 459, ML 506, ML 508 AND ML 537) as being resistant against the insect pests and MYMBV.

The greenish black aphid, *Aphis craccivora* Koch was reported from Kanpur in Uttar Pradesh (Srivastava and Singh, 1976), Punjab (Chhabra *et al.*, 1986a) and Orissa state (Raju and Panda, 1983) in India causing damage to mungbean and blackgram. *Aphis gossipi Glover, A. craccivora* and *Myzus* persicae Sulz were found to transmit the mosaic-mottling virus in mungbean (Rama Krishna *et al.*, 1973). *A. craccivora, Acyrthosiphon pisum* Harris and *A. sesbaniae* David were reported to transmit Iranian strain of YMV from Iran (Kaiser, 1979). Detailed biology of *A. craccivora* was studied by Sri Kanth and Lakkundi (1988) on mungbean and urdbean at Bangalore and revealed that the life cycle of aphid on urdbean was completed earlier (10.34 $\pm$ 4.11 days) than on mungbean (12.47 $\pm$ 3.09). Li Dux *et al.* (1994) studied the bionomics of *A. craccivora* in mungbean and found that it was closely related to the crop growth period.

Three species of leaf hoppers *Empoasca kerri*, Pruthi and *E. terminalis* Distant are known to damage mungbean and urdbean in Haryana, Delhi, India (Chaudhary *et al.*, 1980; Dhuri and Singh, 1983), while *E. moti* and *E. terminalis* have been recorded as serious pest of these crops at Ludhiana, India (Chhabra *et al.* 1981a). Nath and Singh (1998) reported that Jassids *(Cicadellids)* first appeared at the seedling stage of the crop and maintained their population up to pod formation. Prakash and Rao (2000) reported three Jassid species *viz. Empoasea kerri Pruthi, E. moti* Pruthi and *E. terminalis* Dist. which suck the leaves; the former species is prevalent throughout India, while the later is of regular occurrence in north India.

In mungbean and urdbean, four species of *Riptortus* viz. *R. fuscus* (Fab.), *R. linearis* (Fab.), *R. padestris* (Fab.) and *R. Pilosus* (Thunb) have been reported to cause damage in India and Philippines (Dhuri and Singh 1983; Sehgal and Ujagir 1988; Litsinger *et al.* (1988); Sharma and Dutta 1997). *Clavigralla gibbosa* (Spinola) has been reported to cause extensive damage in India to mungbean (Sehgal and Ujagir, 1988). Damage by *Chauliops fallax* Scott. to leaves, flowers and pods have been reported from Madhya Pradesh (Rawat and Sahu, 1973) and Uttar Pradesh (Lal, 1985) in India. *Nezara virudula* (Linn.) has been reported to cause damage to developing pods of mungbean in Madhya Pradesh, India (Singh and Rawat, 1982) and in Philippines (Litsinger *et al.*, 1988). This pest has also been reported as destructive pest of mungbean in India, Philippines and Taiwan that damage pods and seeds. Highly toxic saliva is injected into the plant tissue as it feeds. Pods may fail to develop or seed may be shrivelled from the punctures. Necrotic areas often develop at the feeding sites (Talekar and Chen, 1983). Damage by *N. virudula* is a major cause of flower drop and drying of pods during summer in India (Sehgal and Ujagir, 1988). Sharma and Dutta (1996) studied population dynamics of *Nezara viridula* (L.) on summer mungbean and found it to be appearing during 3rd and 4th week of April at the pod formation stage. Peak numbers of *N. viridula* coincided with the period of pod maturity, and as the crop reached full maturity, pest numbers declined. There was a positive correlation between the minimum temperature and number of *N. viridula.* Natural enemies of *N. viridula* included the predatory thomisid *Thomisus* sp. and the *Scelionid Trissolcus* sp. Comparison of the population trends of pod sucking bugs on two summer green gram varieties was done in which the mean *R. linearis* population on AAU 34 during 1994 was 9.83 bugs/10 sweeps, while the level was 11.87 bugs/10 sweeps during 1995. The population levels of both the bugs were higher on AAU 34 than on SG-1 variety, with AAU 34 having stouter and taller plants with more branches and bigger leaves and thus forming a larger canopy than SG-1, hence providing a more favourable micro climate to the bugs for survival and multiplication (Sharma and Dutta, 1997). *Riptortus pedestris* L. and *Nezara* sp. was reported from Vamban (Anonymous, 2002).

Piezodorus hybneri Gmelin and *P. rubrofusciatus* Fab. the pentatomid bugs have been reported to damage leaves and pods of mungbean in Thailand and India, respectively, whereas *P. hybneri* is an economically important pest of mungbean and soybean (Poonsavasde and Sukonthabhirom, 1990; Singh and Sehgal, 1993). The bugs normally suck the sap from the developing pods causing shrivelled, deformed and undersized pods and seeds.

In mungbean, mites belonging to family Tetranychidae are common and damage the crop by sucking the cell saps of leaves. Mites are of minor importance, but a severe infestation of *Tarsonemus* sp. has been reported by Singh and Chauhan (1984) from Hisar, India. *Tetranychus tilarius* Linn. *T. urticae* Koch, *T. cinnabarinus* (Boisd.) and *T. neocaldonicus* Andre have been reported to cause leaf and pod damage in mungbean in Tamil Nadu, Andhra Pradesh, India and South East Asia (Babu and Rajasekaran, 1988; COPR, 1981). *Polyphagotarsonemus latus* Banus has been reported to cause damage in mungbean at Ludhiana, India (Chhabra *et al.* 1993a) and in SE Asia (Sepswadi *et al.* 1990) by sucking sap of leaves of the crop.

Insect Pests of Bud, Flower and Pod Stages

The adult of blister beetle *Mylabris pustulata* Thunb feeds on the flower on mungbean crop at Pantnagar (Sehgal and Ujagir, 1988), whereas Chhabra *et al.* 1993a, also found this beetle to feed on the anthers of mungbean and urdbean crops in Punjab, India. *M. phalerata* Pallos has been reported to damage flowers of urdbean in Delhi (Dhuri and Singh 1983) and mungbean at Pantnagar (Singh and Sehgal, 1993).

Prakash and Rao (2000) reported blister beetle distribution throughout India, and both the spp. *viz. M. pustulata* and *M. phalerata* are prevalent and they are flowers feeders. *Mylabris* sp. reported damaging mungbean crop during *Rabi* 2001-02 (Anonymous, 2002), whose damage initiated from 6th week and persisted upto 9th week (7.3) at the time of senescence set in at Vamban; interestingly the beetles population was in increasing trend from its initiation (1.7 beetles/5 plants).

While considering bruchid in mungbean and urdbean, the major three species viz. *Callosobruchus chinensis* (L.) *C. analis* Fab., and *C. maculatus* are prevalent and have been reported to damage the pods and grain in field in India (Pruthi and Singh 1950; Gujar and Yadav 1978; Chhabra *et al.* 1993a). These species also damage mungbean pods in Taiwan (Talekar 1988) and Philippines (Litsiger *et al.* 1988).

Singh and Singh (1990) tested the response of various cultivars of urdbean to field infestation of pulses beetle and found that JU 78-3, RU-2, T-9 and JU 78-27 were free of *C. chinensis* infestation. Likewise, Dharmasena and Subasinghe (1986) studied seed characteristics contributing resistance to *Callosobruchus* spp. and found varieties with small seeds and glossy seed coats to be associated with higher degree of resistance than large seeds with a dull surface. Sarkar *et al.* (1991) screened sixteen genotypes of mungbean to test the resistance against *Callosobruchus maculatus* on the basis of the number of healthy seeds (resistance), number of eggs deposited (oviposition preference) and total emergence holes for 50 seeds. All three characters were highly heritable, but ovipositional preference was slightly lower in heritability than the other two. Resistances were negative function of adult survival, showing the importance of this character for selecting bruchid-resistant varieties. Haque *et al.* (1992) screened 150 genotypes for susceptibility to *Callosobruchus maculatus* at 30°C and 70% RH. The genotypes SBG 3 and UH 83-2 were the least susceptible (3.0-3.5% weight loss) whereas PU-26 and UHB 2-26 were the most susceptible. Talekar and Lin (1992) also characterized mungbean resistance line against *C. chinensis.* Raghunathan and Ram (1992) screened mungbean varieties and found lowest feeding damage and adult emergence with cv. PDM-14. The work of Singh *et al.* 1994; Singh *et al.* 1995; Negi *et al.* 1997; Verma 1998; Singh and Rine, 2000; Yadav *et al.* 2000; Singh and Pandeey 2001; Srivastava and Kaur (2002) deserve special mention who screened a number of genotypes of different pulses against this dreaded stored pest.

Larvae of Lycaenids, *Euchrysops* spp. have been reported to damage pods of mungbean in Tamil Nadu (Jayaraj 1988) and Uttar Pradesh, India (Lal *et al.,* 1980) and in North America (COPR 1981). Pods of mungbean and urdbean crops have been reported to be damaged by *Lampides boeticus* (L.) in India (Chhabra *et al.* 1980; Dhuri and Singh, 1983). Damage to flowers and pods of mungbean by *Exelastis atomosa* (Wlsm) has been reported from Uttar Pradesh (Lal *et al.* 1980). Pod borer *Maruca* (*testulalis*) *vitrata* is widely distributed in tropics. Eggs of this pest are deposited on petals and sepals; the larvae feed on the flowers and bore into the pods of *vigna spp.* crops. Young larvae of *M.* (testulalis) *vitrata* may cause flower bud shedding or destroy reproductive parts of flowers in mungbean in India (Chhabra *et al.*, 1980; Sehgal and Ujagir, 1988), in Philippines (Litsinger *et al.* 1988) and in urdbean in India (Dhuri and Singh, 1983).

The other pests of bud flower and pods like *Ostrinia furnacalis* Guenae, *Etiella behrii Zeller* and *E. zinckennela* have been reported to web flowers and feed on pods of mungbean in Philippines. (Litsinger *et al.* 1988). *Helicoverpa amigera* has been reported to feed on buds, flowers, pollen or immature pods of mungbean in India and is a serious pest at post flowering stage in Philippines (Chhabra *et al.* 1980; Chaudhary and Sharma 1982; Litsinger *et al.* 1988; Sehgal and Ujagir, 1988).

PEST MANAGEMENT

The mungbean and urdbean pulses of *vigna* group are by and large cultivated in the subsistence farming systems of hot, humid and semi-arid environments of tropical and subtropical parts of the globe. Besides the management of the scarce available resource for crop production, the farmers have to fight the on-slought of a large and diverse pest fauna and complex of plant disease pathogens. Despite the availability of improved cultivars, which promise yield potentials of >1–2. t ha^{-1}, the farmers' average on-farm yields are often < 0.4 t ha^{-1}

For the successful cultivation of mungbean and urdbean crops, it is imperative to have some efficient management system with the farmers against the damaging effects of the large and diverse insect pest fauna. Under the new concept of IPM of harmful insect pest population, we must exploit the cultural practices, host plant resistance lines against different insect pests and disease pathogens, the endemic natural enemy fauna and options of biological control, and above all need based insecticides used to keep pest population below the economic threshold levels (ETL). Efforts should also be made to develop suitable and precise forecasting modules against the major pests of mungbean (Mathur *et al.,* 1997; Yadav and Singh, 2006). Almost similar objectives were put forth in an earlier review of mungbean pests in India by Sehgal and Ujagir (1988), but a lot more new information that has become available on different component strategies of IPM is documented here.

Cultural Control

The cultural practices evolved by farmers have sustained cultivation of *vigna* spp. crops even under minimum input conditions and often in semi-arid environments. These practices which help destroy different stages of major insect pests or prevent them from causing economic injury include tilling and cultivating relatively higher seed rate to compensate for seedling mortality, regularized irrigation and fertilizer application, low risk cropping schemes, manipulation of crop phonology by adjustments in sowing/harvesting dates and destruction of crop residues. (Asthana and Kim, 1996).

Manipulation in date of sowing in urdbean helped reducing the population of galerucid beetle and jassid *Empoasca kerri* (Saxena *et al.* 1973). Inter-cropping of mungbean with non-host plants like sorghum and pearl millet decreased the incidence of jassids and other major pests populations, which were high in mungbean and urdbean intercropping at Delhi. Similarly, inter-cropping of urdbean with pigeonpea delayed the appearance of green jassid, green bug and membracid (Singh and Singh, 1977). Relatively higher seed rate and synchronous planting of mungbean by the farmers in Philippines, together with crop rotation of grain legumes with rice and/or corn adopted by the farmers, have reduced damage due to flea beetles, major pests and/or soil pests (Litsinger *et al.* 1978b). In India too, mungbean sown at optimum time had reduced damage due to jassids, whitefly, galerucid beetle, pod borers and MYMV (Lal, 1983). Dasgupta and Chaudhary (1985) observed that intercropping of mungbean with urdbean had reduced MYMV incidence. Similarly, growing of trap crop, clean cultivation and crop rotation and destroying alternate host, *Solanum nigrum* had minimized beanfly, *Ophiomyia phaseoli damage* in India (Ramakrishanan, 1986).

Application of potash at 50 kg ha^{-1} reduced the pod borer infestation and increased yield by 36% (Ramakrishnan, 1986). Dhuri *et al.* (1986) observed that urdbean intercropped with sesame, pigeonpea, mungbean and sorghum had influenced the incidence of galerucid beetle, jassid, whitefly, semi-looper, Bihar hairy caterpillar and pod borders. It was reported that there was reduction in jassid population in urdbean-sorghum and urdbean-sesame intercropping; however, intercropping with mungbean and pigeonpea further enhanced the population of this pest. Similarly, intercropping urdbean with sorghum and pigeonpea delayed the appearance of semi-looper, *Plusia orichalcea.* Intercropping with sorghum brought about significant reduction in the incidence of pod borers in the urdbean. Mungbean variety CO4 intercropped with cotton variety MCU 10 registered a minimum incidence of leaf hopper (*Empoasca kerri*) and thrips (Janagarajan, 1987). Maize, sorghum and pearl millet when grown as barrier crops for mungbean, significantly reduced the population of whitefly, *B. tabaci* and incidence of MYMV and resulted in higher yield (Ravindrababu, 1987).

Different combinations of carbofuran 3G and phorate 10G along with nitrogen and phosphorus fertilizer significantly reduced the leaf damage due to *M. obscurella* and *Anomis flava* (Gatoria and Singh 1988). Handpicking of egg masses and young larvae play an important role in controlling the Bihar hairy caterpillar (*D. obliqua*), the banded beetle (*M. phalerata*), the brown bug (*C. gibbosa*) and the tobacco caterpillar (*S. litura*) (Lal, 1987).

Thus, manipulation of dates of sowing and harvest to escape peak pest population; intercropping preferably with non-legume hosts; handpicking of egg masses of hairy caterpillars; proper soil preparation techniques and application of high dose of fertilizers together with carbofuran treatment had helped achieve optimum control of harmful insect pest population.

As regards cultural control measures of bruchid, time and frequency of harvesting, removal of alternate host plants, multiple cropping lines, intercropping beans with maize reduced the abundance of the bruchid probably because of the shade given by the maize canopy (Labeyrie and Maison, 1957). Besides, field infestation may result from bruchids migrating from infested seeds in adjacent granaries to the ripening crop (Prevett, 1961 and Caswell, 1973). Therefore, infested seeds in the warehouses or granaries should be treated or disposed of before the planting season.

High insect proliferation rates in the tropics make good storage hygiene particularly important (De Lima, 1979a) when farmers in Nigeria after harvest leave the pods in heaps or harvest the whole plant and remove the pods. Later pods are open to attack by bruchids. Storage in the granaries immediately after harvest gave physical protection and considerably reduced infestation (Caswell, 1968).

Host Plant Resistance (HPR)

The success of HPR to insect pests, as IPM strategy depends on the efficacy of other available control options. Litsinger *et al.* (1978a) had suggested that in Philippines, priority in breeding for insect resistance in mungbean be given to flea bettles, leaf hopper and stemflies, as these are difficult to control by chemicals. HPR studies in India have largely focused on mungbean and urdbean resistance to key pests like whitefly, jassids, thrips and lepidopteran pod borers (Chhabra and Kooner 1988; Chhabra *et al.*, 1988). Genotypes resistant or least susceptible to major insect pests of mungbean and urdbean at glance are depicted in Table 8.2.

Table 8.2: Mungbean/urdbean genotypes resistant/least susceptible to major insect pests.

Pests	Mungbean	Urdbean
Stemfly	V 2396, V 3495, V 4281, V 2335, V 2533, COGG 407, -408, -424, KM 2, PUSA 107, -16, -8731, 8733, MH 81-7, MG 212, MUG 154, -249, G 65, NARP I, SML 82, -32, -69, -100, -134, -154, -168, -169, -178, -26, 269.	UG 218, PDU 1, -5, Sel 37, COGB 305, ACC 305, Trichy, M 10, NP 7, CO 2- 100, TU 94-2, UPU 95-1, AKU 7, LBG 707, -689, -174, -694, UG 737, AKU 7.
Thrips	MG 815, -331, -423, -218, -368, G 65, MUG 212, UPM 83-6, -82-4, -84-4, M 81, MU 9161, PUSA 119, K 851, NARP 280, SML 7, -69, -77, -99, 100, -97, -103, -110, -112, -168, -130, -191, -258, -264, 84047, PDM 30, -36.	PDU 5, KB 63, PDU 88-23, -2, -5, UG 567, DU 4, T9, UH 80-4, UH 90-9, UH 80-7.
Jassids	Phule M 9037, -9505 PDM 40, PDM 84, -143.	————
Pod Borers	ML 131, OUM 6, PUSA 104, -105, -119, PDM 84-139, PDM 85-199, PDM 11, ML 5, -370, PDM 84-115, PDM 54, AB 19, PS 933, PDM 91-249, LGG 458, MGG 538, MG 316, GM 83-13, PDM 84, PDM 84-143, PMB 27, PUSA 104, -117, ML 131, -332, P 102, -103, JRVM 133, NPRG 3, B 24-88, MGG 341, ML 5, -131, PUSA 9531, PB 11, PMB 34, RMG 353, K 851, PUSA 9531, RMG 353, ML 5, PMB 27, PMB 14, LGG 460.	V 2135, 3274, -3276, -3229, -3277, -3812, -3949, -3281, -4411 Sarla, UG 737, PLU 557, TAU 1, -2
Galerucid Beetle	LGG 470, -471, -475, M 2, PDM 54, ML 131,	LAB 697, -611, -702, Pant U 35,
(Flea Beetle)	OVM 7, B 24-8-4, B 3-8-8, B 12-4-4, HY 4-3, TT 9-E, PDM 11-80	UPU 95-1, TU 94-2, KU 92-1, BG 12/ 15, Sarla, PDU 2.
Bruchids	V 2709, V 2802	VM 2011, 2164

Screening of mungbean resistant cultivars against stemflies at Asian Vegetable Research and Development Centre (AVRDC), Taiwan, has led to the identification of 3 AVRDC, accessions, V 2396, V 3495 and 4281 as resistant to *M. sojae, O. centrosematis* and *O. phaseoli* (Talekar 1990). Balboa (1972) reported Mungong Tagalog as most resistant cultivars of mungbean in Philippines against *O. phaseoli,* and two AVRDC accessions V 1360 and V 2816 were reported as least susceptible in Indonesia (Tengkano and Soekarna 1976), whereas V 2535 and V 2533 were identified as moderately resistant in China (Lin, 1981). Gunasekaran *et al.* (1977) screened 160 urdbean cultivars against *O. phaseoli* and identified No. 55 highly resistant to the fly and also to pod borer, *Catechrysops cnejus.* High trichome density on stems and leaves, purplish small diameter stem and small unifoliate leaves contributed to resistance against *Ophiomyia* spp. in mungbean (Talekar and Lee 1988), whereas a positive correlation was found between the thickness of mungbean leaf and stemfly infection. This implied that stemflies prefer to eat or lay eggs in thick leaf varieties (Lin and Rose, 1976).

Extensive research programme on HPR studies at Ludhiana, India showed that out of about 600 mungbean genotypes screened against whitefly–MYMV complex, jassids and pod borers (*L. boeticus, M. testulalis* and *H. armigera*), ML 1, ML 3, ML 5, ML 170, ML 194, ML 195, ML 235 and 15127 were identified as resistant. (Chhabra and Kooner 1980; Chhabra *et al.* 1980). While studying the mechanism of resistance, Chhabra *et al.* (1981b) found the resistant cultivars having higher contents of phenols, free amino acids and non-reducing sugars at the green leaves stage than the 'infestor'. Three genotypes ML 337, ML 423 and ML 428 were resistant not only to whitefly, jassids and MYMV, but also to three pod borer complex, and the basis of resistance was antibiosis (Chabbra *et al.*, 1988).

Field screening of mungbean breeding material during different years, by Chhabra and Kooner (1992a, 1992b) let to the identification of ML 484, MH 309 and PDM 84-146 out of 67 and ML 370, ML 459, ML 506, ML 508 and ML 536 out of 48 test entries as resistant to whitefly MYMV complex and jassids. In further screening of mungbean material, Chhabra and Kooner (1993) contributed four more genotypes ML 267, ML 326, ML 372 and Pusa 326 as resistant to insect pests and MYMV.

In summer mungbean cultivars SML 77, SML 99, SML 100, SML 103, SML 112, Pusa 107, UPM 82-4 and UPM 83-6 have been identified as resistant to thrips at Ludhiana, India (Chhabra and Kooner 1988; Chhabra *et al.* 1993b).

Out of > 475 urdbean accessions, Chhabra and Kooner (1981) and Chhabra *et al.* (1984), identified 4-5-2, LU 15, LU 178, LU 190, LU 196, LU 330, LU 397, LU 426, LU 436, UL 29,UL 257, UG 119, UG 187, UG 302, UG 407 and Pant U 30 as resistant to whitefly–MYMV and jassids. Chhabra *et al.* (1986a), in the field testing of 30 urdbean genotypes, identified LU 335, LU 274, LU 332, LU 470 as moderately resistant to *A. craccivora* and M 1-1 as highly resistant to aphids.

ECOLOGICAL APPROACH

Every effort is being made to raise yield of pulse by adopting modern agricultural practices viz., use of high-yielding varieties, heavy manuring and assured irrigation in order to meet the growing demands of pulses. The ecological approach as a component to the pest management is also an imperative and indispensable component like developing forecasting modules against the major pests on different hotspots in varying agro-ecosystem by studying pests population fluctuation in relation to agro-eco-factors (Kalaiyarasan and Palanisamy, 2004; Khan and Ramamurthy, 2004; Reddy and Kumar, 2005 and Yadav and Singh, 2006).

Chaudhary and Senapati (2004); Badjena and Mandal (2005) found positive association of thrips with temperature during summer. Yadav and Singh (2006) reported that a unit increase in maximum daily temperature caused an increased of 1.97 thrips/plant at Kanpur and 2.7 thrips/plant at Etawah. A unit increase in minimum temperature caused an increase of 1.30 jassids/plant at Kanpur. The peak incidence of YMD (64.5–78.7%) average temperature raised from 29.0–30.5°C and RH 52.52–72.0% during rainy season was recorded at both the location while during summer average temperature ranged from 31.0–38.5°C along with average relative humidity ranging between 20.50–59.2% was more conductive for the spread of YMD transmitted by whitefly (*B. tabaci*).

Active bruchid beetles lay eggs on soft and smooth pods of mungbean (Doria and Raros, 1977, Dias and Yadav, 1988). Talekar and Lin (1981) reported that resistance in mungbean accession 2011 was due to dense hairs on pods and *C. chinensis* adults avoided alighting on these pods. They found that the hardiness of seed coat in mungbean was behind the resistance against *C. chinensis.* Rathore and Chaturvedi (1997) postulated that combination of several factors could be responsible other than texture alone for preference for oviposition by bruchids.

Seed protein of *Dolichos lablab, Lablab purpureum* and *Rhyncocea cana* revealed the presence high level of non-nutritional chemicals such as viclins, α-amylase inhibitors and lectins in quantities that may impart resistance to *C. maculatus* (Fory *et al.*, 1996; Ignacimuthu *et al.*, 2000). Dongre *et al.* (1996) evaluated cultivated and wild accessions of mungbean and urdbean and resistance to *C. maculatus* was observed in wild progenitor of urdbean, *vigna mungo* var. *silvestris.* Cross of wild type with urdbean accession TU1 revealed the presence of two dominant duplicate genes designated as *Cmr1* and *Cmr2*. Resistance to *C. chinensis* in an accession of TC 1966 of *Vigna radiata* var. *Sublobata* was reported by Fujii and Miyazaki (1987), and resistance was found to be controlled by a single dominant gene,-*R* (Kitamura *et al.,* 1988).

Varieties/genotypes reported to possess low to moderate level of resistance bruchids were reported *viz.,* IGG2 and YKR 415/2 against *C. maculatus* in mungbean (Eduard and Gynathilagaraj, 1994); Ganapathy and Janarthanan (1995); and urdbean genotypes *viz.,* Vamban 1 against this pest was reported by Edward and Gunathilagaraj (1994).

As regards resistance accessions of wild progenitors, V 2709, V 2802, *Vigna sublobata* Acc. TC 1966 in mungbean and VM 2011 (pod resistance) and VM 2164 in urdbean have been reported by Talekar, 1996.

Biological Control

Studies on natural enemies have been limited to the identification of natural control agents. Extensive research will, therefore, be required before these could be exploited for the control of insect pests of mungbean and urdbean.

Out of the 123 natural control agents, 88 are parasitoids, 28 predators and 7 pathogens recorded on whitefly, jassid, aphid, stink bug, green bug, beanfly, red hairy caterpillar, Bihar hairy caterpillar, tobacco caterpillar, pod borers and bruchids.

Bemisia tabaci (Genn.) predation of immature stages by *Brumus* sp., *Chrysopa* spp., *C. cymbela* Banks, *Cybocephalus* sp. and parasitization of nymph and pupa by *Eretomocerus masii Silvi Prospaltella flava* Shaffee *and Geocoris tricolor* Fab. have been reported from India and Philippines (Nair 1986; Lal 1987; Litsinger *et al.* 1988). Adult mortality due to NPV and fungus, *Poecilomyces farinosus* was reported by Nene (1973a, 1973b).

While considering *Empoasca kerri* Pruthi, five nymphal predators viz. *Chrysopa cymbela, Camponotus* sp., *Acantholepis simplex* and *Coccinella septumpunctata* and a nymphal and adult predator, *Distina albida* were reported from India (Nayar *et al.* 1976; Dhuri and Singh 1983; Faleiro and Singh 1990) and two egg parasitoids, *Gonatocerus* spp. and *Oligosita* sp. have been recorded from Philippines (Litsinger *et al.* 1988).

As regards *Aphis craccivora* Koch, predation of nymph and adult by 15 predators viz. *Brumus suturalis, Chilocorus nigritus, Coccinella repanda transveralis, Harmonia arcuata, Ischiodan javana, I. scutellaris, Menochiles sexmaculatus, Vernia* sp., *Cyclosa insulana, Lencauge celebesiana* and *Theridion* sp., *Chrysopa carnea, C. septumpunctata, M. Sexmaculatus, Xantogramma scutellare,* was reported by Khan and Hussain (1965), Patel *et al.* (1976) Raychaudhari *et al.* (1979). Parasitizations of nymphs and adults by *Diaenetiella rapae, Praon* spp. and *Thrioxys indicus* have been reported by Stary and Ghosh (1978) and Sinha and Singh (1980). On *Piezodorus hybneri* Genelin, two egg parasitoids *Telenomus Telenomus* sp. and *Trissolucus* sp. have been reported in Thailand (Poonsavasde and Sukonthabhirom, 1990). As regards *Nezaa viridula* (Linn.), it is parasitized by a nymphal parasitoid, *Trissolucus basalis,* which has been reported from Australia (Turner, 1983).

So far as the *Melanagromyza sojae* (Zehntner) is concerned, its two larval/pupal parasitoids *Afrostilba* sp. and *Opius oleracei,* have been recorded on mungbean in India by Singh (1982). While on *Ophiomyia phaseoli,* parasitization of larva and pupa by 14 parasitoids viz. *Plutarchia* sp., *Chrysomotomyia douglasi, Hemiptarsensus* sp. *H. agromyzae, H. biclavus, Opius oleracei, Poycystomyia benefica, Sphegigaster agromyzae, Crytoprymna* sp. *Diaentiella rapae, Eurytoma* spp. *E. poloni, Halticoptera* sp. and *Parattrigonogastra rugosa* from Australia, Egypt, Malaysia and Philippines has been reported by Ooi (1973), Kleinschmidt (1970), Abdul Nasr and Assem (1968), Otanes (1918).

NPV has been reported to cause larval mortality in *A. moorei* (Nene 1973b) and four parasitoids, *Apanteles* spp. *Carcelia evolens, Sturania inconspicuella* and *Tachina* (*Exorisia*) fallax and a predator *Mermis indica* have been recorded as larval biocontrol agents (Lal, 1987; Dhuri and Singh 1983; Mathur and Singh, 1998).

On *Helicoverpa armigera* (Hubner), Lazarevic (1971) recorded a larval parasitoid, *Disophrys nigricornis* in Sudan. Bhatnagar *et al.* (1982) reported 27 species of Bethylids, Braconids, Ichneumonides, and Tachinids and Trichogrammatids parasitoids from the states of south central India. These workers recorded 5 egg parasitoids, 2 egg/larval parasitoids, 14 larval parasitoids and 4 larval/pupal parasitoids. These workers further reported 19 species of insect predators of eggs and larvae of *H. armigera.* A bacterial biocontrol agent *Bacillus thuringiensis* has been reported to cause larval mortality (COPR 1981). Singh and Mathur (1998-99) while working on life table studies found a high degree of death by parasitoid are pathogen on *H. armigero.* A predator *Ovomermis albicans* has been reported by Usha and Singh (1978) in India on *Maruca* (*testulalis*) *vitrata.*

On *Spilosoma* (=*Diacrisia*) *obliqua* (Walker), Krishnamurthi and Usman (1954) reported *Apanteles creatonoti*, parasitizing larvae. These workers also recorded two egg parasitoids, *Trichogramma chilonis* and *T. minutum,* and 9 species of larval parasitoids, *A. flavips, A. jayanagorensis, A. obliqua, A. ruidus, A. ater, A. vitripennis, Charops* sp. and *Drino* sp. *S. obliqua* (Kalra *et al.* 1962; Shetgar *et al.* 1990).

While accounting *Spodoptera litura* (Fab.), a larval parasitoid *Bracon hebetor,* an egg and larval parasitoid *Chelonus blackburni* and pupal parasitoid *Tetrastichus israeli* have been reported from Tamil Nadu by Udhyakumar and Jayaraj (1987). Krishnaiah *et al.* (1984) reported NPV causing larval mortality. Mohan *et al.* (2000) reported *Gryon* sp. a potential parasitoid on *Spodoptera litura* in Kanpur.

On storage legume pest, *Callosobruchus* spp., an egg parasitoid *Chaetostricha mukerji* and larval parasitoid *Bruchobius laticeps of C. chinensis* have been reported by Pruthi and Singh (1950). Biological control of bruchids has yet received very little attention. It is recommended that possibilities of classical biological control are assessed. Among the natural enemies of bruchids, braconids, Trichogrammalid, Eulophid, Eupelmid Torymid, Pteromalid, Eurytomalid and mite are the major, which kill various stages of the pest, need their inoculation, augmentation, inundation and conservation (Huis, 1991).

Botanical Control

Use of botanical control under field condition is not so promising as under storage condition. The pulse beetle, *Callosobruchus* spp. is the major insect pest of mungbean and urdbean and other pulses causing substantial damages in the storage. In recent past, the use of plant materials and vegetable oils has acquired an important approach for the control of pests. Few significant contributions have been made so far using plant materials (Pandey and Singh, 1995; Thawsatine, 1987; Rajpakse *et al.*, 1998; Ram Singh *et al.*, 2001 and Singh *et al.*, 2004). The vegetative oils have been reported to inhibit oviposition, which results in the reduction of multiplication of bruchids (Invigaro, 1990 and Reddy *et al.*, 1994). Ginger rhizome powder and neem leaf powder have been reported to be effective in reducing the egg laying of beetle (Echendu *et al.*, 1998 and Rajpakse *et al.*, 1998); Singh *et al.*, 2004 found Bidang leaf powder (*Embelia ribes*) to be significantly more effective on mungbean than neem leaf powder (*Azadiracta indica*).

The treatment with ester and neem oil prevents the mungbean seeds upto 75 days after treatments (Ketker *et al.*, 1987; Reddy *et al.*, 1994 and Yadav *et al.*, 2004).

Chemical Control

In general, farmers due to its quick response favour this option of control. Evidence of substantial increase in yield by the use of chemical insecticides in controlling insect pests of the crops is on records with several research publications and reports globally.

Chemical Control of Insect Pests at Root and Seedling Stages

In mungbean and urdbean, *Ophiomyia phaseoli* cause extensive damage at seedling stage in India, Philippines, Taiwan, Indonesia, Malaysia, Thailand and African countries. Soil application of granular insecticides *viz.* aldicarb, carbofuran, disulfoton and phorate at 1.0 kg, 1.5 kg and 2.0 kg along with seed treatment with carbofuran @ 2 g and 4 g per 100 g of seed gave effective control of fly and substantial increase in yield (Roongsook *et al.* 1973; Su *et al.* 1976; Vyas and Saxena 1980; Gupta and Singh 1984b; Lal 1985; Catipon 1986; Sehgal and Ujagir 1988). In the spray formulation, carbaryl, chlorpyriphos, dimethoate, endosulfan, malathion, mavinphos, monocrotophos and thiometon were identified to be very effective for the control of beanfly (Harnoto 1976; Ibrahim 1977; Akinbohungbe, 1982; Mehrotra *et al.* 1989). A comprehensive review on the chemical control of agromyzid flies giving details of presowing application of insecticides, seed treatment, soil treatment and post-sowing application of insecticides have been dealt by Talekar (1990).

Chemical Control of Insect Pests at Vegetative Stage

(i) Foliage feeder

For the control of galerucid beetle, whitefly and jassid on mungbean and urdbean, soil application of aldicarb, carbofuran, disulfoton, and mephospholon at the time of sowing gave significant control of these pests (Chaudhary *et al.* 1981; Yein and Singh 1981; Vyas and Saxena 1983; Singh and Gupta 1986). Similarly, seed treatment of mungbean with carbofuran (4%) followed by disulfotan and phorate granules at 1.5 kg a.i. ha^{-1} proved consistently effective against *M. obscurella* (Vyas and Saxena 1981). Sprays of chlorphyriphos, dimethoate, and monocrotophos reduced the damage caused by *M. obscurella* (Yadav and Yadav, 1979; Chander and Singh 1989a). Endosulfan sprays 6 and 8 weeks after sowing gave the successful control of the beetle and semilooper, *Anomis flava* (Yein and Singh, 1981). Dust and emulsion formulations of fenvalerate rendered effective control of *S. obliqua* on mungbean and urdbean (Chhabra and Kooner, 1987).

Vekaria and Vyas (1985) evaluated the bioefficacy of certain insecticides for the control of *H. armigera* on mungbean and reported that fenvalerate, monocrotophos, endosulfan, quinalphos and phosalone alone or in combination with miraculon were effective in controlling the pest. Gupta and Singh (1988) reported that phorate 10G, mephoslon 5G, disulfoton 5G, carbofuran 3G and aldicarb 10G at 2 kg a.i. ha^{-1} at sowing, reduced population of leaf feeding caterpillar. Dust formulation of quinalphos 1.5%, sevithion 4%, carbaryl 5%, BHC 10% and parathion 2% were effective in controlling foliage flower-feeding weevil, *Apion collaris* (Kapoor *et al.,* 1984).

(ii) Sucking pests

Effective control of whitefly, *B. tabaci* was obtained by the application of granular insecticides aldicarb 10G, disulfoton 5G, phorate 10G at 2 kg a.i. ha^{-1} and two sprays of monocrotophos 0.04% also resulted in good control of *B. tabaci* and reduced the incidence of YMV (Murugesan and Chelliah 1978; Chaudhary *et al.* 1981; Sharma and Verma, 1982). Mixing 2% orchid oil with monocrotophos enhanced the efficacy of whitefly control (Vadodaria and Vyas, 1987), whereas Singh *et al.* (1981) reported the adverse effect of oil application alone. Mathur and Singh (1998) reported the granular effect of phorat or carbofuran as basal application with two foliar spray of either with Mythyl-o-demeton or Monocrotophos or Cypermethrin suppress YMV intensity significantly.

Sprays of monocrotophos, endosulfan, fenitrothion, chlorfen-vinphos, malathion, phasalone, vamidothion, dimethoate at the concentrations of 0.075%, 0.07%, 0.05%, 0.04% and 0.03% were found effective by Sidhu and Sandhu (1971); Vyas and Saxena (1983); Yein and Singh (1981); Gatoria and Singh (1984, 1988) for the control of jassids, *E. kerri, E. moti, A. biguttula, A. devastans.* Galceron and dimethoate were most effective and gave maximum kill of mite *Polyphagotarsonemus latus* (Banks) (Gupa and Dhooria 1973; Chhabra *et al.,* 1993a).

In summer mungbean effective control of thrips, *M. distalis* has been obtained by the use of systemic insecticides dimethoate, monocrotopho-s, oxydemeton methyl and phosphamidon, malathion and endosulfan as per spray by Chhabra and Kooner (1985a, 1985b, 1986), Chhabra *et al.* (1986b). Damage of thrips has been reported to be reduced with a combination of aldicarb at 1 kg a.i. ha^{-1} and spray of monocrotophos 0.04% (Chander and Singh, 1989b).

Mathur and Singh (1998) reported that the combination of phorate granules as basal application with two foliar sprays of cypermethrin or methylo-demeton at 35 and 50 days after sowing was most effective in reducing the population of thrips and jassids.

Chemical Control of Insect Pests at Flowering and Podding Stages

Bioefficacy of fenvalerate, monocrotophos, endosulfan, quinalphos and phosalone alone or in combination with miraculon were effective in controlling the pod borers *C. ptychora, M. vitrata* and *H. armigera* on mungbean. The highest yield was obtained in the treatment with fenvalerate 0.01% + miraculon (Vekaria and Vyas, 1985). Whereas, spray application at flower initiation with endosulfan 0.07%, dimethoate 0.03%, methyldemeton 0.025%, monocrotophos 0.04%, deltamethrin 0.01%, fenvalerate 0.004% also gave significant control of pod borer complex in mungbean (Bhalani and Parsana, 1986; Bhat *et al.,* 1988; Chander and Singh, 1989b). Srivastava *et al.* (1984) and Sukul *et al.* (1987) found synthetic pyrethroids, permethrin, cypermethrin and deltamethrin equally effective for the control of pod borers. Combination of disulfoton of 1 kg a.i. ha^{-1} as basal application followed by endosulfan 0.07% spray too reduced the pod borer complex damage in mungbean and urdbean (Sundararaju and Rangarajan, 1987a). Dusting with endosulfan 4%, folidol 2% and phosalone 4% was also effective against pod borers and burchids on mungbean and urdbean crops (Rajendran *et al.,* 1981). Chlorpyriphos and cynofenphos were found superior in reducing *Etiella zinckenalla* infestation on mungbean in Indonesia (CRIA, 1971), whereas Litsinger *et al.* (1980) found that Decis at 0.05 kg a.i. ha^{-1} applied either twice at 30 and 42 days after emergence (DAE) or three times at 30, 40 and 50 DAE gave excellent control of post flowering insects on mungbean in Philippines.

Combination of Insecticides and Fertilizers for the Control of Insect pest Complex of Mungbean and Urdbean

Application of phorate 10G, carbofuron 3G and aldicarb 10G at 1.5 kg a.i. ha^{-1} alone or in combination with fertilizers in mungbean reduced jassid and whitefly population significantly (Yein and Singh, 1981; Gatoria and Singh, 1984). Similarly, application of carbofuron 3G alone or in combination with fertilizers reduced the incidence of galerucid beetle, *M. obscurella* (Gatoria and Singh, 1988) and disulfoton 5G at 1 kg a.i. ha^{-1} or endosulfan 0.07% in combination with fertilizers or alone were effective in controlling pod borer in mungbean (Sundararaju and Rangarajan, 1987b).

Phytotonic effect of carbofuran soil application and seed treatment of greengram significantly improved plant growth by controlling the pest complex in India and gave maximum green fodder and dry matter yields (Patel and Srivastava, 1990).

Effect of Soil Application and Seed Treatment of Insecticides on Germination

For the control of beanflies, sucking pests like whitefly, jassids, aphids, thrips and mites, presowing application or seed treatment is quite common. Work done on this aspect has revealed that persistence of carbaryl, a soil insecticide for the control of insect pests in mungbean, at a concentration of > 30 ppm, markedly reduced the germination and decreased survival of plants

(Gupta and Mishra, 1970). Similarly, seed treatment of mungbean with insecticides carbofuran, monocrotophos, dimethoate, phosphamidon and formothion at 1, 3 5 a.i. on 100 g of seed, considerably reduced germination percentage in all the treatments of dimethoate and with 5% a.i. of monocrotophos (Pawar and Chowan 1979). Venkataramaiah *et al.* (1980) found that the blackgram seedlings treated with DDT showed retarded growth of the seedlings. Presowing soil application of HCH and Aldrin at 50 kg ha^{-1} also reduced the plant stand of mungbean (Verma *et al.*, 1983).

As regards bruchid, its established infestation can only be controlled in bulk and bag storage in moderate to large quantities by fumigation with safe chemicals to eliminate the infestation. Treatments kill all stages of brunchids. EDB (ethylene dibromide) @ 3 ml/q with an exposure period of 7 days proved very effective. Ampules are available and farmers can use them easily. The only prerequisite is that the space treated should be made airtight by placing mud on the lid or cover gunny bags with thick polythene sheet. When quantities more than 2 tonnes to be fumigated, use AlP (Aluminium phosphide). AlP, available in tablets, is highly effective in controlling bruchids in longer space, should be used under expert guidance. To reduce the risk of above fumigants and to utilize rural based technology, the biogas produced at the village level can be used for the control of bruchids. Ensure airtight store before fumigation. Exposure period of 7 days provide 100% kill of eggs, grubs and adults (Mohan *et al.*, 1989).

Since no single method of control is effective for pests control, it is imperative to adopt holistic approach by integrating individual components to evolve an effective, efficient, cost effective and safe IPM strategy.

Future Agenda for Thrust Areas

Concept of Integrated Pest management (IPM) implies effective use of as many control measures as are compatible to suppress pest populations below damaging levels and optimize yields with minimum disruption of or damage to the environment. Much debate exists on the status of IPM for the *Vigna* spp. corps in India, Africa and elsewhere in the tropics, but little efforts have been made to implement the viewpoints that have often emerged. The possible reasons often put forth are that not sufficient information is available on various components necessary for the formulation of IPM programme. However, this review clearly shows that considerable knowledge already exists concerning pest management systems in mungbean and urdbean to begin to formulate IPM strategies. Modifications to IPM approaches can be made later on as our knowledge of HPR, ETL and EIL for different pests in the agro-ecosystem of the locality increases. It is to be underlined here that IPM strategies need to be initiated urgently to counteract the disastrous effects of possible overdependence on chemical pesticide to obtain quick control of different pests. Thus, results and experience gained so far in the *Vigna* spp. crops, the following further agenda for IPM is suggested.

Information on the extent of damage by major pests of *Vigna spp.* crops in different agro-ecological zones need to be fully documented to focus our research on pest management and HPR studies on the key pests in each area. For this, an ideal survey can be planned coupled with methodology combining systematic sampling with statistics.

Manipulation of date of sowing, seed rate and crop rotation have given significant reduction in the lepidopteran and coleopteran foliage feeders and soil pests. Promoting the right sort of cropping system's diversity by raising companion and/or barrier crops such as sorghum, pearl-millet, maize, cassava to minimize the population of jassid, flower thrips, spotted pod borer has given a lead to identify many other compatible crops and other methods of cultural control, not only to reduce the incidence and damage by the pests and diseases but also to improve the economy of farmers.

Studies on HPR against key endemic pests and the biochemical/biophysical basis, can make rapid progress in developing insect-resistant cultivars. Resistant markers in mungbean and urdbean in India for beanfly, whitefly–YMV complex, thrips and jassids are known. Plant breeders may choose the best parents for their crosses to consolidate insect resistance in high- yielding genotypes. Development of more genetic materials with multiple resistance or at least moderate resistance for key pests of the area should be kept on research priorities.

Although certain biocontrol agents have been identified for use in the control of some major pests like whitefly, jassid, aphids, *M. vitrata,* gram pod borer, hairy caterpillars which attack *Vigna* spp. crops. Among them *Chrysops* spp., *Apanteles* spp., *Trichogramma* spp. *Bacillus thuringiensis* and NPV are suitable candidates for improving the biological control of some of these pests (Table 8.2). Its potential should be assessed for other major pests in the areas/ zones of the country where they are virulent.

Research on botanicals and plant resources with pest suppressing qualities that are within the reach of farmers in rural areas should be intensified. This will require identification of more such plants, evaluation of extracts for insecticidal properties and application methodology.

Studies on proper sampling methodology, spatial distribution pattern ETL, EIL, life tables and forecasting modules of key endemic pests of *Vigna* spp. crops are needed to generate data base for judicious need base application of eco-friendly chemical insecticides and to minimize risks of insecticide resistance and pest resurgence.

REFERENCES

Abul-Nasr, S. and M.A.H. Assem (1968). Studies on biological process of the beanfly, *Melanagromyza phaseoli* (Tryon). *Bull. Soc. Entomol. Egypt.* **52:** 283-295.

Agyen-Sampong, M. (1978). Pests of cowpea and their control in Ghana. Pages 85-92 in Pests of Grain Legumes; Ecology and Control [SR Singh, HF van Emden, TA Taylor (Eds.)].

Ahmad, R., Yadava, C.P. and Lal, S.S. (1998). Evaluation of spray schedule for the control of insect pest of mungbean. *Ind. J. of Pul. Res.,* **11**(2): 146-148.

Akingbohungbe, A.E. (1982). Seasonal variation in cowpea crop performance at Ile-Ife, Nigeria and the relationship to insect damage. *Insect Sci. Appl.* **3:** 287-296.

Allen, D.J., Anno-Nyako, F.O., Ochieng, R.S. and Ratinam, M. (1981). Beetle transmission of cowpea mottle and southern bean mosaic viruses in West Africa. *Trop. Agric.* (Trinidad) **58:** 171-175.

Amoako-Atta, B., Omolo, E.O. and Kidega, E.K. (1983). Influence of maize, cowpea and sorghum intercropping systems on stem/pod borer infestations. *Insect Sci. Appl.* **4:** 47-57.

Anonymous (1976). Outbreak of pests and diseases. Quarterly Newsletter, FAO Plant Protection Committee for the South-East Asia and Pacific region, **19**(3): 5-6.

Anonymous (1993). Project Director's Report, *Kharif* Pulses Group Meeting, AICPIP (ICAR), QUAT. Bhubaneswar, 12-14 May 1993, Annexure III & IV.

Anonymous (2001). Annual reports Kharif, AICRP on MULLaRP (ICAR), IIPR, Kanpur, pp. 96.

Anonymous (2002). Annual Report (Kharif) All India Coordinated Research Project on MULLARP (India Council of Agricultural Research) IIPR, Kanpur, pp. 92-109.

Anonymous (2004). Low cost technology for safe storage of pulses. National Agricultural Technology Project, Research Achievements (1999-2003). Indian Institute Pusles Research, Kanpur, pp. 49-52.

Asthana, A.N. and Kim, Doo Hwan (1996). Recent advances in Mungbean Research. P. 219. Indian Society of Pulses Research, IIPR, Kanpur-208024.

Atwal, A.S. and Dhaliwal, G.S. (1986). Polyphagous pests (In) Agricultural Pests of South Asia and their Management: Kalyani Publishers, Ludhiana, New Delhi: 359-376.

Babu, S.C. and Rajasekaranm B. (1988). Economics of integrated pest management in mungbean production: research of a linear programming model. Pages 343-349 in Mungbean Proceedings of Second International symposium, Bangkok (S. Shanmugasundaram Ed).

Balboa, F.C. (1972). Comparative resistance of five varieties of mung, *Phaseolus aureus* Roxb., to the attacks of the beanfly, *Agromyza phaseoli Coq. Araneta Research Journal* **19:** 66-79.

Bashir, M., Iqbal, M.S, Ghafoor, A., Ahmad, Z. and Qureshi, A.S. (2002). Variability in cowpea germplasm for reaction to virus infection under field condition. *Pakistan J. Botany,* **34**(1): 47-48.

Bhalani, P.A. and Parsana, G.J. (1986). Chemical control of gram pod borer, *Heliothis armigera* Hubner infesting greengram (*Vigna radiata* L.). *Pesticides* **20**(1): 24-25.

Bhat, N.S., Manjunatha, M., Raju, G.T.T., Nagbhushan, G.G. and Deshpande, V.P. (1988). Chemical control pod borer complex of greengram (*Vigna radiata*). *Indian J. Agric. Sci.,* **58**(11): 864-866.

Bhatnagar, P. and Dahia, B. (2005). Reaction of mubgean and urdbean genotypes against insect-pests at yellow mosaic virus. *Indian J. Pulses Res.,* **18**(1): 90-91.

Bhatnagar, V.S., Lateef, S.S., Sithanantham, S., Pawar, C.S. and Reed, W. (1982). Research on *Heliothis* at ICRISAT. Pages 385-396 in Proceedings of the International Workshop on *Heliothis* management, 15-20 November 1981, ICRISAT Centre, Patancheru, Andhra Pradesh, India.

Bindra, O.S. and Jakhmola (1967). Incidence and losses caused by some pod-infesting insects in different varieties of pigeonpea. *Indian J. Agric. Sci.,* **37**(3): 177-186.

Borah, R.K. (1995). Effect of synthetic pyrethroids and organo phosphorus insecticides on the incidence of white fly, *Bemisia tabaci* (Genn.) and yellow mosaic virus in greengram, *Vigna radiata* (L.) Wilczek. *Indian J. Virology,* **11**(1): 75-76.

Caswell, G.H. (1968). The storage of cowpea in the Northern States of Nigeria. *Proc. Agric. Soc., Nigeria,* **5:** 46.

Caswell, G.H. (1973). The storage of cowpea. *Samaru Agric. Newsl.,* **15:** 73-75.

Caswell, G.H. and Akibu, S. (1980). The use of primiphos methyl to control bruchids attacking selected varieties of stored cowpea. *Trop. Grain Legume Bull.* **17/18:** 9-11.

Catipon, E.M. (1986). Mungbean Plant Industry Production Guide (41), Ministry of Agriculture and Food, Manila, Philippines.

Chalfant, R.B. and Gaines, T.P. (1973). Cowpea curuculionid: Correlation between chemical composition of southern pea and varietal resistance. *J. Econ. Entomol.* **66:** 1011-1013.

Chalfant, R.B., Leigh, T.F., Renwick, J.A., Messina, F.J. and Schalk, J.M. (1985). Entomological research on cowpea pests in the USA. Pages 267-273 in Cowpea Research, Production and Utilization. (S.R. Singh, K.O. Rachiee Ed.), London.

Chander, S. and Yesbir Singh (1989a). Control of galerucid beetle *Madurasia obscurella* Jacoby infesting greengram, *Vigna radiata* (L.) Wilczek. *Indian J. Entmol.* **51**(2): 185-191.

Chander, S. and Yesbir Singh, (1989b). Control of thrips *Megalurothrips distalis* and pod borer *viz. Euchryrops cnejus* Fab. and *Maruca testulalis Geyer* infesting greengram, *Vigna radiata* (L.) Wilczek, *Indian J. Entomol.* **51**(2): 131-135.

Chandrakar, H.K. and Srivastava, S.K. (2000). Efficacy and economics of some insecticides to pod borer complex of urdbean *Vigna mungo* (L.) Hepper. Environment and Ecology, **18**(4): 1014-1016.

Chaudhary, J.P., Singh, A.K. and Yadav, L.S. (1981). Effect of disulfoton granules on pest complex, crop growth and yield of green-gram. *Indian J. Entom.* **43**(4): 369-372.

Chaudhary, J.P. and Sharma, S.K. (1982). Biology of gram pod borer, *Heliothis armigera* (Hubnr) in the Haryana State. *Bull. Entomol.* **22**(1/2): 101-112.

Chaudhary, J.P., Yadav, L.S., Poonia, R.S. and Rastogi, K.B. (1980). Some observation on field population of *Empoasca kerri* Pruthi, a jassid, pest on mungbean crop in Haryana, Haryana *Agric. Univ. J. Res.* **10**(2): 250-251.

Chhabra, K.S. (1992). Advances in management of insect pests of *Vigna* species. Pages 178-186 in Proceedings National Symposium on New Frontiers in Pulses Research and Development, (J.N. Sachan ed.), *Directorate of Pulses Research*, Kanpur, India.

Chhabra, K.S., Kooner, B.S. and Brar, J.S. (1980). Resistance behaviour of mungbean, *Vigna radiata* (L.) W. cultivars against insect-pests complex and yellow mosaic virus. *Indian J. Ecology,* **7**(2): 276-280.

Chhabra, K.S., Kooner, B.S., Sharma, A.K. and Saxena, A.K. (1988). Sources of resistance in mungbean (*Vigna radiata)* to insect pests and mungbean yellow mosaic virus. Pages 308-314 in Mungbean Proceedings of the Second International Symposium, Bangkok. (S. Shanmugasundaram ed.).

Chhabra, K.S., Kooner, B.S., Sharma, A.K. and Saxena, A.K. (1993b). Role of allomones in varietal resistance in summer mungbean against thrips, *Megalurothrips distalis* (Karny). In proceedings of the International Symposium on Pulses Research, 2-6 April 1994, New Delhi, India (In Press).

Chhabra, K.S., Brar, J.S. and Kooner, B.S. (1981a). Jassid species recorded on green gram, black gram and red gram in Punjab. *Pulse Crops Newsletter,* **1**(2): 65.

Chhabra, K.S. and Kooner, B.S. (1980). Sources of whitefly, *Bemisia tabaci* G. and yellow mosaic virus resistance in mungbean, *Vigna radiata* W. *Tropical Grain legume Bulletin* No. **19**: 26-29.

Chhabra, K.S. and Kooner, B.S. (1985a). Losses in summer mungbean due to insect pest in Punjab. *Indian J. Ent.* **47**(1): 103-105.

Chhabra, K.S. and Kooner, B.S. (1985b). Problem of flower shedding caused by thrips, *Megalurothrips distalis* (Karny) on summer mungbean, *Vigna radiata* (L.) Wilczek and its control. Tropical *Pest Management,* London, **31**(3): 186-188.

Chhabra, K.S. and Kooner, B.S. (1986). Comparative efficacy of different insecticides in controlling thrips, *Megalurothrips distalis* (Karny) on summer mungbean, *Vigna radiata* (L.) Wilczek. Legume Research **9**(1): 39-42.

Chhabra, K.S. and Kooner, B.S. (1986-92). Research on *Kharif* pulses, Consolidated Reports on Entomology, All India Coordinated Pulses Improvement Project, Punjab Agricultural University, Ludhiana.

Chhabra, K.S. and Kooner, B.S. (1987). Susceptibility of different stages of Hairy caterpillar *Diacrisia obliqua* (Walker) to some insecticides. *Pesticides* **21**(6): 22-24.

Chhabra, K.S. and Kooner, B.S. (1988). Varietal resistance in mungbean [*Vigna radiata* (L.) Wilczek] to Thrips (*Megalurothrips distalis* Karny). TVIS NEWS (The Tropical vegetables Information Service), **3**(2): 10-11.

Chhabra, K.S. and Kooner, B.S. (1992a). Sources of resistance in mungbean against major insect pests and yellow mosaic virus. Legume Research **14**(4): 175-184.

Chhabra, K.S. and Kooner, B.S. (1992b). Response of some promising mungbean genotypes towards whitefly, jassids and mungbean yellow mosaic virus. In Proceedings of National Symposium on 'Recent Advances in Integrated Pest Management', PAU, Ludhiana, 12-15 October 1992, IV, **13:** 136.

Chhabra, K.S. and Kooner, B.S. (1993). Screening of mungbean germplasm against insect pests and yellow mosaic virus. *Indian J. Pulses. Res.* **6**(1): 69-75.

Chhabra, K.S., Kooner, B.S., Saxena, A.K. and Sharma, A.K. (1981b). Effect of biochemical components on the incidence of insect pest complex and yellow mosaic virus in mungbean. Crop Improv. **8**(1): 56-59.

Chhabra, K.S., Kooner, B.S. and Sharma, A.K. (1986b). Check flower drop in summer mungbean by controlling thrip, *Megalurothrips distalis* (Karny). *Indian Farming,* **36**(9): 25-26.

Chhabra, K.S., Kooner, B.S. and Cheema, H.S. (1980a). Incidence of *Aphis craccivora* Koch on grain legumes in the Punjab. Pages 23-28 in Aphidology in India, Proceedings of the National Symposium, 2-4 Nov. 1985, Agartala (B.K. Agarwala ed.).

Chhabra, K.S., Kooner, B.S. and Verma, M.M. (1993a). Insect pests of pulses identification and control manual Co-published by Punjab Agricultural University, Ludhiana and Directorate of Pulses Research, ICAR, Kanpur, pp. 88.

Chhabra, K.S., Kooner, B.S., Saxena, A.K. and Sharma, A.K. (1984). Influence of biochemical components on the incidence of insect pest complex and yellow mosaic virus in blackgram. *Indian J. Entomol.* **46**(2): 148-156.

Chhabra, K.S., Kooner, B.S. and Sharma, A.K. (1986b). Check flower drop in summer mungbean by controlling thrip, *Megalurothrips distalis* (Karny). *Indian Farming,* **36**(9): 25-26.

COPR (1981). Pest control in tropical grain legumes, Center for Overseas Pest Control, Overseas Development Administration, London, pp. 127-206.

CRIA (1971). Mungbean research: Insecticidal research. In Annual Report for 1971. Central Research Institute for Agriculture, Bogor, Indonesia.

Dalaya, V.P., Rajput, S.G. and Awate, B.G. (1986). Occurrence of thrips (*Thrips* spp.) on greengram in Northern Maharashtra and its control. *Pesticides,* **20**(8): 18-19.

Daoust, R.A, Roberts, D.W. and Das Neves, B.P. (1985). Distribution, biology and control of cowpea pests in Latin America. Pages 251-266 in Cowpea Research, Production and Utilization, (Singh, S.R., Rachie, K.O. eds.), London.

Dasgupta, B. and Chauhdury, A.K. (1985). Use of inter-varietal intercropping to minimize the yellow mosaic virus of urd and mungbeans. *Indian J. Plant proct.* **3**(1): 100-101.

De Lima, C.P.F. (1979b). A review of the use of physical storage procedures in East Africa : aspects for improvement and extension. *Proceedings of the Second International Working Conference on Stored Product Entomol.,* pp. 10-16, Sept. 10-16, 1978. Ibadan, Nigeria.

Dhakshinamoorthy, G. and Selvanarayanan, V. (2002). Evaluation of certain natural products against pulse beetle, *Callosobruchus maculatus* (Fab.) infesting stored green gram. *Insect Environment,* **8**(1): 29-30.

Dharmasena, C.M.D. and Subasinghe, S.M.C. (1986). Resistance of mung (*Vigna radiata*) to callosobruchus spp. (Col: Bruchidae). *Trop. Agric.,* **142:** 1-6.

Dhuri, A.V. and Singh, K.M. (1983). Pest complex and succession of insect pests in blackgram, *Vigna mungo* (L.) Hepper, *Indian J. Ent.* **45**(4): 396-401.

Dhuri, A.V., Singh, K.M. and Singh, R.N. (1986). Effect of intercropping on population dynamics of insect pests of blackgram, *Vigna mungo* (L.) Hepper. *Indian J. Ent.* **48**(3): 329-338.

Dias, C.A.R. (1987). Ecological studies on *Callosobruchus maculatus* Lab., *Callosobruchus chinensis* Linn. And *Callosobruchus analis* Fab. Infesting legumes in India, Ph.D. Thesis, I.A.R.I., New Delhi, India.

Dias, C.A.R. and Yadav, T.D. (1988). Incidence of pulse beetles in different legume seeds. *Indian Journal of Entomology,* **50:** 457-461.

Dongre, T.K., Pawar, S.E., Thakare, R.G. and Harwalkar, M.R. (1996). Identification of resistant sources to cowpea Weevil (*Callosobruchus maculatus* (F.) in black gram (*Vigna nungo* Var. mungo). *J. Stored Products Research* **32**(3): 201-204.

Doria, R.C. and Roros, R.S. (1975). Varietal resistant of mung to the bean weevil, *Callosobruchus chinensis* (Linn.) and some characteristics of field infestation. *Philippines Entomology,* **2:** 399-409.

Echendu, T.N.C., Njoku, B.O., Oli, E., Odurukwe, S.O. and Ene, L.S.O. (1988). Preliminary investigations into the use of ginger, Neem and Gaswewnut shell liquid (CNSL) to reduce damage caused to stored cowpea by *Callosobruchus maculatus.* Proceeding of the first *National Ginger Workshop Umuadike Nigeria,* 124-128.

Edward, Y.S.J.T. and Gunathilagraj, K. (1994). Relative susceptibility of greengram, black and cowpea cultivars to *Callosobruchus maculatus* (F.) in the field. *Madras Agricultural J.,* **81**(10): 575-577.

Faleiro, J.R. and Singh, K.M. (1990). Pest relationship on summer cowpea crop in Delhi. *Indian J. Ent.* **52**(4): 711-712.

Fory, L.F., Finhardi-Filho, F., Quintero, C.M., Osborn, T.C., Cardona, C., Christopeels, M.J. and Mayer, J. (1996). Alpha amylase inhibitors in resistance to common beans to the Mexican bean Weevil and the bean weevil (Coleoptera: Bruchidae). *J. Economic Entomology,* **89**: 204-210.

Fujii, K. and Miyazaki, S. (1987). Infestation resistance of wild legumes (*Vigna sublobata*) to azukibean weevil, *Callosobruchus chinensis* (L.) (Coleoptera: Bruchidae) and its relationship with cytogenetic classification. *Applied Entomology* and *Zoology,* **22**: 229-230.

Ganapathy, M. and Janarthanan, R. (1995). Phenetic resistance in blackgram (*Vigna mungo* L. Hepper) against *Callosobruchus maculatus* Fabricius. *Journal of Entomological Research,* **19**(2): 119-122.

Gaswell, G.H. and Akibu, S. (1980). The use of pirimiphos methyl to control bruchids attacking selected varieties of stored cowpea. *Trop. Grain Legume Bull.* **17/18** : 9-11.

Gatoria, G.S. and H. Singh (1984). Effect of the insecticidal and fertilizer application on jassid complex in greengram, *Vigna radiata* (L.) Wilczek. *J. Entomol. Res.* **8**(2): 154-158.

Gatoria, G.S. and Singh, H. (1988). Effect of the insecticidal and fertilizer application on chewing insect pests of greengram, *J. Res. Pb. Agril. Univ.* **25**(2): 225-232.

Gujar, G.T. and Yadav, T.D. (1978). Feeding of *Callosobruchus maculatus* (Fab.) and *Callosobruchus chinensis* (Linn.) in green gram. *Indian J. Ent.* **40:** 108-112.

Gunasekaran, G.R., Kumaresan, D., Balasubramanian, G., Jayaraj, S. and Marappan, P.V. (1977). Relative susceptibility of blackgram varieties to stemfly and pod borer. *Madras Agric. J.* **64**: 56-57.

Gupta, B.N. and Saxena, H.P. (1971). Chemical control of pests of moong (*Phaseolus sureus* Roxb.) leads to spectacular rich plant growth and high yield *Proceedings of the 5th Workshop* (Kharif pulses), H.A.U., Hissar, 18-19 March 1971, pp. 118-119.

Gupta, D.S. and Mishra, R.C. (1970). Carbaryl affects germination of seeds. *Indian J. Entomol.,* **32**(4): 396-397.

Gupta, P.K. and Singh, J. (1984b). Control of *Ophiomyia phaseoli* Tyron in greengram with granular insecticides. *Indian J. Agric. Sci.,* **54:** 321-324.

Gupta, S.K. and Dhooria, M.S. (1973). Occurrence of *Hemitarsonemus latus* (Banks) on blackgram and its control in Punjab. *Curr. Sci.,* **39**(9): 401-402.

H.O., T.H. (1967). The beanfly, *Melanagromyza phaseoli* Coq. and experiments on its control. *Malaysian Agri. J.,* **46:** 149-157.

Hameed, S.E. and Dubey, S.L. (1993). Losses due to insect pests in North Bihar. Pages 136-146 in Special issue Vol.I. *Indian J. Ent.,* 1983.

Hill, D.S. (1993). Agricultural insect pests of tropics and their management. Cambridge Univ. Press, pp. 746.

Honda, Y., Iwaki, M., Thongmeearkam, P., Deema, N. and Srithongchai, W. (1982). Blackgram mottled virus occurring on mungbean and soybean in Thailand. *JARQ* **16**(1): 72-77.

Huis, A. van (1991). Biological methods of bruchid control in the tropics: A review. *Insect Sci. Applic.,* **12** No. 1/2/3, pp. 87-102.

Hurje, R.V., Kulkarni, G.N., Shashidhara, S.D. and Vyakarnahal, B.S. (1990). Effect of insecticide and fungicide treatment on cowpea seed quality, *Seed Research* **18**(1): 90-92.

Hussain, M.H. and Abdel-Ali, V.A.I. (1982). Toxicology of some compound against the cowpea bettle, *Callosobruchus maculatus* (Fab.) (Coleoptera: Bruchidae). *International Pest Control,* **24:** 12-17.

Ibrahim, A.G.B. (1977). Increased production of leguminous crops through control of beanfly. Pages 301-309 in Foods and Agriculture Malaysia 2000, Faculty of Agriculture, University Pertanian, Malaysia.

Ignacimuthu, S., Janarthanan, S. and Balachandran, B. (2000). Chemical basis of resistance in pulses to *Callosobruchus maculatus* (F.) (Coleoptera: Bruchidae) *J. Stored Products Research,* **36:** 89-99.

Islam, W., Ahmad, E.N. and Joarder, O.I. (1984). Timing and extent of damage caused by insect pests of greengram in Bangladesh. *Crop Prot.* **31**(3): 343-348.

Jackai, L.E.N. and Daoust, R.A. (1986). Insect pests of cowpeas, *Ann. Rev. Entomol.* 95-119.

Jackai, L.E.N., Singh, S.R., Raheja, A.K. and Wiedijk, F. (1985). Recent trends in the control of cowpea pests in Africa. Pages 233-234 in Cowpea Research, Production and Utilization, (Singh, S.R., Rachie, K.O. Ed.), London.

Jagginavar, S.B., Kululkarni and Lingappa, S. (1990). A note on seasonal abundance of cowpea pod borer complex. *Karnataka J. Agric. Sci.,* **3**(3-4): 272-275.

Jagruti, Pattnaik, Kole, C. and Pattnaik, J. (2002). Detection of protein marker for screening of MYMV resistant mungbean genotypes, *Indian J. Genetics Plant Breeding,* **62**(1): 77-78.

Janagarajan, A. (1987). Effect of intercropping on the incidence of in cotton and sorghum Page 13 in Proc. Reg. Res. Council, Southern Zone, Tamil Nadu Agric. Univ., Aruppukkottai.

Jayaraj, S. (1988). Non-chemical management methods of some key mungbean pests. Pages 366-371 in mungbean. Proceedings of the Second International Symposium, Bangkok, (S. Shanmugasundaram ed.).

Kaiser, W.J. (1979). Natural infection of cowpea and mungbean by alphalpha mosaic virus in Iran. *Plant Dis. Rep.* **63**(5): 414-418.

Kalra, A.N., David, H. and Srivastava, T.N. (1966). *Drino* (Prosturmia *sp.* (Tachinidae: as a parasite of *Diacrisia obliqua W. Indian J. Entomol.* **28:** 556-557.

Kalra, V.K. (1985). Record of two new species of *Apanteles* parasitizing of *Spilosoma obliqua* (Wks.) *Indian J. Plant Protec.* **12**(2): 155.

Kapoor, K.N., Dhamdhere, S.V., Singh, O.P., Mishra, U.S. and Deole, J.Y. (1984). Evaluation of some insecticidal dust against Alcidodus collaris infesting greengram, *Vigna radiata* (L.) Wilczek, with records of its alternate host. *J. Entomol. Res.* **8**(1): 17-19.

Karel, A.K. and Maligna, Y. (1980). Leafhopper and aphid resistance in cowpea varieties. *Trop. Grain Legume Bull.* **20:** 10-11.

Ketkar, C.M., Schmuttere, R. and Ascher, K.R.S. (1987). Use of tree derived non-edible oils as surface protectants for stored legumes against *Callosobruchus maculatus* and *C. chinensis.* Natural pesticides from the neem tree (*Azadirachta indica, A. juss*) and other tropical plants proceeding of the 3rd International Neem Conference Nairobi Kenya, 10-15 July, 1986, p. 532-542.

Khamala, C.P.M. (1978). Pests of grain legumes and their control in Kenya. Pests of grain Legumes: Ecology and Control. London/New York: Academic. pp. 454.

Khan, M.Q. and Hussain, M. (1965). Role of coccinellid and syrphid predators in biological control of groundnut aphid. *Indian Oilseeds Journal,* **9:** 67-70.

Kitamura, K.M., Ishimoto, M. and Sawa, M. (1988). Inheritance of resistance to infection with Azuki Bean weevil in *Vigna sublobata* and successful incorporation to *V. radiata. Japanese J. Breeding* **38:** 459-464.

Kleinschmidt, R.P. (1970). Studies of some Agromyzidae in Queensland. Queensl. *J. Agric. Anim. Sci.* **27:** 341-384.

Kooner, B.S. and Chhabra, K.S. (1980). Pests of pulse crop and their control. Pages 132-141 in Breeding methods for the improvement of pulse crops, (Gill, K.S. ed.), Punjab Agricultural University, Ludhiana, India.

Kooner, B.S., Chhabra, K.S., Sekhon, H.S., Dhingra, K.K. and Cheema, H.S. (1983). A new deformity in summer mungbean, *Vigna radiata* (L.) wilczek. *Pulses Newsletter,* **3**(1): 40-42.

Krishnaiah, K., Ramakrishnan, N. and Reddy, P.C. (1984). Further trials on control of *Spodoptera litura* Fab. By Nuclear Polyhedrosis Virus on blackgram and groundnut. *Indian J. Plant. Prot.* **12**(1): 81-83.

Krishnamurthi, B. and Usman, S. (1954). Some insect parasites of economic importance notes in the Mysore State. *Indian J. Entomol.* **16**: 327-344.

Labeyrie, V. and Maison, P. (1957). Techniques culturals concernant la preservation du haricot contre la bruche *Acanthoscelides obtectus* Say, *Rev. Zool. Agric.,* **10:** 1-18.

Lal, S.S. (1983). Studies on the feasibility of integrated control of major insect pests attacking *Vigna radiata.* Summary Ph.D. thesis, Kanpur University, Kanpur, pp. 1-10.

Lal, S.S. (1985). A review of insect pests of mungbean and their control in India. *Trop. Pest Management* **31**(2): 105-114.

Lal, S.S. (1987). Insect pests of mung, urd, cowpea and pea and their management. Pages 185-201 in Plant Protection in field crops. 9M. Veerabhadra Rao and S. Sithanantham eds.).

Lal, S.S., Yadav, C.P. and Dias, C.A.R. (1980). Important pests of pulse crops in India and their management. AICPIP, Kanpur, Publication No. 2.

Lambrides, C.J. and Imrie, B.C. (2000). Susceptibility of mungbean varieties to the bruchid species *Callosobruchus maculatus* (F.). *C. phaseoli* (Gyll.) *C. chinensis* (L.) and *Acanthoscellides obtectus* (Say.) (Coleoptera: Chrysomelidae). *Australian J. Agri. Res.,* **51**(1): 85-89.

Lateef, S.S. and Reed, W. (1983). Review of crop losses caused by insect pests in pigeonpea internationally and in India. In: Crop Losses due to Insect Pests, Vol. II. Special Issue, *Indian J. Entomology.* The Entomological Society of India, Hyderabad Branch, India, pp. 284-291.

Lawn, R.J. and Ahn, C.S. (1985). Mungbean *Vigna radiata* (L.) Wilczek, *Vigna mungo* (L.) Hepper). Pages 854-623. in Grain Legume Crops. (RJ Summer field and E.H. Roberts Eds), Collins London.

Li, D.X., Tiang, J. and Zhang, Z.Y. (1994). A study of the bionomics and control techniques of *Aphis craccivora* (Koch) on mungbeans. *Plant Protection,* **20**(6): 4-5.

Lin, C.S. (1981). Resistance of beans to insects. Chinese Journal of Entomology **1**(2): 1-6.

Lin, C.S. and Rose, R.I. (1976). Screening of mungbean varieties for resistance to beanfly and study for resistance mechanism. *Plant Protection Bulletin* (ROC), **17**(4): 14.

Litsinger, J.A., Barrion, A.T., Badong, J.P., Gadela Cruz, C., Canapi, B.L. and Apostol, R.F. (1988). Food web, yield loss and chemical control of insect pests of wetland rice based mungbean in the Philippines. In Mungbean Proceedings of the Second International Symposium, CA MacFoy and ZT Dabrowski. 1984. Preliminary studies on cowpea resistance to *Aphis craccivora* Kock (Hom: Aphididae). *Z. Angew. Entomol.* **97:** 202-209.

Litsinger, J.A., Quirino, C.B., Lumban, M.D. and Bandong, J.P. (1978b). The grain legume pest complex of rice based cropping system at three locations in the Philippines. Pages 309-320 in pests of grain legumes ecology and control. (S.R. Singh, H.F. Van Emden and A.A. Taylor Eds.), Academic Press, London.

Litsinger, J.A., Price, E.C., Herrera, R.T., Bandong, J.P., Lumban, M.D., Quirino, C.B. and Castillo, M.D. (1978a). Introducting pest control technology to a low management crop: The mungbean example in the Philippines. Pages 183-191 in Proc. The First International Mungbean Symposium, 16-19 August 1977, Los Banos.

Litsinger, J.A., Bandong, J.P. and Paragna, F. (1980). Field evaluation of foliar insecticides for post flowering mungbean insects. Pangasinan, Philippines, 1977-78. *Insecticide and Acaricide Tests* **5:** 137-138.

Liu, Xuming, Jin, Dasheng, Cheng, Xuzhen, Wu, Xiaofel, Wang, Suhua, Liu, X.M., Jin, D.S., Cheng, X.Z., Wu, X.F. and Wang, S.H. (1998). Preliminary evaluation on their resistance to *Callosobruchus chinensis* (L.) in mungbean germplasm. *Crop Genetic Resources,* No. **2:** 35-37.

Manjunath, M., Bhat, N.S., Raju, G.T.T., Raju, M. and Viraktamoth, M. (1988). A new predatory ant *Aenactus pachycerus* Smith (Hymenoptera: formicidae) on wire worm, *Agriotes* sp. (Coleoptera: Elateridae). *Curr. Sci.*, **57**(B): 7449.

Marimuthu, T.S., Subramanian, C.L. and Miohan, R. (1981). Assessments of yield losses due to yellow mosaic infection in mungbean. *Pulse Crops Newsletter*, **1:** 104.

Mathur, Y.K. and Singh, S.K. (1998). Field evaluation of various insecticides against major pest of mungbean, Abst. National symposium on management of biotic and abiotic stresses in pulse crops held at I.I.P.R. Kanpur, 26-28 June 1998, pp. 86.

Mathur, Y.K. and Singh, S.K. (1998). Some important of natural enemies of *Amsacta moorei* Butler in Rajasthan, *Indian J. Ecol.* **25**(1): 77-78.

Mathur, Y.K., Singh, S.K. and Kaushal, P.K. (1997). Precasting model of the key pests of mungbean in Uttar Pradesh, India. Abst. International Conference on Ecological Agriculture organized by Indian Ecological Society & CRRID, held at Chandigarh from November 15-17, 1997, pp. 89-90.

Mehrotra, P., Singh, S.D. and Kaul, C.K. (1989). A note on the chemical control of *Melanagromyza obtusa* Malloch on cowpea. *Indian J. Entomol,* **51**(3): 355-356.

Mohan, Krishna, Bisen, R.S. and Singh, S.K. (2000). New record of *Gryon sp. on Spodoptera litura* Fab. from Kanpur, Uttar Pradesh. *Insect Environment* **6**(3): 100-101.

Murugesan, S. and Chelliah, S. (1978). Effect of yellow mosaic infestation of the host greengram on the biology of *Bemisia tabaci. Genn. Entomon.* **3**(1): 41-43.

Nair, M.R.G.K. (1986). Insects and mites of crops in India. Indian Council of Agricultural Research, New Delhi, pp. 408.

Natarajan, N. and Subbarao, P.V. (1992). Suitable time of sowing and insecticide for the management of stemfly (*Ophiomyia phaseoli*) in blackgram (*Vigna mungo*). *Madras Agricultural J.* **79**(2): 119-122.

Nath, P. and Singh, A.K. (1998). Effect of intercropping of groundnut with millets and pigeon pea on the relative incidence of insect pests. *Ann. Pl. Prot. Sci.* **6**(2): 151-154.

Nayar, K.K., Ananthakrishnan, T.N. and David, B.V. (1976). General and Applied Entomology, Mc Graw Hill, New Delhi, pp. 581.

Ndoye, M. (1978). Pests of cowpea and their control in Senegal. In Pests of Grain Legumes: Ecology and Control. London/New York: Academic, pp. 454.

Negi, R.S., Srivastava, Meera and Saxena, M.M. (1997). Egg-laying and adult emergence of *Callosobruchus chinensis* on green gram (*Vigna radiata*) treated with pongam oil, *Indian J. ent.,* **59**(2): 170-172.

Nene, Y.L. (1972). A survey of viral diseases of pulse crops in Uttar Pradesh. Final Technical Report of PL 480 *Ad hoc Project* 88-108.

Nene, Y.L. (1973a). Viral diseases of some warm weather pulse crops in India. Plant Disease reporter, **57:** 463-467.

Nene, Y.L. (1973b). Note on a fungal parasite of *Bemisia tabaci.* Genn., a vector of several plant viruses. *Indian J. Agric. Sci.* **43**(5): 514-516.

Nilakhe, S.S. and R.B. Chalfant (1982). Cowpea cultivars screened for resistance to insect pests. *J. Econ. Entomol.* **75:** 223-227.

Ooi, P. (1973). Some insect pests of greengram. *Phaseolus aureus: Malaysian Agricultural J.* **49**: 131-142.

Otanes, F.Q. (1918). The beanfly *Philipp. Agric.*, **7:** 2-31.

Pandey, N.K. and Singh, S.C. (1995). Effect of neem leaf powder on survival and mortality of pulse beetle *Callosobruchus Chinensis* (L.) infesting blackgram Uttar Pradesh. *J. Zoology,* **15**(3): 162-164.

Panwar, V.P.S. (1995). Agricultural insect pests of crops and their management. Kalyani Pubs., Ludhiana, Punjab, pp. 286.

Parihar, D.R. (1979)., Outbreak of Kutra, *Amsacta moorei* pest in Rajasthan desert. *Ann. Arid Zone,* **18:** 140-141.

Patel, M.B. and Srivastava, K.P. (1990). Phytotonic effect of insecticides on cowpea *V. unguiculata* (Linn.) and green gram *V, radiata* (Linn.). *Indian J. Entomol,* **52**(4): 583-588.

Patel, R.C. and Patel, J.R. (1983). Crop losses due to insect pest in Gujarat State. Pages 27-32 in Special issue Vol. I, *Indian J. Ent.,* 1983.

Patel, R.C., Yadava, D.N. and Patel, J.R. (1976). Natural control of groundnut aphid, *Aphis craccivora* Koch. in Central Gujarat. Current Science **45:** 34-35.

Patnaik, N.C., Mohanty, J.N. and Chhabra, K.S. (2000). (Eds), Prakash, Anand and Rao, Jagadiswari. Insect Pests of Pulses and Oilseeds and their management. Applied Zoologists Research Association, CRRI, Cuttack 0753006, India pp. 11-27.

Pawar, C.S. (1983). Cropping Entomology, Report of Work 1982-83, ICRISAT, Patancheru PO, Andhra Pradesh, India, pp. 1-21.

Pawar, V.M. and Chowan, K.M. (1979). Effect of systemic insecticides on germination and seedling development in greengram (*Vigna aureus* Hepper). *Indian J. Entomol.* **41:** 405-408.

Perrin, R.M. (1978). Varietal differences in the susceptibility of cowpea to larvae of the seed moth, *Cydia ptychora* (Meyrick) (Lepidoptera : Tortricidae). *Bull. Entomol. Res.* **68:** 47-56.

Pohelman, J.M. (1991). The mungbean Mohan Primlani Publication for Oxford & IBH Publishing Co. Pvt. Ltd., New Delhi, pp. 1-375.

Poonsavasde. P. and Sukonthabhirom, S. (1990). Evaluation of the scelionid egg parasitoid for controlling *Piezodorus hybneri* Gmelin. In: Proceedings of the Mungbean Meeting 90, Chiang Mai, Thailand: 277-282.

Prakash, A. and Rao, (2000). Insect Pests of Pulses and Oilseed and their Management Edt. C.R.R.I. Cuttack, pp. 1-259.

Prevett, P.F. (1961). Field infestation of cowpea (*Vigna unquiculata*) pods by beetles of the families Bruchidae and curculionidae in Northern Nigeria. *Bull. Entomol. Res.* **52:** 635-645.

Pruthi, H.S. and Mohan Singh, (1950). Pests of stored grains and their control. Special No. *Indian J. Agric. Sci.,* **18**(4): 18-23.

Rachie, K.O. (1985). Introduction, Cowpea Research, Production Utilization, xi-xii, Page 460 in Cowpea Research, Production and Utilization. (Singh, S.R. and Rachie, K.O. Eds.). John Wiley & Sons, New York.

Raheja, A.K. (1985). Present status of research in cowpea entomology at IAR (Institute of Agricultural Research), Ahmadu, Bello University, Zaria, Research Report: 1-23 (Personal Communication).

Rajapaksa, R.H.S. and Jayasena, K.W. (1989). Field pest problem of mungbean, *Vigna radiata* in Southern region of Sri Lanka, *Entomon.* **14**(12): 159-164.

Rajendran, R., Gopalan, M. and Janarthanan, R. (1981). A note on the control of pod borers in blackgram and green gram with dust formulations. *Pesticides* **15**(10): 35-36.

Rajpakse, R., Senanayake, S.G.J.N. and Ratnasekra, D. (1998). Effect of five botanicals on oviposition adult emergence and mortality of *Callosobruchus maculatus* (Fab.) (Coleoptera: Bruchidae) infesting cowpea (*Vigna unguiculata* L. Walp.). *J. Entomologial Research,* **22**(2): 117-122.

Raju, A.K. and Panda, N. (1983). Observations on the biology of *Aphis croecivora* Koch. On five varieties of green gram. *Indian J. Agric. Sci.,* **53:** 868-869.

Ram Ujagir (1997). Insect pests of peas and their management. In-Integrated Pest and Disease Management. Eds. Upadhyay, R.K. *et al.* A.P.H. Publishing Crop, New Delhi: 210-233.

Ramakrishnan, C. (1986). Management of early season pests of pulses. Pages 120-124 in Pest and Diseases Management: Oilseeds, Pulses, Millets and Cotton, (Jayaraj ed.). Tamil Nadu, *Agril. Univ.,* Coimbatore.

Raman Goud, T. (1993). Sesamum and sunflower pests and their management. Presented in summer institute on advances in Entomology: Oilseeds & Pulses, May 17 to June 5, 1993. Compd. of Lectures 255-267.

Raman, K.V., Singh, S.R. and van Emden, H.F. (1982). Mechanism of resistance to leaf hopper damage in cowpea *J. Econ. Entomol.,* **73:** 484-488.

Ramkrishnan, G., Konadaswamy, T.K., Damodaran, A.P.S. and Ayjaroo, R. (1973). Studies on new mosaic viruses occurring on *Phaseolus aureus* Roxb. *Madras Agric. J.* **60**(7): 465-468.

Rao, S.S., Rao, S.N. and Azam, K.M. (1990). Occurrence of *Tetranychus urticae* Koch on greengram (*Vigna radiata*). *Indian J. Entomo,* **52**(1): 158-159.

Rathore, Y.S. and Chaturvedi, S.K. (1997). Developmental pattern of *Callosobruchus chinensis* (L.) on chickpea seeds of advanced breeding lines. *Indian J. Pulses Research,* **10**(2): 180-184.

Rathore, Y.S. and Sharma, V. (2002). Management of bruchid infestation in pulses. ISPRD, 2002. Pulses for Sustainable Agriculture and Nutritional Security. Proceeding of National Symposium (Eds. Ali, Masood; Chaturvedi, S.K. and Gurha, S.N.). *Indian Institute of Pulses Research,* Kanpur-208024, pp. 111-124.

Ravindrababu, R. (1987). Studies on mungbean yellow mosaic virus disease. Ph.D. Thesis, Tamil Nadu Agril. Univ., Coimbatore, pp. 191.

Rawat, R.R. and Sahu, H.R. (1973). A study on the Bionomics of *Chauliops fallax* Scott (Heteroptera, Lygacidae) at Sehore (Madhya Pradesh). *J. Bombay N. History Soc.* **70**(3): 475-479.

Rawat, R.R. and Kausik, V.K. (1983). Crop losses due to insect pests in the north east Madhya Pradesh. Pages 73-83 in Crop losses due to insect pests. Special issue, Vol.I. *Indian J. Ent.,* **1983:** 73-83.

Ray Chandhari D.N., Dutta, S., Agrawal, B.K., Raha, S.K. and Ray Chaudhury, D. (1979). Some parasites and predators of aphids in north east India and Bhutan 11. *Entomon.* **4**(2): 163-166.

Reddy, M.U., Bharti, S.R. and Reddy, D.D.R. (1999). Efficacy of some vegetable oils as protectants against the pulse beetle (*Callosobruchus chinensis*) in green gram (*Phaseolus aureus*) during storage, *Indian J. Nutrition Dietetics,* **36**(10): 436-442.

Reddy, V.S., Babu, T.R., Hussaini, S.H. and Reddy, B.M. (1994). Effect of edible and non-edible oils on the development of pulse beetle, *Callosobruchus chinensis* L. and on viability of mungbean seeds. *Pest Management and Economic Zoology,* **2**(1): 15-17.

Reed, W., Lateef, S.S., Sithanathan, S. and Panwar, C.S. (1989). Pigeonpea and chickpea insect identification Hand Boo, *ICRISAT Information Bulletin* No. 26, pp. 51.

Robertson, G.A. (1981). Response of mung to mechanical reduction of leaf area at flowering. Pulse Crops Newsletter, **1**(4): 37-38.

Roongsook, D., Sitchawat, T., Knapp, F.W. and Tiemtaisong, A. (1973). The effect of phorate and malathion on beanfly control and on yield of mungbean. *Thai. J. Agric. Sci.,* **6:** 283-286.

Sahoo, B.K. and Patnaik, N.C. (1994). Insect pests of greengram and blackgram in the south central region of Orissa with notes on their seasonal activity. *Orissa J. Agric. Res.* **7**(Suppl.): 74-76.

Sakamoto, Y., Sasaki, H. and Kikuta, H. (1983). Ecological studies on the subfamily Apioninae injurious to leguminous crop in Japan. *I.J. Coll. Dairying,* **10**(1): 153-160.

Sarkar, T.K., Gayen, P. and Somchoudhury, A.K. (1991). Some genetic parameters of resistance to *Callosobruchus maculatus* (F.) (Coleoptera: Bruchidae) in mungban, *Vigna radiata* (L.) Wilczek. *J. Entomological Research,* **15**(1): 36-39.

Saxena, H.P (1983). Losses in blackgram due to insect pests. Pages 294-297 in crop losses due to insect pests. In special issue Vol.II. *Indian J. Ent.*

Saxena, H.P. (1973). Major pest problems in pulses. Pages 191-194 in Pest Management and Plant Protection Training, New Delhi, Deptt. of Agriculture, Ministry of Agriculture and Irrigation.

Saxena, H.P. (1978). Pests of grain legumes and their control in India. Pages 15-23 in Pests of Grain Legumes; Ecology and Control, (SR Singh, H.F. van Emden and T.A. Taylor Eds.). Academic Press, London.

Saxena, H.P., Singh, Y. and Phokela, A. (1973). Effect of date of sowing on pest incidence. Page 115 in Proceedings IX Workshop (Kharif Pulses), New Delhi.

Sehgal, V.K. and Ram Ujagir, M. (1988). Insect pest management of mungbean in India. Pages 315-328 in mungbean Proceedings of the Second International Symposium, Bangkok, (S. Shanmugas-undaram Ed.).

Sehgal, V.K., Y.S. Rathore and Ram Ujagir (2000). Insect pest management in mungbean and urdbean in India. Proceeding of National Symposium on "Management of Biotic and Abiotic Stresses in Pulse Crops" held at Kanpur, June 26-28, 1998, pp. 87-92.

Sethuraman, K., Manivannan, N. and Natrajan, S. (2001). Management of yellow mosaic disease of urdbean using neem products, *Legume Res.,* **24**(3): 197-199.

Sharma, S.R. and Verma, A. (1982). Control of yellow mosaic of mungbean through insecticides and oils. *J. Entomol. Res.* **6**(2): 130-136.

Shetgar, S.S., Bilapate, G.G., Patil, V.V. and Londhe, G.M. (1990). A note on the natural enemies of Bihar hairy caterpillar, *Spilosoma obliqua* (Walker). *Indian J. Entomol.* **52**(1): 158.

Sidhu, A.S. and Sandhu, A.S. (1971). Control of jassid infesting moong crop with new insecticides. Pages 138-142 in Proceedings, Fifth Workshop on Pulse Crops, New Delhi, ICAR.

Singh, P., Bhadauria, N.S. and Jakhmola, S.S. (2004). Efficacy of different plant materials as powder form against pulse beetle *Callosobruchus maculatus* in green gram. *Indian J. Entomology,* **66**(4): 364-365.

Singh, B.B., Gupta, S., Singh, R.A., Singh, S.K. and Singh, K.K. (2006). *Mung aur Urd Ki Unnat Kheti.* Technical Bulletin. *Indian Institute of Pulses Research,* Kanpur, India, pp. 1-20.

Singh, B.K. and Sehgal, V.K. (1993). Insect fauna of mungbean and effects of crop protection on field incidence of major pests and grain yield at Pantnagar in Northern India. Crop Protection, U.K. (Personal Communication).

Singh, C.P. and Pandey, M.C. (1995). Studies on the biology of Amsacta moorei *Butler on groundnut. J. Insect Sci.,* **8**(2): 207-208.

Singh, D.P. and Singh, B.B. (1994). A new variety of mungbean–Narendra mung-1. *Indian Farming,* **43**(11): 3-4.

Singh, J. and Gupta, P.K. (1986). Relative efficacy of some granular insecticides against galerucid beetle (*Madurasia obscurella* Jacoby) an important pest of greengram. *Indian J. Entomol.* **46**(2): 221-231.

Singh, K.M. and Singh, R.N. (1977). Succession of insect pests in mungbean and blackgram under dryland conditions at Delhi. *Indian J. Entomol.* **39:** 365-370.

Singh, O.P. and Rawat, M.M. (1982). Some observations on the greengram on the green stink bug *Nazara viridula* (L.) on summer greengram crop at Jabalpur, Madhya Pradesh. *Sci. Cul.* **48**(3): 108-109.

Singh, R.A., Gurha, S.N., Mishra, D.P. and Gangal, L.K. (1981). Role of systemic insecticides in augmenting yield with reduced yellow mosaic virus incidence in mungbean. *Indian J. Plant Prot.* **8**(2): 167-169.

Singh, Ram and Chauhan, R. (1984). Screening greengram genotyples against *Tarsonemus mite. Indian J. Agric. Sci.* **54**(8): 669-670.

Singh, Ram, Basant Singh and R.A. Verma (2001). Effect of different indigenous plant products as grain protectant against, *Callosobruchus Chinensis* Linn. on pea. *Indian J. Entomology,* **63**(2): 179-181.

Singh, S. (1982). Ecology of Agronomyzidae (Diptera) associated with leguminous crops in India. Memoirs of the School of Entomol. No. 8, St. John's College, Agra, India. pp. 126.

Singh, S.K. and Mathur, Y.K. (1999). Life table studies on *Helicoverpa armigera* Hubner on gram. Rabi programme report Deptt. of Entomology, CSAUA&T, Kanpur pp.

Singh, S.K. and Mathur, Y.K. (1998-99). Life table studies *Helicoverpa armigera* Hub. On gram Rabi Research Technical Programme, Deptt. of Entomology, CSAUA&T, Kanpur, pp. 39-44.

Singh, S.K. and Rai, A.B. (2005). Is the reproductive physiology of Adjuki bean Weevil, *Callosobruchus chinensis* (L.) governed by ageing? *Indian J. Entomology,* **67**(1): 1-6.

Singh, S.K. and Yadav, C.P. (2002-03). Bio-ecology of *Callosobruchus* spp. and identification of sources of resistance in stored chickpea. *Annual Report*, IIPR, Kanpur, pp. 36-38.

Singh, S.R. (1977). Cowpea cultivars resistant to insect pests in world germplasm collection. Trop Grain Legume Buyll. **9:** 3-7.

Singh, S.R. (1978). Resistance to pests of cowpea in Nigeria. Pest of Grain Legumes: Ecology and Control London/New York Academic. pp. 454.

Singh, S.R. and Jackal, L.E.N. (1985). Insect pests of cowpeas in Africa: Their life cycle, economic importance, and potental for control. Page 217-231 in cowpea research, production and utilization, (Singh, S.R., Rachie, K.O. ed.), London.

Singh, S.R. and Taylor, T.A. (1978). Pests of grain legumes and their control in Nigeria. Pages 99-111. In Pests of Grain Legumes: Ecology and Control, London/New York: Academic pp. 454.

Singh, T.B. and Singh, R. (1980). Studies on the bionomics of *Trioxys indicus* S&S. a parasitoid of *Aphis craccivora* Koch – the area of discovery of the parasitoid. Zeithchrift fur Ang. Ent. **89:** 173-178.

Singh, T.V.K., Singh, K.M. and Singh, R.N. (1990). Groundnut pest complex: III. Incidence of insect pests in relation to agroclimatic conditions as determined by graphical super imposition technique *Indian J. Ent.* **52**(4): 686-692.

Singh, T.V.K., Singh, K.M. and Singh, R.N. (1990). Groundnut pest complex IV: Regression studies to determine the associationship between jassid and thrips and weather parameters. *Indian J. Ent.* **52**(4): 693-701.

Singh, V.N. and Pandey, N.D. (2001). Growth and development of *Callosobruchus chinensis* (Linn.) on different gram varieties. *Indian J. Ent.,* **63**(2): 182-185.

Singh, V.N., Pandey, N.D. and Singh, Y.P. (1994). Effectiveness of vegetable oils on the *Callosobruchus chinensis* (Linn.) infesting stored grains (Chickpea). *Indian J. Ent.,* **56**(3): 216-219.

Sinha, M.M., Yadav, R.P. and Kumar, A. (1975). Outbreak of the Bihar hairy caterpillar, *Diacrisia obliqua* Walker in North Bihar. *Entomologists Newsletter,* **5:** 97.

Sinha, M.M., Yadav, R.P. and Kumar, A. (1979). Multidirectional approach for pest management in arhar, *Cajanus cajan* in Bihar. *Pesticides* **13**(11): 14-16.

Sinha, M.M., Yazdani, S.S., Kumar, A. and Singh, R. (1985). Observations on the seasonal occurrence of insect pests of greengram in Northern Bihar, *Bull Entomol.* **28**(2): 210-211.

Sinha, T.B. and Singh, R. (1980). Studies on the bionomics of *Trioxys indicus* S&S a parasitoid of *Aphis craccivora* Koch–the area of discovery of the parasitoid. *Zeitchrift fur Ang. Ent.* **89:** 173-178.

Sivaprakasam N., Balasubramanian, G., Venkatesan, S. and Chandra-sekaran, J. (1988). Control of *Spodoptera litura* F. with nuclear polyhedrosis virus and chloropyriphosaon cowpea. *Madras Agric. J.* **75**(9-10): 366-368.

Speswasdi, P., Pitaksa, S., Chareonrak, T., Phapoom, V. and Heuel-Rof, B. (1990). Crop loss assessment for major mungbean pests in rice based cropping system. Pages 259-269 in proceedings of the mungbean meeting 1990, Chaing Mai, Thailand.

Srikanth, J. and Lakkundi, N.H. (1988). Instar period, fecundity and longevity of cowpea aphid. *Aphid craccivora* Koch. On seven leguminous hosts. *J. Aphidology,* **2**(1/2): 18-19.

Srivastava, A.S., Awasthi, B.K. and Nigam, P.M. (1977). Incidence of *Myllocerus maculotus* Desb. On leguminous crops (Coleoptera: Curculionidae). *Labdev J. Sci. Technol.* **15**B: 215-216.

Srivastava, K.M. and Singh, L.N. (1976). A review of the pest complex of *Kharif* pulses in UP. PANS **22:** 333-335.

Srivastava, K.P., Gajblije, V.T., Jain, H.K. and Agnihotri, N.P. (1984). Efficacy of synthetic pyrethroids against the pod borer, *Euchrysops cnejus* Fab. and their residue in greengram, *Vigna radiata* (L.) and cowpea, *Vigna unguiculata* (L.) *Walp. J. Entomol. Res.* **8**(1): 36-41.

Srivastava, Meera, and Kaur, A. (2002). An evaluation of efficacy of extracs of plant *Peganum harmala* against pulse beetle *Callosobruchus chinensis. Indian J. Ent.,* **64**(2): 138-147.

Stary, P. and Ghosh, A.K. (1978). Further records of aphid parasitoid (Hym. Aphididae) from Meghalaya, *India, Oriental Insects* **12**(1): 77-80.

Su G.Y., Kung, S.P. and Rose, R.I. (1976). Control of beanflies in mungbeans and asparagus beans, 1975. *Insecticide and Acaricide Tests,* **1:** 53.

Subha Rao, P.V., Rangarajan, A.V. and Azeer, A.B. (1974). Record of new host plants for some important crops pests in Tamil Nadu. *Indian J. Entomol.,* **36:** 227-228.

Sukul, P., Handa, S.K. and Singh, K.M. (1987). Comparative efficacy of synthetic pyrethroids against pod borer complex, crop growth and yield of greengram. *Indian J. Entomol.* **49**(3): 408-413.

Sundaraju, D. and Rangarajan, A.V. (1987a). Insecticidal control of pests of greengram and blackgram. *Pesticides* **21**(12): 18-19.

Sundaraju, D. and Rangarajan, A.V. (1987b). Effects of insecticides in combination with fertilizer in controlling yellow mosaic disease and pod borer of greengram. *Pesticides,* **21**(12): 20-21.

Talekar, N.S. (1988). Biology, damage and control of bruchid pests of mungbean. Pages 329-342 in Mungbean Proceedings, Second International Symposium, Bangkok (S. Shanmugasundaram Ed.).

Talekar, N.S. (1990). Agronomyzid flies of food legumes in the Tropics. Wiley Eastern Limited, New Delhi, India, pp. 297.

Talekar, N.S. (1996). Sources of resistance to major insect pests of mungbean in Asia. 40-50. *In:* Recent advances in mungbean research (Asthana, A.N. and Kim, D.H. Eds). *Indian Society of Pulses Research,* IIPR, Kanpur, India.

Talekar, N.S. and Chen, B.S. (1983). Seasonability of insect pests of soybean and mungbean in Taiwan. *J. Eco. Entomology,* **76:** 34-37.

Talekar, N.S. and Lin, C.L. (1992). Characterization of *Callosobruchus chinensis* (Coleoptera: Bruchidae) resistance in mungbean. *Journal of Economic Entomology,* **85**(4): 1150-1153.

Talekar, N.S. and Lin, C.P. (1992). Mechanism of *Callosobruchus chinensis resistance in mungbean. Journal of Economic Entomology,* **85:** 1150-1153.

Talekar, N.S. and Lin, Y.H. (1981). Two sources with differing modes of resistance to *Callosobruchus chinensis* in mungbean. *J. Economic Entomology,* **74**(5): 639-642.

Telekar N.S. (1990). Agromyzid flies of food legumes in the Tropics. Wiley Eastern Limited, New Delhii, India pp. 297.

Telekar N.S. and Lee, Y.H. (1988). Biology of Ophiomyia centrosematis (Diptera: Agromyzidae), a pest of soybean. *Ann. Entomol. Soc. Am.*, **81**(16): 938-942.

Tangkano, W. and Soekarna, D. (1976). Screening of mungbean (*Phaseolus radiata*) varieties for resistance to the beanfly *Agromyza phaseoli* Coq. Pages 39-47 in Annual Report Central Research Institute for Agriculture, Ministry of Agriculture, Bogor, Indonesia.

Taylor T.A. (1978). *Maruca testulalis:* an important pest of tropical grain legumes. Pages 193-200 in Pests of Grain Legumes: Ecology and control. (S.R. Singh, H.F. van Emden and T.A. Taylor Eds.). Academic Press, London.

Thapa, R.B. and Timsina, J. (1990). Incidence of the beanfly, *Ophiomyia phaseoli* (Tryon) in mungbean and soybean varieties after late rice. *J. Institute of Agriculture and Animal Science,* No. **11:** 83-88.

Thawstine, S. (1987). Effect of neem on the development of cowpea weevil *Callosobruchus maculatus* (Fab.) on some leguminous seeds. *Munoz. Nueva Ecija* (Philiphines) **7:** 9-11.

Thontadarya T.S. and Hiremath, I.G. (1983). Losses due to insect pests in Karnataka. In Crop losses due to insect pests. Special issue, *Indian J. Entomol.,* **1:** 43-58.

Turner, J.W. (1983). Influence of plant species on the movement of *Trissolucus basalis woolaston* (Hymenoptra: Scelionidae–a parasite of Nezara viridua. *J. Aust. Ent. Soc.,* **22**(3): 271-272.

Udhaya Kumar, A. and Jayaraj, S. (1987). Influence of host plants on certain parasites of *Spodoptera litura* (F.), *J. Biol. Control.*

Ujagir, Ram, Singh, R.K. and Ujagir, R. (1999). Management of insect pests of mungbean by insecticides using seed treatments, soil and foliar application. *Indian J. Pulses Research,* **12**(1): 82-91.

Usua E.J. and Singh, S.R. (1978). Parasites and predators of the cowpea pod borer, *Maruca testulalis* Gey. (Lep. Pyralidae). *Nigerian J. Entomology,* **3:** 100-102.

Vadodaria, M.P. and Vyas, H.N. (1987). Control of whitefly, *Bemisia tabaci* (Gennadius) and its impact on yellow mosaic virus (YMV) in greengram, (*Vigna radiata* (L.) Wilczek) and the grain yield. *Indian J. Agril. Res.* **21**(1): 21-26.

Venkataramaiah, C., Reddy, A.J. and Chandrashekharan, V. (1980). On DDT–induced radicular proliferation in *Phaseolus mungo* Linn. *Current Science,* **49**(6): 242-243.

Verma, A.N., Bhanot, J.P. and Khurana, A.D. (1983). Effect of presowing soil application of BHC and aldrin on plant stand in wheat and *moong,* Haryana Agricultural University Journal of Research, **13**(1): 150-152.

Vyas, H.N. and Saxena, H.D. (1980). Bio-efficacy of some synthetic insecticides on seed treatment and soil application to control stemfly, *Ophiomyia phaseoli* Cogi (Diptera: Agromygzidae) in pea, *Pisum sativum. Pesticides* **14**(1): 12-13.

Vyas, H.N. and Saxena, H.P. (1981) Toxicological and persistence studies of some insecticides in relation to pest complex and plant growth of greengram, *Vigna radiata* (L.) Wilczek and fieldpea, *Pisum sativum* L. *Pulse Crops Newsletter,* **1:** 98.

Vyas, H.N. and Saxena, H.P. (1983). Bioefficacy of some insecticides against *Empoasca kerri* Pruthi in greengram, *Vigna radiata* (L.) Wilczek. Pesticides **17**(5): 15-17.

Yadav, A.S., Bhaduria, N.S. and Jakhmola, S.S. (2004). Efficacy of edible non-edible vegetable oils against pulse beetle, *Callosobruchus maculatus* (Fab.) in green gram. *Indian J. Entomology,* **66**(4): 365-366.

Yadav, D.K. and Singh, S.K. (2006). Forecast model of major insect pests of mungbean. *Ann. Pl. Protec. Sci.* **14**(2): 323-328.

Yadav, L.S. and Yadav, P.R. (1979). Avoidable losses due to insect pests in pulses. *Military Farm J.* **11:** 17-19.

Yadav, T.D., Singh, S., Jha, A.N. and Yadav, S. (2000). Host suitability of chickpea cultivars to *Callosobruchus chinensis. Indian J. Ent.,* **62**(2): 203-204.

Yein, B.R. and Singh, H. (1981). Effects of some agricultural chemicals on the control of chewing insect pests of greengram. *J. Res. Punjab Agril. Univ.,* **18**(1): 30-36.

Yein, B.R. and Singh, H. (1982). Effect of pesticides and fertilizers on the population of whitefly and incidence of yellow mosaic virus in greengram. *Indian J. Agric. Sci.,* **52**(12): 852-855.

9

Integrated Management of Soybean Insect Pests for Sustainable Agriculture

Debjani Dey and Jaydeep Halder

Division of Entomology, Indian Agricultural Research Institute, New Delhi–110 012

Soybean (*Glycine max* (L.) Merr. is grown in 83.69 million hectares worldwide with an annual production of 189.34 million tonnes (2003). USA tops the list in both acreage and production. The other major soybean producing countries are Brazil, Argentina, Indonesia, China, Paraguay, Canada and Italy. During the last two and a half decades, there has been a phenomenal increase in area under soybean cultivation in India as it has gained paramount importance in India not only for good quality oil (20-22%) but also for high quality protein (42-45%) as well. At present, India occupies fifth place in the world in soybean cultivation with a production of 8.25 million tonnes from an area of 7.71 million hectares during 2005-2006 (http://agricoop.nic.in/Agristatistics.htm) and ranks third among the nine edible oilseeds cultivated in India, after groundnut and rapeseed-mustard. However, the present national average yield of about 1.210 tonnes/ha is much below the world average (2.5 tonnes/ha). It contributes 25-30% to India's oilseeds production. Madhya Pradesh produces nearly 70% of soybean while Maharashtra ranks second (Plate 9.2). The other major soybean producing states are Rajasthan, Karnataka and Uttar Pradesh (Plate 9.1).

About 275 insect species are known to be associated with soybean crop right from germination to harvest (Singh *et al.*, 1989). Amongst these, a dozen cause economic damage and are serious in different parts of the country. However, the severity of these principal pests is not the same at all places. Therefore, to meet the pest management challenges faced by soybean producers, an effort has been made to identify the regional differences in the pest complex and accordingly approach is focussed towards crop protection products and crop production tactics that are important for the economic management of key pests of soybean in the various soybean growing areas.

The major pests infesting soybean crop at various growth stages in India include Stemfly (*Melanagromyza sojae*), Girdle beetle (*Oberiopsis brevis*), Gram pod borer (*Helicoverpa armigera*), Leaf miner (*Aproaerema modicella*), Whitefly (*Bemisia tabaci*), Bihar hairy caterpillar (*Spilosoma obliqua*), Tobacco caterpillar (*Spodoptera litura*), Green semi-looper (*Chrysodeixis acuta*) and Brown semi-looper (*Mods undata*), the details of which have been enumerated below.

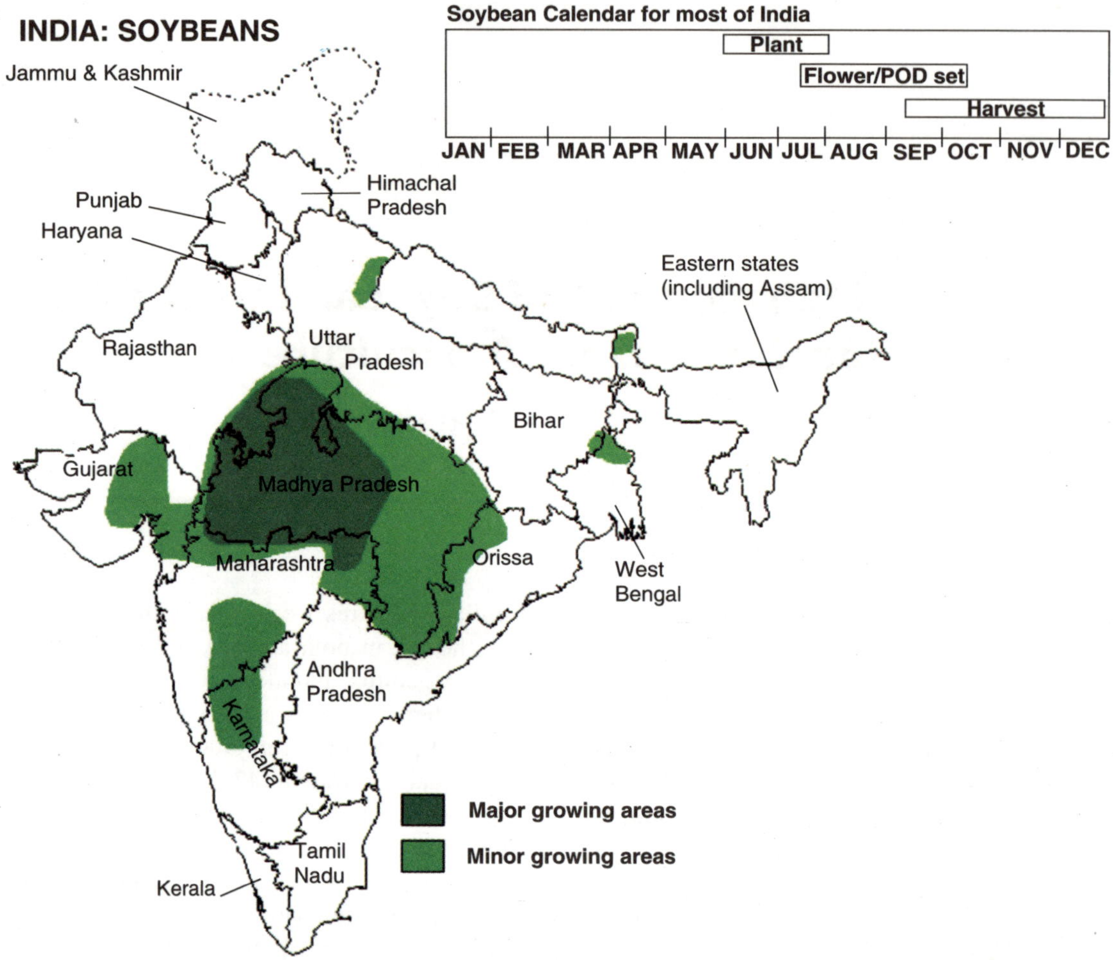

Plate 9.1 : Major and minor soybean growing areas of India.

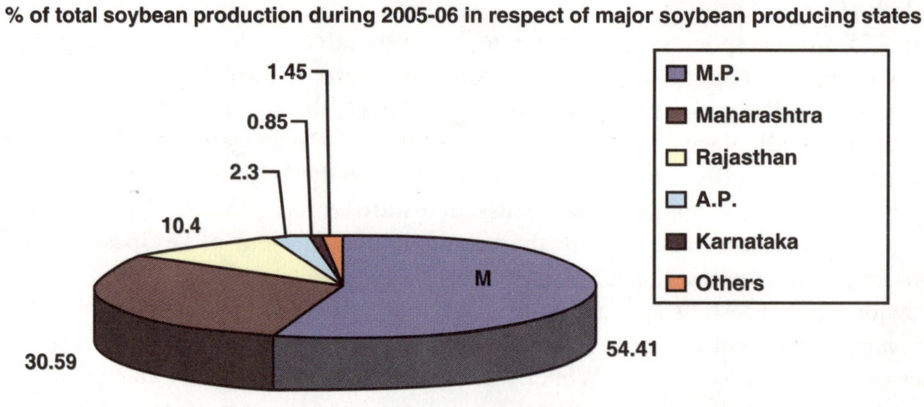

Plate 9.2 : Per cent soybean production during 2005-2006.

WHITEFLY (*Bemisia tabaci* Gennadius: Aleyrodidae, Hemiptera)

These have been termed as whitefly because of the whitish appearance of the adults. The white fly is observed throughout the year on different crops, however, maximum population however, is recorded during the wet season (July to September) when both temperature and relative humidity are quite high.

Distribution: Maharashtra, Bihar, Punjab, Uttar Pradesh, West Bengal and other parts of North India.

Diagnosis: The adult white fly is approximately 1.0 mm long, yellow bodied insects with whitish grey wings, which are densely covered with a waxy powder (Fig. 9.1). The newly emerged nymphs are oval shaped and light yellow in colour, but later change to pale greenish yellow.

Fig. 9.1 : Adult of white fly, *Bemisia tabaci.*

Life cycle: A single fertilized female lays on an average 38 to 106 yellow eggs on the lower surface of leaves. The eggs are minute, less than 0.2 mm long, stalked and remain attached to the underside of the leaves. Yellow colour of the egg changes to dark brown before hatching in 1 or 2 days. They feed approximately for 7 to 14 days before moulting into a pupal stage. The pupae are oval and black in colour. Adults emerge in 8 to 14 days.

Damaging stage of the insect: Both nymphs and adults.

Damage symptoms: The infestation starts right from unifoliate stage of the plants and continues till the crop remains green. The nymphs as well as adults suck the sap from the leaves resulting in lowering the vitality and thus causing stunted growth of the plants. They also secrete honey-dew, which when deposited on the leaves result in the development of a sooty mould on the leaves, thus hampering photosynthesis. They also transmit yellow mosaic disease, a serious disease in the plains of northern India, resulting in yellow patches on the leaves (Fig. 9.2). Due to heavy attack of the insect, up to 80% grain yield loss has been reported. Gill and Singh (1986) reported that a population of 21 flies per plant during the first week of June resulted in yellow vein mosaic disease incidence of 100 per cent by the end of July with a very high yield loss.

Fig. 9.2 : YMV disease symptoms due to whitefly damage.

STEMFLY *(Melanagromyza sojae* M. agromyzidae, Diptera)

Stemfly is a very serious pest throughout India. During kharif it is found to infest 90–100 per cent of plants and during rabi about 60 per cent. It remains active from sowing (June – July) to harvesting (October).

Distribution: Throughout India.

Diagnosis: The adult flies are shining black and about 2 mm long (Fig. 9.3). The maggot is white in colour and remains inside the stem.

Life Cycle: Adults lay eggs either on cotyledonous leaves or first trifoliate leaves. The eggs are light yellow in colour, remain embedded in the leaf tissues and hatch in 2-7 days. The maggots remain inside the stem and start feeding on the leaf tissues, start moving towards the stem and reach the pith region of the stem in 2-3 days, where they feeds and develop. They undergo 3 to 4 moults. The larval period lasts for about 10 to 15 days. While feeding on the

Fig. 9.3 : Adult of stemfly, *Melanagromyza sojae.*

pith they make tunnels, which can be seen by splitting open the stem. Pupation occurs inside the tunnels. The pupae are barrel-shaped and somewhat dark yellowish brown in colour. Pupal period lasts for 7 to 10 days. The life cycle is completed in about 19 to 32 days, and the adult fly escapes through the exit holes made by the feeding maggots earlier.

Damaging Stage of the Insect: Only larvae cause damage by burrowing inside the stem.

Damage Symptoms: Damage due to this pest starts right from germination and continues throughout the plant growth; the early stage of crop growth (up to 7-10 days after germination) being the most vulnerable. In the early stage of plant growth, initially symptoms of withering and drooping of leaves with browning of leaf surfaces are observed. Then due to feeding on the stem pith, tunnels are formed resulting in either drying or stunted growth of plants (Fig. 9.4). Agarwal and Pandey (1961) observed an infestation of 10 per cent during September and 22 per cent during October–November. Infestation due to stemfly results in about 20% reduction in yield. Bhattacharya *et al.*, 1986, studied the correlation between tunnel lengths due to *M. sojae* with yield characteristics and reported a perfect negative correlation between tunnel length and different yield parameters. Kundu and Mehra (1989) observed that in an early soybean variety, Harosoy-63, a per cent tunnel length of more than 26 per cent affected all the yield components, namely pod number, pod width, grain number, grain weight and pod height leading to significant yield reduction. Venkatesan and Kundu, 1994, determined the economic injury levels due to stemfly as 6.40 and 11.23 per cent stem tunnelling with quadratic and linear relationship, respectively. Further they also observed a significant negative correlation between the stemfly infestation and grain yield.

GIRDLE BEETLE (*Oberiopsis brevis* Swed.: Cerambycidae, Coleoptera)

Fig. 9.4 : Stem tunnelling due to stemfly damage.

Fig. 9.5 : Adult of girdle beetle, *Oberiopsis brevis.*

This beetle inflicts serious damage especially during rainy season (July-August) to soybean crop in Central and Southern India.

Distribution : Throughout India.

Diagnosis : Medium sized beetle around 8-10 mm long with a pair of long antennae. The anterior half of the forewing is orange-brown, while the posterior half is black in colour (Fig. 9.5).

Life Cycle : Adult female before egg laying makes 2 parallel rings or girdle on the petiole, stem or base of any of the trifoliate leaves, usually the central one. A slit is made by cutting the stem between the two girdles by the female. Eggs are laid through the pit. A female lay singly from 8 to 72 elongate eggs in the pith of the stem, petiole or stalk of a leaflet. Eggs hatch in about 4-8 days. The larvae is yellowish white, smooth and apodous with a swollen prothoracic segment. It gradually tunnels towards the main stem and hollows it up to the base of the plant. One grub lives in a plant at any one time. Full grown larvae measures from 19 to 22 mm in length. Larval period lasts for about 32 to 65 days in which case the adults emerge to start the next generation in the same season. Otherwise, the grubs require 248 to 308 days including larval diapause period (from October to June-July) to develop into a pupa. The diapause of the grub breaks after rains in June/July, and the grub develop into pupa from which adult emerges out in 8 to 11 days.

Damaging Stages of the Insect: Both grub and adults.

Damage Symptoms: The girdle formation is the most characteristic symptom of damage due to this insect (Fig. 9.6). As a result of girdle formation, the portion above the girdle

Fig. 9.6 : Girdle formation and drying of the stem and leaves above the girdle.

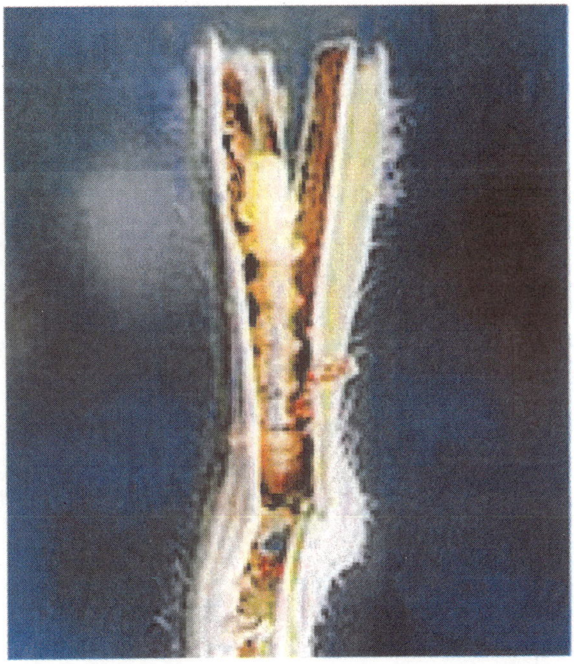

Fig. 9.7: The grub of girdle beetle inside the stem of soybean.

withers and dries up ultimately. The girdled part also dries up later and breaks off. The grub (larva), which is smooth and yellowish, after entering into the stem, makes a tunnel up to the base of the plant resulting in drooping of leaves and flowers and ultimately the stem breaks off during the late stage of plant growth (Fig. 9.7). Heavy infestation can result in about 50 per cent loss in yield (Singh and Singh, 1966).

GREY WEEVIL (*Myllocerus undecimpustulus* Faust : Curculionidae, Coleoptera)

The weevil is active from August to October.

Distribution : Widely distributed throughout India.

Diagnosis : The adult beetles are greyish in colour. The head is prolonged into a snout and bears the mouth parts at the tip (Fig. 9.8). When disturbed the adult beetles fall to the ground and remain motionless.

Life Cycle : Female beetles lay dirty white eggs in batches in soil. Hatching takes place in 3 to 5 days and the grub stage lasts for about 25 to 30 days. The apodus 'C' shaped creamy white to yellowish-brown grubs with reddish-orange or shiny brown head are found in the region of freshly formed roots. The pupa is creamy white and is lodged in an earthen cell at a depth of 2.5 to 7.0 cm. Pupal period lasts for 7 to 8 days. The life history from egg to adult is completed within 35 to 45 days.

Damaging Stages of the Insect: Grubs and adults.

Damage Symptoms: The weevil starts feeding on the margin of leaves with irregular cuttings and then proceeds inward making holes on leaf surfaces. This affects the dry weight

Fig. 9.8 : Adult of grey weevil, *Myllocerus undecimpustulus* causing damage to soybean plant by its feeding.

of stem and pod size leading to 10.2 per cent loss in grain yield. Grubs feed on the roots, while adults attack the foliage of the seedlings as well as grown up plants.

BLUE BEETLE (*Altica cyanea* Web.: Chrysomelidae, Coleoptera)

Distribution: North and Central India.

Diagnosis: Adults are shining dark blue beetles about 5 mm long with black antennae.

Life Cycle: Full grown grubs are 8 to 10 mm long and dark brown in colour. Pupae are 12 to 16 mm long, orange-yellow when freshly formed and gradually darken and finally become light to dark brown.

Damaging Stages of the Insect: Grubs and adults.

Damage Symptoms: The grubs feed on epidermis of leaves, and when serious, they leave behind only the network of veins. Adults nibble and feed on the leaf margins.

TOBACCO CATERPILLAR (*Spodoptera litura* Fab.; Noctuidae, Lepidoptera)

Distribution: Central India.

Diagnosis: The moth is stout, dark with wavy white markings on the forewings while the hind wings are white with a brown margin. Early instar caterpillars are green with black head but become pale greenish brown with dark markings when full grown. Some have transverse and longitudinal grey and yellow bands with rows of dark spots (Fig. 9.9).

Life Cycle: The female moth lays eggs in a mass on the surface of leaves and covers them with brown hairs. The eggs hatch within 3 to 7 days. The larvae on hatching feed on the chlorophyll of the leaves in a gregarious manner, but in later instars they spread to different plant parts. Feeding results in complete defoliation. A full grown caterpillar measures 25 mm in length. Larvae pass through 5 to 6 instars and the larval period lasts for 14 to 20 days. Pupation takes place in the soil. Adult emerges in 7 to 10 days. Life cycle is completed in 30 to 37 days.

Fig. 9.9 : Adult of tobacco caterpillar, *Spodoptera litura.*

Damaging Stage of the Insect: Caterpillars

Damage Symptoms: Only caterpillars cause damage by feeding on leaves. The type of injury inflicted to plants is like that of Bihar hairy caterpillar. The full-grown larvae devour leaves completely and also damage developing pods (Fig. 9.10) resulting in severe reduction in grain yield. In the case of severe infestation, the entire crop is devoured and thus lost completely (Fig. 9.11).

Fig. 9.10 : Caterpillars of tobacco caterpillar feeding on pods of soybean.

Fig. 9.11 : Complete defoliation of the plants due to feeding by larvae of tobacco caterpillar.

GRAM POD BORER (*Helicoverpa armigera* Hubner: Noctuidae, Lepidoptera)

This pest is active from July to October.

Distribution: It is a serious pest in Madhya Pradesh, Uttar Pradesh and southern parts of India.

Diagnosis: The adults are medium sized moths with light yellowish to brown in colour. Distinct black dots are present in the forewings while hind wings are lighter in colour (Fig. 9.12). The young larvae are usually green with longitudinal stripes of various hues of brown, black, pink, brown, yellowish, dark green with a few sparse hairs on the body.

Fig. 9.12 : Adult of gram pod borer, *Helicoverpa armigera*.

Life Cycle: A female moth lays 1219 to 1554 yellowish-green eggs singly on leaves, which hatch in 3 to 4 days. The upper region of the plant is most preferred for oviposition and about 7 times as many eggs are laid on the leaf surface as on the rest of the plant. The young larvae feed on the leaves and developing pods. Fully developed larvae measure about 30 mm and the larval period is completed in 20 to 25 days. Finally larvae pupate in the soil. The pupal period lasts for about 9 to 13 days and the life cycle is completed in 31 to 35 days.

Damaging Stage of the Insect: Caterpillars.

Damage Symptoms: Only caterpillars are destructive. Initially the larvae feed on leaves resulting in defoliation, but at later stages (flowering stage onwards) they feed on flowers, immature and mature pods and developing seeds (Fig. 9.13). Heavy infestation of this insect can result in about 90% grain yield loss.

BIHAR HAIRY CATERPILLAR (*Spilosoma obliqua* Walker: Arctiidae, Lepidoptera).

This is a sporadic pest and causes damage during August-October in different parts of India.

Distribution: Northern India, 'Tarai' hills and plains of Uttarakhand and Uttar Pradesh, Karnataka, 'Khasi' hills of Meghalaya, Bihar, Orissa, West Bengal and Assam.

Diagnosis: The adult moths are light yellowish-brown in colour having pink abdomen with black dots on the dorsal and lateral surfaces. Hind wings also have black dots on the anterior margin.

Fig. 9.13 : Larvae of *H.armigera* feeding on soybean pods.

Life Cycle: A female moth lays 500 to 1200 pale greenish eggs arranged in closely packed batches of 400-700 on the lower surface of the leaves. With increase in temperature, the duration of all the stages and number of larval moults decreases. The life from egg to adult averages 35 to 45 days.

Damaging Stage of the Insect: Caterpillars.

Damage Symptoms: The infestation of the crop starts in patches. At times it becomes very serious. The early instar larvae (first to third) feed gregariously on the leaf surface resulting in skeletonization of the entire leaf. The leaves appear light brownish in colour and these patches are easily detectable in the field. Later instar larvae feed singly and defoliate the entire plant.

Besides the aforesaid, various minor pests also attack soybean crop during various seasons in one or the other region of the country. The minor pests which are of some importance are listed in Table 9.1.

Table 9.1 : List of minor pests associated with soybean from various parts of India.

Scientific & Common name	Order and family	Distribution	Reference (s)
Foliage feeders			
Chrotogonus sp, Oxya sp, Attractomorpha sp (Surface grass hoppers)	Orthoptera, Acrididae	M.P.	Saxena, 1972
Plusia chalycytes, P. orichelcea (Semi looper)	Lepidoptera, Nocuidae	Rajasthan, U.P.	Bharadwaj and Balla, 1976
Triphosa sp. (Looper)	Lepidoptera, Geometridae	H.P.	Bharadwaj and Bhalla, 1976
Amsacta spp., *Estgmene lactinea*	Lepidoptera, Arctiidae	M.P.	Rai *et al.*, 1973
Herpetogramma sp.	Lepidoptera, Pyralidae	H.P.	Bharadwaj and Bhalla, 1976
Anarsia ephippias (Leaf webber)	Lepidoptera, Gelechidae	M.P.	Kundu and Srivastava, 1992
Nacoleia dieminalis (Leaf folder)	Lepidoptera, Pyralidae	M.P.	Kundu and Srivastava, 1992
Euproctis xanthorthoea (Tussock caterpillar)	Lepidoptera, Lymantridae	M.P.	Kundu and Srivastava, 1992
Zizera lysimon, Polymmatus bacticus	Lepidoptera, Lycaenidae	—	Kundu and Gupta, 1985, Gangrade, 1976
Popillia gemma	Coleoptera, Scarabaeidae	M.P.	Kashyap and Adlakka, 1971
Galerucida bicolor, Haltica cearulescens (Flea beetle)	Coleoptera, Alticidae	Simla hills (H.P.)	Sharma and Bhalla, 1964
Root feeders			
Cydia ptychora	Lepidoptera, Totricidae	M.P.	Singh and Jakhmola, 1983
Aulocophora atripennis, A. foveicollis (Pumpkin beetle)	Coleoptera, Chrysomelidae	U.P. and Delhi	Singh, 1969; Kundu and Srivastava, 1992
Holotrichia consanguinea (white grub)	Coleoptera, Melolonthidae	Madhya Pradesh	Singh and Singh,1994
Sap suckers			
Nipaecoccus vastator (Mealy bug)	Hemiptera, Pseudococcidae	M.P.	Gujrati and Singh, 1985
Riptortus pedestris, *Cletus bipunctata* (Pod bug)	Hemiptera, Coreidae	U.P.	Singh, 1969
Tricentrus sp. (Cowbug)	Hemiptera, Membracidae	H.P.	Bharadwaj and Bhalla, 1976
Pyrilla perpusilla (Sugarcane leaf hopper)	Hemiptera, Fulgoridae	India and Sri Lanka	Kundu and Srivastava, 1992
Chuliops nigresecens,	Hemiptera, Lygaeidae	U.P., M.P.	Singh, 1969

Table 9.1 : Contd.

Scientific & Common name	Order and family	Distribution	Reference (s)
C.fallax (Podbug) *Callitettix versicolor,* *Cosmpcarta dorsimaculata,* *C.bispecularis,* *Poophilus costalis* (Frog hopper)	Hemiptera, Cercopidae	H.P.	Kashyap and Adlakka, 1971 Bharadwaj and Bhalla, 1976 Kashyap and Adlakka, 1971
Nezara viridula, *Coptosoma cribraria,* *Dolycoris indicus* (Potato green bug)	Hemiptera, Pentatomidae	U.P.	Singh, 1969, Kundu and Srivastava, 1992
Ayyaria chaetophora	Thysanoptera, Thripidae	—	Ananthakrishnan, 1971
Tetranychus telarius, Red spider mite	Tetranychidae, Acarina	Madhya Pradesh	Singh *et al.*, 1990
Flower feeders			
Mylabris macilenta, *Epicauta mannerheim* (Blister beetle)	Coleopteran, Meloidae	Bihar, U.P.	Singh, 1969, Kashyap and Adlakka, 1971, Kundu and Srivastava, 1992
Cordylocera livida	Coleopteran, Cantharidae	H.P.	Bharadwaj and Bhalla, 1976
Homoneura picta	Diptera, Lauxaniidae	H.P.	Bharadwaj and Bhalla, 1976
Internal feeders			
Nupsera nitidor	Coleoptera, Cerambycidae	Bihar	Dutt, 1915, Kashyap *et al.*, 1980

H.P: Himachal Pradesh, M.P. : Madhya Pradesh; U.P: Uttar Pradesh

INTEGRATED PEST MANAGEMENT

To save the soybean crop from ravages of insect pests, it is very essential to adopt suitable, need-based and quick result-oriented control measures, keeping in mind the resurgence of pests as a result of development of pest resistance to insecticides, minimum environmental pollution as well as less harm to useful biotic agencies such as parasitoids, predators and pollinators. For this, an integrated approach should be adopted to suppress the pest population to a desired level through proper and judicious use of suitable control measures at the right time (Hammond *et al.*, 1991). Prior to understanding soybean insect IPM comes the need to know the crop plant and its growth throughout the season. Many decisions that a grower or IPM practitioner will make depends upon the type of soybean and the plant's growth stage. Growth stage is the most important criterion because the relationship between insect injury and crop damage is dependent on the stage when the injury occurs. Research has shown that injury during the vegetative stages is usually not as detrimental to the plant as that during reproductive stages. Also knowledge of the threshold of economic damage leading to economic loss is highly essential as it will enable a more judicious use of an insecticide. However, it is still advisable to confirm the presence of an insect pest before making any insecticide application. Population monitoring of insect pests is an important task which involves visual examination of soybean plants every 3 to 5 days during the soybean growing season to obtain information on population dynamics of all pests and natural enemies and to estimate damage to the plants. Integrated pest management involves various methods of control, viz., mechanical, cultural and use of both biopesticides and chemical pesticides.

Mechanical Control

Ploughing of fields during summer helps eliminate soil pests either by excessive heat or by exposing them to predatory birds. Bhattacharya and Bhatia (1980) reported that hoeing of crop at fortnightly interval with a soil turning hoe like "Akola hoe" exposed the grubs of *Holotrichia consanguinea,* which are then picked up by predatory birds. Collection and destruction of developing stages (egg/larva/pupa) of serious pests and infested plants from the fields will help in reducing the pest incidence. Use of light traps or pheromone traps in and around the fields helps capture adults of Bihar hairy caterpillar, tobacco caterpillar, gram pod borer, stemfly, girdle beetle and white grubs. Yellow sticky traps attract whiteflies, jassids, aphids, thrips, flea beetles, etc. By using such traps in the fields, first appearance of the pest and pest densities can be monitored for taking timely measures to achieve cent per cent control.

Cultural Control

(i) Use of resistant varieties

Some strains or varieties of a crop possess genetic resistance to attack of certain pests. These resistant varieties can be cultivated without change in normal practice of cultivation and these can also be incorporated into insect pest management practices. Resistant varieties are being continuously developed by the various Agricultural Research Institutes, State Agricultural Universities and State Agriculture Departments of that particular region.

(ii) Early sowing

Early sowing (last week of June) helps reduce the incidence of stemfly and leaf miner, whereas the incidence of girdle beetle can be minimized by sowing in July (after monsoon break) (Chandel and Gupta, 1995). Use of resistant varieties can be effective mean for adoption by farmers to attain the targeted yield of soybean (Chiang and Talekar, 1980). Gain and Kundu (1986) suggested that soybean should be cultivated in Delhi region in the beginning of July to avoid higher bean stem borer. Higher activity of the stem borer was favoured by optimum temperature (24.9–25.2°C), higher per cent relative humidity and high rainfall and these abiotic factors had high positive correlation with per cent stem tunnelling. Early sowing (20th June–5th July during kharif and first fortnight of November during rabi) had minimum pest incidence and recorded highest grain yield during both the seasons, while sowing during August (Kharif) and second fortnight of December (Rabi) recorded maximum incidence of most of the pests (Sontakke and Mishra, 1994). Satish Kumar and Bhattacharya (1988) observed that variation in row spacing showed marked effect in stemfly attack on 30 DAS. Plots having 30 and 45 cm spacing in row to row were infested more by *Melanogromyza sojae* as compared to plots having row to row distance of 60 and 75 cm. Singh and Gangrade (1977) reported that varieties planted on 1st June probably reached a vulnerable stage of growth when the girdle beetles emerged from diapause.

(iii) Trap crop

Some crops that serve as alternate food plants for an insect pest can be grown around soybean field a little earlier to divert or lessen infestation on soybean crop. Wherever possible Lucerne crop, an important fodder crop, can be grown to trap the leaf miner infestation, a serious pest of soybean. Lady's finger, maize can be grown to divert the infestation of grey weevils; crops like chillies, bittergourd, cowpeas are also food plant for girdle beetle, a serious pest of soybean. These crops can be grown as a trap crop to lessen the infestation of girdle beetle in soybean (Kundu and Srivastava, 1992).

Biological Control

Some parasitoids, predators, virus, bacteria and fungal pathogens have been reported controlling the major insect pests of soybean. Parasitoids, viz., *Trichogramma* sp, *Apanteles* sp., *Tetrastichus* sp., *Bracon* sp., *Elasmus* sp., *Iphiaulax* sp., *Spegigaster* sp. *Goniozus* sp. etc. help in reducing the population of egg, larvae or pupae of stemfly, girdle beetle, leaf miner, gram pod borer, Bihar hairy caterpillar, tobacco caterpillar, semiloopers, etc. A predatory mite (*Tyopholodromus* sp.), predatory bug (*Amyotea malabarica*) and green lace wing (*Chrysopa carnea*) have been found predating on pupae of whitefly, nymphs and adults of green bug and full-grown grubs of girdle beetle respectively. Pentatomid bug predators, viz. *Nebis* sp. and *Cantheconidia furcellata* keep the population of larvae of green semilooper and tobacco caterpillar, respectively under control (Higley and Boethel, 1994). Shetgan and Tembhre (1984) documented that larvae of leaf miner, *Stomopteryx subsecivella* Z. were parasitised by *Microchelonus* sp, *Apanteles* sp, *A. litae* var. *operculella* Nixon, *Stenomesioideus ashmedi* and *Goniozus sp* during the second week of August. The parasitisation reached its peak (84 per cent) during the last week of August. Bhattacharya (1976) reported that during September to early November, a Chalcid fly, *Brachymeria* sp. and an Ichneumonid, *Xanthopimpla punctata* Fab. were found parasitizing the pyralid leaf rollers *viz., Lamprosema indicata* and *L. dimemevalis*. The parasitism by both the parasitoids was to the tune of 11.1 per cent. A number of predatory spiders were also seen moving about the soybean plants, though the exact nature of their effect could not be ascertained. Except the first instar nymphs of *Amyotea (Asopus) malabarica* (Fabricius), all other stages fed on both nymphal and adult stages of green painted bug, *Nezara viridula* Linn. (Singh, 1973). Gujrati and Gangrade, 1973 recorded the following hymenopterous parasitoids, *Xanthopimpla predator* (Fabr.), *Echthromorpha agrestoria* (Sweal) and *Brachymeria lasus* (Walker) and two dipterous pupal parasitoids viz., *Exorista sorbillans* (Wield) and *Thelaria macropus* (Wield) on soybean brown semi-looper (*Mods undata*). Singh and Gangrade, 1975 reported that larvae of *D. obliqua* were parasitized by three hymenopterous and four dipterous parasitoids in the field with percentage parasitizations being 23.4 and 5.71% respectively. Out of three hymenopterous parasitoids two, *viz., Apanteles obliqua* (Walker) and *A. glomeratus* were identified. Five Sarcophagids *viz., Sarcophaga nisera* Walker, *Sarcophaga sp.* var *Kankawensis* Baranov, *Goniophthalmus halli* Mensil, *Carcelia corvinoides* Wulp and *Sarcophaga sp.* also attacked the larvae of *D. obliqua* during various season. Two reduviid predators, *Rhincoris fuscipes* (Fabricius) and *Scarda annulipes* Reuben fed on 1-3 larvae per day during their life of 21-29 days. Two Pentatomids *viz., Cantheconidia furcellata* Wolff and *Andrallus spinidens* Fab. were also found to be effectively controlling the larvae of *D. obliqua*. Various entomopathogenic fungi, *Nomurea rileyi* (Farlow) wiped out several colonies of young larvae of *D. obliqua.*

Though these parasitoids and predators, to some extent, help in reducing the pest problem, they are inadequate to suppress the pest menace to a desired level. Further, innundative release of biotic agencies is not feasible on a large scale unless cheap and easy mass rearing technologies of parasitoids and predators are developed.

Biopesticides as a Tool in Pest Management

Biopesticides include pathogens like virus (nuclear polyhedrosis virus–NPV), bacteria *(Bacillus thuringiensis–Bt*), fungi *(Beauveria bassiana–Bb*) and pesticides of plant origin. For their successful utilization, their application on the plants must coincide with the emergence of the pest (just after hatching of young ones, the most suitable and susceptible stage of insect to microbial pesticides). The NPV of gram pod borer and tobacco caterpillar is very effective in controlling the larvae of these pests. The application of *Bt* on soybean plants helps in controlling the larvae of tobacco caterpillar, gram pod borer and other lepidopterous pests, whereas *Bb* is effective in minimizing the larval population of gram pod borer. Since these biopesticides are stomach poisons, the young larvae (first instar) as a result of feeding on the treated plants can be killed in early stages of their development. Thus, the pest population can be minimized. Being economical, environment-friendly and due to their easy application methodology, they can be considered as supplemental tool in pest management. However, every care should be taken as some biopesticides such as NPV are host-specific and easily disintegrated when they are exposed to sunlight.

In outbreak of certain insect pests, the mechanical, cultural and biological control measures are of limited value to quickly suppress the insect pest population. Under such conditions, chemical control measures are preferred for quick results.

CHEMICAL PESTICIDES AS COMPONENT OF PEST MANAGEMENT

Seed coating or furrow treatment with systemic insecticides holds promise, as they not only protect the developing seedlings but also later stages of the crop against various insect pests. These treatments are not only economical but also fit better into the usual farming practices with an added advantage of least disruptive effect on the environment and the useful fauna. Phorate 10 G applied @ 1.0 kg a.i./ha once in furrows at sowing and once as side dressing 3 weeks after germination was most effective against stemfly infestation (Anonymous, 1998). Investigations carried under All India Coordinated Research Project on Soybean revealed that phorate 10 G as furrow treatment at sowing @ 10 kg/ha was quite effective in controlling insect pests of soybean in early stage of crop growth (Anonymous, 1999, Chaudhary, *et al.*, 1981). Kundu (1991), reported that monocrotophos (0.04%), malathion (0.05%), endosulfan (0.105%) and cartap hydrochloride (0.05%) were most effective against second instar larvae of *Spilosoma obliqua.* Among various insecticides tested, quinalphos (0.05%), fenvalerate (0.01%) and monocrotophos (0.04%) were highly effective against leaf miner and also gave maximum yield (Singh and Singh, 1983). Singh and Singh (1991) reported that chloropyriphos (0.04%) was most effective followed by phosphamidon (0.03%) in reducing the number of jassid nymphs and adults. Maximum net return was achieved in case of phosphamidon followed by chlorpyriphos. Depending on the severity of incidence of pests like stem fly, girdle beetle and leaf feeding insects, 1 or 2 foliar spray applications of either Endosulfan 35 EC 0.07% (2 ml/litre of water) or Triazophos 40 EC 0.04% (1 ml/litre of water) or Quinalphos 20 AF 0.035% (1.8 ml/litre of water) or Monocrotophos 36 SL 0.05% (1.5 ml/litre of water) at 15 days interval helps in reducing the infestation (Singh and Singh, 1994). Seed treatment with Thiamethoxam 70 WS @ 3 g/kg seed was found to be very effective in controlling stemfly infestation and yellow mosaic incidence transmitted by whitefly (Siddiqui and Trimohan 2000).

Srivastava *et al.,* 2007 reported that diflubezuron @ 600 g/ha was very promising in suppressing the soybean defoliators (*viz.,* S. *obliqua, Thysanoplusia orichalcea*) population followed by 400 g/ha and 800 ml/ha of lufnuron. In case of whitefly, which is a vector of yellow vein mosaic disease, nearly cent per cent control can be achieved by using little higher concentration of insecticide, as damage by the vector is not density dependent (even one fly can infect many plants).

During the last two and a half decades, soybean cultivation in India has gained paramount importance not only as a source of good quality oil but for protein as well. At present although India occupies the fifth place in the globe in soybean cultivation, yet the present national average yield is much below the world average. The major soybean producing states are Madhya Pradesh, Maharashtra, Rajasthan, Karnataka and Uttar Pradesh. Although temporal changes in the arthropod complex are common within all regions, the major pests infesting soybean crop at various growth stages in India include Stemfly (*Melanagromyza sojae*), Girdle beetle (*Oberiopsis brevis*), Gram pod borer (*Helicoverpa armigera*), Leaf miner (*Aproaerema modicella*), Whitefly (*Bemisia tabaci*), Bihar hairy caterpillar (*Spilosoma obliqua*), Tobacco caterpillar (*Spodoptera litura*), Green semi-looper (*Chrysodeixis acuta*) and Brown semi-looper (*Mods undata*). To save the soybean crop from ravages of insect pests, it is very essential to adopt suitable, need-based and quick result-oriented control measures, keeping in mind the resurgence of pests as a result of development of pest resistance to insecticides, minimum environmental pollution as well as less harm to useful biotic agencies such as parasitoids, predators and pollinators. For this, an integrated approach should be adopted to suppress the pest population to a desired level through proper and judicious use of suitable control measures at the right time.

REFERENCES

Agarwal, N.S. and Pandey, N.D. (1961). Bionomics of *Melanogromyza phaseoli* Coq (Diptera:Agromyzidae) *Ind. J. Ent.* **23**(4): 293-298.

Ananthakrishanan, T.N. (1971). Thrips (Thysanoptera) in agriculture, horticulture and forestry–diagonosis, bionomics and control. *J. Scient. Industr. Res.* **30:** 113-146.

Anonymous, (1998). Proceedings Twenty-eighth Annual Workshop, All India Coordinated Research Project held at Tamil Nadu Agricultural University, Coimbatore, Tamil Nadu during May 6-8, 1998.

Anonymous, (1999). Proceedings Twenty-ninth Annual Workshop, All India Coordinated Research project held at Jawaharlal Nehru Krishi Vishwa Vidyalaya, Jabalpur, Madhya Pradesh during May 10-12, 1999.

Bhattacharjee, N.S. (1976). Record of parasites and predators of soybean leaf roller in India. *Ind. J. Ent.* **38**(4): 383-384.

Bhattacharya, A.K. (1984). Screening of soybean germplasm with special reference to lossess caused by major pests (1983-1984). Annual Research Report, department of Entomology, G.B. Pant University of Agriculture and Technology, Pantnagar, India

Bhattacharya, A.K. and Ratore, Y.S. (1977). Survey and study of the bionomics of major soybean insects and their chemical control. G.B. Pant University of Agriculture and Technology, *Research Bulletin* 107: pp. 324.

Bhattacharya, A.K., Rathore, Y.S., Chaudhary, R.R.P., Sri Ram, and Rathore, R.R.S. (1986). Correlation between degree of damage caused by *Melanogromyza sojae* (Zehnter) and yield of soybean. *Ind. J. Ent* **48**(1): 1-6.

Bhattacharya, N.S. (1979). Annual Report: All India Coordinated Research Project on Soybean. Indian Agricultural Research Institute, New Delhi, India, pp. 32

Bhattacharya, N.S. and Bhatia, H.P. (1980). Infestation of white grub, *Holotrichia consanguinea* Blanchard on soybean. *Ind. J. Ent.* **42**(2): 321-324.

Bhradwaj, S.P. and Bhalla, O.P. (1976). Record of soybean pests in Himachal Pradesh *Ind. J. Ent.* **38**(3): 286-289.

Chandel, Y.S. and Gupta, R.K. (1995). Effect of sowing dates on the incidence of insect pests of soybean (*Glycine max*) *Ind. J. agric. Sci* **65**(8): 624-625.

Chaudhary, R.R.P., Bhattacharya, A.K. and Rathore, R.R.S. (1981). Use of systemic insecticides for stemfly, *Melanagromyza sojae* (Zehnt.) and whitefly, *Bemisia tabaci* Genn. *Indian J. Ent,* **43**(2): 223-225.

Chiang, H.S. and Talekar, N.S. (1980). Identification of sources of resistance to beanfly and two other agromyzid flies in soybean and mungbean. *J. Econ. Ent.* **73:** 197-199.

Dutt, H.L (1915). The soybean stem borer; *Agric. J. Dept. Agric. Bihar and Orissa,* **3:** 52-56.

Dutt, N. and Pal, P.K. (1982). Role of mandibular length of *Obereopsis brevis* Swed in selecting the site of oviposition and damage in soybean, *Glycine max* (L.) Merr. *J. Ent. Res.* **6**(1): 37-47.

Gain, D. and Kundu, G.G. (1986). Seasonal incidence of the bean stem miner, *Melanogromyza sojae* (Zehnter) in soybean at Delhi (India) *J. Ent. Res.* **10**(2): 152-154

Gangrade, G.A. (1976). Insects of soybean. Technical bulletin, Directorate of Research service, JNKVV, Jabalpur, Madhya Pradesh, India, pp. 88.

Gangrade, G.A., Singh, O.P. and Matkar, S.M. (1975). Soybean yield losses in response to damage by varying level of three lepidopterous larvae. *Ind. J. Ent.* **37**(3): 225-229.

Gill, K.C and Singh, H.R. (1986). A note on the incidence, estimation of losses and symtomology of yellow mosaic virus on soybean, *Glycine maxi* Linn. *Ind. J. Ent.* **48**(4): 524-526.

Gujrati, J.P. and Gangrade, G.A. (1973) Record of parasites of *Mocis undata* Fab. *Ind. J. Ent.* **35**(2): 168.

Gujrati, J.P. and Singh, O.P. (1985) Biology of leaf webber, *Anarsia ephippias* Meyrick, in Madhya Pradesh. *Ind. J. Agric. Sci.* **55**(5): 383-384.

Hammond, R.B., Higgins, R.A., Mack, T.P., Pedigo, L.P. and Bechinski, E.J. (1991). Soybean pest management, pp. 341-472. *In: CRC Handbook of Pest Management in Agriculture,* 2nd Edit. D. Pimentel, Ed. CRC Press, Inc. Boca Raton, FL.

Higley, L.G., and Boethel, D.J. (1994). *Handbook of Soybean Insect Pests. Entomol. Soc. Amer.* pp. 136.

Kashyap, N.P. and Adlakka, R.L. (1971). New records of insects pests of soybean crop. *Ind. J. Ent.* **33**(4): 467-468.

Kashyap, N.P., Hameed, S.F. and Vaidya, D.N. (1980). Preliminary studies on the varietal susceptibility of soybean to *Nupsera nitidor* var *atripenis. Trop. Grain Legume Bull.* **13/14**: 50-52.

Kogan, M. and Herzog, D.C. (1980). *Sampling Methods in Soybean Entomology.* Springer-Verlag, New York. pp. 587.

Kundu, G.G. (1991) Control of Bihar hairy Caterpillar, *Spilosoma obliqua* (Walker) 0n soybean. *Ind. J. Ent.* **53**(3): 491-493.

Kundu, G.G. and Goswami, K. (1985). Evaluation of soybean germplasms for resistance to the stem fly, *Melanogromyza phaseoli* (Tryon) *J. Ent. Res.,* **9**(1): 106-108.

Kundu, G.G. and Gupta, S.L. (1985) *Zizera lysimon* Hubner (Lycaenidae: Lepidoptera) as a pest of soybean, *Glycine max* (L) Merr. in India. *Ind. J. Ent.* **9**(1): 106-108.

Kundu, G.G. and Mehra, R.B. (1989). Determination of economic injury level of stemfly, *Melanogromyza sojae* (Zehnter) in soybean. *Ind. J. Ent.* **51**(4): 434-439.

Kundu, G.G. and Srivastava, K.P. (1992). Soybean pests in India and their management, Today and Tomorrow's Printers & Publication, New Delhi, India.

Nair, M.R.G.K. (1995). Insects and mites of crops in India. Publications and information division, Indian Council of Agricultural Research, New Delhi pp. 55-56.

Rai, P.S., Reddy, K.V.S. and Govindan, R. (1973). A list of insects pests of soybean in Karnataka state, *Curr. Res.* **2**(11): 97-98.

Satish Kumar and Bhattacharya, A.K. (1986). Effect of intercropping of soybean with blackgram and maize on incidence of insects. *Ind. J. Ent.* **50**(2): 226-237.

Satish Kumar and Bhattacharya, A.K. (1989) Influence of different row spacing on the incidence of some insects pests of soybean. *Ind. J. Ent.* **51**(1): 1-7.

Saxena, D.K. (1972). Arthropod pests of soybean in Madhya Pradesh and their control *Pesticides* **6**: 17-19.

Sharma, A.N. (1994). Control of stemfly and sucking insects of soybean through seed treatment. *J. Maharastra. Agric. Univ* **19**(2): 310-311.

Sharma, P.I. and Bhalla, O.P. (1964). Survey study of insect pests of economic importance in Himachal Pradesh *Ind. J. Ent.,* **26:** 238-239.

Shetgan, S.S. and Tembhare, U.T. (1984). Occurrence of natural enemies on soybean leaf miner and relative susceptibility of some soybean varieties to its attack *J. Maharastra Agric. Univ.* **9**(2): 218-219.

Singh, D.P. and Singh, K.M. (1966). Certain aspects of bionomics and control of *Obera brevis* S. *J. Sci. Tech* **4**(3): 174-177.

Singh, J.P. (1969). Soybean in Pantnagar, *Tech Bull.,* **6**: 69-77.

Singh, K.J. and Singh, O.P. (1983). Toxicity of some insecticides against leaf miner, *Aproaerema modicella* (Deventer) infesting soybean in Madhya Pradesh, India, *Ind. J. Pl. Protection,* **11:** 127-129.

Singh, K.J. and Singh, O.P. (1985). A study on the loss of soybean crop due to the attack by parakeet, *Psittacula crameri* (S) *Baratiya Krishi Anusandhan Patrika* **2**(2): 77-80.

Singh, K.J. and Singh, O.P. (1991) Efficacy and economics of some effective insecticides against jassida, *Apheliona maculosa* Dist. infestation on soybean. *J. Insec. Sci.* **4**(2): 187-189.

Singh, O.P. and Gangrade, G.A. (1974). Biology of *Diacrisia obliqua* Walker (Lepidoptear:Arctiidae) on soybean and effect of loss of chlorophyll on pod and grain *JNKVV Res. J.* **8**(2): 86-89.

Singh, O.P. and Gangrade, G.A. (1975). Parasites, predators and diseases of larvae of *Diacrisia oblique* Walker (Lepidoptera:Arctiidae) on soybean *Curr. Sci.* **44**(13): 481-482.

Singh, O.P. and Jakhmola, S.S. (1983). New record of *Cydia ptychora* Meyrick as a pod borer pest of soybean in Madhya Pradesh (India), *Ind. J. Agric. Sci.* **53**(7): 615-616.

Singh, O.P. and Singh, K.J. (1990) Efficacy and economics of some emulsifiable concentrate insecticides against *Melanogromyza sojae* (Zehnter) and *Chrysodeixis acuta* (Walker) on soybean. *J. Insect Sci.* **3**(1): 77-82.

Singh, O.P. and Singh, K.J. (1994) Record of white grubs, *Holotrichia consanguinea* (Blanchard) infestation and its effect on yield of soybean in Madhya Pradesh. *J. Insect Sci.* **7**(1): 100-101.

Singh, O.P., Singh, K.J. and Kapoor, K.N. (1990). Seasonal incidence and chemical control of red spider mite, *Tetranychus telarius* Linn. on soybean in Madhya Pradesh, India, *Ind. J. Ent.* **52**(1): 57-62.

Singh, O.P., Verma, S.N. and Nema, K.K. (1989) Insect pests of soybean in India, International Book Distributors, Dehradun, pp. 281.

Singh, Z. (1973). New records of a predatory bug, *Amyotea (Asopus) malabarica* (Fabricius) on *Nezara viridula* Linn. In India. *Ind. J. Ent.* **35**(2): 169-170.

Sontakke, B.K. and Mishra, P.R. (1994). Impacts of different sowing dates/seasons on the incidence of insects pests and yield losses in soybean. *J. Insect Sci.* **7**(1): 84-87.

Srivastava, K., Rahman, S.M.A. and Ram, S. (2007). Evaluation of insect growth regulators against soybean defoliators. *Annals of pl. Prot.* **15**(2): 294-298.

Srivastava, S.K. and Srivastava, B.K. (1988). Varietal resistance and toxicity of insecticides against leaf folder, *Lamprosema indicata* Fab. on soybean. *Pesticides*, **XXII**(12): 45-47.

Venkatesan and Kundu, G.G. (1994) Yield infestation relationship and determination of economic injury level of stemfly, *Melanogromyza sojae* (Zehnter) infesting soybean. *J. Ent. Res.* **18**(3): 265-270.

10

Modified Atmosphere: An Alternative to Control Insects in Storage

Chitra Srivastava and A. Cecily

Division of Entomology, I.A.R.I., New Delhi–110 012

The significant development in post-harvest technologies has been the fast approaching reduction in chemicals used for storage pest control. Fumigation, as a technology, is widely used for disinfestation of durable foodstuffs. Usually, fumigation gives a quick, low cost and effective solution to control insect/pest attack in storage commodity. But increasing restriction to use fumigants in many parts of the world is due to the development of insect resistance, environmental pollution, health hazards both for the public and workforce, carcinogenicity of some fumigants, effects on treated products like production of residues. Widely used fumigant phosphine has developed resistance against most of the storage insects. Ozone depletion problem has led to the phase-out of methyl bromide in developed countries by the year 2005 and worldwide by 2015 under the terms of the Montreal protocol. Thus, there is significant search for the alternatives to methyl bromide. In particular, the search for non-chemical methods of insect control has increased in intensity. Additionally, a public interest has arisen about the pesticide residues in food and the harmful influence of residues on the environment.

The development of new, friendlier chemicals specifically for the stored product pest control has become prohibitively expensive. These constraints have led to the realizations that the emphasis is rapidly shifting to IPM with chemical means of control as a last resort. However, in practice, chemical control still plays a dominant role, with phosphine fumigation as the mainstay of the grain storage industry. Widespread experience has proven that chemicals applied at commercial levels develop resistance by insects. Phosphine is a very useful fumigant, but it is slow acting and insects in various countries have developed resistance (Navarro *et al.*, 2002).

In addition to the poisonous gases that are used for pest control, the normal gases of the atmosphere can be altered or modified to achieve control. The use and manipulation of natural components of the atmosphere, e.g., oxygen, nitrogen and carbon dioxide, is referred as "controlled" or "modified" atmosphere storage. Controlled atmosphere (CA) techniques are widely used in the storage of perishable commodities such as fruit, vegetables, cut flowers, etc. to retard ripening and reduce spoilage from microorganisms. Also, they will control some insects in these products. The most extensive use of controlled atmospheres for insect control is in grain and similar commodities. Here the atmospheres are modified by removing the life-supporting oxygen or by adding high levels of carbon dioxide.

Although the principle of modified atmosphere storage has been used since antiquity, e.g., in hermetic storages, a number of procedures have been developed in recent years to replace the normal atmosphere of storage for the purpose of controlling pest organisms. In many respects the practice of using modified atmospheres for insect control is closely related to fumigation. Gas-tight enclosures are required, many of the procedures are closely related and the problems are often similar to those found in fumigation. When carbon dioxide is used, it is applied as a fumigant and it functions in a similar way. Treatment of stored food commodities with controlled atmospheres, i.e., oxygen-deficient, nitrogen-rich or CO_2-rich (hypercarbia) atmospheres, has been extensively studied as an effective alternate control measure to conventional fumigation. CO_2 is a more effective fumigant than nitrogen because insect respiration is affected while displacing oxygen. Nitrogen kills insects only by displacing oxygen to levels less than one per cent in air. CO_2 has been studied as a fumigant in bulk cereals at inter granular levels more than 60 per cent in volume. Increase in CO_2 concentration at a constant R.H. and temperature usually produced increased mortality with decreased exposure time.

Controlled atmosphere technology can fulfil a specific niche where use of other fumigants is unacceptable, such as in treating organic foods. Controlled atmosphere is limited by the long exposure times required to produce complete mortality, which is similar to those required for phosphine fumigation. This technology can be adopted where cheap source of nitrogen or CO_2 is available and the storage structure is well sealed.

Controlled atmosphere procedures are an appropriate substitute for the fumigation of some commodities because the gases involved do not leave harmful residues and often the atmospheres provide superior conditions to normal storage in air. In some cases, the two procedures may be used in a complementary way to increase effectiveness of a treatment; carbon dioxide increases the toxicity of a number of fumigants to insects.

Since sound pest management should, where possible, promote programme that integrate appropriate control procedures and minimize the use of toxic chemicals and many of the requirements for fumigation are similar to those needed for controlled atmospheres.

BASIC REQUIREMENTS

Controlled atmosphere systems depend on either depletion of oxygen to asphyxiate organisms or the addition of carbon dioxide to act directly and kill them. In these treatments the new atmospheres are maintained for an adequate period to kill all stages of the organism, and they should have no adverse effect on the commodity. To achieve this, the treatment requires:

- a storage structure capable of containing the gas;
- a source of suitable gas or a means of producing the required atmosphere;
- a method of maintaining the atmosphere for the required period of time;
- a method of aerating to remove the altered atmosphere after the treatment.

It should be noted that the controlled atmospheres, which are toxic to pest organisms, are also dangerous to humans and precautions are necessary to ensure that no one is exposed to them without special protection. Although nitrogen itself is non-toxic to humans, the absence of oxygen or the presence of high levels of carbon dioxide is lethal.

LOW OXYGEN ATMOSPHERES

Hypoxic atmosphere loosely applies to any atmosphere containing an O_2 concentration of less than 21%. Oxygen-deficient atmospheres are produced by flushing a storage with nitrogen to displace the normal nitrogen-oxygen atmosphere. Liquid nitrogen from tanks may be used as a gas source (Banks and Annis, 1977). Exothermic inert atmosphere generators that consume the oxygen to leave principally nitrogen have been tested and show promise for insect control. These generators burn propane or other hydrocarbon fuel to give an atmosphere of less than 1 percent oxygen with about 10 per cent carbon dioxide and 89 percent nitrogen. Oxygen can also be removed by the metabolic activity of microorganisms and insects in hermetic storages, thus producing an atmosphere where insects cannot survive. For complete insect control, the level of oxygen must be maintained below 1.2 per cent for one week, at temperatures above 35°C, or more than 24 weeks at 15°C. Upon exposure to hypoxia, insects typically exhibit a suite of behavioral changes which increase local oxygen concentrations. Immediate changes include increased spiracular opening frequencies (Krafsur and Graham, 1970), increased ventilatory movements and increased activity (Heslop et al., 1963). As the period of severe hypoxic increases, terrestrial insects become quiescent, decrease aerobic metabolic rate and switch to anaerobic metabolism (Wegener, 1993).

CARBON DIOXIDE ATMOSPHERES

Insects are generally killed more rapidly by carbon dioxide than by the lack of oxygen. The carbon dioxide gas is applied to storages from a vessel of liquid carbon dioxide with appropriate vaporizers and pressure regulators to control flow rate. Carbon dioxide in the form of dry ice has also been used for the treatment of grain in freight containers (Sharp and Banks, 1980).

GAS TIGHTNESS REQUIREMENTS

Structures used for controlled atmosphere treatments must have a high degree of gas tightness for the process to be effective and economical. They must be of sound construction and suitably modified so that gross gas loss through apertures, such as ventilators, open eaves and imperfections in the fabric is prevented. Changes in temperature, atmospheric pressure and wind forces can have a pronounced effect on gas loss from the storage structure. Gas is added to maintain concentrations at the required level in storage over the entire exposure period. One of the main reasons that gas loss occurs is the diurnal temperature variation in the head space of storage structures. This is less in concrete structures than in unprotected metal bins. Diurnal temperature fluctuations can be reduced by insulation, painting with a highly reflective white paint or by placing a false roof on the bin to leave a ventilated air space next to the permanent roof.

In structures other than welded steel, e.g., concrete, bolted or riveted metal, sealing of the entire fabric of the structure may be necessary. Bolted metal bins can be sealed by treatment of each lap joint and bolt location with silicon rubber sealant, thixotropic acrylics or by application of liquid envelope, "cocoon" type, polyvinyl chloride coatings. Concrete surfaces may also need to be coated with a good sealant that will prevent yes loss and protect the concrete from high levels of carbon dioxide. In bins that are very tightly sealed, some precautions are necessary to avoid unusual stresses on the structure caused by external or internal pressure changes. To prevent such changes in a bin that is sealed for a controlled atmosphere treatment, a pressure relief valve must be installed (Banks and Annis, 1977).

Controlled atmosphere storage can be viewed in two phases—the "purge phase" where the normal atmosphere is replaced with the prescribed atmosphere and the "maintenance phase" where the atmosphere is maintained for the desired period of time (Banks and Annis, 1977).

Liquid nitrogen supplied directly from a road tanker is passed through a heat exchanging facility where it is vaporized and brought to ambient temperature. It is then passed through a flow meter (e.g. "Rotameter") into the gas introduction system of the bin. However, when gas in introduced directly into the bin at the walls, an increased input rate may be less efficient at lowering the oxygen tension. In cases where a low efficiency distribution system is used, pockets of air may remain which are not purged directly but in which the oxygen is removed through the slower diffusive and convective forces. In these instances a slower purge rate allows these processes to occur and is not wasteful of gas.

Gas input must continue uninterrupted until the headspace has been reduced to about one percent O_2. Purging in the grain mass occurs generally by the passage of a sharp front through the grain and direct displacement of the interstitial air. In the headspace where free gas mixing occurs, decay is exponential. When the headspace has reached one percent O_2, the purge may be terminated, the top vent and introduction ports closed and the maintenance input of gas commenced.

MAINTENANCE PHASE

After termination of purging, gas input at a lower rate is continued to maintain the atmosphere. The input rate required is determined by the bin capacity, its degree of gas tightness and the weather. At present it is best to adjust the maintenance rates by systematically reducing the flow of gas until the atmosphere is just maintained. At the correct rate, with grain temperatures exceeding ambient ones, the oxygen tension at the base of the bin will rise slightly during the day and fall to about 1 per cent at night. The headspace tension should remain low. If the grain temperature is below ambient, or if the atmosphere contains more than 3 per cent CO_2, effects will be seen in the headspace, not at the base.

Sitophilus oryzae is one of the most tolerant grain pests in low oxygen atmospheres, and if there is any doubt about what species are present, the exposure for this pest should be used. In commodities where the pests are *Tribolium* spp. and *Oryzaephilus* spp. only, shorter exposure times, half of those given, may be used. Currently available data for *Rhyzopertha dominica* are insufficient for an exact recommendation, but it is considerably more susceptible than *S. oryzae* and should thus be controlled by the exposures given. Exposure times should be based on the minimum temperature within the grain, not average temperatures.

CO_2 dosages should be based on the response of *S. oryzae* pupa and *T. castaneum* adults and larvae. However, where *Trogoderma* grubs are likely to be present, CO_2 exposure should be set for those species, which are more tolerant to high CO_2 concentration.

GRAIN STORAGES

Three procedures for establishing high concentrations of carbon dioxide in large silo-type bins have been tested by Jay (1980). These are based on introducing the gas at the top of a filled storage, at the bottom of the storage or with the grain stream during filling. A procedure for introducing the gas at the base of the storage with subsequent recirculation has been found to be effective in controlling a natural infestation of insects and is deemed to be commercially feasible (Wilson *et al.*, 1980).

For the latter treatment, liquid carbon dioxide supplied in cryogenic tankers is vaporized in a heat exchanger and the gas is diluted with air to give approximately 80 per cent carbon dioxide and 20 per cent air. The gas stream is maintained above 30°C with a super heater and introduced at the base of the bin through a 75 mm diameter iron pipe. When the carbon dioxide concentration at the top of the bin reaches a constant level, gas input is stopped and the atmosphere so established is recirculated through a 50 mm diameter plastic duct leading from the bin apex to the introduction port at the base. With the gas tightness standards specified above, this procedure will give a concentration of carbon dioxide >70 per cent initially and this can be maintained at > 35 per cent for ten days without requiring additional gas.

TERMINATION OF TREATMENT

After a controlled atmosphere treatment is terminated, the modified atmosphere is gradually displaced by entry of air from outside. The rate at which this happens depends on a number of factors, such as gas tightness of the storage, contents of the storage and the weather, and it may be increased by making larger openings or by active ventilation with a blower. A 2000 tonne bin with 5 cm (2 in) diameter openings at the top and base may reach acceptable oxygen levels of 16 per cent from the established 1 per cent in less than two weeks (Banks and Annis, 1977). Care should be taken to avoid structural damage from the reduced pressures caused by rapid emptying of a tightly sealed storage. Steel bins have been severely damaged by rapid removal of grain without sufficient venting.

SAFETY PRECAUTIONS

Adequate precautions should be taken when working in areas close to controlled atmospheres or on entering storages that have been treated to avoid any harmful effects. Nitrogen atmospheres containing less than 14 per cent oxygen or more than 5 per cent carbon dioxide may be dangerous to human life. Personnel entering a nitrogen atmosphere containing less than 10 per cent oxygen may collapse without warning and become unconscious. Carbon dioxide produces respiratory discomfort, lightheadedness and nausea, and unconsciousness may occur in less than five minutes in 9 per cent CO_2. In any case, where unconsciousness or respiratory distress occurs through exposure to a controlled atmosphere, the victim must be taken immediately to fresh air.

Portable oxygen monitors or carbon dioxide analysers should be available on the work site as well as airline or self-contained breathing apparatus. Gas masks with canisters provide no protection against low oxygen or high carbon dioxide atmospheres. All enclosed working areas close to controlled atmospheres should be well ventilated and gas reservoirs should be kept outside, if possible (Banks and Annis, 1977).

PROBLEMS ASSOCIATED WITH MODIFIED ATMOSPHERE STORAGE

In sealed systems, moisture may migrate and condense to produce problems, particularly in grains having more than 12 per cent moisture content. High carbon dioxide or low oxygen atmospheres can inhibit mould and toxin formation and preserve germination under moist storage conditions. However, spoilage will occur once the modified atmosphere is removed unless the grain is dried or processed within a short time (Banks, 1981). Once the modified atmosphere is removed, the commodity is open to infestation unless otherwise protected.

Grain kept under a modified atmosphere may develop taint, at higher moisture levels (> 16 per cent) this taint may be difficult to remove (Banks, 1981). High levels of carbon dioxide may cause structural problems in reinforced concrete storages by reducing alkalinity of the steel.

CHOICE OF TREATMENT

When a controlled atmosphere procedure is going to be used, the choice between using the nitrogen (oxygen deficiency) method or the carbon dioxide method may be determined by several factors. The amount of time available for the treatment, the suitability of the structure for holding the gas, the availability of the gases or equipment for producing them and the overall cost of the operation are all important. Carbon dioxide generally kills insects faster than nitrogen (oxygen deficiency) and requires a less stringent standard of gas tightness (Jay, 1980). The comparative cost of controlled atmosphere treatments will depend on the availability and cost of the gases, the amount required and transportation costs, as well as the costs of equipment and labour.

FUMIGATION, CONTROLLED ATMOSPHERES AND FORCED AERATION

The choice of treatment to be used for controlling insect infestations will vary with the type of problem and the relative merits of the treatment. Fumigants can usually be used effectively in storages prepared for controlled atmosphere treatments, as the standards of gas tightness that are required are generally more stringent for the controlled atmospheres. Also, the quantity of material required, the application costs and the exposure time are usually much less for fumigants than for controlled atmosphere treatments. On the other hand, controlled atmosphere procedures avoid the use of chemical pesticides, leave no harmful residues and can provide superior storage of grain. A combination of fumigation with controlled atmosphere procedures may have some potential, as the effectiveness of fumigants is enhanced by carbon dioxide.

In addition to controlled atmospheres, the forced aeration of grain is closely allied with insect control procedures and may be an integral part of a pest management programme. Calderon (1972) suggested that "a sensible use of ambient or chilled air for aeration of grain offers new possibilities in many parts of the world for preservation of grain without (or with very little) use of chemicals." The use of fumigants and controlled atmospheres together with forced drying procedures and aeration should be considered as complementary conservation methods that form part of an overall pest management programme.

PHOSPHINE(PH_3) AND CARBON DIOXIDE(CO_2)

Kashi and Bond (1975) showed that 4-64 per cent concentrations of carbon dioxide potentiated the toxic action of phosphine. However, Desmarchelier and Wohlgemuth (1984) pointed out that though low concentrations of carbon dioxide aided the toxicity of phosphine against some stored product pests e.g., *T. castaneum*, the action against other pests like eggs and pupae of *Sitophilus* spp. was not changed. Rajendran and Muthu (1989) studied the action of PH_3 and CO_2 on eggs of *T. castaneum* and found that there was an additive effect of these mixtures. The higher susceptibility of the eggs to PH_3-CO_2 mixture has been attributed to their sensitivity towards CO_2. The results indicate that a formulation releasing a high concentration of CO_2 and a low concentration of phosphine may prove more useful in improving the efficacy of the latter.

EFFECTS OF CO_2 AND TEMPERATURE

In cases where disinfestation of commodities is required, the possibility of using CO_2 at temperatures raised to levels that will not affect the commodity should be considered. The combined effects of high CO_2 at optimal elevated temperatures against insects have shown a corresponding increase in insect mortality with increase in temperature (Adler, 1994).

The influence of various CO_2 concentrations on *Oryzaephilus* spp. has shown that the eggs are the most resistant stage at 40°C and 90 per cent CO_2 for 6 hrs exposure period to obtain a LT_{99} value. Tests with *E. cautella* showed that the pupae was the most resistant stage when exposed to 90 per cent CO_2, with a LT_{99} value of 17 hrs at 35°C and only 3 hrs at 45°C. The adult was the most sensitive stage at 35°C requiring only 4 hrs of exposure (Navarro *et al.*, 2002).

In the UK, the bolted metal cab bins of 1000 tonnes capacity, held under a low oxygen atmosphere can be disinfested at a fuel cost of less than 40 pounds a tonne using a self-cooling burner. The atmosphere applied continuously to the bin contained less than 1 per cent oxygen and upto 12 per cent CO_2. Interruptions in maintaining the 1 per cent oxygen atmosphere throughout the bin due to high winds can be tolerated provided they do not exceed 16 hrs. It indicates that although very low level of oxygen (<1%) is required to kill insects, somewhat higher levels (upto 3%) can be effective at preventing breeding and reinfestation (Conyers and Bell, 1996).

The susceptibility to controlled atmospheres is greater at higher temperatures due to enhanced respiratory demand. Outside the optimum range of low and high temperatures can add stress to controlled atmospheres treatment (Mbata and Phillips, 2001). The greater efficiency at higher concentration of CO_2 and low temperatures could be related to the higher solubility of CO_2 in tissues at low temperatures (Yacoe, 1986). Thus, temperature had a significant effect on the metabolic response of *P. stulana* to reduced O_2 concentrations.

INFLUENCE OF PRESSURE ON CO_2

The new process of using carbon dioxide under high pressure offers an exciting and novel alternative to conventional CO_2 application. The high pressure CO_2 treatment combines most of the advantages of CA technology and at the same time addresses its serious drawback, i.e., it requires extremely short durations for lethal exposure, within the range of few minutes. This renders this method promising for quarantine treatment and rapid disinfestation of valuable products. Survival of any stage of species tested, i.e., *S. oryzae, R. dominica, T. castaneum* and *Corcyra cephalonica* can be completely suppressed by CO_2 treatment (>90%) released at a high pressure of 30 kg/cm^2 for 5 minutes. The rice moth was the most susceptible and its pupae succumbed to the lethal atmosphere even at the lowest dosage of 5 kg/cm^2 pressure and exposure period of 5 minutes. Efficacy of pressurized CO_2 on Indian meal moth eggs showed that very young eggs were most tolerant, and 100 per cent mortality was achieved within 40 minutes at 20 bar under CO_2. One to four-day-old eggs required about 20 minutes of exposure to 20 bar for complete kill.

EFFECTS OF LOW PRESSURES AND TEMPERATURE

For *T. castaneum, E. cautella* and *P. interpunctella,* the egg stage was the most resistant to the low pressure and the time needed to obtain LT_{99} was 22, 45 and 49 hrs, respectively. For *Tribolium,* the larva and the adults were the most susceptible stages. The influence of low pressure on *E. cautella* pupa with that of low oxygen concentration, and deduced that the partial pressure of oxygen has decisive effect on insect mortality. The effect of temperature and exposure time on mortality of stored product insects exposed to low pressure reveals that insect mortality under such conditions is predominantly a result of oxygen deficit and not due to physical pressure effects (Navarro et al., 2002).

The lethal effects of a sudden or slow increase and decrease of carbon dioxide pressure as a pest control technique were tested on adult tobacco beetles *Lasioderma serricorne.* A quick build-up and decay of the carbon dioxide pressure of 20 bar within 2 seconds resulted in higher mortalities than a slow increase and slow decrease within 2 minutes. When the pressure increase was slow and the decrease rapid beetle mortality was higher, as when the increase was rapid and the decrease slow. It appeared that the decrease is more efficient than the increase for quickly obtaining mortality.

The effects of CO_2-high pressure methods in which sudden pressure losses lead to two mechanisms of action: the physiological action of CO_2 at high concentrations in target sites; and the physical action of high pressure gas in tissues of the insects causing expansion as the gas is suddenly released to the atmosphere (Riudavets *et al.*, 2002).

BIOCHEMICAL MODE OF ACTION OF CO_2

Triacyl glycerols form an important energy source for insects. These undergo hydrolysis to produce fatty acids, which release further energy by oxidation. During exposure of *Tribolium* pupae to anoxia, lipids did not provide an energy source, however, as soon as insects returned to air, they were utilized in quantity. The accepted belief is that carbohydrate reserves in insects are utilized by the glycolytic pathway during anaerobic metabolism.

Friedlander and Navarro (1979) examined the influence of CO_2 on TG metabolism in *Ephestia cautella* pupae to determine whether water production from TG metabolism forms a regulatory mechanism to recompensate for increased water loss due to the influence of CO_2 on spiracular openings. They found that TG metabolism was not sufficient to compensate for water losses under conditions of high CO_2 concentration and low ambient humidity. Glycogen, which is also a carbohydrate reserve in insects, was also measured by the same authors and it was shown that glycogen utilization increased during hypercarbia.

The utilization of glycogen is accentuated under hypoxia, which is associated with the production of lactate (Adler, 1994). However, after 24 hr, lactate production decreased to almost zero, suggesting the inhibition of this metabolic pathway. It was suggested that the inhibition of glycolysis at low lactate levels is associated with a low energy yield, and therby in atmospheres that combine anoxia and hypercarbia, the lethal action is enhanced under the influence of CO_2. Although without O_2, glycolysis is dominant, in a few cases, the breakdown of certain amino acids can be utilized for the production of energy conserving adenosine triphosphate (ATP). Also, nutritive compounds such as lipids, proteins and carbohydrates can be metabolized with a maximum energy yield in the presence of O_2 but the energy yield of anoxic metabolism is comparatively low. This anaerobic metabolism gives rise to acidic compounds. Therefore, in the application of MAs, a particular interest exists in determining the levels of O_2 and CO_2 or their combination that can prevent oxidative metabolism and thereby cause insect death.

Donahaye and Navarro (2000) reported that *Tribolium* spp. selected for resistance to HCC showed greater quantities of TG, and its lower rate of metabolism enable it to survive longer under hypercarbia, and this appears to be an important mechanism of resistance. The same results were found in psocids selected for resistance to carbon dioxide by Wang and Zhao (2003).

The effect of CO_2 on the main NADPH-producing systems in *E. cautella* pupae, namely the malic enzyme and 6-phosphogluconic acid dehydrogenase, a key enzyme in the hexose monophosphate pathway was investigated by Friedlander *et al.* (1984). It was found that both enzymes were sensitive to CO_2. At high levels of CO_2, the malic enzyme becomes fully reversible, the reaction being driven towards pyruvate production by coupling the malic enzyme to the ubiquitous lactic acid dehydrogenase. Phosphogluconic acid dehydrogenase was largely inhibited at high levels of CO_2. The combined effect results in NADPH production becoming dependent on available malate. High CO_2 affects the ATP production by the influence on insects' membrane systems.

RESISTANCE DEVELOPMENT TO CA

Extensive use of CO_2 in insect control programmes might be expected to favour the selection of resistant strains. Investigations with fig moth have shown that CO_2 affected several metabolic sites, which might facilitate resistance development in the insect. To elucidate the genetic potential of insects to develop resistance to CO_2 alone, Bond and Buckland, 1979, investigated the response of *S. granarius* to selection. They found that the adults produced a 3.3 fold increase in resistance at 42 per cent CO_2 in seven generations, and a 1.8 fold increase in resistance at 75 per cent CO_2 in four generations. It was found that *S. oryzae* has the genetic potential to develop resistance to CO_2. The O_2 and R.H. at the tested level had minor were of minor influence on the tolerance obtained and the action of CO_2 was dominant.

All stages of a strain of *T. castaneum* selected for resistance in the adult stage to a high CO_2 content atmosphere and of an unselected strain showed significant differences to the treatments. Similarly, the strain selected for resistance to low oxygen atmosphere in the adult stage also showed partial resistance.

The results of biochemical assays carried out by Wang *et al.*, 2000 showed that the activities of carboxyl esterase and superoxide dismutase *in vitro* increased the selection of psocids. The enhanced CarE and antioxidation enzyme activities could reduce the effects of these toxic products on insects, resulting in increased insect resistance to CO_2. Therefore, grain storage and warehouse operators should be aware that CA treatments can induce the esterase enzymes, which in turn can promote selection of pest strains resistant to CA treatments, and the rates of development of CA resistance are similar to those recorded for laboratory induced resistance to fumigants. Currently, not all aspects of the mode of action of CO_2 on insects are elucidated.

COMPARISON BETWEEN LABORATORY STUDIES AND FIELD USE

Generally, when a field test using CO_2 is conducted on infested grain, higher mortality is obtained in the natural population than would be expected in laboratory studies. This may be due to the difference between the gas tightness used in laboratory studies, which is not the case in field. Moreover, in laboratory studies on removal of CO_2 exposure, there is rapid desorption from the grains, whereas in field situation the grain may not be moved or aerated for several days or weeks after the treatment is concluded, and there would be a gradual desorption of the carbon dioxide from the grain into the surrounding atmosphere (Jay 1980).

CONCLUSION

The sole alternative to fumigation for in-storage insect control, which also offers a diversity of applications and toxic-residue-free treatment, is the CA method. CA can be generated from liquefied gas, bulk transported in road tankers, or delivered from cylinders for small-scale treatments. The alternative approach on-site generation of CA. possibilities of on-site generation from biologicial sources, including assisted hermetic storage by external biogeneration of CA as well as conventional hermetic storage can be done. The CO_2–air atmosphere is usually more toxic than oxygen deficient ones.

To enhance technology transfer, the extension activities in developing countries should be encouraged. Priorities for future research on protection of stored products should involve:

1. alternatives for treatment of durables for reducing fumigant emissions;
2. revision of dosage rates to avoid overdosing;
3. improvement of gas tightness of enclosures; and
4. reducing the frequency of treatment by preventing or reducing reinvasion of pests subsequent to fumigation.

CA technology is the most potential alternative to MB for disinfestations of durable commodities, but still requires considerable development in both laboratory and field before they can be introduced into routine disinfestations systems.

REFERENCES

Adler, C.S. (1994). A comparison of the efficacy of CO_2 rich and nitrogen rich atmospheres against the granary weevil, *S. granarius. Proc. of the 6th Intl. Working Conf. on Stored Product Production,* Canberra, 11-15.

Banks, H.J. and Annis, P.C. (1981). Treatment of bulk grain in freight containers with CO_2 while in transit as an insect control measure : a commercial trial using bulk malt, Canberra, Commonwealth Scientific and Industrial Research Organization, Division of Entomology, Report No. **26:** p. 13.

Banks, H.J. (1981). Effects of controlled atmosphere storage on grain quality : a review. *Food Technol. Aust.,* **33:** 335-340.

Banks, H.J. and Annis, P.C. (1977). Suggested procedures for controlled atmosphere storage of dry grain. Canberra, Commonwealth Scientific and Industrial Research Organization, Division of Entomology. Technical Paper No. **13:** p. 23.

Bond, E.J. and Buckland, C.T. (1979). Development of resistance to carbon dioxide in the granary weevil. *J. Econ. Entomol.*, **7:** 770-771.

Calderon, M. (1972). Aeration of grain - benefits and limitations. EPPO Bull., **6:** 83-94.

Conyers, S.T. and Bell, C.H. (1996). The effects of periods of increased oxygen on the mortality of adult *Cryptolestes ferrugineus, S. granarius* and *O. surinamensis* caused by exposure to 1% oxygen in a nitrogen atmosphere. *J. Stored Prod. Res.*, **32:** 187-194.

Daniel, V., Hanlon, G. and Maekawa, S. (1993). Eradication of insect pests in museums using nitrogen. *Waac Newsletter*, **15**(3): 15-19.

Desmarchelier, J.M. and Wohlgemuth, R. (1984). Response of several species of insects to mixtures of phosphine and carbon dioxide. *In*: Ripp, B.E. (ed.) *Controlled atmosphere and fumigation in grain storage.* pp. 75-81.

Donahaye, E.J. and Navarro, S. (2000). Comparisons of energy reserves among strains of *T. castaneum* selected for resistance to hypoxia and hypercarbia and the unselected strain. *J. Stored Prod. Res.*, **36:** 223-234.

Friedlander, A. and Navarro, S. (1979). The effect of CA on carbohydrate metabolism in the tissue of *E. cautella. Insect Biochem.*, **9:** 79-83.

Friedlander, A., Navarro, S. and Silhacek, D.L. (1984). The effects of CO_2 on NADPH production in *Ephestia cautella* pupae. *Comp. Biochem. Physiol.*, **77:** 839-842.

Heslop, J.P., Price G.M. and Ray, J.W. (1963). Anaerobic metabolism in the housefly, *Musca domestica* L. *Biochemistry J.* **87:** 35-38.

Jay, E.G. (1980). Methods of applying carbon dioxide for insect control in stored grain. Washington, D.C., U.S. Department of Agriculture, SEAAATS-13. p. 7.

Kashi, K.P. and Bond, E.J. (1975). The toxic action of phosphine : role of carbon dioxide on the toxicity of phosphine to *Sitophilus qranarius* (L.) and *Tribolium confusum. J. Stored Prod. Res.,* **11:** 9-15.

Ke, D. and Kader, A.A. (1992). Potential of controlled atmospheres for postharvest insect disinfestations of fruits and vegetables. *Post harvest News and Information*, **3:** 31-37.

Krafsur, E.S. and Graham, C.L. 1970. Spilacular responses of *Aedes* mosquitoes to carbon dioxide and oxygen. *Annals of the Entomological Soc. America* **63:** 691-696.

Mbata, G.N. and Phillips, T.W. (2001). Effects of temperature and exposure time on the mortality of stored-product insects exposed to low pressure.

Navarro, S., Finkelman, S., Sabio, G., Isikber, A., Dias, R., Rindner, M. and Azrieli, A. (2002). Enhanced effectiveness of vacuum or CO_2 in combination with increased temperatures for control of storage insects. *Proc. of the 8th Intl. Working Conf. on Stored Product Production*, USA, 818-822.

Rajendran, S. and Muthu, M. (1989). The toxic action of phosphine in combination with some alkyl halide fumigants and carbon dioxide against the eggs of *T. castaneum. J. Stored Prod. Res.*, **25**(4): 225-230.

Riudavets, J., Gabaraa, R., Castane, C., Alomar, O., Pons, M.J. and Sanchez, J. (2002). The efficacy of CO_2 treatments (under pressure or by modified atmospheres) for pest control in stored products. *Proc. of the 8th Intl. Working Conf. on Stored Product Production,* USA, 832-834.

Sharp, A.K. and Banks, H.J. (1980). Disinfestation of stored durable foodstuffs in freight containers using carbon dioxide generated from dry ice. Canberra, *First International Conference on Technology for Development.*

Storey, C.L. (1973). Exothermic inert-atmosphere generators for control of insects in stored wheat. *J. Econ. Entomol.*, **66:** 511-514.

Wang, J.J. and Zhao, Z.M. (2003). Accumulation and utilization of triacylglycerol and polysaccharides in *Liposcelis bostrychophila* selected for resistance to carbon dioxide. *J. Appl. Ent.*, **127:** 107-111.

Wang, J.J., Zhao, Z.M. and Tsai, J.H. (2000). Resistance and some enzyme activities in *Liposcelis bostrychophila* in relation to CO_2 enriched atmospheres. *J. Stored Prod. Res.,* **36:** 297-308.

Wegener, G. (1993). Hypoxia and post hypoxic recovery in insects: physiological and metabolic aspects, In: Hochachka, P.W., Lutz, P.L., Rosenthal, M., Sick, T., van den Thillard,G (Eds), Surviving Hypoxia—Mechanism of control and Application. CRC Press. Boca Raton, FL pp. 417-432.

Wilson, A.D., Banks, H.J., Annis, P.C. and Guiffre, V. (1980). Pilot commercial treatment of bulk wheat with CO_2 for insect control: the need for gas circulation. *Aust. J. and Agric., Anim. Husb.*, **20:** 618-624.

Yacoe, M. E. (1986). Effect of temperature, pH, and CO_2 tension on the metabolism of isolated hepatic mitochondria of the desert iguana, *Dipsosaurus dorsalis. Physiological Zoology* **59:** 263-272.

11

Current Modified Three-tier Seed Production of Sugar Cane

Vijai Singh, S.K. Awasthi and B.B. Joshi

Indian Institute of Sugarcane Research, Lucknow–226 002

IMPORTANCE OF THREE-TIER SUGAR CANE SEED

It is known to every one that a good start is a half done. It is true in case of availability of healthy seed of vegetative propagated crops viz., sugar cane and potato where 19-20% crop produce is lost due to diseases alone. Healthy sugar cane or potato seed is developed in phased multiplication. It is referred as three-tier seed in case of sugar cane (Singh, 1983). Thermo-therapy is essentially given at the first phase for raising breeder seed followed by multiplication as foundation and certified seed in II and III tier. Degenerated seed is replaced with the healthy seed of improved varieties (Agnihotri, 1996). Replacement of seed from three-tier sugar cane production has achieved perceptible increase in productivity of the crop.

Diseases are produced by the interaction of host with pathogen under favourable environment. Sufficient level of resistance is difficult to breed due to heterozygous and polyploid nature of sugar cane and the environment remains beyond human control. The pathogens also overcome the resistance in due course of time by mutation and adaptation. It is, therefore, the option of keeping away of the pathogen from host that was utilized for control of sugar cane diseases (Singh, 1983). Several diseases have been kept under control by freeing the setts from pathogens. Sugar cane diseases alone may cause yield loss up to 19–20% each year (Agnihotri, 1996). In India, annual economic loss has been estimated to vary from Rs. 225 to 250 crores. The loss tends to increase prices of sugar under national and international market and the growers shift to other alternative crops due to uneconomical returns. Sugar cane is planted in larger area during spring (Feb.-March) in sub-tropical India. The problem of water scarcity during summer coupled with diseases and insect-pests may cause havoc in sugar cane. Waterlogging leads to predisposition of cane for secondary spread of red rot, a dreaded disease (Agnihotri, 1996). Heat therapy has been utilized since ages to control seed-borne pathogens in many crops including sugar cane (Joshi, 1954, Singh, 1967, 1973). Moist hot air treatment (MHAT) at 54°C for 2:30 h in >90% RH was found highly effective in the control of control sett-borne diseases of sugar cane (Singh, 1983). Regular surveillance for diseases and rouging of affected plants was recommended in all the 3 years of multiplications. The major diseases viz., red rot incited by *Colletotrichum falcatum;* smut by *Ustilago scitaminea;* wilt by *Fusarium sacchari;* ratoon stunting by *Clavibacter xyli* sub sp. *xyli;* grassy shoot by phytoplasma and mosaic by virus are completely or partially controlled by MHAT. The pathogens and/or virulent strains get introduction to newer areas due to farmer's quest of using new varieties. The pathogens escape in the form of incipient infection in the seed cane.

RSD and SCMV are important throughout the world since their presence in seed stocks goes unnoticed. Severely affected varieties invite the infections of other deadly diseases viz., red rot, GSD and smut, and ultimately the diseased cultivars are phased out from cultivation. The decline in yield was progressive due to SCMV in sugar cane.

THREE-TIER SEED TECHNOLOGY

Several methods of heat treatment viz., hot water, hot air, aerated steam and MHAT were tried, but the later method was established to be highly efficient, effective and economical over hot water, hot air and aerated steam. Penetration of heat in the sett was superior in case of MHAT at 54°C for 2:30 h at > 90% RH. A machine was developed at IISR, Lucknow, where 6-7 q seed can be treated each time. Sufficient seed for one ha crop could be treated in two days by running the machine @ 8 h/day (Singh, 1983).

FREEDOM FROM SEED-PIECE TRANSMISSIBLE DISEASES

MHAT treatment is helpful in developing GSD, RSD, smut, leaf-scald free seed material of sugar cane. Incipient infections of shorter duration of red rot and wilt are also eliminated from cane stalks. Deep seated red rot and wilt infections are partly affected. SCMV is beyond control. Rather mosaic symptoms appear very prominently due to MHAT. Such diseased plants are rouged out easily (Agnihotri, 1996).

The incidence of severe strains of SCMV should not be above 5%, failing which the seed needs to be rejuvenated. Sugar cane Grassy shoot is incited by a phytoplasma (Srivastava *et al,* 2006). The pathogen is transmissible through leafhopper vector, *Deltocephalus vulgaris* Dash and Viraktamath (Singh *et al,* 2002). MHAT at 54°C for 2:30 h in >90% RH completely eliminates the pathogen from infected setts (Shukla and Singh, 1990). Germination of the buds from MHAT treated seed was improved by 15-20%. Ratoon stunting disease is one of the important pathogens causing degeneration in the varieties. The pathogen is transmissibility cutting knifes besides the seed. In several varieties, discernible symptoms of RSD are not produced but the pathogen may be eliminated and to get good vigor and yield. In several cultivars viz., Co 527, Co 658, Co 997, Co 66-11 RSD was inactivated. *Colletotrichum falcatum* Went produces red rot. Pith portion of inter node turns reddish brown where white patches (spots) can easily be seen by splitting the cane. Laminar, spindle and midrib infections are also produced in susceptible varieties. Probably thick walled and dormant resting structures present on the nodes could not be inactivated by heat therapy leading to variations in isolation of the pathogen from affected tissues. Smut posed a threat to Co 740 and Co 1158 in larger areas. Smut symptoms consist of black whip-like structure covered with very thin, white and quickly breakable membrane. Several workers have claimed the success of heat therapy by hot water against and by MHAT. The diseased buds fail to germinate after thermotherapy. Wilt is incited by *Fusarium sacchari.* Yellowing and withering of all the crown leaves followed by drying of whole cane is produced by F. in 75-100% plants in Co 0121, Co A 89085 and hybrid 57 NG 159. *Trichoderma, Pseudomonas, Bacillus* and other antagonist's species may be useful to manage wilt. Typical symptoms of SCMV are produced on leaves. Different shades of mild to severe mosaic patterns are formed on the leaf. It is transmissible by aphids in non-persistent manner. *Ropalosiphum maidis, Melanaphis sacchari* and *M. indosacchari* are among the most efficient vectors infesting sugar cane at several places. The infestation of SCMV- A, B, F and SC has been recognized in India. Diseased-free seed may be obtained by meristem tip culture. Rejuvenation of seed stocks from SCMV infection was reported by several workers (Roth, 1969).

DEVELOPMENT OF GENETICALLY PURE VARIETY

Sugar cane crop is propagated through vegetative cane stalks. The mixture of varieties is not detectable easily by the growers. Breeder scientists are supposed to play a major role in maintaining the purity of variety by observing morphological, biochemical and biotechnological characters.

Grow Out Test (GOT)

Genetic purity of seed is determined by GOT. The characterized traits including morphological, qualitative and quantitative descriptors are ascertained and compared. Important characters of stalk, node, root primordial region, leaf size, number, hairiness, droopiness, ligules shape and size are recorded and compared by the breeders. Data on cultivation of crop, its storage and quality are also determined and compared. GOT is highly dependent on environment.

Isozyme Pattern Test

Isozyme pattern was utilized for testing the genetic purity of a variety. The marker has some limitations. DNA region coding for soluble proteins can be sampled, but that may lead to partial genetic changes in plants.

DNA Marker Tests

DNA markers are independent of environment and considered as quick and precise alternative to GOT. DNA markers are based on sequence variations determined between parental lines. PCR amplifications of specific genomic region are compared. Different molecular markers viz., RAPD (randomly amplified polymorphic DNA); AFLP (amplified fragment length polymorphism); STMS (sequence tagged microsatellite sites) or SSR (simple sequence repeat) and ISSR (inter simple sequence repeat) have been utilized in various crops for testing the seed purity. SSR has been picked up as more reliable, easily reproducible and co-dominant in nature. SSR amplified products when visualized for the single strand conformation may depict the variations due to changes in nucleotide sequences because single strand conformation polymorphism depends on differential motilities of single-stranded DNA molecule due to conformational differences. Single-stranded conformation polymorphism PCR analysis was found useful to differentiate closely related genotypes of sugar cane. Though DBT through national certification system has guided for utilization of ISSR markers for testing the genetic fidelity, but according to literature, SSR is more robust, highly polymorphic, reproducible, co-dominant and available in public domain and may be free of cost. SSR is recognized as the best option for establishing the genetic purity of tissue culture raised plants. DBT through all India coordinated research project has framed guidelines for promotion of genetically pure seed cane by tissue culture technique.

MODIFICATIONS IN THREE-TIER SEED PRODUCTION UNDER IPR REGIME

Sugar cane crop was earlier not covered under seed act. The steps were: (i) Foundation seed, (ii) Certified seed, and (iii) Commercial seed during 3 years phased multiplication. But now the crop is covered under seed act where 3 years multiplications are referred as: (i) Breeder seed, (ii) Foundation seed, and (iii) Certified seed (Ann, 2007). The standards have been precisely fixed and are maintained for supplying certified seed to the growers (Table 11.2). Comparative step-wise modifications are given in Table 11.1. Genetic purity of seed cane is imperative for breeders to protect: (i) intellectual property through plant variety protection; (ii) an uniform appearance with sustained agronomic performance that meets the demands of farmers, processors and consumers; (iii) varieties with stable genetic identities can be created so that agronomic performance can be as predictable as possible given unpredictable environmental fluctuations; and (iv) breeders can more completely and precisely characterize and measure genetic diversity so that genetic resources can be more thoroughly evaluated and utilized for creation of improved varieties. As per procedure recommended by the brainstorming group on developing standards for tissue culture raised planting material of sugar cane at VSI, Pune, 2006, the breeder seed should be raised after MHAT treatment followed by tissue culture

Table 11.1: Details of current modified three-tier sugar cane seed.

Old concept	Modified new concept
I tier was called foundation seed Steps required: a. Selection of healthy cane of known variety. b. MHAT at 54°C for 2:30 h at 90% humidity. c. Applications of prophylactics. d. Surveillance for diseases and insects. e. Rouging of diseased plants. Major responsibility—Pathologist and Breeder	I tier is called breeder seed Steps required: a. Selection of genetically tested pure variety. b. MHAT at 54°C for 2:30 h at 90% humidity. c. Applications of prophylactic. d. Surveillance for diseases and insects. e. Rouging of diseased plants. Major responsibility—Breeder and Pathologist
II tier was called certified seed Steps required: Certified seed was not in legal sense under seed act a. Maintenance of purity of variety. b. Disease freedom is maintained. c. Prophylactics are needed. d. Rouging of diseased plants.	II tier is called foundation seed Steps required: Certified seed is now being produced in third year a. Maintenance of purity of variety. b. Disease freedom is maintained. c. Prophylactics are needed. d. Rouging of diseased plants.
III tier was called commercial seed It multiplied with all precautions for disease freedom under the supervision of cane development personnel's and supplied to growers for further use.	III tier is called certified seed Disease free crop is certified for genetic purity and disease freedom. It is covered under seed act. The standards of seed cane are maintained as per norms and disease escapes are not allowed In any case above the permissible limits.

technique may be tagged with specific golden yellow label. At least 15 plants should be indexed for freedom from diseases viz., red rot, smut, SCGSD, SCMV and SCYLV. The standards prescribed by government of India (Vyavhare, 2007) are strictly followed. Genetic purity of tissue culture raised plants should be maintained uniform up to 99.99%. No variation in purity is allowed over 0.01% in case of tissue culture grown plants. Necessity of fast multiplication through tissue culture: There are 425 sugar factories operative for manufacturing sugar in India. Sugar cane crop is grown on an average of 40,00,000 ha land. Considering 50% area under ratoon and 50% in plant, an average of 20,00,000 ha land is planted with fresh seed each year. Approximately 25,00,000–30,00,000 t healthy seed would be required to cover 1/5 land of plant cane each year. Demand of high sugar, high yielder and new varieties is increasing and it is difficult to fulfil the same with conventional planting and multiplication ratio of 1:10 or 1:15. Seed can be multiplied with space transplanting programme (STP) of single bud to obtain higher multiplication ratio of 1:30 or 1:40. But tissue culture technique seed cane could be multiplied at least 1:120. The highest multiplication ratio may be attained with tissue culture and burgeoning demand can be meet out. The technology also takes care of freeing stocks from viral disorders. Once the healthy seed is available in sufficient quantity, subsequent planting of seed may be with STP or conventional.

Table 11.2: Limits of disease and insect escapes for different tier seed.

Disease/pest	Days	Breeder	Foundation	Certified
Red rot	40-60	0.00	0.00	0.00
	120-130	0.00	0.00	0.00
	At harvest	0.00	0.00	0.00
Smut	40-60	0.01	0.01	0.01
	120-130	0.01	0.01	0.01
	At harvest	0.01	0.01	0.01
Wilt	At harvest	0.01	0.01	0.01
GSD	120-130	0.05	0.05	0.05
	At harvest	0.01	0.01	0.01
Leaf scald	120-130	0.01	0.01	0.01
	At harvest	0.01	0.01	0.01
Top borer	120-130	5.00	5.00	5.00
	At harvest	5.00	5.00	5.00
Internode borer	At harvest	10.00	10.00	10.00
Stalk borer	At harvest	20.00	20.00	20.00
Gurdaspur/Plassy borer	At harvest	5.00	5.00	5.00
Mealy bug/scale	At harvest	5.00	5.00	5.00

Limitations of Tissue Culture

Some extra facilities are required for tissue culture work viz., washing, media, inoculation and growth chamber, transfer area and greenhouse with nursery beds, trained personnel for clone selection, explant culture, virus indexing, maintaining clones uniformity, multiplication cycles, and testing quality of plants. A power back-up for uninterrupted light supply is essentially needed. Phenotypic instability has also been observed in tissue culture in different crop plants including sugar cane.

AGRONOMICAL GENERAL CONSIDERATIONS FOR THREE-TIER SEED

- It should contain sufficient moisture (65%) at the time of planting.
- Immature upper half of the cane is preferred for better germination.
- Nitrogen and phosphorus requirement of seed cane are 10-20% higher than other commercial cane. Split doses of N are preferred. Last dose may be given in October.
- Seed material may be taken from up land field with disease-free or with minimum incidence of seed piece transmissible diseases.
- Care should be taken to remove infected canes and setts at the time of planting. Good, sound and healthy looking seed may be preferred.
- Prophylactic viz., seed treatment with 0.2% carbedazim may be given to remove secondary invaders. The seed become clean after MHAT and that may invite soil-borne saprophytes and parasites.
- Application of chlorpyriphos or heptachlor 1 kg a.i./ha in furrows after sett placement may control termite and root borer early shoot borer and white grubs.
- Survey of seed plot is required at 15 days interval to monitor the incidence of disease and insect-pests. The surveillance would facilitate removal of diseased plants/clumps, if any.
- Secondary infections of viral and phytoplasmal diseases through insect vectors are beyond the control of MHAT. Spraying of systemic insecticides at 0.02% at 15 days interval would prevent the entry of aphid and leafhopper carriers of the diseases.
- Irrigation of seed plot at 15 days interval during April-June in subtropics or as required according to area may be given. Fertilizer may be applied thrice. Higher doses of N (25%) and P (50%) are needed for seed crop. P may be given at planting. N may be applied thrice at planting, tiller stage and late in the season (Oct.) after the monsoon. Post-monsoon irrigations are also recommended in Nov., Dec. and Feb. for maintaining moisture in seed cane.

REPLACEMENT OF HEALTHY SEED OF DIFFERENT VARIETIES DEVELOPED THROUGH THREE-TIER SEED

The technology was demonstrated to various sugar mills and farmers' fields of UP, Bihar, West Bengal, Punjab, Haryana, Rajasthan, Orissa, AP, Karnataka, TN, Maharashtra and Gujarat. Disease-free seed of Co 1148, CoJ 64, CoS 767 and CoS 802, CoLk 8001 and CoLk 8102 was developed at IISR, Lucknow. Development workers from different states were trained for the technology. The methodology of MHAT and vigorous crop raised after treatment was also shown to the farmers from various states of India. It was also popularized in neighbouring countries viz., Bangladesh, Sri Lanka and Pakistan. South African, Nigerian and Ethiopian trainees of 4-month training course were also exposed to the success of 3-tier seed programme. MHAT machine was further developed through private entrepreneurs and selected sugar factories of UP, Bihar, Punjab, Karnataka and Maharashtra were encouraged to purchase MHAT machine on subsidized rates from them. State Governments compensated the farmers by allowing subsidies on I, II and III year seed crop. CoJ 64, a high sugar variety from Punjab was successfully grown and covered larger areas of Punjab, Haryana and UP due to back-up of thermotherapy, though it was moderately susceptible to red rot. Recently, during 2002-07, healthy seed of Co Pant 84212, CoS 8436, CoS 94257, CoS 95255, CoS 96268, CoSe 65422, and CoSe 92423 was developed at IISR, Lucknow and supplied to progressive farmers of different sugar mills of Ghaghara Ghat, Nanpara Bahraich, Faizabad, Sitapur, Basti and Hardoi.

REFERENCES

Agnihotri, V.P. (1996). Current sugar cane disease scenario and management strategies. In: Presidential address. *Indian Phytopath.* **49:** 109-126.

Ann. (2007). Proc. *National seminar on sugar cane seed production and certification.* VSI, Pune, (Eds.) AS. Patil, pp. 78.

Joshi, N.C. (1954). Hot water treatment of setts for the control of red rot and smut diseases. *Indian Sugar.* **4:** 1-2.

Roth, G. (1969). Elimination of virus disease-streak by thermotherapy and tissue culturing. Proc. *South African Sugar Technol. Assoc.* **43:** 6671.

Shukla, U.S. and Singh, V. (1990). Inactivation of grassy shoot disease mycoplasma in diseased setts by heat treatment. *Indian Phytopath.* **43:** 561-562.

Singh, K. (1967). Heat therapy of sugar cane. *Indian sugar.* **17:** 1-5.

Singh, K. (1973). Hot air therapy against red rot of sugar cane. *Plant Dis.* Reptr. **57:** 220-222.

Singh, K. (1983). Seed-piece transmissible diseases of sugar cane and the three-tier seed programme. In: *Recent Advances in Plant Pathology,* (Eds.), A Husain, B.P. Singh, K. Singh and V.P. Agnihotri, Print house India, pp 331-341.

Singh, V., Baitha, A. and Sinha, O.K. (2002). Transmission of grassy shoot disease of sugar cane by a leaf hopper (*Deltocephalus vulgaris* Dash & Viraktamath). *Indian J. Sugarcane Technol.* **17:** 60-63.

Srivastava, S., Singh, V., Gupta, P.S., Sinha, O.K. and Baitha, A. (2006). Nested R assay for detection of sugar cane grassy shoot phytoplasma in the leaf hopper vector, *Deltocephalus vulgaris:* a first report. *Plant Pathology* **55:** 25-28.

Vyavhare, D.S. (2007). Seed act rules and seed certification of sugar cane. In: Proc. National seminar-sugar cane seed production and certification, (Eds.) AS. Patil, VSI, Pune, Feb, 22-23, 2007, pp. 34-41.

12

Eco-friendly Management of Potato Pest

V.K. Chandla, Sonu Kashyap and Anupam Sharma

Division of Plant Pathology, Division of Crop Improvement, Central Potato Research Institute, Shimla–171 001, H.P.

The potato is presently grown in 1.35 hectares in our country, which accounts for approximately 1% of total area of country under cropping. This is supposed to be the only food crop in country that fulfills almost all the valuable criteria for a good food and offers a potential for increasing food production and income for the country. India produces almost 25 million tons of potato annually, but the losses in this crop due to diseases, pest, nematodes, weeds, and other causes have always been a limitation. So for high productivity, application of chemical pesticides has always been a first choice. But ever increasing resistance of pest to pesticides and residue remains of these pesticides leads to a major concern about human health and other environmental hazards. Therefore, at present the eco-friendly management of pest is thrust area of research in economically sound pest control techniques. A great diversity of pests attacking potato exists in India due to different agro-climatic zones and their feeding habits. Potato pests are grouped into sucking pest or sap feeders, foliage feeders or defoliators, storage pests, soil pest, nematodes and other non insect-pest.

SUCKING PEST OR SAP FEEDERS

Aphids

Distribution, Nature of Damage and Population Dynamics

Though the direct damage caused by aphids is marginal compared to most other pests damaging the potato crop, yet they are extremely important being vectors of many potato viruses. Both the adults and nymphs suck the leaf sap. Aphid species colonizing on potato are *Myzus presicae* Sulzer, *Aphis gossypii* Glover, *Rhopalosiphum rufiabdominalis* Sasaki, *Aphis fabae* Scopoli and *Rhopalosiphoninus latysiphon* Davidson. Both *M. presicae* and *A. gossypii* transmit potato virus Y (PVY) and potato leaf roll virus (PLRV) (Nagaich et al., 1970; Singh et al., 1982). Among these, *M. persicae* is a potent vector of many other viruses in potato crop.

Fig. 12.1. *Myzus persicae.*

M. persicae, a most potential vector has been studied under different potato growing regions of the country. It starts appearing on potato crop in the North-Western plains of India from the second week of November and in the Eastern plains in early December, reaching the critical level (20 aphids per 100 compound leaves) by the end of December in the former and during 1-2nd week of January in the later area. In the North-Western higher hills, i.e., from Kashmir to Kumaon and in North-Eastern hills, the critical level generally reaches by the end of July. In the plateau region of Maharashtra, aphid infestation remains almost negligible in Kharif crop, while on Rabi crop it appears in the second week of November and critical level reaches by the third week of December. *M. persicae* has a complex life cycle due to different morphs and methods of reproduction. These morphs include winged, wingless parthenogenetic forms, oviparous females, males, etc. Reproduction is of both the types, i.e., asexual and sexual. In the case of asexual reproduction, winged viviparous females start appearing on potato crop and other secondary hosts from the middle of November onwards in the Indo-Gangetic plains. These forms reproduce parthenogenetically and give birth to young ones that pass through four nymphal instars (Verma, 1985; Verma & Chauhan, 1993a).

Cultural and Mechanical Management

Growing potato crop by adopting the seed plot technique with the following precautions, i.e., planting of clean (virus free) potato seed purchased from certified/reliable sources; growing of seed crop during aphid free/low aphid periods; maintaining an isolation of 50 m for seed crop; timely rouging of virus infected plants well before they touch each other; haulm (foliage) cutting of crop after desired maturity of crop but before the aphids cross the critical level of 20 aphids/100 compound leaves and (Verma et al., 1993a); Regular cutting of foliage regrowth so that the aphids could not build up on them.

Biological Management

Natural Enemies

Over 24 predators and 22 parasitoids reportedly attack Aphid, *M. persicae* (Singh, 1988). However, very few of these are effective. Among the predators, *Allograpta favana* Wiedemnn (Pruthi, 1946); *Episyerphus balteatus* De Geer (Gosh et al., 1985); *Ischiodon scutellaris* Fabr.; *S. scripta* Linnaeus; *Sphaerophoria indiana* Bigot (Pruthi, 1946); *Leucopis fumidilarva* Tanas (Saxena, 1981) and *Menochilus sexmaculatus* Fabr. are effective and among the population of parasitoids: Aphelinus sp. and *Aphidius colemani* Viereck have been reported to be promising. At Shimla, Aphelinus sp has been found to parasitize 100 per cent *M. persicae* under glasshouses conditions. Similarly, on *A. gossypii*, over 36 predators and 21 parasitoids have been recorded (Singh, 1988). In Karnataka, Acolemani parasitized upto 70% aphids in certain potato fields (Trivedi, 1988). At Shimla, one Aphelinus has showen its maximum capability by parasitizing 100% of *M. persicae* alone (Verma et al., 1976a). Seven species of coccinellids, two of syrphids and a chryopid were recorded on potato around Farrukhabad, Agra and Meerut districts of Uttar Pradesh. Among these, *Cooccinella septempuncata* Linn. and *M. sexmaculatus* Fabr. were found predominant (Mishra et al., 1995). Ten species of cocccinella, two chrysopids and three predaceous bugs were recorded feeding on aphids in Shimla during May-October (Anon, 1984).

Five entomopathogenic fungi were reported against aphids in Shimla (Singh, 1988).

Leafhoppers

Distribution, Nature of Damage and Population Dynamics

The leafhoppers damage the potato crop both as vector as well as direct pest. The species like *Alebroides nigroscutulatus* Dist. and *Seriana equata* Singh damage the crop as vectors of

Fig. 12.2. Leafhopper damaged potato plants.

phytoplasmal diseases, viz. purple top roll (PTR) and marginal flavescence (MF). These are predominantly present at/around Shimla from May to August. The adults are about 3 mm in length and greenish yellow during summer months but acquire reddish tinge in winter. The winged adults fly away at slightest disturbance and are attracted towards light. The pest remains active throughout the year and adults migrate to potato crop during October from its alternate habitat (Chandla, 1986).

Amrasca devastans Dist., *Empoasca devastans* Dist., *Amrasca beguttula beguttula* Isida cause direct damage to potato crop as a pest and is widely distributed over all the potato growing regions in India. However, it regularly causes serious damage to early planted potato crops normally grown in Western Uttar Pradesh, Haryana and Punjab. On the other hand, it is graded as sporadic pest in Maharashtra and Karnataka but causes severe damage to kharif potatoes. Both the adults and nymphs are responsible for causing the damage by sucking the sap from the foliage and depleting the cell content. The first symptom is black spot at the tip of leaf and/or end of each lateral vein let. Often the entire margin of infested leaf may roll upward, which turns brown and looks as if scorched by fire or drought. These brown margins continue to increase in width until a narrow strip of leaflet along the midrib remains green while the rest of the portions shrivel and die. Normally the older leaves below the growing tips burn first (Mishra et al., 1995; Verma et al., 1993b). Upon heavy infestation, every leaf succumbs rapidly and dies much before the normal development of tubers is completed. On an average, the pest is responsible for 90-100% foliage damage. This foliage damage is popularly known as 'hopper burn' and leads to 20-30% yield losses in early-planted potato crops in the western Gangetic plains.

Cultural and Mechanical Management

The practice of adopting suitable crop rotations with non-host crops in pest prone areas and providing proper isolation to potato crop from susceptible hosts are the most useful management components for protecting the potato crop from *A. devastans*. Delayed planting undertaken on 25th September for early crop and during middle of October for the main crop reduces the incidence and impact of the pest.

Biological Management

Natural enemies

A few species of Cocccinella beetles are often seen feeding on immature stages of leafhoppers.

Broad Mite

Distribution, Nature of Damage and Population Dynamics

The broad mite, *Polyphagotarsonemus latus* Banks *Hemitarsonemus latus* Banks, also known as tarsonemid mite, tropical mite, yellow tea mite, causing tambera in potato (or murda in chillies), is widely distributed. Damage due to broad mite is serious in Maharashtra and Karnataka during kharif season and western UP and Punjab on early-planted rabi potatoes. Besides, mite damage on potato crop has been observed around Gwalior (MP) and in Kangra valley (HP).

Fig. 12.3. Polyphagotarsonemus latus.

Both the adults and nymphs of this mite feed on the lower surface of leaves by sucking the cell sap. The infestation normally starts from the top leaves that gradually proceeds downward. Because of its attack, the lower sides of infested leaves first turn oily, then reddish and finally bronze coloured. In case of severe infestation, the leaves are short and leathery. The pest is responsible for 80-100% foliage infestation, causing 12 to 60% yield losses. High humidity and moderate temperatures are quite condusive for the multiplication of this mite and consequent damage.

This mite generally appears in the 3rd week of October on potato crops planted at/around 15th September in western Gangetic plains and is hyperactive during November and early December. However, it causes little damage on main crop (planted after 15th October) and spring crop (planted during the last week of December). In peninsular India, this mite generally appears in early August on kharif crop and is active during August-September. In Maharashtra and Karnataka its attack on rabi potato crop is generally seen towards crop maturity, i.e., during February-March.

Cultural and Mechanical Management

Suitable crop rotations with non-host crops like wheat in pest prone areas and providing proper isolation to potato crop from susceptible hosts like chillies and brinjal. Delayed planting undertaken in last week of September for early crop and during middle of October for the main crop reduces the incidence and impact of the pest.

OTHER MINOR SUCKING PESTS OR SAP FEEDERS

Distribution and Nature of Damage

Some of the minor sucking pest of sporadic nature infesting potato crop are green bug, *Nezara viridula* Linn.; whiteflies, *Bemasia tabaci* Genendius and *Trialeurodes vaporariorum* Westwood;

thrips, *Thrips hawaiensis*, *T. palmi*, *Megalurothrips distalis* and Scirtothrips sp. and bulb mite, *Rhizoglyphus echinopus* F. and R. The damage is caused both by the nymphs and adults. They suck the sap from the tender shoots and thus affect plant vigour and yield (Mishra et al., 1995). Besides, *B. tabaci* is reported to be an efficient vector of Gemini viruses causing apical leaf curl disease in potato crop in the North Western plains of India. The bulb mite, however, feeds on tubers both in the country stores and standing crop.

Management practices suggested for other sucking pests also found effective against these pest.

FOLIAGE FEEDERS OR DEFOLIATING PEST

Defoliating Caterpillars

Distribution, Nature of Damage and Population Dynamics

Several leaf eating caterpillars such as semiloopers, *Plusia orichalcea* F. and P. ni Hb.; lucerne caterpillar, *Spodoptera exigua* Hb.; gram pod borer, *Helicoverpa armigera* Hb.; Bihar hairy catterpillar, *Spilosoma* (Diacrisia) *obliqua* Walk, and brinjal shoot and fruit borer, *Leucinodes orbonalis* Guen. have been reported to feed on potato foliage from different regions of the country. Among these, *H. armigera* and *P. orichalcea* are quite important.

In the plains, caterpillars of *H. armigera* migrate from chickpea to potato in the spring season and feed on potato foliage. In the hills of Northern India, the moths appear in large numbers by the end of March on ornamental plants and females lay eggs singly on the lower side of leaves. On hatching, the caterpillars feed on potato foliage. *P. orichalcea* caterpillars cause severe damage to foliage in the summer potato crop in Meghalaya and to spring crop in Punjab. It is also a polyphagous pest having a wide host range including pea, cabbage, cauliflower, radish, turnip, carrot, celery, indigo, linseed, etc.

Semilooper *Plusia orichalcea*

Fig. 12.4. *Plusia orichalcea* on potato leaves.

Cultural and Mechanical Management

Handpicking of larvae (caterpillars) and their destruction if undertaken in time is the cheapest and best method of control. This seems quite effective, especially, if the population of these pests is not very high. Adoption of a suitable crop rotation with cereal crops and clean cultivation are some other effective yet economical control methods.

Biological Management

Natural enemies

These polyphagous lepidoterous pests damaging potato crop have a long list of natural enemies on their key host plants. There are important records of 11 species of parasitoids from Bangalore on *H. armigera* (Diwakar & Pawar, 1979). Among ichnemonids, *Campoletis chlorideae* Uchida, Eriborus sp. and *Xanthopimia punctata* Fabr., in braconids, Bracon hebetor Say, *B. greeni* Ashm. and Apanteles sp., from bathylid Goniozu (Parasierola) sp., in trichogrammatid Trichogramma chilonis Westwood; in tachinids *Carcelia illota* Curq, *Palexorista laxa* Curr. and *Goniophthalmus halli* Mesnil and among mermithid Hexmerimis sp. have been found to be effective.

Three hymenopterons, *Chelones narayani* Gupta, *Horogene fenestralis* Opius and Ecphoropsis sp. have been reported to parasitise the larvae of *H. armigera* in Bihar. The predators of *H. armigera* and *P. orichalcea* including Chrysopa sp. feed on the eggs, while *Micromius nilghiriousm* Novas, *Hemerabiuin onontanus* Kimmins, *Henricohania spinosa* Dist. and *Sphedanolestes variabillis* were found feeding on the larvae in Shimla hills (Saxena & Singh, 1982).

Virus

A nuclear polyhyedrosis virus on *P. orichalcea* and *H. armigera* from Bihar were recorded during February-March (Singh, 1974).

Bacteria

Spraying the crop with a bioinsecticide, i.e., *Bacillus thuringiensis* (Bt) Berliner @ 109 spore/ ml on observing an initial damage. Conservation and an early release of the natural enemies, i.e., before the damage by caterpillars is not significant.

Epilachna Beetles

Distribution, Nature of Damage and Population Dynamics

Henosepilachna ocellata Redtenbacher causes severe damage to the potato crop in the higher hilly regions of Northern India. In the years of severe attack, the crop gets completely defoliated (Kapoor, 1950). The activity and damage of another species, i.e. *H. vigintioctopunctata* Fabr. is restricted to plains and mid-hills. Epilachnids are reported to feed on a wide range of host plants including potato, tomato, brinjal, datura, and many other Solanum sp. (Lall, 1964). Both the adults and grubs voraciously feed on the leaves by scrapping the chlorophyll resulting in pre-mature destruction of foliage.

Fig. 12.5. *Henosepilachna ocellata* grubs on potato leaf.

Biological Management

Natural enemies

A eulophid egg parasitoid, *Tetrastichus ovulorum* Ferriere, was recorded on *F. vigintioctopunctata* from Bangalore and Ranchi (Krishnamurti, 1932; Lall, 1964). A chalcid parasitoid, Ugna menoni Kerrich was recorded on *H. vigintioctopuncta* from Andhra Pradesh (Azam et al., 1974) and *H. ocellata* from Shimla hills (Saxena and Singh, 1982). The parasitisation of *H. vigintioctopunctata* in the field by *Pediobius foveolatus* Crawford has been reported to be as high as 77 per cent. Two reduviids, Coranus sp. and Henricoclania sp. nr. spinosa Dist. and a nabid, *Aptus mussooriensis* Dist. have been found feeding on the grubs and adults of *H. ocellata* at Shimla (Saxena and Singh, 1977) and *Cantheconidea fusicellata* Wolf. from Kerala on *H. vigintioctopunctata.* (Shiella and Abraham, 1981).

MINOR DEFOLIATING PESTS

Distribution, Nature of Damage

There are many minor or sporadic defoliating pests of localized importance recorded on potato crop from different parts of the country. Important among them are the flea beetle, *Psylloides plana* Maulik which is prevalent in North-western hills; weevil, *Myllocerus subfasciatus* Guerin in Nilgiris and blister beetle, *Epicauta hirticornis* Hagg in Shimla hills; army worm, *Mythimna seperata* Walk. in North West plains. All of these, damage the potato crop by feeding on its foliage and tender shoots.

Cultural and Mechanical Management

In the initial stage of attack, collect and mechanically destroy affected leaves as also the different stages of the pest.

STORAGE PEST

Potato Tuber Moth

Distribution, Nature of Damage, Losses and Life Cycle

The potato tuber moth, *Phthorimaea operculella* Zeller, (Lepidoptera: Gelechiidae) (PTM) is a native of South America and is the most obnoxious pest of potato both in fields and the country stores. In areas where the pest is left unmanaged, the losses to potatoes kept in the country stores may be as high as 70 per cent. Due to inadequate cold storage capacity, raising more than one crop in a year and non-adoption of recommended control measures, PTM has the status of a serious/key pest. Prolonged dry and hot weather is quite conducive for PTM multiplication. PTM larvae damage the crop foliage and also stems, exposed tubers in field and stored tubers in the country stores with estimated losses to the tune of 26 and 56 per cent, respectively (Trivedi, 1990). In infested tubers, the feeding tunnels are packed with black excretory pellets and the larvae are inside the tunnels.

The female moth lays eggs on the underside of leaves or on exposed tubers near the eye. The young larvae are grey-yellowish white with brown head. The full-grown larvae come out of the foliage/tubers and pupate in greyish silken cocoons often covered with soil for camouflage. The foliage feeding larvae (caterpillars) are green in colour, while those feeding on tubers are pinkish. Pupation occurs either in dried leaves, soils or in cracks and crevices of country stores. The duration of life cycle varies from 20 to 30 days at the optimum temperature of 20 to 25°C. Egg laying is inhibited and moths die at 36°C and no emergence takes place at temperature below 13°C and above 40°C. In a year there are 8 to 9 generations in northern plains, 10 to13 generations in Peninsular India and 11 generations in North-eastern hills of India (Chandla, 1986).

Cultural and Mechanical Management

Planting seed tubers at a depth of 10 cm as against the traditional planting depth of 6 cm reduce its damage to a great extent (Akhade et al., 1970). The fields should be ridged after 6 to 7

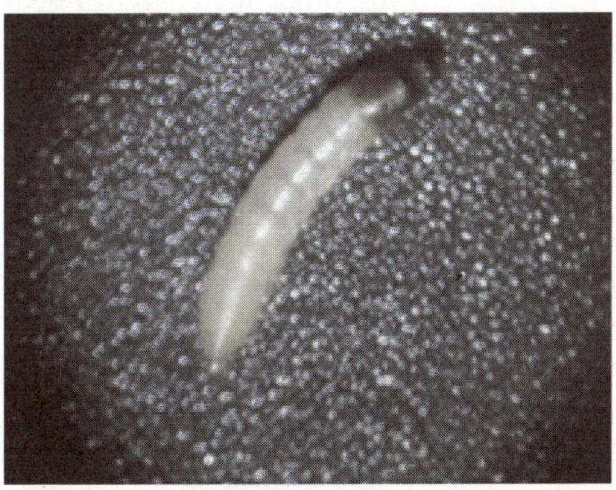

Larvae

Adult

Fig. 12.6. *Phthorimaea operculella.*

weeks of planting so that the tubers are burried at least 25cm below the soil surface. In areas where PTM population remains quite high and severe tuber damage is expected, ridging should be done twice so that the tubers are not exposed at any time for egg laying and infestation. Timely and adequate irrigations minimize soil cracking and thereby reduce the risk of tuber exposure to PTM attack or their laying eggs. This problem is quite common in areas where potato crop is taken in heavy soils.

Harvested tubers must be removed from the field as early as possible and should not be kept overnight in the field. Leftover tubers, after harvest, should also be collected. All the plant debris including the weeds belonging to family solanaceae should also be collected and destroyed. The crops like tomato, tobacco, chillies and brinjal should not be grown in the vicinity of potato fields, particularly in PTM prone areas (Lal, 1993).

Varietal resistance remains the first line of defence in the control of insect pests. Although immunity to PTM has not been identified in potato cultivars so far, yet various levels of genetic resistance/tolerance have been reported in several potato cultivars. An Indian potato hybrid, QB/A 21-29 was found to be tolerant to PTM. Store healthy (PTM free) potatoes in cold stores. Covering of healthy tubers stored in country stores with 2-3 cm thick layers of chopped dried leaves of either of Lantana, Soapnut, Neem, Eucalyptus or Eupatorium. Do not leave the harvested potatoes in the field overnight and never cover them with infested potato foliage. Collect the leftover tubers after harvesting for table purpose or destroy them. Also remove the volunteer plants from the potato fields.

Biological Management

Natural enemies

In Karnataka, Indigenous parasitoids like *Chelonus curvimaculatus* Cameron, *Bracon gelechiae* Asheamd, Apanteles sp., *Pristomerus vulnerator* Panzer and several other braconids have been found in abundance and are reported to cause 4-17% parasitisation under field conditions (Nair & Rao,1972). A number of exotic parasitoids were also introduced into India between 1970 and 1980. Of these, *Copidosoma koehleri* Blanchard, an exotic egg/larval parasitoid, gave 28.4-60.8% parasitization in Maharashtra (Dalaya and Patil, 1973). On the other hand, field releases of *Bracon hebator* Say registered 12% parasitisation of the larvae under field conditions in Bangalore (Divakar and Pawer, 1979). Melanips sp. and Diadgma molliplum (Holmgren) caused 2.5 to 5.0% parasitisation in Shimla hills (Saxena and Raj, 1980). Besides, *Orgilus jennieae* Muesebeck, *Apanteles subandinus* Blanch, *Chelonus blackburni* Cameron and *C. kellieae* Cameron, the ovo-larval parasitoids were also tested for their effectiveness in suppressing PTM population at Rajgurunagar (Maharashtra). Of these, *C. jennieae* and *A. subandinus* registered 59.6 and 16.6% parasitization, respectively (Chaudhary et al., 1983).

Botanical

Natural insect repellents like leaves, stems, flowers and fruits of various plants possessing repellency properties. Three plant species, viz. *Lantana camara* Linn., *Eucalyptus globulus* L. Herit. and Minthostachys sp. either dried/shredded leaves or powder form are effective in preventing PTM damage to potatoes stored for four months in country stores. In India, out of the several plant species tested against PTM, *Lantana aculeata* L. provided the best protection by reducing PTM damage from 70% to below the 5%. Likewise, sprout damage was also reduced from 45% to below 3%. For this purpose, fresh leaves of lantana are chopped, dried and spread in 2 cm thick layers both below and above the stored potato. Dried powders of five different plants, rhizomes of *Acorus calamus*, leaves of *Melia azedarach*, ripen berries of *Piper longum*, leaves of *Prunus persica* and ripen fruit of *Lindera neesiana* were tested for the control of potato tuber moth at laboratory. Three concentrations — 0.05%(w/w), 0.5%(w/w) and 5%(w/w) — were used. All the concentrations of *L. neesiana* and *A. calamus* were found effective in PTM control. *L. neesiana* showed 66.7%, 70.0% and 83.3% and *A. calamus* 56.7%, 66.7% and 70% adult mortality (.Niroula and Vaidya, 2004).

Four natural plant oils (Margorum, Cardamon, Rosemary and Terpintin) were tested against different stages of PTM. The 0.02 and 0.05% concentrations of cardamon oils exhibited the best reduction in percentage of eggs hatchability (67.47 and 86.74%). Dusting potato tuber by 1.5% concentrations of cardamom and rosemary oils elicited the lowest percentage of larval penetration, pupation and adult emergence. (Moawad and Ebadah, 2007).

Sex pheromones

PTM sex pheromones consisting of trans-4, cis-7-ol-tridicadien-1 acetate (PTM-1) and trans-4, cis-7-cis-10-tridectrien-1 acetate (PTM-2) in ratio 0.4 mg PTM 1 + 0.6 mg PTM-2 are commercially available from International Potato Center Lima, Peru, Laboratory for Research on Insecticides, Marijkeweg 22,6700 PG, Wageningen, Netherlands and Bhabha Atomic Research Centre, Trombay, Mumbai. These sex pheromones are commonly used for monitoring of field population, detecting the occurrence, facilitating proper use of insecticides and mass trapping of PTM for direct control (Chandla et al., 1986). These sex pheromones can be used effectively in funnel, water or sticky traps. For monitoring purposes, two traps per field should be installed in the crop 50 metre apart. The traps should be checked at 7-day intervals. The number of traps to be used for direct control depends upon the population density of moths, for example, in Lima, population above 2000 moths/trap/day requires 42 traps/ha (Raman & Booth, 1983).

Bacteria

Bacillus thuringiensis Berliner, a Gram positive, flagellated, rod-shaped bacterium, which produces a para sporal crystal during sporulation, has also been reported to be effective in controlling PTM in Peru, Tunisia, and India (Amonlar et al., 1979; Chandla et al., 1993; Ranjekaret et al., 2003).

Virus

A baculovirus has also been collected in Peru and reported as effective against PTM. In this case, the diseased larvae become opaque and milky white. Such larvae do not move violently when disturbed and also fail to pupate. This virus has been reported quite effective in suppressing the PTM populations both in fields and stores. This virus remains effective up to a period of 120 days under storage condition (Chandla et al., 1993; Lal, 1993). Treat potatoes with the dust of granulosis virus before storing the potatoes. For this purpose a mixture in the ratio of 1.0 kg of talc powder + 20 GV infected larvae + 1.0 litre of water + 2 ml of triton should be prepared and used @ 5 kg of mixture/ ton of potatoes.

CIPC

CIPC (Isopropyl N-(3-chlorophenyl) carbamate) is a sprout suppressant commonly used on ware potatoes in Country stores. But beside suppression of sprouts. CIPC was found to reduce PTM infestation. After the period of 60 days of incubation only 2-6% tuber damage was observed in the country stores (Chandla et al., 2003).

SOIL PEST

Cutworms

Distribution, Nature of Damage and Population Dynamics

Cutworms are polyphagous insects of cosmopolitan distribution. Five species, viz. *Agrotis ipsilon* (Hfn), *A. segetum* Schiff., *A. flammatra* Schiff., *A. interacta* Wlk. and *A. spinnifera* (Hb.) occur on potato crops in India. The first two species are widely distributed in plains and hilly regions, respectively. The larvae (caterpillars) of cutworms damage the crop by cutting the young plants at the base and later on by feeding on shoots and leaves. After tuberisation, they feed on tubers by making deep and irregular galleries in them, thus reducing the market value of infested tubers. In badly infested fields, tuber damage may vary from 12-40 per cent (Chandla, 1983). The moths are generally not found in North Indian plains between May to September, but reappear during October and remain active till March-April. A possibility of migration of moths from the plains to hills during summer months cannot be ruled out and mass appearance of moths occurs during April in the hills.

Cultural and Mechanical Management

Cutworms either aestivate during summer months or hibernate during winters in the soil while completing their life cycles. Therefore, deep ploughing of potato fields during summer months in the plains exposes the immature stages to high temperature and predatory birds (Chandla,

Fig. 12.7. *Agrotis segetum* on potato tuber.

1985; Mishra & Agarwal, 1988; Mishra et al., 1995). Similarly, deep ploughing of fields during autumn in the hills also minimises cutworm's population.

Light traps installed in/around potato fields attract the adults of cutworms, and helps in mass collection and destruction of the moths.

Biological Management

Botanical

Garlic as intercrop with potato was found to be effective in minimizing cutworm damage in potato crop at Shimla.

Natural enemies

Broscus punctatus Dist and *Liogryllus bimaculatus* Linn have been reported to be the parasitises of cutworm larvae (Fletcher, 1916). *Macrocentrus collaris* Spin, *Netelia ocellaris* Thomson, *Coelichneumon* sp. nr *truncatulus* Thomson, *Periscepsia carbonaria* Panzer and *Turanogonia chinensis* Wiedemann also parasitise *A. ipsilon* and *A. segetum* under natural field conditions (Chandla et al., 1989; Singh, 1993).

Bacteria

Bacillus thuringiensis Berliner is a well-known biopesticide. Spraying the crop and ridges with this biopesticide (Bt @ 109 spores/ml) gives a good control (Mishra et al., 1995).

Fungus

Entomogenous fungus, *Metarrhizium anisopliae* Meld. is a best known fungal control of cutworm (Mishra et al., 1995).

Nematodes

Entomophilic nematode, Stinernema (Neoaplectana) sp. are also well known as dominant regulatory factors for cutworm populations from various parts of the country (Mishra et al., 1995).

White Grub

Distribution, Nature of Damage and Population Dynamics

White grubs are distributed throughout the country. The two major species affecting potato crop are Lachnosterna (Brahmina) coriacea Hope L. (B.) longipennis Blanch. and Melolontha indica Blanch. The damage to potato is caused by the grubs only which feed on rootlets, roots and tubers. The first stage grub can survive on the organic matter present in the soil but roots are preferred as and when encountered. The damage to potato tubers is mainly caused by the second and third instar grubs, which make large, shallow and circular holes on them. The white grubs infested tubers, therefore, have poor market value. These grubs

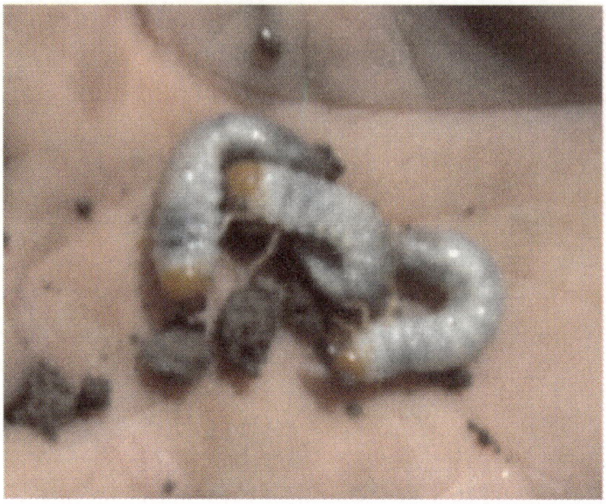

Fig. 12.8. *Brahmina coriacea.*

damage the tubers without any symptoms on the foliage. Thus, the farmer remains unaware of the damage to tubers until harvest. The damage is more severe in the years following high adult activity.

In Shimla hills, yield losses to potato crop have been reported to be as high as 85% (Mishra and Chandla, 1989). The adults start emerging from the soil immediately after good monsoon showers at dusk during June-July. The emerging beetles could be seen settling on bushes/trees such as acacea, peach, plum, apricot, pear, apple, etc. for mating feeding. Later on the mated females re-enter the soil for egg laying at dawn. All the larval stages live in the soil from July-November and initially feed upon roots of living plants, ingesting at the same time large quantities of soil and dead organic matter and afterward feed upon tubers. On completion of development, these grubs construct earthen cell in the soil in which they undergo either hibernation or aestivation prior to pupation during December to April-May. Emergence of adults starts with the onset of first monsoon showers in June onwards (Chandla, 1986). White grubs seem to be periodic pests, attacking potato crop irregularly from year to year. The major factors influencing the development of damaging numbers of grubs are soil moisture and rainfall. In general, in years with normal or above normal rainfall, grub populations increase. Well-maintained land next to ornamental plants favoured by the adults seems to be more commonly attacked.

Cultural and Mechanical Management

In the endemic areas, autumn ploughing in hilly areas not only exposes the grubs to adverse conditions (low temperature) but the exposed grubs also become prey of the birds. The use of nitrogenous fertilizers, especially ammonia and urea, at higher doses kill the first instar grubs. The light traps may also be used for collecting the beetles during night. The beetles can also be collected by shaking or jerking the host plants during night. The fallen beetles should be collected and destroyed by putting them either in kerosinized water or by burning. The host trees of adults (beetles) should be lopped or pruned.

Water Management

Practically all white grub species require moist soil for their eggs to hatch. The young larvae are also very susceptible to desiccation. In areas which stand some moisture stress, do not water in July and early-August when white grub eggs and young larvae are present. On the other hand, moderate grub infestations can be outgrown if adequate water and fertilizer is applied in August through September and again in May when the grubs are feeding.

Host Plant Modifications

Certain species of scarab adults prefer specific host plants. Where beetles are common, do not plant roses, grapes and lindens around high maintenance turf areas. May/June beetles prefer oaks and the green June beetles feed on ripening fruit such as peaches. The fine and tall fescues are not as severely attacked as perennial rye grass.

Biological Management

A number of bio-control agents are known to manage white grubs in different parts of the country. These should be encouraged either by conserving the existing populations or by introducing and establishing the known bio-control agents obtained from new localities.

Botanical

In India, out of the several plant species tested against grubs, Neem seed kernal and Eucalyptus globulus in powder form or in solution made in water are effective.

Natural enemies

Some of the potential parasites scoliids, *Campsomeris collaris* F., *Scolia aureipennis* Lepeletier and *S. pustulata* Magr. are known to be effective against white grubs.

Bacteria

Bacillus sp is a well-known biopesticide. Spraying the crop and ridges with this biopesticide *Bacillus thuringiensis* (Bt @ 109 spores/ml) gives a good control (Mishra et al., 1995).

Several strains of the bacterium, *Bacillus popilliae*, have been found that attack white grubs. However, the commercial preparation of this bacterium is extracted from Japanese beetle grubs and are most active against this species. This bacterium is picked up by feeding grubs and it causes the body fluids to turn a milky-white before grub death. Fresh bacterial preparations should be used and three to five years are needed to provide lasting controls.

Fungus

Entomogenous fungus, *Metarrhizium anisopliae* Meld is best known fungal control of white grubs (Mishra et al., 1995). *Beauveria brongniartii* Sacc. is also very effective against white grubs.

Nematodes

Products containing microscopic, parasitic nematodes may be purchased from centres to help control white grubs. These nematode products are mixed with water and drenched in soil. In general, the treatment area should be irrigated before the application of the nematodes, as nematodes move by swimming through the soil water. Entomophilic nematode, Stinernema (Neoaplectana) sp. and Heterorhabtitis are also well known as dominant regulatory factors for grubs populations from various parts of the country (Mishra et al., 1995).

THE MINOR SOIL PESTS

The minor soil pests recorded on potato crop are wire-worm, Agroitis spp; Termites, Odontotermus obesus Ramb and Eremotermus spp; Lunate fly or bulbfly, Eumerus spp.; Red ant, *Dorylus orientalis* Westwood; mole cricket, *Gryllotalpa africana* P. de Beauv., earthworm, *Allodobophora caliginosa* Savingny and *A. foetida* Savingny and field rats, *Nesokia indica* Gray and *Bandicota bengalensis* Gray. All these pests are sporadic in nature and cause damage to potato tubers at different intensities in different parts of the country.

NEMATODE PESTS OF POTATO

Potato Cyst Nematode (PCN)

Distribution, Nature of Damage, Population Dynamics and Biology

Andes mountainS of South America, the native place of potatoes, are also considered as the original home for the PCNs. The nematode cysts must have got introduced into Europe with the breeding material brought for late blight resistance in 1850s and would have spread throughout the world with the improved potatoes developed in Europe at that time. Presently, PCNs are prevalent in about 60 countries. The yellow cyst nematode, *Globodera rostochiensis* is relatively widespread (being reported from 58 countries), while *G. pallida* is less prevalent (present only in about 27 countries). The latter is able to adopt under subtropical climatic

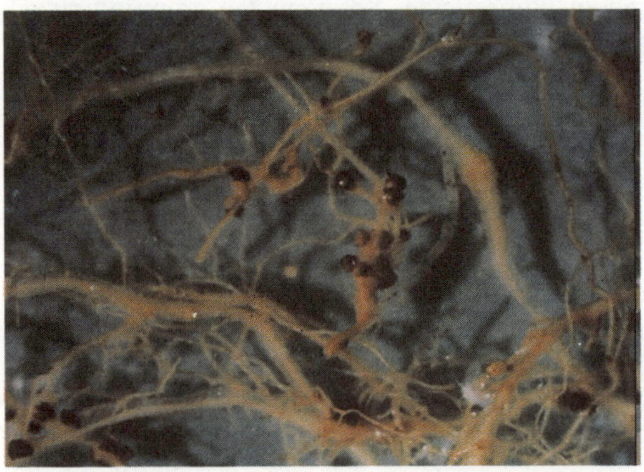

Fig. 12.9. *Globodera rostochiensis* on roots of potato plant.

conditions showing a wide genetic variability that even exists in the place of origin of potato, Andes Mountains.

In India, Dr. F.G.W. Jones first detected the PCNs from a field at Vijayanagaram state Farm in Ootacamund; Nilgiris hills (Seshadri & Sivakumar, 1962). These were probably introduced from British Islands because these fields contained European weeds. Later on, their occurrence was also noticed from Kodaikanal hills. Extensive surveys conducted earlier in other major potato growing areas of Assam, Karnataka, Himachal Pradesh, Punjab and Uttar Pradesh indicated the absence of PCNs in these states. Realising the potential danger of PCNs to potato cultivation in the country, the Destructive Insect Pest Act 1919 was amended by the Tamil Nadu Government in 1971 to ensure strict checking of movement of potatoes grown in infested fields/areas prior to their marketing.

Species and Patho-types: Detailed studies carried out with several populations of PCNs collected from major potato growing localities of Ootacamund revealed that Rol of *G. rostochiensis* and Pa2 of *G. pallida* were the most prevalent pathotypes. The other prevalent pathotypes were Ro2 and Ro5 of the former and Pal and Pa3 of the later species.

Symptoms

PCN infestation initially appears in patches and then spreads in the entire fields within 4-5 years through irrigation water, wind, agricultural implements, humus and movement of seed material grown from the infested fields. The infested plants lose vigour, become sickly in appearance with stunted growth, dull appearance and unhealthy foliage. Infested plants wilt during the sunny and bright days and require more water per unit weight compared to the healthy ones. Besides, they accumulate more calcium but efficiently use N.P. and K. The total photosynthesis per plant is significantly reduced because of reduced leaf area. This leads to drastic reduction in yield by 65 to 71% from summer and autumn crops, respectively (Prasad, 1993).

Cultural and Mechanical Management

The potato cyst nematodes are the most competent plant parasitic nematodes exhibiting a highly specialised survival mechanism. They are restricted to a limited host range belonging to family Solanacae and have got themselves distributed along with the hosts by adapting to the surroundings. They cannot be eradicated completely once established in a given locality, hence they can only be managed by adopting judicious and effective IPM strategies.

Growing non-host crops and following effective crop rotations at least for a year with any non-solanceous vegetable such as beetroots, cabbage, carrots, cauliflower, French beans, garlic, radish, turnips, etc. during autumn brings down the cysts population to a great extent. Presently, the farmers of Nilgiri follow Potato-Cabbage-Carrot rotations for the best management of PCNs. This rotation also helps in maximum utilization of nutrients applied.

Resistant varieties

High degree of PCN resistance occurs in *Solanum ajanhurri*, *S. bulbocastanum*, *S. gandarlassi* and *S. tuberosum* sub species andigena, *S. vernei* and *S. spegazzinii*. Of a large population produced from PCN resistant *S. tuberosum* ssp. andigena clones and *S. multidissectum*, a selection

(No. 3246) was obtained from Scottish Plant Breeding Station, Edinburgh, and U.K. along with some commercial varieties. Three hybrids had a high degree of resistance. However, further screening against PCNs showed that initially resistant genotypes proved susceptible because of presence of two species of PCNs with several pathotypes. Later germplasm screening revealed high degree of resistance in several clones of tuber bearing Solanum sp.

Even though *S. vernei* clone Vtn 62-33-3 obtained from Netherlands had a high degree of resistance, it was highly susceptible to late blight, hence crossed with Kufri Jyoti to obtain several desirable genotypes. Of these, one selection bearing No. SON 110 possessed desirable yield characters, late blight and PCN resistance. This hybrid clone was released as cv. Kufri Swarna in 1985 that today occupies about 40 per cent of potato area in Nilgiris. It performs well even under drought conditions in Nilgiri hills. No doubt many other vertically resistant hybrids are available now; their vertical resistance can be broken easily by quick adoption of both late blight and PCN.

Biological Management

Fungus

There are thirteen fungi that are able to infect the cysts, out of which frequency of occurrence was maximum for Dactylaria (31.6%) followed by *Aspergillus* spp. (14%), *Humicola grasea* and three other non-sporulating fungi in the frequency range of 7.2 to 1.75% in Nilgiris. *In vitro* parasitization of eggs by these fungi indicated maximum mortality by Dactylaria sp (82.4%) followed by *Penicillum* spp (80.6%), *Aspergillus* spp (68%) and *Humicola grisea* (60%). Other fungi (excepting *Fusarium* spp) parsitize the eggs of PCN to varying extents. The fungus *P. lilacinus,* an effective bio-agent against nematodes, does not occur in Nilgiris.

Root-Knot Nematode

Distribution, Nature of Damage, Population Dynamics

The root-knot nematodes (RKNs) causing root galls are the well-known and most popular nematode pests of potato crop. These are prevalent in all parts of the world, particularly in the

Fig. 12.10. *Meloidogyne incognita* damage on potato tuber.

subtemperate, subtropical and tropical regions affecting almost all the agricultural crops including potatoes.

Root knot nematodes occur in all the potato growing countries of the world and from all the potato growing zones of our country. They are considered an important pest of many commercial crops but certainly not potato in India.

Symptoms

Root-knot nematode infests the roots and tubers of potato plants leading to the hindrance in normal functions of nutrient and water uptake/translocation. Infested plants are dwarfed, chlorotic and have stunted foliage and poor root development. The galls on roots are small and often go unnoticed, but tubers have warty growth or rough surface that reduces their market value (Thriumalachar, 1951).

The 2nd stage juveniles hatch from the eggs and first infest the young roots. They form giant cells through which the nematodes extract nourishment from the plant cells. Later on the giant cell and development of nematode within the roots results in the formation of root-knots or galls. The female larvae gradually enlarge and undergo four moults before forming a pear-shaped structure. The male nematodes, however, retain their thread-like appearance. Males come out freely from the root system and mate with females to fertilize the eggs. The females are sedentary, depositing 300 to 400 eggs into a gelatinous matrix usually adhering to the root galls. These eggs hatch readily and invade the fresh roots. (Prasad, 1993).

In Shimla hills, *Meloidogyne incognita* completes its life cycle in 25-30 days during summer (April-September) but takes about 65 to 100 days in winter (October-March). In the hills normally two generations are completed by the time of tuberization, therefore, tuber infestation is invariably observed even if the inoculum pressure is low in fields. On the other hand, tuber infestation in plains remains low mainly because the crop duration is short and fresh roots are not always available. This situation many a times leads to the conclusion that the root knot nematodes are absent in such localities although they are present in the roots. Further, hot summers in plains reduce the soil populations of RKNs.

The eggs and larvae can survive for more than 100 days even in the absence of the hosts during summer months at Shimla hills. This could be one of the reasons for continuation of higher inoculum levels causing higher infestation in subsequent crops. There is 42.5 per cent yield reduction upon 100 per cent tuber infestation. RKNs infested show higher weight loss compared to uninfected tubers during storage. Although post storage performance of infested tubers is normal, the number of sprouts on RKNs infested tubers are fewer. Root knot nematodes have got their role in spreading and development of bacterial wilt and early blight (Bombawala et al., 1982).

Cultural and Mechanical Management

The basic principle of RKN management programme is to achieve an increased yield of quality produce with reduced nematodes population.

Deep ploughing and sun drying of fields during summer months help in drying of infective stages of larvae. Planting of potato crop during the 2nd week of October in autumn and in early January for the spring can limit RKN infestation on tubers. Burning of trash in field before planting helps not only by sterilizing the soil but also enriching it, though possible only for small holdings. Growing one row of trap plants like *Tagetes patula* and *T. erecta* (African marigold) in between 2 or 3 rows of potatoes improves the crop performance by reducing RKN infestation. RKN are being polyphagous in nature having a wide host range, yet a few crops are free from *M. incognita* infestation allowing crop rotation with a non-host, like maize, wheat, millets, beans, etc. for reducing RKN infestation. Potatoes harvested from RKN infested field should not be used as seed for subsequent crop. Movement of the soil and water from infested fields to RKNs free fields should also be avoided. Besides, the field should be kept free from weeds as RKNs have a wide host range and most of the weeds help them.

Host Resistance

Host resistance is the most practical approach for the management of RKN. An intervarietal hybrid HC-294 possesses resistance against RKNs, which inhibits the formation of giant cell in the roots.

Biological Management

A large number of biotic agents such as fungi, bacteria, predacious nematodes, protozoan, etc. are reported for managing the RKNs. However, an efficient use of these bio-agents has not yet been possible.

REFERENCES

Akhade, M.N., Tidke, P.M. and Patkar, M.S. (1970). Control of potato tubemoth (*Gnorimoschema operculella Zell.*) in Deccan Pleatue through insecticides and depth of planting. *Indian J. Agri. Sci.* **40:** 1071-1076.

Amonkar, S.V., Pal, A.K., Vijaylakshmi, L. and Rao, A.S. (1979). Microbial control of potato tuber moth (*Phthorimaea operculella Zell.*). *Indian J. Expetl. Biol.* **17:** 1127-1133.

Anonymous (1984). Quinnquennial review team reports, AICRP on biological control of pests and weeds. *Tech. Doc. No.* **6:** 141.

Azam, K.M., Aziz, S.A. and Ali, M.M. (1974). A new record of *Ugna menoi* Kerrich as parasite of Epilechna *vigintioctopunctata* (F) in India. *Curr. Res.* **3:** 88.

Bombawala, O.M., Bedi, P.S., Sharma, N.K. and Sharma, S.K. (1982). *B.B.* Nagaich *In: Developing Countries.* Indian Potato Association CPRI, Shimla. pp. 425-427.

Chandla V.K., Singh, B. and Chandel, R.S. (2003). Management of Potato tuber moth in countary stores with CIPC. *J. Indian potato Assoc* **30**(1-2): 153-154.

Chandla, V.K. (1983). Extent of damage by pests. B.B. Nagaich (Eds.). *In: potato production storage and Utilization.* Central Potato Research Institute, Shimla. pp. 425-427.

Chandla, V.K. (1985). Potato pest and their management. *Indian farming* **36**(12): 31-32.

Chandla, V.K. (1986). Insect pest complex of potato crop in Shimla hills and their management. *Ph.D Thesis.* Department of entomology–Apiculture, Dr. Y.S. Parmar University of Horticulture and Forestry, Solan (H.P.) pp. 94.

Chandla, V.K., Mishra, S.S. and Verma, K.D. (1989). Natural enemies of Agrotis species in *Shimla hills* (Abstract). Presented in "Xth Annual., General Body Meeting of IPA., held at HAU, Hisar on 20th Aug. 1989.

Chandla, V.K., Mishra, S.S., Sharma, D.C. and Kashyap, N.P. (1993). Potato tuber moth menece in Himachal Pradesh and its management (Abstract). *J. Indian potato Assoc* **20**(1): 60-61.

Chaudhary, R., Trivedi, T.P. and Raj, B.T. (1983). Field evaluation of some exotic parasitoid of Potato tuber moth (*Phthorimaea operculella Zell.*). *Indian J. Ent.* **45:** 504-506.

Dalaya, V.P. and Patil, S.P. (1973). Laboratory rearing and field releases of *Copidosoma koehleri* Blachard, an exotic parasite for the control of *Gnorimoschema operculella* Zeller. *Res. J. Mahtma phule Agri. Univ* **4:** 97-107.

Diwakar, B.J. and Pawar, A.D. (1979). Field recovery of *Chelonus blackburni* and *Bracon hebtor* from Potato tuber moth. *Indian J. Plant Pro.* **7:** 214.

Fletcher, T.B. (1916). Annotated list of Indian crop pests. *Rep Agri. Res. Inst and Col. Pusa* 1915. pp. 58-77.

Ghosh, D., Nath, D. and Chakrabarti, S. (1985). Predators and parasites of aphids (Hom., *Aphididae*) from North-West Himalaya: Ten species of syrphids (Diptera-*Syrphidae*) from Garhwal range. *Entomon* **10**(4): 301-303.

Kapoor, A.P. (1950). A note on *Epilachna ocellata* Redt. (Coleoptera: Coccinellidae) with description of three species hitherto confused with it. *Rec. Indian Mus.* **48:** 17-19.

Krishnamurti, B. (1932). The potato epilachna beetle, *Epilachna vigintioctopunctata* (Fabr.). *Bull. Dept. Agric. Mysore Ent. Ser.* **9:** 16.

Lal, L. (1993). Potato tuber moth-Bionomics and Management. *In: Advances in Horticulture-potato*, Eds. Chandha, K.L. and Grewal, J.S. Malhotra Publishing House, New Delhi. **7:** 591-602.

Lall, B.S. (1964). Vegetable pests. In: *Entomology in India.* (Eds). N.C. Pant. *Entomological Soc. India*. pp. 184.

Misra, S.S. and Agarwal, H.O. (1988). Potato pests in India and their control. *Trop. Pest manag* **35:** 199-209.

Misra, S.S. and Chandla, V.K. (1989). White grubs infesting potatoes and their management. *J. Indian Potato Assoc.* **16**(1&2): 29-33.

Misra, S.S., Chandla, V.K. and Singh, A.K. (1995). Potato pests and their management. *Technical Bulletin No.* 45. Central Potato Research Institute, Shimla. pp. 57.

Moawad, S.S. and Ebadah, I.M.A. (2007). EbadahImpact of Some Natural Plant Oils on Some Biological Aspects of the Potato Tuber Moth, *Phthorimaea operculella*, (Zeller) (Lepidoptera : Gelechiidae). *Res. J. Agri. Biol. Sci.* **3**(2): 119-123.

Nagaich, B.B., Verma, K.D. and Upreti, G.C. (1970). Hereditary variation in the ability of *Myzus persicae* to transmit potato leaf roll and virus Y. *Final Tech. Rept. PL 480 Scheme*, CPRI.

Nair, R. and Rao, V.P. (1972). Results of the survey for natural enemies of potato tuber moth, *Phthorimaea operculella* Zeller (Lepidoptera : Gelechiidae) in Mysore state and the parasites reared from Maharashtra state. *Tech. Bull. Commonwealth Inst. Biolcontro.*, **15:** 115-130.

Niroula, S.P. and Kamini Vaidya (2004). Efficacy of Some Botanicals against Potato tuber moth, *Phthorimaea operculella* (Zeller, 1873) Our Nature **2:** 21-25.

Prasad, K.S.K. (1993). Nematodes-distribution, biology and management. *In: Advances in Horticulture: Potato.* Eds. K. L. Chadha and J. S. Grewal. Malhotra Publishing House, New Delhi. **7:** 635-647.

Pruthi, H.S. (1946). Report of the Imperial Entomologist. *Abrigd. Sci. Rep. Agric Res. Inst.* New Delhi. pp. 64-71.

Raman, K.V. and Booth, R.H. (1983). Evaluation of technology for integrated control of potato tuber moth in field and storage. *Technology Evaluation Series No. 10.* International Potato Center, Lima, Peru. pp. 18.

Ranjekar P.K., Aparna Patankar, Vidya Gupta, Raj Bhatnagar, Jagadish Bentur and Ananda Kumar, P. (2003). Genetic engineering of crop plants for insect resistance *Cur. Sci.* **84:** 321-329.

Saxena, A.P. (1981). *Leucopis fumidilarva* Tanas (Diptera: Chamaemycides), a new predator of aphids and leaf hoppers on potato crop. *J. Indian potato. Assoc* **8**(2): 96-98.

Saxena, A.P. and Raj, B.T. (1980). Studies on the biological control of potato pests. *Proc. III. Workshop on Biol. Control of crop pests and weeds.* PAU, Ludhiana, October 27-30, pp. 121-125.

Saxena, A.P. and Singh, V.P. (1982). Natural enemies of potato pests in India. *In: Potato in developing countries,* Ed.B.B. Nagaich. Central Potato Research Institute, Shimla. pp. 349-355.

Saxena, A.P. and Singh, V. (1977). Predacious bugs on *Epilachna ocellata* Redt. (Coccinellidae: Coleoptera). *J. Indian Potato Assoc.* **4:** 70.

Seshadri, A.R. and Sivakumar, C.V. (1962). Golden nematode of potatoes (*Heterodera rostochiensis* Woll.), a threat to potato cultivation in the Nilgris. *Madras Agric. J.* **49:** 281-288.

Sheila, M.K. and Abraham, G.C. (1981). *Cantheconidea furcellata* (Wolf.) (Pentatomidae:Hemiptera), as a predator of *Henosepilachna vigintioctopunctata* (Fabr.) *Agric. Res. J. Kerala* **19:** 142.

Singh, M.N., Khurana, S.M.P., Nagaich, B.B. and Agarwal, H.O. (1982). Efficiency of *Aphis gossypii* and *Acyrthosiphon pisum* in transmitting potato viruses X and leafroll. *In: Potato in Developing countries.* (Eds.). Nagaich, B.B., Shekhawat, G.S., Gaur, P.C. and Verma, S.C. Central Potato Research Institute, Shimla. pp. 289-293.

Singh, S.P. (1974). Observations on the biology of *Trichoplusia orichalcea* Fabr. on potato. *Indian. J. Plant Prot.* **2**(1&2): 127-128.

Singh, S.P. (1988). Natural enemies of aphids. *Fourth National Symposium of Aphidology*, CPRI, Shimla. Octoder **4-6:** pp. 66.

Singh, S.P. (1993). Biological control of potato pests. *In: Advances in Horticulture: Potato. Vol. 7.* Malhotra Publishing house, New Delhi, 11006 (Eds.). Chadha, K.L. and Grewal, J.S. pp. 615-634.

Thirumalachar, M.J. (1951). Root-knot nematode in potato tubers in Shimla. *Curr. Sci.* **20:** 104.

Trivedi, T.P. (1988). Natural enemies of potato pests in Karnataka. *J. Indian Potato Assoc.* **15**(3&4): 159-160.

Trivedi, T.P. (1990). Spatial distribution, phenology, life table and assessment of losses due to potato tuber moth. *Phthorimaea operculella* (Zeller) Ph.D. Thesis of University of Agriculture sciences. Banglore (India). pp. 182.

Verma, K.D. (1985). Potato aphids with special reference to *Myzus persicae* (Sulzer). *Proc. 2nd Nat. Symp. on recent trends in Aphidological studies.* (Eds). S.P. Kurl pp. 39-46.

Verma, K.D. and Chauhan, R.S. (1993a). The life cycle of potato vector, *Myzus persicae* (Sulzer). *Curr. Sci.* **65:** 488-89.

Verma, K.D., Misra, S.S. and Saxena, A.P. (1976a). A role of *Aphelinus* sp. in the natural control of *Myzus persicae* on potato. *J. Indian Potato. Assoc.* **3**(1): 40.

Verma. K.D., Parihar, S.B.S. and Khan, I.A. (1993b). Field reaction of some potato varieties/germplasm lines to leafhopper, Empoasca fabeae (Harris). *J. Indian Potato. Assoc.* **20:** 49.

13

Forgotten Pollinator: Protecting the Bees from Pesticide Poisoning

Seema Singh Thakur

Biocontrol Laboratory, Department of Entomology, College of Agriculture, G.B. Pant Univ. of Agri. & Tech., Pantnagar–263 145, Uttarakhand

Pesticides are substances used to eliminate unwanted pests. Insecticides rid us of unwanted insects. Despite significant achievements in the field of non-conventional methods of pest control, Indian agriculture still depends on chemical pesticides. However, the harmful effects of the pesticides against beneficial insects continue to be a serious problem. The role of insect pollinators, particularly the honeybee in increasing crop productivity, is well documented. It is estimated that about 90 per cent of the pollination in various cross-pollinated crops such as oilseeds, vegetables, fruits and field crops is done by honeybees alone. In addition to pollinating crops, bees are the only source of useful products like honey and beeswax. Unfortunately, honeybees are insects that are greatly affected by insecticides. There are several ways by which honeybees can be killed by insecticides. One is direct contact of the insecticide on the bee while it is foraging in the field. The bee immediately dies and does not return to the hive. In this case the queen, brood and nurse bees are not contaminated and the colony survives. The second more deadly way is when the bee comes in contact with an insecticide and transports it back to the colony, either as contaminated pollen or nectar or on its body. The main symptom of honeybee pesticide kill is large numbers of dead bees in front of the hives. Another symptom is a sudden loss of the colony's field force. After a honeybee pesticide loss, the colony may suffer additionally from brood diseases and chilled brood.

CAUSES OF PESTICIDE POISONING

Most bee poisoning occurs when insecticides are applied to crops during the blooming period. It may also occur due to:

- drift of toxic pesticides onto adjoining crops or weeds that are in bloom;
- contamination of flowering cover crops when orchards are sprayed;

- insecticidal dusts adhere to foraging bees and ultimately become packed with the pollen onto the hind legs. Penncap-M and Sevin are especially dangerous because they may be stored with pollen and kill newly emerged workers the following season;
- bees drinking or touching contaminated water on foliage or flowers; and
- bees collecting contaminated pollen or nectar;

TYPICAL SYMPTOMS OF PESTICIDE POISONING

The most apparent indication of serious pesticide poisoning is the sudden loss of adult bees. This loss is characterized by the appearance of many dead and dying adult bees and sometimes pupae at the colony entrances. However, in many instances, the bees are lost in the field before returning to the colony. If the pesticide is brought back to the hive by the foragers, the nurse bees are killed when they feed on contaminated honey or pollen and the brood will exhibit symptoms of neglect or poisoning. The symptoms of poisoned honey bees often depend on the class of pesticide involved (Table 13.1).

- The most common symptom of bee poisoning is the appearance of excessive numbers of dead bees in front of the hives.
- Another common symptom is lack of foraging bees.
- Aggressiveness in bees may be caused by most pesticides.
- Stupefaction, paralysis and abnormal activities of bees are commonly caused by chlorinated hydrocarbons and organophosphorus insecticides.
- Regurgitation of the honey stomach contents is often caused by poisoning with organophosphorus insecticides.
- Bees may perform abnormal communication dances on the horizontal landing board at the hive entrance while under the influence of insecticide poisoning.
- Disorganized behaviour patterns may lead to lack of recognition of affected field bees by guard bees.
- Many bees poisoned with Sevin or Dieldrin slow down and appear as though they had been chilled; such bees may take two to three days to die. Beekeepers familiar with sevin poisoning quickly learn to recognize the crawlers that move about in front of the hive but are unable to fly. Dead brood in or in front of the hive is typical of sevin, microencapsulated methyl parathion (Penncap-M), or arsenical poisoning.
- One forager returning to the hive with a load of contaminated pollen or nectar can cause extreme agitation and death of a number of bees. Several such foragers can seriously disrupt and damage the colony.
- Often, the queen is superseded because of the agitation of the workers, possibly aggravated by a reduction in the secretion of queen substance. Queens may be affected, especially by slow acting materials such as arsenicals, Sevin and Microencapsulated methyl parathion (Penncap-M) which may be taken into hive with pollen. Queens may behave abnormally; for instance, lay eggs in a poor pattern. Severely weakened or queen less colonies will not live through the following winter.

Table 13.1. Typical symptoms of pesticide poisoning.

Class of pesticide	Examples of pesticides	Symptoms of poisoning
Organo-phosphorous	Parathion, methyl parathion, dimethoate (Cygon), naled (Dibrom), mevinphos (phosdrin), diazinon, dichlorvos (Vapona), monocrotophos (Azodrin), malathion, tepp, phorate (thimet), phosphamidon (Dimecron), chlorpyyrifos (Dursban)	Regurgitation (bees are wet); disorientation, lethargy. Perhaps distended abdomen, erratic attempts to clean selves, tumbling about, paralysis, ultimately death. Wings held away from body but usually remaining hooked together. High percentage of poisoned bees die at colony.
Chlorinated hydrocarbon	Aldrin, chlordane, DDT, dieldrin, heptachlor, lindane, toxaphene	Erratic movements, abnormal activities, trembling. Hindlegs dragging as if paralyzed and wings held away from body but usually remaining hooked together. High percentage of poisoned bees die in field as well as at the colony.
Carbamate	Carbaryl (Sevin), carbofuran (Furadan), aminocarb (Matacil), dimethilan (Dimetilane), mexacarbate, methomly (lannate, Nudrin	Aggressiveness, erratic movements; then inability to fly, stupefaction as though chilled; followed by paralysis, morbidity, death. Most bees usually die at colony. Queens often cease egg laying; hive bees initiate supersedure, rearing queens before egg laying resumes
Dinitrophenyl	Dinocap (karathane), DNOC, DNOCHP, DNOSBP	Similar to symptoms of chlorinated hydrocarbons but often accompanied by regurgitation of substances in digestive tract, as is typical of organophosphorous pesticides. Most affected bees usually die at colony.
Botanical	Pyrethrum, allethrin, and the pryethroid: resmethrin, bioresmethrin, bioethanomethrin, and cismethrin; also nicotine, rotenone, ryania, sabadilla	Perhaps regurgitation from highly toxic pyrethroids, together with erratic movements; then inability to fly and stupefaction, followed very soon by paralysis, moribundity, and death. Bees often die between foraging area and colony.

STRATEGIES TO PROTECT POLLINATORS

Plant Growth Stage

Almost all cases of bee poisoning result from insecticides being applied to blooming crops or being allowed to drift onto blooming crops or weeds. Insecticide applications usually are not recommended for blooming crops. When the label permits insecticide application to blooming crops, growers and applicators need to communicate with beekeepers and exercise all reasonable measures to lower the risk to bees.

Location of Apiary/Distance from Treated Fields

The location of your apiary is probably the most important factor in eliminating the risk of pesticide poisoning. The farther colonies are away from fields or orchards that are treated with pesticides, the better chance the bees have against pesticide poisoning. Establish apiaries at least 4 miles from crops being treated with toxic materials and subjected to drift if moving is impossible, covering colonies with a well-ventilated screen to restrict honey bee flight during peak foraging hours may be your only course of action. However, this method has dangerous side effects and could lead to higher mortality than would have occurred from pesticide exposure. Colonies may have difficulty controlling their hive temperature when confined and can easily overheat, so care must be taken. Providing water inside the screen will allow the bees to reduce temperatures. Do not keep the colony covered for more than two days. Covering the colonies with large wet burlap sacks and providing them with shade is another method for preventing overheating. Most honeybees forage within 1-2 miles of their colony, but they can go up to 5 miles when nectar and pollen are scarce near the hive. Honeybee mortality usually decreases as the distance of colonies from treated fields increases. (The most severely damaged colonies are usually those closest to the application site.) Most foraging activity occurs within one to two mile radius of the hive; however, during periods of pollen or nectar shortage, bees forage at greater distances, and colonies up to five miles from the treated area can be injured.

Relative Toxicity of the Chemical

Pesticides vary in their toxicity to honeybees. Most fungicides and herbicides are not toxic to honeybees and can be used without endangering them. Some insecticides and miticides are not hazardous to bees and can be applied without risks of bee injury. For example, *Bacillus thuringiensis* (sold under the brand names: Dipel, MVP, Thuricide, Biobit, etc.) is a biological insecticide that is not toxic to bees. Insecticides that are moderately toxic to bees can be applied when bees are not actively foraging. Synthetic pyrethroid insecticides will kill the bees present during application, but then, many of them exhibit repellency to bees. Bee losses caused by pyrethroids are usually moderate. Insecticides that are highly toxic to honeybees cannot be applied to blooming crops when bees are present without causing serious injury.

Choice of Formulation

Different formulations of the same insecticide often vary considerably in their toxicity to bees. Granular insecticides generally are not hazardous to honeybees. Dust formulations (seldom used today on commercial field crops) are typically more hazardous than emulsifiable concentrates, because they adhere to the bee's body hairs and are carried back to the beehive. Wettable powder and flowable formulations essentially dry to a dust-like form and also can be carried to the hive by foragers. Likewise, microencapsulated insecticides can be collected by foragers along with pollen and carried back to the beehive. Honeybee exposure to insecticides that kill foraging bees in the field reduces honey production, but colonies recover as young bees emerge. Exposure to dust, wettable powder, flowable and microencapsulated insecticide formulations can cause severe losses of both foraging bees and hive bees. In the worst cases, toxins may remain active in the hive for several months and prevent colonies from recovering from the injury.

Method of Application

The method of application can also change the risk of pesticide poisoning. Aerial applications (Fig. 13.1) have the highest potential risk for causing bee kills. Most bee kills occur when the

Fig. 13.1. Aerial spraying.

pesticide drifts or moves from the target area into the apiary or onto crops attractive to the bees. The outcome of drift can be catastrophic. Spraying during windy days greatly increases the risk of drift. Using granular formulations, soil treatments or equipment that confines the spray to the intended target can help reduce the risk of drift from pesticides. When the target crop is not blooming or is unattractive to bees, insecticide drift can cause significant bee poisoning if drift reaches adjacent flowering crops or weeds. In general, sprays should not be applied if wind speed exceeds 10 mph and favours drift toward flowering crops or weeds. While pesticides should not be sprayed over colonies or allowed to drift onto them, drift over colonies rarely causes bee poisoning. When evaluating potential drift hazards, focus on drift to blooming plants.

Residual Action

Residual activity of an insecticide is an important factor in determining its safety to pollinators. An insecticide that degrades within a few hours can generally be applied with minimal risk when bees are not actively foraging. Applying insecticides with extended residual activity – more than 8 hours – when bees are not actively foraging will not prevent bee injury if bees visit the crop during the period of residual activity. Insecticides with extended residual activity merit extra precaution to prevent bee exposure. Penncap-M (microencapsulated methyl parathion) with a residual period of 8 days and Furadan F (flowable carbofuran) with a residual period of 14 days fall into this classification.

Time of Application

Timing of application is related to all the previously mentioned factors. Again, the most critical factor is to control pests prior to bloom or after bloom is complete, when possible. Evening application of a short-residual insecticide can greatly reduce the potential for bee injury.

Temperature

Temperature can have a substantial effect on the bee poisoning hazard. If temperatures after treatment are unusually low, insecticide residues can remain toxic to bees for much longer than if normal temperatures prevail. Conversely, if high temperatures occur during late evening or early morning, bees may forage during these times.

Lack of Communication

Reducing insecticide injury to honeybees requires communication and cooperation between beekeepers, farmers and applicators. It is important that beekeepers understand cropping and pest management practices used by farmers near their apiaries. Likewise, insecticide applicators should be aware of apiary locations, have a basic understanding of honeybee behaviour and learn which materials and application practices are hazardous to bees. It is unlikely that all bee poisonings can be avoided entirely; however, in most cases, bee losses can be prevented by knowing the hazards and maintaining effective communication.

STEPS APPLICATORS CAN TAKE TO REDUCE RISKS TO HONEYBEES

Do Not Treat Fields in Bloom

Be especially careful when treating crops are highly attractive to bees. Insecticide labels carry warning statements about application during bloom. Always read and follow the label. Examine fields and field margins before spraying to determine if bees are foraging on flowering weeds. Where feasible, eliminate blooming weeds by mowing or tillage prior to insecticide application. It is also important to be aware that many plants offer pollen and nectar only for a few hours

Fig. 13.2. Do not treat fields in bloom.

each day, and that fields should be scouted for bees at the same time of day as the anticipated insecticide application.

Choose Short Residual Products and Low Hazard Formulations

If insecticides must be applied during the flowering period to save a crop, select the least hazardous option. Notify local beekeepers as far in advance as possible. Avoid spray drift. Give careful attention to the position of blooming crops and weeds relative to wind speed and direction. Changing spray nozzles or reducing pressure can increase droplet size and reduce spray drift.

Apply Insecticides When Bees are not Foraging

Some insecticides can be applied in late evening or early morning (i.e., from 8 p.m. to 6 a.m.) with relative safety. In the case of corn, bees collect pollen from tassels in the early morning and are not present in the afternoon or evening hours. Short residual materials applied from late afternoon until midnight do not pose a bee hazard in corn fields if blooming weeds are not present.

Adjust Spray Programs in Relation to Weather Conditions

Reconsider the timing of insecticide application if unusually low temperatures are expected that night. Cool temperatures can delay the degradation process, and cause residues to remain toxic to bees the next day. Cease applications when temperatures rise and bees re-enter the field in early morning.

Contact Local Beekeepers and Obtain Locations of Bee Yards

It requires that apiaries be clearly identified with name, address and phone number of the beekeeper. Identification may appear on one or more colonies, or a separate sign may be posted in the apiary. If colonies are present in an area that will be sprayed with a bee-toxic insecticide, contact beekeepers in time for them to protect or move the colonies. Many pesticide applications pose minimal risk to bees, and beekeepers may choose to accept some risk rather than move colonies.

Read the Pesticide Label

Carefully follow listed precautions with regard to bee safety.

Maintain Bee Forage Areas

Providing forage in uncultivated areas will benefit both wild and managed bees. Many other beneficial insects, such as parasitic wasps, will also benefit from the availability of nectar and cover. Intensive agriculture often increases bee dependence on cultivated crops for forage. Encouraging bee forage plants in wild or uncultivated areas will reduce bee dependence on crop plants that may require pesticide treatments.

STEPS BEEKEEPERS CAN TAKE TO PROTECT THEIR COLONIES

Choose Low Hazard Apiary Locations

Do not locate bees adjacent to crops that are likely to be sprayed with an insecticide. Choose apiary sites that are relatively isolated from intensive insecticide applications and not normally subjected to drift. Keep bee hives out of low areas. Establish holding yards of honey bee colonies at least four miles from orchards being treated with insecticides. Keep hives out of fields treated with the more hazardous insecticides for at least 36 hours after the application. Tests have shown that about 90 per cent of bee mortalities occur within 24 hours after application.

Know the Risks

Many crop pests can be controlled without endangering bees. Attend crop pest management training sessions to stay informed about crop pests and control measures used by growers and applicators. These sessions also provide an opportunity to establish communication links with growers and applicators. Learn about pest control problems and programs so that you can develop mutually beneficial agreements with growers concerning pollination service and prudent use of pesticides.

Maintain Positive Working Relationships with Applicators

Risk management decisions can best be made when both parties understand each other's needs. Establish a communication link prior to the spray season rather than during peak activity when both parties are busy.

Register Apiaries and Post Identification

Register your apiaries with the State Department of Agriculture or any other agency that assists applicators seeking to contact beekeepers. Post your name, address and telephone number in a conspicuous place in your apiaries. Mark colonies of bees that are next to orchards or fields that may be treated. Post your name, mailing address, email address, and phone number in printing large enough to be read at some distance in all apiaries so you can be contacted readily to move the colonies before hazardous insecticides are applied.

Be Prepared to Protect Colonies if Necessary

Frequently, pests can be controlled without putting honeybees at risks. If pest control measures are necessary but carry unacceptable risks, know the options for protecting colonies and be prepared to implement them. These options include:

- Briefly confine bees to their hive with wet cloth when products with short residual activity are applied. This measure is feasible only feasible if a small number of colonies are involved and if the confinement period is brief and early in the morning. Caution! This measure can result in colony injury due to overheating and should be used with care.
- Disrupt foraging activity temporarily when short residual materials are applied by removing colony covers and offsetting boxes so as to increase colony exposure. This will result in a temporary reduction in foraging. Most bees will remain in the hive to protect their stores and maintain nest conditions in the exposed nest. After a few hours to one day, colonies will adjust to the change and resume foraging. This method is safer than confining colonies but is not recommended if bees are located in or adjacent to fields that will be treated.
- Move bees to another location at least four miles from the treated area when highly toxic products with extended residual activity are applied to blooming crops. Moving populous colonies during hot weather can result in considerable bee mortality and should be avoided if possible. Moves should be made in the early morning hours when temperatures are cool and the bees are least active.

Provide Educational Resources to Growers and Applicators

There are many ways to alleviate bee poisoning. Often, severe losses can be avoided by relatively simple modifications of pest control programmes. Talk with growers and applicators about how to reduce bee injury, and provide them with reference materials on protecting bees when you establish apiaries in their areas.

What the Grower Can Do

- Mow or beat down orchard cover crops before applying sprays hazardous to bees. Treatment with 2,4-D is the best way to remove dandelion blooms. This is especially important in relation to the first cover spray on apples, which is applied during a critical foraging period when bees will fly several miles to obtain pollen and nectar from even a few blooms of dandelion, mustard, etc.
- Blossom-thinning sprays have not been hazardous to bees in orchards. However, Sevin used as a fruit thinner 15 to 25 days past full bloom of apples is highly hazardous if cover crop blooms become contaminated. When insect pests have been damaging a crop every season, use a preventative program of early season application before the insect populations increase or before foliage growth and weather conditions reduce the effectiveness of insecticides.
- Learn the pollination requirements of the crops you raise. Application of insecticides hazardous to bees on these crops, or driving beekeepers out of your area by the use of insecticides on other blossoming crops will likely cause lower crop yields.
- Learn about the beekeeper's problems with the poisoning of bees and enter into mutually advantageous agreements with him or her to best produce bee-pollinated crops.

CONCLUSION

It is unlikely that all bee poisonings can be avoided entirely; however, in most cases, bee losses can be prevented by knowing the hazards and maintaining effective communication. By applying suitable preventive measures we can reduce the risk of pesticide poisoning to bees who are the friends of mankind since thousands of years as besides pollinating crops they also produce honey and beewax. So it is imperative to save these pollinators from the hazards of pesticides. There are many ways to alleviate bee poisoning. Often, severe losses can be avoided by relatively simple modifications of pest control programmes. Talk with growers and applicators about how to reduce bee injury, and provide them with reference materials on protecting bees when you establish apiaries in their areas.

14
Prospects of Biocontrol Agents against Chickpea Wilt

Sonika Jamwal[1], Anamika Jamwal[2] and J.S. Jamwal

Department of Plant Protection, Allahabad Agricultural Institute-Deemed University, Allahabad–211 007

India grows a variety of pulse crops under a wide range of agro-climatic conditions and is recognized globally as a major player in pulses production contributing 2.5% to the global production (Ali and Kumar, 2005). Chickpea (*Cicer arietinum* L.) is the most important Rabi pulse crop cultivated in India covering 7.01 million hectares with a production of 6.12 million tones, which is a significant contribution to pulse production, accounting 30.1% of the total area and 38.5% of pulses production (Asthana and Mishra, 1999). Legumes, in general, contain albumin, globulin, prolamin, gluten and free nitrogen (Singh *et al.*, 1992). Generally, chickpea is grown under rainfed situations, but responds to variable irrigation (Chundawat *et al.*, 1976). Pulses are an important source of protein in vegetarian diets. Being a leguminous crop, chickpea possesses root nodules, which fix atmospheric nitrogen, a process requiring energy, but add up to 30 kg N/ha to the soil, improving soil fertility, and thus is not dependent on synthetic nitrogen (Ahlawat *et al.*, 1997, 1998). Out of the total biological nitrogen fixed on global basis, per year more than 75% (about 40 × 10^6 tonnes per year) is contributed by the pulse-rhizobium association (Kush and Mishra, 1981).

DISEASES OF CHICKPEA

Chickpea is prone to several diseases, viz. wilt, collar rot, dry root rot, ascochyta blight, grey mould, rust, etc. Wilt is a serious disease of chickpea caused by *Fusarium oxysporum* f. sp. *ciceri* causing considerable damage to the crop. It is most common disease of eastern part of Uttar Pradesh. Wilt is also a major constraint to the chickpea cultivation in India and in the Mediterranean region. Losses due to wilt may be 10-40% (Singh, 2000).

Present Address
1. Dryland Research Sub-Station, SKUAST-Jammu, Rakh Dhiansar–181 133 J&K
2. Division of Plant Pathology, SKUAST-Jammu–180 002 J&K

Fusarium Wilt

F. oxysporum f. sp. *ciceri* is one of the most common fungi occurring in all types of soils throughout the world (Burgess, 1981). *F. oxysporum* f. sp. *ciceri* forms three kinds of asexual spores viz. micro-conidia, macro-conidia and chlamydospores (Sharma, 2000).

Fusarium spp. are most prominent among the vascular wilt fungi. Wilt occurs as a result of the presence and activities of the pathogen in xylem vessels of the plant. There are two theories regarding the wilt: toxin theory (Gaümann, 1953, 1954, 1957) and plugging theory (Ludwig, 1952).

Mode of Infection

The mycelium of *Fusarium* invades the vascular tissue and finally blocks the xylem vessels that adversely affects the translocation, produces some toxic secretion in the vessels of the host which might also be the cause of wilting. The mycelium consists of branched septate, often colourless hyphae, which turns brown at maturity and produces toxic secretion in the xylem vessels that ultimately blocks the vessels and results into wilting of the host plant. The pathogen is a soil inhabitant, and between crops it survives in the infected plant debris in soil as mycelium and sometimes on contaminated farm equipment. When healthy plant grows in contaminated soil, the germ tubes of the spores or the mycelium penetrates root tip or enters through wounds or at the point of formation of lateral root. The mycelium advances through the root cortex intercellularly, and when it reaches the xylem vessels, it enters them through the pits. The mycelium then remains exclusively in the vessels and travels through them mostly upward towards the stem of the plants. While in the vessels, branched mycelium produces micro-conidia, which germinate at the point where their upward movement is stopped the mycelium penetrates the upper wall of the vessels and more micro-conidia produced in the next vessels. The mycelium also advances laterally into the adjacent vessels, penetrating them through the pits, vessels clogging by mycelium, spores, gels, gums, tyloses and cruising vessels by proliferation in adjacent parenchyma cells, is responsible for the breakdown of the water economy of the infected plants.

Symptomatology

The older leaves become greyish yellow between the veins, and then die and fall to the ground. The heart leaves though still attached to the plant, curl inwards. Collar region is discoloured and infected plants can be uprooted easily and the lateral roots weakened by infection, remain in soils. Vascular cells are discoloured. A transverse section of the root shows some brown vascular rings. Mature plants are seldom killed but yield loss can nevertheless be considerable. The damage may kill seed plants.

Disease Management

Control of plant disease, particularly those which are soil-borne in nature, through biological agents has now become an integral part of plant disease control measures mainly due to the hazardous effect of chemicals, which are not a permanent solution for soil-borne diseases. Only recently, the importance of biological control of wilt in chickpea has been realized. Relatively little work has been done in this direction. But the studies made at Pantnagar and All India Coordinated Research projects on chickpea centres have clearly indicated that bio-control agents like *Trichoderma viride*, *Trichoderma harzianum Gliocladium virens*, *Bacillus subtillis* and *Pseudomonas fluorescens* have tremendous potential for the control of soil-borne pathogens.

The study on wilt of chickpea was conducted at the A.A.I.D.U. Allahabad. Various bio-control agents, chemicals and plant extracts were tested against chickpea. Chemicals inhibited the test fungus more effectively as compared to the biocontrol agents, followed by the plant extracts. But the application of chemicals alone is not an answer to eradicate the soil-borne diseases because chemicals are lethal also to non-target soil microflora and disturb the ecological equilibrium of soil.

(1) It increases the risk of human disease.
(2) It decreases the number of defenders as natural enemies
(3) It has temporary effects.

For avoiding these ill effects, biocontrol agents are used that are eco-friendly, non-poisonous and do not adversely affect the crop.

Biological Control

Biological control has been defined by Baker and Cook (1974) as the "reduction of inoculum density and disease producing activity of the pathogen or parasite in its active or dormant state, by one or more organisms, accomplished naturally or through manipulation of the environment host or by mass production of one or more antagonists". According to Lupton (1984), biological control is "accelerating or diverting evolutionary process in order to obtain genotypes adaptable to man's needs are a most important example of the application of biological control to agricultural and horticultural crops." Being safe and non-hazardous, biological methods of plant disease management have gained popularity in the past few years.

The important genera of fungi exploited as biocontrol agents are the species of *Trichoderma, Gliocladium, Aspergillus, Chaetomium, Penicillium, Neurospora, Catenaria, Paecilomyces, Glomus* etc. (Chaube *et al.,* 2001).

Known Biocontrol Agents

Fungi and bacteria with known biocontrol activity against soil-borne fungal pathogens, including fungal biocontrol agent *Trichoderma viride* and bacterial biocontrol agent *Pseudomonas fluorescens* were used in screening tests against *Fusarium oxysporum* f. sp. *ciceri.*

Fungal Biocontrol Agents

Among fungal biocontrol agents, species of *Trichoderma* have been very widely explored and recommended against many soil-borne fungal pathogens viz. *Sclerotium rolfsii, Rhizoctonia solani, Pythium* spp., and *Fusarium* spp. (Hader *et al.,* 1979a; Sivan and Chet, 1986).

USE OF FUNGAL ANTAGONISTS

In India, a larger number of highly effective antagonistic fungi have been identified and found effective "*in vitro*" as well as "*in situ*" condition, but rarely in the farmer's field. Talc based formulation of *T. viride* for dry seed treatment of groundnut, chickpea, cotton, green gram and pigeon pea to control *Macrophomina phaseolina* has been developed (Jeyarajan *et al.*, 1994) and distributed to farmers to cover about 20,000 ha annually. It has been found effective in controlling pepper wilt, capsule rot of cardamom and root diseases of tea (Jeyarajan and Angappan, 1998). *T. viride* is a fungal bio-fungicide. This is used for seed and soil treatment for suppression of various diseases causing fungal pathogens. Seeds of chickpea were dressed with formulation of *T. viride* @ 4 g/kg of seed. Seed treatments were kept in shade for few hours before sowing to dry.

Characteristics of *Trichoderma viride*

Conidiophores hyaline, much branched, non-verticillate, may occasionally display a pyramidal arrangement, phialides single or in-group, hyaline, flask-shaped and inflated at the base, attached to conidiophores at the right angles. Conidia (phialospores) hyaline, 1-celled, borne in small terminal clusters 3 μm in diameter, round or ellipsoidal in shape, smooth or rough walled and grouped in sticky heads at the tips of the phialides. The colour of the conidia is mostly green. *T. viride* belongs to class Hypomycetes and family Moniliaceae (Hawks worth *et al.*, 1983), hyphae septate, hyaline, conidiophores, phialides and conidia are observed. *T. viride* produces chlamydospores.

Possible Mechanism Involved in *Trichoderma* Antagonism

Trichoderma spp. are common fungi in almost every soil and are antagonistic to other fungi. Many isolates produce volatile and non-volatile antibiotics (Dennis and Webster, 1971 a, b). The ability to produce antifungal substances varies with the isolate, even within the same species aggregates (Dennis and Webster, 1971 a).

When no effect of *Trichoderma* can be directly observed or detected, but its controlling activity in natural soil is significant, the possibility of competition (Alexander, 1982) between the biocontrol agent and the pathogen should be considered (Cook and Baker, 1983). Showing successful biological control, the antagonistic *Trichoderma* was found to be a mycoparasitism (Boosalis, 1964; Hadar *et al.*, 1979a; Harman *et al.*, 1980, Chet and Baker, 1981; Chet and Elad, 1982, Elad *et al*, 1983b).

After recognition and attachment, *Trichoderma* grew along or coiled around the hyphae of the phytopathogenic fungus (Elad *et al.*, 1983c; Harman *et al.*, 1980; Henis and Chet, 1975; Wells *et al.*, 1972). Sites of interaction were photomicrographed with infrared film. Bright regions indicating infrared irradiation were observed, apparently due to the high parasitic activity occurring there (Elad *et al.*, 1983e).

The specificity of *Trichoderma* could not be explained by the different activity of these enzymes, since nonantagonistic *Trichoderma* isolates also produce lytic enzymes, although fewer (Elad *et al.*, 1982a). This may, therefore, support the idea that recognition is an important factor in the specificity of *Trichoderma*.

The activities of the enzymes were tested not only in liquid mixed cultures, but their specific effect was also observed using different microscopic techniques. Interaction sites were stained by fluorescein isothiocyanate-conjugated lectins or Calocofluor White M2R New. The appearance of fluorescence indicated the presence of localized cell wall lysis at points of interaction between the antagonist and its host (Elad *et al.*, 1983c).

Mode of Action

The mode of action of *Trichoderma* spp. includes all the attributes of a successful antagonist such as competition for space and nutrition on the same energy source as the pathogen, production of enzymes (glucanase and chitinase) to lyse fungal cell walls, antibiotic and inhibitory metabolite production (Dennis and Webster 1971 a,b,c), predatory effect or direct parasitic action against the pathogen and indirect effect resulting in modification of the environment adverse to the pathogen. Several species of *Trichoderma* have been extensively studied for their biological control against fungal plant pathogens (Sivan and Chet, 1986; Harman, 1991). The antifungal abilities of these beneficial microbes have been known since 1930s and there have been extensive efforts to use them for plant disease control. Species of *Trichoderma* are strongly antagonistic to other fungi. The exact nature of relationship is still not clear, but it appears that they kill other fungi with a toxin and then consume them using a combination of lytic enzymes. This suggests they are actually microbial predators. *Trichoderma* spp. are common in soil (especially waterlogged soil) and decaying plant materials.

Use of bacteria for management of plant diseases and yield improvement increased steadily since the mid-1960s (Baker, 1987).

Characters of *Pseudomonas fluorescens*

Pseudomonas are straight or slightly curved rod 0.5-1.0 µm in diameter and 1.5-5.0 µm in length. They do not produce prosthecae and are not surrounded by sheath. No resting stages are known. They are gram negative and motile by one or several polar flagella.

Bacterial Biocontrol Agent

Pseudomonas fluorescens is a biocontrol agent used against wilt of chickpea. *P. fluorescens* make up a diverse group of bacteria that can generally be distinguished from other *Pseudomonas* by their ability to produce water-soluble yellow green pigment. They are typical Gram-negative chemoheterotrophic motile rods with polar flagella and are grouped in r-RNA homology group I (Palleroni *et al.*, 1973).

Use of Bacterial Antagonists

In India, work on the use of bacterial antagonists for plant disease control started in the 1950s and gained momentum in the 1980s. Firstly *Bacillus subtilis* was used to control *Fusarium* wilt in pigeon pea (Vasudeva *et al.,* 1958). Fluorescent pseudomonads, especially the *P. fluorescens*, have been studied by various workers for their efficacy against diseases of major crops such as cotton, chickpea, pigeonpea, etc. Potential isolates of *P. fluorescens* have been identified. Their efficacy alone, or in combination with other agents, like *Trichoderma* spp., has been studied against wilt diseases of pigeonpea, sunflower, etc. Formulation of *P. fluorescens* with various carrier materials like talc, peat, vermiculite, lignite, kaolinite, etc. has been developed with the talc-based ones giving good results (Vidhyasekaran and Muthamilan 1995).

Suppression of plant pathogens by antagonistic bacteria and their subsequent aid in promoting plant growth is less understood. Stress is now upon isolating antagonists from the right kind of environment. It is advocated to choose a healthy source (e.g., roots) from a disease prone environment for isolating PGPR (Campbell, 1989). The application of microorganisms as agents for biocontrol of plant disease in agriculture is now considered an important alternative to the use of chemicals fungicides. PGPR including certain *Pseudomonas* spp. have been shown to effectively control fungal pathogens like *Fusarium oxysporum, Sclerotium rolfsii* (Hebber *et al.*, 1992), Ganesan and Gnanamanickain, 1987, Vasantha Devi *et al.*, 1989). *Sclerotium rolfsii* Sacc. has a host range of more than 500 plant species (Punja 1985). It causes pre and post emergence damping off/collar rot and wilt of sunflower and is considered as one of the economically important disease of sunflower.

Mode of Action

P. fluorescens suppress by various modes of action, namely competition for nutrients and space, antibiosis by production of various antibiotics, siderophores and lytic enzymes; other mechanisms include production of hydrogen cyanide and degradation of toxins. The bacterium produces pyoverdin and several secondary metabolites with antibiosis properties, e.g. HCN, Pyrolnitrin and 2-4 diacetyl pholoroglucinol (DAPG) and siderophores.

Antagonistic Metabolites Produced by Fluorescent *Pseudomonas*

Fluorescent *Pseudomonas* produces a variety of secondary metabolites with antagonistic characteristics, most of which are nitrogen-containing heterocyclic compounds. Among these are phenazines (pyocyanin and pyrrolnitrine type compounds) and indole derivatives (Leisinger and Margraff, 1966). Other secondary metabolites with antagonistic characteristics are unusual amino acids and peptides, including the iron chelating agents, some of which act as iron carriers (siderophores). Some of the *Pseudomonas* secondary metabolites, however, may act as phytotoxins. The production of both strong siderophores and other antagonistic metabolites is of particular interest in the search for microbial agents for crop production. Pyoverdine and pseudobactin, water-soluble compounds with pronounced and strong siderophore properties produced by fluorescent pseudomonas, have been chemically characterized. Pseudobactin was described as the siderphore of plant growth promoting *P. fluorescens-putida* isolate B10 and claimed to be involved in the mechanism of plant growth promotion (Kloepper *et al.,* 1980 a). Pyoverdine was already detected by Elliot in 1958 and was shown to have an important function in microbial iron transport by Meyer and Abdullah (1978).

Findings at A.A.I.D.U.

T. viride significantly inhibited the growth of *F. oxysporum* f. sp. *ciceri* in solid as well as liquid media by dual culture method. *T. viride* in both the ratios (1:1 and 2:1) resulted in reduced growth of the pathogen. Parasitism and excessive branching of hypha was observed due to *T. viride*. This might be due to the exhibition of antibiosis by *T. viride* under "*in vitro*" tests. Similar findings have been reported by Singh *et al*., (2003) and Sonawane and Pawar (2001). In the "*in situ*" studies Padomodaya (1996) found *T. viride* to be highly inhibitory in suppressing the growth of *F. oxysporum* f. sp. *ciceri* by radial growth and was found to be significantly superior to other isolates after seven days. Under *in vitro* studies, Somasekhara (1998) evaluated twelve *Trichoderma* isolates against *Fusarium udum* wherein *T. viride*, *T. harzianum* and *T. koningi* were found effective.

Pseudomonas fluorescens significantly inhibited the growth of *F. oxysporum* f. sp. *ciceri* in solid as well as liquid media resulting in reduced growth of the pathogens. *P. fluorescens* might have suppressed the pathogen by various modes of action such as competition for nutrients and space, production of various antibiotics, siderophores and lytic enzymes. Other mechanisms include production of hydrogen cyanide and degradation of toxins. The growth of *F. oxysporum* f. sp. *ciceri* was inhibited by *P. fluorescens* on PDA in solid as well as in liquid medium. Percent inhibition was maximum with *P. fluorescens* on PDA. The bacterium produced pyoverdin and several secondary metabolites with antibiosis properties, e.g., hydrocyanic acid, pyrolnitrin and 2-4 diacetyl phloroglucinol (DAPG) siderophores.

Effect of Biocontrol Agents on Wilt Incidence

Seed treatment with talc-based formulation of *T. viride* and *P. fluorescens* (4g/kg seed) not only controlled the wilt incidence, but also increased the number of nodules/plant and maximum grain yield in chickpea (*Cicer arietinum).* The seed treatment completely eradicated the disease and the antagonist moved to the rhizosphere and survived well in it controlling chickpea wilt. It showed profound effect over *F. oxysporum* f. sp. *ciceri* causing chickpea wilt. Many researchers have found species of *Trichoderma* effective against wilt diseases of various crops (Gupta *et al.*, 1988; Zaidi, *et al.*, 1991; Karthikeyan and Karunanithi, 1996; Hemani and Bhatnagar, 1996; Ushamalini *et al.*, 1997; Zaman *et al.*, 1998).

Effect of Antagonist on Plant Growth under Field Conditions

Seed treatment with talc-based formulation of both *T. viride*, *P. fluorescens* has been found effective in increasing the growth of chickpea, i.e., shoot length, root length, shoot weight and root weight production. Talc based formulation of *F. oxysporum* f. sp. *ciceri* + *P. fluorescens* was more effective. Moreover fluorescent *Pseudomonas* produced plant growth promoting substances such as auxins and gibberellins and enhanced the growth of plant and its yield. Enhancement of growth in chickpea plant conforms to the findings of many workers (Defago *et al.,* 1990; Vidyasekaran and Muthamilan, 1995; Kumar, 1999; Rangeshwaran and Prasad, 2000).

CONCLUSION

Biocontrol agents inhibited *Fusarium* wilt. Seed should be dressed with biocontrol agents @ 4 g/kg of seed. Wilt incidence decreased by the use of biocontrol agents; shoot length, root length, shoot weight and root weight also increased due to biocontrol agents.

Different methods for disease management are being developed day by day in India. But it has been realized that a single method of disease control is not at all satisfactory. Therefore, time has come to employ all the suitable methods available so as to keep the disease level below economic threshold, a strategy termed as Integrated Disease Management.

Crop losses vary from year to year in different areas. These variations are primarily due to weather factors and intensive cultivation using high-yielding crop varieties and indiscriminate use of pesticides, which are very toxic for controlling the soil-borne diseases. Pesticides are very hazardous to our flora and fauna. Therefore, to combat crop losses due to diseases is really a challenging task for scientists as well as farmers in India.

Use of pesticides is obviously an immediate option in this sphere, but over dependence and injudicious application of pesticides is a serious concern to our environmentalists, scientists as well as farmers in India. Alternatively, the diseases can be safely controlled by the use of biocontrol agents along with other diseases management practices.

REFERENCES

Ahlawat, I.P.S., Om Prakash and Sain, G.S. (1997-98). Scientific crop production in India, Gram, Importance and utility. pp. 150.

Alexander, M. (1982). *Introduction to Soil Microbiology.* Wiley, New York, pp. 467.

Ali, M. and Kumar, S. (2005). Pulses yet to see a breakthrough. *The Hindu Survey of Indian Agriculture.* 54-56.

Asthana, A.N. and Mishra, J.P. (1999). Pulse–a little impetus needed. *The Hindu Survey of Indian Agriculture* Kasturi and Sons Ltd. National Press Kasturi Building, Chennai, pp. 61-65.

Baker, K.F. (1987). Evolution concepts of Bio-control of plant pathogens. *Ann. Rev. Phytopathol.* **25:** 67-85.

Baker, K.F and Cook, R.J. (1974). Biological control of Plant Pathogens., Freeman, San Fransisco.

Boosalis, M.G. (1964). Hyperparasitism. *Ann. Rev. Phytopathol.*, **2:** 363-376.

Burgess, L.W. (1981). General ecology of the fusaria. *Fusarium* Diseases, Biology and Taxonomy, (Eds. Nelson, P.E., Toussoun, T.A. and Cook, R.J.), University Park PA, USA. pp. 225-235.

Campbell, R. (1989). Biological control of Microbial Plant Pathogens. pp 50-51 Cambridge University Press, Cambridge U.K., pp. 218.

Chaube, H.S., Vishwakarma, S.N., Kaur, A., Akhtar, J. and Anurag, (2001). Biological control of Plant diseases. *Indian Farmers Digest.* 23-29.

Chet, I. and Baker, R. (1981). Isolation and biocontrol potential of *Trichoderma hamatum* from soil naturally suppressive of *Rhizoctonia solani. Phytopathology,* **71:** 286-290.

Chet, I. and Elad, Y. (1982). Prevention of plant infection by biological means. *In: La Selection des Plantes. Bordeaux (France). Colloq. l'INRA,* **11:** 195-204.

Chundawat, G.S., Sharma, R.G. and Shekhawat, G.S. (1976). Effect of nitrogen, phosphorus and bacterial fertilization on the growth and yield of gram in Rajasthan. *Indian J. Agron.* **21:** 127-130.

Cook, R.S. and Baker, K.F. (1983). The Nature and Practice of Biological Control of Plant Pathogens. *The American Phytopathological Society,* St. Paul, MN, pp. 539.

Defago, G., Berling. C.H., Berger. U., Hass. D., Kahr. G., Keel. K., Vaisarel. C., Thner, P. and Wuthricl, B. (1990). Suppression of black root of tobacco and other root diseases by strains of *Pseudomonas fluorescens* potential application and mechanism. In Biological control of Soil Borne Plant Pathogen [Ed. Horn by D,) CAB International Wellingboard, Oxon, U.K. pp. 93-108.

Dennis, C. and Webster, J. (1971a). Antagonistic properties of species-groups of *Trichoderma.* I. Production of nonvolatile antibiotics. *Trans. Br. Mycol. Soc.* **57:** 25-39.

Dennis, C. and Webster, J. (1971b). Antagonistic properties of species-groups of *Trichoderma.* II. Production of volatile antibiotics. *Trans. Br. Mycol. Soc.* **57:** 41-48.

Dennis, C. and Webster, J. (1971c). Antagonistic properties of species-groups of *Trichoderma* III. Hyphal interactions. trans. *Trans. Br. Mycol. Soc.* **57:** 363-369.

Elad, Y., Chet, I. and Henis, Y. (1982a). Degradation of plant pathogenic fungi by *Trichoderma harzianum. Can. J. Microbiol.* **28:** 719-725.

Elad, Y., Barak, R., Chet, I. and Henis, Y. (1983b). Ultrastructural studies of the interaction between *Trichoderma* spp. and plant pathogenic fungi. *Phytopath. Z.* **107:** 168-175.

Elad, Y., Chet, I., Boyle, P. and Henis, Y. (1983c). Parasitism of *Trichoderma spp.* on *Rhizoctoina solani* and *Sclerotium rolfsii.* Scanning electron microscopy and fluorescence microscopy. *Phytopathology.* **73:** 85-88.

Elad, Y., Sadovski, Z. and Chet, I. (1983e). Detection of mycoparasitism by infrared photomicrography. *Microb. Ecol.,* **9:** 185-187.

Elliot, R.P. (1958). Some properties of pyoverdine, the water-soluble pigment of the pseudomonads. *Appl. Microbiol.,* **6:** 241-246.

Ganesan, P. and Gnanamanickam, S.S. (1987). Biological control of *Sclerotium rolfsii* sacc. in peanut by inoculation with *Pseudomonas fluorescens. Soil Biol. Biochem.* **19:** 35-38.

Gaumann E. (1953). Some problems of pathological wilting in plants. *Advan. Enzyzmo,* **11:** 401-435.
Gaumann E. (1954). Toxins and plant diseases. *Endeavour.* **13:** 198-204.
Gaumann E. (1957). Fusaric acid at a wilt toxin. *Phytopathology.* **47:** 342-357.
Gupta, R.P., Katiyat, R.P. and Singh, D.P. (1988). Seed treatment with Bavistin and Rhizobium and its effect on wilt incidence, nodulation and yield of chickpea. *Pesticides–Bombay* **22:** 9-10.
Gupta, S., Srivastava, A.K. and Arora, D.K. (1996). Plant growth promotion and *Fusarium* wilt suppression by herbicides tolerant *Pseudomonas* strains. Proceeding of the second international weed control congress Copenhagen Denmark 1-**4:** 369-374.
Hader, Y., Chet, I. and Henis, Y. (1979a). Biological control of *Rhizoctonia solani,* Damping off with wheat bran culture of *Trichoderma harzianum. Phytopathology.* **69:** 64-68.
Haral, A.K and Konde, R.K. (1983). Biological management of soil borne plant pathogens effects on nodulation and dry matter of chickpea. *J. Maharashtra Agric. Univ.* **8:** 112-113.
Harman, G.E., Chet, I. and Baker, R. (1980). *Trichoderma hamatum* effects on seed and seedling disease induced in radish and pea by *Pythium* spp. or *Rhizoctonia solani. Phytopathology.* **70:** 1167-1172.
Harman, G.E., Jin. X., Stasg. T.E., Peruzzotti. G., Leopold, A.C. and Taylor, A.G. (1991). Production of conidial biomass of *T. harzianum* for biological control. *J. Biol. Control.* **1:** 23-28.
Hawksworth, D.L., Sutlon, B.C. and Ainsworth, G.L. (1983). Ainsworth and Bisky's dictionary of the fungi commonwealth Mycological Institute, Kew, Survey, **7:** 445.
Hebbar, K.P., A.G. and Dart, P.J. (1992). Rhizobium of maize antagonistic to *Fusarium moniliforme,* a soil-borne fungal pathogens, isolation and identification *Soil Biol. Biochem.* **24:** 979-987.
Hemani, B. and Bhatnagar, H. (1996). Influence of environmental conditions on antagonistic activity of *Trichoderma* spp. against *Fusarium udum. Indian J. Mycol. Pl. Pathol.* **26:** 58-63.
Henis, Y. and Chet, I. (1975). Microbial control of plant pathogens. *Adv. Appl. Microbiol.* **19:** 85-111.
Jeyarajan, R., Ramakrishnan. G., Dinakaran, D. and Sridar, R. (1994). Development of products of *Trichoderma viride* and *Bacillus subtilis* for biocontrol of root rot diseases, *In: Biotechnology in India* (Ed. Diwedi). *Bioved Res. Soc., Allahabad.* pp. 25-36.
Jeyarajan, R. and Angappan, K. (1998). Mass production technology for fungal antagonists and field evaluation, pp. 48-56. *In: Biological suppression of Plant Diseases, Phytoparasitic Nematodes and Weeds* (Eds. Singh, S.P. and Hussaini, S.S.), Project Directorate of Biological Control, Bangalore, India, pp. 284.
Karthikeyan, A. and Karunanithi, K. (1996). Influence of organic amendments on the intensity of *Fusarium* wilt of banana. *Pl. Dis. Res.* **2:** 180-181.
Kloepper, J.W., Leong, J., Teintze, M. and Schroth, M.N. (1980a). Enhanced plant growth by siderphores produced by plant growth promoting rhizobacteria. *Nature* **286:** 885-886.
Kumar, B.S.D. (1999). Fusarial wilt suppression and crop improvement through two rhizobacterial strains in chickpea growing in soils infested with *Fusarium oxysporum* f. sp. *ciceri. Biol. Fert. Soils.* **29:** 97-91.
Kush, A.K. and Mishra, A.K. (1981). Wonders of Rhizobium. *Indian Farming.* **31:** 84-86.
Landa, B.B., Navas-cortes, J.A. and Jimenez-Diaz, R.M. (2004). Integrated management of *Fusarium* wilt of chickpea with sowing date, host resistance and biological control. *Phytopathlogy.* **94:** 946-960.
Leisinger, T. and Margraff, R. (1966). Secondary metabolites of the fluorescent pseudomonads. *J. Appl. Bact.* **29:** 470-489.
Ludwig, R.A. (1952). "Studies on the physiology hydromycotic wilting in the tomato plants" Mac Donald coll. *Tech. Bull.* **20:** 1-140.
Lupton, (1984). "The Biology of Symbiotic Fungi" John Wiley and Sons, London, New York, pp. 282.
Meyer, J.M. and Abdullah, M.A. (1978). The fluorescent pigment of *Pseudomonas fluorescens*: biosynthesis, purification and physicochemical Properties. *J. Gen. Microbiol.* **107:** 319-328.
Padomodaya, B. and Reddy, H.R. (1996). Screening of *Trichoderma spp.* against *Fusarium oxysporum* f. sp. *lycopersici* causing will of Tomato. *Indian J. Mycol. Pl. Pathol.* **26:** 266-270.
Palleroni, N.J., Kuniseuoa, R., Contopoulou, R. and Dondoroff, M. (1973). Nucleic acid homologies in genus *Pseudomonas. Int. J. Sys. Bacterial.* **23:** 333-339.
Punja, Z.K. (1985). The biology, ecology and control of *Sclerotium rolfsii. Annal. Rev. Phytopathol.* **23:** 97-127.
Rangeshwaran, R. and Prasad, R.D. (2000). Isolation and evaluation of rhizospheric bacteria for biological control of chickpea wilt pathogens. *J. Biol. Control.* **14:** 9-15.
Rangeshwaran, R., Wasnikar, A.R., Prasad, R.D., Anjula, N. and Sunanda, C.R. (2002). Isolation of entophytic bacteria for biological control of wilt pathogens. *J. Biol. Control.* **16:** 125-133.
Sharma, O.P. (2000). Deuteromycotina. Textbook of fungi. pp. 263.
Sharma, O.P. (2000). Textbook of fungi. pp. 269-271.
Sharma, O.P. (2000). Toxins produced by fungus. Textbook of fungi pp. 27.

Shukla, P., Singh, R.R. and Mishra, A.N. (1981). Search for best seed dressing fungicides to control chickpea wilt. *Pesticides* **15**(7): 15-16.

Singh, B.K., Mukesh-Srivastava and Udit-Narain (2003). Evaluation of bio-agents against *Fusarium oxysporum* f. sp. *ciceri* causing chickpea wilt. *Farm-Science J.* **12:** 48-49.

Singh, R.S. (2000). Management of the disease. *Plant Disease* **5:** 481-482.

Singh, S.K., Rahman, S.J., Gupta, B.R. and Kalha, C.S. (1992). Integration of pesticides application schedules for disease and insect pest management in chickpea under dry land conditions. *Indian J. Pl. Protec.* **2:** 158-161.

Sivan, A. and Chet, I. (1986). Biological control of *Fusarium* spp. in cotton, wheat and muskmelon by *Trichoderma harzianum. J. Phytopathol.* **118:** 39-47.

Somasekhra (1998). Evaluation of *Trichoderma* isolates and their antifungal extract as potential biological control agents against pigeon pea wilt pathogen. *Current-Research,* Univ. of Ag. Sci. Bangalore. **27:** 158-160.

Sonawane, S.S. and Pawar, N.B. (2001). Studies on biological management of chickpea wilt. *Journal of Maharashtra Agril. Univ.* **26:** 215-216.

Ushamalini, C., Rajappan, K. and Kousalya, Gangadharan (1997). Control of cowpea wilt by non-chemical means. *Pl. Dis. Res.* **12:** 122-129.

Vasantha Devi, T., Malarvizhi, R., Sakthivel, N. and Gananamanilkam, S.S. (1989). Biological control of sheath blight of rice in India with antagonistic bacteria. *Plant and Soil* **11a:** 325-330.

Vasudeva, R.S., Jain, A.C. and Nema, N.G. (1952). Investigations of the inhibitory action of *Bacillus subtilis* on *Fusarium udum* Butler, the fungus causing wilt of pigeon pea (*Cajanus cajan*). *Ann. App. Biol.* **39:** 229-238.

Vasudeva, R.S., Subbaiah, T.V., Sastry, M.L.M. Rangaswamy, G. and Ayenger, R.S. (1958). Bulbiformin, an antibiotic produced by *Bacillus subtilis. Ann. App. Biol.* **46:** 336-345.

Vidyasekaran, P. and Muthamilan, M. (1995). Development of formulations of *Pseudomonas fluorescens* for control of chickpea wilt. *Pl. Dis.* **79:** 782-786.

Wells, H.D., Bell, D.K. and Jawarski, C.A. (1972). Efficacy of *Trichoderma harzianum* as a biological-control for *Sclerotium rolfsii. Phytopathology,* **62:** 442-447.

Zaidi, S.B.I., Khan, M.L. and Saxena, S.K. (1991). Effect of fungicides on mycoflora of chickpea seeds. *Indian Phytopath.* **44:** 399-395.

Zaman, M.A., Saleh, A.K.M. and Mehman, G.M.M. (1998). Seed-borne fungi of mustard and their control with indigenous plant extracts. *Bangladesh J. Pl. Pathol.,* **13:** 25-28.

15

Antifungal Activity of Essential Oils of *Ocimum sanctum* on *Rhizoctonia solani* and *Fusarium oxysporum*

Deb Prasad Ray

Fertiliser Control Laboratory, Paschim Medinipur, West Bengal, India

The light yellow coloured essential oil has been extracted by steam distillation from the Tulsi (*Ocimum sanctum*) plant leaves yielded 0.45%. The oil was subjected to TLC, GLC and NMR analysis. The terpenes, α-pinene, limonene, phellandrene, linalool, Eugenol, methyl eugenol, Caryophyllene oxide and two Sesquiterpene alcohols have been identified from the oil. The screening of the essential oil for bioassay was done against the two fungal species *Rhizoctonia solani* and *Fusarium oxysporum*. The oil exhibited excellent fungitoxicity against the test fungi showing complete inhibition of their mycelial growth. The minimum inhibitory concentration of ocimum oil was found to be 0.025% against *R. solani* and 0.050% against *F. oxysporium*. The oil was fungistatic in nature at its respective minimum inhibitory concentration against both the test fungi, but turned fungicidal at higher concentration.

Plants like every living being in earth have the defense mechanisms. Plant chemicals or botanicals are often referred to as either primary or secondary metabolites. Primary metabolites because of their roles in basic cell metabolisms are needed for physiological development of a plant, whereas secondary metabolites are produced as a result of defense mechanism evolved by the plant against various natural enemies. Several higher plants rich in secondary metabolites have shown success in plant disease control and are proved to be harmless and non-phytotoxic unlike chemical fungicides. In the present study, an effort has been made to study the fungicidal property of the essential oil obtained from Tulsi (*Ocimum sanctum*) plant.

Tulsi (*Ocimum* sp.) is a very much well-known plant of family Labiatae and has found respectable place in Ayurvedic system of medicine for treatment of wide range of diseases. The essential oil obtained from the leaves of the plant is a rich source of terpenes. The study of population of this species raised from seeds revealed that the species is highly heterozygous and possesses a rich genetic variability for aroma chemicals and important morphological characters (Khosla *et al.*, 1980). The seeds are diuretic, tonic and antiseptic and are used for treatment of constipation and piles (Gupta, 1992). The plant has unique germicidal and antiseptic activity (Nene *et al.*, 1968).

The essential oils are obtained through steam distillation of leaves and terminal branches of *Ocimum sanctum* plant. The oil percentage was calculated on moisture free basis. The chemical analysis of the oil was carried out by means of thin layer chromatograpy (TLC), gas liquid chromatography (GLC), column chromatography and NMR in the way described by Sobti *et al.*, (1977, 1978). The GLC was studied on a Hewlett Packard 5890 A Series-II fitted with SE-30, 2 m long column at a temperature ambient of 110°C and FID using N_2 as the carrier gas. The presence of various compounds in the oil was confirmed on the gas chromatogram by comparison of the retention times with those of the reference samples. The compounds of the chromatogram were thus identified.

In vitro tests were conducted to evaluate the fungitoxicity of the plant against *R. solani* and *F. oxysporium*. The fungitoxicity was calculated following *Dixit et al.* (1976) and recorded in terms of percentage of mycelial inhibition of growth of test fungi. The fungi were grown on Potato Dextrose Agar (PDA) poured Petri plates and the inoculum of 4 mm diameter was used to inoculate the Petri plates treated with three different concentrations. Approximately 30 mL of potato dextrose agar medium having temperature between 60-70° C after autoclave was mixed with the oil in four concentrations of 0.1, 0.050 and 0.025 and 0.010 % (v/v) (Srivastava *et al.*, 1993). The oils were mixed with the medium thoroughly before pouring into the sterile Petri plates of 90 mm diameter. In the same way, the three concentrations were mixed and poured into the sterile Petri dish in triplicates. After 48 hrs of incubation, these plates were inoculated with equal size of inoculum of 4 mm diameter of each of the fungi. All the inoculated plates were incubated at 25 ± 1°C. The final observations were recorded after sixth days. The mycelial growth of the fungus in treated (T) and control (C) Petri plates was measured diametrically in three different directions. From the mean growth of the above readings per cent inhibition (I) was calculated. The minimum inhibitory concentration of Ocimum oil was determined by the poisoned food technique (Ray *et al.*, 2002) by using the following formula:

$$\text{Per cent omjobotopm} = \frac{C-T}{T} \times 100$$

where, C = Diameter of control treatment

T = Diameter of test compound

The per cent inhibition was converted to corrected per cent inhibition (CPI) using the following formula:

$$\text{CPI} = \frac{\text{Per cent inhibition} - \text{C.F.}}{100-\text{C.F.}} \times 100$$

where, $$\text{C.F.} = \frac{C-T}{T} \times 100$$

C.F. = Correction factor

X = Diameter of Petri dish

ESSENTIAL OIL

The essential oils are obtained through steam distillation of leaves and terminal branches of *Ocimum sanctum* plant. For extraction of oil, the plant samples were then taken into a Clevenger apparatus and extracted in dichloromethane for 12 hrs. To make moisture free, the solvent extract was dried over anhydrous sodium sulphate bed and subsequently evaporated under vacuum. The light yellow coloured essential oil with pleasing spicy odour from the Tulsi (*Ocimum sanctum*) plant leaves yielded 0.45%.

CHEMICAL ANALYSIS OF THE OIL

The oil on TLC examination showed the presence of eugenol, methyl eugenol and sesquiterpenes as major constituents. The amount of eugenol was estimated by shaking 5 ml of oil with 5% KOH in a Erlenmeyer flask, which was found to vary from 9 to 55%. The eugenol free oil was extracted and examined by TLC in two different solvent systems. The compounds identified in the eugenol free oil by CoTLC were eudesmol, linalool, caryophyllene oxide, and methyl eugenol.

A total 11 peaks were recorded in gas chromatograph. The compounds along with their percentages are given in Table 15.1. Eugenol was the major constituent of the oil contributed 55.26%. The next highest constituents were sesquiterpene alcohols (16.39%). This was followed by methyl isoeugenol (10.22%), caryophyllene (8.26%) and linalool (5.23%). Other compounds present in the oil were in traces such as phellandrene (1.24%), limonene (0.71%) and α-pinene (0.38%). Three compounds were unidentified from the peaks. In the investigated oil, eugenol was the major constituent that was quite similar to the result reported by Khosla *et al.*, 1980 in other species of *Ocimum* oil. However, the variation of other constituent in the chemical composition of the oil may be due to the change in climatic condition in which the plants are grown.

Table 15.1. Compounds identified from the essential oil of *Ocimum sanctum* by gas chromatography.

Peak No.	Compound identified	Percentage present in the oil
1.	α-pinene	0.38
2.	limonene	0.71
3.	phellandrene	1.24
4.	unidentified	0.36
5.	linalool	8.23
6.	eugenol	55.26
7.	methyl isoeugenol	10.22
8.	caryophyllene	5.26
9.	unidentified	0.76
10.	unidentified	1.19
11.	sesquiterpene alcohol	16.39

During the screening of the essential oil against the two fungal species, the oil exhibited a very good fungitoxicity against the test fungi showing complete inhibition of their mycelial growth. In Table 15.2, the per cent growth inhibition of *Rhizoctonia solani* and *Fusarium oxysporium* and their corrected per cent of growth inhibition has been presented. The oil was fungistatic in nature at its respective minimum inhibitory concentrations against both the test fungi, but turned fungicidal at higher concentration. The *in vitro* fungal growth was carried out at the level of concentration ranging from 0.010% to 0.500%. The initial growth in the Petri dish was quite slow at the low concentration. From the experiment it was revealed that the minimum inhibitory concentration of *Ocimum* oil was found to be 0.025 ppm against *R. solani* and 0.050 ppm against *F. oxysporium*. In higher concentration the growth was rarely seen, which revealed the fungitoxic nature of the *Ocimum* oil. The result was quite similar to the observation made by Ray *et al., (*1999*)*. From the oil of Tagetes against the Test fungi *R. solani*. However, *Ocimum* oil was found to be more effective at the higher concentration. This may be correlated to the chemical constituent of the oil. The sesquiterpenes and eugenol are found to be fungistatic at lower concentration and fungitoxic at the higher concentration. This result emphasizes the preparation of antifungal formulation from the *Ocimum* oil.

Table 15.2. Per cent growth inhibition of *Rhizoctonia solani* and *Fusarium oxysporium* while tested against the essential oil of *Ocimum sanctum*.

Conc (%)	**Per cent growth inhibition**		**Correction factor**	**Corrected per cent age of inhibition**	
	Rhizoctonia solani	***Fusarium oxysporium***		***Rhizoctonia solani***	***Fusarium oxysporium***
0.010	20.00	10.50	6	14.89	4.79
0.025	30.24	17.21	6	25.79	11.93
0.050	40.25	25.32	6	36.44	20.55
0.100	78.12	48.56	6	76.72	45.28
0.500	98.32	85.36	6	98.21	84.43

The findings of the present study revealed that the oil exhibited that the oil of *Ocimum sanctum* may constitute an ideal storage of fungitoxicant for protection of seed and soil-borne diseases. Since the plant grows luxuriantly all over India, its essential oil would provide an indigenous, cheap and renewable source of plant protection in place of synthetic chemicals, most of which exhibit adverse effects. Further studies on the shelf-life and evaluation of other fungal species may find a wide arena into the pest control world.

ACKNOWLEDGEMENTS

The author is thankful to the Agricultural Chemist, Fertilizer Control Laboratory, Midnapore and Joint Director of Agricultural (Chemistry), Government of West Bengal for their kind helps during the experimental work.

REFERENCES

Dixit, S.N., Tripathi, S.C. and Upadhyay, R.R. (1976). The Antifungal substance of rose flower (*Rosa indica*). *Economic Bot.* **30:** 371-374.

Gupta, S.C. (1992). Population study and improvement of yield characters in *Ocimum sanctum* Linn. *Indian Perfumer.* **36**(4): 289-292.

Khosla, M.K., Pushpagandhan, P., Thapa, R.K. and Sobti, S.N. (1980) Search for New Aroma Chemicals from *Ocimum* species III: Studies on genetic variability for essential oil and other allied characters of South American species *O. micranthum* Wild. *Indian Perfumer* **XXIV**(3): 148-152.

Nene, Y.L., Thapliyal, P.N. and Krishan Kumar, (1968). Screening of some plant extracts for antifungal properties. *J. Sci. Tech*. **63**(4): 226-228.

Ray, D.P. (1999). Phytochemical insect antifeedant and antinemic screening of marigold (*Tagetes erecta* L.) flower. M.Sc. Thesis, PG School, Indian Agricultural Research Institute, New Delhi, India.

Ray, D.P., Singh, R.P. and Kumar, B. (2002). Composition and antifungal activity of African marigold flower. *Bangladesh J. Agri. Res.* **27**(4): 709-717.

Sobti, S.N., Pushpagandhan, P. and Atal, C.K. (1977). *Indian Perfumer* **21**(1): 13-16.

Sobti, S.N., Pushpagandhan, P., Thapa, R.K., Aggarwal, S.G., Vashist, V.N. and Atal, C.K. *Lioydia* **41**(1): 50-55.

Srivastava, S., Naik, S.N. and Maheswari, R.C. (1993). *In vitro* studies on antifungal studies of Palmarosa and Eucalyptus oils. *Indian Perfumer.* **37**(3): 277-279.

16

Application of Biocontrol Agents for Management of Soil-borne Pathogens

Anamika Jamwal, Sonika Jamwal, V.S. Verma and J.S. Jamwal

Sher-e-Kashmir University of Agricultural Sciences and Technology, Jammu–180 002, J&K

During the past 20 years more attention than ever has been given to the development of biological methods to control plant diseases. Indeed the concern for food of high quality, without residues of pesticides and for a sustainable agriculture that will preserve the fertility of soil and eco-friendly environment has stimulated research dealing with biological control. Seedlings of economically important crop plants are attacked by various soil-borne pathogenic fungi, such as *Fusarium oxysporum*, *Rhizoctonia solani*, *Sclerotium rolfsii, Pythium* spp., *Phytophthora* spp. and others causing diseases like seed rot, seedling rot, stem rot, etc. These diseases are often termed as pre- and post-emergence damping off, seedling blights and wilt. Disease problem is one of the major constraints for achieving higher production in agricultural crops. Every year these diseases result in considerable economic losses not only in India but throughout the world. India loses about 30% of its crops due to pests and diseases each year. The damage due to these diseases is estimated to be Rs. 60000 crores annually. Currently, the most recommended and common method for reducing or suppressing the soil-borne pathogens is the use of chemical fungicides. Chemical control of diseases require continuous use of fungicides, most of which have now been banned by WHO due to their severe toxicity and recalcitrant nature. However, quite often the indiscriminate and unscientific use of chemicals like insecticides, fungicides and herbicides, which are being used regularly, has led to many problems such as development of pathogen resistance to fungicides, mutations of soil micro-flora, inability of seed treated fungicides to protect the roots of mature plants, rapid degradation and high capital investment of chemicals and has given impetus to alternative methods (Cook and Baker, 1983 and Bezbaruah *et al*., 1995a, b). Such an increase in chemical degrader populations disturbs microbial diversity, which leads to deterioration of soil fertility. Considering these limitations one approach to overcome this problem is to use beneficial, naturally occurring and eco-friendly and environmentally safe microorganisms as biocides, alone or in integration with other practices (Lumsden and Locke, 1989). Biological fungicides or biofungicides are commercial products that are composed of microorganisms such as fungi, bacteria or actinomycetes that suppress plant diseases. These biological control organisms (BCO) are formulated as powders for seed treatments, in granular form for soil incorporation and as suspension for root drenches and foliar sprays. Biological control has gained momentum in last few years which was defined by Cook and Baker (1983) as the reduction of the amount of inoculum or disease producing activity of a pathogen accomplished by or through one or more organisms other than man. Most approaches involving biocontrol are directed against the pathogens, and in the early stages of the development of the concept of biocontrol, mostly the soil-borne pathogens were studied (Baker and Cook, 1974; Bruehl, 1975; Papavizas and Lumsden, 1980). The biocontrol agents can be used for the management of rot diseases at comparatively cheaper cost with long-term antagonistic effects against the pathogens because these biocontrol agents multiply and remain near the root zone of the plant giving protection to these plants throughout the growth period. Some of the microbial taxa that have been successfully commercialized and currently marketed are EPA-registered biopesticides in the United States including fungi belonging to the genera *Ampelomyces, Candida, Coniothyrium*, and *Trichoderma* and bacteria belonging to the genera *Agrobacterium, Bacillus, Pseudomonas* and *Streptomyces* (United States Environmental Protection Agency). The future success of the biological control industry will depend on innovative business management, product marketing, extension education and research (Mathre *et al.*, 1999 and Biological Control Working Group, 1998). Several biological control agents have been commercialized or have been registered for commercial field trials (Lumsden *et al.*, 1995). However, a major obstacle encountered in the area of biological control is the inconsistencies in performance of biocontrol products (Lumsden and Lewis, 1989). The objective of this chapter is to describe the mechanism of action of antagonistic microorganisms and biological control of soil borne pathogens by use of two biocontrol agents: the fungus *Trichoderma* and the bacterium *Pseudomonas*.

BIOCONTROL AGENTS

The use of fungal and bacterial biocontrol agents for reducing soil-borne pathogens is receiving much attention these days not only because they provide protection against major pathogens, but also due to the fact that many of them enhance plant growth resulting in better yield. Biocontrol agents include all classes/groups of organisms existing in an ecosystem. Maximum emphasis for developing biocontrol agents has invariably gone to fungal and bacterial biocontrol agents. The important genera of fungi studied as biocontrol agents are *Trichoderma, Gliocladium, Fusarium, Aspergillus Chaetomium, Neurospora, Paecilomyces, Catenaria, Penicillium, Dactylella, Arthobotrys*, etc. Among these *Trichoderma* have been widely explored and recommended as biocontrol agents against many soil-borne pathogens (Raghuchander *et al.*, 1997). Use of bacteria for the management of plant pathogens and yield improvement has increased steadily since the mid-1960s (Baker, 1987). *Pseudomonas* is an effective bacterial antagonist for the management of soil-borne diseases. Biological control of plant diseases, especially soil-borne pathogens, has been the subject of much research in the last two decades.

Essential Characteristics of Effective Biocontrol Agents

1. Speed/mobility to prevent pathogen in developing resistant structures.
2. Longevity, enough to protect plant during its vulnerable period, whatever that may be.

3. Environmental tolerance to sustain activity under different soil and climatic conditions.
4. Mode of action for physical contact and chemical nature of killing component varies from pathogen to pathogen.

While using natural enemies, it is important to have fast growing biocontrol organisms in the fields which can eventually make the conditions hostile and unfavourable for the pathogen proliferation.

Major Advantages of Biocontrol Agents

Biocontrol agents are preferred over chemical pesticides for the following reasons:

1. Biocontrol agents are usually less harmful than conventional pesticides.
2. Biocontrol agents generally affect the target pests and closely related organisms in contrast to the broad spectrum conventional pesticides that are non-discriminative and may affect useful microorganisms, besides insects, birds and mammals.
3. Eco-friendly as they do not pollute environment and can be used along with organic manure.
4. Lower cost and longer efficacy than fungicides.
5. Do not lead to development of resistance in plant pathogens.
6. No phytotoxic effects.
7. Enhance plant growth, increase yield and net profit.
8. Effective suppression of the fungal pathogen before it causes economically important damage to a crop.

The likelihood for successful development of commercially viable fungicides is maximized once all these criteria are fulfilled.

Importance of Biocontrol Agents

Biological control really developed as an academic discipline during the 1970s and is now a full-grown science supported in both the public and private sector (Baker, 1987). In nature every ecosystem exists in a balance. Growth and multiplication of each organism depends on the food-chain. The natural microorganisms found in the diversity are selected for studying its biology for mass multiplication and utilization to check the target pest and disease. Bacteria and fungi are highly versatile, i.e., they can carry out almost all known biological reactions. They are also specific in their action, do not leave any residue, safe and economical. Biological control of diseases is one of the important means for checking disease problem in almost all agro-ecological situations. Bioagents are living organisms that can intervene in the life cycle of diseases in such a way that the crop damage is minimized. These bioagents can be conserved, preserved and multiplied under laboratory conditions for field release, and when they are introduced in the field, they are capable of bringing down the targetted disease population below economic threshold level (ETL). The use of environmentally sound disease management technologies, based on biological antagonists, is being encouraged now as part of Integrated Pest/ Disease management programmes.

Biocontrol can be divided into three strategies or approaches:

1. **Regulation of the pathogen population:** Biocontrol agents are used to regulate the pathogen population at or below an acceptable threshold.
2. **Exclusionary system of protection:** Beneficial micro organisms are used as a living barrier that excludes infection.
3. **Self defence:** Resistance mechanisms in the host plant prevent or suppress disease damage.

Commercially Available Biocontrol Products

The following is the list of commercially available products of *Trichoderma* and *Pseudomonas* for the biocontrol of plant pathogen and plant growth promotion involving the induction of plant host defences (Table-16.1).

MECHANISM OF ACTION OF BIOCONTROL AGENTS

Biocontrol organisms suppress disease by different mechanisms such as competition, enhanced nutrient uptake, antibiosis, antagonism or by inducing host resistance. It is now possible to understand the mechanisms by which an antagonist can limit either the density or the activity of the pathogen. The knowledge of the mechanism of bioagents should help to identify the environmental conditions favourable for application of biocontrol and to improve efficiency and consistency of biocontrol methods. The biocontrol agents have to be chosen in conjunction with their mode of action, which should not exclude each other, but on the contrary, show a synergistic effect.

Mechanisms of Biological Control Activity of *Trichoderma*

Trichoderma spp. have evolved numerous mechanisms that are involved in attacking other fungi and enhancing plant and root growth. These mechanisms include mycoparasitism (Haran, *et al.,*1996), competition for space and nutrients (Elad *et al.,*1999), production of inhibitory compounds (Sivasithamparam and Ghisalberti, 1998), inactivation of the pathogen's enzymes and induced resistance (Yedidia *et al.,* 1999; Kapulnik and Chet, 2000). The mechanisms are described as under:

Direct Action

Microbial antagonism

Microbial antagonism is the direct interaction between two microorganisms that share the ecological niche. There are three types of direct interaction, viz. mycoparasitism, antibiosis, and competition for nutrients. This type of antagonism occurs mostly during the saprophytic phase of pathogens and contributes in reducing the inoculum potential of the pathogen in soil and at the root surface.

Mycoparasitism

Mycoparasitism has long been considered as an important mechanism of action of biocontrol

Table 16.1. Biocontrol products of *Trichoderma* and *Pseudomonas* commercially available in the United States.

Biocontrol agent	Product	Supplier	Target Pathogen/Disease	Crop	Application
Trichoderma harzianum	Root Shield, Plant Shield, T-22 Planter box	Bioworks Inc. 122 North Genese St. Geneva, USA	*Pythium* spp, *Rhizoctonia solani, Fusarium* spp.	Trees, shrubs, all ornamentals, cabbage, tomato cucumber	Granules mixed with soil or potting medium, powder mixed with water and added as a soil drench
Pseudomonas aureofeciens	Bioject Spot Less	Eco Soil Systems, Inc. 10740 Thommint Rd., San Diego, USA.	Dollarspot, Anthracnose, *Pythium aphanidermatum*	Turf and others	Overhead irrigation, can only be used with the Bioject Automatic Fermentation system
Pseudomonas syringae	Bio-save 10LP, 110	Ecoscience Corp., 153 Sable Palm Dr., Longwood, FL 32 779.	*Botrytis cinerea, Penicillium* spp, *Mucor pyroformis, Geotrichum candidum*	Pome fruit, citrus, cherries and potatoes	Pellets added to water to produce liquid suspension, postharvest application to fruit as drench, dip or spray
Pseudomonas fluorescens A506	Blightban A506	NuFarm Inc., 1-708-754-3330	*Erwinia amylovora* and russet-inducing bacteria	Almond, apple, apricot, blueberry, cherry, strawberry, tomato	Bloom time spray of the flower and fruit peach, pear, potato
Pseudomonas chlororaphis	Cedomon	BioAgri ABP.O. Box 914 Uppsala, Sweden	Leaf stripe, net blotch, Fusarium sp, spot blotch, leaf spot and others	Barley and oats; potential for other cereals	Seed treatment

mainly by fungal biocontrol agents. Findings reported during the last few years indicate that mycoparasitism is one of the main mechanisms involved in the antagonism of *Trichoderma* as a biocontrol agent. The parasitic activity of strains of *Trichoderma* spp. has been extensively studied and plays a major role in the antagonism expressed against *Rhizoctonia solani* (Weindling, 1934; Elad *et al.*, 1983). The events leading to mycoparasitism are complex and occur as follows: *Trichoderma* strains detect other fungi and grow tropically downwards (Chet *et al.*, 1981), remote sensing is at least partially due to the sequential expression of cell wall degrading enzymes. Different strains can follow different patterns of induction, but the fungi apparently always produce low levels of an extracellular exochitinase. Diffusion of this enzyme catalyses the release of cell wall oligomers from target fungi, and this in turn induces the expression of fungitoxic endochitinase which also diffuse and begin attack on the target fungus before contact is actually made. Once the fungi come into contact, *Trichoderma* spp. attach to the host and can coil around it and form appressoria on the host surface. Attachment is mediated by the binding of carbohydrates in the *Trichoderma* cell wall to lectins on the target fungus (Inbar *et al.*, 1996). Once in contact, *Trichoderma* degrades cell wall of the target fungi by the secretion of different lytic enzymes. This process (mycoparasitism) limits growth and activity of plant pathogenic fungi. *Trichoderma* attaches to the host hyphae via coiling, hooks and appressoria like bodies and penetrate the host cell wall by secreting lytic enzymes. Direct entry of *Trichoderma* hyphae into lumen of the target fungus occurs. There are at least 20-30 known genes, proteins and other metabolites that are directly involved in this interaction, which is typical of the complex systems that are used by these fungi in their interactions with other organisms (Harman *et al.,* 2004). These enzymes have been reported mainly in isolates of *Trichoderma harzianum* (Sivan and Chet, 1989; Lorito *et al.,* 1993; Lorito *et al.,* 1996b). *T. harzianum* is known to produce relatively high concentrations of cell wall degrading enzymes such as α-1, 3- glucanases and different chitinolytic enzymes. These enzymes have been suggested as the key enzymes in mycoparasitism (Elad *et al.,* 1982; Cherif and Benhamou, 1990). Several enzymes have been purified and characterized and their ability to inhibit the germination of spores and elongation of the hyphae belonging to the pathogenic fungi has been shown *in vitro* (Lorito *et al.,* 1998). Whether parasitism is one mode of action that can be deliberately associated with other modes of action to improve efficacy of biological control has not yet been investigated.

Antibiosis

Antibiosis is the antagonism resulting in the release or production of antibiotics and some other metabolites by antagonistic organisms that are toxic to other microorganisms and inhibit their growth. Many isolates of *Trichoderma* spp. produce volatile and non-volatile antibiotics (Dennis and Webster, 1971 a,b). Trichodermin was described as an antibiotic produced by *Trichoderma* (Godtfredsen and Vangedal, 1965). Howell and Stipanovic (1983) isolated and described an antibiotic gliovirin from *Gliocladium* (*Trichoderma*) *virens* that was strongly inhibitory to *Pythium ultimum* and *Phytophthora* species. The species of *Trichoderma* are also known to produce a number of antibiotics, such as trichodermin, trichodermol, harzianum A and harzianolide (Barbosa *et al.*, 2001) and secrete a number of antibiotics including pyrones, isocyanates, peptides, peptaibols and trichothenes (Ghisalberti and Sivasithamparam, 1991; Schirmbock *et al.,* 1994).

But whether or not these antifungal metabolites are relative to biocontrol under field conditions is still a point of speculation (Howell and Stipanovic, 1983; Ghisalberti and Sivasithamparam, 1991; Wilhite *et al.,* 1994; Howell and Stipanovic, 1995).

Competition for Space and Nutrients

Competition for food and space is one of the mechanisms of biological control activity of *Trichoderma* spp. against phytopathogenic fungi. *Trichoderma* species are generally considered to be aggressive competitors (Samuels, 1996). Competition for carbon has also been involved in the determinism of antagonism expressed by different strains of *Trichoderma* spp. against several plant pathogens, especially *Fusarium oxysporum* (Sivan and Chet, 1989). Rhizosphere competence by biocontrol agents is important in the mechanism of competition, especially with seed treatments. This is the zone where protection against pathogens is critical (Harman, 1992). It is important for control of root diseases because a biocontrol agent cannot compete for space and nutrients if it is unable to grow in the rhizosphere (Howell, 2003). *Trichoderma* species, either added to the soil or applied as seed treatments, grow readily along with the developing root system of the treated plants (Ahmed and Baker, 1987; Harman, 2000).

Indirect action

Induced systemic resistance

More and more studies are devoted to the resistance induced in the host plant by application of biocontrol agents. The induced localized or systemic resistance is an important component of plant disease control by *Trichoderma* spp. However, different mechanisms might be responsible for biocontrol caused by different strains, in different plants and with different pathogens. In fact, *Trichoderma* spp. in general have been found to be highly resistant to a variety of toxins and xenobiotic compounds, including antibiotics produced by other microorganisms, plant antimicrobial compounds and chemical fungicides (Harman, 1996). The molecular basis of this resistance, which makes this fungus an active colonizer of toxic environments and a strong competitor has been partially elucidated with the recent discovery that *Trichoderma* strains produce a set of ATP-binding cassette (ABC) transporters. These ATP-dependent permeases mediate the transport of many different substrates through biological membranes, and overexpression of ABC-transporter genes decreases the accumulation of toxicants in *Trichoderma* cells. In *Trichoderma* spp., ABC transporters have recently been shown to be important in many processes. These include resistance to environmental toxicants that are produced by soil microflora or introduced by human activity (e.g., fungicides and heavy metal pollutants) and secretion of antibiotics and cell wall degrading enzymes that are necessary for the establishment of a compatible interaction with a host fungus or for the creation of a favourable microenvironment. Three classes of compounds are produced by *Trichoderma* strains, which induce resistance (biochemical elicitors of disease resistance in plants are now known). These are proteins with enzymatic or other functions, homologues of proteins encoded by the avirulence (Avr) genes and oligosaccharides and other low molecular weight compounds that are released from fungal or plant cell walls by the activity of *Trichoderma* enzymes.

Mechanisms of Biocontrol Activity of *Pseudomonas*

Pseudomonas spp. suppress the pathogens by various modes of action, namely, competition for nutrients and space, antibiosis by production of various antibiotics. Other mechanisms include production of hydrogen cyanide and degradation of toxins (Defago *et al.*, 1990). The mechanisms are explained as under:

Direct action

Antibiosis

Antibiosis is a very common phenomenon responsible for the biocontrol activity of many organisms, including *Pseudomonas* spp. developed as biocontrol agents. Fluorescent Pseudomonas suppress the pathogens by antibiosis through the production of various antibiotic substances such as 2,4-diacetylphloroglucinol, phenazine-1-carboxylic acid, oomycin A, oxychlororaphine, pyoluteorin, HCN and pyocyanine. A given strain of biocontrol agent may produce several types of antifungal compounds effective against certain species of fungal pathogens. The introduction of *Pseudomonas fluorescens* that produce the antibiotic 2,4-diacetylphloroglucinol reduce the suppression of various soilborne pathogens (Weller *et al.,* 2002). The production of phenazines and 2-4 – diacetylphloroglucinol by fluorescent *Pseudomonas* was shown to be the primary mode of antagonism against *Gäeumannomyces graminis* var. *tritici* (Vincent *et al.,* 1991 and Keel *et al.,* 1992). A single antifungal metabolite generally does not account for all the antagonistic activity of a biocontrol agent (Loper and Lindsow 1993). Therefore, it may be very useful to associate several strains of biocontrol agents producing different types of antifungal metabolites to improve the efficacy or enlarge the activity spectrum of biological control.

Competition for nutrients and space

Competition for nutrients and space is considered as one of the classical mechanisms of biological control. It seems to be an important mechanism of biocontrol, but it is difficult to prove experimentally. *Pseudomonas fluorescens* produces yellow green fluorescent pyoverdines and act as siderophores. The best known of these pyoverdines is pseudobactin, a hexapeptide cyclized through a fluorescent quinoline derivative with a hydroxamic iron chelating group derived from ornithin, aspartic acid, and aniline residues (Leong, 1986). Competition for iron is one of the modes of action by which fluorescent Pseudomonads limit the growth of pathogenic fungi and reduce incidence or severity. Under conditions of iron stress, these bacteria synthesize siderophores, which show a higher affinity for Fe^{3+} than fungal siderophores and are responsible for increased plant growth by PGPR (Kloepper *et al.,* 1980). The production of siderophores chelates the iron available in the soil and makes it unavailable to pathogen, thus the pathogen dies for want of iron. Iron starvation also prevents the germination of fungal spores, actively multiplies in the root zone, enters into the vascular system and protects the plant. Siderophores produced by *Pseudomonas* species have been implicated in the biological control of damping off of cotton (Loper 1988, Laha *et al.*, 1992) and in suppression of several vascular wilts caused by *Fusarium oxysporum* (Kloepper *et al.*, 1980). Competition for nutrients and space play an important role in the efficiency of a biocontrol agent even if another mode of action has been investigated.

Indirect action

Induced systemic resistance

Fluorescent pseudomonads suppress the pathogens indirectly by inducing plant-mediated defense reactions inducing systemic resistance (ISR) by enhancement of plant defense capacity. Fluorescent pseudomonads activate ISR against various fungal, bacterial and viral diseases (Lin *et al.,* 1995a, b). Colonization of bean root by fluorescent bacteria has correlation with induction of pathogenesis related proteins (RP proteins). Maurhofer *et al*., 1994 reported that ISR by *Pseudomonas fluorescens* strain CHAO against tobacco mosaic virus in tobacco was associated with accumulation of B-1-3 glucanases and endochitinase. ISR classically occurs when an inducing agent is applied prior to challenge inoculation with a pathogen resulting in reduced disease in comparison to the non-inoculated control. The fluorescent pseudomonads selected for their biocontrol activity and growth promoting capacity have been shown to induce systemic resistance in the plant. (Kloepper *et al*., 1996). The first evidence was given by Van Peer *et al*., 1991, who demonstrated that root colonization of carnation by a strain of fluorescent *Pseudomonas* sp. resulted in an accelerated and increased accumulation of phytoalexins in the stem of carnation after inoculation with *Fusarium oxysporum* f. sp. *dianthi*. Induced resistance is not exclusive from other modes of action and may exert a complementary effect on microbial antagonism. Indeed, direct antagonism usually limits the saprophytic growth of the pathogen, resulting in a decreased number of infection sites, and induced resistance limits the growth of the pathogen during its parasitic phase inside the plant.

The control resulting from an association of several modes of action is generally more effective and consistent than the control provided by a single mode of action. It would be interesting to associate several modes of action in combining several microorganisms for biological control.

METHOD OF APPLICATION OF BIOCONTROL AGENTS

Different methods are commonly used for delivering biological control agents that are described as under:

Seed Treatment

Seed treatment is an alternative approach to introduce *Trichoderma* and *Pseudomonas* into the soil. The application of biocontrol agent to seed as coating is the most common method of application. The biocontrol agent is suspended in water and applied as slurry.

Soil Application

For soil application, biocontrol agents are either applied in furrows or at the base of the plants or spread over in entire field, preferably with the FYM.

Foliar Spray

The most efficient way to utilize antagonists, is that they should be added to soil close to the seed. Several attempts have been made to apply the antagonists directly to the seed furrow e.g., by adding the antagonists to the gel used for fluid drilling of vegetables seeds.

Biological Control of Soil-borne Diseases by Fungal Antagonists

Several reports have been published on biological control of a wide range of soil-borne fungal pathogens by *Trichoderma* spp. and found effective in the laboratory, green house and experimental field conditions. A talc-based formulation of *Trichoderma viride* for dry seed treatment of groundnut, gingelly, sunflower, cotton chickpea, black-gram, green-gram and pigeonpea to control *Macrophomina phaseolina* has been developed (Jeyarajan *et al.,* 1994). It has been found effective in controlling pepper wilt, capsule rot of cardamom and root diseases of tea (Jeyarajan and Angappan, 1998). Species of *Trichoderma* have also been used similarly to provide protection to seeds against *Pythium* spp. and *Rhizoctonia solani* during germination. Seedling roots, corms, bulbs tubers, etc. can also be treated with spore or cell suspension of such antagonists. The technique of integrating *Trichoderma* seed treatment with SMP (Solid Matrix Priming) was applied for registration in U.S.A. for biological control of seed rot and seedling decay (Harnam and Taylor, 1990). Seed treatments of *Trichoderma viride, Trichoderma harzianum* and *T. koningii* reduced damping off cabbage in nursery (Roy *et al.,* 1998). In field trials, application of *Trichoderma* not only reduced the incidence of damping off and wilt caused by *Pythium* and *Fusarium* in tomato, chilli and brinjal, but also enhanced the growth, i.e., shoot length, root length and dry matter production (Manoranjitham *et al.,* 1999, 2000, 2001; Ganavel and Jayraj, 2003; Ramesh and Korikanthimath, 2004). In fact, the antifungal abilities of these beneficial microbes have been known since 1930s, and there have been extensive efforts to use them for plant disease control since then (Samuels, 1996), Weindling (1932, 1934) and Weindling and Fawcett (1936) perhaps were the first investigators to demonstrate the potential of *Trichoderma* spp. to control plant diseases. Different species of *Trichoderma* like *T. harzianum, T. hamatum, T. viride, T. longibrachiatum, T. koningii, T. pseudokoningii* effectively inhibit soil-borne pathogens viz., *Fusarium oxysporum, Pythium aphanidermatum* and *Rhizoctonia solani* causing damping off in *Acacia nilotica, Albizia lebbeck, Dalbergia sissoo, Prosopsis julifora* and *Leucaena leucocephal*. Among all of these, *T. viride* recorded the maximum inhibition of all the three pathogens and also recorded the highest seed germination and seedling survival in *Acacia nilotica*. (Sanjay and Kaushik, 2001). Karpagavalli and Ramabadran (2001) showed that *Trichoderma harzianum* and *T. viride* antagonistic to *Pythium aphanidermatum* produced cellulolytic enzymes in fungicide amended medium and had profound influence on the damping off incidence, shoot and root length and dry matter production of tomato seedlings.

Biological Control of Soil-borne Diseases by Bacterial Antagonists

In India, work on the use of bacterial antagonists for the plant disease control started in the 1950s and gained momentum in the 1980s. Fluorescent pseudomonads especially *Pseudomonas fluorescens* has been studied by various workers for their efficacy against diseases of major crops such as cotton, chickpea, pigeonpea, etc. Formulations of *Pseudomonas fluorescens* with various carrier materials like talc, peat, vermiculite, lignite, kaolinite, etc. have been developed with talc based ones giving good results (Vidhyasekaran *et al.,* 1997). Strains of *Pseudomonas fluorescens* and *Pseudomonas putida* applied to potato seed pieces, sugarbeet and radish are reported to increase the yield of potato by 5-33%, sugarbeet by 4-8 tonnes per hectare, and root weight of radish by 6-144% (Weller, 1988). Commercial formulations of *Pseudomonas fluorescens* are now available for the control of damping off, root rots, stem rot of various important food crop plants grown in greenhouses, nurseries, etc. The biological control efficacy of *Pseudomonas fluorescens* was tested extensively and determined to be consistent and reliable when used as prescribed for the control of damping off in nursery application. Preliminary results have shown that seed treatment and soil application of *P. fluorescens* with combination of *Trichoderma* or alone reduced damping off caused by *Pythium aphanidermatum* in tomato both in pot and field trials (Manoranjitham and Prakasam, 1999; Manoranjitham *et al.,* 2000). In field trials, *P. fluorescens* effectively reduced damping off disease in *Acacia nilotica*, *Azadirachta indica*, *Dalbergia sissoo* and *Prosopsis juliform* caused by *Fusarium oxysporum* f.sp. *dalbergiae, F. moniliforme*, *Rhizoctonia solani* and *Pythium aphanidermatum* (Kaushik *et al.,*2002). A group of rhizosphere bacteria (rhizobacteria) that exerts a beneficial effect on plant growth is referred to as plant growth promoting rhizobacteria or PGPR (Scroth and Hancock, 1981). Currently *Pseudomonas* spp. are receiving much attention as PGPR because of their multiple effects on plant growth promotion. PGPR are believed to improve plant growth by colonizing the root system and pre-emptying the establishment or suppressing deleterious effects of rhizosphere microorganisms on the root. (Scroth and Hancock, 1981). Inoculating planting material with PGPR presumably prevents or reduces the establishment of pathogens (Suslow, 1982).

MICROBIAL ASSOCIATIONS FOR BIOLOGICAL CONTROL

The efficacy and consistency of biological control can be improved by using microbial associations. It is necessary to evaluate the compatibility of the microorganisms under various environmental conditions and to develop production and formulation processes leading to a commercial product easy to handle and apply.

Compatibility between *Trichoderma* and *Pseudomonas*

One of the best examples of a microbial association used to improve efficacy and consistency of biological control is provided by the association of *Trichoderma* and *Pseudomonas*. For an additive effect, the microorganisms to be associated must be fully compatible. They must establish together in the rhizosphere of the host plant without excluding each other by competition or antibiosis. Hubbard *et al.* (1983) reported that a strain of *Trichoderma hamatum* applied as conidia to pea seeds was effective in controlling seed rots caused by *Pythium* spp. in some soils, but not in other soils. This antagonism between fluorescent pseudomonads and *T. hamatum* was controlled by iron availability in the soil and addition of ferrous oxalate to soil permitted *T. hamatum* to protect pea seeds. *In vitro* experiments confirmed that extracellular compounds produced by fluorescent pseudomonads under conditions of Fe-deprivation were responsible for inhibition of *T. hamatum* and suggested that pseudomonads inhibit *T. hamatum* through the production of siderophores. Dandurand and Knusden (1993) failed to demonstrate any beneficial effect of the association of strain of *P. fluorescens* with a strain of *T. harzianum* in controlling *Aphanomyces* root rot of peas although the presence of the bacteria stimulated the hyphal growth of the fungus developed from the coated seeds.

Having selected independently, two efficient microorganisms do not guarantee that their association will result in an increased beneficial effect. Therefore, more attentions should be given to the natural associations of microorganisms in order to select antagonists that could be used in combinations.

CONCLUSION

Biological control of plant diseases, especially those induced by soil-borne pathogens, is presently restricted to a limited number of commercial preparations based on a single antagonistic microorganism. Biocontrol agents have a complementary beneficial effect on soil-borne pathogens through several modes of action. Since a single antagonist does not possess all the possible modes of action, it would be very interesting to associate several antagonists to improve the efficacy and consistency of biological control. Indeed, the selection of compatible antagonist requires a good knowledge of the modes of action of the antagonists and also their ecological behaviour in different soils and in the rhizosphere of several plants. Increased demand for organic produce has enlarged the market for biocontrol products. The field of plant pathology will contribute substantially to making the 21st century the age of biotechnology by the development of innovative biocontrol strategies. The efficiency and productivity in modern agriculture has traditionally been dependent on the use of chemicals to control pests. Reliance on such singular control has created formidable problems. The use of biocontrol strategies offers several advantages over the current control measures primarily relying upon broad spectrum pesticides. Biocontrol may not totally replace chemicals, but the judicious use of biocontrol agents can significantly enhance our lives, the environment, and agriculture productivity. Biocontrol agents have distinct advantages in being compatible with most of the agricultural practices and, hence, can be successfully utilized as a part of total crop management practices or, broadly, as a component of the agro-ecosystem management.

REFERENCES

Ahmad, J.S. and Baker, R. (1987). Competitive saprophytic ability and cellulytic activity of rhizosphere- competent mutants of *Trichoderma harzianum Phytopathol.* **77:** 358-362.

Baker, K.F. (1987). Evolving concepts of biological control of plant pathogen. *Ann. Rev. Phytopathol.* **25:** 67-85.

Baker, K.F. and Cook, R.J. (1974). Biological control of plant pathogens. Freeman, San Francisco.

Barbosa, M.A.G., Rehn, K.G., Menezes, M. and Mariano, R.R. (2001). Antagonism of *Trichoderma* species on *Cladosporium herbarum* and their enzymatic characterization.*www.google.com*

Bezbaruah, B., Bora, T.K. and Saika, N. (1995a). *Indian J. Microbiol.* **35**(1): 55-60.

Bezbaruah, B., Bora, T.K. and Saika, N. (1995b). *Indian J. Agri. Sci.* **65**: 578-583.

Biological Control Working Group (1998). Alternative paradigms for commercializing biological control: Workshop white paper, on line, Rutgers Workshop Organizing Committee and Experimental Station Committee on Policy.

Bruehl, G.W. (1975). Biology and control of soil borne plant pathogens. *American Phytopathological Society St. Paul.*

Cook, R.J. and Baker, K.F. (1983). The nature and practice of biological control of Plant Pathogens. *American Phytopathol. Soc. St. Paul, Minn.* pp. 539.

Chet, I., Harman, G.E. and Baker, R. (1981). *Trichoderma hamatum*: Its hyphal interactions with *Rhizoctonia solani* and *Pythium* spp. *Microb. Ecol.* **7:** 29-38.

Cherif, M. and Benhamou, N. (1990). Cytochemical aspects of chitin breakdown during the parasitic action of a *Trichoderma* spp. on *Fusarium oxysporum* F. spp. *radicis-lycopersici. Phytopathology.* **80:** 1406-1414.

Dandurand, L.M. and Knudsen, G.R. (1983). Influence of *Pseudomonas fluorescens* on hyphal growth and biocontrol activity of *Trichoderma harzianum* in the spermosphere and rhizosphere of pea. *Phytopathol.* **83:** 265-270.

Defago, G., Berling, C.H., Burger, U., Hass, D., Kahr, G., Keel, K., Voisard, C., Wirthner, P. and Wuthrick, B. (1990). Suppression of black root rot of tobacco and other root rot diseases by strains of *Pseudomonas. fluorescence.* Potential application and mechanism. In : Biological control of soil borne Plant Pathogen. *CAB International Wellingabord oxon*, U.K. 93-108.

Dennis, C. and Webster, J. (1971a). Antagonistic properties of speices groups of *Trichoderma.* II. Production of volatile antibiotics. *Trans. Br. Mycol. Soc.*, **57:** 41-48.

Dennis, C. and Webster, J. (1971b). Antagonistic properties of species groups of *Trichoderma.* 3. Hyphal interaction. *Trans. Brit. Mycol. Soc.* **57:** 25-39.

Elad, Y., Chet, I. and Henis, Y. (1982). Degradation of plant pathogenic fungi by *Trichoderma harzianum. Canadian J. Microbiol.* **28:** 719-725.

Elad. Y., Chet, I., Boyle, P. and Henis, Y. (1983). Parasitism of *Trichoderma* spp. on *Rhizoctonia solani* and *Sclerotium rolfsii.* Scanning electron microscopy and fluorescence microscopy. *Phytopathology.* **73:** 85-88.

Elad. Y., David, D.R., Levi, T., Kapat, A. and Kirshner, B. (1999). *Trichoderma harzianum* T-39 mechanisms of biocontrol of foliar pathogens, *In:* Modern fungicides and antifungal compounds – II (Lyr, H., Russell, P.E. Dehne, H.W. and Sisler, H.D. Eds.) *Andover, Hants, UK. Intercept.* pp. 459-467.

Gnanavel, I. and Jayaraj, J. (2003). Integration of soil solorization and *Trichoderma viride* pers; Fr. to control damping off of tomato *(Lycopersicum esculentum Mill). J. Biological control.* **17**(1): 99-101.

Ghisalberti, E.L. and Sivasithamparam, K. (1991). Antifungal antibiotics produced by *Trichoderma* spp. *Soil Boil. and Biochem.* **23**: 1011-1020.

Godtfredsen, W.O. and Vangedal, S. (1965). Trichodermin, a new sesquiterpene antibiotic. *Acta. Chem. Scands,* **19:** 1088.

Haran, S., Schickler, H., Oppenheim, A. and Chet, I. (1996). Differential expression of *Trichoderma harzianum* chitinases during mycoparasitism. *Phytopathology.* **86:** 980-985.

Harman, G.E. (1992). The development and benefits of rhizosphere competent fungi for biological control of plant Pathogens. *J. Plant Nutrition.* **15:** 835-843.

Harman, G.E. (1996). *Trichoderma* for biocontrol of plant pathogens: From basic research to commercialization products. *In:* Cornell Community Conference on Biological control.

Harman, G.E. (2000). The myths and dogmas of biocontrol: changes in perceptions derived from research on *Trichoderma harzianum* strain T-22. *Plant Disease.* **84:** 377-393.

Harman, G.E., Howell, C.R., Viterbo, A., Chet, I. and Lorito, M. (2004). *Trichoderma species*—opportunistic avirulent plant symbiots. *Nature Reviews Microbiology.* **2:** 50-54.

Harnam, G.E. and Taylor, A.G. (1990). In biological control of soil-borne plant pathogens (Ed. Hornby, D.) CAP Wallingford, U.K. 415-526.

Howell, C.R. and Stipanovic, R.D. (1983). Gliovirin, a new antibiotic from *Gliocladium virens* and its role in the biological control of *Pythuim ultimum. Canadian. J. Microbiolol.* **29:** 321-324.

Howell, C.R. and Stipanovic, R.D. (1995). Mechanisms in the biocontrol of *Rhizoctonia solani*-induced cotton seedling disease by *Gliocladium virens*: Antibiosis. *Phytopathology.* **85:** 469-472.

Howell, C.R. (2003). Mechanisms employed by *Trichoderma* species in the biological control of plant diseases: The history and evolution of current concepts. *Plant Disease.* **87:** 4-10.

Hubbard, J.P., Harman, G.E. and Hadar, Y. (1983). Effect of soil borne *Pseudomonas* spp. on the biological control agent, *Trichoderma hamatum,* on pea seeds. *Phytopathology* **73:** 655-659.

Inbar, J., Menendez, A. and Chet, I. (1996). Hyphal interacation between *Trichoderma harzianum* and *Sclerotinia sclerotium* and its role in biological control. *Soil Biol. Biochem.* **28:** 757-763.

Jeyarajan., R., Ramakrishnan, G., Dinakaran, D. and Sridar, R. (1994). Development of products of *Trichoderma viride* and Biocontrol of root diseases, In: Biotechnology in India. *Bioved Research Society. Allahabad.* pp. 25-36.

Jeyarajan, R. and Angappan, K. (1998). Mass production technology for fungal antagonists and field evaluation, pp. 48-56. *In:* Biological suppression of plant diseases Phytoparasitic Nematodes and Weeds (Eds. Singh, S.P. and Hussaini, S.S.), *Project Directorate of Biological control, Bangalore, India*, pp. 284.

Kapulnik, Y. and Chet, I. (2000). Induction and accumulation of PR proteins activity during early stages of root colonization by the mycoparasite *T. harzianum* strain T-203. *Plant Physiol. Biochem.* **38:** 863-873.

Karpagavalli, S. and Ramabadran, R. (2001). Effect of fungicides and *Trichoderma* species on cellulolytic enzyme production, damping off incidence and seedling vigour of tomato. *Plant disease research.* **16**(2): 179-185.

Kaushik, J.C., Sanjay, Arya and Gupta, P.P. (2002). Management of damping off disease in forest nurseries by soil solarization and biocontrol agents. *Forage Research.* **28**(1): 29-31.

Keel, C., Schinder, U., Maurhofer, M., Voisard, C., Laville, J. and Burger, U. (1992). Suppression of root diseases by *Pesudomonas fluorescens* CHAO; Importance of the bacterial secondary metabolite 2, 4-diacetylphloroglucinol. *Mol. Plant Microbe Interact.* **5:** 4-13.

Kloepper, J.W., Leong, J., Teintze, M. and Schroth, M.N. (1980). Enhanced plant growth by siderophores produced by plant growth promoting rhizobacteira. *Nature.* **285:** 885-886.

Kloepper, J.W., Zehnder, G.W., Tuzun, S., Murphy, J.F., Wei, G., Yao, C. and Rapuch, G. (1996). Toward agricultural implementation of PGPR- mediated induced systemic resistance against crop pests, *in Advances in Biological Control of Plant Diseases* (Tang, W., Cook, R.J., and Rovira, A., Eds.), *Proceedings of the International Workshop on Biological Control of Plant Diseases, China Agricultural University Press, Beijing,* May 22-27, pp. 165-174.

Laha, G.S., Singh, R.P. and Varma J.P. (1992). *Indian Phytopathol.* **45:** 412-415.

Lin, L., Kloepper, J.W. and Tarun, S. (1995a). *Phytopathology.* **85:** 695-698.

Lin, L., Kloepper, J.W. and Tarun, S. (1995b). *Phytopathoglogy* **85:** 843-847.

Loper, J.E. (1988). Role of *Fluorescent* siderophore Production in biological control of *Phythium ultimum* by a *Pseudomonas flurescens* strain. *Phytopathology.* **78**: 166-172.

Loper, J.E. and Lindsow, S.E. (1993). Roles of competition and antibiosis in suppression of plant diseases by bacterial biological control agents. In Pest Management: Biologically based Technologies (Lumsden, R.D. and Vaughn, J.L., Eds.), *American Chemical Society, Washington, DC,* pp. 173-180.

Lorito, M., Harman, G.E., Hayes, C.K., Broadway, R.M., Troncoso, T., Woo, S.L. and di Pietro, A. (1993). Chitinolytic enzymes produced by *Trichoderma harzianum*: antifungal activity of purified endochitinase and chitobiase. *Phytopathology.* **83:** 302-307.

Lorito, M., Woo, S.L., Ambrosio, M.D., Harman, G.E., Hayes, C.K., Kubicek, C.P. and Scala, F. (1996). Synergistic interaction between cell wall degrading enzymes and membrane affecting compounds. *Molecular plant microbe interactions,* **9:** 206-213.

Lorito, M., Woo, S.L., Fernandez, I.G., Colucci, G., Harman, G.E., Pintor-Toro, J.A., Filippone, E., Muccifora, S., Lawrence, C.B., Zoina, A., Tuzun, S. and Scala, F. (1998). Genes from mycoparasitic fungi as a source for improving plant resistance to fungal pathogens. *Proc. Nat. Acad. Sci.,* USA, **95:** 7860-7865.

Lumsden, R.D., Lewis, J.A. and Fravel. D.R. (1995). Formulation and delivery of biocontrol agents for use against soil borne pathogens. *In:* Biorational Pest Control Agents (Hall, F.R. and Barry, J.W. Eds.), *Am. Chem. Soc. Washington, DC,* pp. 166-182.

Lumsden, R.D. and Lewis, J.A. (1989). Selection, production, formulation and commercial use of plant disease biocontrol fungi, problem and progress, *In:* Biotechnology of fungi for improving plant growth (Whipps, J.M. and Lumsden, R.D. eds.). *Cambridge University Press, Cambridge, UK,* pp. 171-190.

Lumsden, R.D. and Locke, J.C. (1989). Biological control of damping off caused by *Pythium ultimum* and *Rhizoctonia solani* with *Gliocladium virens* in soilless mix. *Phytopathology.* **79:** 361-366.

Manoranjithan, S.K. and Prakasm, V. (1999). Biological control of damping off disease of tomato. *South Indian Horticulture.* **47**(1/6): 302-303.

Monoranjitham, S.K., Prakasam, V. and Rajappan, K. (1999). Effect of antagonists on *Pythium aphanidermatum* (Edson) Fitz. and the growth of chilli seedlings. *J. Biological Control.* **13**(1/2): 101-106.

Manoranjithan, S.K., Prakasm, V., Rajappan, K. and Amutha, G. (2000). Control of chilli damping off using bioagents. *J. Mycology and Plant Pathoglogy.* **30**(20): 225-228.

Manoranjitham, S.K., Prakasam, V. and Rajappan, K. (2001). Biocontrol of damping off of tomato caused by *Pythium aphanidermatum. Indian Phytopathology.* **54**(1): 59-61.

Mathre, D.E., Cook, R.J. and Callan N.W. (1999). From discovery to use: Traversing the world of commercializing biocontrol agents for plant disease control. *Plant Dis.* **83:** 972-983.

Maurhofer, M.P., Hase, J.P., Meuwly, P., Metraux, and Defago, G. (1994). *Phytopathology.* **84:** 139-146.

Papavizas, G.C. and Lumsden, R.D. (1980). Biological control of soil-borne fungal propagules. *Ann. Rev. Pythopathol.* **18:** 389-412.

Raghuchander, Rajappan, K. and Sarniyappan, R. (1997). Evaluating methods of application of biocontrol agent in the control of mungbean root rot-*Indian Phytopathology* **50:** 229-234.

Ramesh, R. and Korikanthimath, V.S. (2004). Fungal and bacterial antagonists for the management of damping off and wilt in brinjal. *Indian J. Plant Protection.* **32**(2): 80-84.

Roy, S.K., Das, B.C. and Bora, L.C. (1998). Non pesticidal management of damping off of cabbage by *Rhizocotonia solani* Kuhn. *J. Agri. Sci. Soc. North-East India.* **11**(2): 127-130.

Sanjay, Arya and Kaushik, J.C. (2001). Efficacy of *Trichoderma species* and *Gliocladium virens* as biocontrol agents against damping-off of forest nurseries. *Plant-Disease-Research.* **16**(1): 46-51.

Schirmbock, M., Lorito, M., Wang, Y.L., Hayes, C.K., Arisan-Atac, I., Scala, F., Harman, G.E. and Kubicek, C. (1994). Parallel formation and synergism of hydrolytic enzymes and peptaibol antibiotics, molecular mechanisms involved in the antagonistic action of *Trichoderma harzianum* against phytopathogenic fungi. *App. Env. Microbiol.* **60:** 4364-4370.

Schroth, M.N. and Hancock, J.G. (1981). Selected topics in biological control. *Ann. Rev. Microbiol.* **35:** 453-476.

Sivan, A. and Chet, I. (1989). The possible role of competition between *Trichoderma harzianum* and *Fusarium oxysporum* on rhizosphere colonization. *Phytopathology.* **79:** 198-203.

Sivasithamparam, K. and Ghisalberti, F.L. (1998). Secondary metabolism, *Trichoderma* and *Gliocladium.* In: *Trichoderma* and *Gliocladium* (Kubicek, C.P. and Harman, G.E. Eds.). Taylor and Francis Ltd. London, pp. 139-191.

Suslow, T.V. (1982). Role of root colonizing bacteria in plant growth. *In:* Phytopathogenic prokaryotes (Mount, M.S. and Lacy, G.H. Eds.). Academic Press London. pp. 187-223.

United States Environmental Protection Agency. Termss: Biopesticides. Online. *Office of Pesticides Programs.*

Van Peer, R., Niemann, G.J., and Schippers, B. (1991). Induced resistance and phytoalexins accumulation in biological control of *fusarium* wilt of carnation by *Pseudomonas* sp. strain WCS 417 r. *Phytopathology* **81:** 728-734.

Vidhyasekaran, P., Sethuraman, K., Rajappan, K. and Vasumathi, K. (1997). Powder formulations of *Pseudomonas fluorescens* to control pigeon-pea wilt. *Biological Control.* **8:** 166-171.

Vincent, M.N., Harrison, L.A., Brackin, J.M., Kovacevich, P.A., Mukerji, P., Weller, D.M., and Pierson, E.A. (1991). Genetic analysis of the antifungal activity of a soil-borne *Pseudomonas auerofaciens* strain. *Appl. Env. Microbiol.* **57:** 2928-2934.

Weindling, R. (1932). *Trichoderma lignorum* as a parasite of other soil fungi. *Phytopathology.* **32:** 837-845.

Weindling, R. (1934). Studies on the lethal principle effective in the parasitic action of *Trichoderma lignorum* on *Rhizoctonia solani* and other soil fungi. *Phytopathology.* **24:** 153-170.

Weindling, R. and Fawcett, H.S. (1936). Experiments in the control of *Rhizoctonia* damping off of citrus seedlings. *Hilgardia,* **10:** 1-16.

Weller, D.M. (1988). Biological control of soil-borne plant pathogens in the rhizosphere with bacteria, *Annual rev. Pythopathology* **26:** 379-407.

Weller, D.M., Raajimakers, J., McSpadden Gardener, B., and Thomashow, L.M. (2002). Microbial populations responsible for specific soil suppressiveness to plant pathogens. *Ann. Rev. Phytopathol.* **40:** 309-348.

Wilhite, S.E., Lumsden, R.D. and Straney, D.C. (1994). Mutational analysis of gliotoxin production by the biocontrol fungus *Gliocladium virens* in relation to suppression of *Pythium* damping off. *Phytopathology.* **84:** 816-821.

Yedidia, I., Benhamou, N. and Chet, I. (1999). Induction of defense responses in cucumber plants (*Cucumis sativus L.*) by the biocontrol agent *Trichoderma harzianum. App. Env. Microbiol.* **65:** 1061-1071.

17

Insect Pests of Stored Products and their Control with Plant Products

M.O. Ashamo

Department of Biology, Federal University of Technology, P.M.B. 704, Akure, Nigeria
Email: oashamo@yahoo.com

Although chemical method is the most effective means of controlling pests, it has a lot of shortcomings such as high mammalian toxicity, high level of persistence in the environment, health hazards, residual effects, and pest resistance amongst others. One solution to these problems might be to totally replace synthetic chemicals with compounds, which occur naturally in plants. Most of these plant products are cheap, safe, readily available, edible and eco-friendly. They have been found to be very effective against many of the stored product pests in Nigeria. This paper highlights the different plant products that have been tested against insect pests of stored products, their management, the different types of formulations available and future prospects.

Nigeria has a population of about 140 million. This number is increasing and will continue to do so for some years more and therefore food production ought to be increased substantially to meet this ever-growing population. As the population increases, the available land for cultivation reduces as a result of urbanization, industrialization and provision of amenities for the people. The surplus food produced needs keeping facilities and care during storage. Efforts are being made at various levels to reduce post-harvest losses so that more food can be made available to the ever-increasing population. The availability of food commodities in countries like Nigeria is in most cases seasonal. To cope with current and future demands, adequate storage and preservation of food during the period of abundance is very important. Primarily proper storage ensures food availability throughout the year, preserves them in good quality, prevents attack/ infestation by pests—insects, microorganisms, rodents, birds etc., and helps to achieve market stability.

The stored product pests attacking commodities in Nigeria are many and varied, but emphasis in this chapter was given to insect pests (Table 17.1). The nutritive requirement of stored products insects are much the same as those of vertebrates, especially man, therefore food commodities with the highest nutritive values for man are also those most susceptible to pest damage (Odeyemi and Daramola, 2000).

Table 17.1 1Some important insect pests of stored products and commodities attacked.

Insect pest	Common name	Commodities attacked
Lasioderma sericorne (Fabricius)	Cigarette beetle	Cocoa, tobacco, oilseeds, cereals and cereal products, processed food.
Araecerus fasciculatus (Degeer) maize,	Coffee weevil	Stored coffee, cocoa beans, dried cassava chips, spices. etc.
Rhyzopertha dominica (Fabricius)	Lesser grain borer	Whole cereals, dried cassava chips
Prostephanus truncatus (Horn)	Larger grain borer	Maize, dried cassava chips
Dinoderus porcellus (Lesne)	Grain borer	Cereals, dried yam chips
Callosobruchus maculatus (F.)	Cowpea beetle	Legumes especially cowpea, pigeon pea, lentil, etc.
C. *subinnotatus* (Pic.)	Cowpea beetle	Bambara groundnut *(Voandzeia subterranean)*
Caryedon serratus (Olivier)	Groundnut borer	Groundnuts, oilseeds
Necrobia rufipes (Degeer)	Copra beetle	Dried copra, oilseeds, palm kernel, fish and fish meal
Cryptolestes ferrugineus (Stephens)	Rust-red grain beetle	Grain and grain products, nuts, oil cakes, dried fruits, dates, cowpea
Sitophilus zeamais Motsch.	Maize weevil	Cereals, especially maize, dried cassava
S. oryzae (L.)	Rice weevil	Rice and wheat
Balanogastris kolae (Desbrochers)	Kola weevil	Nuts of *Kola accuminata*
Sophrorhinus gbanjaensis D & T	Kola pod borer	Nuts of *Kola nitida, K. accuminata,* wild *Kola* species
Trogoderma granarium Everts	Khapra beetle	Oilseeds, pulses, damaged cereals
Dermestes maculatus Degeer	Hide beetle	Dried fish of genus *Clarias* and *Tilapia,* hides
Carpophilus hemipterus (L.)	Dried fruit beetle	Dried fruits, ripe fruits, cereals, oilseeds, etc.
Oryzaephilus surinamensis (L.) spices	Saw-toothed grain beetle	Cereals, cereal products, copra, nuts, dried fruits,
O. *mercator* (Fauvel)	Same as above	Same as above
Ahasverus advena (Waltl)	Foreign grain beetle	Groundnuts, copra, cocoa beans
Tribolium castaneum (Herbst) cocoa,	Rust red flour	Cereal, cereal products, groundnuts, nuts, coffee, beetle spices
Sitotroga cerealella (Olivier)	Angoumois grain moth	Whole cereal grains
Ephestia (Cadra) cautella (Wlk)	Tropical warehouse moth	Cereal flours, cereal products, nuts, cocoa, dried fruits, spices
Corcyra cephalonica (Stainton)	Rice moth	Cereals, cereal products, oilseeds, cocoa beans, dried fruits, spices
Plodia intyerpunctella (Hubner)	Indian meal	Grains, milled cereal products, nuts, dried fruits, moth chocolate
Dasyses rugosella Stainton	Yam moth	Yam tubers *(Dioscorea* spp)
Euzopherodes vapidella Mann	Yam moth	Yam tubers *(Dioscorea* spp)

Chemical method is the most effective means of controlling both field and storage pests. Insecticides are the most powerful tools available for use in pest management. They are highly effective, rapid in curative action, adaptable to most situations and flexible in meeting changing agronomic and ecological conditions (Olaifa *et al.,* 1987). Despite the dramatic successes recorded in the control of insect pests using synthetic insecticides, it has a lot of shortcomings such as high mammalian toxicity, high level of persistence in the environment, health hazards, residual effects on non-target organisms and pest resistance (Sighamony *et al.,* 1986). These have necessitated the use of other control measures, which are eco-friendly with little or no negative impact on the environment and are non-toxic to mammals especially man. One solution to these problems might be to totally replace synthetic chemicals with compounds, which occur naturally in plants. It is well known that plants manufacture and store various chemical substances, which protect them from attack by insects, bacteria, fungi and viruses (Fraenkel, 1969). These plant products can and have been exploited to protect stored products from insect attack. Vegetable oils, plant powders and extracts have been used to reduce post-harvest losses of cereals and grain legumes and other products in Nigeria (Agbakwuru *et al.,* 1978; Ivbijaro and Agbaje, 1986; Ofuya, 1986; Lale, 1992; Odeyemi, 1998; Adedire and Lajide, 1999; Ashamo, 2004).

This paper reviews the present status in the management of insect pests of stored products with plant products in Nigeria.

PLANT PRODUCTS USAGE IN NIGERIA

Traditionally, many different types and parts of plants are used against stored product pests. Plants are rich sources of bioactive organic chemicals with substances toxic to insects. Plants are known to produce a diverse range of secondary metabolites such as alkaloids, glucosides, terpenoids, polyacetylenes, flavonoids, unusual amino acids, etc. Many of these chemicals/ compounds are medicinal for humans and also protect plants from pests and pathogens. These chemicals repel approaching insects, deter feeding and oviposition on the plants, disrupt behaviour and physiology of insects in various ways and even prove toxic to different developmental stages of many insects. Long before the advent of synthetic insecticides, plant parts, i.e., leaves, fruits, seeds, roots, etc. were used to protect agricultural crops and stored products from attack by insects and non-insect pests. Decades before rural farmers were known to place neem leaves amongst stored grains to keep insects away. Some apply dry, ground or whole chilli pepper, *Capsicum* spp into their grains to repel insects and protect against insect damage. Plant ashes have been applied to bruises or wounds on yam tubers to protect against insects and prevent rotting and wood ash can be admixed with raw dried grains at small farm level and household level (Olaifa *et al.,* 1987). Also rice and maize grains are rubbed with vegetable oil to make them more attractive to buyers and prevent insect attack. The wood ash, fine sand, clay dust, quick lime admixed with dried grains cause invisible injury to the protective wax layer of stored food pests leading to dehydration. As these substances fill in the spaces between the grains, locomotion is restricted with regards to finding partners for reproduction. Respiration difficulties also result. Wood ash is very useful, however, it may affect the taste of grains and a considerable large amount is needed, for example, 300-400 g of ash per kg of grains (Odeyemi and Daramola, 2000). The admixture of vegetable oils of groundnut, castor and palm oil to stored grains has reduced oviposition and larval growth in storage insect pests (Schoonhoven, 1978; Odeyemi, 1991; Lale, 1992; Adedire, 2002, Ashamo, 2005). Farmers are generally able to obtain the oil required by grinding the seeds themselves. About 5 ml oil per kg is required for effective protection of cowpea. The oil is active against eggs and larvae. Other plants also used are known to have toxic, growth regulating and repellent effects. For example, *Lippia adoensis* and *Monodora tenuifolia* were effective in the control of *S. zeamais* (Olaifa *et al.,* 1987; Odeyemi, 1993).

Table 17.2 shows some of the plant products tested against stored products insects in Nigeria.

Table 17.2. Some plant products tested against stored products insects in Nigeria.

Plant	Common name	Family	Parts used	Target insect species
Azadirachta indica	Neem	Meliaceae	Leaves, bark, fruit, seed	1,2,3,4,5,6,7,12*
Melia azedarach	China berry	Meliaceae	Leaves, seed	1,2,3,7
Ocimum basilicum	Basil	Labiataceae	Leaves	1,2
Capsicum frutescens	Chilli pepper	Solanaceae	Leaves, fruits	1,2,3,4,12
Capsicum annum	Chilli pepper	Solanaceae	Fruit	1,2,3,4,12
Zingiber officinale	Ginger	Zingiberaceae	Leaves, corn	1,2,7
Khaya senegalensis	Mahogany	Meliaceae	Wood ash	2
Eugenia aromatica	Kanafuru	Myrtaceae	Fruit	1,2,3
Piper umbellatum	'Musoro'	Piperaceae	Seed	1,2,3
Piper guineense	Black pepper	Piperaceae	Seed	1,2,3,4,5,7,8,11
Erythrophleum guineense	'Epo obo'	Caesalpinaceae	Bark	1,2
Aframomum melegueta	Alligator pepper	Zingiberaceae	Fruit, seed	1,2,3,4
Hyptis sauvolens	Curry leaves	Labiataceae	Leaves	1,2,3
Allium cepa	Onion	Liliaceae	Dried skin	1,2,3,4,12
Allium sativum	Garlic	Liliaceae	Scale	1,2,3,4,12
Tetrapleura tetraptera	'Aidan'	Mimosaceae	Fruit	2,3
Uvaria afzelli	Pepper fruit	Annonaceae	Fruit	1,2,3
Myristica fragrans	Nutmeg	Myristicaceae	Seed	1,3,7
Citrus sinensis	Sweet orange	Rutaceae	Peel	1,2,3,4,5,9,12
Hura crepitans	Sandbox	Euphorbiaceae	Seed	1,2
Monodora tenuifolia	Pepper fruit	Annonaceae	Seed, Fruit	1,2,3
Eichhornia crassipes	Water hyacinth	Pontederaceae	Leaves	9
Dennettia tripetala	Pepper fruit	Annonaceae	Seed	1,2,5,8
Jatropha gossypifolia	Wild cassada	Euphorbiaceae	Seed	2
Arachis hypogaea	Groundnut	Leguminosae	Seed	1,2,3,4,10
Parkia clappertoniana	Locust bean	Mimosaceae	Seed	2

* 1=*Callosobruchus maculatus*, 2=*Sitophilus zeamais*, 3=*Sitophilus oryzae*, 4=*Dasyses rugosella*, 5=*Dermestes maculatus*, 6=*Rhyzopertha dominica*, 7=*Tribolium eastaneum*, 8=*Necrobia rufipes*, 9=*Corcyra cephalonica*, 10= *Trogoderma granarium*, 11=*Oryzaephilus mercator*, 12=*Euzopherodes vapidella*

NEEM PRODUCTS

Neem, *Azadirachta indica* A. Juss is indigenous to India and was introduced to West Africa between 1918 and 1921. It is rapid growing, drought resistant and provides many useful by-products including timber, firewood, animal fodder, fertilizer/mulch and chemicals (easily prepared by subsistence farmers) of great potential in repelling or killing field or post-harvest pests (Donli and Sule, 1998). The tree is central to the afforestation programmes of West African countries including Nigeria. Almost every part of the tree is bitter and tests with different insect species indicate maximum repellency by seed and seed products. All parts of the neem tree possess insecticidal activity, but seed kernel is the most active. Neem bark, leaf, fruit and oil as well as extracts with various solvents especially ethanol have been found to exhibit activity against insect pests ranging from ovicides, repellents, deterrents, antifeedants, ovipositional deterrents, chemo~terilants, metabolic inhibitors. The extracts, oil and dust of leaves and crushed seeds of the neem tree, *A. indica,* are potent against storage insects (lvbijao, 1983; Makanjuola, 1989). Azadirachtin is the most important biologically active component of neem and showed phagorepellent and toxic effects when incorporated into diets of different stored products insect species.

Management of *Callosobruchus maculatus* (F.) with Plant Products

Callosobruchus maculatus (F.) is a major primary pest of *Vigna unguiculata* (cowpea), *Lens culinaris* (lentils) and *Vigna radiata* (green gram). It also attacks other pulses such as *Cicer arientinum* (chick peas) and *Voandzeia subterranean* (bambara groundnut). The control of stored product insects using plant materials is a common practice among small-scale farmers in Nigeria. Many plant derivatives appear to have insecticidal, ovicidal and larvicidal properties against C. *maculatus*. Neem seed and leaf powders *(A. indica)* were toxic to *C. maculatus* effecting high mortality within 48 hrs and discouraged oviposition. Progeny emergence was inhibited. Neem kernel effectively preserved cowpea for eight months against *C. maculatus* (Sowunmi and Akinnusi, 1983). The leaf extract of neem was more toxic than extract from the bark. *Piper guineense* Schum and Thonn and *Capsicum* spp (lvbijaro and Agbaje, 1986) caused adult mortality in *C. maculatus*. Dry chilli pepper fruits, *C. annum,* onion scale leaves *Alliun cepa* and *Citrus* peels afforded some protection against damage by *C. maculatus* (Ofuya, 1986; Lale, 1992; Don-Pedro, 1989; Adedire and Lajide, 1999). Other plant materials have also been used against *C. maculatus* (Makanjuola, 1989; Lale, 1995; Ofuya and Lagunju, 1998). Oil coating is effective in controlling *C. maculatus* and other stored products insects (Don-Pedro, 1989; Lale and Abdulrahaman, 1999; Ajayi and Lale, 2001; Adedire, 2002; Adedire *et al.,* 2003). Regardless of the source of oil, all gave protection from bruchid attack. The main effect of the oil coating is its ability to prevent hatching of bruchid eggs, the oil acting as an ovicide. The oil, however, did not affect adversely the viability, the palatability, the cooking quality and the physical appearance of legume seeds.

Management of *Sitophilus* spp with Plant Products

Sitophilus zeamais Motschulsky predominantly attack maize, while *S. oryzae* (L.) attacks rice (paddy) and wheat. Makanjuola (1989) evaluated various extracts of neem for the control of *S. oryzae, S. zeamais, C. maculatus* and an apionid *Cylas puncticollis.* All the extracts tested resulted in a significant reduction in adult emergence of *S. oryzae.* Seed kernel, leaf, flower and bark from the tree were used in powdered and extract form. However, the seeds appeared more effective than the leaf. The powders of *A. indica, Melia azedarach* L., *Capsicum frutescens* L., *Zingiber officinale* Rosc, *Eugenia aromatica, Aframomum melegueta* Schum, *Uvaria afzelli, P. umbellatum, Allium cepa* were toxic to *S. zeamais.* There was a significant reduction in adult emergence, grain damage and grain weight loss. Weevil mortality was also very high (Adedire and Ajayi, 1996; Yusuf *et al.,* 1998). Petroleum ether extracts and oils of plants such as *Piper guineense, A. melegueta, Monodora tenufolia* Benth, *Arachis hypogaea* L., *Lipia adoensis* Hoschst, *Elaeis guineensis* Jacq had insecticidal activity against *S. zeamais* with high weevil mortality and greatly reduced adult emergence. The extracts did not have adverse effect on water absorption capacity and viability of the seeds (Odeyemi, 1993; Ashamo and Odeyemi, 2001a). The nutmeg oil, *Myristica fragrans* (Houtt.) was also effective against *S. oryzae* (Ashamo, 2005).

Management of Moth Pests of Stored Yam Tubers

Yam tubers, *Dioscorea* spp are attacked by moths in storage. The moth pests include *Dasyses rugosella* Stainton (Dina, 1977; Iheagwam and Ezike, 1989; Ashamo, 2005), *Euzopherodes vapidella* Mann (Dina, 1976; Ashamo and Odeyemi, 2001 b; Akinneye and Ashamo, 2006) and *Decadarchis minuscula* Walsingham (Echendu and Alozie, 1990). Several plant powders and oils including *Capsicum annum, C. frutescens, A. cepa, A. sativum, P. guineense* have been found to be effective against *E. vapidella* and *D. rugosella* (Ashamo, 2004; Ashamo and Akinneye, 2004).

TYPES OF FORMULATIONS USED

Plant products used in the management of stored products are obtained from leaves, flowers, fruits, seeds, bark and stems of plants (Lale, 1995). Seeds and fruits, however, appear to be the commonest and perhaps the most important sources of natural chemicals for management of stored product pests (Ivbijaro, 1983; Ivbijaro and Agbaje, 1986; Lale, 1995).

Powders

These are prepared by harvesting the appropriate plant parts, which are then sundried and pulverized by means of pestle and mortar, an electric kitchen mill or a coffee blender. The required quantity is admixed with appropriate quantity of commodity prior to storage. This is perhaps the most convenient method and easy to apply, the commodity remains clean compared to those treated with plant oils and the moisture content of the grains is not increased as it is if aqueous solutions are used (Lale 1992; 1995).

Oils

These are usually extracted from flowers, fruits, seeds or roots by means of organic solvents and used very much like emulsifiable concentrates. Both fatty and essential oils are used for the protection of stored products against insect pests. The plant oils have relatively high efficacy against virtually all stages in the life cycle of the insects (Agbakwuru *et al.,* 1978; Ivbijaro and Agbaje, 1986; Lale, 1992; Ashamo and Odeyemi, 2001a). However, there are some limitations in the use of plant oils: (a) fatty acids reduce the cleanliness and visual appeal of treated commodities; (b) some plant oils, especially non-edible fatty oils such as neem seed oil, have colours and taste that may taint commodities intended for human consumption; and (c) some plant oils affect seed germination (Sighamony *et al.,* 1986; Lale, 1995).

Aqueous Solutions

These are crude plant extracts obtained by using water as a solvent. Aqueous solutions are employed more frequently in the spraying of growing crops (Olaifa *et al.,* 1987) and their use as protectants of stored products is rather restricted. This is probably because their application increases the moisture content of stored commodities (Lale, 1995).

Mode of Action

Plant products have been reported to act as *de facto* contact insecticides against both adults and immature stages especially eggs and larvae of several species of stored product insects. The essential oils are highly lipophilic and have the ability to penetrate the cuticle of the insect. Some fatty vegetable oils kill insects by flooding their spiracles, thus causing rapid asphyxiation

(Su *et al.,* 1977; Schoonhoven, 1978; Don-Pedro, 1989; Lale, 1992; Lale, 1995). Some plant products act as attractants. Some of the plant products inhibit/reduce oviposition, while others acts as repellents and antifeedants against stored product insects (Lale, 1995). Powdered neem seeds as well as the pure active ingredient, azadirachtin, are among the most potent repellents so far screened (Sighamony *et al.,* 1986; Lale, 1995).

CONCLUSION

There is an abundance of luxuriant vegetation in the tropics, especially Nigeria, which have not been fully exploited. The plant products presently in use have not been properly formulated and application rates have not been standardized, although both volumetric and gravimetric methods are reported in literature. Although powders are measured on the basis of their weight and oils in terms of volume, a uniform standard should be developed that farmers can easily use. The active ingredients of the plant products in use should be determined to reduce the bulk of plant products used in storage. Repellent and antifeedant plant products should be developed since the insect pests do not have to contact the stored commodities before they are affected.

REFERENCES

Adedire, E.O. (2002). Use of nutmeg, *Myristica fragrans* (Houtt.) powder and oil for the control of cowpea storage bruchid, *Callosobruchus maculatus* Fabricius. *J. Plant Diseases and Prot.,* **109**(2): 193-199.

Adedire, C.O. and Ajayi, T.S. (1996). Assessment of the insecticidal properties of some plant extracts as grain protectants against the maize weevil, *Sitophilus zeamais* Motschulsky. *Nigerian J. Ent.,* **13:** 93-101.

Adedire, C.O. and Lajide, L. (1999). Toxicity and oviposition deterrency of some plant extracts on cowpea storage bruchid, *Callosobruchus maculatus* Fabricius. *J. Plant Diseases and Prot.,* **106:** 647-653.

Adedire, C.O., Adebowale, K.O. and Dansu, O.M. (2003). Chemical composition and insecticidal properties of *Monodora tenuifolia* seed oil (Annonaceae). *J. Trop. Forest Prod.,* **9**(1&2): 15-25.

Agbakwuru, E.O.P., Osisiogu, I.U.W. and Ugochukwu, E.N. (1978). Insecticides of Nigerian vegetable origin 11, some nitroalkanes as protectants of stored cowpeas and maize against insect pests. *Nigerian J. Sci.,* **12:** 493-504.

Akinneye, J.O. and Ashamo, M.O. (2006). Some aspects of the biology of the yam moth, *Euzopherodes vapidella* Mann (Lepidoptera: Pyralidae), a pest of stored yam tubers in Nigeria. *J. Ent. Res.,* **30**(1): 59-63.

Ashamo, M.O. (2004). Effects of some plant powders on the yam moth, *Dasyses rugosella* Stainton (Lepidoptera: Tineidae). *Bioscience Res. Comm.,* **16**(1): 41-46.

Ashamo, M.O. (2005). Integration of varietal resistance and nutmeg, *Myristica fragrans* (Houtt.) oil in protecting post harvest infestation by *Sitophilus oryzae* (L.) in rice. *J. Ent Res.,* **29**(4): 259-263.

Ashamo, M.O. and Odeyemi, O.O. (2001a). Protection of maize against *Sitophilus zeamais* Motsch. using seed extracts from some indigenous plants. *J. Plant Diseases and Prot.,* **108**(3): 320-327.

Ashamo, M.O. and Odeyemi, O.O. (2001b). Effects of rearing temperature on the fecundity and development of *Euzopherodes vapidella* Mann (Lepidoptera: Pyralidae), a pest of stored yam. *J. Stored Prod. Res.,* **37:** 253-261.

Ashamo, M.O. and Akinneye, J.O. (2004). Toxicity of powders of some tropical plants against the yam moth, *Euzopherodes vapidella* Mann (Lepidoptera: Pyralidae). *Nigerian J. Exptal. Applied. Bio.,* **5**(1): 63-68.

Dina, S.O. (1976). Observations on *Euzopherodes vapidella* Mann (Lepidoptera: Pyralidae) infesting yam tubers in Ibadan, Nigeria. *Nigerian J. Ent.,* **2:** 35-41.

Dina, S.O. (1977). Occurrence of *Dasyses rugosella* Stainton (Lepidoptera: Tineidae) on stored yam tllbers in Nigeria. *Trop. Sci.,* **19:** 113-117.

Donli, P.O. and Sule, B.S. (1998). Host range of Oriental yellow scale insect *(Aonidiella orientalis* Newstead) attacking neem as a background to its control. *ESN Occassional Publ.,* **31:** 23-31.

Don-Pedro, K.N. (1989). Effect of fixed vegetable oils on oviposition and adult mortality of *Callosobruchus maculatus* (F.) on cowpeas. *Int. Pest Control* **31:** 34-37.

Echendu, T.N.C. and Alozie, S.O. (1990). Susceptibilities of some yam cultivars to damage by *Decadarchis minuscula* Walsingham (Lepidoptera: Tineidae) in South East Nigeria. *Beitrage Trop. Landwirtschaft Veterinamed.,* **28:** 63-67.

Fraenkel, G. (1969). Evaluation of our thoughts on secondary plant substances. *Entomologia Expt. Appl.,* **12:** 473-486.

Iheagwam, E.U. and Ezike, C.I. (1989). Biology of the moth, *Dasyses rugosella* Stainton, a storage pest of yams *(Dioscorea* spp) (Lepidoptera: Tineidae). *Deutsche Entomol. Zeith.,* **36:** 381-386.

Ivbijaro, M.F. (1983). Toxicity of neem seed, *Azadirachta indica* A. Juss to *Sitophilus oryzae* (L.) in stored maize. *Prot. Ecol.,* **5:** 353-357.

Ivbijaro, M.F. and Agbaje, M. (1986). Insecticidal activities of *Piper guineense* Schum and Thonn, and *Capsicum* species on the cowpea bruchid, *Callosobruchus maculatus* (F.). *Insect Sci. Appl.,* **7:** 521-524.

Lale, N.E.S. (1992). A laboratory study of the comparative toxicity of products from three spices to the maize weevil. *Postharvest Bio. Tech.,* **2:** 61-64.

Lale, N.E.S. (1995). An overview of the use of plant products in the management of stored products *Coleoptera* in the tropics. *Postharvest News Inf.,* **6:** 69N-75N.

Lale, N.E.S. and Abdulrahaman, H.T. (1999). Evaluation of neem *(Azadirachta indica* A. Juss) seed oil obtained by different methods and neem powder for the management of *Callosobruchus maculatus* (F.) (Coleoptera: Bruchidae) in stored cowpea. *J. stored Prod. Res.,* **35:** 135-143.

Maina, Y.T and Lale, N.E.S. (2004). Efficacy of integrating varietal resistance and neem *(Azadirachta indica* A Juss) seed oil for the management of *Callosobruchus maculatus* infesting bambara groundnut in storage. *Nigerian J. Ent.,* **21:** 94-103.

Makanjuola, W.A. (1989). Evaluation of extracts of neem *(Azadirachta indica* A. Juss) for the control of some stored products pests. *J. Stored Prod. Res.,* **25:** 231-237.

Odeyemi, O.O. (1991). Control of Khapra beetle, *Trogoderma granarium* Everts in decorticated groundnut with vegetable oils. *Applied Ent. Phyto.,* **58:** 31-38.

Odeyemi, O.O. (1993). Insecticidal properties of certain indigenous plant oils against *Sitophilus zeamais* Mots. *Applied Ent. Phyto.,* **60:** 19-27.

Odeyemi, O.O. (1998). Feeding and oviposition inhibiting action of powdered extracts of *Dennettia tripetala* and *Aframomum melegueta* on *Corcyra cephalonica* (Stainton) and *Ephestia cautella* Wlk. *African J. Sci.,* **2:** 32-44.

Odeyemi, O.O. and Daramola, A.M. (2000). *Storage Practices in the tropics,* vol. 1. Food Storage and Pest Problems. Dave Collins Publications, Akure, Nigeria. pp. 253.

Ofuya, T.I. (1986). Use of wood ash, dry chilli, pepper fruits and onion scale leaves for reducing *Callosobruchus maculatus* damage in cowpea seeds during storage. *J. Agric. Sci. Camb.,* **107:** 467-468.

Ofuya, T.I. (1990). Oviposition deterrence and ovicidal properties of some plant powders against *Callosobruchus maculatus* in stored cowpea *(Vigna unguiculata)* seed. *J. Agric. Sci. Camb.,* **115:** 343-345.

Ofuya, T.I. and Lagunju, S. (1998). Potential of using trona (Urao) in the control of *Callosobruchus maculatus* (F.) (Coleoptera: Bruchidae) infesting cowpea seeds in storage in Nigeria. *Entomon* **23:** 219-225.

Olaifa, J.T., Erhun, W.O. and Akingbohungbe, A.E. (1987). Insecticidal activity of some Nigerian plants. *Insect Sci. Appl.,* **8:** 221-224.

Schoonhoven, A.V. (1978). Use of vegetable oils to protect bean from bruchid attack. *J. Econ. Ent.,* **71:** 254-256.

Sighamony, S., Annes, I., Chandrakala, T. and Osmani, Z. (1986). Efficacy of certain indigenous plant products as grain protectants against *Sitophilus oryzae* and *Rhyzopertha dominica. J. Stored Prod. Res.,* **22:** 21-23.

Sowunmi, O.L. and Akinnusi, O.A. (1983). Studies on the use of neem kernel in the control of stored cowpea beetle, *Callosobruchus maculatus* (F.). *Trop. Grain Leg. Bull.,* **27:** 28-31.

Su, H.C.F. (1977). Insecticidal properties of black pepper to rice weevils and cowpea weevils. *J. Econ. Ent.,* **70:** 18-22.

Yusuf, S.R., Ahmed, B.I., Chaudhary, J.P. and Yusuf, A.U. (1998). Laboratory evaluation of some plant products for the control of maize weevil *(Sitophilus zeamais* Mots.) in stored maize. *ESN Occassional Publ.,* **31:** 203-213.

18

Eco-friendly Management of Plant Parasitic Nematodes in Vegetable Crops

V.K. Singh

Division of Plant Pathology, Sher-e-Kashmir University of Agricultural Sciences and Technology DLRSS, Dhiansar, Jammu–181 133 (J&K)

Vegetable crops are prone to several biotic and abiotic stresses during their growth period. Plant nematodes constitute one of the major causes of crop losses all over the world. Sasser and Freckman (1987) have indicated that annual crop losses due to plant parasitic nematodes on worldwide basis exceed \$100 billion. Root-knot nematodes are major pests of vegetables belonging to leguminous cruciferous, solanaceous and several other root and bulb crops like onion, garlic, carrot, radish, etc. The yields of Okra, tomato and brinjal 27.3 per cent losses respectively due to *Meloidogyne incognita* infestation @ 3-4 juveniles/g soil under field conditions (Bhatti and Jain, 1977). Even 50 per cent reduction in potato yields caused by *Globodera rostochiensis* cannot be predicted by inspection of the upper plant. Also, a 50 per cent cabbage yield reduction caused by sugarbeet cyst nematode, *Heterodera schachtii* in microplots cannot be assessed accurately by inspection alone. A large number of plant parasitic nematodes have been found associated with vegetable in India, but pathogenicity has not been proved for all. Most of the commonly occurring vegetable nematodes belong to genera—*Meloidogyne, Globodera, Heterodera, Rotylenchulus. Pratylenchus, Tylenchorhynchus, Ditylenchus, Radopholus, Belonolaimus, Dolichodorus, Hemicydiophora, Pratylenchus, Paratrichodorus, Longidorus, Nacobbus, Helicotylenchus, Rotylenchus, Trichodorus, Xiphinema,* etc.

IMPORTANT NEMATODES ON VEGETABLES

Root-Knot Nematode (*Meloidogyne* spp.)

Root-knot nematode is the most important pathogen of vegetable crops. Almost all the vegetable crops have been found to be susceptible to this nematode to a varying degree. The four major species are *Meloidogyne incognita. M. javanica, M. arenaria* and *M. hapla.* These four species account for more than 95% of the root-knot nematode population world over (Sasser and Carter. 1985). The above ground symptoms of the infected plants include yellowing of leaves, stunting of shoot and reduced size of various shoot parts. Symptoms of the disease appear on the underground parts of the plants. Infected roots swell at the point of invasion and develop into the typical root knot galls that are two to several times as large in diameter as the healthy root. The root system is galled. Several infections take place along the same root, and the developing galls give the root a rough, clubbed appearance. Usually, infected roots remain smaller, show various stages of necrosis, and develop rotting, particularly late in the season. The male and female root-knot nematodes are easily distinguishable morphologically. The males are worm-like and about 1.2 to 1.5 mm long by 30 to 36 μm in diameter. The females are pear-shaped and about 0.40 to 1.30 mm long by 0.27 to 0.75 mm wide. A life cycle is completed in 25 days at 27°C, but it takes longer at lower or higher temperatures.

Cyst Nematodes (*Heterodera* spp. *and Globodera* spp.)

Various species of *Heterodera* and *Globodera* are popularly known as the cyst nematodes. The round cyst nematode, known as the golden nematode, is particularly severe on potato. The name golden nematode of potato was proposed in 1944 by Chitwood. The above ground symptoms resemble those associated with root damage and include stunting of shoot, yellowing of foliage, etc. The severity depends upon density of nematodes attacking roots. The mature female bodies are found attached to roots by their head end embedded almost in the stele. The site of feeding is modified into a syncitium similar to that found in case of *Meloidogyne.* On the surface of infected roots, white to brown bodies of females can be discerned with naked eyes. The intensity of body colour depends upon the maturity stage of the young female or cyst. In case of cyst, nematode matrix with eggs may also be found attached to the posterior region of the female body. The male is worm-like, about 1.3 mm long by 30 to 40 μm in diameter. Fully developed females are lemon-shaped, 0.6 to 0.8 mm in length and 0.3 to 0.5 mm in diameter. Approximately, 21 to 24 days are required for the completion of life cycle of this nematode. The life cycle process of *Globodera* is similar to the other cyst forming species, with one life cycle being completed in 5-7 weeks. However, those which do not emerge can remain in viable condition inside the cysts even for 7-8 years.

Lesion Nematode (*Pratylenchus* spp.)

Lesion nematodes occur in all parts of the world. Amongst vegetables, tomato is an important host for this nematode. The plants show chlorosis, stunting and general lack of vigour resulting into wilt. The plants form patches or zones in the field. The roots show necrosis and lesions, which become ideal for infection of other microorganisms. Both male and female of this nematodes are worm-like, 0.4 to 0.7 mm long and 20 to 25 μm in diameter. They are migratory, endoparasitic nematodes. The life cycle of the various species of *Pratylenchus* is completed within 45 to 65 days. Generally, temperature and host are the most important factors, which have profound influence on the build-up of the nematode populations.

Reniform Nematode (*Rotyenchulus reniformis)*

Reniform Nematode is a pest of a large number of vegetables. In India, it is considered as pest of great significance after root-knot nematode. The nematode is a semi-endoparasite and remains attached to roots. Infected plants shows stunting in growth with reduced and discoloured root system. Damage during pre- and post-emergence of seedlings leads to reduction in germination and crop stand. The life cycle is generally completed in about a month. Soil pH is also an important factor affecting reproduction of the nematode. Soil moisture and temperature have profound influence on infectivity, penetration and the biology of the nematode.

Stem and Bulb Nematode (*Ditylenchus* spp.)

This nematode commonly attacks onion, garlic, potato, pea, carrot, etc. The symptoms of injury differ from different plants and different parts of the same plant. Infected seedlings become twisted, enlarged and deformed, leading to death of the plant. The plants become stunted. The nematode feeds on stem leaves and bulbs and is rarely found in soil. The nematode is 1.0 to 1.3 mm long and about 30 μm in diameter. The total duration of life cycle ranges from 19-23 days at 15°C with four moults and four Juveniles stages, the first moult being within the egg (Yuksel 1960).

CULTURAL CONTROL

The cultural practices include crop rotation, fallowing, flooding, sanitation, ploughing during summer season, mulching, inorganic fertilizer, spacing of plants in the field, time of sowing, resistant varieties, etc.

In crop rotation, for the control of nematodes, susceptible (host) crops are rotated with non-host crops. Flooding is reported to reduce incidence considerably. The period of flooding appears to vary with several factors such as kind of soil, season, etc. Warm conditions are said to reduce the period of lime required for control. Juveniles (larvae) are more easily killed by flooding than eggs. The period of flooding needs to be worked out for each condition. Fallow periods in cropping sequences can also reduce nematode populations.

The summer ploughing 2-3 times during hottest period of the year help to expose nematodes to the drying action of sun and wind and reduce the population. Haque and Prasad (1982) reported drastic reduction of plant parasitic nematodes and significantly increased the yield with three deep ploughings. Sanitation term covers a wide range of cultural practices including weed control, crop residue destruction and disinfestation of farm equipment before moving it from heavily infested fields to uninfested fields. In monocultures, eliminating the weed hosts can be important in reducing the populations of plant parasitic nematodes. Although manipulation of sowing dates is not very effective for nematode control, there are several reports where early planting of potato, sugar beet and tomato has been found to reduce the infestation of *Globodera rostochiensis, Heterodera schachtii* and *Meloidogyne incognita.* The nematodes can also be controlled by mulching with different available waste organic materials such as straw, grass clipping, weeds, or plastic mulches around the crops. The solar heat can be trapped and conserved using transparent polythene mulch. In summer, soils polarization effectively controls the diseases caused by nematodes (Sharma and Nene 1985, Sharma 1985a) as well as many other soil-borne diseases. It induces microbial processes that can contribute to the control of pathogens (Katan, 1981). Prasad and Yaduraju (1993) indicated that 30 days solarization is enough to achieve 95% control of nematodes. Prasad *et al.*, (1995) found about 90% kill of root-knot and reinform nematodes when solarised for 30 days during May-June months.

The addition of inorganic fertilizers alone without organic manure usually increase the nematode population and disease intensity. NH_4-N reduces the disease incidence while NO_3-N may increase the same. Particular forms as well as dose and proportion of NPK may also reduce or increase the incidence of disease. Use of tolerant/resistant varieties is the most practical approach for the management of nematode diseases. Crop cultivars resistant to phytonematodes can be the most useful and cheapest means of nematode control for the small-scale farmers.

BIOLOGICAL CONTROL

Nematodes have many natural enemies including fungi, bacteria and predacious nematodes. Certain fungi capture and kill nematodes in the soil. *Arthrobotrys* sp. *Dactylaria* sp., *Dactylella* sp., *Catenaria* spp., and *Trichothecium* spp., are the genera most commonly represented. The biological control of root-knot on various vegetable crops under greenhouse conditions by using several fungi and bacteria has been reported by Pandey and Trivedi (1992); Zucker man *et al*., (1993); Bandyopadhyay *et al*., (2001), Singh *et al*., (2001). Some fungi capture nematodes, by adhesion, but many employ specialized devices that include networks of adhesive branches, stalked adhesive knobs, non-constricting rings and constricting rings. The surface of the nematode is penetrated and the fungus hyphae grow throughout the nematode body, digesting and absorbing its contents. Under favourable conditions, large numbers of nematodes may be captured and killed especially by those fungi that form adhesive networks or hyphal loops.

A bacterial parasite of nematodes, *Pasteuria penetrans,* has received much attention and research effort in recent years. *P. penetrans* is probably the most specific obligate parasite of nematodes, with a life cycle remarkably well adapted to parasitism of certain phytonematodes. Applications of *P. penetrans* in several greenhouse and microplot tests have consistently suppressed the nematode induced root galling and egg production with consequent enhanced plant growth Walia (1994); Walia and Dalal (1994); Gogoi and Neog (2005), and Anil Kumar *et al.* (2006).

Nematicidal plants with roots containing nematicidal substances have been investigated. These toxic substances reduce the population level of some nematode species. African marigolds (*Tagetes* spp.), asparagus, crotalaria, mustard and several cruciferous plants have been reported to produce toxic substances.

Soil amendments with green manure, compost, oil cakes of neem, mahua, mustard, groundnut, cotton, linseed, karanj and saw dust, etc. have been found to reduce nematode populations. Neem, karanj or groundnut cakes incorporated into soil at the rate of 1,500 kg/ha gives good control of plant parasitic nematodes and could be practised wherever possible. Apart from encouraging the multiplication of natural enemies like nematode trapping fungi, the decomposition products of these organic amendments are toxic to nematodes. Use of organic manures is of great value since the decomposition products and promotion of natural enemies decrease nematode populations Singh and Sitaramaiah (1973), Singh and Singh (1991), Singh and Singh (1992), Alam (1986), Alam (1990), Goswami and Swarup (1971), Haseeb *et al*., (1984a), Mankau (1961), Mishra and Prasad (1974), Singh and Singh (2001). The amount of oilcakes or any organic matter to be incorporated depends on various factors like soil type and texture, crop to be planted, the predominant nematode fauna of the soil and the amount of soil moisture present in the soil. In trap crops, when grown in infested soils, the nematodes penetrate into the root system and start multiplying. Before the nematodes complete their life cycle, the plants are uprooted and destroyed.

CHEMICAL CONTROL

Nematicides are used to protect highly susceptible varieties of vegetable from nematode attack. The most important consideration in the control of nematode by chemical is the cost/return ratio. Chemicals are thus recommended for highly valuable or cash crops. The older nematicides are mostly fumigants, which are applied to soil. These are seldom recommended because of their hazardous nature and high toxicity to non-target organisms. Recently, large number of non-fumigant and systemic nematicides are available that are safe on plant. DBCP (Nemagon) has been found very effective for standing crops against root-knot nematodes, but its use has been suspended due to adverse effect on human beings. Carbofuran 3G (Furadon), Phorate lOG' (Thimet), Fenamiphos (Nemacur), Fensulphothion (Dasanit) have been recommended for a variety of crops. The dose and method of application would vary with crop. Chemicals and Phytoalexins that have implications in the nematode control (Prasad and Swarup 1986), synthetic pheromones to disrupt nematode reproduction Bone (1987), and chemicals that directly interfere with nematode chemoreceptors Jansson (1987) should be exploited for nematode population management. Prasad (1990) studied the control by seed pelleted with carbofuran (40 flowable) and aldicarb sulfone (75 wp) @ 3 and 6% W/w. The chemical control of root knot nematode on tomato under greenhouse conditions by using various chemicals has been reported by Singh and Singh (1991), Singh and Singh (1998).

REGULATORY CONTROL

Numerous attempts have been made to prevent the introduction of nematodes into countries or provinces by means of quarantine. Quarantines are established by legislative action in parliament, etc., and usually give quarantine authorities power to make and enforce regulations to accomplish the purpose. Such regulations usually prohibit bringing infected plants into protected areas where similar crops might become infected.

PHYSICAL CONTROL

Soil treatment with heat for a short period of time kills most of the nematodes. Steam treatment or hot water dips (at 45-55°C for 5-120 min.) and heat treatment of plant materials particularly the bulbs infected with endoparasitic nematodes are suitable methods but are applicable only in capital-intensive crops, as in the greenhouses or for seedlings in the nurseries. Electricity, radiation and X-rays or γ-rays and ultrasonics like heat, these measures can only be recommended as control measure in capital-intensive crops or in certified seed production programmes. Potato tubers could be freed from *G. rostochiensis* by ionizing irradiation. Some nematologists have reported control of nematodes by osmotic pressure. But this is an impractical and uneconomical proposition. These aspects were reviewed by Jenkins (1960), Jonesh (1978), Mass (1987), Southey (1965) and Southey (1978) also reported the role of physical factors in the soil for regulating nematode population.

INTEGRATION OF CONTROL MEASURES

Utilization of the best combination of available management strategies for the pest complex at hand (nematodes, insect pests, disease organisms, weeds, etc.) constitutes an integrated crop protection system. Resistant cultivars, crop rotation, pesticides, and sanitary and cultural practices can all be employed to the best possible advantage. An integrated management strategy prevents the excessive build-up of any single nematode, insect, or disease population and minimizes the development of pest resistance to any single tactic. Integrated pest management systems require flexibility and depend upon the specific pest problem and locally available management options. A fixed set of recommendations may keep a pest complex in check for a limited period of time, but as the pest population shifts, recommendations will also have to change. Therefore, system development takes into account many factors including the species and race of pests present, the availability of resistant host plants, the longevity of the pest, and the crops, cropping systems, and climate of the geographical region. The end result is a management strategy tailored to fit the unique circumstances of each pest situation. AICRP (1992) proposed an integrated approach by combining cultural methods with organic amendments, chemicals and inorganic fertilizers to reduce the population of pathogenic nematodes and enhance the yield of vegetable crops. Application of urea @ 23.8 kg/ha and neem cake @ 200 kg/ha in combination with carbofuran @ 1 kg a.i/ha in root-knot infested field gave an enhanced yield of tomato (32.2%, brinjal (36.7%) and reduced the galls by 0.2 to 0.9 times over untreated crop (AICRP, 1992). The control of plant parasitic nematodes by combined application on various vegetable crops has been reported by Jain and Bhatti (1985), Singh and Singh (1991), Reddy (1992), Singh (2006) and Singh (2006).

REFERENCES

AICRP, (1992). Achievements (1977-87). All India coordinated Research Project on Nematode Pests and their Control. pp. 16-26.

Alam, M.M. (1986). Possible use of weeds and soil amendment for the management or root-knot and stunt nematodes attacking egg plant. *Agricultural wastes,* **16:** 97-102.

Alam, M.M. (1990). Control of plant parasitic nematodes with organic amendments and nematicides in nurseries of annual plants. *J. Bangladesh Acad. Sci.* **14:** 107-113.

Anil Kumar, Walia, R.K. and Kapoor, A. (2006). Efficacy *of Pasteuria penetrans* as seed treatment in controlling *Meloidogyne Javanica* on three succeeding crops. *J. Biological Control (accepted).*

Bandyopadhyay, P., Kumar, D., Singh, V.K. and Singh, K.P. (2001). Eco-friendly management of root-knot nematode of tomato by *Arthrobotrys oligospora* and *Dactylaria brochopaga. Indian J. Nematol.* **31**(2): 153- 156.

Bhatti, D.S. and Jain, R.K. (1977). Estimation of loss in Okra, tomato and brinjal yield due to *Meloidogyne incognita. Indian J. Nematol.* **7**(1): 37-47.

Chitwood, B.E. (1944). The golden nematode of potatoes. U.S. *Dept. Agr. Circ.,* 875.

Goswami, B.K. and Swarup, G. (1971). Effect of oil cake amended soil on the growth of tomato and root-knot nematode population. *Indian Phytopath.* **24:** 491-494.

Gogoi, B.B. and Neog, P.P. (2005). Efficacy of *Pasteuria penetrans* against *Meloidogyne incognita* on green gram under field conditions. *Indian J. Nematol.* **35:** 80-81.

Haseeb, A., Alam, M.M. and Khan, A.M. (1984a). Control of plant parasitic nematodes with chopped plant leaves. *Indian J. Plant Pathol.* **2:** 180-181.

Jain, R.K. and Bhatti, D.S. (1985). In nematol network newsl. **2:** 17-18.

Jenkins, W.R. (1960). Control of nematodes by Physical methods. In (Eds) In: Sasser and W.R. Jenkins. Nematology chapel Hill, North Carolina.

Jones, F.G.W. and Krempston, R.A. (1978). *In:* Plant Nematology, (Ed.) JF. Soc. The HMS London, pp. 440.

Katan, J. (1981). Solar heatings (solarization) of soil for control of soil borne pests. *Annual review of phytopathology*, **19:** 211-236.

Mankau, R. (1963). Effect of organic soil amendments on nematode population. *Phytopathology.* **53:** 881-882.

Mass, P.W. (1987). Physical methods and quarantine. *In:* Principles and practices on nematode control of crops. Academic press, Australia, 265-279.

Mishra, S.D. and Prasad, S.K. (1974). Effect of soil amendments on nematodes on the crop yields. *Indian J. Nematol.* **4:** 1-19.

Pandey, R. and Trivedi, P.C. (1992). *Indian phytopath.* **45**(1): 134-135.

Prasad, D. (1990). Control of root-knot nematode *Meloidogyne arenaria* on groundnut by chemical seed pelleting. *Current nematol.* **1:** 31-34.

Prasad. D. and Yaduraju, N.T. (1993). Solarization of filed soil for effective management of nematodes and weeds. Nat. Sym. ecofriendly approaches Mgmt. pests, dis. Ind. Effluents, Dec. 20-22, 67 (Abstr.)

Rao, M.S. and Reddy P.P. (1992). Prospects of management of root-knot nematode on tomato through the integration of biocontrol agent and botanicals. Sem. on current trends in diagnosis and management of plant dis. IIHR Bangalore, pp. 10.

Sasser, J.N. and Carter C.C. (1985). An advanced treatise on *Meloidogyne.* Vol I: Biology and control, North Carolina state university graphics pp. I-422.

Sasser. J.N. and Freckman, D.W. (1987). A world perspective on nematology: the role of society. *Vistas of Nematology*, 7-14.

Sharma, S.B. (1985a). Nematode diseases of chickpea and pigeonpea. Pulse pathology progress report No. 43 Patancheru, A.P. (India) International crops research institute for the semi arid tropics. pp. 103.

Sharma, S.B. and Nene, Y.L. (1985). Effect of presowing solarization on plant parasitic nematodes in chickpea and pigeonpea fields. *Indian J. Nematol.* **15:** 277-278.

Singh, R.S. and Sitaramaiah, K. (1973). Res. Bull. No. 6, G.B. Pant University of Agril. Sci. & Tech., pp. 1-289.

Singh. K.P. and Singh V.K. (1991). Nematicidal nature of Arjun bark *Terminelia arjuna* on cyst nematode *Heterodera cajani* of pigeon pea. *New Agriculturist* **2**(1): 77-78.

Singh K.P. and Singh V.K. (1992). Terminelia arjuna leaf powder reduces population density of *Heterodera cajani. Interrnational Pigeonpea Newsletter* **16:** 17-18.

Singh. K.P. Bandyopadhyay, P. and Singh, V.K. (2001) Predation and performance of two predacious fungi for control of root-knot nematode of brinjal. *Mycopathological Research* **39**(2): 95-100.

Singh, V.K. and Singh, K.P. (1991). Effect of various nematicides on the control of root knot nematode, *Meloidogyne javanica* on tomato. *Indian Biologist.* **23**(2): 49-50.

Singh, V.K. and Singh, K.P. (1991). Population management of root-knot nematode *Meloidogyne incognita* on cauliflower. *Indian biologist.* **23**(2): 51-52.

Singh, V.K. and Singh K.P. (1998). Effect of granular nematicides on the biology of *Heterodera cajani* of pigeonpea. *Indian J. Nematol.* **28**(2): 168-173.

Singh, V.K. and Singh, K.P. (2001). Effect of some medicinal plant leaves on the biology of *Heterodera cajani. Indian J. Nematol.* **31**(2): 143-147.

Singh, V.K. (2006). Management of root-knot nematode, *Meloidogyne incognita* infecting cauliflower. *Indian J. Nematol.* **36**(1): 126.

Singh V.K. (2006). Management of root-knot nematode infecting tomato. *Indian J. Nematol.* **36**(1): 127.

Southy, J.F. (1965). Physical method of control. In: plant nematology (Ed.). J.F. Southey, GDI Min. Agr. Fish Food, HMSO, London, pp. 323-332.

Southey, J.F. (1978) Regulation controls. *In:* plant nematology (Ed.) J.F. Southey GDI. Min Agr. Fish food, IIMSO, London pp. 326-332.

Walia, R.K. (1994). Assessment of nursery treatment with *Pasteuria penetrans* for the control of *Meloidogyne javanica* on tomato in greenhouse. *J. Biological Control.* **8:** 68-70.

Walia, R.K. and Dalal, M.R. (1994). Efficacy of bacterical parasite, *Pasteuria penetrans* application as nursery soil treatment in controlling root-knot nematode *Meloidogyne javunica* on tomato in greenhouse. *Pest Management and Economic Zoology* **2:** 19-21.

Yuksel, H.S. (1960) Observations on the life cycle of *Ditylenchus dipsaci* on onion seedlings. *Nematologica* **5:** 289-296.

19

Use of Biotechnological Tools in Integrated Pest Management

Seema Singh Thakur, A.K. Karnatak and D.C. Karnatak

Department of Entomology, College of Agriculture, G.B. Pant University of Agri. & Tech. Pantnagar–263 145, Uttarakhand

Biotechnology has contributed much to the field of insect pest management so far through development of transgenic plants and other novel eco-friendly products to control insects. Recognition of the value and relevance of biotechnological applications in insect control is gaining momentum. *Integrated Pest Management* is today a widely accepted strategy to reduce overdependence on chemical insecticides and their potentially negative environmental and economic effects. Biotechnology has considerable potential to contribute to sustainable biological elements of IPM. Biotechnology for insect pest management has to some extent been an early by product of the acquisition of biotechnological know how, which will have more substantial implications for agriculture than simply improved IPM.

Agriculture in India accounts for nearly 65% of the country's employment, 26% of the total GDP and nearly 20% of total export earning and supplier of raw material to major industries. Agriculture is not only the backbone of Indian economy and food security but also a way of life, a tradition and anchor of overall livelihood opportunity for about 700 million of our one billion population. Agriculture, therefore, is and will continue to be central to all strategies for planned socio-economic development of the country. India's population is expected to reach approximately 1.5 billion by 2050. It is estimated that around 300 million (roughly 30%) of India's population is suffering from malnutrition. Despite major advances in agriculture and strong growth in food production in the latter part of the twentieth century, hunger and famine are still common in many parts of the country. The application of biotechnology techniques within the agriculture sector can potentially improve food security by raising crop tolerance to adverse weather and soil conditions, by enhancing adaptability of crops to different climates and by improving yields, pest resistance and nutrition, particularly of staple food crops.

Genetic modification of crops through genetic engineering is one such method allowing individual characters (gene, factor or trait) to be transferred into crop plants.

Advantages of Transgenics in IPM

- Less use of conventional insecticides thus, safe to natural enemies.
- Reduction in traditional spray of insecticides.
- No resurgence or outbreak of pests.
- Slow development of pest to economic level.
- Horizontal transfer of genetic information from one species to another.
- Insecticidal genes are single dominant, multiple genes often code similar trait.
- Total pest control input has been reduced by 57% in cotton and 90% in bollworm control.
- Increase in worker safety as well as safe to non-target organism and human beings.
- Reduction in environmental pollution as there is no drift and no groundwater pollution.
- Transgenic plants eliminate the problem of shelf-life and field stability faced by insecticide formulation.
- Since toxin will always be there, so no need to monitor the plant.
- Provides protection to those parts which are not protected with insecticides like borers and bollworms.
- Cost of application will be nil.

Disadvantages

- Development of resistance to GE crops, though it may be slow but pest will remain continuously in the toxic environment.
- There may be no reduction in the use of insecticides.
- Little attention to ETL on GE crop.
- Increase in population of non-target pests that will require additional sprays.
- Adopters are not using cultural control practices—longer crop rotations.
- Adult lepidopterans are pollinators, and their elimination may reduce the number as pollinator. This would become more serious if the toxin is expressed in nectar and pollen.
- Uncertainty about the stability of genes. Expression of gene may be tissue specific, wound specific or at particular growth stage.
- Additional insecticide sprays were needed due to Bt cotton failing to control bollworm in Eastern Texas.
- Twenty-seven per cent yield reduction and lower copper level in leaves in Bt cotton.
- Aphids sequestered the Bt toxin apparently affecting coccinellids in potato.
- High cost of production.
- Expression of undesirable phenotypic traits.

Steps Involved in Genetic Engineering

The steps involved in genetic engineering are selection of plant gene where introduction would be of positive value, identification and isolation of such gene, transfer into plant cell and regeneration of complete plants from transfer cells/tissue.

Criteria for Selection of Genes for Resistance against Insect Pests

- Select and transfer genes that are more effective in the management of pests than traditional HPR genes.

- Better understanding of mode of action of new gene is essential.
- Resistant gene must be environmentally benign and safe.
- Must combine with HPR gene to provide more effective control.

METHODS OF GENETIC ENGINEERING IN PLANTS

1. Vector-Mediated Gene Transfer

(i). Agrobacterium is a Gram-negative soil bacterium. In addition to carrying chromosomal DNA, it carries Ti plasmid and is the natural genetic engineer of plants. During infection, Ti plasmid transfers a portion of its DNA (T DNA) into plant cell and becomes integrated into the chromosomal DNA of plant. T DNA segment of Ti plasmid carries a number of genes encoding enzymes for the synthesis of phytochromes, which stimulate the growth and division of the plant tissues resulting into the formation of tumour. Foreign gene inserted within the T-region of the Ti plasmid are transferred to and stably integrated into the plant genome.

(ii). DNA viruses as vectors: The genomes of cauliflower mosaic virus (CaMV) and geminivirus (Tomato golden mosaic virus, TGMV) are double-stranded DNA, which make these viruses as potential transformation vector.

2. Vectorless Gene Transfer

(i) Direct uptake of DNA

In the presence of PEG (polyethylene glycol), plant protoplast is capable of taking DNA.

(ii) Electroporation

A high voltage current is applied in a pulsed manner, which creates tiny holes in the plant cell membrane. These holes are large enough for DNA molecule to diffuse into the cell. The cell recovers from electric shock and can be generated into whole plant.

(iii) Microinjection

Injection of DNA directly into the plant cell nucleus or intact plant organ has been used to develop transgenic plants.

(iv) Microprojectile bombardment (Biolistic approach)

Microparticles (l-3μ) of gold or tungsten are coated with DNA and are accelerated (1400 feet/sec) by accelerator into meristematic tissues, pollen, etc. Thus, DNA is transferred into plant nuclear genome.

In IPM programme, resistant plants are one of the most acceptable tools that are eco friendly, non-hazardous to non-target organisms, no residue problems, no deleterious effect on natural enemies of insects, no pest resurgence and outbreak of secondary pests as has been seen with the use of insecticides. First insect control protein genes of *Bacillus thuringiensis* were introduced in tobacco and tomato.

Resistance Genes from Microorganisms

- ***Bt* toxin gene:** Bt gene codes for Bt toxin, which differs in insecticidal spectrum. The insect toxicity of Bt resides in a large crystal protein (delta-endotoxin). They are converted into active form upon ingestion by the susceptible insect, thus killing the insect by disrupting the ion transport across the brush border/membrane.
- **Cholesterol oxidase** (CO): This protein is present in streptomyces culture filtrate and is toxic to boll weevil larvae.

Resistance Genes from Higher Plants

At present there are two groups of plant derived genes that are used to confer insect resistance on crop plants by retarding growth and development in insect.

- **Proteinase inhibitors:** Plants contain peptidases acting as protease inactivating proteins (PIPs). The different proteinases are serine, cysteine, aspartic acid and metalloproteinase. They catalyse the release of amino acids from dietary protein.
- **Cowpea trypsin inhibitor gene (CpTI) :** From cowpea, *Vigna unguiculata* is most active inhibitor. This gene produces antimetabolite substances that provide protection against insects, Bruchids, etc.
- **Alpha-amylase inhibitor:** This is isolated from *Phaseolus vulgaris* and is very effective against Bruchids. The inhibitor protein blocks the larval feeding in the midgut as it blocks the function of amylase, which is responsible for digestion of starch.
- **Lectins:** They are glycoproteins. Lectins from snowdrop, *Galanthus nivalis* showed activity against aphids.

Resistance Genes from Animals

Genes involved are serine proteinase inhibitors from mammals and tobacco horn worm. They inhibit proteolysis in lepidopteran. Bovine pancreatic inhibitor (BPTI), alpha-antitrypsin (alpha AT) and spleen inhibitor (SI) are promising insect resistance protein.

Table 19.1: Transgenic plants carrying Bt gene for insect resistance.

S. No.	Plant	Gene (s)	Target insect
1.	Rice	Cry 1A(b), Cry 1A(o)	*Chilo suppressalis, Cnaphalocrocis medinalis, Scirpophaga incertulas*
2.	Tomato	Cry 1A(c)	*Manduca sexta*
3.	Poplar	Cry 1A	*Lymantria dispar*
4.	Tobacco	Cry 1A(b)	*Helicoversa virescens, H. zea, M. sexta*
5.	Cotton	Cry 1A(b), Cry 1A(c)	*H. zea, Pectinophora gossypiella, Spodoptera exigua, Trichoplusia ni*
6.	Sugar cane	Cry 1A(b)	*Diatraea saccharalis*
7.	Potato	Cry 3A, Cry 1A(b)	*Leptinotarsa decemlineata*

What has Biotechnology brought to IPM

In order to understand whether biotechnology is making a positive contribution to IPM, it is useful to review the pest control biotechnologies presently available or in development. One focus of biotechnological research has been on improving natural enemies of pests as pest control agents. This has focused principally on pathogens of insect pests and their use as formulated biological pesticides. Emphasis has been placed on bacteria and viruses, largely because they are better understood and more easily manipulated, as opposed to fungi, protozoa, nematodes and arthropod predators.

Research on bacteria has concentrated on *Bacillus thuringiensis* (Bt), which is already widely used as a biopesticidal formulation to control caterpillar and beetle pests of crops and flies, which are disease vectors. Research has focused on increasing the host range and virulence of Bt by combining genes with different host specificities and properties. Research is also underway on improvement of *Xenorhabdus* spp., the bacteria responsible for the mortality of nematode-infected hosts. Here the emphasis is on stabilizing and improving the virulence of these bacteria. Insect viruses also have a market in their natural form as biopesticides, mostly against caterpillar pests of forestry and field crops. Biotechnological research has focused on engineering of certain viruses to express genes whose toxins kill faster than the wildtype viruses.

The second principle area of biotechnology for pest control has been the development of crop varieties resistant to pests and diseases. This has concentrated on incorporating insect and virus resistance into the plant genome. In addition, modification of the genome of plant-associated microorganisms has been followed as a strategy to confer insect resistance to plants. By far the most extensive application has been the incorporation of genes, which produce various *Bt delta endotoxins* into crops, primarily to confer resistance against caterpillars and beetle pests. In 1996, the first commercial transgenic varieties of cotton, maize and potatoes were released in the USA. Research is also conducted on incorporating genes for plant or microbially derived compounds, which affect insect pests, such as trypsin inhibitase and cholesterol oxidase. Overall, the application of biotechnology to IPM has so far been quite conservative. It has focused largely on improvement of existing crop protection products or technologies by use and manipulation of viral and bacterial genomes. Thus, engineered viruses and bacteria are variants of existing biopesticidal formulations of the same species. Some transgenic plants or plant-associated bacteria seek to improve on what can already be achieved less effectively by topical application of particular microorganisms, particularly Bt. The technical objectives of many of these manipulations are directed at improving the performance of an engineered product relative to its "wildtype" competitor by:

- broadening the target spectrum of the product;
- increasing the speed of action of the product;
- improving the delivery of the product to the pest.

The rationale has much to do with competition with existing products and the rapid acquisition of large markets. To consider how they may succeed at this, it is useful to examine the future of the relevant areas of pest control, namely biopesticides and host plant resistance.

BIOTECHNOLOGY IN BIOPESTICIDE DEVELOPMENT

Today, there are over a hundred commercial biological control products in the market, and many more are locally produced and supplied for particular productions systems. However, most commercial biological control has focused on insect pathogens because of their relative ease of mass production and their capacity to be used in the same manner as formulated chemical insecticides.

Bt has been the principal target of product development. As a product, *Bt* is valuable in IPM systems because it is much less harmful to predators and parasites than broad spectrum chemical insecticides. Therefore, it can be substituted for chemical products in "insecticide treadmill" situations and will allow the recovery of natural enemy populations. Like many biopesticides, it is often less effective on its own than a highly potent chemical product.

However, in an IPM system, where it is used only when needed and it conserves natural enemies, its impact is augmented by the action of those natural enemies (which is free of charge to the farmer) and it can be both more economical and sustainable.

A second problem facing *Bt* is the risk of resistance. Where Bt is still used as a single technology solution, like its chemical predecessors, it is sprayed regularly and a range of insect pests are now developing resistance.

Bt's third problem is that it lacks the most desirable biological property of a biological control agent: its ability to reproduce and perpetuate itself in crops.

MYTHS AND REALITIES OF *Bt* COTTON

- **Myth:** Transgenic cotton will pollinate other species planted nearby or be carried over long distances with unknown consequences.
- **Reality:** Trials show that in only 2.1% of cases, do the *Bt* cotton's pollen travel across non-*Bt* cotton and that too not beyond 2 metres. Also, the pollination is incompatible for cross-pollination with near relatives other than cotton in any Indian conditions.
- **Myth:** Since *Bt* cotton has an in-built resistance against bollworm, a pest that is the major threat to cotton, it might also kill other insects, which may be advantageous.
- **Reality:** Bt proteins have no detrimental effects on the survival and production of collmbolla, a beneficial species of insects. *Bt* can be harmful to useful pests but it has proven to be much better than conventional pesticides.
- **Myth:** Cattle or animals that gaze may be adversely affected.
- **Reality:** Studies in the Industrial Toxicology Research Centre Lucknow, established that Bt cotton is safe to mammals, birds and fishes. The data on allergy studies on rats show no adverse effect.
- **Myth:** *Bt* cotton ensures that seeds die with each year's crop forcing the farmer to buy afresh every year. (Referred to as terminator gene).
- **Reality:** There is no terminator gene in Indian *Bt* cotton. The seeds have already progressed to six generations.

The most significant downside of *Bt* cotton is the threat of development of resistance by insects and pests. This threat can be countered by the use of "Refugia" to prevent or delay the development of resistance.

"Refugia" is a method when non-*Bt* cotton is allowed to grow along with *Bt* Cotton so that the pests can thrive on the non-*Bt* cotton and breed with the pests that may have developed resistance, hence reducing the risk of resistance.

BENEFITS OF *Bt* COTTON

In addition to direct control of insects during crop growth of cotton and economic benefits that allow cultivars to produce cotton crop more profitably, *Bt* cotton also provides some secondary benefits.

1. *Bt* Cotton: Pesticides, Labour and Health

In South Africa, a study reveals a direct cost benefit for *Bt* growers per hectare per season due to a reduction in the number of insecticide applications. Cost savings emerged in the form of lower requirements for pesticide, but also important were reduced requirements for water and labour. The reduction in the number of sprays was particularly beneficial to women who do some spraying and children who collect water and assist in spraying. The increasing adoption rate of *Bt* cotton appears to have a health benefit measured in terms of reported rates of accidental insecticide poisoning. These appears to be declining as the uptake of *Bt* cotton increases. *Bt* cotton gives higher yield than non-*Bt* varieties and generated greater revenue. Seed costs for *Bt* cotton were double than those of non-*Bt*, although pesticide costs were lower.

2. Reduced Fuel Usage, Air Pollution and Related Waste

During farm operations large amount of fuel especially diesel is used. Around three litres of diesel or petrol is used to spray one ha of crop of 70 days old (when first peak of bollworm appears on cotton). Further field trips are reduced as the number of sprays is reduced. Further burning of every litre of diesel/petrol fuel releases 1.67 kg CO_2. Hence, air pollution is averted by using *Bt* cotton varieties as 60-80 per cent insecticide application is reduced.

3. Increase in Population of Beneficial Insects

It is a known fact that due to application of broad spectrum insecticides, many non-target, beneficial insects are also eliminated, but *Bt* cotton is toxic to only some lepidopteran insects and does not possess adverse effect on non-lepidopteran, beneficial fauna.

4. Safety to Farm Workers, other Birds and Animals

Reduced application of broad spectrum insecticides due to *Bt* cotton has proven safe for farm workers. Further, spraying of broad spectrum pesticides has taken lives of nesting birds, butterflies and wild animals visiting fields during night-time. There is a huge saving on water, which is used as a diluent for pesticides.

BIOTECHNOLOGY AND CROP RESISTANCE

Engineering genes for *Bt* toxins into plants is an ingenious method of delivering these toxins to pests which might naturally avoid them, such as insects that feed inside plants. From an IPM perspective, this technology has more similarities to plant resistance breeding than biopesticide development.

Pros and Cons of the Technology in Agriculture

- ***Allergies:*** Eating certain transgenic foods has occasionally led to the development of allergies. A study reported in the *New England Journal of Medicine* in 1998 showed that people consuming transgenic soyabean intended as animal feed developed certain allergic reactions. Transgenic crops marketed for human consumption have not been linked directly to causing widespread allergies.
- ***Monarch butterfly mortality:*** Recent studies reported in *Science* and *Oecologia* journals suggested that pollen from the transgenic *Bt* corn may be fatal to monarch butterflies, which feed on milkweed coated with com pollen. Scientists have confirmed the mortality of monarch butterflies exposed to Bt com pollen under both laboratory and field conditions. A study published in *Nature* (1999) indicated that secretions from remains of *Bt* com adversely affected certain other soil-borne non-target insect species.
- ***Soil erosion:*** Herbicide-tolerant crops are used for post-emergent weed control in no-till farming. This reduces the need for tillage, which in turn reduces soil erosion and nutrient losses. A potential environmental risk comes with this technique because it involves total vegetation control, which may affect an ecosystem's species diversity.
- ***Herbicide-resistant weeds:*** The evolution of "superweeds" capable of resisting herbicides as a result of using herbicide-tolerant crops has been a topic of public interest. Such a phenomenon is highly unlikely because there are more than 12 distinct groups of herbicides displaying different sites and modes of action. The popular herbicide, Roundup, is only one member of one of these herbicide groups. Therefore, the likelihood of a weed becoming resistant to all known herbicides is almost impossible. The same argument and counter-argument are true for "escapes" of transgenic plants into the surrounding habitat.
- ***Food supply and pesticide usage:*** Transgenic crops may provide increased profits to the farmer while providing cheaper and more nutritious food. Genetic engineering also has helped make crops available that could not otherwise tolerate adverse environmental conditions (drought, cold, high salt levels in the soils, etc.). Such crops are capable of resisting pests, generating higher yields, and producing food with high nutrient content. They are considered an effective means of dealing with pest problems while reducing production costs.
- ***Gene flow:*** The flow of transgenes into other organisms through pollution (termed "genetic pollution") may pose unknown risks to the ecosystem. Once these genes are released, it is difficult to recall them. Proponents of the technology claim that this form of gene pollution is similar to introducing alien species into an ecosystem. The ecosystem is dynamic, and human interference caused it to change throughout history. Some organisms have become a nuisance although they were originally introduced as a biological control agent (e.g., Japanese ladybug).
- ***Crop diversity:*** Since genetic engineering focuses on crops with certain highly desirable traits, genetic diversity within the crop could be diminished. This can make crops more susceptible to natural calamities such as disease outbreaks. Such problems also have been encountered with hybrids generated by traditional breeding techniques.
- ***Faster breeding:*** A possible benefit of transgenic crops or animals is that they can be bred for desirable traits very precisely and much faster than when traditional methods are used.

The related disadvantage is that because the actual "breeding" in genetic engineering is carried out under laboratory or sterile conditions, the implications under field conditions may not be fully understood until a problem arises.

- ***Nutrient levels:*** Certain transgenic crops (e.g., "golden rice" capable of synthesizing the precursor of vitamin A) are capable of producing higher amounts of nutrients and vitamins, which could have a great impact on solving nutrition problems in heavily populated and underdeveloped countries.
- ***Resistant microbes:*** Although there is no evidence, there are claims that transgenic crops may lead to the release of resistant strains of microbes into the environment by plants. This has been contradicted by proponents of the technology who state that such risks are highly unlikely compared to similar releases from medical or veterinary practices.
- ***Pharmaceuticals:*** Researchers are testing transgenic plants that are valuable to farmers and consumers now. They are capable of producing vaccines, pharmaceuticals, and other materials used in the medical industry. But their ability to safely contain such products has been questioned.
- ***Disease pathogens:*** Plant pathogens capable of causing cancers *(Agrobacterium tumefaciens)* are used sometimes to carry the novel genes into the transgenic plants. There is a low likelihood that such pathogens could recombine with their equivalents in the plant and cause new plant diseases. Other techniques (e.g., "gene gun") can be used to insert genes into the DNA of an organism, which may rule out the possibility of such recombining.

FUTURE PERSPECTIVES/CONCLUSION

The future of biotechnology has much promise. Biological control and host plant resistance stand out as elements of IPM, which are potentially self-renewing and available to all farmers, rich or poor. Biotechnological innovations, which improve the persistence or efficiency of these biological processes, for instance by improving survival or transmission rates of pathogens, or facilitating broadly based quantitative crop resistance to pests, will be valuable to the sustainable IPM of the future. Beyond these areas, biotechnology has considerable potential application in improving mass production technologies for natural enemies of pests, and for improving diagnostic systems, which allow scientists to recognize desirable plant genes and natural enemies, and which allow farmers to recognize potential pest problems before they cause damage. In its next generation, and with the benefit of the IPM experience, biotechnology stands to contribute greatly to sustainable pest management.

20

Management of Rice Diseases through Induced Resistance

Chinmay Biswas, S.K. Biswas and S.S.L. Srivastava

PDCSR (ICAR), Modipuram, Meerut–250 110
Department of Plant Pathology, CSAUA&T, Kanpur–208 002, U.P.

Systematic management of plant diseases dates back to 1885 with the advent of Bordeaux mixture by Milardet. Thereafter, a number of contact and systemic fungicides have been reported over the years for the management of various plant pathogens, of which many have become obsolete and many are still being used with excellence. However, the ecological hazards caused by the chemical fungicides have prompted scientists to search for other avenues of crop protection. Most of the conventional chemical and biological plant protection procedures, as well as the use of disease-resistant cultivars, tend towards the direct control of pests and diseases by their elimination. But these practices generally and particularly in case of monoculture crops like rice, raise problems. The risk of developing pesticide resistance in plant pathogens is enhanced by the long-term use of the same chemical. The resistance genes of the plants can also be overcome by changes in the pathogens. Therefore, the situation demands development of new strategies for plant protection, which can supplement the use of resistance breeding and chemical control. Of late, to this endeavour, researchers are trying to develop induced resistance against plant pathogens. Induced resistance refers to heightened resistance in a plant towards pathogens as a result of a previous treatment with a pathogen, an attenuated pathogen or a chemical that is not itself a pesticide (Deverall and Dann, 1995). Induced resistance may be localized at the site of the inducing treatment or it may be systemic and thereby effective in all or some parts of the plant distant from the site of induction. There is enhancement in resistance in response to an extrinsic stimulus without a known alteration of the genome. The protection is based on the stimulation of defense mechanisms by metabolic changes that enable the plants to defend themselves more efficiently.

The major rice diseases against which induced resistance have been reported are rice blast, brown spot, sheath blight and bacterial blight. The inducing agents could be either biotic or abiotic. Resistance has been reported to be induced through pre inoculations with non-pathogens, less virulent or avirulent races, biocontrol agents and pretreatments of the plants with naturally occurring metabolites or chemically defined substances. The induction of resistance in rice against various diseases has been discussed below under different sub-heads based on the nature of the inducing agents.

PRE-INOCULATION WITH NON-HOST PATHOGENS

Pre-inoculation of rice plants with some pathogens of other plants can induce resistance.

Manandhar *et al.* (1998) reported that isolates of *Bipolaris sorokiniana,* a non-rice pathogen could substantially reduce leaf blast caused by *Magnaporthe grisea,* when applied 24 h or more before the pathogen and induced resistance was suggested to be involved in the suppression of the disease.

Inoculation of rice plants with barnyard grass *(Echinochloa* sp.) pathogens viz. *Alternaria alternata, Curvularia lunata, Helminthosporium gramineum* and *Exserohilum monocera* has been reported to increase the control of rice blast by 53-82% (Huang-Shiwen *et al.*, 2005).

Hassa *et al.* (2000) noticed that inoculation of rice plants with non-host pathogen *Pseudomonas syringae pv. syringae* leads to the activation of defence-related genes and ultimately to induced resistance against the rice blast fungus. One of the molecular determinants of *P. syringae* pv. *syringae* that is recognised by the plant cells and evokes these defence responses is syringolin. The elicitor syringolin, secreted by the bacteria is a novel and unusual peptide consisting of a 12-membered ring formed by the 2 non-proteinogenic amino acids, 5-methyl-4amino-2hexenoic acid and 3,4-dedhydrolysine. Application of syringolin to rice leaves elicited the accumulation of Pir 7b, as well as other defence related transcripts, and induced resistance to *M. grisea* (Waspi *et al.,* 1998).

Pseudomonas syringae isolates from cucumber *(P. syringeae* pv. *lachrymans),* tobacco *(P. syringae* pv. *tabaci),* bean *(P. syringae* pv. *syringae)* and tomato (*P. syringae* pv. *tomato)* were reported to induce systemic resistance against bacterial blight caused by *Xanthomonas oryzae* pv. *oryzal* with reduction in disease lesions by 30%, 35%, 38% and 41% respectively (Jin *et al.*, 1997).

PRE-INOCULATION WITH AVIRULENT/MILDLY VIRULENT ISOLATES

There are many instances where inoculation with avirulent or mildly virulent isolate of a pathogen confers protection against the concerned virulent race.

Sinha and Das (1972) reported that inoculation with spore suspension of a mildly virulent race of *Helminthosporium oryzae* induced considerable resistance in rice cultivars viz. Dular and Latishal to a challenge inoculation with its virulent race. Prior inoculation with the mild race reduced both the number and size of brown spot lesions caused by the virulent race. This effect gradually increased with an increase in the concentration of mild race inoculum up to a certain level (c. 5×10^5 spores/ml of suspension), but rapidly declined thereafter.

It had also been possible to induce considerable resistance in susceptible rice plants (cv. Dular and Dharial) to a virulent race of *H. oryzae* by previously inoculating them with an avirulent race. (Sinha and Trivedi, 1972).

When rice seedlings were preinoculated with weakly virulent isolates of Magnaporthe from *Eleusine indica* and *Setaria viridis* expression of the virulent race was suppressed with reduction of blast lesion numbers by 35–38% (Du-Xin Fa *et al.*, 1996).

Pretreatment with weakly virulent strain of *Xanthomonas oryzae* pv. *oryzae* was reported to induce systemic resistance to the virulent strain. (Zeng-FuHua *et al.,* 1999). Liu-FengQuan and Wang-JinSheng (1998) reported 48.5% control of bacterial leaf blight in rice by applying on avirulent mutant strain Du 728.

ROLE OF PLANT GROWTH PROMOTING BACTERIA

Yang-HaiLian *et al.* (1999) reported that when two strains of nitrogen-fixing bacteria viz. *Enterobacter cloacae* MR 12 and *Aicaligenes paradoxus* R4 were sprayed onto the rice leaves at booting stage, five days prior to the inoculation with *Xanthomonas oryzae* pv. *oryzae,* resistance to the bacterial blight disease was increased by 38% and 27% respectively.

Seed imbibition as well as phyllosphere spray of Methylobacterium was reported to promote plant growth, increase plant height, number of tillers, plant biomass and grain yield. The bacteria also significantly reduced the sheath blight (*R. solani*) incidence. The percentage disease reduction recorded for seed imbibition and phyllosphere spray were 17.8 and 23.5% respectively. Rice plants sprayed with Methylobacterium sp. strain PPFM-Os-07 showed increased presence of PR-proteins and phenolic contents on day one after inoculation. Maximum phenylalanine ammonialyase and peroxidase activity on day 4 and beta-1, 3-glucanase and chitinase activity on day 5 were recorded. (Munnusamy-Madhaiyan *et al.* 2004).

USE OF ANTAGONISTIC MICROORGANISMS/BIOCONTROL AGENTS

Antagonistic microorganisms are generally used in direct control of plant pathogens by virtue of their capacity of hyperparasitism, inhibition, antibiosis, etc., and the phenomenon is called biological control. However, some biocontrol agents are also known to induce resistance upon preinoculation.

1. ***Trichoderma:*** Rice seedlings pre-treated with *Trichoderma harzianum* strain NF 9 exhibited moderate resistance against blast as well as bacterial leaf blight. The activities of phenylalanine ammonia-lyase and peroxidase were found to increase (Nzojiyobiri *et al.*, 2003). Kumawat (2006) reported induction of resistance against brown spot *(Drechslera oryzae)* by preinoculation with *T. viride* with increased levels of protein and phenolic content.

   Ngueko *et al.* (2002) reported that preinoculation of *T. harzianum* prior to inoculation of *M. grisea* led to a decrease in blast severity.
2. ***Serratia marcescens:*** Application of an antagonistic bacterium, *Serratia marcescens* as spray on rice phylloplane and pouring into rhizosphere before inoculation with blast pathogen suppressed the disease. However, the induced resistance was not associated with increased activity of peroxidase, PAL, or chitinase. However, lipoxygenase levels were elevated. (Someya *et al.*, 2002).
3. ***Pseudomonas fluorescens:*** Krishnamurthy and Gnanamanickam (1997) reported 18% suppression of rice sheath blight due to systemic resistance induced by *Pseudomonas fluorescens.*

The Pseudomonas application as a bacterial suspension or a talc-based formulation through seed, root, rest and foliar application either alone or in combination effectively reduced sheath blight disease incidence, promoted plant growth and ultimately increased yields and increases in chitinase and peroxidase activity were noticed. (Nandakumar *et al.*, 2001; Pathak *et al.*, 2004).

Mosa (2002) reported induction of resistance in rice also against blast by *Pseudomonas fluorescens* strain PF5 with increased Peroxidase activity.

Vidhyasekaran *et al.* (2001) reported that severity of bacterial blight caused by *Xanthomonas oryzae* pv. *oryzae* decreased from 6.8 to 1.2 due to pre-treatment with *Pseudomonas fluorescens* Pfl on rice leaves or seeds. In induced resistant leaves, a sharp increase in lignification and activities of Peroxidase, PAL, etc. was observed.

SYNTHETIC ORGANIC COMPOUNDS

A variety of synthetic organic compounds have been tried to induce resistance against plant pathogens and many are successful to combat rice diseases.

1. **Chitosan:** Sinha *et al.*, (1993) reported that seed treatment with low doses (0.002-0.02%) of chitosan, an oligomer of fungal origin induced resistance in many fungal pathogens.

   Enhanced defence against blast pathogen was also observed in rice seedlings treated with chitosans. Chitosans of different molecular weights were found to have different capacities to induce defence responses. Chitosans of low molecular weights were more effective. The induced resistance was associated with a set of defence responses including production of hydrogen peroxide, increased activities of phenylalanine ammonia lyase and chitinase. (Lin-WuLing *et al.*, 2005).
2. **2, 6-dichloro isonicotinic acid:** The synthetic inducer 2, 6-dichloro isonicotinic acid (INA) efficient protected rice leaves from infection by the rice blast fungus, *Magnaporthe grisea.* Lipoxygenase enzyme activity and levels of non-conjugated jasmonic acid (JA) were enhanced in leaves of plants treated with a high dose of INA (100 ppm). (Schweizer *et al.*, 1997).
3. **Acibenzolar-s-methyl:** Application of acibenzolar-s-methyl (ASM) induced resistance in rice against bacterial blight caused by *Xanthomonas oryzae* pv. *oryzae.* Induced systemic resistance persisted upto 3 days in the pre-treated rice plants. Increased phenolic content and accumulation of PR proteins viz., chitinase, beta 1,3-glucanase and thaumatin-like protein were observed in rice plants pre-treated with ASM followed by inoculation with the blight bacterium (Babu *et al.*, 2003).

   Oostendorp *et al.* (2001) reported that acibinzolar-S-methyl induced long-lasting resistance in monocots against a broad spectrum of diseases. ASM is translocated systemically in plants and can take the place of salicylic acid (SA) in the natural systemic acquired resistance (SAR) signal pathway, including the same spectrum of resistance and the same set of molecular marker.
4. **3-Chloro-l methyl-1H-pyrazole-5 carboxylic acid (CMPA):** A pyrazole derivative, 3-chloro-l methyl -lH-pyrazole-5 carboxylic acid (CMPA) exhibited high anti rice blast activity without any significant antimicrobial activity with ED 80 (80% effective dose, the concentration needed for 80% inhibition of disease development) of 0.05 mg/pot. The carboxyl group at 5-position of 1-methyl-lH pyrazole played an important role in the activity and the halogen atom (chlorine/bromine) at 3-position enhanced the activity. CMPA was also reported to induce resistance against rice bacterial blight caused by *Xanthomonas oryzae* pv. *oryzae* (Nishioka *et al.,* 2003, Nishioka *et al.,* 2005).

5. **Benzo (I, 2, 3)-thiadiazole-7 carbothioic acid S-methyl ester (BTH):** Foliar application of BTH significantly reduced the disease severity of rice seedling blast. The optimum concentration of BTH for inducing SAR was 0.5 mmol/litre, the best time of treatment was 7 days before inoculation with *M grisea* and the period of induced resistance could last 14 days or more. BTH increased cinnamyl-alcohol dehydrogenase (CAD), lipoxygenase (LOX) and beta -1, 3-glucanase activities both in treated and untreated rice leaves. (Ge-Xiuchun *et al.* 1999, Ge-Xiuchun *et al.,* 2002).

   Pre-treatment of rice plants with BTH showed resistance also to the bacterial blight. Approximately 50% reduction in disease intensity was observed in plants treated with BTH at 100 micro g/ml (Babu *et al.* 2003, Song-FengMing *et al.,* 2001).

   Zhang-WeiDong *et al.* (2003), reported induction of systemic acquired resistance of rice against sheath blight disease (*Rhizoctonia solani*).
6. **Probenazole:** Probenazole (3-allyl oxy-l, 2-benziso thiazole-l, 1dioxide) is an agricultural chemical primarily used to prevent blast disease in rice. Probenazole treated rice acquires resistance to the pathogen irrespective of rice variety. This chemical is applied prophylactically and is thought to induce or bolster endogenous plant defences. It is also effective against bacterial blight (Saka moto *et al.,* 1999; Oostendorp *et al.,* 2001).
7. **Jasmonic acid:** Transient accumulation of non-conjugated jasmonic acid (JA) was detected in wounded young rice leaves, and it triggered the induction of a number of pathogenesis related genes. The non-treated leaves of wounded plants that emerged after the treatment were more resistant to challenge infection by rice blast fungus, *M. grisea.* Exogenous local application of JA also results in a similar degree a systemic disease resistance as wounding. Thus, jasmonic acid can be applied for the management of rice blast (Schweizer *et al.,* 1998).
8. **Riboflavin:** Averyanov *et al.* (2000), reported that exogenous application of riboflavin and dimethylated derivative, roseoflavin induced resistance in rice plants against blast disease. Both the compounds were applied two days prior to inoculation. However, roseoflavin was a better inducer of disease resistance than riboflavin. The involvement of active oxygen is suggested for the defence mechanism of the flavins.
9. **Thiamine:** It has been demonstrated that in addition to its nutritional value, thiamine (vitamin B1) induces systemic acquired resistance (SAR) in plants. Thiamine treated rice plants showed resistance to fungal, bacterial and viral infections. It induces transient expression of PR genes. The induced resistance lasts for more than 15 days after treatment. Thiamine induces SAR in plants through salicylic acid and Ca^{2+} related signaling pathway. (AhnIlPyung *et al.* 2005).
10. **Salicylic acid:** Although salicylic acid (SA) is an important signal compound of the SAR pathway, its exogenous application has been reported to induce resistance in many instances.

Cai-XinZhong *et al.* (1996) reported that spraying of SA induced resistance to rice blast. The optimum concentration of SA to induce blast resistance was in the range of 0.01-0.1 mM and the required interval between induction treatment and challenge inoculation was 2-4 days. The blast severity was reduced by 59.8% and the induced resistance lasted for 15 days, after which treated seedlings still showed 24.3% disease reduction.

The activities of phenylalanine ammonia lyase (pAL) and peroxidase increased significantly in treated leaves. Increase in lignin content was also observed. Four basic PR proteins with MW 48, 52, 59 and 106.5 KDa were detected (Cai-Xinghong and Zheng-Zhong, 1997).

Liu-FengQuan and Wang-JinSheng (2000) reported that SA could induce resistance also against bacterial blight of rice. The optimum concentration of SA for foliar spray was 10 micro g/ml and for root soaking 5 micro g/ml solution. The disease indices of bacterial leaf blight were reduced by 50.8% and 50.4% respectively and the resistance could be maintained for 15-20 days.

Immunoblot analysis revealed an induction of a 25 KDa protein cross reacting with rice-thaumatin like protein (TLP) antiserum in response to SA pretreated and SA non pre-treated rice plants followed by pathogen inoculation. Increased phenolics content and enhanced activities of some pathogenesis-related proteins, i.e., TLP, chitinase and beta-1, 3-glucanase, were observed in rice plants treated with SA (Babu *et. al.*, 2003).

Foliar spray of SA has also been reported to induce resistance to rice sheath blight caused by *R. solani* (Kandhari and Gupta, 2003).

INORGANIC CHEMICALS

There are several inorganic elements, compounds or salts, which are known to induce acquired resistance in plants.

1. **Hydrogen peroxide:** Seed treatment with 0.5% hydrogen peroxide (H_2O_2) reduced blast severity on rice seedlings. (Mosa, 2002).
2. **Dipotassium hydrogen phosphate (K_2HPO_4):** K_2HPO_4 has been used to induce resistance against rice blast as foliar spray as well as soil drenching. Foliar spray with 50 mM K_2HPO_4 resulted in significant reduction in blast severity. It was reported to increase the activity of peroxidase enzyme in the treated leaves (Mosa, 2002; Manandhar et *al.,* 1998).
3. **Ferric chloride:** Ferric chloride applied as soil drench or foliar spray induces resistance to rice blast and also increases the grain yield (Manandhar *et al.,* 1998).
4. **Nickel nitrate:** Pre-treatment of rice seedlings with nickel nitrate (2.2 m mol/litre) induced resistance to bacterial blight caused by *X. oryzae pv. oryzae.* The lesion length of pre-treated rice leaves was 69.6% shorter than that of the control. The resistance induced by nickel nitrate lasted for about 21 days (Wang-HaiHua *et al.,* 2000).
5. **Silicon:** Although silicon mediated resistance to rice blast is hypothesized to be the result of a mechanical barrier produced from Si polymerisation in plants, there are reports which suggest active role of silicon in induced resistance.

Differential accumulation of glucanase, peroxidase and PR-l transcripts was found to be associated with limited colonization by blast pathogen, *M. grisea* in epidermal cells of rice plants grown in soils amended with silicon. (Rodrigues *et al.,* 2005).

Higher levels of momilactone phytoalexins were found in leaf extracts from plants inoculated with *M. grisea* and amended with silicon (Si^+) than in leaf extracts from inoculated plants not amended with silicon (Si^+) (Rodrigues *et al.,* 2004).

INDUCTION OF RESISTANCE WITH NATURAL PRODUCTS

In addition to biotic agents and synthetic compounds, natural products can also be used for inducing resistance.

1. **Cowdung water extract:** Pre-treatment of rice leaves with 20% cowdung water extract (CDWE) induced resistance against bacterial blight with reduction of disease severity by 50%. Foliar spray of CDWE induced accumulation of phenolics and enhanced the activities of peroxidase, chitinase and beta-1, 3-glucanase. (Sible *et al.,* 2004).
2. **Plant extracts:** Many plant extracts have been reported to be antimicrobial, however induction of resistance due to pre-treatment of plant extracts is rare.

Sateesh-Kagale (2004) reported that leaf extracts of *Datura metel* induce systemic resistance in rice against sheath blight and bacterial blight. It was evident from increased accumulation of PR proteins and other defence related compounds.

Although evidences of induced resistance against rice diseases by various biotic and abiotic agents are continuously being reported, there are considerable variations in mode, time and dose of applications of the inducing agents. Therefore, further research is required to consolidate these findings. In most of the reports, the induced resistance lasts for only 15-20 days even or less. Efforts are to be made for reinduction of resistance so that the induced resistance is maintained for a longer period, otherwise it will be of no practical use for the farmers. In the long run, at advanced stage of research, formulations may be developed and marketed specifically for induced resistance.

REFERENCES

Averyanov, A.A., Lapikova, V.P., Nikolaev, O.N. and Stepanov, A.I. (2000). Active oxygen associated control of rice blast disease by riboflavin and roseoflavin. *Biochem.* Moscow, **65**(11): 1292-1293.

Babu, R.M., Sajeena, A., Samundeeswari, A.V., Sreedhar, A., Vidhyaesekaran, P. and Reddy, M.S. (2003). Induction of bacterial blight *(Xanthomonas oryzae* pv. *oryzae)* resistance in rice by treatment with acibenzolar-S-methyl. *Ann. Appl. Biol.,* **143**(3): 333-340.

Babu, R.M., Sajeena, A., Samundeeswari, A.V., Sreedhar, A., Vidhyasekaran, P., Seetharaman, K. and Reddy, M.S. (2003). Induction of systemic reisistance to *Xanthomonas oryzae* pv. *oryzae* by salicylic acid in *Oryza sativa* (L.) *Zeitschrift. fur. Pflanzenkrankheiten. und. Pflanzenschutz,* **111**(5): 419-431.

Babu, R.M., Velazhahan, R., Vidhyasekaran, P., Seetharan, K. and Sajeena, A. (2003). Induction of resistance to *Xanthomonas oryzae* pv. *ryzae* in rice by benzothiadiazole (BTH). *Acta Phytopathologica Enomologica Hungarica,* **38**,(1/2): 73-78.

FengMing, S., XiuChun, G., Zhong, Z. and Yan, X. (2001). Benzothiadiazole induced systemic acquired resistance in rice against *Xanthomonas oryzae* pv. *oryzae. Chinese. J. Rice Sci.,* **15**(4): 323-326.

FengQuan, L. and JinSheng, W. (1998). A preliminary study on controlling rice bacterial leaf blight with a virulent gene deleted strain DU 728. *Chinese J. Biol. Contr.,* **14**(3): 115-118.

FengQuan, L. and JinShing, W. (2000). Preliminary study on resistance of rice seedling to leaf blight induced by salicylic acid. *Acta Phytophylacica Sinica,* **27**(1): 47-52.

FuHua, Z., Vue Xuan, W. and ZeMin, L. (1999). Resistance to rice bacterial blight *(Xanthomonas oryzae* pv. *oryzae)* induced by biotic and abiotic agents and its relation to active oxygen metabolism in rice seedlings. *Chinese J. Rice Sci.,* **13**(3): 165-169.

HaiHua, W., FuHua, Z., CiSheng, C. and Jian, K. (2000). Resistance to bacterial blight induced by nickel nitrate in rice seedlings. *J. Hunan Agric. Univ.,* **26**(2): 119-121.

HaiLian,Y., XiaoLu, S., Wei, S. and YunShan, W. (1999). Studies on disease resistance induced in rice by plant growth promoting bacteria. *Acta Phytopathologica Sinica,* **29**(3): 286-287.

Hassa, P. Granado, J., Freyde, E., Waspi, U. and Dudler, R. (2000). Syringolin mediated activation of the Pir 7b esterase gene in rice cells is suppressed by phosphatase inhibitors. *Mol. Pl. Microb. Interac.,* **13**(3): 342-346.

IlPyung, A., SooNok, K. and YongHwan, L. (2005). Vitamin B_1 functions as an activator of plant disease resistance. *Pl. Physiol.,* **138**(3): 1505-1515.

Jin, Q.L., Liu, N.Z., Qiu, J.L., Li, D.B. and Wang, J. (1997). A truncated fragment of harpin Pss inducers systemic resistance to *Xanthomonas campestris* pv. *oryzae* in rice. *Physiol. Mol. Pl. Pathol.,* **51**(4): 243-257.

Kandhari, J. and Gupta R.L. (2003). Efficacy of fungicides and resistance inducing chemicals against sheath blight of rice. *J. Mycopathol. Res.,* **41**(1): 67-69.

Krishnamurthy, K. and Gnanamanickam, S.S. (1997). Biological control of sheath blight of rice: Induction of systemic resistance in rice by plant associated *Pseudomonas* spp. *Curr. Sci.,* **72**(5): 331-334.

Kumawat, G.L. (2006). Effect of Trichoderma on *Drechslera oryzae* and its role in defense against brown leaf spot of paddy. M.Sc. Thesis. pp. 78. *C.S.A. Univ. Agric. & Tech., Kanpur.*

Madhaiyan, M. Poonguzhali, S., Senthilkumar, M., Seshadri, S., HeekYung, C., Jinchul, Y., Sundaram, S. and TongMin, S. (2004). Growth promotion and induction of systemic resistance in rice cultivar Co-47 *(Oryzae sativa* L.) by *Methylobacterium* spp. *Bot. Bull. Aca. Sinica,* **45**(4): 315-324.

Manandhar, H.K., Jorgensen, H.J.L., Mathur, S.B. and Smedegaard-Petersen, V. (1998). Resistance to rice blast induced by ferric chloride, dipotassium hydrogen phosphate and salicylic acid. *Crop Protec.* **17**(4): 323-329.

Manandhar, H.K., Jorgensen, H.J.L., Mathur, S.B. and Smedegaard-Petersen, V. (1998). Suppression of rice blast by preinoculation with avirulent *Pyricularia oryzae* and the non-rice pathogen *Bipolaris sorokiniana, Phytopathology,* **88**(7): 735-739.

Mosa, A.A. (2002). Induced resistance in rice against blast disease using abiotic and biotic agents. *Ann. Agric. Sci. Cario,* **47**(3): 9931008.

Nandakumar, R., Babu, S., Viswanathan, R., Raguchander, T. and Samiyappan, R. (2001). Induction of systemic resistance in rice against sheath blight disease by *Pseudomonas fluorescens. Soil Biol. Biochem.,* **33**(4/5): 603-612.

Ngueko, R.B., Ying, S. and Tong, X. (2002). Induced resistance in rice plants by *Trichoderma harzianum* P-I and its efficiency to *Magnaporthe grisea. Chinese Rice Res. Newsletter,* **10**(3/4): 16-18.

Nishioka, M., Nakashita, H. Yasuda, M., Yashida, S. and Yamaguchi, I. (2005). Induction of resistance against rice bacterial leaf blight by 3-chloro-1-methyl-IH-pyrazole-5-carboxylic acid. *J. Pesticide Sci.,* **30**(1): 47-49, 56.

Nishioka, M., Nakashita, N., Suzuki, H., Akiyama, S., Yoshida, S. and Yamaguchi, I. (2003). Induction of resistance against rice blast disease by a novel class of plant activator, pyrazolecarboxylic acid derivatives. *J. Pest. Sci.,* **28**(4): 416-421.

Nzojijobiri, J.B., Tong, X., FengMing, S. and Ying, S. (2003). Resistance induced by *Trichoderma harzianum* NF 9 against *Magnaporthe grisea* and *Xanthomonas oryzae* pv. *oryzae* in rice. *Chinese J. Biol. Contr.,* **19**(3): 111-114.

Oostendorp, M., Kunz, W., Dietrich, B. and Staub, T. (2001). Induced disease resistance in plants by chemicals. *European J. Pl. Pathol.,* **107**(1): 19-28.

Pathak, A., Sharma, A., Johri, B.N. and Sharma, A.K. (2004). Pseudomonas strain GRP3 induces systemic resistance to sheath blight in rice. *International Rice Res. Notes,* **29**(1): 35-36.

Rodrigues, F.A., Jurick, W.M., II., Datnoff: L.E., Jones, J.B. and Rollins, J.A. (2005). Silicon influences cytological and molecular events in compatible and incompatible rice *Magnaporthe grisea* interactions. *Physiol. Mol. Pl. Pathol.,* **66**(4): 144-159.

Rodrigues, F.A., McNally, D.J., Datnoff, L.E., Jones, J.B., Labbe, C., Benhamou, N., Menzies, J.G. and Belanger, R.R. (2004). Silicon enhances the accumulation of diterpenoid phytoalexins in rice : a potential mechanism for blast resistance. *Phytopathology,* **94**(2): 177-183.

Sakamoto, K., Tada, Y., Yokozeki, Y., Akagi, H., Hayashi, N., Fujimura, T. and Ichikawa, N. (1999). Chemical induction of disease resistance in rice is correlated with the expression of a gene encoding a nucleotide binding site and leucine rich repeats. *Pl. Mol. Biol.,* **70**(5): 847-855.

Sateesh Kagale., Marimuthu, T., Thayumanavan, B., Nandakumar, R. and Samiyappan, R. (2004). Antimicrobial activity and induction of sysptemic resistance in rice by leaf extract of *Datura metel* against *Rhizoctonia solani* and *Xanthomonas oryzae* pv. *oryzae. Physiol. Mol. Pl. Pathol.,* **65**(2): 91-100.

Schweizer, P., Buchala, A. and Metraux, J.P. (1997). Gene-expression patterns and levels of jasmonic acid in rice treated with the resistance inducer 2, 6-dichloro isonicotinic acid. *Pl. Physiol.,* **115**(1): 61-70.

Schweizer, P., Buchala, A., Dudler, R. and Metraux, J.P. (1998). Induced systemic resistance in wounded rice plants. *Pl. J.,* **14**(4): 475-481.

Shiwen, H. Jiying, L., Hang, Z., GuiFang, D., ZhouJin, T. and Kuan, L. (2005). Preliminary studies on rice blast control by defence inoculation with barnyard grass *(Echinochloa* spp.) pathogenic fungi. *Chinese J. Rice Sci.,* **19**(4): 384-386.

Sible, G.V., Marimuthu, T., Rabindran, R. and Velazhahan, R. (2004). Induction of resistance in rice against *Xanthomonas oryzae* pv. *oryzae* by foliar application of cow dung water extract. *Acta Phytopathologica Entomologica Hungarica*, **39**(1/3): 15-27.

Sinha, A.K. and Das, N.C. (1972). Induced resistance in rice plants to *Helminthosporium. Physiol. Pl. Pathoi.*, **2:** 401-410.

Sinha, A.K. and Trivedi, N. (1972). Resistance induced in rice plants against *Helminthosporium* infection. *Phytopath,* Z, **74:** 182-191.

Sinha, A.K., Chaudhary, A.K. and Das, A.R. (1993). Chitosan induces resistance in crop plants against their fungal pathogens. *Indian Phytopath.* **46**(4): 411-14.

Someya, N., Nakajima, M., Hibi, T., Yamaguchi, I. and Akutsu, K. (2002). Induced resistance to rice blast by antagonistic bacterium, *Serratia marcescens* strain B2. *J. General Pl. Pathol.,* **68**(2): 177-187.

Vidhyasekaran, P., Kamala, N., Ramanathan, A., Rajappan, K., Paranidharan, V. and Velazhahan, R. (2001). Induction of systemic resistance in rice by *Pseudomonas fluorescens* against *Xanthomonas oryzae pv. oryzae* in rice leaves. *Phytoparasitica,* **29**(2): 155-166.

Waspi, D., Blanc, D., Winkler, T., Ruedi, P. and Dudler, R. (1998). Syringolin, a novel peptide elicitor from *Pseudomonas syringae* pv. *syringae* that induces resistance to *Pyricularia oryzae* in rice. *Mol. Pl. Microb. Interac.,* **11**(8): 727-733.

WeiDong, Z., XiuChun, G., FengMing, S. and Zhong, Z. (2003). Systemic acquired resistance of rice against sheath blight disease induced by benzothiadiazole. *Acta Phytophytacica Sinica,* **30**(2): 717-176.

WuLing, L., XianYang, H., WenQing, Z., Rogers, W.J. and WeiMing, C. (2005). Hydrogen peroxide mediates defence responses induced by chitosans of different molecular weights in rice. *J. Pl. Physiol.,* **162**(8): 937-944.

XinFa, D., ShuYuan, S. and Zhong, Z. (1996). Interactions between Pyricularia isolates from rice and non-rice hosts on nce seedlings. *Acta Phytophylacica Sinica,* **23**(2): 97-101.

XinZhong, C. and Zhong, Z. (1997). Biochemical mechanism of salicylic acid induced resistance to rice seedling blast. *Acta Phytopathologica sinica,* **27**(3): 231-236.

XinZhong, C., Zhong, Z. and FingMing, S. (1996). Effect of salicylic acid on the induction of resistance to rice seedling blast. *Acta Phytopathologica sinica,* **26**(1): 7-12.

XiuChun, G., FengMing, S., YongYe, C. and Zhong, Z. (2002). Changes in activites of defense-related enzymes in rice resistance induced by benzothiadiazole to blast fungus *(Magnaporthe grisea). Chinese J. Rice Sci.,* **16**(2): 171-175.

XiuChun, G., FengMing, S., Zhong, Z. and HaiDan Q. (1999). Systemic acquired resistance to *Magnaporthe grisea* in rice induced by BTH. *Acta Agriculturae Zhejiangensis,* **11**(6): 311-314.

21

Present Situation of Crop Losses Caused by Plant Virus

K.K. Biswas

Plant Virology Unit, Division of Plant Pathology,
Indian Agricultural Research Institute, New Delhi–110 012

The science of plant pathology started because of the result devastating crop losses. Estimate of loss is prerequisite to the rational development of any agricultural research program that has plant protection as a component. Crop loss is attributed to reduction in value and/or financial return due to damage, often measured as the difference between actual yield and attainable yield due to the effects of one or more pathogens or pests. Actual yield is the site specific yield obtained when crops are grown using current production practices at the farm level, and attainable yield is the site specific yield obtained when crops are grown using all available pest management technologies to minimize the biotic stress, i.e., a measure of the genetic potential of a crop genotype at a specific site. The reliable estimates of losses facilitate the objective of identification of relative importance of biotic pests. Crop loss information provides a means for evaluating the potential benefits of implementing any management decision and to develop decision thresholds.

DEGREES OF CROP LOSS

Viruses are certainly responsible for far greater economic losses than is recognized because the nature of losses caused by viruses is very different than that attributed to bacteria, fungi, nematodes. Losses are often more insidious, frequently less conspicuous and therefore they go unnoticed. Many viruses are latent in some of their host species, even in case of adverse affects. Latent infections usually result in plants that grow more slowly, attain a smaller size, produce less fruit, reduce lifespan of perennial crops and act in various other ways that go unnoticed. Only when comparative experiments are conducted with virus free and infected plants, we become aware that looses are occurring. It is conceivable that we sustain greater looses due to chronic and widespread latent infections by hundred of viruses in many different crops genera. Therefore, in perennials crops and those that are vegetatively propagated, the term virus "infections" is more appropriate then the term virus diseases" in order to consider all losses associated with viruses.

For these reasons mentioned above, there have been very few comprehensive surveys undertaken with loss data with many different viruses in many crops. Data presented in some of the few comprehensive surveys are based on estimations, extrapolations or educated guesses. There are several obvious reasons for not having accurate loss data. Among them are variations in losses by a particular virus in a particular crop from year to years, different degrees of losses among regions within the same year, differences in loss assessment methodologies, viral losses are often complicated with other loss factors, variations in definitions of the term 'loss', and some component of loss are virtually unmeasurable. Virus induced symptoms can be confused with disorders due to pesticide toxicities, air pollution, genetic abnormalities, combined infections with the fungi, and mineral excesses or deficiencies.

A few reviews focus on the losses due to plant viruses, in general. Reviewers discussed crop loses by plant viruses in different ways, such as types of losses, relationships of losses to mode of virus transmission, economic costs of virus management, experimental methods to determine losses, estimation of potential and actual losses, predication of crop losses and ways of preventing crops losses.

KINDS OF CROP LOSSES

Crop loss is the measurable reduction/damage in quantity and quality of yield provided through description of categories of yield as well as the many forms of direct and indirect losses caused by plant diseases caused by viruses (Table 21.1).

Table 21.1. Some direct and indirect losses associated with plant virus infections in crops.

Direct damage	Indirect damage
Reduction in growth	1. Yield reduction (including symptomless infection) 2. Crop failure
Reduction in vigour	1. Decreased sensitivity to frost and drought 2. Increased predisposition of pathogens and pests
Reduction in quality and market value	1. Defects of visual attraction: size, shape, colour 2. Reduced keeping quality 3. Reduce consumer appeal like grading, taste, texture and composition 4. Reduced fitness for propagation
Costs of attempting to maintain crop health	1. Cultural hygiene on farm including vector control 2. To produce virus free propagation materials 3. Eradicating programs 4. Breeding for resistance 5. Research, extension and education

Plant Vigour Loss

Loss of plant vigour is one of the more common forms in which viruses cause losses. For example, crinkle virus and some mosaic viruses in severe cases reduce plant vigour in Leguminous, Solanaceous and other many crops, etc; bushy dwarf virus reduce height and diameter of many perennial fruit crops; necrotic spot viruses cause decline in vigour in papaya and stone fruit trees; latent viruses reduce the growth and number of stool bed apple trees; and apple mosaic virus significantly reduce tree growth and number of laterals produced (66,72,14, Biswas and Gupta, 2007). Many other examples of the detrimental effects of latent virus infection become readily apparent only in controlled experiments.

Productive Life Loss

Virus infection also reduces the productive life of crops, which may be blamed on other causes. This happens in annual crops such as potatoes infected with leafroll, but the effect is greater in perennial crops such as orchards where life span is reduced by several years. Leaf curl virus reduces the lifespan of raspberry beds, and latent viruses in citrus and many perennial fruit crops render orchards prematurely unproductive without typical disease symptoms. Tomato ring spot virus in peach trees, cherry leaf roll virus in walnut trees, tristeza virus in citrus, leaf roll viruses in grapevine are examples of viruses that significantly reduce the productive life of perennial crops.

Incompatible Budunion

Grafting and budding failure were often simply explained as incompatibility on the vegetatively propagated fruits, nut and woody ornamental. It is now known that viruses latent in rootstocks or scion variety were responsible in many cases. Apples, pears, citrus and most of the stone fruit were especially affected by this problem. Apple mosaic virus decreased the bud take by 2% to 20% among several cultivars tested in trials. Fan leaf virus reduced the ability to produce rooted cuttings and success in grafting grapevines.

Seed Loss

Viruses are responsible for a wide range of economic losses associated with seed production. Hundreds of viruses infect seed and many are transmitted to resulting seedling. Infected seed may appear normal depending on the virus and the host species. Losses also occur when seed is aborted, discolored, cracked or reduced in size, weight, or viability. Neergaard described many ways in which viruses affect seeds of various crops. For example, germination of wheat seed is reduced by barley yellow dwarf virus, as is apple seed from infected trees, and lettuce seed infected with mosaic virus. Seed was smaller when wheat and barley were infected with BYDV or when soybeans were infected with mosaic virus.

Growth Habit Loss

Viruses also reduce yield in a number of unusual ways by affecting growth habit of their host crops. Viruses in strawberry reduced runner production; pox virus in plums caused 40% to 100% fruits to drop before maturity among several cultivars, as did mosaic virus in apple trees. This virus was also responsible for less branching in young apple trees. Leaf roll virus reduced the number of stems in potato plants, while para crinkle virus produced tubers of less uniform size. Therefore, infected plants compete more successfully with virus free adjacent plants, which results in yield loss.

Loss in Combination with other Pathogens

Viruses in combination with other kind of pathogens can result in more adverse effect. For example, pea enation mosaic virus reduced yield of broad bean seed by 6%, its aphid vector alone reduced yield by 50%, but both reduced yield by 93%. Barley yellow dwarf virus enhanced mildew in oat, ergot disease in wheat. Wheat streak mosaic infection was associated with an increase in *Helminthosporium sativum* in wheat. Certain viruses in potato plants cause greater susceptibility to *verticillium dahliae*, late blight. Leaf blight fungus, *Alternaria* was more severe on sugar beets infected with beet western yellows virus. Apple trees with latent viruses were more susceptible to collar rot disease.

Plant Susceptibility

Virus infection can increase plant susceptibility to low temperatures. Leafroll virus in grapevines made them more susceptible to spring frosts, and wheat with BYMV did not survive the winter. When growing under heat or drought stress, fanleaf infected vines were killed. Pome and stone fruit crops are more susceptible to frost damage when trees are infected with viruses.

Other Adverse Effects

Viruses are also responsible for some unusual effects on crops such as greater nutrient requirements of fruit trees and reduced coloration in grapes, which adversely affect fresh market sales and red wine color. Sugar content can also be reduced by viruses. Necrotic ring spot virus in peach trees and beet western yellows virus reduced sugar content of the harvest. Leaf roll virus in grapevines reduced sugar content, which in turn affected raisin production to the extent that some crop was not harvested. Certain viruses in cereal crops reduced milling qualities, other examples could be given on some of the unusual effects that viruses have on plants or the quality of their harvest that affect economics of production.

ESTIMATIONS OF SOME NATURAL CROP LOSS CAUSED BY PLANT VIRUSES

Virus diseases affect most of the economic crops and are responsible for far greater economic losses of crop productivity and quality not only in India but also worldwide. The extent of losses depends on many factors and may vary widely. Some viruses under appropriate conditions may infect a plant without producing any obvious signs of disease (latent), whereas others may lead to rapid death of the whole plant (adverse affects). Between these extremes, a wide variety of diseases can be produced. The estimation of crop losses caused by important viral diseases is mentioned in Table 21.2.

Table 21.2. Important viral diseases in India and the crop losses by them.

Crop	Virus	Yield reduction
Apple	Apple mosaic	30-40%
Banana	Banana bunchy top	Upto 100%
Bean	Bean golden mosaic virus	20-40%
Bhindi	Bhindi yellow vein mosaic virus	50-80%
Blackgram	Mungbean yellow mosaic virus	50-80%
Cassava	Indian cassava mosaic virus	18-25%
Chilli	Chilli mosaic virus	94%
Cotton	Leaf curl	68-71%
Cucurbit	Cucumber mosaic virus	50-75%
Groundnut	Groundnut bud necrosis	>80%
Maize	Sugarcane mosaic virus	32%
Melons	Zucchini yellow mosaic	40-50%
	watermelon mosaic	40-50%
Mungbean	Mungbean yellow mosaic virus	21-70%
Orange	Citrus tristeza virus	10K trees
Papaya	Papaya leaf curl virus	Upto 100%
Pigeonpea	Pigeonpea sterility mosaic virus	>80%
Potato	Potato virus X	10-20%
	Potato virus A	10-20%
	Potato virus S	10-20%
	Potato virus M	15-20%
	Potato virus Y	40-80%
	Potato virus X + Potato virus A	40%
	Potato virus X+ Potato virus Y	40-100%
	Potato leaf roll	7-16% (Autumn)
		39-60% (Spring)
Rice	Rice tungro virus	
Sugar cane	Sugar cane mosaic virus	30-40%
Sunflower	Sunflower necrosis virus	12-17%
Tomato	Cucumber mosaic virus	30-60%
	Tomato leaf curl	28-92%

USE OF CROP LOSS KNOWLEDGE

Crop loss knowledge is a common approach for managing disease in a given field, season and context, and for continuing the phytopathological research including the research priorities in a particular area. The practical usefulness of crop loss knowledge is for getting real benefit of control against the cost of control. Knowledge about the crop loss is essential to make short term as well as long-term strategic decisions. The short-term strategic decisions include the choice of crops, cultivars and genotypes, and the choice of tillage, etc. They also include other choices, such as the timing of the crop establishment, the type of the seed treatment used, sowing density and spatial pattern of the crop plant, etc. The long-term strategic decisions enable decision makers to identify and anticipate where IPM is needed for a given disease, and when the disease does not represent a threat to a crop.

REFERENCES

Agrios, G.N. (1990). Economic Considerations. Pages 1-22 *In:* Plant Viruses, Vol. II, Pathology. Mandahar, C.L. ed. CRC Press.

Anonymous. (1999-2003). Annual Report, Indian Agricultural Research Institute, New Delhi.

Anonymous. (2000). Diseases and pest of Potato-A manual. Central potato research Institute (Paul Khurana, S.M. ed.), Shimla, HP.

Ahlawat, Y.S. (1997). Viruses, greening bacterium and viroids associated with citrus (citrus species) decline in India. *Indian J. Agri. Sci.* **67:** 51-57.

Basu, N. and Giri, B.K. (1993). The essentials of Viruses, Vectors and Plant Diseases. Wiley Eastern limited, 4835/24 Ansari Road, Dayaganj, New Delhi–110002.

Bhargava, K.S. (1975). Sugarcane Mosaic- retrospect and prospects. *Indian Phytopathol.* **28:** 1-11.

Biswas, K.K., Pant, R.P., Das, D.R., Pun, K.B., Chowdhuri, A. and Ahlawat, Y.S. (2004). Incidence, detection and management of chilli mosaic disease syndrome in capsicum annum cv Kalimpong local in the Hills of West Bengal. National Symposium on Molecular Diagnostics for the Management of Viral Diseases, pp. 89-90 (Abs.).

Biswas, K.K., Pun, K.B., Pant R.P. and Ahlawat, Y.S. (2005). Mosaic disease in chilli (*Capsicum annuum*) cv Kalimpong Local in Darjeeling Hill of West Bengal and its management. *Indian Phytopath.* **58:** 456-461.

Biswas, K.K. and Chowdhury, A. (2005). Rai mosaic disease in rai (*Brassica juncea*) and Radish (*Raphanus sativus*) in the Himalayan hilly regions of West Bengal. *J. Mycopathol Res.* **46:** 263-266.

Biswas, K.K., Das, D.R. and Ahlawat, Y.S. (2005). Viral diseases of fruits, vegetables and spices in North Eastern Himalayan region of India and their management. *In:* Crop Protection-Management strategies (Prasad, D. Ed.), Daya Publishing House, New Delhi-110035, pp. 225-246.

Biswas. K.K. and Gupta, A. (2007). Management of important plant viruses in Indian agriculture. In: Sustainable Pests Management (Prasad, D. ed.), Daya Publishing House, New Delhi-110035, pp. 208-138.

Bridddon, R.W. and Markham, P.G. (2000). Cotton leaf curl viral disease. *Virus Res.* **71:** 151-159.

Bos, L. (1982). Crops losses caused by viruses. *Crop Protection* **1:** 263-282.

Campbell, A.I. (1965). The effect of some apple viruses on clonal root stock production. Zast. Bilja No. 85-88: 261-265.

Campbell, A.I. (1980). The effects of viruses on the growth, yield and quality of three apple cultivars on healthy and infected clones of four rootstock. *Acta. Hortic.* **114:** 185-191.

Cisar, G., Brown, C.M. and Jedlinski, H. (1982). Effect of fall or spring infection and sources of tolerance of barley yellow dwarf of winter wheat. *Crop Sci.* **22:** 474-478.

Dasgupta, I., Malathi, V.G. and Mukherjee, S.K. (2003). Genetic Engineering for Virus Resistance. *Current Science* **84:** 341-354.

Geske, S. and Riesselman, J. (1988). Barley yellow dwarf and wheat streak mosaic in small grains. Montguide. MT State Univ. Ext. Service Publ. 8802. 3 pages.

Gibbs, A. and Harrison, B. (1976). Ways of preventing crop losses. Pages 219-231 *In:* Plant Virology, The principals. Halsted Press.

Gill, C.C. (1980). Assessment of losses on spring wheat naturally infected barley yellow dwarf virus. *Plant Dis.* **64:** 197-203.

Hull, R. (2002). "Matthews' Plant Virology." (4th Ed.). Academic Press. A Harcourt Science and Technology Company, Harcourt Place, 32 Jameston Road. London, NW1 7BY, UK.

Khetrapal, R.K., Maisonneuve, B., Maury, Y., Maury, Y., Chalhoub, B., Dinant, S., Lecoq, H. and Varma, A. (1998). Breeding for resistance to plant viruses. *In: Plant Virus Disese Control*, (Hadidi *et al.*, Eds.), APS Press, pp. 14-32.

Lee, R.F. and Rocha-Pena. (1992). Citrus tristeza virus. Pages 226-249. *In:* Plant Diseases of International Importance, Vol III. (Kumar, J., Chube, H.S., Singh, U.S. and Mukhopadhyay, A.N. Eds.). Prentice Hall, NJ.

Madden, L.V., Pennypacker, S.P., Antle, C.E. and Kingsolver, C.H. (1981). A Loss Model for Crops. *Am. Phytopathol. Soc.* **71:** 685-689.

Madden, L.V. (1983). Measuring and Modeling Crop Losses at the Field Level. The *Am. Phytopathol. Soc.* **73:** 1591-1596.

Millikan, D.F. (1955). The influence of infection by ring spot virus upon the growth of one year old Montmorency nursery trees. *Phytopathology* **45:** 565-566.

Mink, G.I. (1993). Pollen and seed transmitted viruses and viroids. *Annu. Rev. Phytopathol.* **31:** 375-402.

Neergaard, P. (1977). Seed borne viruses. Pages 71-117. *In:* Seed Pathology, Vol I. Neergaard, P. ed. Halsted Press.

Nemeth, M. (1986). Virus, mycoplasma and rickettsia disease of fruit trees. Martinus Nijhoff Publ. pp. 841.

Nutter, Forrest W., Teng, Paul, S. and Royer Matthew H. (1993). Terms and Concept for Yield, Crop Loss, and Disease Thresholds. The *Am. Phytopathol. Soc.*, February, Special Topic. 211-215.

Paul Khurana, S.M. and Thind, T.S. (2002). Important biotic diseases of potato and their control. *In: plant Pathology at cross road* (eds. C. Sen, M.K. Dasgupta, S.).

Satapathy, M.K. and Anjaneyulu, A. (2001). Rice tungro epidemic disease epidemic: analysis. Discovery Publishing House, New Delhi.

Singh, B.R. Gupta, S.P. Pandey, R. and Tripathi, D.P. (1990). Response of some chilli lines to mosaic and leaf curl diseases. *Indian J. Virol.* **6:** 113-115.

Singh, S.J. (2003). Virus and phytoplasma diseases of papaya, passion fruit and pineapple. Kalyani Publishers, Noida.

Singh, S.J. (1996). Virus and mycoplasma diseases of papaya. *In:* Singh, S.J. (ed) advances in diseases in fruit crops in India, Kalyani publishers, Ludhiana, pp. 275-296.

Stouffer, R.F. and Lewis, R.H. (1969). The present status of stem pitting in Pennsylvania. *Plant Dis. Rptr.* **53:** 429-434.

Varma, A. and Sinha, S.K. (1992). Sustainable development through long term biotechnological alternatives in agriculture. Pages 44-57. *In:* Proceedings of the International seminar on Impacts of biotechnology on agriculture and food in Developing countries. Committee on Science and Technology in Developing Countries (COSTED), Madras.

Varma, A. and Malathi, V.G. (2003). Emerging geminivirus problems: A serious threat to crop production. *Ann. Appl. Biol.*, **142:** 145-164.

Verma, H.N. and Prasad, V. (1985/86). Virus diseases in papaw (Papaya). *Rev. Tropical. Plant. Path.* **2:** 311-327.

Walkey, D.G.A. and Payne. C.J. (1990). The reaction of two lettuce cultivers two mixed infection by beet western yellow virus, lettuce mosaic virus and cucumber mosaic virus. *Plant Pathol.* **39:** 156-160.

22

Present Scenario of Management Strategies of Plant Viral Diseases

K.K. Biswas, Sumita Kumari and Avijit Tarafdar

Plant Virology Unit, Division of Plant Pathology,
Indian Agricultural Research Institute, New Delhi–110 012

Increasing quantities of plant protection chemicals employed and their wide and uncritical application at a time by modern spraying and dusting machineries has had some unfortunate effects. The ultimate aim of plant pathologists is to evolve suitable control measures so that the loss due to diseases may be prevented and the economic values of the crops rose, but it is true that the excessive and indiscriminate use of chemicals has led to various environmental and health hazards. One example in this respect was that a notorious large-scale campaign in the USA to eliminate the fire ant by the use of chloro-organic insecticides had a limited effect on the pest, but resulted in the death of the several wild, domestic and other non-target organisms and ultimately illness of the man. Matter came to a lead in 1962 with the publication of Rachel Carson's "Silent Spring", which drew attention to the abuses in the use of the agro-chemicals to control pests, diseases and weeds. Then, the global community has expressed a willingness to reduce or stop its reliance on chemical pesticides and to move recently towards "organic farming".

ORGANIC FARMING

Presently the rapid growth of organic farming has been among the most remarkable changes in global agriculture. The organic movement began as a reaction of agricultural scientists and farmers against the industrialization of agriculture. Advances in biochemistry, plant breeding and engineering in the early 20th century led to profound changes in farming. By the use of hybrid seeds, large-scale irrigation, fertilization, and pesticides, fields grew in size and cropping became specialized to make efficient use of machinery and reap the benefits of the green revolution. The organic farming promotes and enhances biodiversity, biological cycles and soil biological activity based on minimal use of off-farm inputs and management practices that restores, maintains and enhances ecological harmony. Enhancing soil health is the corner stone of organic farming. A variety of methods are employed including crop rotation, green manure, cover cropping, application of compost, and mulching. Organic farmers also use certain processed fertilizers such as seed meal, and various mineral powders such as rock phosphate and greensand, a naturally occurring form of potash.

Organic pest control against targets pests, weeds and diseases involves the cumulative effect of many techniques allowing for an acceptable level of pest damage encouraging or even introducing beneficial organisms. These techniques generally provide benefits in addition to pest control, soil protection and improvement, fertilization, pollination, water conservation, etc. and these benefits are both complementary and cumulative in overall effect on farm health. Many farm studies concluded that organic farming produces the same as conventional methods, but consumes less energy and contains no pesticide residues. In other cases the farm studies showed an average of 20% lower organic yields over conventional, along with 50% lower expenditure on fertilizer and energy, and 97% less pesticides. In some cases it has been observed that organic yields were 95-100% of conventional yields. Organic methods often require more labor, providing rural jobs but increasing costs to urban consumers.

All aspects of organic farming and organic food are under debate by environmentalists, food safety advocates, various consumer protection, social justice and labor groups, small independent farmers to agribusiness, food consumers and government agricultural policies. Controversy centers on the overall value and safety of chemical agriculture, with organic farming popularly regarded as the "opposite" of modern, large-scale, chemical-based farming, but also of the value of organic farming's solutions. The following topics may be argued in organic farming, viz. pesticides, environment, food contamination and quality, human health, soil conservation, crop rotation, genetically modified organisms, shifting from chemical to ecofriendly methods of pest control, multiple diseases and host resistance.

CULTURAL CONTROL

There are many ways to control diseases caused by plant viruses. However, unlike control of fungal diseases, use of chemical pesticides is not available in practice for viral pathogens. Numerous chemicals have been reported to control viral multiplication, movement and formation of diseases symptoms caused by viruses, but none of them has been adopted for practical purpose in crop production systems. The single case where chemical treatment of virus infections has a practical value is the preparation of healthy propagative materials from totally infected lines of crops such as garlic, potato and other herbaceous ornamental crops. The chemicals are applied to meristem tip or callus cultures and often heat treatment is done leading to virus free lines.

Virus diseases may also be avoided or controlled by planting certified virus free material or vegetatively propagated crops such as potatoes and ornamental crops, perennial trees, by testing seed lots for the absence of seed transmitted viruses, by using cultural practices designed to avoid transmission or mechanically transmitted viruses, and by seeking to prevent transmission of vector transmitted viruses by controlling the vector of its access to the host plant. All these methods can make a valuable contribution to practical disease control in specific favorable situations, but they do not offer the whole for all situations.

CROSS PROTECTION

Cross protection was first demonstrated by McKinney in 1929 by inoculating tobacco plant with a "green" strain of tobacco mosaic virus (TMV) to protect systemic infection by another severe strain of TMV inducing yellow mosaic symptoms. It was also of potential interest for protecting plants against viruses in the field. Cross protection has not only been observed to occur between viruses, it was also demonstrated to occur between viroids or plant virus satellites. Many mild strains have been utilized as cross protecting strain in many crops in field level experimentally or commercially. An ideal mild strain to be used in the field for cross protection should have some features—it should induce milder symptoms than isolates commonly encountered in the field; should be fully systemic and invade all host tissues; should be genetically stable and not evolve towards more severe forms at times; should not easily disseminated by means of vectors; should provide protection towards wider range of challenging severe isolates; and the protective inoculum should be easy to produce, store and provide to farmers, etc.

Table 22.1: Viruses for which mild strains have been used for cross protection.

Virus	Origin of mild strains	Crops controlled
Cocoa swollen shoot	Field isolates	Cocoa*
Citrus tristeza	Field isolates	Citrus**
Arabis mosaic	Field isolates	Grapevine*
Papya ring spot	Mutagenesis	Papaya**, Squash*
Zucchini yellow mosaic	Greenhouse variants	Squash**, cucumber**
Soybean mosaic	Cold treatment	Soybean*
Tomato mosaic virus	Mutagenesis, field isolates &	
Heat treatment	Tomato**	

*: Experimental **: Commercial

Cross protection proved to be a very efficient and popular method to control *Tomato mosaic virus* (ToMV). Mild strains were obtained either from naturally occurring strains or by culturing a severe strain in tissue held above 34°C. The mild strain MII-16, which was extensively used in Western Europe was obtained by nitrous acid mutagenesis followed by single local lesion selection. Cross protection against ToMV in tomatoes has been widely used in Europe, Canada, New Zealand and Japan. In France, a subculture of MII-16 was used commercially from 1972 to 1983. Preliminary experiments and pilot tests demonstrated that it did not reduce the yields but only slightly delayed fruits setting. It provided very efficient protection against ToMV severe strain; however, the risk like synergism with other viruses limited the use of ToMV crosses protection.

Citrus tristeza virus (CTV) is prevalent in many regions of citrus growing countries in the world. It causes very important losses, particularly in certain scion root combination where it reduces fruit sizes and yield causing stem-pitting, declining and premature death of the tree. It has been currently used for grapefruit in South Africa and Australia and for Perra Sweet oranges in Brazil. More than 8 million Pera Orange trees were cross protected in 1980 and more than 50 millions trees in 1993. Mild isolates in Brazil, Australia and South Africa were isolated from the trees noticed as showing no or very discrete symptoms in areas where all others trees were severely infected. In the areas where server stem pitting isolates are widespread and the brown citrus aphid is epidemic, cross protection has been used as a control strategy. Cross protection presumably requires a certain minimum degree of relationships between the protecting and challenge isolates at least in certain areas of viral genome. Better knowledge to more accurately select the effective and mild protecting isolates is needed. Defective interfering RNAs have recently been associated with CTV infection may also have applications for cross protection.

Cross protection has been successfully used in the biocontrol of several cases of *Cucumber mosaic* (CMV) infection. Cross protection can be obtained using a sat-RNA free mild strain. Deploying this technique, a CMV pseudo recombinant was assembled, characterized and used as a cross protection agent in Italy and called CMV-ST. The CMV-ST was tested worldwide and had no deleterious effect on growing plants or environment. Cross protection has been favorably exploited in the management of viroids. When tomato plants with a prior infection by a mild strain of *Potato spindle tuber viroids* (PSTVd) are challenged-inoculated with a severe strain, symptom development by the challenged strain is delayed by 48 days. More recently it has been demonstrated that tomato plant pre-inoculated with non-necrogenic satellite RNA of CMV (CARNA-5) produced milder symptoms, when challenged inoculated with a severe strain of PSTVd.

DISEASE RESISTANCE

The recent advances made in molecular biology with plant transformation and regeneration technologies have widened the scope of conferring resistance to viruses in plants. For an effective viral disease control, the breeding for disease resistance through conventional breeding methods is always have most important role side by with recently developed novel approach for breeding for disease resistance. Extensive work has been done on breeding for resistance against plant viruses. The resistant varieties if available are an effective strategy for minimizing the losses caused by the plant viruses. This is largely due to intrinsic properties of viruses, which do not permit their control by any simple physical and chemical control. The molecular biology has facilitated different steps of conventional methods of breeding for disease resistance and also helped in incorporating novel resistance genes. The development of various molecular markers in breeding resistance has also shortened the selection process.

GENETIC ENGINEERING

Genetic engineering is the branch of biotechnology that modifies genetic structure of the crop or pest genome, which is accomplished either by gene deletion or blocking or by transferring gene from one organism to another. Successful genetic improvement of crop plants by methods developed through biotechnology has resulted in introducing new genes for virus resistance into existing commercial crop varieties. These new genes are derived from the viruses themselves in a concept referred to as plant derived resistance (PDR). The organisms/crops that have been modified by one of this gene modifying technique are referred to as genetically modifying organisms/crops (GMOs/GM crops). Sometimes it also called genetically engineered organisms/ crops (GEOs/GE crops). The term "transgenic crop" is applied when the genetic modification involves transfer of genes from one organism to another. These genes that are transferred are called transgenes. Thus, identification of genes conferring pest resistance, and then transferring these genes to crop plants is performed by the genetic engineering techniques. Since the first report of this strategy to protect tobacco and tomato plants against tobacco mosaic virus (TMV) infection by engineering the viral coat protein gene into plants, there have been many reports of transgenic plants have been resistant to viral diseases reviewed. Success has been obtained in field level in crops like wheat against *Wheat streak mosaic* (WSMV), citrus against CTV and *Citrus psorosis* (CPV), Squash against CMV, *Zucchini yellow mosaic* (ZYMV) and *Watermelon mosaic* II (WMV II) and Papaya against *Papaya ring spot* (PRSV).

Table 22.2: Transgenic crop resistance to viruses using pathogen derived genes.

Crop	Transformation with the gene	Against the virus
Citrus	CP gene P23 gene, RdRp gene, Replicase	CTV
Squash	Replicase gene	CMV
Potato	CP gene, Replicase gene	PVX
	CP gene	PVX, PVY, PLRV
Sugarcane	CP gene	SCMV
Maize	CP gene	MDMV, MCMV
Rice	CP gene	RSV, RTSV
Wheat	CP gene	WSMV
Apricot	CP gene	PPV
Cantaloupe	CP gene	ZYMV*,WMV2*,CMV*
Grape	CP gene	GCMV, GFLV, ToRSV
Muskmelon	CP gene	ZYMV*
Papaya	CP gene	PRV*
Plum	CP gene	PPV
Squash	CP gene	ZYMV*, WMV2*
Pepper	CP gene	TSWV
Tomato	CP gene	ToMV* YMV* CMV*, TYLCV*
Potato	CP gene	Yes
Lettuce	CP gene	LMV, TSWV
Pea	CP gene	PEMV
Cucumber	CP gene	CMV*
Sugarbeet	CP gene	BNYVV
Peanut	CP gene	TSWV
Soybean	CP gene	BPMV
Bean	CP gene	BPMV

Coat Protein Gene

Use of viral coat protein gene as a transgene for obtaining virus resistant plants is one of the most important successes achieved in the plant biotechnology. Coat protein-mediated resistance (MR) was demonstrated for the first time a decade ago (Powell-Abel *et al.*, 1986) by demonstrating of virus resistance in transgenic plants utilized the tobacco mosaic virus (TMV) coat protein (CP) gene expressed in tobacco plants.

Table 22.3: Comparative performance of transgenic virus resistance plants.

Host	Transgene	Yield increase (%)
Tomato	TMV CP	40
Tomato	CMV satellite	14
Potato	PVX + PVY CP	38
Squash	CMV + ZYMV + WMV2 CP	97
Squash	ZYMV + WMV2 CP	90
Squash	ZYMV CP	77
Papaya	PRSV CP	90

Replicase Gene

In addition to CP genes, other genetic elements from plant viruses are also being tested in transgenic plants for the ability to confer resistance to virus infection or disease. The expression of replicase (RNA dependent RNA polymerase, RdRp) native genes and modified version has been shown to confer resistance to virus infection. The modified proteins are thought to be defective in functions critical for the normal replication or spread of virus in the plant. Both, the expression of native viral replicase genes and modified replicase genes have resulted in transgenic plants demonstrating high levels of resistance to virus infection or disease.

Movement Protein Gene

These are the virus encoded proteins essential for cell-to-cell movement of plant viruses. These proteins have been shown to modify the gating function of plasmodesmata, thereby allowing the viral particles or their nucleoprotein derivatives to spread to adjacent cells. This phenomenon was first used to engineer resistance against TMV in tobacco by producing modified MP, which is partially active as transgenes. The conferred resistance is believed to be based on the competition between wild-type viruses encoded MP and the preformed dysfunctional MP to bind to the plasmodesmatal sites. The above resistance was also effective against distantly related or unrelated viruses. As, for example, resistance against TMV can be achieved in tobacco by using the MP derived from *Brome mosaic virus*, suggesting the functional conservation of this protein among several viruses. This protein can be made from single MP gene as in tobamovirus, whereas it is made by a set of three overlapping genes, known as triple-gene-block (TGB), in case of potex-Carla virus. The resistance is dependent upon the interaction of viral derived and the transgene derived MPs.

Satellite RNA

Satellite RNAs are sub viral RNA. They can be distinguished from the viral genome by their dispensability for normal viral propagation. The classical definition of satellite is a linear or circular RNA that requires a helper virus to supply proteins for replication, but shares little sequence relatedness with the helper virus and is not required for the accumulation of the helper virus. The replications of satellite viruses are completely dependent on enzymes encoded by the helper virus. Thus, their amplification is limited to confined cells. It is nearly exclusively associated with plant viruses. Satellites are sequences that are either mostly or completely unrelated to any large contiguous segment of the helper virus. They can cause large alteration in normal disease progression induced by their helper virus, and are thus fascinating tools for controlling viral diseases. Satellite RNAs are divided into two groups—ones that generally encode a single non- structural protein are known as large satellite RNAs, and other with no functional coding capacity called small satellite RNAs. Satellite viruses are those, which are capable to encode their own coat protein.

Defective Interfering Viral Nucleic Acid

Defective interfering (DI) viral nucleic acid is also a type of sub viral RNA. It is derived mainly from the genome of the helper virus by one or more premature termination and reinitiating events that infrequently results in a functional template capable of further amplification. DI RNAs are associated with many plants and nearly all animal RNA viruses. These viruses are missing some or all the viral genes required for the movement, replication and encapsidation, but contain all cis acting element necessary for replication by RdRp of the parental virus. Most DI RNAs moderate the symptoms of their helper viruses, but some of them exacerbate the severity of symptoms as in DI RNAs of *Broad bean mottle virus* and *Turnip crinkle virus* (TCV). No single mechanism is accepted to account for the DI RNA interference with the accumulation of their helper virus and corresponding symptom attenuation. Many mechanisms of DI RNA symptom modulation have been proposed.

1. Most logical and widely used explanation for the reduction of helper viral genomic RNA levels when associated with DI RNAs is a competition between the DI RNA and the viral genome for limited factors required for replication; reduced virus levels in turn lead to attenuated symptoms.
2. Alternative model proposed suggesting that a link between DI RNA accumulation and viral gene expression leads to symptom attenuation.
3. Activation of PTGS is also believed to be a mechanism of symptoms modulation.

Post Transcriptional Gene Silencing

Post transcriptional gene silencing (PTGS) is an adaptive sequence specific RNA degradation system of any organism that plays a role in control of aberrant, unwanted excess of foreign RNA intracellularly in a homology dependent manner. It is prevalent in various forms of life of plant, fungus and invertebrate animals. This activity could be present constitutively to help normal development or it may be induced in response to cellular defense mechanisms against pathogens. According to recent models, PTGS is activated by double-stranded (ds) RNA and results in cleavage of the RNA into 21-26 nucleotides long double-stranded (ds) small interfering

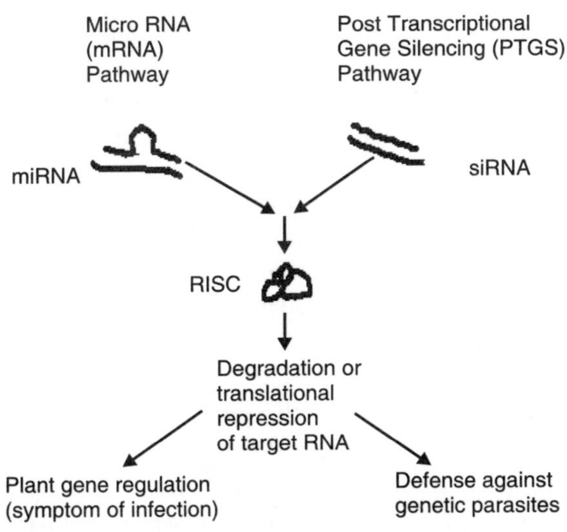

Fig. 22.1. Schematic diagram showing micro RNA (miRNA) and post transcriptional gene silencing (PTGS) pathway.

(si) RNAs by an Rnase III like enzyme called DICER. The generated siRNA are associated with an enzyme complex (RNA-induced silencing complex; RISC) and guide the RISC to degrade any RNA having sequence homology with the inducer dsRNA. This complex (RISC) is made of different cellular factors, namely RNA dependent RNA polymerase (RdRp), RNA-helicase, translation elongation factor, RNAse, etc. This degradation process initiating from a concerned cell having the elicitor RNA spreads later within the entire organism in a systemic fashion. This process is generally regarded to evolve as a plant defense mechanism against invading viruses containing either DNA or RNA genomes. In addition to the cell-autonomous defense function, PTGS in plants generates mobile signals that instruct target RNA degradation at a distance (systemic PTGS). When the viral RNA is either elicitor or target of PTGS, the degradation mechanism is known as virus induced gene silencing (VIGS). The VIGS come into play when plant recovers from initial viral infection.

CONCLUSION

The development of transgenic plants resistant to viral diseases requires an effective expression cassette for the expression of viral transgenes. The expression cassette consists of a transcriptional promoter and terminator, which control the production of viral gene messenger RNA in transgenic plants. Generally constitutive promoters that result in high levels of gene expression have achieved greatest success. The cauliflower mosaic virus (CaMV) 35S promoter is a single or double copy version (enhanced 35S) and is used most frequently, although several other promoters have also been used successfully. In cases where a virus is restricted to certain cells or tissues, the use of tissue specific promoters may be considered.

Wherever possible, major crop varieties should be considered for transformation to allow for immediate use of transgenic plants. In cases where target varieties cannot be transformed, other varieties can be used to introduce genes of interest, following by breeding into commercial varieties. The CP-MR and replicase-MR are the two PDR strategies most likely to be commercially available in the near future, especially in vegetable and grain crops. Viral genes should be selected based on analogy with previous performance experience. In most cases, CP would be considered first because there are many examples in which CP has conferred higher resistance to viruses from various groups. The coat protein gene is generally well characterized and therefore easily cloned. Following expression in plants, the coat protein can be detected using readily available antisera generated to virions. In the case of luteoviruses, CP has conferred only weak to moderate resistance, so the use of other genes such as replicase should be considered. For virus groups where high resistance has not yet been achieved, basic research must be done to select the best resistance genes or modifications of commonly used genes.

REFERENCES

Antignus, Y., Vunsh, R., Lachman, O., Pearlsman, M., Maslenin, L., Hananya, U. and Rosner, A. (2004). Truncated Rep gene originated from tomato yellow leaf curl virus-Israel [Mild] confers strain specific resistance in transgenic tomato. *Annals of Applied Biology* **144:** 39-44.

Asad, S., Hars, W.A.A., Bashir, A., Zafar, Y., Malik, K.A., Malik, N.N. and Lichtenstein, C.P. (2003). Transgenic tobacco expressing Gemini viral RNAs are resistant to the serious viral pathogen causing cotton leaf curl disease. *Archives of Virology* **148:** 2341-2352.

Biswas, K.K. and Varma, A. (2001). Evaluation of resistance in blackgram (*Phaseolus mungo*) to variants of mungbean yellow mosaic geminivirus. *Indian Agril Sci.*, **71:** 215-218.

Biswas, K.K., Das, D.R. and Ahlawat, Y.S. (2005). Viral diseases of fruits, vegetables and spices in North Eastern Himalayan region of India and their management. *In:* Crop protection–Management strategy (Prasad, D. Ed.), Daya Publishing House, New Delhi-35 pp. 225-226.

Biswas, K.K. and Gupta. A. (2007). Management of important plant viruses in Indian agriculture. *In:* Sustainable pest management (Prasad, D. Ed.), Daya Publishing House, New Delhi-35 pp. 208-238.

Boogaart, T., Maule, A.J., Davies, J.W. and Lomonossoff, G.P. (2004). Source of target specificity associated with the recovery against *Pea seed-borne mosaic virus*, infection mediated by RNA silencing in pea. *Molecular Plant Pathology* **5:** 37-43.

Febes, V.J., Niblet, C.L., Lee, R.F. and Moore, G.A. (2003). Characterization of grapefruit plants (*Citrus paradisi* Macf.) transformed with *Citrus tristeza* clostero virus genes. *Plant Cell Rep.*, **21:** 421-428.

Fitch, Maureen, M.M., Manshardt, R.M., Gonsalves, D., Slightom, J.L. and Sanford, J.C. (1992). Virus Resistant Papaya Plants Derived from Tissues Bombarded with the Coat Protein Gene of Papaya Ringspot Virus. *Bio/Technology* **10:** 1466-1472.

Dasgupta, I., Malathi, V.G. and Mukharjee, S.K. (2003). Genetic engineering for virus resistance. *Curr. Sci.* **84:** 341-54.

Domínguez, A., Guerri, J., Cambra, M., Navario, L., Moreno, P. and Pena, L. (2000). Efficient production of transgenic citrus plants expresseing the coat protein gene of citrus tristeza virus. *Plant cell Reporter* **19:** 427-433.

Ghorbel, R., Lopez, C., Fagoaga, C., Moreno, P., Navaro, L., Flores, R. and Pena, L. (2000). Transgenic citrus expressing the citrus tristeza virus p. 23 protein exhibit viral like symptoms. *Mol Plant Pathol.* **2:** 27-36.

Golemboski, D.B., Lomonssoff, G.P. and Zatlin, M. (1990). Plant transformed with tobacco mosaic virus no structural gene sequence are resistant to virus. *Proc. Nat. Aca. Sci., USA*, **87:** 6311-6315.

Grumet, R. (1994). Development of virus resistant plant via genetic engineering. *Plant Breeding reviews* **12:** 47-49.

Gutierrez, M.A., Luth, D. and Moore, G.A. (1997). Factor affecting *Agrobacterium* mediated transformation in Citrus and production of sour orange (*Citrus aurantium* L.) plants expressing the coat protein gene of citrus tristeza virus. *Plant cell reporter* **16:** 745-753.

Gonsalves, D. and Garsney, S.M. (1989). Cross protection techniques for control of plant virus diseases in the tropics. *Plant Disease* **73:** 592-597.

Gonsalves, D., Chee, P., Provvidenti, R., Seem, R. and Slightom, J.L. (1992). Comparison of coat protein-mediated and gentically-derived resistance in cucumbers to infection by cucumber mosaic virus under field conditions with natural challenge inoculations by vectors. *Bio. Technology,* **10:** 1562-1570.

Hafner, G.J., Harding, R.M. and Dale, J.L. (1995). Movement and transmission of banana bunchy top virus DNA component one in bananas. *J. General Virology,* **76:** 2279-2285.

Harriron, B.D., Mayo, B.D. and Boulcombe, D.C. (1987). Virus resistance in transgenic plant that expresses Cucumber mosaic virus satellite RNA. *Nature* **328:** 799-801.

Hull, R. (2002). "Matthews' Plant Virology." (4th Ed.). Academic Press. A Harcourt Science and Technology Company, Harcourt Place, 32 Jameston Road. London, NW1 7BY, UK.

Khoury, J., Singh, R.P., Boucher, A. and Coombs, D.H. (1988). Concentration and distribution of mild and severe strains of potato spindle tuber viroid in cross-protected tomato plants. *Phytopathology* **78:** 1331-1336.

Khetrapal, R.K., Maisonneuve, B., Maury, Y., Maury, Y., Chalhoub, B., Dinant, S., Lecoq, H. and Varma, A. (1998). Breeding for resistance to plant viruses. *In: Plant Virus Disese Control*, (Hadidi *et al.*, Eds.). APS Press, pp. 14-32.

Lawson, E.C., Kaniewski, W., Haley, L., Rosman, R., Newell, C., Sanders, P. and Turner, N. (1990). Engineering resistance to mixed virus infection in a commercial potato cultivar: resistance to potato virus X and potato virus Y in transgenic Russet Burbank. *Bio/Technology* **8:** 127-134.

MacDiarmid, R. (2005). RNA silencing in productive virus infections. *Ann. Rev. Phytopathol.* **43**: 523-44.

May, G.D., Afza, R., Mason, H.S., Wiecko, A., Novak, F.J. and Arntzen, C.J. (1995). Generations of transgenic Banana (*Musa acuminata*) plants via Agrobacterium-mediated transformation. *Biotechnology,* **13:** 486-492.

Miller, N., Sulistyowati, E. and Dietzgen, R.G. (2003). Cucumber mosaic virus infection transiently breaks DsRNA–induced transgenic immunity to potato virus Y in tobacco. *Molecular Plant-Microbe Interactions* **16:** 936-944.

Montasser, M.S., Tousignant, M.E. and Kaper, J.M. (1991). Satellite-mediated protection of tomato against cucumber mosaic virus: I. Greenhouse experiments and simulated conditions in the field. *Plant Disease* **75:** 86-92.

Powel-Abel, P., Nelson, R.S., De, B., Hoffman, N., Rogers, S.G., Fraley, R.T. and Beachy. (1986). Delay of disease development in transgenic plant exprees the tobacco mosaic virus coat protein Gen, *Science* **232:** 738-743.

Scorza, R., Ravelonandro, M., Malinowski, T., Cambra, M. and Minoiu, N. (2003). Potential use of transgenic plums resistant to Plum pox virus field infection. *Acta Horticulture* **622:** 119-122.

Simon, A.E., Rossinck, M.J. and Havelda, Z. (2004). Plant virus satellite and defective interfering RNAs: new paradigms for a new sentury. *Ann. Rev. Phytopathol.* **42:** 415-37.

Singh, S.J. (2003). Virus and phytoplasma diseases of papaya, passion fruit and pineapple. Kalyani Publishers, Noida.

Smith, G.R., Ford, R., Frenkel, M.J., Shikla, D.D. and Dale, J.L. (1992). Transient expression of the coat protein of sugarcane mosaic virus in sugarcane protoplasts and expression in *E. coli. Arch. Virol.* **125:** 15-23.

Tapfer. M. (2002). Risk assessment of virus-resistant transgenic plants. *Ann. Rev. Phytopathol.* **40:** 467-91.

Vigne, E., Komar, V. and Fuchs, M. (2004). Field safty assessment of recombination in transgenic grape veins expressing the coat protein gene of *Grapevine fanleaf virus. Transgenic Research* **13:** 165-179.

Yang, Z.N., Ingelbrecht, I.L., Louzada, E., Skaria, M. and Mirkov, T.E. (2000). Agrobacterium–mediated transformation of the commercially important grapefruit cultivar Rio Red (*Citrus paradisi* Maca.). *Plant Cell Reports,* **19:** 1203-1211.

Young, Y., Sherwood, T.A., Patte, C.P., Hiebert, E. and Polston, J.E. (2004). Use of *Tomato yellow leaf curl virus* (TYLCV) *Rep* gene sequences to engineer TYLCV resistance in tomato. *Phytopathology* **94:** 490-496.

23

Nematophagous Fungi in Biocontrol of Plant Parasitic Nematodes

Satyandra Singh and S.C. Dhawan

Division of Nematology, Indian Agricultural Research Institute, New Delhi–110 012

Plant parasitic nematodes are very active creatures and important agents (pests) of destruction of agriculturally important crops. They are one of the most important biotic stresses hampering the production of many crops. They infect almost all cultivated crop plants and also act as vectors of many plant viruses. In association with other organisms they form disease-complex and enhance the disease severity in one or other way. Several chemical nematicides were quite effective in reducing the damage due to these microscopic worms and increasing the yield in many crops during 60 and 70s. However, most of the nematicides have been withdrawn from the market because of environment and health concerns. As a result of general awareness over chemical pesticides, there has been a shift in the development of alternative methods of nematode control including search for biological control agents (BCAs) of nematodes.

Root-knot, cyst forming, reniform, ear-cockle, burrowing, lesion and ecto-parasitic nematodes are distributed throughout the world. According to Sasser and Freckman (1987), the estimated overall annual yield loss of the world's major crops due to damage by plant parasitic nematodes is 12.3%. The avoidable yield loss due to *M. incognita* in India is 31-71 reported from Indian states—Assam, Gujarat, Karnataka and Rajasthan (Sharma *et al.* 2002). In India, extensive studies on crop loss assessment has not been conducted. Different workers reported various estimated crop losses on different crops ranging from 4% to total crop failure due to important plant parasitic nematodes in India (Jain, 2003).

The term biological control of nematodes in its true sense can be defined as the reduction of nematode population density in its active or dormant state by introducing one or more organisms naturally or through manipulations of environment, host or antagonists or mass production of any antagonists. There are many biological agents such as nematophagous fungi, bacteria, etc. which have been found to attack plant parasitic nematodes in soil and reduce their population. A biological change in the pest environment can be achieved by one of the four ways, namely, (a) the regular release of large numbers of parasitic/predatory fungi; (b) introduction and establishment of parasitic/predacious fungi of the nematode pest from an epidemic area; (c) manipulation in environment so as to increase the number of parasitic/ predacious fungi of the pest that are already present; and (d) integration of these biological means with other options including chemical means of control. There are still enough diseases with no good chemical control and no host resistance, so there are still plenty openings for biological and integrated control without competing directly with chemicals (Cook, 1993).

Exploitation of these enemies for practical use in the management of plant parasitic nematodes seems to be practically demanding and is a relevant approach in the present context. Of the microorganisms, fungi hold an important position and some of them have shown great potential as biocontrol agents. Fungi utilize many different sources of carbon, some living and some dead. That nematodes can be colonized by fungi led to the possibility that the interaction might be used to control pest nematodes of plants and animals. Closer examination of the interaction has led to the belief that only partial control of nematodes is possible and that in a practical sense, populations of pests may be suppressed not eliminated. In the case of plant and animal parasitic nematodes, fungi may be used to regulate nematode populations in conjunction with other processes.

The organisms of the fungal family include mushrooms, rusts, smuts, puffballs, truffles, morels, molds, and yeasts, as well as many less well-known organisms (Alexopoulos *et al.*, 1996). More than 70,000 species of fungi have been described; however, some estimates of total numbers suggest that 1.5 million species may exist (Hawksworth, 1991; Hawksworth *et al.*, 1995). Within their varied natural habitats, fungi are usually the primary decomposer organisms present. Fungi are filamentous plants that lack chlorophyll and develop as fine threads or hyphae. These elongate branches may fuse to produce a mycelium (a network of hyphae). Individual hyphae that comprise a mycelium may vary from 3 to 10 microns in diameter. In some low nitrogen environments, several independent groups of fungi have adaptations such as nooses and sticky knobs with which to trap and degrade nematodes and other small animals. A number of references on fungal ecology are available (Carroll and Wicklow, 1992; Cooke and Whipps, 1993; Dix and Webster, 1995). Both fungi and nematodes have the same niche.

HISTORY

The history of biological control of nematodes can be divided into two phases. The first phase, prior to 1975 refers to descriptive research. Nematode trapping fungi are not a recent discovery, *Arthrobotrys oiligospora* was described initially by Fresenius in 1852, and Woronius (1870) reported that the organism forms networks of hyphal slings of loops. However, the purpose served by these networks remained unknown until 1888 when Zopf become the first to observe a fungus trap a living animal. Other workers also observed this phenomenon, but there were few additions to our knowledge of nematode traping fungus until 1937 when the classical work of Dreschler was published, who reported that predaceous fungi will forever be associated and proceed further by Duddington (1955) who initiate work on biological control.

Antagonists of nematodes, especially nematode-trapping fungi, were considered interesting wonders of nature with a little relevance to agriculture. Nematicides were inexpensive and effective and little research effort was devoted to studying the effect of antagonists on nematode population in soil.

The work of Lohde, 1874, on nematode destroying fungi was the beginning when he observed the endoparasite, *Harposporium aggullulae* over a century ago. Since then many organisms have been found to feed on nematodes, but most interest has concentrated on fungi, especially those that trap their hosts in specialized traps (Kerry, 1981).

The second phase that began in 1977 with the loss of DBCP, which is continuing in the present, is typified by attempts to replace nematicides with antagonists. This phase will be remembered for replacement of chemical nematicides with antagonists. Thus far, few of these efforts have resulted ineffective biological control may or may not be achieved. Our greatest need of today is for a sound, in depth biological information on how organisms, populations and communities operate in the soil. After some initial failure in the biological control of plant parasitic nematodes (Duddington, 1960), interest has been revived by the demonstration that some species of endoparasitic fungi prevented *Heterodera avenae* populations from increasing and even reduced the damage caused by root-knot nematodes (Jatala *et al.* 1979, Kerry *et al.* 1982, Stirling and Mankau, 1978).

The next phase will be typified by basic investigations of organisms, population and community ecology. There are many biological agents such as nematophagous fungi, bacteria, etc. which have been found to attack plant parasitic nematodes in soil and reduce their populations. They are undoubtedly important in controlling numbers of nematodes of agricultural importance. Since the original description of activities of *Arthrobotrys oligospora* by Zopff in 1888, significant activities in the field of nematophagous fungi was initiated worldwide. Linford (1973) made the first attempt to harness nematophagous fungi for the control of root-knot nematode and he found some measurable success in controlling root-knot nematode by adding *Monacrosporium ellipsospora* (*Dactylella ellipsospora*) when chopped tops of pineapple were amended in the infested field. Since that time many other researchers started working with different soil amendments. Duddington *et al.* (1961) decreased the number of cereal cyst nematodes (*Heterodera avenae*) larvae invading oat seedlings by green manure treatments of chopped cabbage leaves. And this was followed by various workers on different crops with different antagonistic fungi and occupy a significant place in the literature.

ISOLATION AND ESTIMATION

Nematophagous fungi were isolated by spreading a small quantity (1-2 g) of soil or organic matter on weak agar medium (Duddington, 1955, Tolmsoff, 1959, etc.) The sequence of organisms that developed after incubation at 25°C was observed. Multiorganism growth can be observed including common fungi. Fast growing fungi that produce adhesive network traps develop first in the succession of the fungi in which endoparasitic fungi appear, if at all only after several weeks incubation (Barron, 1977). The number of fungi can be increased by baiting nematodes, usually *Panagrellus redivivus* (Wyborn *et al.* 1969). Mankau (1975) and Mc Culloch, 1977, found more fungal species and reduced variability when fungal propagules from larger soil samples are extracted by sieving and floatation technique on agar plates before sprinkled. A quantitative assessment of the number of nematophagous fungi can be obtained by using serial dilution of soil suspensions on water agar and most probable number analysis (Eren and Pramer, 1965). Soil dilution plates have number of demerits in the estimation of those fungi, which exist as mycelium and small quantity. They are slow growing and require the presence of nematode hosts for producing traps and sporulation (Johnson and Curl, 1972). Barron (1969) used low speed centrifugation to separate trapping fungi and endoparasites that produce small conidia before spread over water agar, the larger spores of trapping fungi being discarded with soil particles. Sugar floatation was the best method for isolating endoparasitic fungi (Stirling and Kerry, 1983), although some fungi may not develop from nematodes extracted in this way (Mankau, 1980).

Cook (1962) used **agar disc technique,** in which small agar discs are mounted on microscopic slides and buried in soil. Later, these slides are retrieved, washed, and examined. Trapped nematodes can be counted directly under the microscope. However, there is no way of enumerating the fungi.

According to Nicolay and Sikora, 1998, the **method for examination of nematode trapping fungi** is as follows:

- Hundred g of soil is suspended in 150 ml of water.
- The suspension is poured onto a 250 μm aperture sieve and washed with as little water as possible through the sieve into a beaker.
- The filtrate is agitated for 10 sec, then poured onto a 100 μm aperture sieve and washed through the sieve into a beaker.
- The resulting suspension is centrifuged at 650 g for 2 min.
- The supernatant is transferred to clean centrifuge tubes, while the sediment containing the spores of nematode-trapping fungi is mixed with 1 ml of water. 1-1.5 ml of the soil suspension is pipetted into a 5-6 mm wide groove cut into the middle of water agar Petri dish (9 cm diam.) by means of two joined scalpel.
- 3000 chrysoidin stained nematodes are added as bait in 0.6 ml of water.
- The supernatant of the first centrifugation step containing the spores of endoparasitic fungi is centrifuged again at 4000 g for 30 min.
- The sediment is resuspended and pipetted into the grooves of water agar. Endoparasitic fungi are best observed by cutting out a cross-shaped groove, which increases the contact of nematode with the fungal spores. In addition, it allows a 2 ml soil suspension to be used per plate.
- Three thousand nematodes are pipetted onto the soil in the grooves for detection of endoparasites. Adding a small amount of silver sand to the grooves alleviates nematode penetration into the agar.
- The plates are kept at room temperature, and parasitized specimens examined immediately after addition and at 2 days intervals for two weeks.

A semi-selective medium has been developed by Kerry and Bourne (2002) for isolation of *P. chlamydosporia* from soil, roots and nematodes; which is useful for reisolating the fungus after large numbers of propagules are added to the soil. However, in naturally infested soils, numbers of propagules are too small to be reliably isolated. The medium was modified slightly to contain corn meal agar, NaCl, rose Bengal, antibiotics, chloramphenicol, chlortetracycline and streptomycin sulphate, fungicide, carbendazim and Thiabendazole, and Triton X-100. The fungicides and antibiotics suppress the growth of other fungi and bacteria respectively, and NaCl and Triton X-100 reduced the rate of colony growth so that the colonies remain distinct even after 2 weeks at 20°C by which time they have sporulated and can be identified. Rose Bengal is also an antimicrobial agent and slows colony growth.

To isolate the fungus from eggs, crush the egg masses using a hand-held homogeniser or cyst crusher. It is better to use mechanical means rather than sodium hypochlorite solution, as the bleach affects the permeability of the egg membrane and may affect the growth of the fungus. Plate the eggs onto water agar + antibiotics and incubate at 25°C for 2-4 days. Transfer each colony growing out of an egg to corn meal agar (CMA) to obtain a pure culture.

Fungi can be isolated from adult females by dissecting the females out of roots, squashing them and plating them onto Petri dishes containing water agar + antibiotics. After incubation at 25°C for a few days, fungi will grow out of infected females and can be transferred to plates of CMA. Petri dishes of semi-selective medium must be stored and incubated in the dark, as exposure to sunlight causes inhibition of fungal development due to degradation of the rose Bengal.

Protocol to Quantify Nematophagous Fungi in Soil

- This is an outline and modification of methods developed by Jaffee *et al.* (1996).
- Prepare quarter-strength cornmeal agar (CMA/4)[1] amended with 0.1 g/L of streptomycin sulfate after the CMA/4 is cooled down to approximately 50°C (does not burn your hand).
- Pour the agar medium into 9 cm diameter or 6 cm diameter Petri dishes. Prepare 15 plates of CMA/4 for each sample.
- Subsample 100 g of soil from each sample placed in a 250-mL beaker, and add 200 mL sterile distilled water. Stir the suspension vigorously for 3 minutes.
- Conduct 10-fold dilution of the well-mixed soil suspension twice. This will provide three soil dilutions: 0.5, 0.05, and 0.005 g of soil/mL H_2O.
- Each dilution will be plated on 5 replicated Petri dishes. Each Petri dish will receive 100 ml of soil dilutions. Thus, there will be a total of 15 plates for each sample with 5 plates each of 0.05, 0.005, and 0.0005 g of soil/plate.
- Add approximately 1000 *Steinernema glaseri* (a large and long-lived entomophilic nematode) in 0.1 mL distilled water to each plate as bait for nematode-trapping fungi. This will increase the chances of detecting nematode-trapping fungi.
- Infective juveniles of *S. glaseri* can be obtained from infected *Galleria mellonella* according to the protocol by Woodring and Kaya (1988) and stored at 10°C.
- Observe the entire surface of each Petri dish 3 weeks after plating under a dissecting microscope with 50-140 magnification.
- Record the presence and absence of nematode-trapping fungi in species using the Cooke and Godfrey key (1964) for each Petri dish.
- Use a most-probable number table (Alexander, 1982).

TYPES OF NEMATOPHAGOUS FUNGI

Nematophagous (nematode destroying) fungi are extremely common in soil comprising more than 200 species of taxonomically diverse fungi, which share the ability to attack living nematodes (juveniles, adults and eggs) and use them as nutrients. While many of the traps forming and egg parasitic fungi can survive in soil saprophytically, the endoparasitic fungi are mostly dependent on nematodes as nutrients (obligate parasites).

The fungi involved in biocontrol of nematodes are **endoparasitic, predaceous** and **opportunistic fungi**. Assessment of potential of endoparasitic and predacious fungi in biocontrol has not received much attention except *Hirsutella rhossiliensis* and *Arthrobotrys oligospora.* Opportunistic fungi are gaining more importance in biocontrol programs than the previous two groups. These fungi

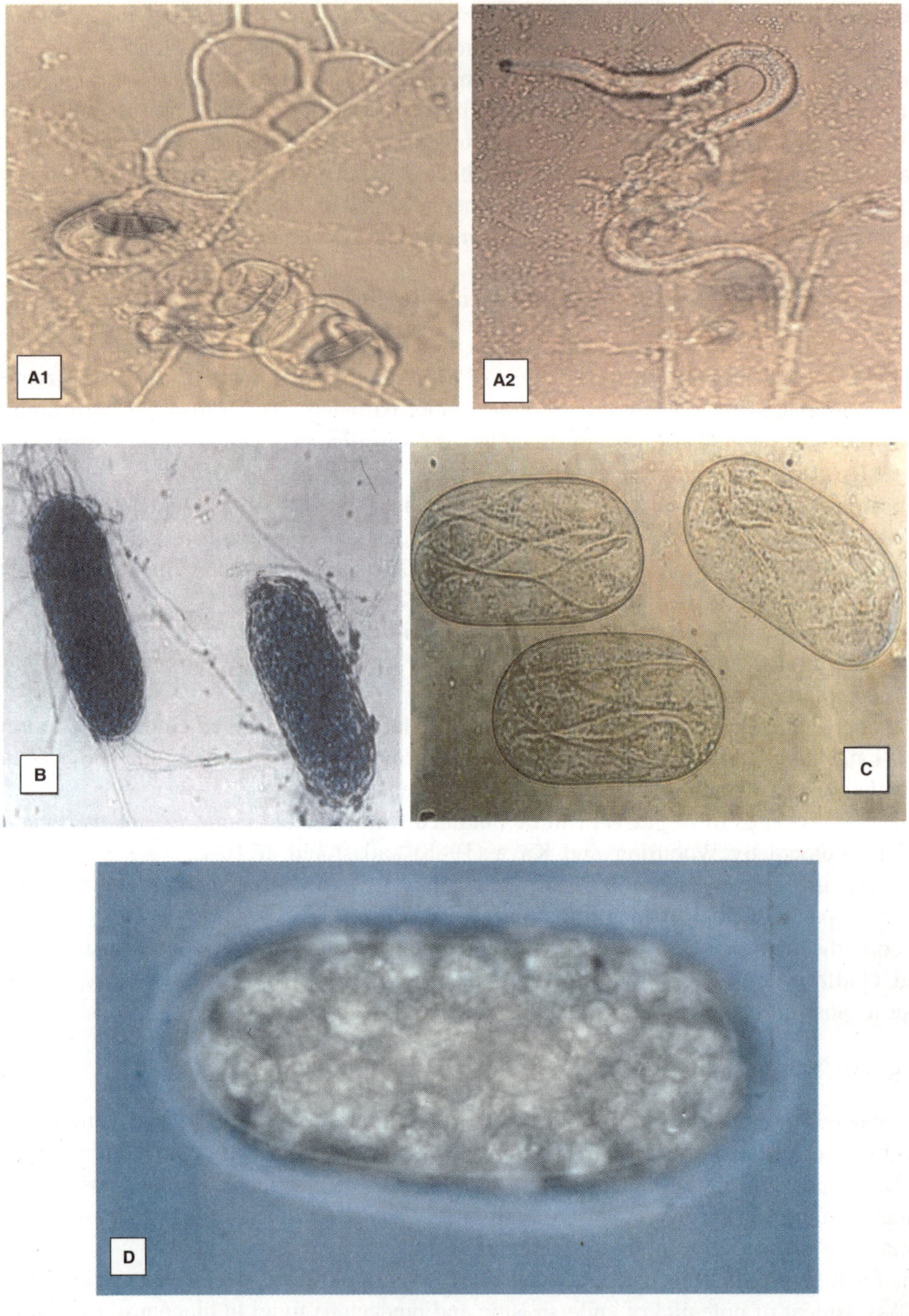

Fig. 23.1 A1. Three-dimensional rings of *A. oligospora*; **A2.** Nematode trapped in the rings. **B.** *M. incognits* eggs parasitized by *P. lilacinus*; **C.** Healthy eggs of *M. incognita*; **D.** *M. incognita* egg containing spores of *P. chlamydosporia*.

are capable of colonizing nematode reproductive structures and deleteriously affecting them. The fungi frequently encountered are *Cylindrocarpon, Exophiala, Paecilomyces, Pochonia*, etc. *Paecilomyces* and *Pochonia* are capable of preventing egg hatching and colonize eggs by hyphal penetration (Fig. 23.1 B & D). Other opportunistic fungi, namely, A*cremonium, Chlamydosporium, Gliocladium, Trichoderma*, etc. were also found to check nematode pathogenesis.

The ability to capture nematodes is connected with a specific developmental phase of the fungal mycelium. The trapping fungi have developed sophisticated hyphal structures, such as hyphal nets, knobs, branches or rings in which nematodes are captured by adhesion or mechanically. Trapping fungi might have a greater effect on ectoparasitic nematodes that are always in soil exposed to attack, than as endoparasites that are protected within roots for much much of their active life. The endoparasites, on the other hand, attack nematodes with their spores, which either adhere to the surface of nematodes or are swallowed by them. Examples of first group are *Arthrobotrys* spp. such as *A. oligospora, A. conoides, A. musiformis* and *A. superba*, which all form three-dimensional adhesive nets, whereas *A. dactyloides* uses constricting rings to capture nematodes mechanically by the swelling of the ring cells (Fig. 23.1 A-1 and A-2). Adhesive branches and adhesive knobs appear in the genus *Monacrosporium. M. haptotylum* (*Dactylaria candida*) produce both adhesive knobs and non-constricting rings (Table 23.1).

Table 23.1. Infection structures of some nematophagous fungi.

Infection structure	**Species**	**Target prey**	**Taxonomic classification**
Adhesive nets	*Arthrobotrys oligospora, conoides, musiformis, A. superba*	Free living and vermiform plant- parasitic nematodes	Deuteromycetes
Adhesive branches	*Monacrosporium gephyrophagum*	——Do ——	Deuteromycetes
Adhesive knobs	*M. ellipsospora* *M. haptotylum*	——Do ——	Deuteromycetes
Adhesive knobs and Adhesive spores	*Nematoctonus concurrens*	——Do ——	Basidiomycetes
Adhesive spores	*N. leiosporus* *Drechmeria coniospora* *Hirsutella rhossoliensis*	——Do —— ——Do —— ——Do ——	Basidiomycetes Deuteromycetes Deuteromycetes
Adhesive hyphae	*Stylopage hydra* *Cystopage cladospora*	——Do —— ——Do ——	Zygomycetes Zygomycetes
Ingested spores	*Harposporium anguillulae* *Catenaria anguillulae* *C. vermicola* *Heptoglossa dickii*	——Do —— ——Do —— ——Do —— ——Do ——	Deuteromycetes Chtridiomycetes Chtridiomycetes Oomycetes
Toxic droplets Appresoria	*Pochonia chlamydosporia*	Eggs and females of root-knot and cyst nematodes	Deuteromycetes
Fungal hyphal tips without appresorial formation	*Paecilomyces lilacinus*	——Do ——	Deuteromycetes

Among the endoparasites, *Drechmeria coniospora*, *Hirsutella rhossoliensis*, *Haploglossa dickii* and *Catenaria anguillulae* infect nematodes with their spores and spend vegetative lives inside infected nematodes. The genus *Nematoctonus* captures nematodes with both adhesive traps and adhesive spores and thus constitute a link between the two groups. The egg-parasitic fungi e.g., *Pochonia chlamydosporia* (previously *Verticillium chlamydosporium*) use appresorium to nematode eggshells, while *Paecilomyces lilacinus* uses fungal hyphal tips to enter into nematode eggshell (Fig. 23.1B). The fungi that parasitize eggs can limit nematode multiplication in the field and seem most suitable for development as biocontrol agents. Opportunistic fungi exist as a number of different isolates, which vary in their pathogenicity and biology, and these differences between populations are common and increase the problem in selecting a virulent organism that will retain its characteristics in mass productions.

ECOLOGY

In vitro studies of nematophagous fungi are dominant in literature. They kill large number of nematodes but result in non-sterile soil are not encouraging. In controlled conditions, fungus has no competition, no option of food source and the host has no escape, exaggerate the effect of the parasite on nematode population.

Nematophagous fungi are widespread throughout the world and have been isolated from an extremely wide range of material, from agricultural soil to Antarctic lichens (Grey and Duff, 1982; Grey *et al.*, 1982). Nematophagous fungi, *Arthrobotrys oilogospora* was found in aerial parts of palnts infected by *Ditylenchus dipsaci* (Goody, 1938) and in seed galls caused by *Anguina tritici* (Dhawan *et al.* 1980). Nematophagous fungi that survive in the rhizosphere may be more effective in parasitizing plant parasitic nematodes than those that occur throughout the soil.

Certain biotic and abiotic factors regulate the growth and abundance of nematophagous fungi. Biotic factors include host tissues, root rhizosphere and competitive/antagonistic microflora, whereas abiotic factors include organic amendments, temperature, moisture, pH and soil texture.

Nematode trapping fungi are different in their requirement of temperature for growth and trapping nematodes (Cayrol and Brun, 1975; Mankau, 1964 and Feder, 1963). Most trap formation of these fungi occurs at the temperature range of 10-30°C, but the most optimal range was found as 20-25°C. These fungi remain active and can capture the nematodes at the pH range of 5-8, but the pH of 6-6.5 was found to be best for its activity. 25°C was the best temperature for growth of antagonistic fungi i.e., *A. oligospora*, *P. chlamydosporia* and *P. lilacinus*. These were very sensitive at temperature 5 and 45°C as none of the fungi was able to grow further. It was found that growth of *A. oligospora* was excellent compared to *P. chlamydosporia* and *P. lilacinus* at 25°C. Sandy loam soil is best for these fungi. Maximum production of cfu's of *A. oligospora*, *P. chlamydosporia* and *P. lilacinus* was reported in hill brown soil (sandy loam) followed by alluvial and black soil. (Dhawan *et al.,* 2004) Excessive moisture in the form of rainfall hampers trap formation. Heavy metals like cadmium and lead have inhibitory effect on the trap formation of these fungi. Magnesium also inhibits trap formation. But the elements like phosphorus, zinc, manganese and molybdenum have stimulatory effect on the trap formation.

Effect of chemical fungicides is not encouraging. Except chlorothalonil and myclobutanil, almost all fungicides have deleterious effect on *Arthrobotrys* spp. The integration of chemical control with the activity of nematode trapping fungi appears possible. Mankau and Imbriana

(1971) and Mitsui (1972) showed that certain nematode trapping soil fungi tolerated upto 700 ppm ethylene dibromide in culture media, suggesting that nematocides may be metabolized by the fungi. The fungi tolerated direct exposure to dosages considerably higher than nematicidal field rates, suggesting that nematicides do not alter whatever biological control is contributed by predacious fungi.

The mycelial growth and trap formation, which are essential features before predation, require energy which becomes available through the carbon sources of organic matter. Thus, the addition of organic amendments to soil is usually followed by a short period in which the activity of nematode trapping fungi increases. It leads to the higher rate of nematode capture. Also, intense microbial activity is initiated by the addition of organic matter, which may lead to food competition resulting in forced nematode trapping activity. A further increase in the nutritional level may reduce the competition and thus also nematode trapping activity. It has been postulated, therefore, that the total amount applied should not be excess of 100-200 kg/ha. The addition of organic amendments to soil also enhances the population of free living nematodes. Ultimately trapping is diverted to these nematodes, and so plant parasitic nematodes escape trapping. These fungi also exhibit the phenomenon of mycostasis (i.e., failing to germinate in the soil at the same rate as observed under lab conditions). This phenomenon is associated with active microbial population and may be removed by addition of a nutrient source in the form of organic amendments.

Obviously, complex interaction predacious fungi decomposing organic substance, and the remainder of the soil microbial population determine the final level of each in the soil. Organic amanedment can be given in the form of dried alfalfa and grape leaves, chickpea pod waste with sugar solution, subabool leaf powder, undecomposed materials, composts, manures, etc. of the four substrates (wheat straw, FYM, chickpea pod waste and subabool leaves). Dhawan *et al.* (2005) reported that the growth of all the three antagonistic fungi, *A. oligospora*, *P. chlamydosporia* and *P. lilacinus* was found maximum in chickpea pod waste. The maximum growth of these three fungi on chick pea pod waste powder + 5% sugar substrate was witnessed by *A. oligospora* (13.30 × 0^9 cfu/g), *P. chlamydosporia* (13.53 × 0^9 cfu/g), and *P. lilacinus* (16.96 × 0^9 cfu/g) spore count. Monosaccharides and disaccharides were suitable for fungal growth. For most isolates, D-glucose, D-mannose, sucrose and trehalose were superior to pectin and soluble starch among the polysaccharides and lactic acid among the organic acids. Both ethanol and methanol could accelerate growth of most isolates but not biomass. (ManHong and XingZong, 2006)

PREPARATION AND METHOD OF APPLICATION

BCAs can be easily and inexpensively mass cultured in nutrient agar/broth medium, which is a great advantage to allow effective commercial production. They can be conveniently applied both as inundative (granular formulation can be incorporated directly to soil or through water irrigation system) and inoculative agents (can be applied as seed treatment in directly seeded crops or as seedling root dip in transplanting crops before transplantation). Their universal applicability is an inherent advantage, so they can be applied in temperate as well as tropical environment. They are compatible to be applied with fungicide and nematicides (Nagesh and Jankiram, 2004; Dhawan and Satyendra Singh, 2008). The persistence and survival of these BCAs varies from few weeks to months.

The antagonistic fungus, previously isolated from agriculture soil was raised in PD broth contained in the conical flasks and incubated at 25 ± 2ºC for 21 days. Thereafter, 300 ml of broth with fungus was taken and mixed thoroughly. Fungal biomass was mixed with 1 kg of talc powder as a carrier material to bulk up the material. To this 20 g fine powder of chickpea pod waste and carboxymethyl cellulose (2% w/w) was added to the carrier material as carbon source and sticking agent, respectively. All the contents were remixed manually and dried under shade. The material after drying was packed in polythene bags for further biomanagement trials. The concentration of chlamydospores was estimated in diluted samples using a haemocytometer. The spore load of *P. chlamydosporia* was maintained at 2.3 × 10^8 cfu/g. Granular formulations of BCA's are prepared by encapsulating different quantities of fungal bioagents in alginate.

The following method of application has been tried a) Seed treatment b) Root dip treatment c) Soil application. Seed treatment with combined formulation of *Pseudomonas fluorescens* (2 × 10^8 cfu/g) and *Paecilomyces lilacinus* (2 × 10^6 cfu/g) at the rate of 10 g/kg of seed and nursery beds treatment with this formulation at the rate of 50 g and neem cake 500/sq.m showed that vigorous seedling of tomato and capsicum can be obtained. This treatment also showed least amount of infestation by root-knot nematodes. Treated seedlings when transplanted in the main field resulted in the significant reduction of root-knot nematodes and increased the yield of these crops (Rao, 2007).

A.oligospora can be applied in soil by four possible means—Encapsulating them with seed, dipping seedling roots into a suspension of the organism at transplanting, direct incorporation into soil and application through irrigation system. Soil application of *A.oligospora* is most effective in suppressing the population of root-knot nematode than seed treatment. As in case of soil application, the fungus is well distributed in root rhizosphere having more access in trapping nematodes compared to seed treatment where the fungus remains localized near the seed surface.

POTENTIAL AS BIOCONTROL AGENTS

Nematophagous fungi are extremely common in soil, where they are known to capture, kill and digest free living and plant parasitic nematodes. Since the original description of the activities of *Arthrobotrys oligospora* by Zopff in 1888, significant activities in the field of nematophagous fungi was initiated worldwide. Linford (1937) made the first attempt to harness nematophagous fungi for the control of root-knot nematode, and found some measurable success in controlling root-knot nematode by adding *Monacrosporium ellipsospor*a (*Dactylella ellipsospora*) when chopped tops of pineapple were amended in the infested field. Since that time many other researchers started working with different soil amendments. Duddington *et al.* (1961) decreased the number of cereal cyst nematode (*Heterodera avenae*) larvae invading oat seedlings by green manure treatments of chopped cabbage leaves. The mode of action of these amendments is probably to stimulate the microbial activity, which in turn increase the number of nematodes and their nematophagous fungi.

After 1950, there have been several investigations on nutritional requirements of nematode-trapping fungi. Knowledge of these requirements is necessary for an understanding of ring formation or morphogenesis in these fungi, and ultimately it may be the key to regulating their saprophagous or predacious behaviour in soil. The carbon and nitrogen requirements have also been investigated. Realisation is that predacious activity is more intense in soil after the addition of sucrose than after the addition of chopped cabbage tissues. Olthof and Estey (1966) found that low levels of dextrose and ammonium nitrate did not reduce the radial growth of *A. oligospora* but greatly reduced its predacity on nematodes. All net forming fungi grew well on polysaccharides than did the ring formers. The predacity of *A. oligospora* was less when nitrogen was supplied as ammonium than when it was supplied as nitrate. *A. robusta* predacity was greatest on media deficient in biotin and with nitrate as the nitrogen source.

Diversity in infection structures has been studied in nematode-trapping, endoparasites, and egg-parasitic fungi. Nematode trapping fungi present a high diversity not only in respect of taxonomic distribution but also in trapping structures formed, which depends on species or even strains of species as well as on environmental conditions, both biotic and abiotic. The most important biotic factor is living nematode causing two-fold relationships: first, the nematode may induce the formation of structures in which they are later captured, and second, after invasion of the nematodes by the fungus they serve as an additional food source. A similar diversity also exists in endoparasites. *D. coniospora* form large numbers of conidia in comparison to production of hyphal material. In a single infected nematode, *D. coniospora* may produce as many as 10,000 conidia, while the endoparasite *H. rhassoliensis*, which sporulates singly, produces 100 to 1000 conidia per infected nematode. Both fungi develop adhesive bud on their conidia with which they infect the nematode. The fungi that parasitize the non-motile stages of nematodes, i.e., eggs, use a different strategy. Hyphae of *Pochonia chlamydosporia* and other fungi grow towards the eggs and appresoria are formed on the hyphal tips, which penetrate the eggshell. The fungi then digest the contents of eggs.

Information has also been generated on the potential of strains of nematophagous fungi as potential biological control agents. Mankau and Xuiying (1985) evaluated different isolates of *Monacrosporium ellipsospora. In vitro* tests exhibited considerable variability in response to *Meloidogyne* eggmasses. Isolate 585-1 and 597 formed sticky knob traps and ensnared larvae so rapidly. In greenhouse experiment, the fungus reduced galling by 49 and 42% respectively in the 10 g and 5 g fungal treatments. The biocontrol efficiency of *Arthrobotrys dactyloides* in reducing the population of *M. javanica* was reported by Galper *et al.* (1995). However, excellent control of root-knot nematodes of tomato following the application of granular formulations of *A. dactyloides* in pot and field was reported by Stirling and Smith (1998). Two strains (ORS 18692 S7 and ORS 18692 S5) of *A. oligospora* isolated from Senegal trapped 98% of *M. mayaguensis* claimed by Duponnois (1995).

There is a renewal interest in using nematode-trapping fungi, partly due to an increased knowledge on the biology of these fungi and partly due to better methods of formulating and applying fungal biocontrol agents to soil. One way to improve the control potential of nematophagous fungi would be to use genetic engineering to increase the pathogenicity and survival of introduced fungus. Using genetic transformation it was possible to generate mutants of nematode-trapping fungus *A. oligospora* over expressing a protease gene (PII). Mutant containing additional copies of the P II gene developed a higher number of infection structures and had an increased speed of capturing and killing nematodes (Ahman *et al.*, 2002). The detection of proteinase production is an important aspect for selecting some biological control agents, because of the role these enzymes play in the infectious process. Studies were conducted to modify a method for detecting proteinases of *Pochonia chlamydosporia* in solid medium supplemented with gelatin using different dyes and measuring the gelatin degradation halo, the colony growth and the proteinase activity indicator. Trypan blue at 0.03% was the most efficient alternative with a good visualization of the degradation halo and the least effect on the fungal colony growth (Peteira *et al.*, 2006).

A wide range of nematophagous fungi infect sedentary nematodes such as eggs of cyst and root-knot nematodes (Kerry, 1988; Stirling and West, 1991) and also of *Tylenchulus semipenetrans* (Walter and Kaplan, 1990). In general, fungi colonizing eggs and cysts are more numerous than those parasitizing females (Gintis *et al.*, 1983). Among egg parasitic fungi, *Paecilomyces lilacinus* and *Pochonia* (*Verticillium*) *chlamydosporia* are most studied species.

Discovery of *P. lilacinus* as an effective parasite of *M. incognita* and *Globodera pallida* eggs was reported by Jatala *et al.* (1979). The proliferated hyphal branches penetrate the eggs of *Meloidogyne* species. In cysts of *Globodera* species the fungus penetrates the vulva. After entering the cysts the fungus grows saprophytically on the mucilaginous body content surrounding the eggs during or before its infection in the eggs. In all cases, eggs in the early embryonic developmental stages prior to gastrulation process are more vulnerable to infection. Once the hypha is in contact with the egg, a series of ultrastructural changes occurs in the eggshell due to exogenous metabolites and the chitinolytic activities of the fungus. Experimental evidence indicated that *P. lilacinus* penetrated eggs of *Meloidogyne* at a much faster rate than it did those of *Globodera* and *Nacobbus* species. Field application of *P. lilacinus* in several on-farm trials in Peru provided evidence to proving its efficiency in controlling *M. incognita* and *T. semipenetrans* Guevara *et al.*, 1984; Herrera *et al.*, 1984; Salas and Jatala, 1980). Encouraging results were found with the use of *P. lilacinus* as biological control agents of golden nematode of potato (Saifullah and Gul, 1991) and root-knot nematode on chickpea (Zaki and Maqbool, 1991).

The fungus *P. chlamydosporia* infect by producing an appresorium laterally or at the tip of hyphae growing across the egg surface (Dunn *et al.*, 1982). Typically, penetration of the eggshell is considered to be due to both enzymatic degradation and physical forces. Attempts at using *P. chlamydosporia* as a biocontrol agent are well documented (Kerry, 1990; Stirling and Smith, 1991; Kerry and Bourne, 2002; Wang *et al.*, 2005; Lopes *et al.*, 2007). Most researchers have reported very good efficacy in greenhouse trials but less convincing performance from field trials (Stirling and Smith, 1998; Tzortzakakis, 2000; Verdejo-Lucas *et al.*, 2003). For successful establishment of *P. chlamydosporia* in non-pasteurized field soil, several applications might be necessary because its potential often drops within a few weeks after introduction (Kerry *et al.*, 1993; Bourne *et al.*, 1996; Stirling and Smith, 1998; Viaene and Abawai, 2000).

Commercial formulations are available in the world market as Royale 300, royale 350 from France. USA developed the formulation like NemAttack, 066026, 066027 and 066028 from *A.oligospora*. Russia developed Nematofazin-BL from *A. oligospora*. Biopesticide has not been listed as PAN bad actor chemical and cholinesterase inhibitor. There is no available weight of evidence regarding acute toxicity, carcinogenic property, endocrine disruptor, development or reproductive toxin and groundwater contamination by these biopesticides. It does not cause allergy. There is no report of ban, restricted or cancelled in any country. There is no legal restriction for import of these biopesticides and has not been listed in UNEP persistent organic pollutant and WHO obsolete pesticide. US EPA (Environmental Protection Agency) has not registered it as it is totally safe to human beings.

STATUS OF RESEARCH AND INNOVATION IN INDIA

It is quite apparent from literature that in India a very good commencement has been made in the field of biocontrol research and development. Progress in the field of biological control of plant-parasitic nematodes by nematophagous fungi is remarkable and various workers have reported occurrence of large number of these fungi from the country. Till 1960, most of the work pertained to nematophagous fungi was reports on occurrence, new records and taxonomical description of nematophagous fungi isolated from soil and different habitats (Dasgupta *et.al.*, 1964; Sachchidananda and Swarup, 1966; Sachchidananda and Rama Krishnan, 1971: Patil and Pendse, 1976; Dayal and Srivastava, 1978). Subsequently, information of these fungi was generated on soil amendments with organic material to see the effect of this material on nematophagous fungi. Work carried out on this aspect showed that organic material on decomposition releases compounds that were stimulatory for the growth of nematophagous fungi and also helped in improving the plant growth (Singh and Sitaramaiah, 1966; Singh *et al.*, 1967; Khan *et al.*, 1974).

After 1980, research was intensified on the application of biological control agent for suppressing the population of plant-parasitic nematodes both *in vitro* and *in vivo* conditions. Besides, hunt for newer nematophagous fungi and their taxonomical description was continued (Prasad *et al.*, 1994; Prasad and Dayal, 1985; Dhawan *et al.*, 1979; Chowdhury and Dhawan, 1984; Dhawan and Kaushal, 1987; Dayal, 2000). Although large number of nematophagous fungi have been identified, greater focus was restricted to only few, namely, *Arthrobotrys* species (nematode-trapping fungi), *Paecilomyces lilacinus* and *Pochonia chlamydosporia* (egg-parasitic fungi). *A. oligospora* mass multiplied on rice bran at 2% w/w recorded an increase in plant growth characters. The maximum shoot weight was recorded in rice bran substrate 2% w/w. The maximum root length and root weight were recorded in treatments with rice bran substrates at higher doses that are at par with the control. However, there was significant reduction in the number of galls and egg masses per plant in treatment with rice bran at the highest dose (Devi and Thaur, 2007).

Efficiency of trap-formation and its viability under different storage conditions was studied by Dhawan *et al.* (2004). They observed that trap formation was density dependent and the fungus could grow well in sandy loam soil and the optimum temperature for its growth was 25°C.

The commercialization of biocontrol agents and necessary fermentation technologies are the prime requirement of mass production of nematode biocontrol agents. A suitable medium is necessary for commercial fermentation using inexpensive and readily available materials. *Paecilomyces lilacinus* and *Pochonia chlamydosporia* are facultative parasite, which can grow on a wide variety of media. The fungi grow readily in liquid media but colonize solid substrates like bran, straw and cereals. Granular formulation is considered to be most suitable for microorganisms that can be easily applied to soil.

Attempts at using talc-based application of *Arthrobotrys oligospora* as a biological control agent against root-knot nematode are well documented. Sankaranarayanan *et al.* (2000) reported reduction in root-galling and increase in plant growth of sunflower. Dhawan *et al.* (2005) observed significant suppression in root galling and egg masses/plant by talc-based application of *A. oligospora* when added in the soil compared to seed treatment. Kumar and Singh (2006) tested the predacity of 5 isolates of *A. dactyloides* isolated from different locations of India against *Meloidogyne incognita* (J2), *Tylenchorhynchus brassicae* and *Hoplolaimus indicus*. All isolates of *A. dactyloides* captured and killed *M. incognita* and *T. brassicae* but not *H. indicus*. The isolates also differed in their predacity of the first two nematode species. They also reported reduction of root-knots by 5.6 to 45.6%, of females by 44.7 to 72.9%, of egg masses by 44.5 to 51.3% when mass culture of *A. dactyloides* was applied in the *Meloidogyne incognita* infested soil before planting of tomato seedlings.

Besides, nematode-trapping fungus, significant progress has also been done with egg-parasitic fungi. A number of glass house trials on *Paecilomyces lilacinus* have revealed the efficacy of the fungus in suppressing root-knot nematode population on okra (Walia *et al*., 1991; Saikia and Roy, 1994; Dhawan *et al*., 2006), on tomato (Zaki, 1994; Khan and Goswami, 2000; Priya and Kumar, 2006), cowpea (Hasan, 2004) against *Radopholus similis* on banana (Devrajan and Rajendran, 2001). The fungus could survive for 6 months when stored under natural condition in micro-plots and for 12 months when stored in laboratory in soil or calcium carbonate or charcoal at a temperature ranging between –20 to 25°C (Dhawan *et al*., 2006).

Another prospective egg-parasitic fungus is *Pochonia chlamydosporia*, which has gained a reputation of biological control agent against plant-parasitic nematodes, particularly root-knot and cyst forming nematodes. A good progress on this bioagent has been made at two centres namely, Indian Institute of Horticultural Research, Bangalore and the Indian Agricultural Research Institute, New Delhi. Investigations on abiotic and biotic factors have shown that sandy loam soil, optimum temperature of 25°C and soil pH of 7 favour good growth and development of *P. chlamydosporia* (Dhawan *et al*., 2004).

Potentiality of *P. chlamydosporia* as a biocontrol agent of root-knot, cyst forming and citrus nematode has been demonstrated on several crops (Parvatha *et al*., 1999, 1996, Dhawn *et al*., 2004; Naik and Rao, 2004; Shylaja *et al*., 2004; Rao, 2004). Biochemical studies done on tolerant and susceptible tuberose cultivars inoculated with *M. incognita* and *P. chlamydosporia* showed increased activity of peroxidase and phenyl alanine ammonia lyase (PAL) in tolerant cultivar than susceptible cultivar (Rao and Shylaja, 2004). Biotechnological investigations on cloning and sequence analysis of 4 strains of *P. Chlamydosporia* (Vc-1, Vc-2, Vc-3 and Vc-4) have revealed the presence of glutamic acid at position 162 and 180 codon region in Vc-1 and at 120 and 162 in Vc-3. This variability has got the influence on the efficacy of the fungus in inhibiting egg hatching. Variability studies at ITS and IGS regions of rDNA and fingerprinting pattern of DNA and ERIC primers suggested that all four isolates of *P. chlamydosporia* have the same geographical origin (Naik and Rao, 2004).

Attempts made at the Indian Institute of Horticultural Research on the development of bio-pesticides have yielded encouraging results by producing number of biopesticides such as **IIHR-Trichorich** (*Trichoderma viride*), **IIHR-Biovert**, P. *chlamydosporia* and combination formulation containing two bioagents such as **IIHR-Biovert Plus**, a formulation of *P. chlamydosporia* and *P. lilacinus*; **IIHR-Trichovert**, a formulation containing *T. viride* and *P. chlamydosporia*.

A private company-Kumar Krishimitra Bio-products (I) Pvt. Ltd. has come out with a bionematicide traded as **Nemastin**. This product contains spores of nematophagous fungi, which bring about destruction of plant-parasitic nematodes.

CONCLUSION

The soil harbours a variety of microorganisms, some of which are antagonists to plant-parasitic nematodes. It is well known that potential for biological control of plant-parasitic nematodes is great and it needs to be exploited on large scale, but spectacular results have never been achieved while using it under field condition. In order to produce spectacular results, there is a need to explore virulent strains of antagonists that can capture nematodes fast and also adopt to different geographical regions. Advances in the techniques of biotechnology would involve selection or introduction of variants of potential biological control agents with characteristics that enhance their effectiveness. It would also require development of cultural practices that promote the growth of beneficial biota in the soil. There is a vital arena of basic and applied research opportunities in the genetic tailoring organisms for effectiveness in specific environmental and cultural situations.

Quantitative assays are lacking for most antagonists of nematodes. Biological control occurs when a population of a community of antagonists suppresses a population of nematodes. Quantitative assays and models must be developed in order to understand the interaction of pest population and biological control agents. Clearly there are myriad interactions and potential interactions among soil organisms that require detailed study. These complex interactions, if better identified and understood, might be manipulated to reduce the population of plant-parasitic nematodes.

The tactics of introduction, inundation, and augmentation as biological controls of soil-inhabiting nematodes raises a series of implementation challenges. In other cases, the organisms will already be present in the soil and will be augmented by alteration of environment or introduction of food substrate. Application and management of biological control agents should be developed case-by-case, taking into consideration the prevailing cultural and environmental conditions.

FUTURE PROSPECTS

It is quite encouraging that we have been able to isolate a large number of bioagents and demonstrate their efficacy under *in vivo*, *in vitro* and controlled field conditions, but there is a need to reach at farmer's field. Major research priorities must be given to the following areas:

1. Isolation and identification of native biocontrol agents with high competitive saprophytic ability and rhizosphere competence, having biological control potential against more than one pathogen, particularly against fungus and nematode that caused complex diseases based on location specific requirements.
2. Genetic manipulation to increase the ability of nematophagous fungi to control nematode pest through biotechnological approaches.
3. Awareness among farmers/growers about harmful effects of chemical nematicides and benefits of BCAs is essential to develop organic farming in India.
4. Development of Fermentation Technology for mass production and to ensure the delivery of the product to farmer's field.
5. Interaction and compatibility of bioagents with other control components to develop integrated management system.
6. More emphasis should be given to cost-benefit ratios to understand economic factor.
7. Popularization of this environment-friendly technology among farmer's/growers with proper training and guidance.

Future research is, therefore, necessary to develop the methodology and delivery system for the use of suitable biocontrol agent that will be effective and competitive with other existing or introduced parasites in rhizosphere. Integrated nematode management strategies should use biocontrol potentials of nematophagous fungi to reduce dependence on chemical nematicide. Application and management of biocontrol agents should be developed on a case by case basis, taking into consideration the prevailing cultural and environmental conditions. Proper and efficient utilization of biocontrol agents in suppressing nematode population ia a great challenge. The goal can be achieved by more research on this aspect.

REFERENCES

Ahman, J., Olsson, M. and Johansson, T. *et. al.* (2002). Improving the pathogenicity of a nematode-trapping fungus by genetic engineering of subtilis in with nematotoxic activity. *Applied and Environmental Microbiology* **68:** 3408-3415.

Alexander, M. (1982). Most probable number method for microbial populations. pp. 815-820. Page, A.L. (ed.). *In:* Methods of soil analysis, part 2. Chemical and microbiological properties. Agronomy Monograph No. 9 (2nd edition). ASA-SSSA, Madison, Wisconsin.

Alexopoulos, C.J., Mims, C.W. and Blackwell, M. (1996). Introductory Mycology (4th ed.). John Wiley and Sons, New York, USA. p. 868.

Barron, G.L. (1969). Isolation and maintenance of endoparasitic nematophagous Hypomycetes. *Canadian J. Botany*, **47:** 1899-1902.

Barron, G.L. (1977). The nematode destroying fungi. Guclph: Canadian Biological Publications, pp. 140.

Bourne, J.M., Kerry, B.R. and de Leji, F.A.A.M. (1996). The importance of the host plant on the interaction between root-knot nematodes (*Meloidogyne incognita*) and the nematophagous fungus, *Verticillium chlamydosporium* Goddard. *Biocontrol Science and Technology* **6:** 539-548.

Carroll, G.C. and Wicklow, D.T. (1992). The Fungal Community: Its Organization and Role in the Ecosystem. Marcel Deker, Inc., New York.

Catrol, J.C. and Brun, J. (1975). Etude a differentes temperatures, de l'activite predatuce. Revue de Zoologic Agricole et de pathogenic vegetole, **74:** 139-146.

Choudhary, P.N. and Dhawan, S.C. (1984). *Catenaria vermicola* in *Heterodera avenae* nematode—A new record. *Curr. Sci.* **53:** 96.

Cook, R.C. (1962a). Behaviour of nematode-trapping fungi during decomposition of organic matter in soil. *Transactions of the British Mycological Society* **45:** 314-320.

Cook, R.C. (1962b). The ecology of nematode-trapping fungi in the soil. *Ann. App. Bio.* **50:** 507-513.

Cook, R.J. (1993). Making greater use of introduced micro organisms for biological control of plant pathogen. *Ann. Rev. Phytopahology* **31:** 53-80.

Cooke, R.C. and Godfrey, B.E.S. (1964). A key to the nematode destroying fungi. *Trand. Brit. Mycol. Soc.*, **47:** 61-74.

Cooke, R.C. and Whipps, J.M. (1993). Ecophysiology of Fungi. Blackwell Scientific Pub., London, U.K.

Dasgupta, S.N., Shome, U. and Shome, K.S. (1964). Preliminary report on predacious fungi in India. *Curr. Sci.*, **33:** 380-381.

Dayal, R. and Srivastava, S.S. (1978). Studies on predacious fungi V. Some members of Hyphomycetes. *Proceedings of National Academy of Science* **48:** 88-92.

Devarajan K. and Rajendran, G. (2001). Effect of fungus *Paecilomyces lilacinus* on the burrowing nematode, *Radopholus similis* in banana. *Pest Management in Horticultural Ecosystems* **7:** 171-173.

Devi, G. and Thaur, N.S.A. (2007). Efficacy of *Arthrobotrys oligospora* substrate against root-knot nematode (*Meloidogyne incognita*) infecting tomato. *Indian J. Nematology.* **37**(1): 97-99.

Dhawan, S.C., Narayana, R. and Prakash Babu, N. (2004). Biomanagement of root-knot nematode, *Meloidogyne incognita* in okra by *Paecilomyces lilacinus. Annals of Plant Protection Sciences* **12:** 356-359.

Dhawan, S.C., Narayana, R. and Prakash Babu, N. (2004). Efficacy of an egg-parasitc fungus, *Pochonia chlamydosporia* as a potential biocontrol agent of root-knot nematode, *Meloidogyne incognita* on okra. *In: National Symposium on Paradigms in Nematological Research and Biodynamic Farming* at University of Agricultural Sciences, Bangalore from 17-19 Nov., 2004, p. 57.

Dhawan, S.C., Narayana, R. and Prakash Babu, N. (2004). Influence of abiotic and biotic factors on growth of *Paecilomyces lilacinus, Arthrobotrys oligospora* and *Pochonia chlamydosporia* and parasitization of eggs/ trapping of *Meloidogyne incognita* juveniles. *Annals of Plant Protection Sciences* **12:** 369-372.

Dhawan, S.C. and Kaushal, K.K. (1987). *Avena fatua*, wild oat, a new colateral host of *Heterodera avenae* and parasitisation of *Pythium aphanidermatum. Indian J. Nematology,* **17:** 335.

Dhawan, S.C. Jain, K.K. and Swaroup, G. (1980). A nematode trapping fungus of *Anguina tritici* larvae. *Indian J. Nematology*, **9:** 180-183.

Dhawan, S.C., Sumeet, Tripathi, M.N. and Prakash Babu, N. (2006). Mass multiplication and viability of an egg-parasitic fungus, *Paecilomyces lilacinus. Indian J. Nematology,* **36:** 94-98.

Dhawan, S.C. and Singh Satyendra (2008). Compatibility of biocontrol fungus, *Pochonia chlamydosporia* with nematicide and neem cake against root-knot nematode, *Meloidogyne incognita* under microplot on okra. (Unpublished).

Dix, N.J. and Webster, J.W. (1995). Fungal Ecology. Capman and Hall. London, U.K.

Drechsler, C. (1937). *Mycologia*, **29:** p. 447.

Duddigton, C.L. (1955). Notes on technique of handling predacious fungi. *Transactions of the British Mycological Society*, **38:** 97-103.

Duddington, C.L. (1960). Biological control-fungi. In Nematology, Fundamentals and Recent Advances with emphasis on the Plant Parasitic and soil form. Ed. J.N. Sasser, W.R. Jenkins, pp. 461-465. Chapel Hill, UNC Press.

Dudduington, C.L., Everard, C.O.R. and Duthoit, C.M.G. (1961). Effect of green manurung and predacious fungus on cereal cyst eelworm in oats. *Plant Pathology*, **10:** 108-109.

Dunn, M.T., Sayre, R.M., Carell, A. and Wergin, W.P. (1982). Colonization of nematodes eggs by *Paecilomyces lilacinus* (Thom) Samson as observed with scanning electron microscope. *In: Scanning Electron Microscopy*, edited by Johari, O.M. and Albrecht, R.M. pp. 1351-1357. Chicago: Sem Inc.

Duponnois, R. (1995). Biological characteristics and effects of two strains of *Arthrobotrys oligospora* from Senegal on *Meloidogyne* species parasitizing tomato plants. *Biocontrol Science and Technology* **5:** 517-526.

Eren, J. and Pramer, D. (1965). The most probable number of nematode traping fungi in soil. *Soil Science*, **99:** 285.

Feder, W.H. (1963). A comparison of nematode capturing effectieness of five Dactylalla sps at four temperatures. *Mycopathologia Applicata*, **19:** 99-104.

Fresenius, G. (1852). *Beitr. Mykol.* No. 1-2.

Gintis, B.O., Morgan-Jones, G. and Rodriguez-Kabana, R. (1983). Fungi associated with several developmental stages of *Heterodera glycines* from an Alabama soybean field soil. *Nematropica* **13:** 181-200.

Goodey, T. (1938). Observation on the destruction of the stemm eelworm *Anguillulina dipsaci* by the fungus *Arthrobotrys oligospora. J. Helminthology*, **16:** 159-164.

Gray, N.F. and Duff, K.R. (1982). Some preliminary observations on predaceous fungi from Ireland. *Irish Naturalists' J.*, **20**(8): 374-380.

Guevara, E., Jatala, P., Salas, R. and Herrera, E. (1985). Efficiencis de algunos nematicidas y *Paecilomyces lilacinus* en el control de *Meloidogyne* spp.en olivos. *Resumenes ix cong. Peru. Fitopatol.* Huanuco, Perd, p. 63.

Hasan, N. (2004). Evaluation of native strains of *Paecilomyces lila*cinus against *Meloidogyne incognita* in cowpea followed by Lucerne. *Annals of Plant Protection Sciences* **12:** 121-124.

Hawksworth, D.L. (1991). The fungal dimension of biodiversity: magnitude, significance, and conservation. *Mycological Research* **95:** 641-655.

Hawksworth, D.L., Kirk, P.M., Sutton, B.C. and Pegler, D.N. (1995). Ainsworth and Bisby's Dictionary of the Fungi (8th Ed.). CAB International, Wallingford, United Kingdom. p. 616.

Herrera, E., Jatala, P. and Canicoba, R. (1984). Estudio de la dynamics poblacional de *Tylenchulus semipenetrans* Cobb 1913, mediante applicaciones de nematicidas y *Paecilomyces lilacinus* (Thom) Samson. *XXVII conv. Nat. Entomol. Ica*, Peru, p. 22.

Jaffee, B.A., Strong, D.R. and Muldoon, A.E. (1996). Nematode-trapping fungi of a natural shrubland: Tests for food chain involvement. *Mycologia* **88:** 554-564.

Jaffee, B., Philips, R., Muldoon, A. and Mangel, M. (1992). Density dependent host-pathogen dynamics in soil microcosms. *Ecology* **73:** 495-506.

Jain, R.K. (2003). Integrated pest management for plant parasitic nematodes. In Nematode Management in Plants. (Ed. Trivedi, P.C.) Scientific Publishers (India), Jaipur. pp. 293-304.

Jannson, H.B. and Nordbring-Hertz, B. (1980). Interactions between nematophagous fungi and plant-parasitic nematodes: attraction, induction of trap formation and capture. *Nematologica* **26:** 383-389.

Jatala, P., Kaltenbach, R. and Bocangel, M. (1979). Biological control of *Meloidogyne incognita acrita* and *Globidera palida* on potatoes. *J. Nematology*, **11:** 303.

Jatala, P., Kaltenbach, R., Bocanngel, M., Devau, A.J. and Campos, R. (1979). Biological control of *Meloidogyne incognita acrita* and *Globodera pallida* on potato. *J. Nematology,* **12:** 226-227.

Johnson, L.F and Curl, E.A. (1972). Methods for research on ecology of soil borne plant pathogens. *Minneapolis: Burgess*, pp. 247.

Kerry, B.R. and Bourne, J.M. (eds.) (2002). A manual for research on *Verticillium chlamydosporium*, a potential biological control agent for root-knot nematodes. IOBC/OILB, Druckform Gmbh, Darmstadt, Germany.

Kerry, B.R. (1981). Progress in the use of biological agents for control of nematodes. In: Biological control in crop production (BARC Symposium No. 5). (Papavizas, G.C. Editor) Totawa: Allenheld and Osmum, pp. 79-90.

Kerry, B.R. (1988). Fungal parasites of cyst nematodes. *Agri. Ecosys. enviro.* **24:** 293-305.

Kerry, B.R. (1990). An assessment of progress towards microbial control of plant-parasitic nematodes. *J. Nematology,* **22:** 621-631.

Kerry, B.R. and Mullen, L.A. (1981). *Verticillium chlamydosporium* in eggs of *Heterodera avanae. In:* Rothamsted Report for 1980. Part I, p. 159.

Kerry, B.R. Crump, D.H. and Mullen, L.A. (1982). Natural control of cereal cyst nematode, *Heterodera avenae* Woll, by soil fungi at three sites. *Crop Protection,* **1:** 99-109.

Kerry, B.R., Kirkwood, I.A., de-Leji, F.A.A.M., Barba, J., Leijdens, M.B. and Brookes, P.C. (1993). Growth and survival of *Verticillium chlamydosporium* Goddard, a parasite of nematodes in soil. *Biocontrol Science and Technology,* **3:** 355-365.

Khan, A.M., Alam, M.M. and Ahmad, R. (1974). Mechanism of the control of plant-parasitic nematodes as a result of the application of oilcakes. *Indian J. Nematol.* **4:** 93-6.

Khan, M.R. and Goswami, B.K. (2000). Effect of different doses of *Paecilomyces lilacinus* isolate 6 on *Meloidogyne incognita* infecting tomato. *Indian J. Nematology,* **30:** 5-7.

Linford, M.B. (1973). Stimulated activity of natural enemies of nematodes. *Science,* **85:** 123-124.

Lohde, G. (1874). Finige neuen parasitischen Pilze. Tageblatt der 47. Versammlung deutscher Naturforscher und Aerzte in Breslau, 203-206.

Lopes, -E-A., Ferraz, -S., Ferreira, -P-A., Freitas, -L-G., Dhingra, -O-D., Gardiano, -C-G. and Carvalho, -S-L. (2007). Potential of nematophagous fungi isolates to control *Meloidogyne javanica.* Nematologia Brasileira, **31**(2): 78-84.

Mankau, R. and Xuiying, W.U. (1985). Effect of nematode–trapping fungus, *Monacrosporium ellipsospora* on *Meloidogyne incognita* populations in field soil. *Revue Nematology,* **8:** 147-153.

Mankau, R. (1964). Ecological relationship of predacious fungi associated with the citrus nematode. *Phytopathology*, **54:** 1435.

Mankau, R. (1975). A semiquqantitative method for enumerating and observing parasites and predators of soil nematodes. *J. Nematology*, **7:** 119-122.

Mankau, R. (1980). Biocontrol: Fungi as nematode control agents. *J. Nematology*, **12:** 244-252.

Mankau, R. and Imbriani, J. (1969). Uptake of labeled nematicides by nematophogous fungi, *Phytopathology,* **59:** 1039.

ManHong, S. and XingZong, L. (2006). Carbon requirements of some nematophagous, entomopathogenic and mycoparasitic Hyphomycetes as fungal biocontrol agents. *Mycopathologia.* **161**(5): 295-305.

McCulloch, J.S. (1977). A survey of nematophagous fungi in Quensland. *Quensland J. Agri. Animal Sci.*, **34:** 25-34.

Mitsui, Y. (1972). Fungicidal effect of nematicides on nematode trapping fungi. *Japanese J. Nematology,* **1:** 25-29.

Naik, D. and Rao, M.S. (2004). Biotechnological approaches for the management of nematode induced wilt disease complex in capsicum. *In: National Symposium on Paradigms in Nematological Research and Biodynamic Farming* at University of Agricultural Sciences, Bangalore from 17-19 Nov., 2004, p. 140-142.

Nagesh, M. and Jankiram, T. (2004). Root-knot problem in polyhouse roses and its management using dazomat, neem cake and *Pochonia chlamydoporia. J. Ornamental Horti. New Series.* **7**(2): 147-152.

Nicolay, R. and Sikora, R.A. (1998). Improved techniques for the detection of nematophagous fungi and their activity against target nematodes. *Revue de Nématologie*, 1988, Vol. 11, Num. 1, p. 115-116.

Nordbring-Hertz, B., Jansson, H.B. and Freeman, E. *et al.* (1995). Nematophagous fungi. Film. No. C1851 Gottingen, Germany: *Institut fur den Wissenschaftlichen* Film.

Olthof, T.H.A. and Estley, R.H. (1966). Carbon and nitrogen levels of a medium in relation to growth and nematophagous activity of *Arthrobotrys oligospora* Fresenius. *Nature,* **209:** 1158.

Parvaha, R.P. and Rao, M.S. *et al.* (1999). Eco-friendly management of Meloidogyne incognita on tomato by integration of Verticillium chlamydosporia with neem and calotropis leaves. *Zeitschrift fuer Pflanzenkrankheiten und Pflanzenschutz.* **106:** 530-533.

Parvatha Reddy, P., Khan, R.M. and Rao, M.S. (1996). Integrated management of the citrus nematode, *Tylenchulus semipenetrans* using pesticides and parasitic fungus, *Paecilomyces lilacinus. Pest Management in Hortil. Ecosystems,* **2:** 61-63.

Parvatha Reddy, P., Rao, M.S. and Nagesh, M. (1999). Ecofriendly management of *Meloidogyne incognita* on tomato by integration of *Verticillium chlamydosporium* with neem and calotrophis leaves. *Nematologia Mediterranea* 16-20.

Patil, S.D. and Pendse, M.A. (1976). Some endozoic nematophagous fungi from Maharashtra, India. *M.V.M. Patrika* **11:** 27-30.

Persmark, L., Banck, A. and Jansson, H.B. (1996). Population dynamics of nematophagous fungi and nematodes in arable soil: vertical and seasonal fluctuation. *Soil Biology and Biochemistry,* **28:** 1005-1014.

Peteira, B., Esteves, I. and Hidalgo-Diaz, L. (2006). Detection of proteases produced by *Pochonia chlamydosporia* in solid medium. *Revista-de-Proteccion-Vegetal.* **21**(3): 186-190.

Prasad, D., Nagia, D.K., Kumar S. and Saini M.L. (1994). Integrated approach for management of plant-parasitic nematodes in brinjal, *Solanum melongena* L. *Plant Prot. Bulletin,* **46:** 31-3.

Priya, M.S. and Kumar, S. (2006). Dose optimization of *Paecilomyces lilacinus* for the control of *Meloidogyne incognita* on tomato. *Indian J. Nematology,* **36:** 27-31.

Rao, M.S. and Shylaja, M. (2004). Biochemical basis for the resistance in tuberose and carnation cultivars and induced systemic resistance by *Pseudomonas fluorescens* and *Pochonia chlamydosporia* in these crops. *In: National Symposium on Paradigms in Nematological Research and Biodynamic Farming* at University of Agricultural Sciences, Bangalore from 17-19 Nov., 2004, p. 45.

Rao, M.S. (2004). Bio-pesticides for the management strategies of nematodes of horticultural crops. In: *National Symposium on Paradigms in Nematological Research and Biodynamic Farming* at University of Agricultural Sciences, Bangalore from 17-19 Nov., 2004, p. 64.

Rao, M.S. (2007). Production of bio-agent colonized vegetable seedlings for sustainable nematode management. *Acta-Horticulturae,* **752:** 545-548.

Sachchidananda, J. and Ramakrishnan, K. (1971). Nematophagous fungi of agricultural soil. *Mycopathology Mycology and Applied* **43:** 235-241.

Sachchidananda, J. and Swarup, G. (1966). Nematophagous fungi in Delhi soils. *Indian Phytopathology* **19:** 279-285.

Saifullah and Gul, A. (1991). Biological control of golden nematode of potato with *Paecilomyces lilacinus* (Thom) Samson. *Sarhad J. Agri.* **7:** 377-381.

Saikia, M.K. and Roy, A.K. (1994). Efficacy of *Paecilomyces lilacinus* on the reduction of attack of *Meloidogyne incognita* on okra. *Indian J. Nematology,* **24:** 163-167.

Salas, R. and Jatala, P. (1980). Capacidad nematofaga del hongo *Paecilomyces lilacinus* (Thom) Samson para el control del nematodo del mudo *Meloidogyne incognita*. *Resumenes 13 th Conv. Nac. Entomol.* pp. 20-21, Huacho, Peru.

Sankaranarayanan, C., Hussaini, S.S., Kumar, P.S. and Rangeshwaran, R. (2000). Biological control of *Meloidogyne incognita* on tomato by *Verticillium chlamydosporium* on different substrates. *J. Biological Control,* **1491:** 39-43.

Sasser, J.M. and Freckman, D.W. (1987). A world prospective on Nematology. The role of society of Nematology (Eds. Veech, J.A. and Dickson, D.W.) published by SON, USA, 1987. pp. 7-14.

Sharma, S.B., Sharma, H.K. and Pankaj (2002). Nematode problem in India. *In:* Crop Pest and Disease Management-Challenges for the Mellinium. (Ed. Prasad, D. and Puri, S.N.). Jyoti Publishers, pp. 267-275.

Shylaja, M., Rao, M.S. and Rahiman, A.B. (2004). Studies on the development of bio-intensive management strategies of nematode induced wilt disese complex in tuberose (*Polyanthus tuberosa*) and carnation (*Dianthus caryophyllus*) using biocontrol agents. In: *National Symposium on Paradigms in Nematological Research and Biodynamic Farming* at University of Agricultural Sciences, Bangalore from 17-19 Nov., 2004, p. 136.

Singh, R.S. and Sitaramaiah, K. (1966). Incidence of root-knot of okra and tomatoes in oilcake soil. *Pl. Dis. Reptr.* **50:** 668-72.

Singh, R.S., Singh B. and Beniwal S.P.S. (1967). Observations on the effect of sawdust on the incidence of root-knot and yield of okra and tomatoes in nematode infested soil. *Pl. Dis. Reptr.* **5:** 108-24.

Stirling, G.R. and Smith, L.J. (1998). Field tests of formulated products containing either *Verticillium chlamydosporium* or *Arthrobotrys dactyloides* for biological control of root-knot nematodes. *Biological control* **11:** 231-239.

Stirling, G.R. and Mankau, R. (1979). Mode of parasitism of Meloidogyne and other nematode eggs by Dactyllela oviparasitica. *J. Nematology,* **11:** 282-288.

Stirling, G.R. and West, L.M. (1991). Fugal parasites of root-knot nematode eggs from tropical and sub-tropical regions of Australia. *Australasian Plant Pathology,* **20:** 149-154.

Stirling, G.R. and Kerry, B.R. (1983). Antagonists of the ceral cyst nematodes, *Heterodera avenae* Woll. In Australian soil. *Australian J. Experimental Agriculture and Animal Husbandry.* **53:** 201-106.

Tolmsoff, W.F. (1959). The isolation of nematode trapping fungi from Oregon soils, *Phytopathology*, **49:** 113-114.

Tzortzakakis, E.A. (2000). The effect of *Verticillium chlamydosporium* and oxamyl on the control of *Meloidogyne javanica* on tomatoes grown in a plastic house in Crete Greece. *Nematologia Mediterranea* **28:** 249-254.

Verdejo-Lucas, S., Sorribas, F.J., Ornat, C. and Galeano, M. (2003). Evaluating *Pochonia chlamydosporia* in a double- cropping system of lettuce and tomato in plastic houses infested with *Meloidogyne javanica. Plant Pathology,* **52:** 521-528.

Viaene, N.M. and Abawai, G.S. (2000). *Hirsutella rhasiliensis* and *Verticillium Chlamydosporium* as biological control agents of the root-knot nematode Meloidogyne hapla on lettuce. *J. Nematology,* **32:** 85-100.

Walia, R.K., Bansal, R.K. and Bhatti, D.S. (1991). Effect of *Paecilomyces lilacinus* application time and method of controlling *Meloidogyne javanica* on okra. *Nematologica Mediterranea* **19:** 247-249.

Walter, D.E. and Kaplan, D.T. (1990). Antagonists of plant-parasitic nematodes in Florida citrus. *J. Nematology,* **22:** 567-573.

Wang, K.N., Riggs, R.D. and Crippen, D. (2005). Isolation, selection and efficacy of *Pochonia chlamydosporia* for control of *Rotylenchulus reniformis* on cotton. *Phytopathology,* **95**(8): 890-893.

Waronin, M. (1870). Abhandl-Senckenbug Natureforsch. Ges. **7:** 325.

Wyborn, C.H.E., Priest, D. and Duddington, C.L. (1969). Selective technique for the determination of nematophagous fungi in soil. *Soil Biology and Biochemistry,* **1:** 101-102.

Zaki, F.A. (1994). Dose optimization of *Paecilomyces lilacinus* for the control of *Meloidogyne javanica* on tomato. *Nematologia meditterranea,* **22:** 45-47.

Zaki, M.J. and Maqbool, M.A. (1991). *Paecilomyces lilacinus* controls *Meloidogyne javanica* on chickpea. *International Chickpea Newsletter* **25:** 22-23.

Zopf, W. (1888). Zur Kenntnis der Infektions-Krankheiten niederer Thiere und Pflanzen. *Nova Acta Acad. Leop. Carol.* **52:** 314-376.

24

Curry Leaf (*Murraya koenigii*) Chemistry and its Role in Plant Protection

D.P. Ray[1], S. Srivastava[2] and D. Prasad[3]

Division of Agricultural Chemicals, Indian Agricultural Research Institute, New Delhi-110 012

The hazards of synthetic pesticides are well known in today's world. The growing concern about the widespread usage of synthetic pesticides has led to the surge of research in alternative pest control techniques, which resulted in the emphasis on development of newer pesticides of safer attributes.

The plant kingdom is a rich storehouse of biochemicals. The secondary metabolites produced by the plants are merely active against different pests as decided by the nature. The use of botanical extracts for protection of plant and human health is an age old practice. Metabolites from botanical origin are found to be not only free from harmful effects such as environmental pollution, human health hazard, pest and disease resistance, but also serve as lead chemicals that gives a new hope for future pest control agents.

The botanicals are gaining importance for the pest control in agriculture, forestry and health program. Botanicals and biopesticides being eco-friendly are considered as potential alternatives to synthetic pesticides. The secondary metabolites, which are produced in the plant system, have both ecological and evolutionary history. These are produced as a result of biochemical mechanism inside the plant body. But these compounds can deploy for their protection from the external enemies.

Curry leaf has an excellent history of use in the culinary purposes. It is a popular leafy-spice used either as fresh form when available or as preserved as dried or frozen for long-term storage, in Indian cuisine for its characterizing authentic flavour and distinct aroma. The plant scientifically known as *Murraya koenigii* (Subtribe: Clauseneae, Sub family: Aurantoidae, Family: Rutaceae, Order: Rutales) is distributed from South and East Asia to Australia. The plant is most valued on account of the spicy odour and resistance of the leaves, stem bark and root to

Present Address:

[1]Fertiliser Control Laboratory, Midnapore, Krishi Bhavan, Abash, Paschim Midnapore, West Bengal

[2]Indian Institute of Natural Resins and Gums, Ranchi-834010, Jharkhand

[3]Division of Nematology, Indian Agricultural Research Institute, New Delhi-110 012

insect and microbial attack. It is cultivated largely for its fragrant leaves which have also been found to yield essential oils. The essential oil of fragrant plants is well documented and found very effective against various agriculturally important pests (Ray, 1999). Hiremath *et al.*(1996) reported 47 components in the essential oil of curry leaves. The major components of oil identified were β-caryophyllene (30.1%) and β-pinene (28.6%) followed by β-gurjunene (6.3%), sabinene (3.6%), limonene (2.7%), linalool (0.82%) and β-phellandrene (0.18%). Although the plant is well known for its antibiotic, antiflammatory and anticonvulsion activities (Nigam and Purohit, 1961), There is scanty of report other activity. The effort has been made to isolate the chemical constituents of the leaves and their bioactivities were tested against different agricultural pests.

POSITION OF *MURRAYA KOENIGII* IN PLANT KINGDOM

Curry leaf tree can be classified as following in the plant kingdom:

Order:	Rutales
Family:	Rutaceae, Sub-tribe-Clausenae
Subfamily:	Aurantoidae
Genus:	*Murraya*
Species:	*koenigii, paniculata, elongata, euchrestifolia*

DISTRIBUTION

The curry leaf comes from the small deciduous curry leaf tree, or *Murraya*, which is native to southern Asia. The plant is largely cultivated for its fragrant leaves and is common in most parts of India. It is also found in Sri Lanka, Myanmar, Indo-China, South China and Australia. The common species found in India are *Muraraya koenigii* (Linn.) Spreng or Curry leaf tree and *Muraraya paniculata* (Linn.) Jack Syn. *Muraraya exotica* (Linn.) popularly known as Orange Jessamine. The other important species are *Murraya elongate* DC (Native of Myanmar and richest source of monomeric coumarine) and *Murraya euchrestifolia* Hyata (Native of Taiwan and richest source of carbazole alkaloids).

BOTANY OF THE PLANT

The curry leaf plant is a small spreading shrub, about 2.5 metres high. The main stem is dark green to brownish, with numerous dots on it. Its bark can be peeled off longitudinally, exposing the white wood underneath; the girth of the main stem is about 16 cm. Leaves are exstipulate, bipinnately compound, 30 cm long, each bearing 24 leaflets, having reticulate venation; leaflets, lanceolate, 4.9 cm long, 1.8 cm broad, having 0.5 cm long petiole. Flowers are bisexual, white, funnel-shaped, sweetly scented, stalked, complete, ebracteate, regular, actinomorphic, pentamerous, hypogynous, the average diameter of a fully opened flower being 1.12 cm; inflorescence, a terminal cyme, each bearing 60 to 90 flowers; calyx, 5-lobed, persistent, inferior, green; corolla, white, polypetalous, inferior, with 5 petals, lanceolate; length, 5 mm; androecium, polyandrous, inferior, with 10 stamens, dorsifixed, arranged into circles of five each; smaller stamens, 4 mm. long whereas the longer ones, 5 to 6 mm; gynoecium, 5 to 6 mm long; stigma, bright, sticky; style, short; ovary, superior. Fruits are round to oblong, 1.4 to 1.6 cm long, 1 to 1.2 cm in diameter; fully ripe fruits, black with a very shining surface;

pulp, Wistaria blue; the number of fruits per cluster varying from 32 to 80. There is only one seed in each fruit, 11 mm long, 8 mm in diameter, colour spinach green. The plant possesses a pungent curry scent with overtones of anise and citrus.

MEDICINAL USES

Curry leaf has a prolong history of usage in the society and has been used in folk medicine in China and other Asian countries as an analgesic, astringent, antidysenteric, antioxidant, febrifuge, hypolipidemic, hypglycemic, for improvement of vision, to treat night-blindness, and for regulation of fertility. The reported nutritional value of curry leaves have prompted the Tamil Nadu government to include two grams of either curry leaf powder (or drumstick leaf powder) for five days in a week along with the noonmeals served to children of economically weaker section, with an expectation that inclusion will help eradicate "Vitamin A" deficiency among children between 3-5 years (Venkateswaran, 2004)). The leaves, the bark and the roots of *Murraya koenigii* (L.) Spreng is used as a tonic and a stomachic. The bark and the roots are used as a stimulant by the traditional physicians. They are also used externally to cure eruptions and the bites of poisonous animals. The green leaves are stated to be eaten raw for curing dysentery, and the infusion of the washed leaves stops vomiting. Strong odiferous oil occurs in the leaves and the seeds of curry leaf plant. The chemical examination of this oil reported that this essential oil exhibited a strong antibacterial and antifungal activity. Analysis of purified extracts of *Murraya koenigii* (L.) Spreng indicate a significant α-amylase inhibitory activity. Effect of curry leafs on carbohydrate metabolism has been studied using rats as experimental animals and significant hypoglycemic action as been observed. There was increase in the concentration of hepatic glycogen and glycogenesis.

CHEMICAL CONSTITUENTS

The curry leaf or *Murraya koenigii* (L.) Spreng. plant is rich storehouse of diverse and structurally complex bioactive molecules. Although the leaves of the plant are most important source of ingredients, the fruits, bark and root also contain a variety of bioactive principles. The analysis of curry leaves gave assays as detailed in the Table 24.1.

Table 24.1: Composition of curry leaf.

Components	Amount
Moisture	66.3%
Protein	6.1%
Fat (Ether extract)	1.0%
Carbohydrate	16.0%
Fibre	6.4%
Mineral matter	81.0 mg
Iron	3.1mg
Carotene (Vit A)	12.6 IU
Nicotinic acid	2.3 mg
Vitamin C	4 mg/ 100 g

The chemical constituents isolated from various parts of the plant are described hereunder.

Seeds

From the petroleum ether extract of the seeds of this plants a number of carbazole alkaloids have been isolated. Bhattacharya and Chowdhury (1985) isolated 2-methoxy-3-methyl carbazole, a member of the C_{13} skeleton group of compounds (Fig. 24.1).

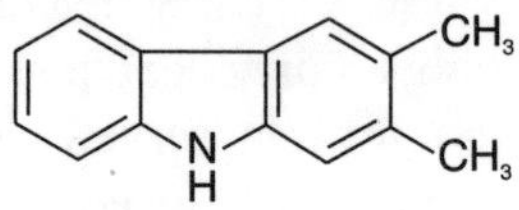

Fig. 24.1. 2-methoxy-3-methyl carbazole alkaloid isolated from seeds of *Murraya koenigii* (L.) Spreng.

Stem Bark

Different solvent extracts of stem bark of the plant, carbazole was predominated group of compounds. Some of the compounds isolated from the solvent extract of stem bark are presented in Table 24.2.

Table 24.2. Compounds isolated from the solvent extract of stem bark of *Murraya koenigii* (L.) Spreng.

Carbon Skeleton Group	Name	Reference
C_{13} skeleton group	Murrayanine (I)	Chakraborty *et al.*, 1965
	Mukoeic acid (II)	Chowdhury *et al.*, 1969
	Mukonine (III)	Chakraborty *et al.*, 1976
	Mukonidine (IV)	
C_{18} skeleton group	Girinimbine (V)	Chakraborty *et al.*, 1964
	Murrayacine (VI)	Chakraborty *et al.*, 1968
C_{23} skeleton group	Mahanimbine	Chakraborty *et al.*, 1966
	Murrayazoline (VII)	Chakraborty and Chowdhury, 1966
	Murrayazolinine (IX)	Chakraborty *et al.*,1973
	Murrayazolidine (X)	Chakraborty *et al.*,1970
	Murrayacinine (XI)	Chakraborty *et al.*,1974
	Mahanimbol (XII)	Rama Rao *et al.*, 1980
	Isomurrayazoline	Bhattacharyya *et al.*, 1982
Other groups	2-methoxy carbazol -3-methyl carboxylate (XIII) and 1-hydroxy-3methyl carbazole (XIV)	Bhattacharyya *et al.*, 1994
	Girinimbilol (XV)	Reisch *et al.*, 1994
	Mahanimbilol	
	Koenigine quinine A	Saha and Chowdhury, 1998
	Koenigine quinine B	

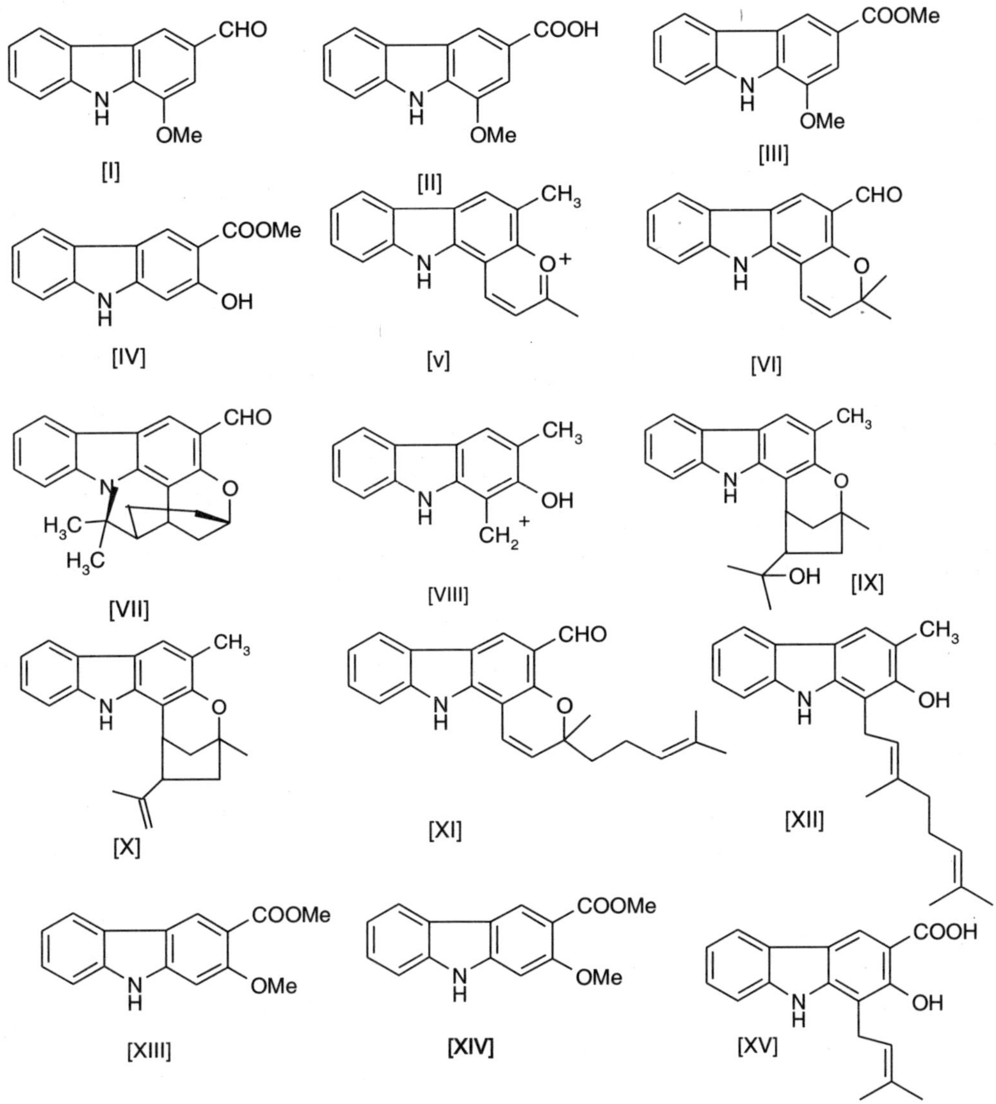

Fig. 24.2. Some carbazole alkaloids of C_{13}-skeleton group isolated from stem bark of *Murraya koenigii* (L.) Spreng.

Roots/Root bark

(i). Members of the C_{23} skeleton group

Roy *et al.,* 1979 isolated mahanimboline (XVI) from the petroleum ether extract of the root bark of *Murraya koenigii* (L.) Spreng. The UV spectrum of the compound was very similar to pyranocarbazole alkaloids like girinimbine and mahanimbine, and thus indicated the presence of the same pyranocarbazole chromophore in mahanimbine.

(ii) Members of the C_{13} skeleton group

Roy *et al.* (1982) isolated two optically active carbazole alkaloids namely Mukoline (XVII) and Mukolidine (XVIII) from the benzene extract of the roots of the plant. Fiebig *et al.* (1985) isolated Koenoline (XIX), a cytotoxic carbazole alkaloid from the chloroform extract of the root bark. Bhattachryya *et al.* (1986) isolated 2-hydroxy -3-methyl carbazole (XX) from the roots of *Murraya koenigii.*

(iii) Members of other group

Srivastava *et al.* (1993) isolated three new compounds, murrayenol, murrayagetin and marmesin-10-rutinoside from the root extract of the plant.

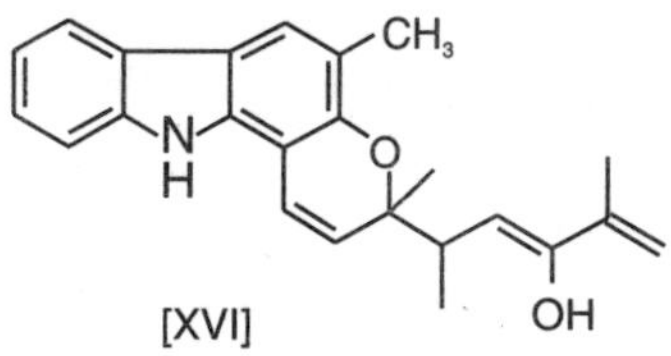

[XVI]

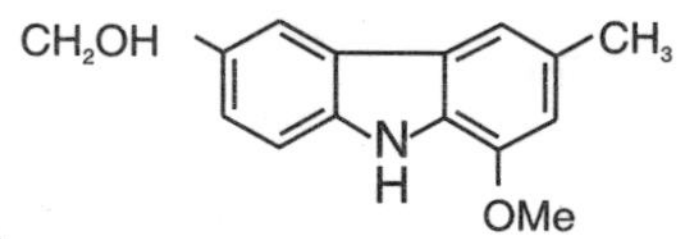

[XVII]

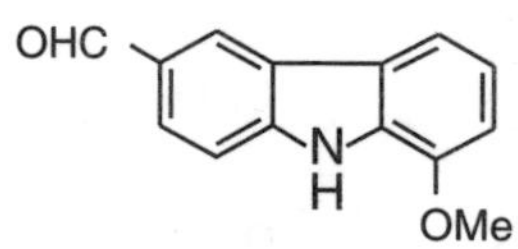

[XVIII]

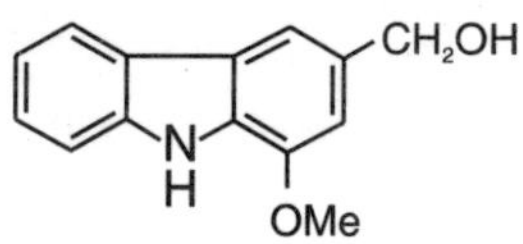

[XIX]

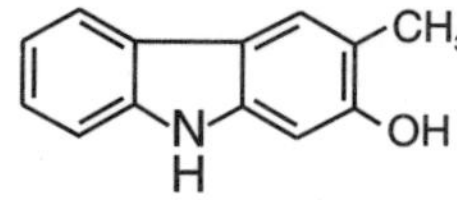

[XX]

Fig. 24.3. Some carbazole alkaloids of C_{13}-Skeleton Group isolated from roots/root bark of *Murraya koenigii* (L.) Spreng.

Fruits

Along with the known carbazole alkaloids, mahanimbine, murrayazolidine, girinimbine, koenimbine and mahanine, two new carbazole alkaloids, isomahanine (XXI) and murrayanol (XXII) were isolated from the fruits of *M. koenigii* by Johannes *et al.* (1992). The structures of isomahanine and murrayanol were confirmed by the formation of 9-methoxymahanimbicine on their methylation and cyclization, respectively.

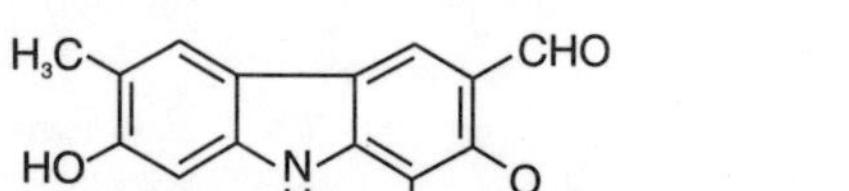

[XXI]

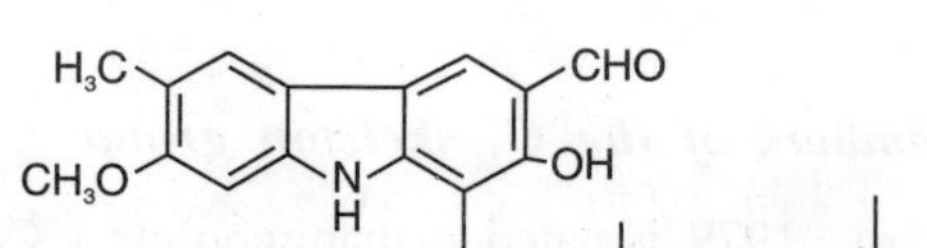

[XXII]

Fig. 24.4. Carbazoles of C_{13}-Skeleton Group isolated from *Murraya koenigii* (L.) Spreng. fruits.

Leaves

Naturally occurring carbazole alkaloids were reported from the plants belonging to the genera *Murraya, Glycosmis* and *Clausena* of the Family Rutaceae. So far the genus *Murraya* has been found to elaborate simple carbazoles with C_{13}, C_{18} and C_{23} skeletons. Although the C_{13} group carbazoles are very rare in the leaves of carbazoles, other carbon groups are common in the leaves. Narasimhan *et al.*(1968) and Kureel *et al.* (1969) isolated a compound of molecular formula ($C_{19}H_{19}NO_2$) from the curry leaves which was found to be optically inactive and identical with koenimbine (XXIII).

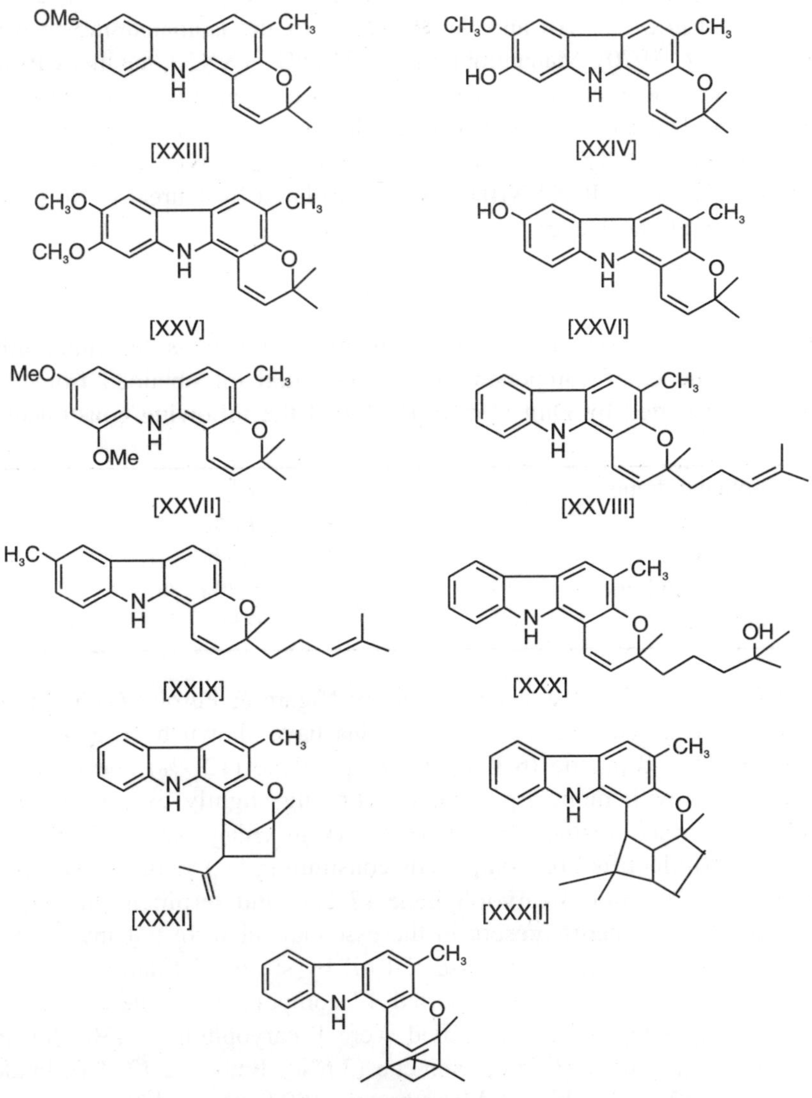

Fig. 24.5. Carbazoles of C_{13}-Skeleton Group isolated from *Murraya koenigii* (L.) Spreng leaves.

Other carbazole alkaloids, e.g., koenigine ($C_{19}H_{19}NO_3$) (XXIV) and koenidine ($C_{20}H_{21}NO_3$) (XXV) were isolated from the leaves by three group of workers [Kureel *et al.* (1969), Joshi *et al.*,1970 and Narasimhan *et al.*(1970)]. Koenine ($C_{18}H_{17}NO_2$) was isolated by Narasimhan *et al.*(1970) from the leaves (XVI). Mukonicine ($C_{20}H_{21}NO_3$) (XVII) was isolated from the alcoholic extract of the leaves of *Murraya koengii* (Spreng.) (Mukherjee *et al.*, 1983).

Mahanimbine ($C_{23}H_{25}NO$) (XVIII), the first member of the C_{23} carbazole alkaloid group was isolated from the stem bark by Chakraborty *et al.*,(1966). Later on it was isolated from the leaves of *Murraya* by Narasimhan *et al.*(1968). Isomahanimbine (XXIX) isolated by fractionation of the mother liquor of the leaf extract after the removal of mahanimbine [Joshi *et al.*, 1970 and Kureel *et al.*, 1970] and the structures of the isomahanimbine was confirmed by synthesis (Kureel *et al.*, 1970). Mahanimbinine ($C_{23}H_{27}NO_2$) (XXX), another carbazole alkaloid isolated from the leaves (Kureel *et al.*, 1970) of *Murraya koengii*. Two groups of workers, Kureel *et al.* (1969) and Dutta *et al.* (1969) isolated the racemate of murrayazolidine ($C_{23}H_{25}NO$) called cyclomahanimbine (XXXI) or curryanine from the leaves of *M. koenigii* Spreng. Bicyclomahanimbine (XXXII OR XXXIII) has been isolated by Kureel *et al.* (1969) from the petroleum ether extract of the leaves.

Essential Oil

The oil extracted from the fresh leaves of this plant is known as 'essential oil' due to its characteristic aroma. The chief components of the essential oil obtained from the leaves by hydrodistillation were isolated by Dutt (1958). He found the following constituents:

dl-phellandrene	4.6%
d-sabinene	9.2%
d-α-pinene	5.5%
β-caryophyllene	26.3%
dipentene	6.8%

Essential oil from curry leaf has been studied by Nigam & Purohit (1961) and Macleod & Pieris (1982). Leaves collected and analysed in China have shown by Liquan *et al.*,(1988) to contain α-pinene (38.4%), β-pinene (6.3%), β-caryophyllene (12.9%) and γ-elemene (10.1%) as the main constituents with monoterpenoids constituting slightly excess of 50% of the oil. However, the plant materials obtained from trees grown in village garden in Penang, Malaysia by Wong and Tie (1993) identified 62 compounds constituting 96.9% of oil with β-phellandrene (24.4%), α-pinene (17.5%) and β-caryophyllene (7.2%) and terpinen (6.1%) as the major constituents. The polar components present in the essential oil were not investigated until few years back. Wagner (1995) subjected the essential oil by steam distillation from the leaves of *Murraya koenigii*. Hiremath *et al.*(1996) reported 47 components in the essential oil of curry leaves. The major components of oil identified were β-caryophyllene (30.1%) and β-pinene (28.6%) followed by β-gurjunene (6.3%), sabinene (3.6%), limonene (2.7%), linalool (0.82%) and β-phellandrene (0.18%). Singh and Maheshwari (1997) showed that super critical fluid extraction using liquid carbon dioxide is convenient and superior in terms of both quantity as well as qualitative properties of the leaves extract as compared to traditional hydrodistillation process.

An exhaustive research on essential oil of curry leaf was carried out by Srivastava *et al.* (2000) Ray and Srivastava (2006), Srivastava *et al.* (2007) Ray and Srivastava (2008), and it was found that that the essential oil comprised of monoterpene hydrocarbons (16%) and sesquiterpens hydrocarbon (80%) with β-caryophyllene, β-gurjunene, β-elemene and β-phellandrene as the major components (Table 24.3). Though the composition of oil is more or less similar, however, composition of the constituents may differ due to the botanical ambiguity or climatic conditions of grown area.

Table 24.3. Chief components of essential oils of curry leaf.

Components	Per cent present
β-pinene	5.84
Sabinene	3.67
β-phellandrene	6.1
Limonene	2.67
Linalool	0.82
β-caryophyllene	29.0
β-gurjunene	21.0
β-elemene	6.9

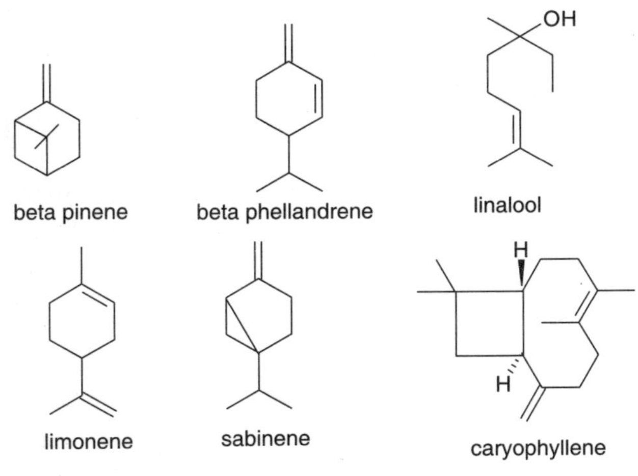

Fig. 24.6. Terpenoids isolated from essential oil of *Murraya koenigii* leaf.

Curry Leaf in Pest Control

There has been a hue and cry during last few decades for the newer molecule for controlling agricultural pests. Exhaustive use of synthetic chemicals leads to a notorious ecological imbalance and several other problems. So, there has been a great concern over identification of the newer plants and plant products.

Plants virtually are nature's chemical factories, providing unlimited natural sources of botanical pesticides. So, more emphasis is laid to obtain pesticides that are specific, non-toxic to man and beneficial organisms; which are biodegradable and less prone to pest resistance, cost-effective and serve as prototype for synthesis of new chemicals of high biological activity. Therefore, attempt has been made by scientists to utilize the leaves of this plant in pest control measures not only on the basis of essential oil and pure carbazole alkaloids present in the leaves alone, but also on the basis of derivatisation of carbazoles.

Carbazole alkaloids are biologically active displaying antibacterial, antifungal, insecticidal, anticancer and antiyeast agent. Therefore, apart from medicinal properties like anti-inflammatory, convulsant and diuretic, etc. shown by this plant, it offers also an opportunity to explore the efficacy of essential oil as a source of insect antifeedant, antinemic, antifungal, larvicidal, insect repellent and insect growth regulatory properties whereas carbazole alkaloids remain as a source of antifungal, antinemic and insect antifeedant principles.

Of the carbazole alkaloids, girinimbine, mahanimbine and murrayanine have antifungal activities (Das *et al.,* 1965). The antibiotic activities of some carbazole alkaloids e.g., glycozoline and related compounds were also evaluated (Chakraborty *et al.,* 1975). Chowdhury *et al.,* 1978 revealed that tetrahydrocarbazoles were more toxic to houseflies than the corresponding carbazoles. Chowdhury and Das (1979) further studied the toxicity of tetrahydrocarbazole, carbazole, glycozoline and glycozolidine along with the extract of toot bark of *Glycosmis pentaphylla* on mosquito larvae.

Antifungal and antibacterial properties of some carbazoles isomeric with glycozoline and girinimbine were studied by Randelia and Patel (1982) but the compounds showed little activity. The authors suggested that symmetrical disposition of the methyl and hydroxyl groups on the carbazole ring is essential for maximum antifungal activity.

Some tetrahydro carbazoles had low insecticidal properties, whereas some substituted 1,2,3,4-tetrahydrocarbazoles were found to be active against *Trypanosoma cruzi* (Chowdhury *et al.,*1978).

Some acidic tetrahydrocarbazoles have been shown to have anti-inflammatory activity (Pecca and Albonico, 1970). Thus, 1-ethyl-8-n-propyl 1,2,3,4-tetrahydrocarbazole-1-acetic acid is a novel anti-inflammatory agent [C1].

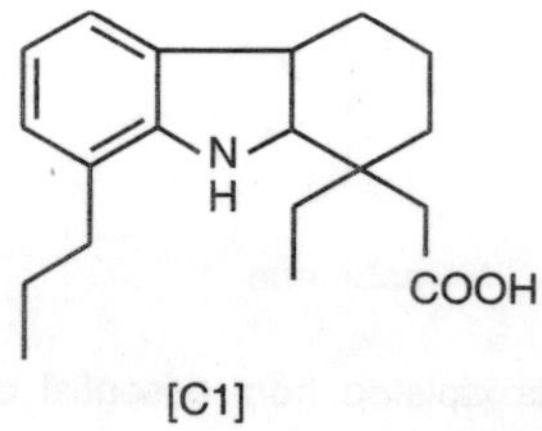

[C1]

6-chloro-1,2,3,4-tetrahydrocarbazole-2-carboxylic acid [C2] was found to be clinically active in the treatment of acute gut.

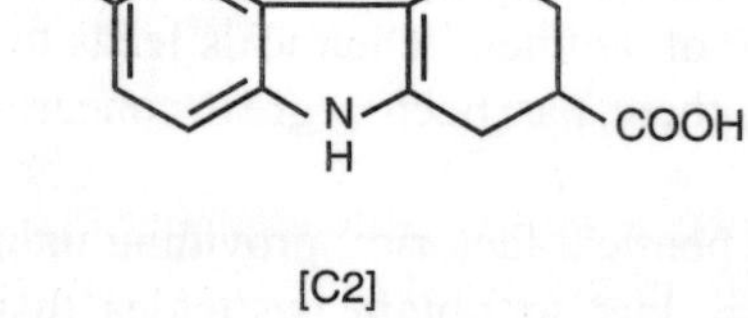

[C2]

3-dimethyl amino-1,2,3,4-tetrahydrocarbazole [C3], which has a modified tryptamine structure prevented amphetamine induced stereotyped behaviour in rats and prevented reserpine induced ptosis in mice (Moorradian *et al.,* 1977).

Another potent analogue was 9-ethyl-N,N-1-trimethyl -1-1,2,3,4- tetrahydro carbazole-1-thanamine [C4].

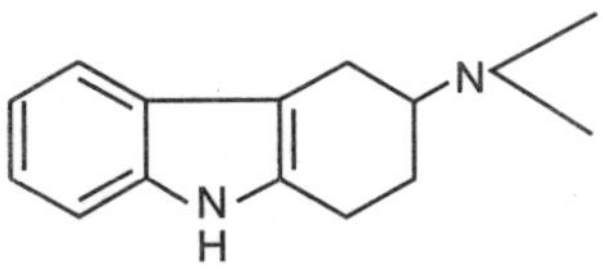

[C3]

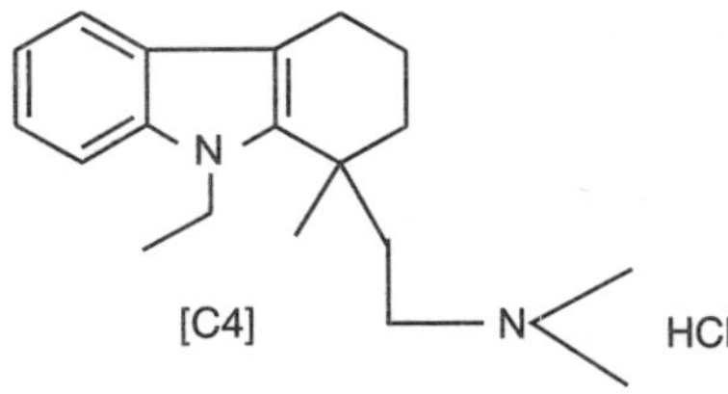

[C4]

N-alkylamino carbazoles [C5 and C6] possess significant anticonvulsant and diuretic activity (Shoeb *et al.,* 1973). Introduction of aminopropyl chain at the N-atom seems to enhance the anticonvulsant activity in combination with CH_3O- at positions 2,3 and 4. Koenoline [C7] exhibited cytotoxic activity against KB cell culture system (Fiebig *et al.,* 1985)

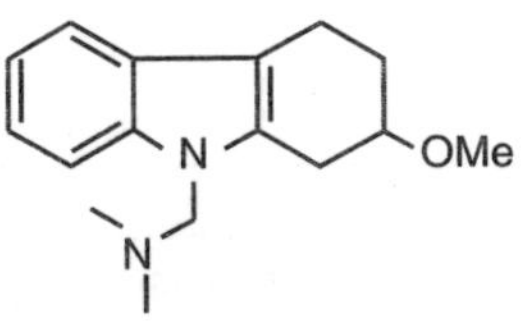

[C5]

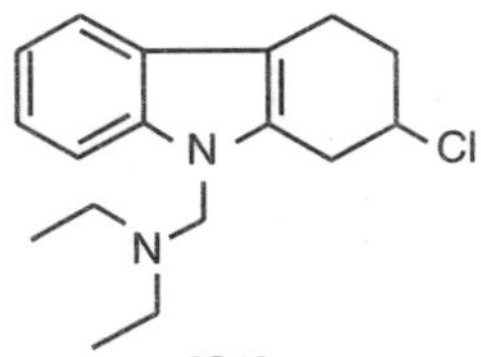

[C6]

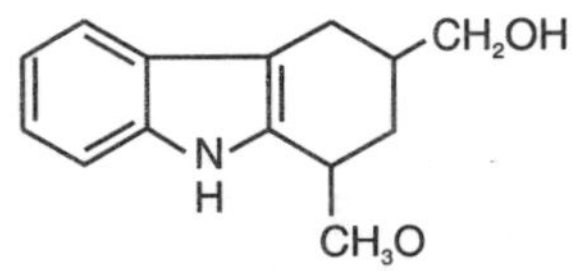

[C7]

Another study on the toxicity of some N-substituted carbazoles with piperine and piperic acid on mosquito larvae (*Culex* sp.) showed that N-methyl, N-acetyl and 2,9-diacetyl carbazoles and also piperine were highly toxic to mosquito larvae but piperic acid whereas n-benzoyl carbazole, N-p-nitrobenzoyl carbazole were not so.

Further investigation of the toxicity of N-substituted carbazoles to mosquito larvae (*Culex* sp) revealed that replacement of the NH-proton by simple groups viz. methyl, acetyl enhanced the larvicidal properties, but the presence of a long chain like the cinnamoyl group did not enhance the toxicity of carbazoles (Das, 1989).

Antibacterial, anti-inflammatory and antifeedant activities of root extract of *M. koenigii* were evaluated (Srivastava and Srivastava, 1993) and the study revealed that antibacterial activity of ethanolic extract of root was found to show positive effect on all the tested organisms e.g., *Vibrio cholerae, Salmonella typhimurium* and *Staphylococcus aureus.* Streptomycine was used as a standard. The anti-inflammatory activity of root extract was evaluated in albino rats. The percentage inhibition of inflammation after 5 hours was calculated, oxyphenbutazone was used as a standard. And the antifeedant activity of the extract was determined against 3rd instar grub stages of brinjal beetle, *Henosepilachna vigintioctopuntata* Fabricus.

Biological Properties of Essential Oil

The terpenes and in particular the lower volatile members of the class have been known from antiquity, having been isolated from plants and used for a variety of human purposes since pre-Christian era.

Their volatility that made them easily discoverable in fragrant plant material and at the same time readily obtainable by simple distillation of leaves lent to term essential oil. Various properties of the oil, such as the fungicidal nature, broad fungitoxic spectrum and absence of phytotoxicity to crop plant suggest that it may be useful as an effective fungitoxicant (seed dresser) against many fungi e.g., *Rhizoctonia solani, Alternaria alternate, Aspergillus niger* and *R. bataticola*, etc.

Kishore *et al.* (1983) extracted the leaves of 22 taxa of Gorakhpur (UP) with distilled water. The extracts were screened for their volatile fungitoxicity. From their experiment a few extracts along with *Murraya* showed 82-95% inhibition of mycelial growth.

(i) Antifungal properties

Various properties of the oil such as the fungicidal, broad fungi toxic spectrum and absence of phytotoxicity to crop plant suggest that it may be useful as an effective fungi toxicant (seed dresser) against many fungi (Table 24.4). Perusal of the data of fungi toxicity of all the concentrations of essential oil against the fungi showed a comprehensive outlook of the total essential oil of the plant.

Table 24.4. Fungal strains controlled by the curry leaf oil.

	Fungal species	Causal organism of
1.	*Rhizoctonia solani*	Damping off and root rot of seedlings, sheath blight of rice
2.	*Alternaria alternate*	Leaf spot
3.	*Rhizoctonia bataticola*	Root rot of seedlings
4.	*Helminthosporium oryzae*	Brown spot of rice, wheat, oat, etc.
5.	*Sclerotium rolfsii*	Collar rot
6.	*Fusarium oxysporium*	Wilt
7.	*Pythium* sp	Wet rot or damping off

During fungitoxicity studies, the minimum inhibitory concentration (MIC) of essential oil of curry leaf was 5000 ppm, which is as effective as other prominent fungicide Agrosan GN, (MIC 5000 ppm) but less efficaceous than Captan (MIC 4000 ppm) (Srivastava *et al.,* 2007). At concentration 1000 ppm of oil, the percentage of inhibition of mycelial growth of *Rhizoctonia solani* is 39.0, while for *Helminthosporium oryzae* it is 46.0. The oil of *M. koenigii* showed a total inhibition of mycelia growth against *R. solani* and *R. bataticola* at 5000 ppm concentration.

The fungitoxicity of hexane and benzene extract of curry against *Rhizoctonia solani, Rhizoctonia bataticola* and *Helminthosporium oryzae* as found by the Ray and Srivastava (2007) has been presented in Tables 24.5 and 24.6. The carbazole alkaloids from the leaf extract and essential oils showed very good antifungal activity against three test fungi. The essential oils had shown fungitoxicity of more than 50% at 2000 ppm concentration level. Between crude extracts, benzene extract was found to be more antifungal than hexane extract. Antifungal activity of pure carbazoles e.g. koenimbine was improved by their derivatisation into corresponding N-methyl koenimbine and N-methyl koenidine. However, hydrogenation of the pure carbazoles into dihydro derivatives caused reduction in the fungitoxicity.

Table 24.5. Fungitoxicity of hexane and benzene extract of curry leaves and purified compunds against *Rhizoctonia solani, Rhizoctonia bataticola* and *Helminthosporium oryzae.*

Extracts/Purified Compound	Concentration (ppm)	Per cent growth inhibition		
		R. solani	*R. bataticola*	*H oryzae*
Hexane	1000	30.0	27.0	32.0
	5000	59.0	57.0	54.0
Benzene	1000	49.0	50.0	52.0
	5000	100.0	100.0	99.0
Koenimbine	1000	48.0	51.0	52.0
	5000	100.0	100.0	100.0
Koenidine	1000	51.0	52.0	49.0
	5000	100.0	99.0	100.0
Mahanimbine	1000	31.0	29.0	30.0
	5000	57.0	55.0	61.0
Isomahanimbine	1000	37.0	31.0	34.0
	5000	62.0	58.0	53.0

Table 24.6. Fungitoxicity of derivatives of pure compounds isolated from the leaves of *M. koenigii.*

Extracts/Purified Compound	Concentration (ppm)	Per cent growth inhibition		
		R. solani	*R. bataticola*	*H. oryzae*
N-methyl koenimbine	1000	50.0	55.0	53.0
	2000	68.0	67.0	70.0
	3000	81.0	79.0	83.0
	5000	100.0	100.0	100.0
N-methyl koenidine	1000	52.0	49.0	47.0
	2000	64.0	62.0	65.0
	3000	78.0	78.0	77.0
	5000	100.0	100.0	99.0
Dihydro koenimbine	1000	38.0	39.0	40.0
	2000	47.0	48.0	46.0
	3000	54.0	57.0	54.0
	5000	68.0	69.0	60.0
Dihydro koenidine	1000	33.0	37.0	38.0
	2000	44.0	49.0	44.0
	3000	53.0	57.0	50.0
	5000	65.0	72.0	61.0

(ii) Antifeedant activities

When a piece of castor leaf (*Ricinus comunis*) dipped in curry leaf oil is kept in front of voracious feeder insects like tobacco armywom (*Spodoptera litura*), it showed reluctance to feed the leaf whereas the ordinary leaf is eaten by the same insect within a minute when given prior to the dipped leaf disc. This result shows the antifeedant property of the essential oil (Ray, 1999). When the insect is exposed to the essential oil, it inhales the essential oil, which affects its nervous system as well as the olfactory tract, thus not allowing it to feed anymore (Ray and Srivastava, 2008). Relentless effort of scientists to find the exact compounds for antifeedancy revealed that the activity of the essential oil is due to β-caryophyllene and β-gurjunene, the two major compounds of the oil.

(iii) Antinemic properties

Plant nematodes are microscopic organisms, generally ranging from 0.5 to 1.0 mm in length. They are sometimes called threadworm, eelworms, roundworms etc. and mostly inhabit the soil in association with the roots of the plants. A few species feed on the aerial parts of plants too, such as stems, leaves, ear heads and grains, etc.

Among the different plant parasitic nematodes, root knot nematode *(Meloidogyne incognita)* is very common in Indian soil and their infestations are easily recognisable by the characteristic root galls they cause. Vegetable crops like tomato, brinjal, bhindi, cucurbits, crucifers, pulses, potato, sugarcane, banana, citrus, etc. are all highly susceptible hosts. Even cereal and millet crops, which are generally not favoured, are attacked under certain conditions and suffer considerable damage. Though a number of insecticides have been treated as nematicides, not a single nematicide has been reported in India so far. Different botanicals pesticides are being used for the control of nematode population. In the laboratory study of essential oil against root knot nematode, it was found that essential oil is very much effective in controlling the population (Ray *et al*, 2000).

The solvent extracts, purified compounds and their derivatives as well as the essential oil of *Murraya koenigii* leaves have been studied by Srivastava *et al.*, (2000) for their antinemic activity aginst root knot nematode, *Meloidogyne incognita* (Table 24.7). The essential oil had shown corrected non-moyility more than 50% at 125 ppm level after 24 hrs exposure to root knot nematode. Between crude extracts, benzene extract was found to be more antinemic than hexane extract. Antinemic activity of pure carbazole alkaloids was improved by their derivatization into corresponding N-methyl derivatives. The chief chemical constituents of essential oil were identified as β-caryophyllene (29.0%) and β-gurjunene (21%).

Table 24.7. Antinemic activity of solvent extracts, purified compounds and essential oil against J_2 of root knot nematode.

Extracts/ Purified Compound/ Essential oil	Dose (ppm)	Corrected Mortality (%) after			LT_{50}
		24h	*48h*	*72h*	
Hexane extract	1000	15.4	42.5	74.4	49.8
	500	8.4	34.9	58.8	63.1
Benzene extract	1000	26.0	46.6	74.7	43.7
	500	19.0	34.1	60.5	61.8
Koenidine	1000	6.0	42.3	65.5	56.3
	500	0.0	38.2	42.3	-
Konimbine	1000	4.5	40.6	62.6	58.7
	500	0.0	37.0	40.6	-
N-methyl koenidine	1000	29.5	51.4	82.3	39.4
	500	21.6	38.2	66.0	54.3
N-methyl konimbine	1000	26.6	48.7	78.6	42.7
	500	19.2	36.2	62.4	48.2
Essential oil	1000	100.0	100.0	100.0	-
	500	88.7	100.0	100.0	-

(iv) Mosquito larvicidal activity of curry leaf essential oil

The essential oil of curry leaf has shown mosquito larvicidal effect and thus proved its importance in the arena of public health. A pioneer work in the Division of Agricultural Chemicals, Indian

Agricultural Research Institute, New Delhi, revealed that when a few drops of oil are added over the stagnant water where mosquito larvae is floating, within 24 h maximum mortality of the *Anopheles* larvae can be observed. Thus, the oil could be employed effectively to check the menace of malaria and other mosquito carrying diseases.

EPILOGUE

In search of the safer plant protection chemicals, plants which has plays a pivotal role in therapeutic uses and in culinary are often screened. It is the way by which neem, pyrethrin, nicotine, ryania and many other plants have been placed in very high position for development of biopesticides. Curry leaf (*Murraya koenigii* (L.) Spreng.) has a very long successful therapeutic history and has been used in folk medicine in China and other Asian countries as an analgesic, astringent, antidysenteric, antioxidant, febrifuge, hypolipidemic, hypglycemic, for improvement of vision, to treat night blindness, and for regulation of fertility. The plant contains carbazole alkaloids, which are fairly effective against insect, nematode and fungi and the essential oil shows excellent activity against this pests which paved the way for development of new eco-friendly molecules. The derivatized product of the carbazoles already proved this concept and recognized the plants ability in the plant protection. It is now necessary to develop the right formulation from the plant extracts. Who knows besides providing delicacy to the Indian cuisine, it will be used equally for plant protection in very near future.

REFERENCES

Bhattacharya, P. and Chowdhury, B.K. (1985). 2-methoxy -3-methyl carbazole from *Murraya koenigii. Indian J. Chem.* **24B:** 452.

Bhattacharyya, L., Roy, S.K. and Chakraborty, D.P. (1982). Structure of the carbazole alkaloid Isomurrayazoline from *Murraya koenigii* Spreng. *Phytochem.* **23:** 471.

Bhattacharyya, P., Maiti, A.K., Basu, K. and Chowdhury, B.K. (1994). Carbazole alkaloids from *Murraya koenigii* Spreng. *Phytochem.* **35**(4): 1085-1086.

Bhattachryya, P., Jash, S.S. and Chowdhury, B.K. (1986). A biogenetically important carbazole alkaloid from *Murraya koenigii* Spreng. *Chem. and Ind.* 246.

Chakraborty, D.P. and Das, K.C. (1968). Structure of Murrayacine. *Chem. Commun.* 967.

Chakraborty, D.P., Barman, B.K. and Bose, P.K. (1964). On the structure of girinimbine, a pyranocarbazole derivative isolated from *Murraya koenigii* Spreng. *Science and Cult.* **30:** 445.

Chakraborty, D.P., Barman, B.K. and Bose, P.K. (1965). On the constitution of Murrayanine, a carbazole derivative isolated from *Murraya koenigii* Spreng. *Tetrahedron,* **21:** 681.

Chakraborty, D.P., Bhattacharyya, A., Islam, A. and Roy, S. (1974). Structure of Murrayacinine, a new carbazole alkaloid from *Murraya koenigii* Spreng. *Chem. and Ind.* 165.

Chakraborty, D.P., Bhattacharyya, P., Roy, S., Guha, R. and Bhattacharyya, S.P. (1976). Minor carbazole alkaloids of *Murraya koenigii* Spreng. Paper presented at the 4th Indo-Soviet Symposium of the Chemistry of Natural Products held in Feb. 1976 at Lucknow (India).

Chakraborty, D.P., Das, K.C. and Bose, P.K. (1966). Structure of mahanimbine a pyranocarbazole derivative from *Murraya koenigii* Spreng. *Science and Cult.* **32:** 83.

Chakraborty, D.P., Das, K.C., Das, B.P. and Chowdhury, B.K. (1975). On the antibiotic properties of some carbazole alkaloids. *Trans. Bose Res. Inst.* **38:** 1.

Chakraborty, D.P., Ganguly, S.N., Maji, P.N., Mitra, A.R., Das, K.C. and Weinstein, B. (1973). Murrayazoline, a carbazole alkaloid from *Murraya koenigii* Spreng. *Chem. and Ind.* 322.

Chakraborty, D.P., Islam, A., Basak, S.P. and Das, R. (1970). Murrayazolidine, the pentacyclic carbazole derivative isolated from *Murraya koenigii* Spreng. *Chem. and Ind.* 593.

Chowdhury, B.K. and Chakraborty, D.P. (1969). Mukoeic acid, the first carbazole carboxylic acid from plant source. *Chem. and Ind.* 549.

Chowdhury, D.N., Basak, S.K. and Das, B.P. (1978). Studies on the insecticidal and antimicrobial properties of some carbazole alkaloids. *Curr. Sci.* **47:** 490.

Chowdhury, D.N. and Das, B.P. (1979). Investigation on the toxicity of some carbazole derivatives and plant extract. *Curr. Sci.* **48:** 344-346.

Das, B.P. (1989). Some investigations of the pesticidal properties of carbazoles. *International Pest Control.* **31:** 144.

Das, K.C., Chakraborty, D.P. and Bose, P.K. (1965). Antifungal activity of some constituents of *Murraya koenigii* Spreng. *Experientia,* **21:** 340.

Dutt, S. (1958). The Indian curry leaf (*Murraya koenigii* Spreng.) and its essential oil. *Indian Soap J.* **23:** 201-206.

Dutta, J., Haque, M. and Chakraborty, D.P. (1969). Thin layer chromatography of carbazole derivatives. *J. Chromatography,* **42:** 555.

Fiebig, M., Pezzuto, J.M., Soejarto, D.D. and Kinghorm, A.D. (1985). Koenoline, a further cytotoxic carbazole alkaloids from *Murraya koenigii. Phytochem,* **24:** 3041.

Hiremath, S.M., Madalgeri, B.B. and Basarkar, P.W. (1996). Essential oil constituents of wild curry leaf (Murraya koenigii Spreng.). *Indian Perfumer,* **40**(4): 110-112.

Johannes, R. and Olar, Goj. (1992). Carbazole alkaloids from the fruits of *Murraya koenigii* Spreng. *Phytochem.* **31**(8): 2877-9.

Joshi, B.S., Kawat, V.N. and Gawad, D.H. (1970). On the structure of girinimbine, mahanimbine, isomahanimbin, koenimbidine and murrayacine. *Tetrahedron,* **26:** 1475.

Kishore, N., Dubey, N.K., Srivastav, O.P. and Sing, S.K. (1983). Volatile fungitoxicity in some higher plants as evaluated against *Rhizoctonia solani* and some other fungi. *Indian Phytopath,* **36:** 724-725.

Kureel, S.P., Kapil, R.S. and Popli, S.P. (1969). Terpenoid alkaloids from *Murraya koenigii* Spreng. *Experientia* **26:** 1055.

Kureel, S.P., Kapil, R.S. and Popli, S.P. (1970). Terpenoid alkaloids and synthesis of Mahanimbine. *Experientia* **26:** 1055.

Liquan Zhu, Lonfeng, Paul, P.H., Hungta Waterman, and Peter, G. (1988). Monoterpene and sesquiterpene rich oil from the leaves of *Murraya* sp.: Chemotaxonomic significance. *Biochem. Syst. Ecol.* **16**(5): 491-494.

Macleod, A.J. and Pieris, N.M. (1982). Analysis of volatile essential oil of *Murraya koenigii* and *Paradonus latifolia. Phytochem.* **21:** 1653-1657.

Moorradian, A., Dupont, P.E., Alava, M.D.A.G. and Pearl, J. (1977). 3-aminotetrahydrocarbazoles as a new series of central nervous system agents. *J. Medizin. Chem.* **20:** 487.

Mukherjee, M., Mukherjee, S., Shaw, A.K. and Ganguly, S.N. (1983). Mukonicine, a carbazole alkaloid from leaves of *Murraya koenigii. Phytochem.* **22:** 2328.

Narasimhan, N.S., Paradhkar, M.V. and Kelkar, S.L. (1970). Alkaloids of *Murraya koenigii.* Structure of Mohanine, koenine, koenigine and koenidine. *Indian J. Chem.* **8:** 473.

Narasimhan, N.S., Pradhkar, V. and Chitguppi, V.P. (1968). Structure of mahanimbine and koenimbine, *Tetrahedron Letters,* 5502.

Nigam, S.S. and Purohit, R.M. (1961). Chemical examination of the essential oil derived from the leaves of *Murraya koenigii* Spreng. *Perfumery and Essential Oil Record.* **52:** 152-155.

Pecca, J.C. and Albonico, S.M. (1970). Tryprocides I. Substituted 1,2,3,4-tetrahydrocarbazoles. *J. Med. Chem.* **13:** 327.

Ramarao, A.V., Bhide, K.S. and Majumder, R.B. (1980). Mahanimbol. *Chem. and Ind.* 697.

Randelia, B.E. and Patel, B.P. (1982). Antimicrobial activity of carbazole derivatives. *Experientia* **38:** 529-531.

Ray, D.P. (1999). Phytochemical, Insect antifeedant and Antinemic screening of Marigold (*Tagetes erecta*) flower, M.Sc. Thesis, Post Graduate School, Indian Agricultural Research Institute, New Delhi-110012, India.

Ray, D.P. and Srivastava, S. (2006). Curry Leaf: The Aromatic Bio pesticides. *J. Interacademicia,* **10**(2): 231-235.

Ray, D.P. and Srivastava, S. (2008). Fungitoxicity and chemical constituent of *Murraya koenigii* (L) Spreng. *J. Interacademicia* (Accepted).

Ray, D.P., Prasad, D. and Singh, R.P. (2000). Chemical examination and antinemic activity of marigold (*Tagetes erecta* L.) flower. *Annals of Plant Protection Sciences* **8**(2): 212-217.

Ray, D.P., Prasad, D. and Singh, R.P. (2001). Marigold: A potential Plant in Pest Management. *In:* Crop pest and Disease Management: Challenges for the Millennium, (Eds. Prasad, D. and Puri, S.N.) Jyoti Publishers, New Delhi, pp. 138-152.

Reisch, J., Adebajo, A.C., Vijeya Kumar, and Aladesanmi, A.J. (1994). Two carbazole alkaloids from *Murraya koenigii* Spreng. *Phytochem.* **36**(4): 1073-1076.

Roy, S., Ghosh, S. and Chakraborty, D.P. (1979). Structure of Mahanimboline *Chem. and Ind.* 669.

Roy, S., Bhattacharyya, L. and Chakraborty, D.P. (1982). Structure and synthesis of Mukoline and Mukolidine, two new carbazole alkaloids from *Murraya koenigii* Spreng. *J. Ind. Chem. Soc.* **59:** 1369.

Saha, C. and Chowdhury, B.K. (1998). Carbazoloquinones from *Murraya koenigii. Phytochem.* **48**(2): 363-366.

Shoeb, A., Anwer, F., Kapil, R.S., Popli, S.P., Dua, P.R. and Dhawan, B.N. (1973). N-alkylaminocarbazoles as potent anticonvulsant and diuretic agents. *J. Medizim. Chem.* **16:** 425.

Singh, R. and Maheshwari, R.C. (1997). Essential oil constituents: Hydrodistillation vs CO_2 extraction. *Indian Perfumer,* **40**(1): 26-29.

Srivastava, S., Prasad, D. and Singh, R.P. (2000). Antinemic studies of the chemical constituents of the leaves of *Murraya koenigii* (L.) Spreng. *Annals of Plant Protection Sciences,* **8**(2): 183-186.

Srivastava, S.K. and Srivastava, D. (1993). New constituents and biological activity of the roots of *Murraya koenigii. J. Ind. Chem. Soc.* **70:** 655-659.

Srivastava, S., Ray, D.P. and Singh, R.P. (2007). Chemical examination and antifungal activity of Curry Leaf (*Murraya koenigii* (L.) Spreng). *Pesticide Research J.* **19**(2): 149-1510.

Venkateswaran, T.V. (2004). Curry Leaf: An Authentic Indian Spice. *Dream 2047,* **6:** 22.

Wagner, W., Kalinowski, H.O. and Jork, H. (1995). Isolation and identification of 11-selinen-4-alpha,7-beta-ol and 10-aromadendranol in the essential oil of *Murraya koenigii* Spreng. *Planta Medica,* **61**(2): 196-197.

Wong, K.C. and Tie, D.Y. (1993). The essential oil of the leaves of *Murraya koenigii* Spreng. *J. Essent Oil Res.* **5:** 371-374.

25

Nicotine to Neonicotinoid: The Most Successful Agrochemicals

Kaushik Majumdar, D.P. Ray[1] and P. Dureja

Division of Agricultural Chemicals, Indian Agricultural Research Institute, New Delhi–110 012

In the history of plant protection, like every invention in the science a little breakthrough made wide open arena for the future. The large kingdom of today's pyrethroides market was initiated with the discovery of pesticidal properties of *Chrysanthemum cinerarifolium*, while the carbamate group of pesticides got the initiation from the compound Physostigmine from a very little known plant *Physostigma venerosum*. Unlike these, the tobacco plant has a long and fascinating history. After its introduction in 1559 to Spain and Portugal from the Americans where it has long been cultivated by the Indians, tobacco reached France and Italy where it came into use mostly for smoking, but since late seventeenth century it has been used as an important insecticide or repellant. As early as 1690, water extracts of tobacco leaves were used to control sucking insects on garden plants. Still at that time the exact active principle was not known. Later it was revealed that the usefulness of tobacco leaf preparation was due to its content of the nicotine alkaloid.

NICOTINE: THE TOBACCO ALKALOID

The principal bioactive alkaloid found in tobacco extract is nicotine along with nornicotine, anabasin and myosinine. Nicotine has been isolated from at least 18 species of *Nicotiana*, among which *tabacum* and *rustica* are most common. It is also found in species completely unrelated to *Nicotiana* as well: *Asclepia syriaca* (milk weed), *Atropa belladonna* (deadly nightshade), *Equisetum arvense* (Horse tails) and *Lycopodium clavatum* (club moss).

In pure state, nicotine (Fig. 25.1) is a colorless, nearly odorless liquid of low volatility. There are several reviews on the chemistry (Fodor and Colasanti, 1985), biosynthesis and metabolism (Leete, 1983; Bush and Crowe, 1989) and insecticidal activity (Metcalf, 1955; Schmeltz, 1971)

Present Address:
[1]Fertiliser Control Laboratory, Krishi Bhavan, Abash, Midnapore, West Bengal–721 102

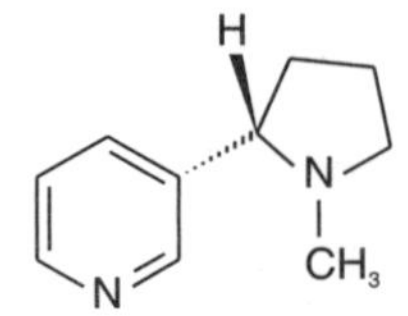

Fig. 25.1. Nicotine molecule.

of nicotine and related tobacco alkaloids. The chemical name of the compound is *l*-3-(1-methyl-2-pyrrolidinyl)-pyridine. The commercial name is Black Leaf-40 containing 40% nicotine sulfate. The sulfate is stable and freely soluble in water and alcohol.

Free nicotine, mainly due to its high volatility, is used for fumigation in greenhouses. In agriculture, nicotine is used as nicotine sulfate, which acts as a stomach poison of leaf-eating insects. Also, the sulfate when spread gradually, releases free nicotine and is useful in controlling insects over extended periods. Addition of alkaline compounds such as soap and calcium caesinate liberates the active free nicotine, which acts as a contact insecticide or as a fumigant against soft-bodied insects.

MODE OF ACTION OF NICOTINE

To obtain a clear insight into the mode of action of nicotine, it can be foretold that they interfere with the neurotransmission through synapsis. Neurotransmission through a vertebrate

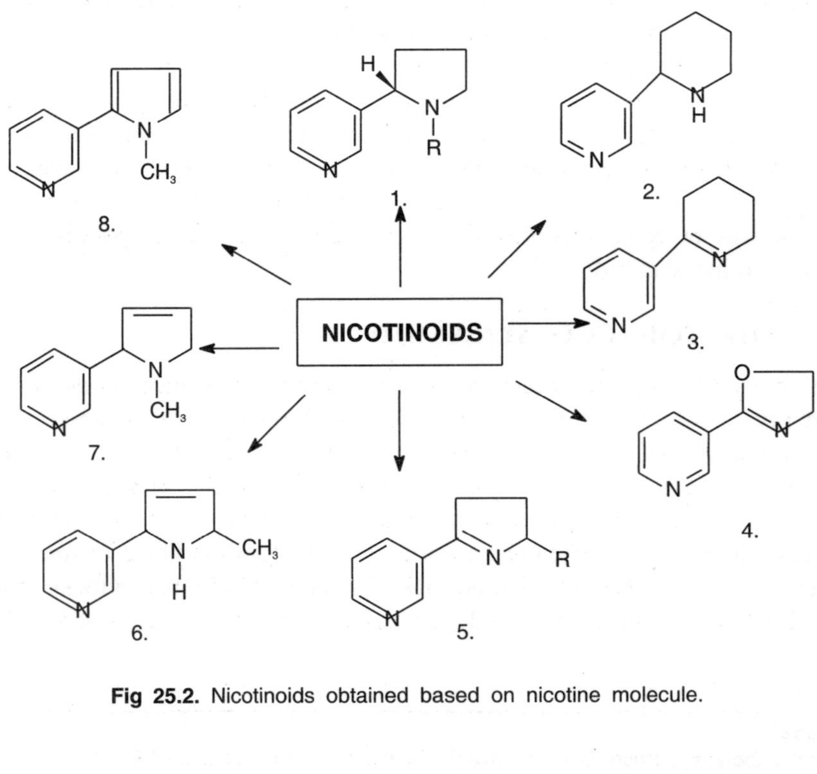

Fig 25.2. Nicotinoids obtained based on nicotine molecule.

nAchR is initiated from the biding of acetyl choline, a neurotransmitter, to the Ach recognition site on the ? subunit. When the electrical signal carried by the presynaptic neuron reaches synaptic end of the cell, Ach is released from the presynaptic vesicle into synaptic cleft. Ach readily diffuses across the cleft to post synaptic membrane, where it binds to the high affinity receptor site on the nAchR receptor.

1. Vertebrate nAchR

The vertebrate nAchR consists of diverse subtypes assembled as five subunits in combinations from nine α (α 1-9), four β (β 1-4), γ, δ and ε subunits (Fig. 25.3). The skeletal muscle or electric ray (Torpedo) subtype is made up of two α1 subunits and one each of β1, d, g (or e in adult muscle) subunits and is best understood relative to the ligand-binding site environment. Neuronal nAchR subtypes expressed in vertebrate brain and ganglia are assembled in combinations of α2-9 and α2-4 and are pharmacologically classified into two branches based on sensitivity to the antagonist α-bungarotoxin (a snake toxin, α-BGT). The α-BGT insensitive branch is made up of subtypes with combinations of α 2–6 and β 2–4 subunits. Of these, the α3β × α5 (X = 2 and/or 4) subtype is mainly found in ganglia and the α4β2 subtype is the most abundant subtype in brain. The α7–9 subunits are involved in the α-BGT sensitive subtypes in brain and ganglia, etc., and the amount of α7-containing receptor is comparable to that of the α4β2 subtype in brain. The specific subunit combinations confer differences in the pharmacological profile among the receptor subtypes, since the drug-binding site is localized at

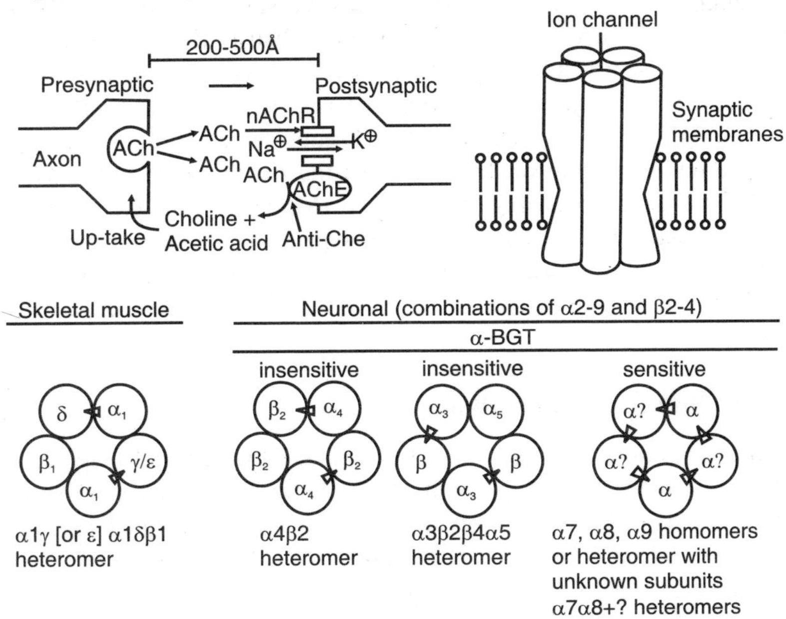

Fig. 25.3. Functional architecture and diversity of the vertebrate nicotinic acetylcholine receptor (nAChR). Top left panel represents neurotransmission through cholinergic synapse. The right is an image of the membrane-associated nAChR. The bottom illustration displays the typical vertebrate nAChR subtypes (cross-section images, seen from top of receptor). The receptor is assembled from five subunits, and different subunit combinations make the receptor subtype. The α8 designater binding site regions for cholinergic agents.

the interface region between subunits. The mammalian nAchR is a target for several potential therapeutic agents for neuropathic disorders, cognitive enhancement and analgesia, leading to a recent interest in the possibility of subtype-selective agents at this receptor (for further reviews, Lindstrom, 1997; Decker and Meyer, 1999; Cordero-Erausquin *et al.*, 2000). In addition, existence of the multiple receptor subtypes should also be considered in the insecticide safety.

2. Insect nAchRs

In the fruit fly, *Drosophila melanogaster*, three genes encoding the ligand binding α-type (Dα1, Dα2 and Dα3) subunits and two for the structural β-type subunits have been identified based on conserved sequence. However, expression of homomeric receptors with either the Dα1 (ALS) or Dα2 (SAD) alone does not generate a functional receptor and the *Drosophila* β-type (ARD and SBD) subunits do not contribute to functional receptor expression. Coexpression of the *Drosophila* α- and β-type subunits in various combinations does not produce any electrophysiological or biochemical response. At present, the functional receptor with ion channel property and/or radioligand binding activity can be generated only when any of the three α-type subunits is coexpressed with the vertebrate (chick or rat) β-type subunit. These results strongly suggest the importance of the β- or non-α-type subunit and the heterooligomeric status of the native *Drosophila* nAchR with possible involvement of unidentified subunit(s) (Bertrand *et al.*, 1994; Lansdell and Millar, 2000; for review, Tomizawa *et al.,* 2000). Interestingly, the antibodies raised against Dα1 and Dα2 subunits can visualize specific regions in the *Drosophila* central nervous system and their distribution patterns are quite similar. Furthermore, it has been demonstrated that both of the above two subunits are functionally co-assembled with the chick β2 subunit in *Xenopus* oocyte within a single receptor complex, suggesting the possibility that the insect nAchR consists of three subunits (Schulz *et al.*, 2000). Recently, the understanding of the nAchR of the peach-potato aphid *Myzus persicae* has rapidly increased. Five genes encoding the α -type (Mpα1-5) and one for the β-type (Mpβ1) subunit have been cloned. Also, nAchR α- and β-type subunit genes have been cloned in the migratory locust, *Locusta migratoria*. Each one of α-type subunit gene has been found in the desert locust *Schistocerca gregaria* or the tobacco hornworm *Manduca sexta* (for review, Tomizawa, 2000).

3. Mechanism of Impulse Transmission

In closed state, the large non-polar residues (L) occluded the channel (Fig. 25.4) by forming a tight hydrophobic ring; binding of Ach could tilt the helices ($H_2\alpha$ helix) to shift the position of the ridges. Pore (Na^+-channel) remains open because it would be lined by negatively charged polar amino acid residues, and specific cations(Na^+) only flow through the channel because of its small size (Na^+:H_2O, radius 0.95E). K^+: H_2O, because of its larger ionic radius (1.33 E) can not flow through the channel.

Cations in the extracellular fluid (Na^+) are present in higher concentrations than in the cytosol of the neuron, and so Na^+ flows through the open gate, down their concentration gradient, thereby depolarizing the postsynaptic cell.

If the total depolarization occurring at the post synaptic membrane is large enough (+30 mV), it triggers a muscle action potential. Entry of Na^+ further depolarizes the membrane and

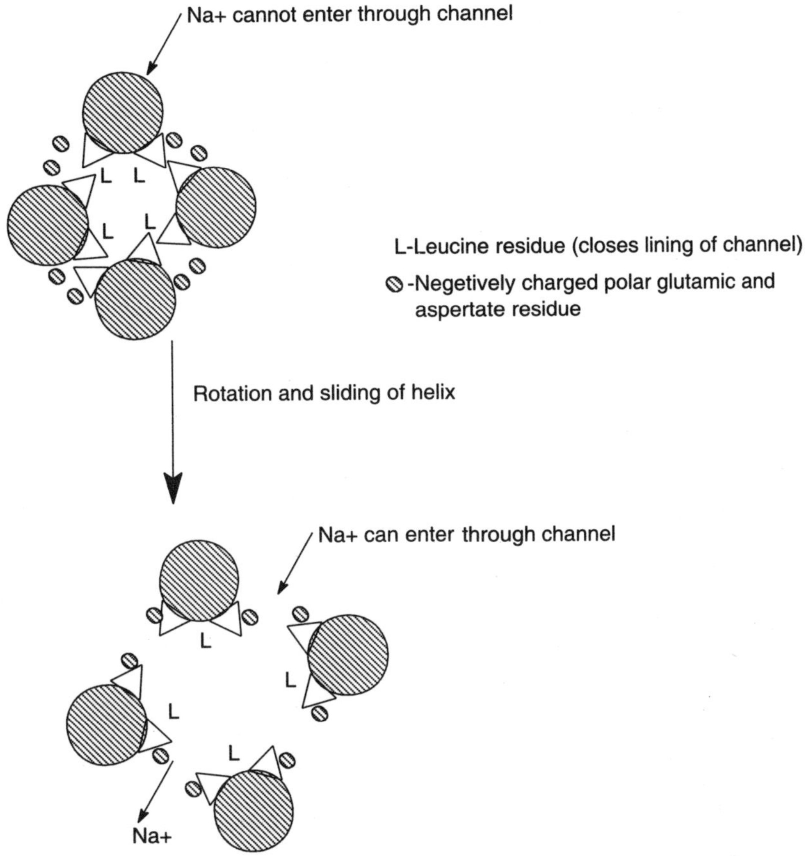

Fig. 25.4. Opening of Na+ channel.

more Na^+ are opened. After the membrane potential attains a value of +30 mV, Ach hydrolyses to choline and acetate thereby lowering the Ach concentration in the synaptic gap, and the Na+ channel gets closed.

WHAT NICOTINE DOES?

Nicotine exerts its effect in both vertebrates and invertebrates by binding to a subset of cholinergic receptors, the nicotinic acetylcholine receptors (nAchRs). Nicotine mimics the action of Ach by interacting with nAchR, i.e., the nAchR fails to distinguish between nicotine and Ach. Hence, the effects produced by nicotine across the cholinergic synapses of ganglia of CNS. Both nicotine and Ach produce stimulation at low concentrations and blockage at high concentrations. The nicotinium ion is attached through positive charge of the pyrrolidine N-atom to the anionic site of nAchR like Ach, but unlike Ach, it is not susceptible to hydrolysis. In actual poisoning by nicotine, Na^+-channel remains open for indefinite period resulting in continuous nerve impulse leading to early twitching, convulsions, followed by paralysis and death (Tomizaoa *et al.,* 1995).

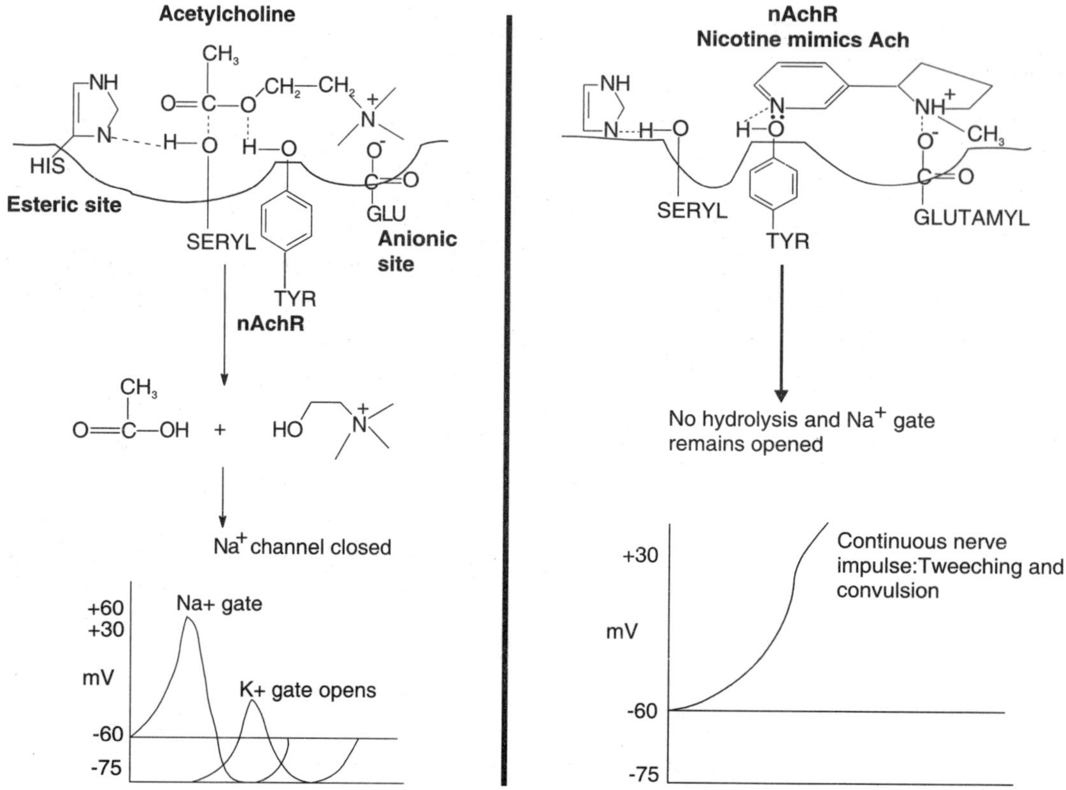

Fig. 25.5. Inhibition of nicotinic acetylcholine esterase by nicotine.

EMERGENCE OF NEONICOTINOIDS

Nicotine has never achieved the prominence that the synthetic insecticides attained due to its expensiveness, lack of commercially applicable synthesis, extreme toxicity to mammals, and limited insecticidal spectrum. In 1979, Soloway had reported some nitromethylene derivatives with highly unusual structure, unique mode of action and interesting biological activity. Chemists of Nihon Bayer Agrochem (Shiokawa, Tsuboi, and Moriya) started to optimize these structures. In 1985, the coupling of the chloropyridyl moiety to the N-nitro substituted imidazolidine ring system enabled the synthesis of highly active insecticide imidacloprid. It is the first commercial example of chloronicotinyl insecticides acting on nicotinic acetylcholine receptors. It is now registered in more than 60 countries as a compound with new or non conventional mode of action to combat highly resistant insect pest (Elbert *et al.,* 1990; Nauen *et al*., 1998). Chloronicotinyl insecticides are getting more and more importance because of the close analogs of imidacloprid, such as Takeda's and Nippon Soda's open chain derivatives, Nitenpyram and acetamiprid, respectively (Tomizawa *et al.* 1995; Yamamoto *et al.* 1995). Later on other analogs come out such as thimethoxam,thiacloprid, clothiamidin and dinotefuran (Table 25.1)

Table 25.1. List of important Neonicotinoids.

Name	Generation	Structure	Year	Commercialized By
Nithiazine	1st Generation	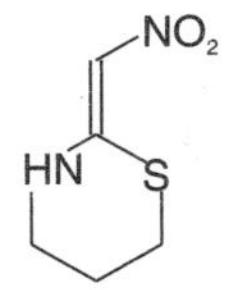	1972	Not commercialised
Imidacloprid	2nd Generation	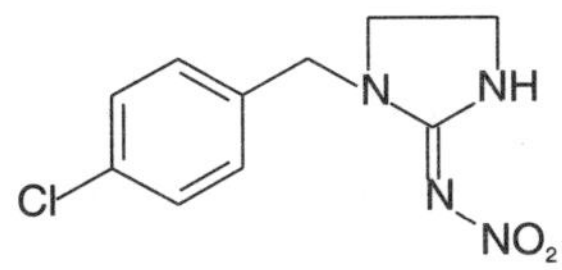	1991	Bayer AG
Nitenpyram	2nd Generation	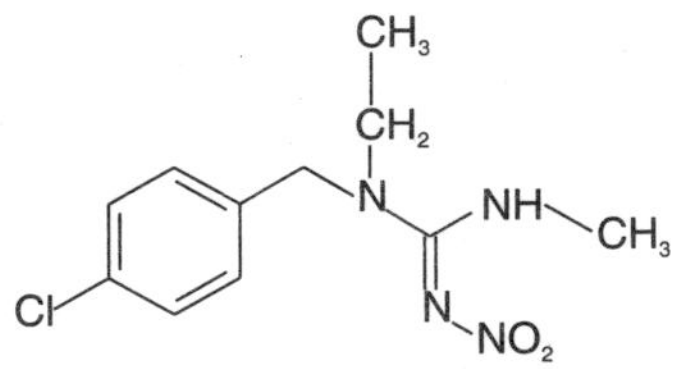	1995	Takeda
Acetamiprid	2nd Generation	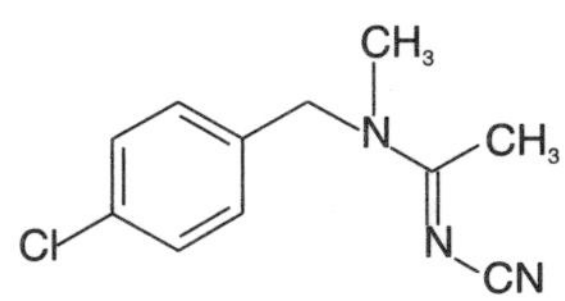	1996	Nippon Soda
Thimethoxam	3rd Generation	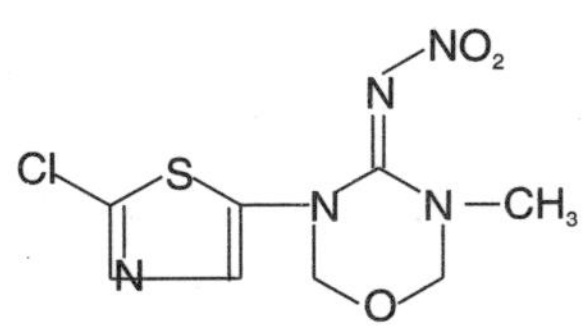	1998	Novertis
Thiacloprid	3rd Generation	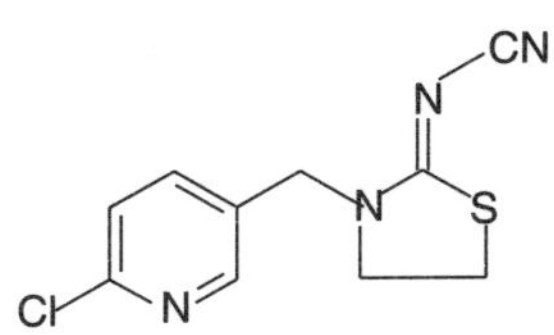	Still not marketed	Bayer
Acetamiprid	3rd Generation	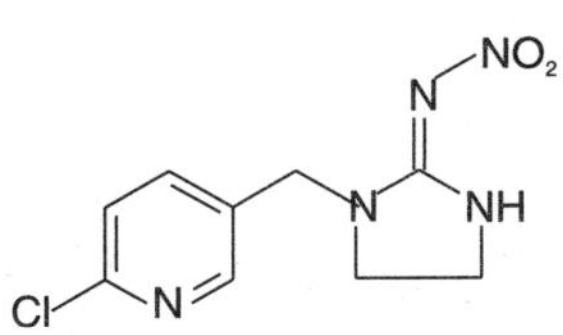	Still not marketed	Takeda, Bayer
Dinotefuran	3rd Generation	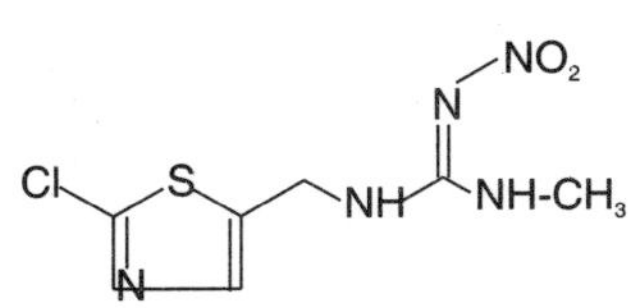	Still not marketed	Mitsui chemicals

All these compounds resembling imidacloprid are collectively called Neonicotinoids. Imidacloprid and other neonicotinoids interact with the acetylcholine binding site of the nicotinic acetylcholine receptors as agonists, which cause excitation and eventually paralysis leading to death as nicotine does.

STRUCTURE ACTIVITY RELATIONSHIP (SAR) OF NEONICOTINOIDS

The structural features responsible for the toxicity of nicotinoids are according to the following analyses:

- Individual ring parts of nicotinoid are of low toxicity (Yamamoto *et al.,*1962)
- Pyridine ring is significant for toxicity (Perry *et al.,*1998)
- N-atoms of pyrrolidine or piperidine ring are separated from pyperidine are separated from pyridine ring by a minimum distance of 4.2 E. It has been recognized that many nAchR ligands share the same feature: the presence of an electron donating group that makes H-bonding to the esteric niche of the nAchR at 5.9 E apart from the positively charged centre, which interacts with the anionic niche of the receptor for nicotine. N-atom of 3-pyridyl group is the electron donor, the N(protonated) atom is the positive centre.
- Neither of the two N-atoms to be quaternary, because, charged N-atoms hinder the penetration through insect cuticle by decreasing lipophilicity.
- pKa value of N-atom of side chain or pyrrolidine ring lies between 7.4-9.0 and are highly toxic to insects, while nicotinoids having pKa less than 5.8 are of low toxicity.
- Then moving to pyridyl derivatives, which were additionally interesting because they represents part of the essential structure of nicotine, a natural antagonist of acetylcholine. The unsubstituted pyridylmethyl derivatives all showed improved activity relative to their corresponding chlorobenzyl parameter. The 3- and 4-pyridylmethyl derivatives were active at 8 ppm. While the 2-pyridylmethyl one was only as active as the unsubstituted benzyl derivative. Since the introduction of a Cl atom enhanced activity in the benzyl series, it was introduced into the pyridine ring of pyridylmethyl-2-nitromethyleneimidazoline.
- 6-chloro compounds are 25 times more active than the parent, while a Cl atom at 2 or 5 position diminishes the activity. The 6-chloro-3-pyridyl methyl derivative was highly active against rice green leaf hopper (Kagabu *et al.,*1992).

In short, the structural requirement for insecticidal activity among 6-chloro-3 pyridyl methyl derivatives can be summarized in Scheme-25.1.

- A methylene group is the best spacer.
- As a ring size, the 5,6 membered ring was best.
- NH and S are best for the part X.
- For part Y, a methylene or amino group were nitromethylene and nitramino.

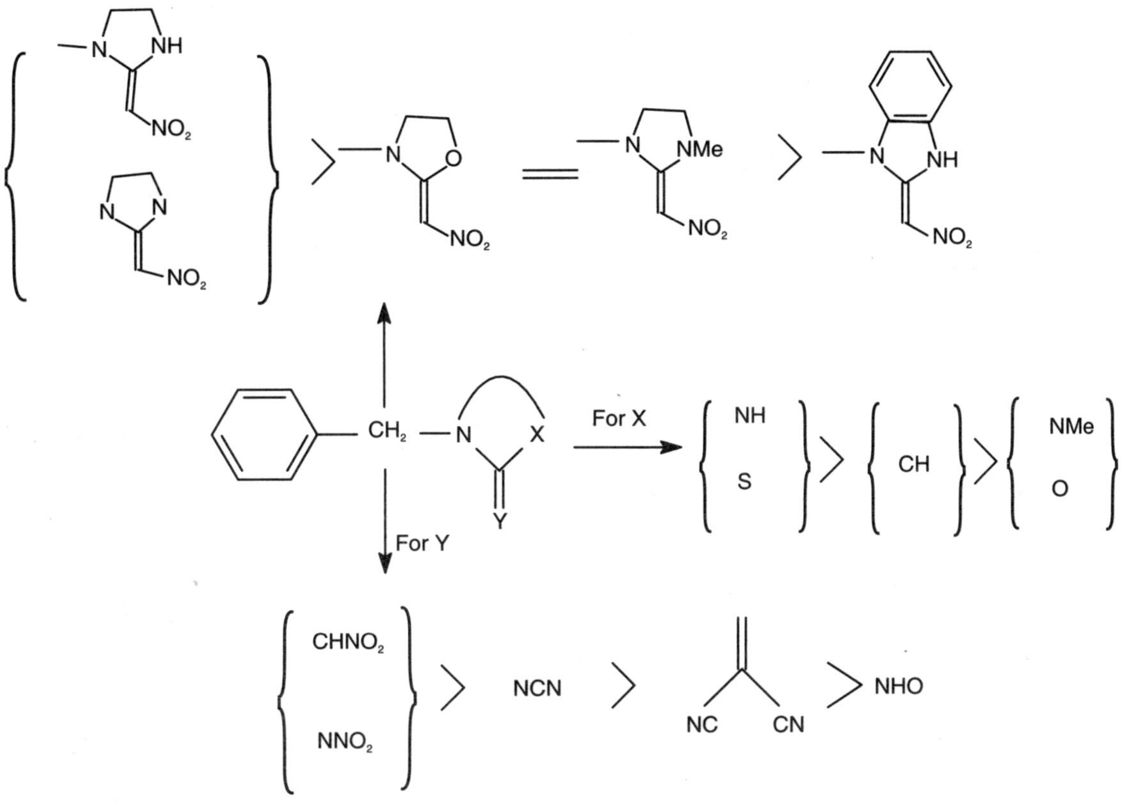

Scheme 25.1. Structure activity relationship of Neonicotinoids.

The term "Neonicotinoids" was derived from the recognition that imidacloprid and related compounds share the same feature with nicotinoids with an additional feature: the partial positive charge on the nitrogen atom conferred by neighboring electron withdrawing group (Fig. 25.6). Based on the essentiality for the higher insecticidal activity of the presence of chlorine atom at 6-position of the 3-pyridyl group, the "chloro-nicotinyl" insecticides was also proposed.

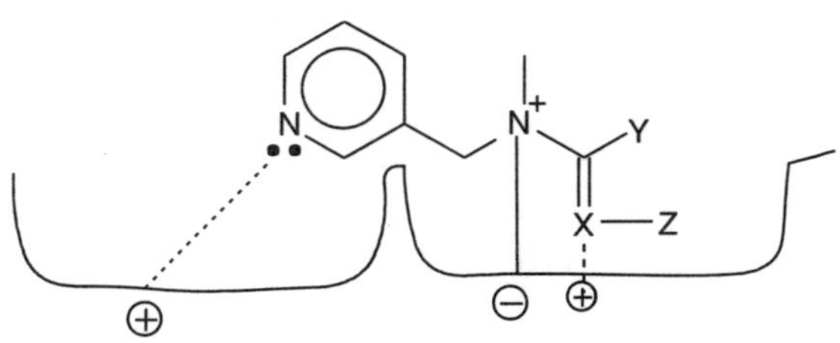

Fig. 25.6. Interaction of Neonicotinoids with nAchR.

FEATURES OF CERTAIN IMPORTANT NEONICOTINOIDS

Among the neonicotinoids, the features of some important ones are discussed below:

1. Imidacloprid [1-((6-chloro-3-pyridinyl)methyl)-N-nitro-2-imidazolidinimine]

It was first registered in US in 1994. It is a systemic insecticide and moves through the plant system. It has low soil persistence, high insecticidal activity and relatively low mammalian toxicity. It is a broad spectrum insecticides killing different orders of insects (Kagabu, 1994).

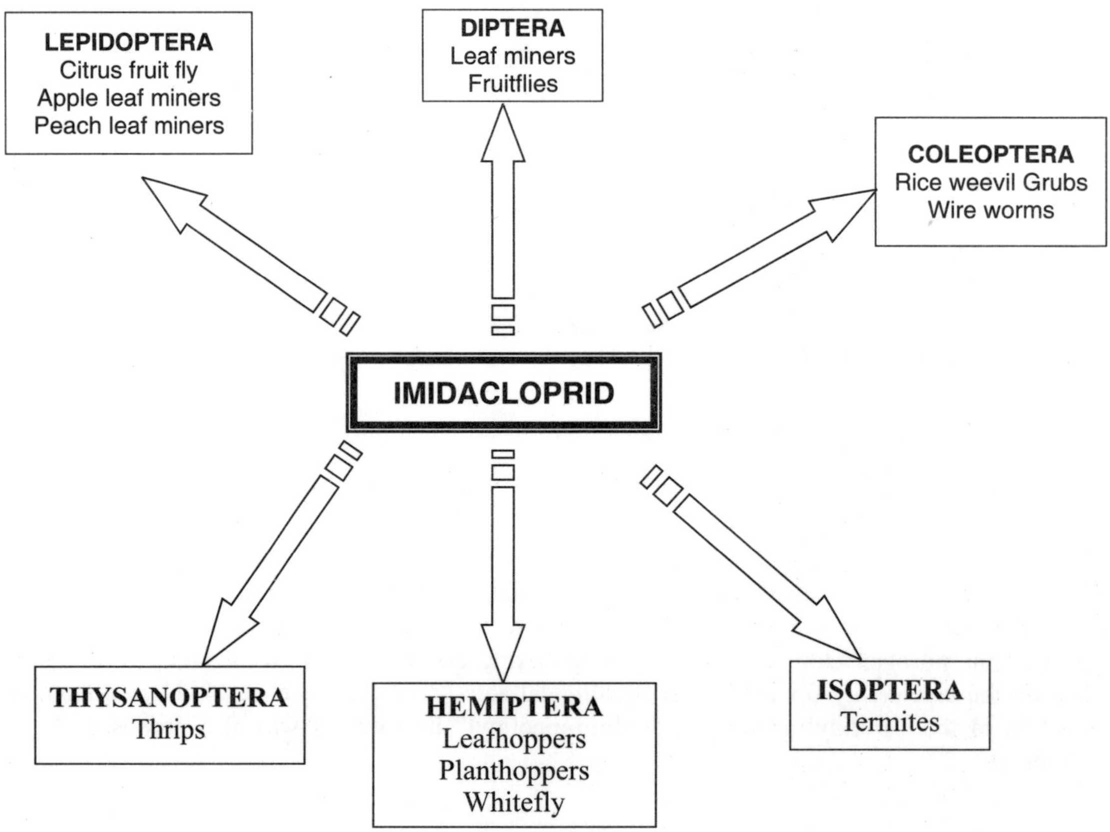

It is marketed as Gaucho®, Confidor®, and Provado®. It is especially useful in seed treatment and soil application. It is also used as a potential termite control agent by the name Premise®. Bee toxicity is its major disadvantage.

Synthesis of Imidacloprid

Scheme-1: It consists of three steps:

1. Synthesis of dimethyl ester of dithiocarbamate:

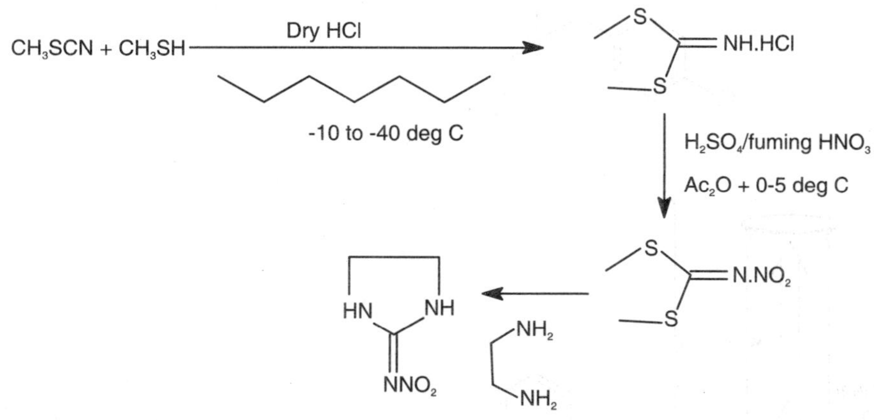

2. Synthesis of second part:

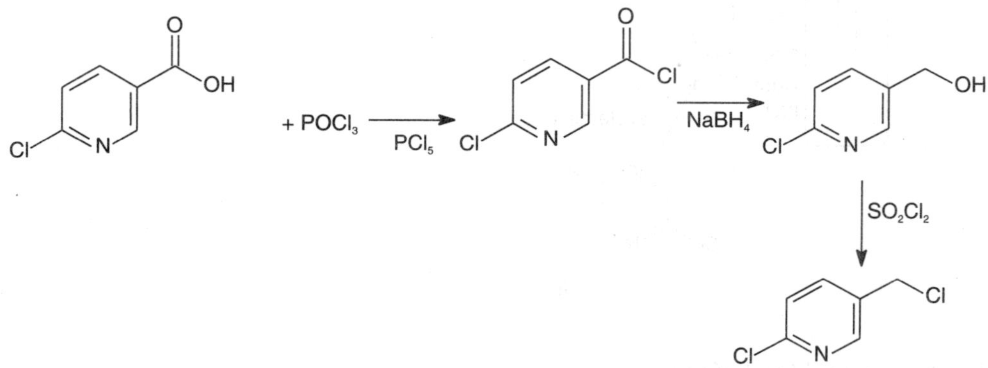

3. Condensation of two moiety:

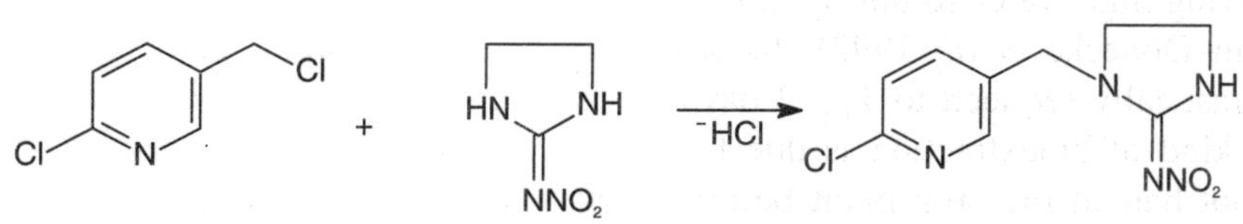

Scheme-2: This method was given by Rallis

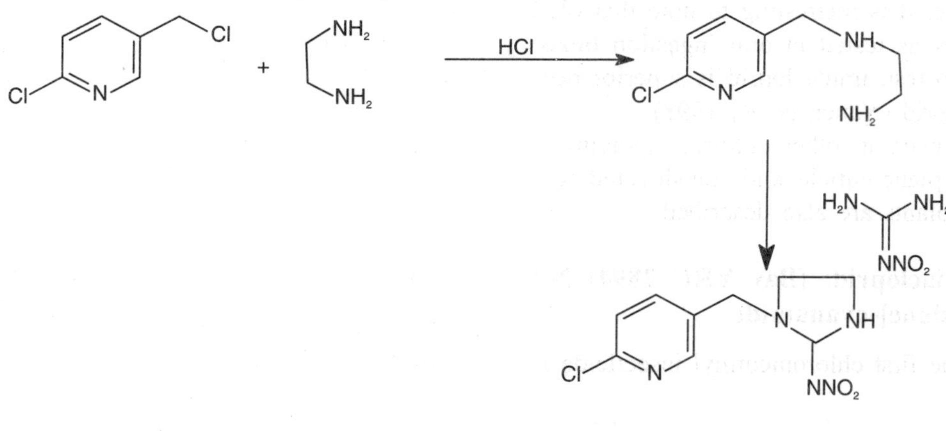

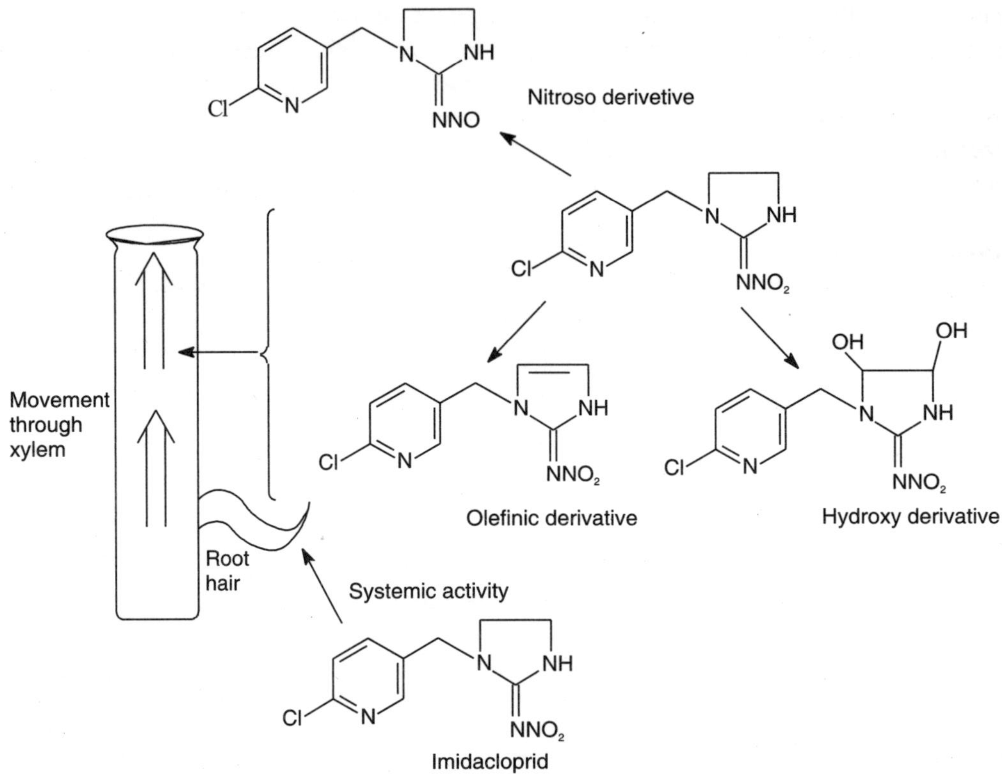

Translocation and Metabolisms

Although having high water solubility, no leaching problems have been encountered in Lysimeter studies (Stein–Dönecke *et al.* 1992). Its solubility in soils is rather high ($T_{1/2}$=150 days), but this value drastically reduced to $T_{1/2}$=2 days by ground cover on the soil (Scholz and Spiteller, 1992). This kind of bioextraction is due to its systemic property. Imidacloprid is readily taken up by the root hair of the crop plant before being metabolized. Imidacloprid at a concentration of 0.3 ppm in soil is sufficient to protect cabbage from *Myzus persicae* for 8 weeks (Elbert *et al.* 1990). It can be converted to olefinic metabolites (Westwood *et al.,* 1998) and nitroso derivative. It is interesting to note that olefinic and nitroso metabolites show more insecticidal properties as tested in oral ingestion bioassays in *Myzus persicae* and *Aphis gossipii.* But in aphid dip test, imidacloprid is superior because lipophilicity of these metabolites is lower than imidacloprid (Nauen *et al.,* 1998).

However, in other cultures, spraying application is also possible. Imidacloprid can enter through plant cuticle and translocated to xylem phloem system. Translaminar properties in various plants are also described.

2. Thiacloprid: (Bay YRC 2894) N-[3-(6-chloropyridin-3-yl-methyl)-thiazolidin-2-ylidene]-cyanamide

It was the first chloronicotinyl insecticide having activity not only against sucking insects such as aphids, white flies and some jassids, but also against weevils, leaf miners and Cydia pomonella in apples and various species of beetles (e.g., *Leptinotarsa decemlineata, antonomus pomorum, Lissorhoptrus oryzophilus, Lema oryzae*) and shows good plant compatibility in all relevant crops. It is registered worldwide under the trade name of Calypso®, but in Japan and Switzerland, it has trade names Bariard® and Alanto®, respectively. The basic formulation for the a.i. is a 480 SC having good compatibility with conventional crop protection products.

Synthetic Pathway

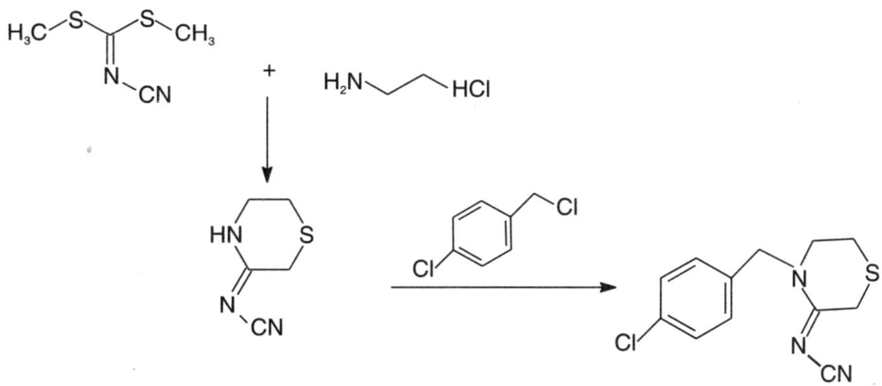

Important features

It is stable towards hydrolysis under the condition of heavy rain. Half life at pH 5, 7 and 9 is above 500 hours. Photolysis in water (buffered in pH 7.0) shows a half life of >100 days. On soil surface, thiacloprid is also under sunlight irradiation. Very safe to the bees.

3. Thiomethoxam

This is the irst and best representative of second generation of neonicotinoid. In 1998, Novertis launched thiomethoxam (Maienfisch *et al*., 1999), a novel neonicotinoid with a unique structure having novel 1,3,5-oxadiazine ring substituted at 5-position by a methyl group and outstanding insecticidal activity. It ia also called the "first of the second generation neonicotinoid" and is the first commercially available neonicotinoid having a 2-chloro-5-thiazolyl heterocycle. This unique combination of structural features is responsible for its outstanding insecticidal activity, e.g., the methyl group at 5-position enhances the activity against sucking pests, whereas 2-chloro-5-thiazolyl heterocycle is essential for improved efficacy against chewing pests (Maienfisch *et al.,* 2001). Lot of research is going on this novel compound. It is effective against aphids, white fly, thrips, leaf hopper and certain beetles. It is marketed as Actara® for foliar and soil treatment and Cruiser® for seed treatment. This compound is more mobile in soil than imidacloprid. One of its formulations Platinum® is registered for use in melons, fruiting vegetables and tuberous vegetables. Pending uses on citrus, brassica, leafy vegetable, barley, sorghum and wheat. Good alternative of OP.

Synthetic Pathway

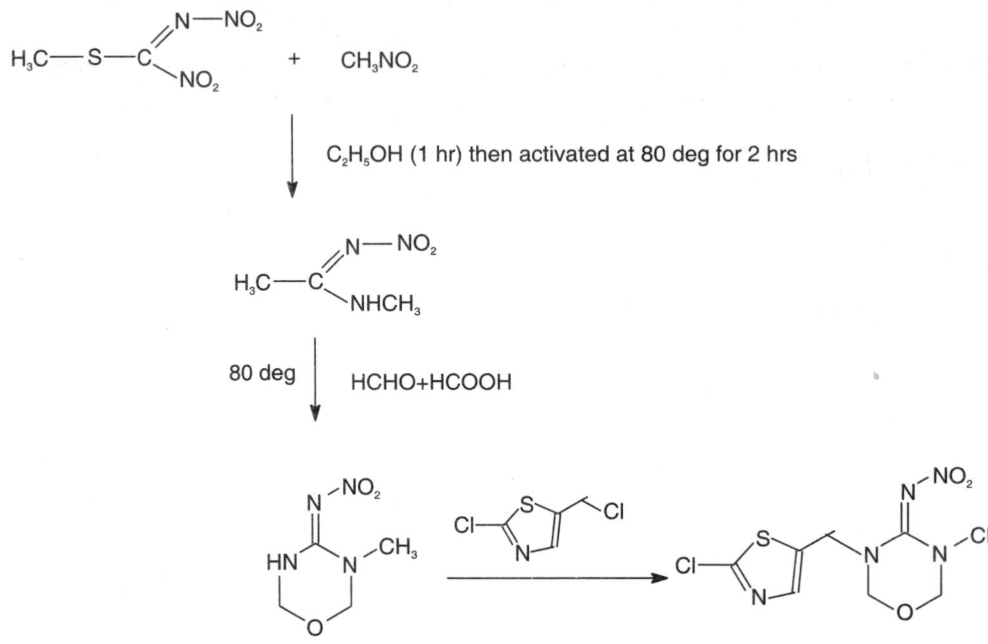

ADVANTAGES OF NEONICOTINOIDS

Besides being highly toxic to insects and relatively less toxic to mammals as compared to nicotinoids, neonicotinoids also relived resistant pressure on other valuable group of insecticides. They can be easily bio extracted because of highly systemic activity. Finally, it has a long lasting effect in plant at low application dose.

EPILOGUE

Nicotine has a long history of pest control. Although with the elapse of time the utility of nicotine has been faded out due to its different complication and toxicity to mammals, its main implication was hidden into a long cherished way to obtain the long lasting safer agrochemicals. It has served as a lead for highly potent synthetic pesticide. Imidacloprid, the first member of the new class of insecticides, neonicotinoides, acts on nAchR to find wide application. Identification of natural lead and modification of structure on the basis of biochemical and biophysical studies has paved a wide road to achieve a group of good chemicals known as neonicotinoides. It is now well accepted that the integration and complementation of this group of chemicals in IPM will improve future crop protection management in the world. Therefore, instead of condemning nicotinoids to posterity, scientists should accept the opportunity to have the neonicotinoides by synthetic moleculer conversion. That is where the morale of the story of nicotinoid to neonicotinoids lies.

REFERENCES

Bertrand, D., Ballivet, M., Gomez, M., Bertrand, S., Phannavong, B. and Gundelfinger, E.D. (1994). Physiological properties of neuronal nicotinic receptors reconstituted from the vertebrate β2 subunit and Drosophila α subunits. *Eur. J. Neurosci.* **6:** 869-875.

Bush, L.P. and Crowe, M.W. (1989). Nicotiana alkaloids. Pages-87-107 in *Toxicants of plant origin. Vol. I. Alkaloids.* eds. Cheeke, P.R. CRC Press, Inc., Boca Raton, Fla., USA. pp. 335.

Cordero-Erausquin, M., Marubio, L.M., Klink, R. and Changeux, J-P. (2000). Nicotinic Receptor function: New Perspective for knockout mice. *Trends Pharmacol. Sci.* **21:** 211-217.

Decker, M.W. and Meyer, M.D. (1999). Therapeutic potential of neuronal nicotinic acetylcholine agonists as novel analgestics. *Biochem. Pharmacol.* **58:** 917-923.

Elbert, A., Overbeck, H., Iwaya, K. and Tsubi, S. (1990). Imidacloprid, a novel systemic nitromethylene analogue insecticide for crop protection. *Brighton Crop protection conferences-Pest and Diseases,* **1:** 21-28.

Fodor, G.B. and Colasanti, B. (1985). In *Alkaloids: Chemical and Biological Perspectives.* Eds S.W. Pelletier, Vol. 3, Wiley, New York pp. 1-91.

Kagabu, S. (1994). Chloronicotinyl Insecticides: Development of imidacloprid. Eighth International Congress of Pesticide Chemistry, Washington, DC, July 4-9, 1994.

Kagabu, S., Morya, K., Shibuya, K., Hattori, Y., Tsuboi, S. and Shikawa, K. (1992). 1-(6-Halonicotinyl)-2-nitromethylene-imidazolidienes as potential new insecticides. *Biosci. Biotech. Biochem.* **56:** 362-363.

Lansdell, S.J. and Miller, N.S. (2000). The influence of nicotinic receptor subunit composition upon against α-bungrotoxin and insecticide (imidacloprid) binding affinity. *Neuropharmacology,* **39:** 671-679.

Leete, (1983). *In:* (Pelletier, S.W. ed.). Alkaloids: Chemicals and Biological properties. VII. J. Wiley & Sons.

Lindstorm, J. (1997). Nicotinic acetylcholine receptors in health and disease. *Mol Neorobiol.* **15:** 193-222.

Maienfisch, P., Brandl, F., Kobel, W., Rindlisbacher, A. and Senn, R. (1999). CGA293343: a novel, broadspectrum neo-nicotinoid insecticide, in *Nicotinoid Insecticides and the Nicotinic acetylcholine receptor,* Eds. Yamamoto I and Cassida JF, Springer-Verlag, Tokeyo, Japan.

Maienfisch, P., Huerlimann, H., Rindlisbacher, A., Gsell, L., Dettwiler, H., Haettenschwiler, J., Sieger, E. and Walti, M. (2001). The discovery of Thiomethoxam: a second generation neonicotinoid. *Pest Manag Sci.* **57:** 165-176.

Metcalf, R.L. (1955). Physiological basis for insect resistance to insecticides. *Physiol. Rev.* **35:** 197-232.

Nauen, R., Tietjen, K., Wagner, K. and Elbert, A. (1998). Efficacy of plant metabolites of imidacloprid against Myzus persicae and Aphis gossypii (Homoptera: Aphididae). *Pestic Sci.* **52:** 53-57.

Perry, A.S., Yamamoto, I., Ishaaya, I. and Perry, R. (1998). *Insecticides in Agriculture and Environment-Retrospects and Prospects.* Springer-Verlag Berlin Heidelberg. pp. 113-116.

Schloz, K. and Spiteller, M. (1992). Influence of groundcover on the degradation of 14C-imidacloprid in soil. Brighton Crop Protection Conferences-Pest and Diseases. **1:** 883-888.

Schmeltz, I. (1971). Nicotine and other tobacco alkalioids. *In: Naturally occurring insecticides;* eds. Jacobson, M., Crosby, D.G. Mercel Dekker, New York, pp. 99-136.

Schulz, R., Bertrand, S., Chamaon, K., Smalla, K-H., Gundelfinger, E.D. and Bertrand, T. (1990). Neuronal nicotinic acetylcholine receptors from Drosophila: two different types of a subunits coassemble within the same receptor complex *J. Neurochem.* **74:** 2537-2546.

Stein-Donecke, U., Fuhr, F., Winenecke, J., Hartwig, J. and Leicht, W. (1992). Influence of soil moisture on the formation of dressing zones and uptake of imidacloprid after seed treatment in winter wheat. Pflanzenschutz Nachrich Bayer. **45:** 327-368.

Tomizaoa, M., Otsuka, H., Miyamoto, T., Eldefrawi, M.E. and Yamamoto, I. (1995). Pharmacological characteristics of insect nicotinic acetylcholine receptor with ion channel and the comparison of the effects of nicotinoids and neonicotinoids. *J. Pestic. Sci.* **20:** 57-64.

Tomizawa, M., Lee, D.L. and Cassida, J.E. (2000). Neonicotinoid insecticides: Molecular features conferring selectivity for insect versus mammalian nicotinic receptors. *J. Agric. Food Chem.* **48:** 6016-6024.

Westwood, F., Bean, K.M., Dewar, A.M., Bromilow, R.H. and Chamberlain, K. (1998). Movement and persistent of 14C imidacloprid in sugarbeet plants following application to pelleted sugarbeet seed. *Pestic. Sci.* **52:** 97-103.

Yamamoto, I., Kamimura, H., Yamamoto, S.S. and Goda, M. (1962). Studies on Nicotinoids as an insecticide *Agr. Biol. Chem.* **26:** 709-716.

26

Genetic Engineering Approaches for Detoxification and Decontamination of Pesticides

Deb Prasad Ray[1], K. Majumdar, S. Srivastava[2] and G. Kulshrestha

[1]*Division of Agricultural Chemicals, Indian Agricultural Research Institute, New Delhi–110 012*
[2]*Indian Institute of Natural Resins and Gums, Ranchi–834 010, Jharkhand*

To meet the global challenges of feeding the world's burgeoning population, increase in agricultural productivity is essential. Increase in the agricultural productivity and gross output of agricultural products necessitates the elimination of losses of harvest due to pests, plant diseases and weeds. It is evident that biological factors lead to losses of about 35% of the world's harvest. So, the demand for pesticides is overwhelming throughout the world. The simple reason behind this is that pesticides make the most practical way to control the pests and diseases. The fact about usage of pesticides is true especially when making projection for meeting the targets set by our planners within the constraints of the dwindling resources.

But indiscriminate use of pesticides raised the eyebrows of environmental scientists around the world. Pesticides have several adverse effects in the environment viz., development of resistance in insect pests, resurgence of insect pests, adverse effect on pollinators, reduces the species diversity, affect the soil microflora and fauna, bioaccumulation of pesticides and finally last but not the least, direct human poisoning. So, there is hue and cry around the world for the detoxification of pesticides, which are already deposited in the environment.

DETOXIFICATION AND DISPOSAL

Detoxification can be defined as getting rid off toxicant's toxicity by chemical or physical means, such as metabolism (mostly catabolism or degradation) of xenobiotics in an organism

Present Address:
[1]Fertiliser Control Laboratory, Midnapore, Paschim Medinipur, Government of West Bengal.

or plant into ultimate harmless products, whereas disposal is discarding the pesticides or pesticide containers.

A significant amount of technology is available for pesticide detoxification and disposal. However, due to diversity in chemical properties of pesticides and their formulated products and quantity and composition of pesticides wastes, which must be treated, no single treatment system can be applied. Some methods are excellent for control of point sources discharges from pesticide production and formulation plant, i.e., (i) incineration, (ii) solvent extraction, and (iii) activated sludge treatment. Others are better suited for small-scale discontinuous disposal situation. For instance, burial, landfills and chemical hydrolysis. Methods of pesticides disposal and detoxification can be divided under four heads. They may be enumerated as in the following table (Table 26.1).

Table 26.1. Methods of pesticides disposal and detoxification.

Methods	Example
Chemical methods	i. Acid hydrolysis
	ii. Oxidation and reduction
	iii. Fixation
	iv. Solvent extraction
	v. Chlorinolysis
	vi. Wet scrubbing
	vii. Irradiation
	viii. Adsorption
	ix. Precipitation
	x. Ion exchange
Physical methods	i. Incineration
	ii. Deep well injection
	iii. Ground burial
	iv. Microwave plasma
Biological methods	i. Landfills
	ii. Soil incorporation
	iii. Activated sludge
	iv. Enzyme treatment
Traditional (Field/household) methods	i. Washing
	ii. Cooking
	iii. Peeling
	iv. Sundrying

Among the various methods enlisted above, the traditional methods are more appropriate for decontamination of pesticides. Tucker and Carson (1985) have compiled some stabilization methods for deactivation of hazardous chemical wastes. These are as follows:

1. Solidification with cement
2. Lime based processing
3. Thermoplastic materials, organic polymers
4. Encapsulation
5. Vitrification (for extremely hazardous or radioactive wastes)

Deactivation offers advantages to the industrial community including lowered exposure to the workers, a decreased need for storage capacity, lower handling and transport costs and small impact on the environment (Chudyk, 1989).

BIODEGRADATION OF XENOBIOTIC COMPOUNDS

Xenobiotic compounds are man-made chemicals that are present in the environment at unnaturally high concentrations. The xenobiotic compounds are either most produced naturally, or are produced at much lower concentration than man. Microorganisms have the capacity of degrading all naturally occurring compounds, which is known as the principle of microbial infallibility proposed by Alexander in 1965. Microorganisms are able to degrade many others. The compounds that resist biodegradation and thereby persist in the environment are called recalcitrants. The xenobiotic compounds may be recalcitrant due to one or more of the following reasons.

- They are not recognized as substrate by existing degradative enzymes.
- They are highly stable, i.e. chemically and biologically inert due to presence of substitution groups like halogens, nitro-, sulphonate-, amino-, methoxy-, and carbamyl- group.
- They are insoluble in water or are absorbed to external matrices like soil.
- They are highly toxic to give rise to toxic products due to microbial cells.
- Inability of the compounds to include the synthesis of degrading enzymes.
- Lack of permease needed for their transport into microbial cells.

Field *et al.* (1993) screened the ligninolytic fungi applicable to the biodegradation of xenobiotics. He reported that under culture condition, which favours lignolysis, many xenobiotics such as some polycyclic aromatic hydrocarbons (PAHs), chlorinated phenols, polychlorinated biphenols (PCBs), pesticide, trinitrotoluenes (TNT and industrial dyes) are oxidized and mineralized to varying extent by white rot fungi (Table 26.2). The initial attack has in many cases been shown to be due to extracellular enzymes of the ligninolytic system (Table 26.3). Purified preparations are able to oxidize a variety of PAH compounds to PAH quinines, pentachloro phenols (PCPs) to tetrachlorobenzoquinone and cleave dichlorodibenzo p-dioxin molecule. Manganese dependent peroxidase can oxidise dichlorophenol to chloro p-benzo-quinones and Manganese (III) in acetate buffer can oxidize PAH compounds to acetoxylated PAH and PAH quinones. However, certain aromatics with high ionization potential (e.g., phenthrene and methoxybenzene) are oxidized to peroxidases.

From the various studies on different microorganisms, it is evident that are particular enzymes responsible for detoxification or decontamination. So, scientists gave emphasis on enzymes. Later it was proved that the particular gene from the genetic sequences of a microorganism is responsible for decontamination and detoxification. From this point, overwhelming research started throughout the world to discover a new arena of genetically modified organisms for detoxification of pesticides.

Table 26.2. Xenobiotic compounds degraded by whole cultures of white rot fungi during lignolytic metabolism.

Group	Xenobitic compounds	Species responsible for degradation
Chlorinated phenol	Pentachlorophenol	*Phanerochate chrysosporium*
		P. sordida
		Pleurotus ostreatus
		Trametes hirsute
		Ceriporiopsis subvermispora
	Trichlorophenol/Dichlorophenol	*Phanerochate chrysosporium*
PCB	Arochlor 1254	*Phanerochate chrysosporium,*
		Trametes versicolor
		Funalia gallica
	Tetrachloropbiphenyl	*Phanerochate chrysosporium*
Dioxin	Dichlorodibenzino p-dioxin	*Phanerochate chrysosporium*
	Tetrachlorodibenzino p-dioxin	*Phanerochate chrysosporium*
DDT		*Phanerochate chrysosporium*
PAHs	Fluorene	*Phanerochate chrysosporium*
	Anthracene	*Phanerochate chrysosporium*
		Trametes versicolor
		Bjerkandera adjusta
	Phenanthrene	*Phanerochate chrysosporium*
		Trametes versicolor
		Chrysosporium lignosum
	Pyrene	*Phanerochate chrysosporium*
Alkyl halides	Lindane	*Phanerochate chrysosporium*
	Chlordane	*Phanerochate chrysosporium*
Nitrotoluenes	Dinitro- and trinitrotoluenes	*Phanerochate chrysosporium*
Choroanilines	*Phanerochate chrysosporium*	*Phanerochate chrysosporium*
		Trametes versicolor
		Chrysosporium lignosum

CONVERSION OF XENOBIOTICS THROUGH DIFFERENT BIOCHEMICAL PROCESSES

Generally, xenobiotics contain functional groups and partial structures not usually found in nature. Many pesticides are bulky complex molecules, and others display hydrophobic properties due to the presence of reduced hydrocarbon fragments. The principal enzymes responsible for the bioconversion of xenobiotics are various lyases and oxidoreductases, specifically hydrolases, oxygenases and various enzymes capable of dehalogenation (Table 26.3).

Table 26.3. Enzymes responsible for xenobiotic degradation.

Xenobiotic Compounds	Enzymes
Chlorophenol, Dichlorophenol, Trichlorophenol, Pentachlorophenol, Chlorocatechol, Anthracene, Pyrene, Benzo [a] pyrene, Benzo [a] anthracene, Perylene, Dibenzo-p-dioxin, Azo dyes	Lignin peroxidase
Dichlorophenol, Chlorocatechol, Nitroaminotoluene, Polymeric dyes, Chlorolignin	Manganese dependent peroxidase
Anthracene, Pyrene, Benzo[a] pyrene, Benzo [a] anthracene, anthanthrene	Mn(III)
Chlorophenol, Chloromethylphenol, Dichlorophenol, Irichlorophenol, Tetrachlorophenol, Pentachlorophenol, Chloroguaiacols, Chloroacetals	Laccase

Hydrolysis

Amide and ester bonds undergo hydrolytic cleavage in anilides, phenylureas, esters of carbamic, thiocarbamic, phosphoric, thiophosphoric, and of other acids. Hydrolases responsible for the cleavage of pesticides are among the best studied groups of enzymes. Most of these hydrolases are extracellular enzymes, except for the cell wall-bound enzymes of *Penicillium* and *Arthrobacter* sp., which hydrolyse barban and propham, respectively. Aryl- and acylamidases, which hydrolyse the amide bond in phenylamides, have been studied and described in sufficient detail. Usually, these enzymes are induced by a broad spectrum of substrates and exhibit low substrate specificity (Engelhardt *et al.*, 1973).

Dehalogenation

Many xenobiotics contain halide atoms bonded to aliphatic or aromatic carbon. Therefore, dehalogenation reactions are important initial steps in the degradation of such xenobiotics. Although molecular mechanisms of dehalogenation are not fully understood and require further studies, certain reaction types are fairly well known: hydrolytic, reductive and oxidative dehalogenations. The latter type includes dehydrohalogenation, i.e., dehalogenation that involves molecular oxygen, and results in the formation of a double bond. A number of enzymes have been isolated, described and characterized as effecting dehalogenation (Goldman *et al.* 1968). These are haloacetate-halidohydrolase, which split off a halogen from halide acetate to form glycolic acid; hydrolases effecting dehalogenation of dichloroacetate; 2-haloacid halidohydrolase. The above enzymes are responsible for the hydrolytic release of a halide from the 2 position of short-chain fatty acids. The enzymes that carry out reductive dehalogenation of aromatic carbon have not yet been characterized, although literature provides a number of examples of the process including DDT, 2,4-D and polychlorophenols. As far as oxidative dehalogenation is concerned, the enzyme, DDT-dehydrochlorinase has been characterized and descriptions of the dioxygenase-mediated dechlorination of picloram and monooxygenase dechlorination of chlorophenoxy acids, chlorobenzoic acids, and chlorophenols are available.

Oxidation

Oxidative processes are very important in the bioconversion of pesticides and other xenobiotics, especially bioconversions that follow hydrolysis and dehalogenation of the parent. In this case, the basic reactions are hydroxylation, followed by cleavage of the aromatic ring, oxidative O- and N-dealkylation, and epoxidation. Certain compounds, e.g., derivatives of aromatic acids, are completely degraded due to the involvement of oxidative enzymes.

Reduction

In general, reductive processes take place under anaerobic conditions in the early stages of degradation. Mention has already been made of reductive dechlorination. Another no less important process is reduction of nitro groups. Enzymes that bring about this process are called nitroreductases. They perform their functions in the presence of NADH, the reaction being stimulated by the reduced form of FAD, as well as Mn and Fe ions. A number of authors reported that these enzymes have low substrate specificity. Thus, the enzymic preparation from *Vellonella alkalescens* was found to catalyse the reduction of nitro groups of some 40 different compounds. The amino products of nitro group reduction may resist further transformation due to the formation of bound residues.

MICROORGANISMS EFFECTING DETOXIFICATION OF XENOBIOTICS

An exhaustive research on microbial degradation of xenobiotics reveals that bacteria are chiefly responsible for the detoxification of xenobiotics. Fungi and yeasts are less important, and microalgae and protozoa appear to be only rarely involved in the degradation of xenobiotics. Among the bacteria, pseudomonads are considered to be the most efficient group involved in the degradation of xenobiotics. *Bacilli* are quite often responsible for processes of pesticide hydrolysis. However, pseudomonads can bring about hydrolysis as successfully as they do dehalogenation, hydroxylation, aromatic ring cleavage and nitro group reduction. Some examples of bioconversion of various compounds effected by pseudomonads are listed in Table 26.4. The bacterial genera *Acinetobacter, Arthrobacter, Rhodococcus* and *Flavobacterium* also deserve mention, since they often bring about various bioconversions of different xenobiotics. Fungal cultures, especially the genera *Aspergillus* and *Penicillium* are often involved in bioconversions of s-triazines. Certain fungi have also been reported to bring about methylation of hydroxy- and amino groups, and metals. Thus, the fungus *Trichoderma virgatum* effected the methylation of pentachlorophenol, whereas *Penicillium notatum*, *Aspergillus niger* and *Scopulariopsis brevicaulis* methylated arsenic derivatives (McBride and Wolfe, 1971).

Table 26.4. Bioconversion of xenobiotics affected by *Pseudomonas.*

Mode of Action	Species	References
Hydrolysis of carbaryl, dichlorvos, diazinon, parathion	*Pseudomonas melophthora*	Matsumura and Boush, 1968
Hydrolysis of parathion	*Pseudomonas stutzeri*	Munnecke and Hsieh, 1976
Dehalogenation of halide acetate	*Pseudomonas* sp.	Goldman and Milne, 1966
Total dehalogenation of DDT, aromatic ring cleavage	*Pseudomonas aeruginosa*	Golovleva *et al.*, 1980 and Golovleva *et al.*, 1981
Total dehalogenation of 3-chlorobenzoate	*Pseudomonas putida*	Golovleva *et al.*, 1984
Oxidative dehalogenation of lindane	*Pseudomonas putida*	Matsumara *et al.*,1976
Reduction of nitro group in 4,6-dinitro o-cresol	*Pseudomonas* sp.	Tewfik and Evans, 1966
Total degradation of 2,4,5-T	*Pseudomonas sepacia*	Chatterjee *et al.*, 1982
Degradation of toluene, xylenes, styrenes, l-methyl styrenes	*Pseudomonas aeruginosa*	Golovleva *et al.*, 1976 Williams and Worsey, 1976

ROLE OF MICROORGANISMS IN DETOXIFICATION OF PESTICIDES

The important role played by microorganisms in the degradation of pesticide residues is well recognized. Microorganisms degrade 25 to 70% of the xenobiotic residues. Literature carries reports on microbial strains capable of degrading such recalcitrant pesticides as DDT, polychlorobiphenyls, dicofol, 2,4,5-T and others (Golovleva *et al.,* 1981; Chatterjee and A.M. Chakrabarty, 1983). Unfortunately, very few xenobiotics can be utilized by pure cultures as the sole source of carbon. Those, which can, are fairly simple compounds such as dalapon, chlorobenzoates, 2,4-D of greater significance are bioconversions of pesticides affected by mixed cultures when the initial degradation steps are brought about by one culture, and metabolites produced are further degraded by other microbial strains. One example is the hydrolysis of parathion by *Pseudomonas stutzeri* to produce nitrophenol, which is further degraded by another strain, *Pseudomonas aeruginosa* (Munnecke and Hsieh, 1976).

Exchange of genetic information by natural microbial populations also appears to be a potent factor facilitating total degradation of xenobiotics. For example, one hypothesis states that phenoxyalkane carboxyherbicides, such as 2,4-D and MCPA, do not accumulate in the environment exactly because plasmids, which several years ago were found to encode their degradation, can be transferred under natural conditions from strain to strain (Don and Pemberton, 1981). Recently, plasmids responsible for biodegradation of polychlorobiphenyls, chlorobenzoates, petroleum, Surfactants and other compounds were discovered. They apparently play an important role in degradation of numerous foreign compounds by the natural microflora.

GENETIC ENGINEERING: THE GREATEST APPROACH OF ALL TIME

Genetic engineering is the modern concept for degradation or detoxification of environmental malaise. Microorganisms are efficient in degrading recalcitrant pesticides. Development of recombinant technology has now made it possible to identify and characterize catabolic genes responsible for their degradation. Several pesticide degrading microbes have been identified and genetically manipulated for the construction of hybrid pathways leading to their complete mineralization, characterization and cloning of catabolic genes for compounds such as HCH, Parathion, 2,4-D and polychlorinated biphenyls (Lal *et al.,* 1995).

Restriction Endonucleases

The discovery of restriction endonuclease gave a major breakthrough in the genetic engineering. Endonucleases are enzymes that produce internal cuts called cleavage in DNA molecules. Many endonucleases cleave DNA at random sites, but a class of endonuclease cleaves DNA only within or near those sites which have specific base sequences. Such endonucleases are known as restriction endonucleases and the sites recognized by them are called recognition sequences or recognition sites. The recognition sequences are different and specific for the different restriction endonucleases or restriction enzymes.

The presence of restriction enzyme was postulated by W. Arbur during 1960s while the first true restriction endonuclease was isolated in 1970. Smith, Nathans and Arbur were awarded the Nobel Prize for physiology and medicine in 1978 for the discovery of endonucleases. Restriction endonucleases are indispensable for DNA cloning and sequencing. They serve as the tools for cutting DNA molecules at predetermined sites, which is the basic requirement for gene cloning or recombinant DNA technology.

Types of Restriction Endonucleases

There are three distinct types of endonucleases

- Type I
- Type II
- Type III

Type I Restriction Endonucleases

These are complex endonucleases and have recognition sequence of about 15 base pair (bp). They cleave the DNA about 1000 bp away from 5'-end of the sequence 'TCA' located within recognition sites or EcoR, EcoB, etc.

Type II Restriction Endonucleases

Type II restriction endonucleases are remarkably stable and induce cleavage either, in most cases, within their recognition sequence or very close to them. More than 350 different types of II endonucleases are known. They require Mg^{+2} ions for cleavage. The first type II endonuclease isolated was Hind II in 1970. Only type II restriction endonucleases are used for restriction mapping and gene cloning in view of their cleavage only at specific sites.

Type III Restriction Endonucleases

Type III restriction endonucleases are intermediate between the Type-I and Type-II enzymes. They cleave DNA in the immediate vicinity of their recognition sites, e.g., EcoP15, Hind FIII, etc. Type I and Type II restriction enzymes are not used in gene cloning.

Table 26.5. Restriction endonucleases, their occurrence and their sites of cuts at nucleotide sequences.

Restriction endonucleases	Source	Recognition sequences
Ava I	*Anabaena variables*	C / Py CG Pu G G Pu GC Py/ C
Bam HI	*Bacillus amyloliquefaciens* H	G / G A T C C C C T A G/ G
Eco RI	*Escherichia coli* RY13	G / A A T T C C T T A A/ G
Hae III	*Haemophilus aegyptius*	GG/ CC CC/ GG
Hind II	*Haemophilus influengae* Rd	G T Py/ Pu A C C T Pu/ Py T G
Hind III	*Haemophilus influengae* Rd	A/ A G C T T T T C G A /A
Mbo II	*Moraxella bovis*	G A A G A N_8/ C T T C T N_7/
Pst I	*Prodencia stuartii*	C T G C A/G G/ A C G T C
Sau 3AI	*Staphylococcus aureus 3A*	/G A T C C T A G/
Sma I	*Serratia marcessens*	C C C/ G G G G G G/ C C C
Xma I	*Xanthomonas malvacearum*	C/ C C G G G G G G C C/ C

Joining of DNA Molecules

DNA fragments can be joined to create artificially recombinant molecules. At present there are three methods for joining of DNA fragments *in vitro*. The first of these capitalize on the ability of DNA ligase to join covalently and anneal cohesive ends produced by certain restriction enzymes. The second depends upon the ability of the DNA ligase from Phage T4 infected *E. coli* to catalyse the formation of phosphodiester bonds between blunt ended fragments. The third utilizes the enzyme terminal deoxynucleotidyl transferase to synthesize homopolymeric 3'-single-stranded trails at the end of fragments.

DNA Ligase

E. coli and Phage T4 encode an enzyme, DNA ligase, which steals single-stranded nicks between adjacent nucleotides in a duplex DNA chain. Although the reactions catalysed by the enzymes *E. coli* and T4 infected *E. coli* are very similar, they differ in their co-factor requirements. When terminal created by the restriction endonucleases that create cohesive ends associate, the joint has nicks, a far base pairs apart in appropriate strands. DNA ligase can then repair these nicks to form an intact duplex. The ligation reaction can be performed so as to favour the formation of recombinants. First the population of recombinant can be increased by performing the reaction at high DNA concentration in dilute solutions. Circulation of linear fragments is relatively because of the reduced frequency of intermolecular reactions. Second, by treating linearised plasmid vector DNA with alkaline phosphate to remove 5'- terminal phosphate groups; both circulation and plasmid dimmer formation are prevented. In this case circularization of the vector can occur only one 5'-terminal phosphate at each join. One nick at each join remains unligated, but after transformation of host bacteria, cellular repair mechanisms reconstitute the intact duplex.

STEPS IN GENE CLONING

Identification and isolation of the desired gene or DNA fragment to be cloned

⇩

Insertion of the isolated gene in suitable vector

⇩

Introduction of the vector into a suitable organism/cell called host (transformation)

⇩

Selection of the treated host cell

⇩

Multiplication/expression/integration followed by expression of the introduced gene in the host construction and characterization of a new cloning vehicle pBR 322.

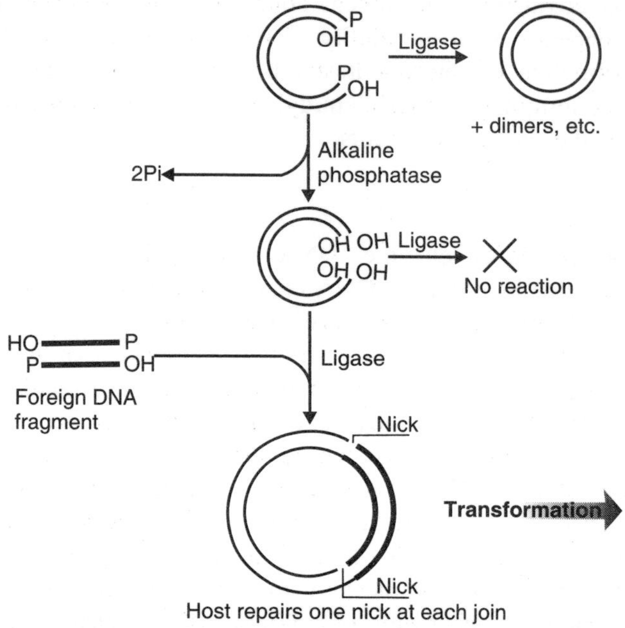

Fig. 26.1. Insertion of a required foreign gene in vector.

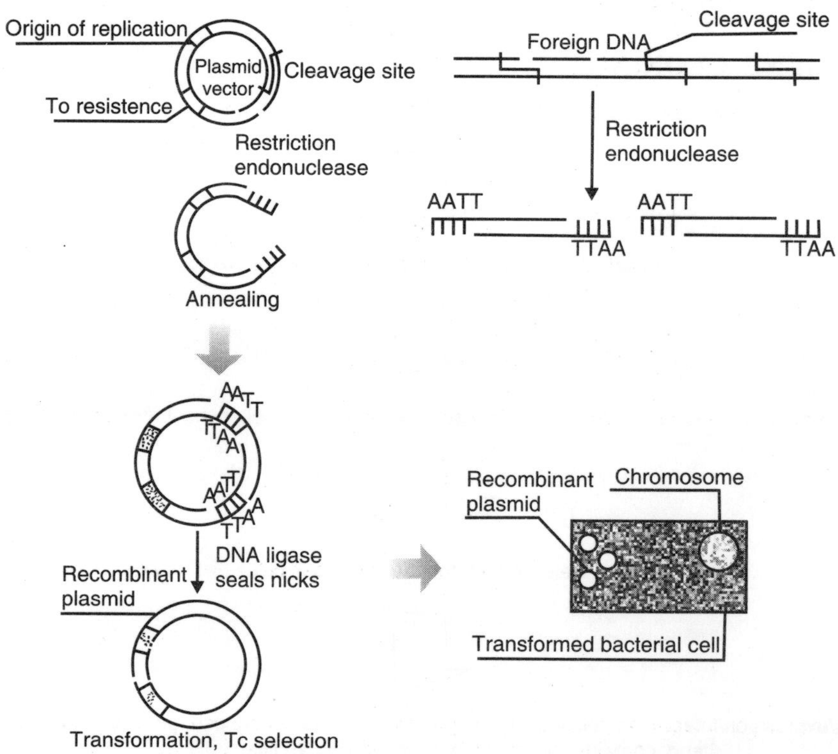

Fig. 26.2. Insertion of a required foreign gene in vector and transformation of it in bacterial cell.

Construction and Characterization of a New Cloning Vehicle pBR322

Although pSC101, Col E1 and RSF 2124 can be used to clone DNA, they suffer from a number of disadvantages. For this reason considerable effort has been expended on constructing, *in vitro*, superior cloning vehicles. Undoubtedly, the most versatile and widely used of this artificial plasmid vector is pBR 322 containing the AP^R and TC^R genes of RSF 2124 and pSC 101, respectively combined with replication element pMB 1 a Col E1 like plasmid.

Plasmid pBR 322 has been completely sequenced. The original published sequence was 4362 base pairs long. Position of the sequence was arbitrarily set between the A and T residues of the EcoR I recognition sequence (GAA TTC). The sequence was revised by the inclusion of an additional CG base pair at position 526, thus increasing the size of the plasmid to 4363 bp. Plasmid pBR 322 is the most widely used cloning vehicle. In addition it has been widely used as a model system for the study of prokaryotic transcription and translocation as well as investigation of the effect of topological changes on DNA confirmation.

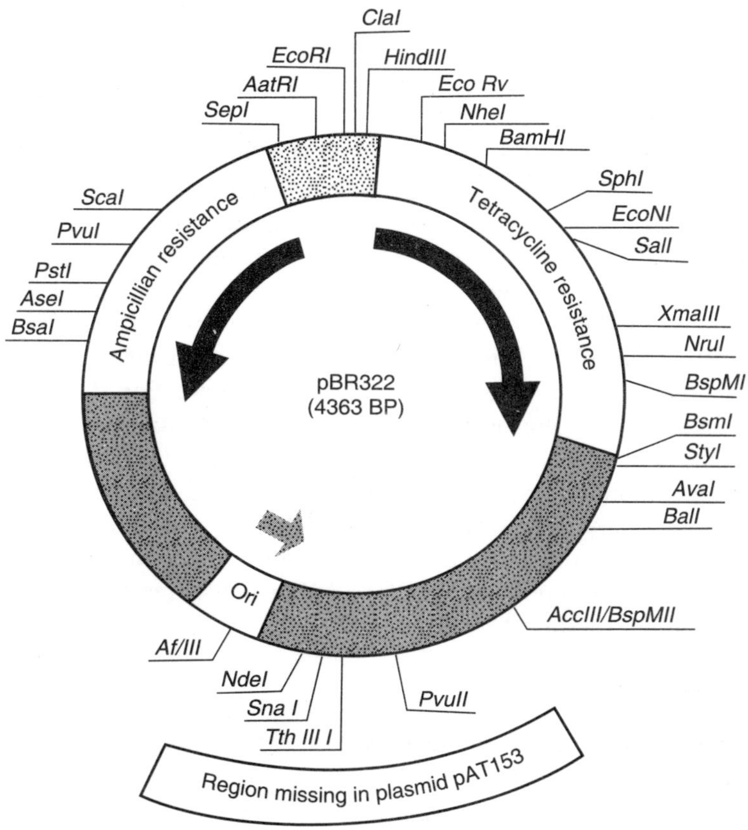

Fig. 26.3. The structure of pBR 322 showing the unique cleavage sites.

Degradation of Pesticides in the Light of Genetic Modification of Microorganisms

The recalcitrant xenobiotics in the environment can cause long-term harm. The development of recombinant DNA technique therefore brought a new hope and made it possible to identify and characterize the catabolic genes responsible for their degradation. Isolation of several pesticide degrading microbes have paved the way for genetic manipulation of microbes for the construction of pathways leading to complete mineralization. Degradation of some of these molecules is presented here.

Degradation of Lindane

Only a few microorganisms are known to degrade the chlorinated hydrocarbon lindane. Ohisa and Yamagachi (1979) reported that lindane undergoes anaerobic degradation by *Clostridium* sp. They also isolated the organism and gave degradative pathway for lindane.

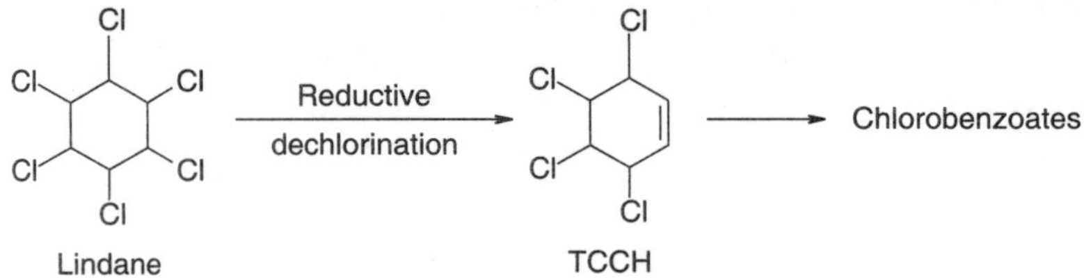

Fig. 26.4. Degradation pathway of lindane.

Lindane is extremely persistent in upland soil. Matsumara *et al.,* 1976 found that *Pseudomonas putida* ATCC 17484 can degrade lindane through hydrochlorination, while Imai *et al.*, 1989 reported that the *P. paucimobilis* SS86 strain degrades the HCH by dehydrochlorination. Under aerobic condition, *Pseudomonas* sp. not only metabolizes the HCH, but also produces thermodynamically more stable β- and α-isomer (Sahu *et al.*, 1991). Singh and Kuhad (2000) studied the degradation of lindane by two rot fungi and reported that *Cyathus bulleri* is more efficient in degradation of HCH than *Phanerochaete sordida*. Imai *et al.* (1991) cloned the specific gene responsible for dehydrochlorination of lindane. For the purpose he used *P. paucimobilis* UT 26 strain and prepared a genomic library by using broad host range cosmid vector pKS13. As much as 2300 clones were screened and only three clones were found to be responsible for degradation of lindane. 500 bp fragments of the DNA are responsible for dehydrochlorination, and only open reading frame of 465 bp was found within 500 bp, which was reported to encode HCH dehydrochrinase, which confers dehydrochlorination yielding 1,2,4-TCS. The gene *Lin A* encoded a polypeptide of 17.3 KDa consisting of 155 amino acid residues were identified from this experiment.

Degradation of Parathion

Munnecke and Hsieh (1976) studied the degradation of parathion by mixed microbial culture.

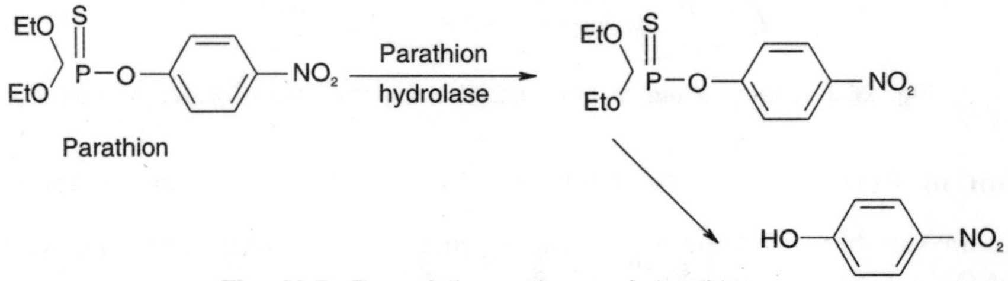

Fig. 26.5. Degradation pathway of parathion.

From their studies the major metabolites were found to be p-nitrophenol (PNP) and diethyl phosphoric acid.

With the development in identification and cloning of genes for parathion hydrolase, the construction of engineered microbes for its detoxification became feasible. In 1982, Serder *et al.,* observed that parathion hydrolase activity in *Pseudomonas dimunta* MG was plasmid encoded, which is pcMS1with 80 kb. Later it was found that *Flavobacterium* sp. ATCC 27551 contained a 39 kb plasmid pPDL2 involved in parathion degradation (Mulbry *et al.,* 1986, 94).

Degradation of 2,4-D

Several strains in multiple genera capable of degradation of 2,4-D have been isolated from soil (Amy *et al.,* 1985). The most thoroughly studied strain among these is *Alcaligens eutrophus* (Den and Pemberton, 1981). For degradation, the genes that encode the degradative pathways for 2,4-D are plasmid encoded. Ghosal *et al.* (1985) isolated the plasmid pJP4 of 80 kb. Don *et al.*, (1985) mapped the catabolic functions of plasmid pJP4 by transposan mutagenesis in a region corresponding to BamH1C fragment. The Hind III B fragment of P EML 159 of *Alcaligens* sp. was then cloned (Amy *et al.,* 1985) into *E.coli* HB101 and demonstrated the release of radiolabelled CO_2 from cultures grown with 2,4-D labelled on the alkyl side chain. This includes the cloning of TFD-monooxygenase. Don *et al.* (1985) reported that plasmid containing the EcoR1B fragment cloned into a curved derivative of JMP 134 enabled the strain to grow 2,4-D as a sole carbon source. The tfd A gene has been cloned as subfragment of the Hind III B fragment of pJP4.

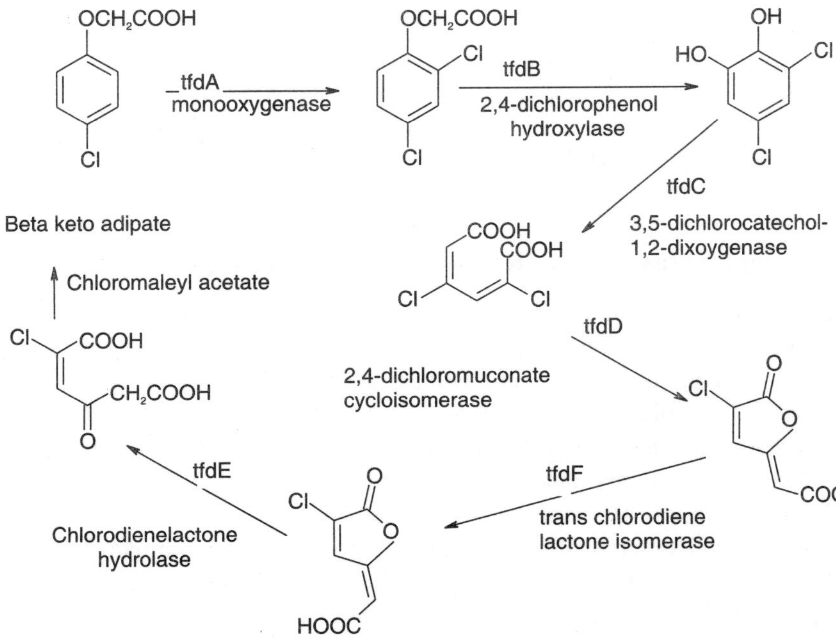

Fig. 26.6. Degradation pathway of 2,4-D.

Degradation of 2,4,5-T

2,4,5-T is poorly biodegradable and persists for a long period in the environment. Karns *et al.* (1984) reported a combination of strong selective pressure and the introduction of plasmid gene pools for the evolution of new pathways, which has allowed the isolation of a strain *Pseudomonas cepacia* that can utilize 2,4,5-T as its sole source of C and energy.

Haugland *et al.* (1990) transferred 2,4,-D degradative plasmid pJP4 from *Alcaligens eutrophus* JMP 134 to *P. cepacia* AC1100. The transformed strains (RHJI) completely metabolized both 2,4-D and 2,4,5-T and efficiently degraded one of these two. Sarodkar *et al.* (1999) isolated a cosmid clone bank of *P. cepacia* AC1100 into mutant PT 88 (CHQ), a hybrid cosmid pus 1 carrying a 25 kb insert, which complemented the mutation of 2,4,5-T (CHQ). The specific region of DNA for restoration of CHQ phenotype was mapped and further identified as to a 4.2 kb segment.

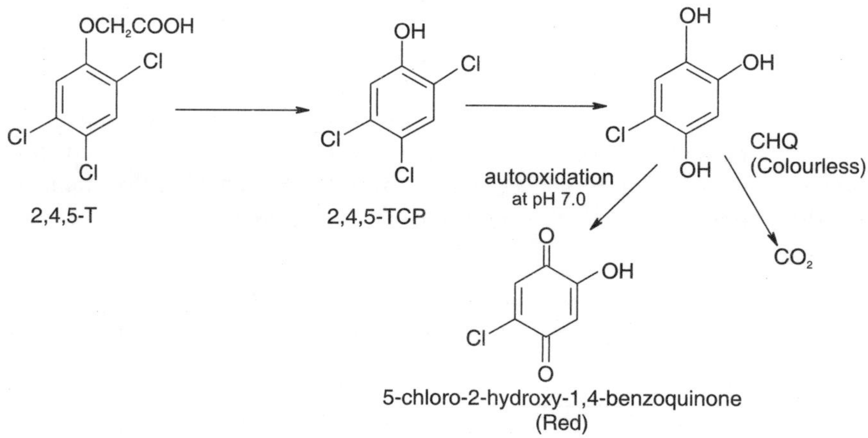

Fig. 26.7. Degradation pathway of 2,4,5-T.

Peroxidases, Phenoloxidases, and Related Oxidoreductases

Plants and microorganisms produce a wide range of oxidative enzymes (e.g., peroxidase, polyphenoloxidase, laccase, and tyrosinase) other than P450s that catalyze the polymerization of various anilines and phenols (Dec and Bollag 2001). For example, peroxidase-mediated pesticide transformations in plants that function similar to P450s include decarboxylation, sulphur oxidation, *N*-demethylation, ring hydroxylation, and aromatic methyl group oxidations (Lamoureux and Frear, 1979). In plants, peroxidase enzymes often function in Phase III metabolism, e.g., formation of bound residues. Horseradish (*Amorocia lapathifolia* Gilib.) roots contain large quantities of peroxidase. Horseradish root tissue has been used to remove 2,4-dichlorophenol from water and was more effective in contaminant removal than the purified peroxidase enzyme (Dec and Bollag, 2001). White rot fungi (*Phanerochaete chryosporium*) offer high potential for xenobiotic transformation because they possess free radical–based lignin degrading systems (lignin peroxidase and manganese-dependent peroxidases) that can degrade a wide range of pollutants such as polychlorinated biphenyls (PCBs) and nitroaromatic explosives (Barr and Augst, 1994). In most instances, polymerization products have reduced toxicity compared with the substrate (Dec and Bollag, 2001), whereas polymerization of 3,4-dichloroaniline (propanil metabolite) by soil microorganisms results in the formation of carcinogenic tetrachloroazobenzene (Pothuluri *et al.* 1991) (Fig. 26.8). Generally, polymerization products are considered to be unextractable humic components (Dec and Bollag, 2000).

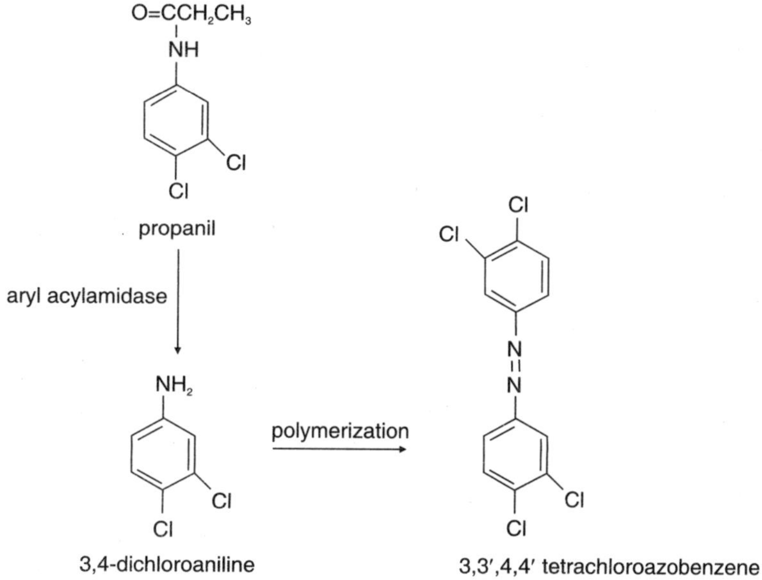

Fig. 26.8. Amide hydrolysis of propanil by aryl acylamidase in plants and microorganisms and subsequent polymerization by microbial peroxidases.

Oxidative Nitroaromatic Transformations

In microorganisms, as opposed to plants, numerous enzymes from many different pathways are capable of oxidizing nitroaromatic compounds, and in many cases the enzymes have been purified, and the genes cloned and sequenced (see review by Zablotowicz *et al.*, 2001). Oxidative reactions that transform various nitroaromatic compounds from several genera of aerobic bacteria have been described (Kadiyala and Spain, 1998; Leung *et al.*, 1997; Zablotowicz *et al.*, 1999). In bacteria, monooxygenases, flavin monooxygenases, and dioxygenases are generally involved in the initial oxidation of nitroaromatic pesticides, e.g., 2,4-dinitrophenol can be metabolized by these three enzymes (Cassidy *et al.*, 1999) (Fig. 26.9). Depending on the compound, nitrite can be released before or after ring cleavage. The flavin monooxygenase from *Sphingomonas* strain UG30 is responsible for initial nitrite removal from the herbicide 4,6-dinitrocresol, but not dinoseb (4,6-dinitro-*o*-*sec*-butylphenol), because of the steric hindrance caused by the bulky butyl group of dinoseb (Zablotowicz *et al.*, 1999). Overall, these diverse microbial nitroaromatic degradative pathways allow for ring hydroxylation, ring cleavage, and subsequent mineralization of several xenobiotics (e.g., nitrobenzene, nitrobenzoic acid, nitrophenols, and nitrotoluene) (Zablotowicz *et al.*, 2001).

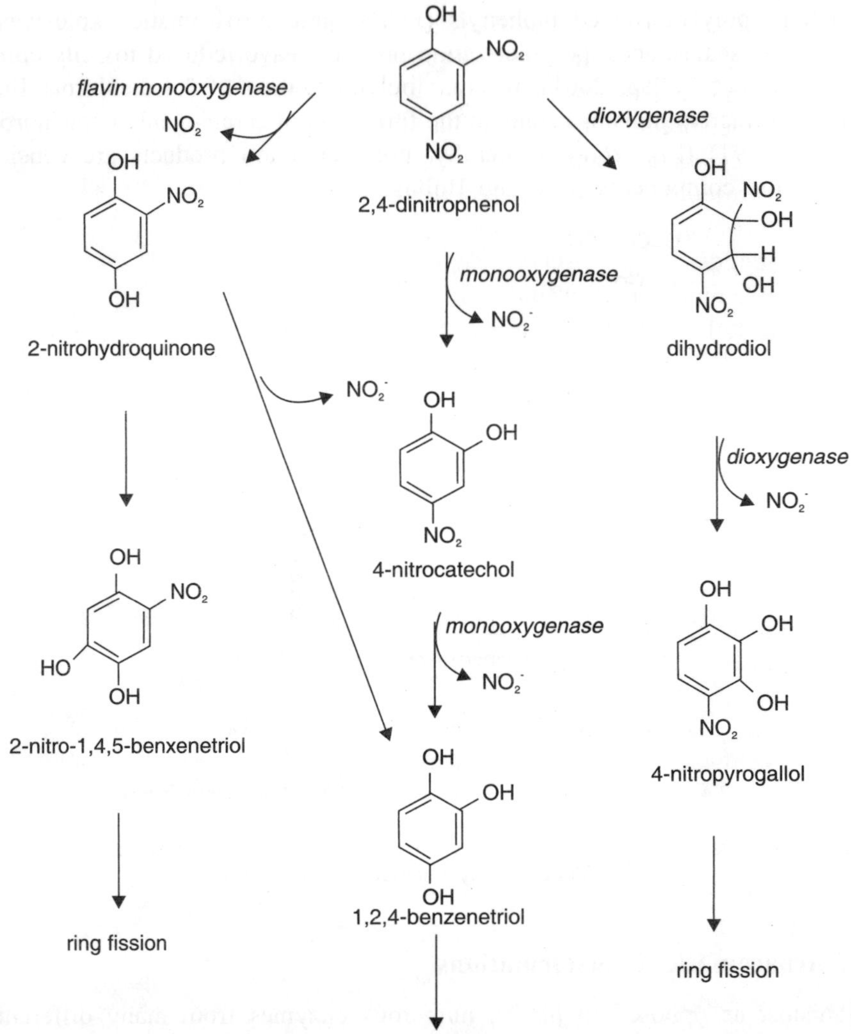

Fig. 26.9. Biotransformation of 2,4-dinitrophenol in *Spingomas sp.* UG30.

Degradation of s-Triazines

Atrazine was traditionally considered to be moderately persistent in soil; however, in the past several years, many bacterial strains representing several genera have been isolated that can completely mineralize atrazine (Sadowsky and Wackett, 2001). A unique operon of genes encoding for *s*-triazine degradation has evolved in areas where this herbicide has been used extensively. The gene regions encoding the first three enzymes of atrazine degradation have been isolated and characterized from *Pseudomonas* sp. strain ADP (Boundy-Mills *et al.*, 1997; de Souza *et al.*, 1995, 1996; Sadowsky *et al.*, 1998). This bacterium mineralizes high concentrations (500 mg L21) of atrazine under both growth and non-growth conditions, using the herbicide as the sole nitrogen source (Mandelbaum *et al.,* 1995). The *atzA* gene encodes atrazine chlorohydrolase, which dechlorinates atrazine hydrolytically to the non-phytotoxic metabolite hydroxyatrazine (Fig. 26.10). The next step in the degradation pathway is hydrolytic removal of the aminoethyl group from hydroxyatrazine by the *atzB* gene product, hydroxyatrazine ethyl amidohydrolase. Finally, the *atzC* gene encodes for another amidohydrolase that converts *N*-isopropylammelide to cyanuric acid. Martinez *et al.* (2001) have recently sequenced the complete catabolic plasmid from strain ADP and have identified three additional genes *atzD, atzE,* and *atzF* encoding for cyanuric acid amidohydrolase, biuret hydrolase, and allophanate hydrolase. Thus, the total genetic basis for the complete atrazine metabolism in strain ADP has now been identified. Many soil bacteria have the capability to mineralize cyanuric acid (Cook, 1987; Cook and Hutter, 1981; Erickson and Lee, 1989; Korpraditskul *et al.,* 1993). Five other atrazine-degrading bacteria with *s*-triazine–degrading genes have been identified with nearly 99% homology to *atzABC* from *Pseudomonas* sp. strain ADP, suggesting that horizontal transfer of atrazine degradation genes may have occurred recently (de Souza *et al.,* 1998a, 1998b). In fact, in *Pseudomonas* sp. strain ADP, the three *atz* genes are on a self-transmissible plasmid pADP-1 (de Souza *et al.,* 1998b). Many lines of evidence suggest the ability of microorganisms to mineralize *s*-triazines developed after the first use of these herbicides in the mid-1950s (Sadowsky and Wackett, 2001). In contrast to bacteria, atrazine and other *s*-triazines are metabolized in plants via *N*-dealkylation by cytochrome P450s, hydrolytic dehalogenation, or displacement of chlorine with glutathione (GSH) (Lamoureux *et al.,* 1998). In microorganisms, there have been no reports of GSH conjugation resulting in dechlorination of s-triazines (Zablotowicz *et al.,* 1994).

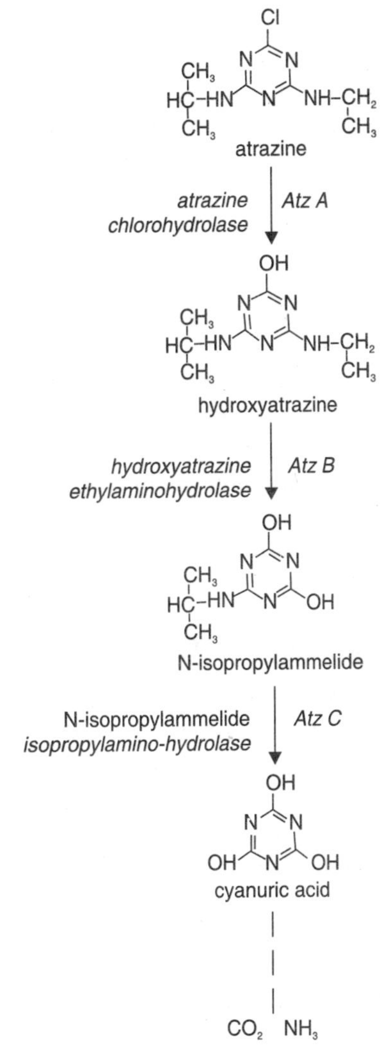

Fig. 26.10. Initial degradation pathway of atrazine by *Pseudomonas* sp. ADP.

PLASMID MEDIATED MINERALIZATION OF NAPHTHALENE, PHENANTHRENE AND ANTHRACENE

Sanseverino *et al.* (1993) described the plasmid mediated mineralization of naphthalene, phenanthrene and anthracene. For this plasmid encoded naphthalene degradation pathway in *P. putida*, PpG7 (NAH7) was used to investigate the role of NAH plasmid encoded pathway in mineralizing phenanthrene and anthracene. Three *Pseudomonas* strains, designated 5R, DFC49 and DFC50 were recovered from a polynuclear aromatic hydrocarbon degrading inoculum from a manufacturer gas plant slurry reactor. From these plasmids pKA1, pKA2 and pK3 approximately 100 kb size were isolated and characterized. By conjugation experiments, these plasmids including NAH7 have been shown to encode the genotype for mineralization of ^{14}C phenanthrene and ^{14}C anthracene as well as ^{14}C naphthalene. One strain, *Pseudomonas fluorescence* 5RL has the complete lower pathway inactivated by transposon insertion in NAHG accumulated a metabolite from phenanthrene and anthracene degradation.

Polychlorinated Biphenyl Degradation

Khan and Walia (1991) reported cloning of genes of *Pseudomonas putida* strains in the plasmid vector pUC19. The resultant hybrid plasmid pAW6194 contained cpbABCD genes on a 9 kb DNA fragment necessary for the catabolism of polychlorinated biphenyls.

These genes were further subcloned on an 8.0 kb Hind III fragment of pAW540. Degradation of 3-chlorobiphenyl, 2,4-dichorophenyl and 2,4,5-trichorobiphenyl into a chloro derivative of benzoic acid was found in *E. coli* harbouring chimeric plasmid pAW540.

Expression of cbpABCD genes of *P. putida* OU83 was driven by its own promoters located on the cloned DNA of pUC19 and was examined by subcloning an 8.0 kb DNA pKK232-8. The resulting recombinant plasmid pAW560 and pAW561 expressed cbpABCD genes and conferred chroramphenicol resistance only in *E.coli* pAW560.

Variations in Chlorbenzoate Catabolism of *Pseudomonas putida*

Brenner *et al.* (1990) reported that *P. putida* is able to utilize a broad range of monochlorinated and trichlorinated benzoates. The involvement of two separate dioxygenases was noted from data on plasmid profiles and DNA hybridization. The benzoates dioxygenase, which convert 3-chlorobenzoate (3-CB), 4-CB and benzoate to the corresponding catechol via reduction of dihydrodiol, was shown to be chromosomally coded. The chlorobenzoate-1,2- dioxygenase that converts ortho-chorobenzoates to the corresponding catechols without the need of functional dioldehydrogenase was shown to be encoded on plasmid pPb 111 (75 kb). Cured strains were unable to utilize o-chlorobenzoates for growth. DNA hybridization data indicated that catabolism of the corresponding chlorocatechols was coded on chromosomal genes. An unique variant of p111 able to grow on 3,5-dichlorobenzoate (3,5-DCB) was obtained by continuous subculturing from media containing progressively lower and higher concentrations of 3-CB and 3,5-DCB, respectively. The low frequency of segregants able to grow on 1,5-DCB, 2,3-DCB and 2,3,5-trichlorobenzoate was evident by lag periods greater than 200 hr. Continued subculture on 3,5-DCB resulted in the formation of new plasmid pPH111 (120 kb), which was homologous to pPB 111.

Catechol Catabolic Genes from *Rhodococcus erythropolis* 1 cp

Eulberg *et al.* (1997) biochemically characterized the muconate and the choromuconate cycloisomerase strain of the chlorophenol utilizing *Rhodococcus erythropolis* 1 cp. Based on the sequences of the N-terminus and tryptic peptides of the muconate cycloisomerase, a fragment of the corresponding gene has been amplified and used as a probe for the cloning of catechol catabolic genes from *Rhodococcus erythropolis.* The clone thus obtained expressed catechol 1,2-dioxygenase muconate cycloisomerase and mucobolactone isomerase activities. Sequencing of the insert on the recombinant plasmid pRER1 revealed that the genes are transcribed in the order Cat A, Cat B, Cat C. Open reading frames downstream of Cat C may have function in carbohydrate metabolism. The predicted protein sequence of the catechol 1,2 dioxygenase was identical to the one from *Arthrobacter* sp. strain mA3 in 59% of the positions. The chlorocatechol 1,2-dioxygenase and the chloromuconate cycloisomerase of Gram-negative bacteria appear to be more closely related to the catechol 1,2-dioxygenase and muconate cycloisomerase of the Gram-positive strains than to the corresponding enzymes of Gram-negative bacteria.

CONCLUSION

The recalcitrant xenobiotics can cause long-term harm in the environment. The development of recombinant DNA technique has brought a new hope to eliminate this nuisance. The advent of recombinant DNA technologies has now made it possible to identify and clone the catabolic genes from different organisms. Isolation of several pesticide degrading microbes has paved the way for genetic manipulation of microbes for the construction of pathways leading to complete mineralization. The gene responsible for degradation of the compounds has been cloned and further attempts are being made to characterize genes responsible for complete mineralization of these molecules. Characterization and cloning of catabolic genes for many pesticides such as parathion, 2,4-D and 2,4,5-T and PCB have already been completed. Further research may lead to the development of other strains with better abilities to degrade or detoxify several other hazardous compounds. General awareness, commercialization of the technology and a compact database in this regard may fulfil the requirement of the need of all.

REFERENCES

Amy, P.S., Schulke, J.W., Frazier, L.M. and Seidler, R.J. (1985). Characterization of aquatic bacteria and cloning genes specifying partial degradation of 2,4-dichlorophenoxy acetic acid. *Appl. Environ. Microbiol.* **49:** 1237-1245.

Barr, D.P. and Augst, S.D. (1994). Pollutant degradation by white rot fungi. *Rev. Environ. Contam. Toxicol.* **138:** 49-72.

Boundy-Mills, K.L., de Souza, M.L., Mandelbaum, R.T., Wackett, L.P. and Sadowsky, M.J. (1997). The atzB gene of Pseudomonas sp. strain ADP encodes the second enzyme of a novel atrazine degradation pathway. *Appl. Environ. Microbiol.* **63:** 916-923.

Bremmer, V., Hernandez, B.S. and Focht, D.D. (1993). Variation in chlorobenzoate catabolism by *Pseudomonas putida* p111 as a consequence of genetic alterations. *Appl. Environ. Microbiol.* **59**(9): 2790-2794.

Cassidy, M.B., Lee, H., Trevors, J.T. and Zablotowicz, R.M. (1999). Chlorophenol and nitrophenol metabolism by *Sphingomonas* sp. UG30. *J. Ind. Microbiol. Biotechnol.* **23:** 232-241.

Chatterjee, D.K. and Chakrabarty, A.M. (1983) *J. Beteriol.* **153:** 532-534.

Chatterjee, D.K., Kilbane, J.Y. and Hakrabarty, A.M. (1982) *Appl. Environ. Microbiol.* **45:** 1697-1700.

Chudyk, W. (1989). Field screening of hazardous waste sites. *Environ. Sci. Technol.* **23:** 503-507.

Cook, (1987). Biodegradation of s-triazine xenobiotics. *FEMS Microbiol. Rev.* **46:** 93-116.

Cook, A.M. and Hutter, R. (1981). s-triazines as nitrogen sources for bacteria degradation of herbicides. *J. Agric. Food Chem.* **29:** 1135-1143.

de Souza, M.L., Seffernick, J., Martinez, B., Sadowsky, M.J. and Wackett, L.P. (1998a) The atrazine catabolism gene atzABC are widespread and highly conserved. *J. Bacteriol.* **180:** 1951-1954.

de Souza, M.L., Wackett, L.P. and Sadowsky, M.J. (1998b) The atzABC genes encoding atrazine catabolism are located on a self-transmissible plasmid in Pseudomonas sp. strain ADP. *J. Appl. Environ. Microbiol.* **64:** 2323-2326.

de Souza, M.L., Wackett, L.P., Boundy-Mills, K.L., Mandelbaum, R.T. and Sadowsky, M.J. (1995). Cloning, characterization, and expression of a gene region from Pseudomonas sp. strain ADP involved in the dechlorination of atrazine. *J. Appl. Environ. Microbiol.* **61:** 3373-3378.

de Souza, M.L., Sadowsky, M.J. and Wackett, L.P. (1996). Atrazine chlorohydrolase from Pseudomonas sp. strain ADP: gene sequence, enzyme purification and protein characterization. *J. Bacteriol.* **178:** 4894-4900.

Dec, J. and Bollag, J.M. (2001). Use of enzymes in bioremediation. Pages 182–193 in J.C. Hall, R.E. Hoagland, and R.M. Zablotowicz, (Eds. Van Eerd *et al.*). Pesticide metabolism x 489 Pesticide Biotransformation in Plants and Microorganisms: Similarities and Divergences. *ACS Symposium Series 777*. Washington, DC: American Chemical Society.

Don, R.H. and Pemberton, J.H. (1981). Properties of six pesticide degradation plasmids isolated from *Alcaligens paradoxis* and *Alcaligens eutrophus*. *J. Bacteriol.* **145:** 681-683.

Don, R.H. and Pemberton, J.M. (1985). Genetic and physical map of 2,4-D degradative plasmid pJP4. *J. Bacteriol.* **161:** 466-468.

Don, R.H., Weightman, A.J., Knackmass, H.J. and Timmis, K.N. (1985). Transposon mutagenesis and cloning analysis of the pathway of degradation of 2,4-D and 3-chlorobenzoate in *Alcaligens eutrophus* JMP 134 (pJP4). *J. Bacteriol.* **141:** 85-90.

Engelhardt, G., Wallenofer, P.R. and Plapp, R. (1973). *Appl. Microbiol.*, **26:** 709-718.

Erickson, L.E. and Lee, K.H. (1989) Degradation of atrazine and related s-triazines. *Crit. Rev. Environ. Control* **19:** 1-14.

Eulberg, D., Ludmila, A. and Scholman, M. (1997). Characterization of catechol catabolic genes from *Rhodococcus erythropolis*. *J. Bacteriol.* **179**(2): 370-381.

Field, J.A., de Jong, E., Feijoo Costa, G. and De Bont, J.A.N. (1993). Screening for ligninolytic fungi applicable to the biodegradation of xenobiotics. *Trends in Biotechnology* **11:** 44-49.

Ghosal, D., You, T.D., Chatterjee, D.M. and Chakraborty, A.M. (1985). Genes specifying degradation of 3-chlorobenzoic acid in plasmids pAC27 and pJP4. *Proc. Nat. Acad. Sci. USA* **82:** 1638-1642.

Goldman, P. and Milne, W.A. (1966) *J. Biol. Chem.* **241:** 5557-5559.

Goldman P., Milne, G.W. and Keister, D.B. (1968) *J. Biol. Chem.* **243:** 428-434.

Golovleva, L.A. and Skryabin, G.K. (1981). Microbial Degradation of Xenobiotics and Recalcitrant Compounds (Leisinger, T., Cook, A. and Hutter, R. Eds.) London, N.Y., 291-298.

Golovleva, L.A., Ilyaletdinov, A.M., Alieva, R.M. and Dzhusupova, D.B. (1984). USSR patent No. 1117316, BI No. 37.

Golovleva, L.A., Zyakun, A.M., Baskunov, B.P., Pertsova, R.N., Skryabin G.K. and Izvestiya, A.N.S.S.S.R. (1980). *Ser. Biol.* **2:** 243-253.

Golovleva, L.A., Golovlev, E.L., Ganbarov, K.G. and Skryabin, G.K. (1976). *Mikrobiologiya,* **46:** 5-10.

Golovleva, L.A., Aharonson, N., Greenhalgh, R., Sethunathan, N. and Vonk, J.W. (1990). The role and limitations of microorganisms in the conversion of xenobiotics *Pure & Appl. Chem.* **62**(2): 351-364.

Golovleva, L.A., Finkelshtein, Z.I., Popovich, N.A., Skryabin, G.K. and Izvestiya, A.N. S.S.S.R. (1981). *Ser. Biol.* **3:** 348-357.

Haugland, R.A., Sangodkar, U.M.X. and Chakraborty, A.M. (1990). Repeated sequences including RS 11100 from *Pseudomonas cepacia* AC1100 function as IS elements. *Mol. Gen. Genet.* **220:** 1295-1297.

Imai, R., Nagata, Y., Fukuda, M. and Yano, K. (1991). Molecular cloning of *Pseudomonas paucimobilis* gene encoding 17 KD polypeptide that eliminates HCl molecules from γ-HCH. *J. Bacteriol.* **173:** 6811-6819.

Kadiyala, V. and Spain, J.C. (1998) A two-component monooxygenase catalyzes both the hydroxylation of p-nitrophenol and the oxidative release of nitrite from 4-nitrocatechol in *Bacillus sphaericus* JS905. *Appl. Environ. Microbiol.* **64:** 2479-2484.

Karns, J.S., Kilbane, J.J. and Chakraborty, A.M. (1984). Microbial biodegradation of 2,4,5-T and Chlorophenols In: Genetic control of environmental pollutants. (Eds. Omen and Hollaeender, A.). Plenum Press, New York, pp. 3-21.

Khan, A.A. and Walia, S.K. (1991). Expression, localization and functional analysis of polychlorinated biphenyl degradation genes cbpABCD *Pseudomonas putida*. *Appl. Environ. Microbiol.* **57**(5): 1325-1332.

Korpraditskul, R., Katayama, A. and Kuwatsuka, S. (1993) Degradation of atrazine by soil bacteria in the stationary phase. *J. Pestic. Sci.* **18:** 293-298.

Lal, R., Dhanraj, P.S. and Saxena, D.M. (1995). Genetic manipulation of microorganisms for degradation of pesticides In: Pesticides, Crop Protection and Environment. (Eds. Walia, S. and Parmar, B.S.) Oxford and IBH publishing Co Pvt. Limited, New Delhi, pp. 263-277.

Lamoureux, G.L. and Frear, D.S. (1979) Pesticide metabolism in higher plants: in vitro enzyme studies. Pages 77-128 in Paulson, (G.D., Frear, D.S. and Marks, E.P. Eds.). Xenobiotic Metabolism: *In Vitro* Methods. *ACS Symposium Series 97*. Washington, DC: American Chemical Society.

Lamoureux, G.L., Simoneaux, B. and Larson, J. (1998). The metabolism of atrazine and related 2-chloro-4,6-bis (alkylamino)-s-triazines in plants. Pages 60-81 in (Ballantine, L.G., McFarland, J.E. and Hackett, D.S. Eds.). Triazine Herbicides: Risk Assessment. *ACS Symposium Series 68*. Washington, DC: American Chemical Society.

Leung, K.T., Cassidy, M.B., Shaw, K.W., Lee, H., Trevors, J.T., Lohmeier-Vogel, E.L. and Vogel, H.J. (1997). Pentachlorophenol biodegradation by *Pseudomonas* spp. UG25 and UG30. *World J. Microbiol. Biotech.* **13:** 305-313.

Mandelbaum, R.T., Allan, D.L. and Wackett, L.P. (1995). Isolation and characterization of a Pseudomonas sp. that mineralizes the s-triazine herbicide atrazine. *Appl. Environ. Microbiol.* **61:** 1451-1457.

Martinez, B., Tompkins, J., Wackett, L.P., Wing, R. and Sadowsky, M.J. 492 x Weed Science 51, July–August 2003, 2001. Complete nucleotide sequence and organization of the atrazine catabolic plasmid pADP-1 from *Pseudomonas* sp. strain ADP. *J. Bacteriol.* **183:** 5684-5697.

Matsumara, F., Benexet, H.J. and Patil, K.C. (1976). Factors affecting microbial metabolism of γ-BHC. *J Pestic Sci.* **1:** 3-8.

Matsumura, F. and Boush, M. (1968). *J. Econom. Entomol.* **61:** 610-612.

Matsumura, F., Benezet, H. and Patil, K.C. (1976). *J. Pestic. Sci.*, **1:** 3-8.

McBride, B.C. and Wolfe, R.S. (1971) *Biochemistry,* **10:** 4312.

Mulbry, W.W., Kearney, P., Nelson, J.O. and Karns, J.S. (1987). Physical comparison of parathion hydrolase plasmids from *Pseudomonas dimunita* and *Flavobacterium* sp. *plasmid.* **18:** 173-177.

Munnecke, D. and Hsieh, D. (1976). *Appl. Environ. Microbiol.*, **31:** 63-69.

Munnecke, D.M. and Hsieh, D.P.H. (1976). Pathway of microbial metabolism of parathion. *Appl. Environ. Microbiol.* **31:** 63-39.

Ohisa, N. and Yamagachi, M. (1978). γ-BHC degradation accompanied by the growth of *Clostridium rectum* isolated from paddy field soil. *Agric. Biol. Chem.* **42:** 1819-1823.

Pothuluri, J.V., Hinson, J.A. and Cerniglia, C.E. (1991). Propanil: toxicological characteristics, metabolism, and biodegradation potential in soil. *J. Environ. Qual.* **20:** 330-347.

Sadowsky, M.J., Tong, Z., deSouza, M. and Wackett, L.P. (1998). AtzC is a new member of the amidohydrolase protein superfamily and is homologous to other atrazine-metabolizing enzymes. *J. Bacteriol.* **180:** 152-158.

Sadowsky, M.J. and Wackett, L.P. (2001). Genetics of atrazine and s-triazine degradation by Pseudomonas sp. strain ADP and other bacteria. Pages 268-282 in (Hall, J.C., Hoagland, R.E. and Zablotowicz, R.M. Eds.). Pesticide Biotransformation in Plants and Microorganisms: Similarities and Divergences. *ACS Symposium Series 777.* Washington, DC: American Chemical Society.

Sahu, S.K., Patnaik, K.K., Sharmila, M. and Sethunathan, N. (1991). Degradation of alpha, beta and gamma HCH by soil bacterium under aerobic conditions. *Appl. Environ. Microbiol.* **199:** 3620-3622.

Sangodkar, U.M.X., Aldrich, T., Haugland, R.A., Johnston, J., Rothmel, R.K., Chapman, P.J. and Chakraborty, A.M. (1989). Molecular basis of biodegradation of chloroaromatic compounds. *Acta Biotechnol* **9:** 301-316.

Sanseverino, J., Applegate, B.M., Henry King, J.M. and Sayler, G.S. (1993). Plasmid mediated mineralization of naphthalene, phenanthrene and anthracene. *Appl. Environ. Microbiol.* **59**(6): 1931-1937.

Serder, C.M., Gibson, D.T., Munnece, D.M. and Lancaster, J.H. (1982). Plasmid involvement in parathion hydrolysis by *Pseudomonas dimunita. Appl. Environ. Microbiol.* **44:** 246-249.

Singh, B.K. and Kuhad, R.C. (2000). Degradation of insecticide lindane by white rot fungi *Cythus bulleri* and *Phanerchaete sordida. Pest Management Science,* **56:** 142-146.

Tewfik, M.S. and Evans, W.S. (1966). Proceedings of the Biochemical Society, *Biochem. J.*, **99**(2): 31-32.

Tu, C.M. (1976). Utilization of degradation of lindane by soil microorganisms. *Arch. Microbiol.* **108:** 259-263.

Tucker, S.P. and Carson, G.A. (1985). Deactivation of hazardous chemical wastes. *Environ. Sci. Technol.* **19**(3): 215-220.

Williams, P. and Worsey, M.J. (1976). *J. Bacteriol.* **125:** 818-828.

Zablotowicz, R.M., Leung, K.T., Alber, T., Cassidy, M.B., Trevors, J.T., Lee, H., Veldhuis, L. and Hall, J.C. (1999). Degradation of 2,4-dinitrophenol and selected nitroaromatic compounds by Sphingomonas sp. UG30. *Can. J. Microbiol.* **45:** 840-848.

Zablotowicz, R.M., Hoagland, R.E. and Locke, M.A. (1994). Glutathione S-transferase activity in rhizosphere bacteria and the potential for herbicide detoxification. Pages 184-198 in (Anderson, T.A. and Coats, J.R. Eds.). Bioremediation through Rhizosphere Technology. *ACS Symposium Series 563.* Washington, DC: American Chemical Society.

Zablotowicz, R.M., Hoagland, R.E., Lee, H., Alber, T., Trevors, J.T., Hall, J.C. and Locke, M.A. (2001). Transformation of nitroaromatic pesticides and related xenobiotics by microorganisms and plants. Pages 194-216 in (Hall, J.C., Hoagland, R.E. and Zablotowicz, R.M. eds.). Pesticide Biotransformation in Plants and Microorganisms: Similarities and Divergences. *ACS Symposium Series 777.* Washington, DC: American Chemical Society.

27
Nematodes as Bioinsecticides against Vegetables Pests

Gitajali Devi

Division of Entomology, ICAR Research Complex for NEH Region, Umium, Barapani, Meghalaya–793 103

Vegetable crops suffer a great loss due to ravages of insect pest, thus causing a serious threat to our agricultural production. Introduction of high yielding hybrids/varieties, off-season production, use of high fertilizer and pesticides and monocropping in vegetable cultivation resulting in development of differential susceptibility to insect attack, cases of minor pests assuming major pest status, resistance problem among insects to insecticides, residue hazards of pesticides and thereby upsetting of balance in nature. Considering all these drawbacks, many components of the IPM concept were developed in the late nineteenth and early twentieth centuries and biopesticides and bioagents were placed as an important tool in pest management strategies. Insect-parasitic nematodes help farmers by providing 'biological control' of soil-dwelling insect pests. These nematodes occur naturally in the soil, and have been found associated with most of the insect orders. There are natural associations between insects and nematodes involving 11 orders and more than thirty families of nematodes and 19 orders, 100 families, more than 250 species of insects. The association may range from a phoretic relationship to obligate entomoparasitism and include host death, sterility, reduced fecundity, delayed development, or aberrant behavior.

TAXONOMY AND BIOLOGY

The phylum Nematoda is divided into two classes: Adenophorea and Secernentia, both of which contain important insect-parasitic nematodes. In the class Adenophorea, parasitic nematodes have been found in the family Mermithidae and in the class Secernentea in the families Steinernematidae and Heterorhabditidae. Most research and development has focused on nematode species in seven families: Mermithidae, Tetradonematidae, Allantonematidae, Phaenopsitylenchidae, Sphaerulariidae, Steinernematidae and Heterorhabditidae. Three genera of nematodes, *Hexamermis* sp., *Steinernema* spp., and *Heterorhabditis* spp are currently receiving the most attention as microbial control agents of insects. Till now about 56 species of *Steinernema,* one species of *Neosteinernema* and 10 species of *Heterorhabditis* have been listed. They are simple roundworms with long, cylindrical shaped bodies, ranging in length from 0.4 mm to 1.1 mm. They are ubiquitous, having been isolated from every inhabited continent from a wide range of ecologically diverse soil habitats including cultivated fields, forests, grasslands, deserts, and even ocean beaches. Although some entomopathogenic nematodes have been isolated from insects naturally infected in the field, they are most commonly recovered from soil by baiting with susceptible insects. The parasitic cycle of nematodes is initiated by the third stage infective juveniles. These non-feeding juveniles locate and invade suitable host insects through natural body openings (i.e. anus, mouth, and spiracles) or even through the cuticle. Once inside the host, infective juveniles invade the hemocoel and release a symbiotic bacterium, which is held in the nematode's intestine. All species of *Steinernema* are associated with *Xenorhabdus* bacteria and all *Heterorhabditis* nematode species are associated with *Photorhabdus* bacteria. The bacteria cause septicemia, killing the host within 24-48 h. The infective juveniles feed on the rapidly multiplying bacteria and disintegrated host tissues. Steinernematid infective juveniles may become males or females, whereas heterorhabditids develop into self-fertilizing hermaphrodites, although subsequent generations within a host produce males and females as well. Under ideal conditions, infective juveniles of Steinernematid emerge from the cadever for 6-11 days and of Heterorhabditids 12-14 days after infection. About 2-3 generations of the nematode are completed within the host cadaver. When food reserves are depleted, nematode reproduction ceases and the offsprings develop into resistant infective juveniles, which disperse from the dead host, and are able to survive in the environment and to seek out new hosts. Insecticidal nematodes use one of two methods to locate their hosts—ambushing or cruising. Ambusher nematode species sit and wait for a suitable host by nictating, i.e., they stand on their tails, waiting to attack a host as it comes by. Since infective juveniles are only a couple of millimeters long, a host that is even a few millimeters away will not be attacked. In addition, these nematodes are most effective in soil near the surface, where spaces between the soil particles are sufficient to allow this nictating behavior. *Steinernema carpocapsae* and *S. scapterisci* are examples of ambusher nematodes. *S. carpocapsae* is sold under the trade names Biosafe, Savior, Millenium, and BioVector 25. Cruiser nematodes actively search out their hosts. Although they can nictate, the infective juveniles also tunnel through the soil looking for their hosts. These cruisers are attracted to carbon dioxide and other chemicals that are exuded by potential hosts. Due to their searching behavior, cruisers are more effective than ambushers against less active insects and those that live deeper in the soil. *Heterorhabditis bacteriophora* and *Steinernema glaseri* are examples of cruiser nematodes. Some species of insecticidal nematodes are intermediate in activity between ambushers and cruisers, including *S. feltiae* and *S. riobravis*. *S. feltiae* is sold under trade name Nemasys, Entonem, X-Gnat, and Magnet. Unlike most insecticidal nematodes, it is infective against insects in soil temperatures below 50°F. *S. riobravis* is sold as BioVector 355 and Devour. In the laboratory, *S. riobravis* has been shown also to be effective against mole crickets. It is active in drier soils than other insecticidal nematodes, but it needs high soil temperatures. It is most effective at temperatures above 95°F. *H. bacteriophora* is commonly called the Hb nematode in trade literature and is sold under the trade names Cruiser, Heteromask, and Gardens Alive Hb Nematodes. This nematode has a shorter shelf-life than most others. The bacteria in this nematode is luminescent, which causes killed insects to glow slightly in the dark. Killed insects turn brick red instead of the medium brown color typical of insects killed by other nematode species.

MASS PRODUCTION, FORMULATION AND APPLICATION

Entomopathogenic nematodes can be mass-produced by *in vivo* or *in vitro* methods. The wax moth, *G. mellonella* larvae are most commonly used to rear nematodes. Using the *in vivo* process, yields have been obtained between 0.5×10^5–4×10^5 infective juveniles per larva, depending on the nematode species. Nematodes are cultured on a crumbed polyether polyurethane sponge impregnated with emulsified beef-fat and pig's kidneys along with symbiotic bacteria. Using this method ("solid" culture), approximately 6×10^5–10×10^5 infective juveniles/g of medium have been achieved. In liquid fermentation technique, costs of production decrease rapidly up to a capacity of approximately 50×10^{12} infective juveniles/month. The current yields of *S. carpocapsae* in liquid culture average at about 2.5×10^5 infective juveniles/g. *S. carpocapsae, S. riobravis, S. scapterisci, S. feltiae, S. glaseri, H. bacteriophora*, and *H. megidis* have been produced successfully in large-scale liquid cultures.

The shelf-life of different nematode-based products varies depending on the formulation, nematode species and temperature. In the simplest type of formulation, the nematodes are impregnated onto moist carrier substrates providing substantial interstitial spaces leading to increased gas exchange. Such carriers include polyether polyurethane sponge, cedar shavings, peat, vermiculite, activated charcoal, calcium alginate, etc. Nematodes held on the sponge need to be hand-squeezed into water before application, whereas from the other carriers they may be applied directly to the soil. The alginate-based *S. carpocapsae* products possess room temperature shelf-life of about 3-4 months, and is gaining acceptability of nematodes in the markets. Another water dispersible granule formulation in which infective juveniles are encased in 10-20 mm diameter granules consisting of mixtures of various types of silica, clays, cellulose, lignin and starches is also gaining acceptability of nematodes in the markets.

Insecticidal nematodes may be applied through sprayers and irrigation systems. Often it is desirable to tank mix one or more inputs to save time and money. Infective juveniles are tolerant of short exposures (2-6 h) to most agrochemicals including herbicides, fungicides, acaricides, and insecticides and therefore, can often be tank-mixed. However, some pesticides can reduce nematode infectivity and survival. Heterorhabditids tend to be more sensitive to physical challenges, including pesticides, than steinernematids. Nematodes are also compatible with most inorganic fertilizers. It is recommended that these sprayer-nozzle screens be removed before application. Nematodes can survive tank agitation. Storage time for the water-suspension formulation may be limited to only a few days under refrigeration due to the oxygen needs of the infective juveniles. The gel formulation is usually soaked in water to remove the infective juvenile-containing gel from the screen matrix. The majority of applications involving the use of entomopathogenic nematodes are in inundative biological control. They have been most efficacious for insects in soil or cryptic habitats where there is protection from rapid desiccation and UV radiation. It is best to apply nematodes to moist soil in the early morning or late evening when air temperatures are between 60 and 85°F. A pre-application irrigation can be applied to moisten the soil and a post-application irrigation can be applied to wash any nematodes on plant surfaces to the soil surface. Although nematodes are generally applied as curative treatments, prophylactic applications to the soil surrounding seedlings and seeds have been advocated.

EFFECTIVENESS

Numerous insect pests of economic importance are encountered in vegetable crop production. Being high value crops, the introduction of biological pest control agents such as entomopathogenic nematodes has stimulated great interest worldwide on both above and below ground pests. Entomopathogenic nematodes are useful against a large number of insect pests of vegetable crops Table 27.1.

Table 27.1. Entomopathogenic nematodes against vegetable crop pests.

Crop	Insect	EPN	Reference	Place
Vegetable	Insect pests of vegetable	Entomopathogenic nematodes	Welch H.E., Briand L.J. 1961	Ontario
	Lepidoptera, Hemiptera, Thysanoptera, Coleoptera, Diptera, mites and snail	*N. feltiae* (*N. carpocapsae*)	Xu J.L. 1983	China
	Spodoptera litura	*N.carpocapsae* (DD-136)	Kondo E., Ishibashi N. 1983	Japan
	cutworms (*Agrotis segetum*), (*A. ipsilon*) (*A. puta*)	*Steinernema*. sp.	Caballero P., Vargas Osuna E., Santiago Alvarez C. 1989	Andalucia and Extremadura regions of Spain
	diamondback moth, greenhouse whitefly *Thrips palmi*, common cutworm, army worm	*S. feltiae* (DD-136)	Yoshihara T. 1991	Japan
	diamondback moth, (*Plutella xylostella*)	*S.carpocapsae* and *H.bacteriophora*	Abdel Razek AS2003	Gharb area of Morocco
Cabbage	cabbage worm [*Artogeia rapae*], purple backed cabbage moth [*Evergestis pallidata* (Hfn.)], cabbage root maggot [*Delia radicum* (L.)]	*H.* sp. and *N.* sp	Finney J.R. 1983	Canada
	black cutworms (*Agrotis ipsilon*)	*N. carpocapsae* (strains Mexican and Kapow), *S.bibionis* [*N. bibionis*], *H.heliothidis*	Capinera J.L., Epsky N.D., Menout G.S., Pelissier D. 1988	Colorado
	cabbage webworm (*Hellula undalis*), cabbage pyralid (*Crocidolomia binotalis*)	*Steinernema carpocapsae* (A11 strain)	Hsiao W.F 1997	Taipei

Table Contd.

Table 27.1. Contd.

Crop	Insect	EPN	Reference	Place
	cabbage butterfly, (*Pieris brassicae*) flies (*Musca* and *Chrysomyia* sp.) red thrips, *Hylopterus pruhi*	*N. carpocapsae* DD-136	Kurashvili B.E., Kakuliya G.A., Gurgenidze T.V. 1987	India
	cabbage worm *Artogeia rapae* (*Pieris rapae*), diamondback moth *Plutella xylostella*, cabbage looper *Trichoplusia ni.*	*Steinernema carpocapsae* All, *S. feltiae* UK, *S. feltiae* 27, and *S. riobrave* 335	Belair G; Fournier Y; Dauphinais N. 2003	
	cabbage maggot [*Delia radicum* (L.)]	*Steinernema feltiae* Filipjev, *S. bibionis* (Bovien), *Heterorhabditis bacteriophora* Poinar, and *H. heliothidis* (Khan, Brooks, and Hirschmann)	Bracken G.K. 1990	
	cabbage maggot	entomopathogenic nematodes	Schroeder P.C., Ferguson C.S., Shelton A.M., Wilsey W.T., Hoffman M.P., Petzoldt C. 1996	
	cabbage root fly (*Delia radicum* L)	*Steinernema feltiae*	Vanninen I; Vainio A; Jaakkola S.1992	
	diamondback moth (*Plutella xylostella*)	*S. carpocapsae* *S. riobravis, S. feltiae*	Ratnasinghe G; Hague NGM 1997	Pakistan
	cabbage root fly (*Delia radicum*)	*Neoaplectana bibionis*	Hommes M. 1988	Germany
	cabbage root maggot, (*Hylemya brassicae*), cabbage fly	entomogenous nematodes Entomopathogenic nematodes	Georgis R., Poinar G.O. Jr., Wilson A.P. 1983. Danilov L.G., Karpova E.V., Shilovskaya T.S. 1995	Russia
Cauliflower	*Pieris rapae crucivora*	*Steinernema feltiae*	Wu H.J., Chow Y.S. 1989	
	cabbage moth (*Barathra brassicae)*	Neoaplectana carpocapsae agriotos	Zhilyaeva I.N., Ezhov G.I., Kondratov E.S. 1973	Russia
	diamondback moth (*Plutella xylostella*)	*Heterorhabditis indicus* [*Heterorhabditis indica*], (*Steinernema glaseri*)	Subramanian S. 2002	India
Cabbage and Japanese radish	*Plutella xylostella*	Entomopathogenic nematodes	Baur M.E., Kaya H.K., Gaugler R., Tabashnik B. 1997	
	lepidopterous insect	*Steinernema feltiae*	Inque T. 1990	Japan

Contd. Table

Table 27.1. Contd.

Crop	Insect	EPN	Reference	Place
Potato	Colorado potato beetle (*Leptinotarsa decemlineata*)	Entomopathogenic nematodes	Welch H.E. 1958	Ontario
	Colorado potato beetle (*Leptinotarsa decemlineata*)	Entomopathogenic nematodes	Fedorko A1971	Hungari
	Colorado beetle	*Neoaplectana carpocapsae*	Salmanov A.A. 1988	Russia
	potato moth (*Phthorimaea operculella*)	*Steinernema carpocapsae* Agriotos	Stepanova N.E., Perera M. 1993	Russia
	Colorado potato beetle, (*Leptinotarsa decemlineata*)	*H.* species *S.* species	Berry RE; Liu J; Reed G1997	
	Colorado potato beetle (*Leptinotarsa decemlineata*)	*Neoaplectana carpocapsae* and *Heterorhabditis* spp	Fodor A., Sáringer G., Georgis R. 1989	Russia
	black cutworms (*Agrotis ipsilon*)	*Steinernema feltiae* [*Neoaplectana carpocapsae*] 'A11'	Georgis R., Wojcik W.F., Shetlar D.J. 1989	
	black cutworms (*Agrotis ipsilon*) (Hufnagel) and *A. segetum*	*Steinernema bicornutum* (isolates PDBC 3.1, PDBC 3.2, and PDBC 2.1), *S. carpocapsae* (PDBC 66.1, PDBC 13.1, and PDBC 6.11), *Heterorhabditis indica* (PDBC 6.71 and PDBC 13.3)	Hussaini SS; Singh SP; Parthasarathy R; Shakeela V2000	India
	black cutworms (*Agrotis ipsilon*) (Hufnagel) and *A. segetum*	entomopathogenic nematodes	Hussaini SS; Singh SP; Parthasarathy R; Shakeela V2001	India
	potato moth (*Phthorimaea operculella*)	*Steinernema feltiae, S. bibionis, S. carpocapsae or Heterorhabditis heliothidis*	Ivanova TS; Borovaya VP; Danilov LG A1994	
	Colorado potato beetle (*Leptinotarsa decemlineata*)	*S. carpocapsae*, 'All' strain	Stewart JG; Boiteau G; Kimpinski J. 1998	New Brunswick and Prince Edward Island

Contd. Table

Table 27.1. Contd.

Crop	Insect	EPN	Reference	Place
Turnip	turnip moth	entomopathogenic *nematode*	Tojo S. 1993	Japan
	turnip moth *(Agrotis segetum)*	*Steinernema carpocapsae*	Yokoo S., Tojo S., Ishibashi N. 1992	Japan
	turnip root fly (*Delia floralis*)	*Neoaplectana carpocapsae*	Hellqvist S., Johansson L., Burman M. 1988	Sweden
	Agrotis ipsilon, *Spodoptera litura*	*Steinernema* (=*Neoplectana*) *glaseri*	Vyas R.V., Yadav D.N 1992	Ludhiana
	Coleopterans, lepidopterans	Entomopathogenic nematodes	Capinera J.L., Epsky N.D. 1992	Caribbean
Sugarbeet	Conorhynchus mendicus	*Steinernema carpocapsae* *S. feltiae*	Akalach M; Wright DJ.,1995	
	sugarbeet weevil (*Temnorrhinus mendicus* Gyll.)	*Heterorhabditis*, strain NH-HL81	Boselli M., Curto G.M., Lazzari G.L. 1991	Italy
	sugarbeet weevil *Temnorhinus* (= *Conorrhynchus) mendicus Gyll.*	*Heterorhabditis* sp. (NL-HL81 strain), *H. bacteriophora* (HP 88 strain) *Steinernema carpocapsae* ('All')	Boselli M., Curto GM, Tacconi R. 1997	Italy
	sugarbeet weevil (*Temnorhinus mendicus*)	*Heterorhabditis* sp., *H. bacteriophora, Steinernema carpocapsae*	Tacconi R., Boselli M., Curto GM, Santi R. 1999	
	sugarbeet weevil *(Temnorhinus mendicus) Gyll*	*Heterorhabditis sp., H. bacteriophora, S.carpocapsae*	Tacconi R., Boselli M., Curto GM., Santi R. 1998	
	sugarbeet weevil *Temnorhinus* (= *Conorrhynchus*) *mendicus* Gyll.	*H.* sp. (NL-HL81 strain), *H. bacteriophora* (HP 88 strain) and *S. carpocapsae* ("All" strain).	Curto GM; Boselli M; Tacconi R; Gwynn RL Smits PH, Griffin C, Ehlers R.U, Boemare N, Masson J.P. 1999	Italy
Carrot	carrot weevil	*Steinernema feltiae, S. bibionis* and *Heterorhabditis heliothidis*	Bélair, G., and G. Boivin. 1985.	Quebec
	carrot weevil, *Listronotus oregonensis* (Le Conte)	*Steinernema feltiae* Filipjev, DD 136 and Listronotus	Boivin G., Bélair G. 1989	
	carrot weevil	*Steinernema* Listronotus strain	Bélair G., Anderson R.V., Boivin G. 1989	

Contd. Table

Table 27.1. Contd.

Crop	Insect	EPN	Reference	Place
	carrot weevil (*Listronotus oregonensis*)	*Steinernema carpocapsae, S. riobrave, S. feltiae, Heterorhabditis megidis, H. bacteriophora*	Miklasiewicz TJ; Grewal PS; Hoy CW; Malik VS2002	Great Lakes
		entomopathogenic nematodes	Philipsen, Holger and Nielsen, Otto, 2003	
Cucumbers	*Acalymma vittatum*	*Steinernema riobravis* [*Steinernema riobrave*]	*Ellers Kirk CD;* Fleischer SJ; Snyder RH; Lynch JP2000	
	Spodoptera litura	*Steinernema feltiae* (*Neoaplectana carpocapsae*)	Choi D. R., Ishibashi N., Tanaka K 1988	Japan
Cucurbits	pickleworm *(Diaphania nitidalis)*	*Steinernema carpocapsae*	Webb SE; Capinera JL 1996,	Florida
Onion	onion maggot (*Delia antiqua*)	*Steinernema feltiae* [*Neoaplectana carpocapsae*], *Heterorhabditis heliothidis*	Choo H.Y., Kaya H.K., Reed D.K. 1988	Korea
	Plutella xylostella Chilo suppressalis	*Steinernema* spp Hanrim Sancheong Pocheon *Heterorhabditis* spp Hamyang	Choo Ho Yul; Lee SangMyeong; Chung BuKeun; Park Yeong Do; Kim HyeongHwan; Choo HY; Lee SM; Chung BK; Park YD; Kim HH. 1995	Korea
	Spodoptera frugiperda	*Steinernema feltiae*	Fuxa J. R., F. Agudelo Silva, and A.R. Richter, 1988	
	Heliothis armigera (Helicoverpa armigera)	*Neoaplectana carpocapsae*	Ghode M.K., Mishra R.P., Swain P.K., Nayak U.K., Pawar A.D. 1988	India
	Heliothis armigera	entomoparasitic nematodes	Glazer I., Navon A. 1989	
	Leafminer (*Liriomyza trifolii*)	entomopathogenic nematodes	Hara, AH; Kaya, HK; Gaugler, R; Lebeck, LM; Mello, CL. 1994.	
	Liriomyza trifolii	*Steinernema carpocapsae*	Harris M.A., Begley J.W., Warkentin D.L. 1990	

Contd. Table

Table 27.1. Contd.

Crop	Insect	EPN	Reference	Place
Sweet potato	sweet potato weevil (*Cylas formicarius*)	*H.bacteriophora*	Jansson R.K., Lecrone S.H., Gaugler R. 1991	Southern Florida
	sweet potato weevil (*Cylas formicarius elegantulus*)	*H.bacteriophora Poinar (= H. heliothidis Khan, Brooks, & Hirschmann) (HP88 strain) and Steinernema (= Neoaplectana) carpocapsae Weiser (All and G-13 strains)*	Jansson R.K., Lecrone S.H., Gaugler R., Smart G.C. Jr. 1990	
	sweet potato weevil	entomopathogenic nematodes	Jansson, RK; Lecrone, SH; Gaugler, R. 1993	
	Cylas formicarius	*Steinernema carpocapsae All strain and Heterorhabditis bacteriophora HP88 strain*	Jansson RK; Lecrone SH 1997	North Carolina
	sweet potato weevil *Cylas formicarius*	*S.carpocapsae, Agriotos, All, Breton, Italian, and Mexican strains; S. feltiae, N27 strain; S. intermedia; H. bacteriophora, HP88, and North Carolina strains; and an unidentified H. sp., FL2122 strain)*	Mannion CM; Jansson RK 1992	
	sweet potato weevil, *Cylas formicarius* (F.)	*H. bacteriophora strain HP88 and H. sp. strain FL2122 (isolated in Florida) S. carpocapsae, S. feltiae strain N-27 and S. glaseri.*	Mannion CM; Jansson RK 1992	
	Cylas formicarius	*Steinernema carpocapsae*, S. *feltiae (All, Mexican and N-27 strains), S. glaseri, Heterorhabditis bacteriophora HP88 strain* and *Heterorhabditis* sp. *FL2122 strain (an undescribed nematode isolated in Florida*	Mannion CM; Jansson R K. 1993	
Beet	*Spodoptera litura*, beet armyworm *Spodoptera exigua*	*Steinernema glaseri. Neoplectana carpocapsae*	Josephrajkumar A; Sivakumar CV 1997	India

Contd. Table

Table 27.1. Contd.

Crop	Insect	EPN	Reference	Place
	cutworm (*Spodoptera litura)*	*Steinernema feltiae, S. bibionis and S. glaseri*	Kaya, H.K. & B.J. Grieve. 1982 Kondo E. 1989	Japan
	Spodoptera litura	*Steinernema feltiae [Neoaplectana carpocapsae], S. bibionis [N. bibionis] and S. glaseri [N. glaseri]*	Kondo E., Ishibashi N1988	Japan
	Spodoptera litura	*Steinernema feltiae [Neoaplectana carpocapsae] (DD-136)*	Kondo E1987	Japan
	Spodoptera litura	*Steinernema feltiae [Neoaplectana carpocapsae] (DD-136)*	Narayanan K., Gopalakrishnan C. 1987	India
Brinjal	brinjal root weevil, (*Myllocerus discolor* Boh)	*Steinernema glaseri and Heterorhabditis sp.*	Prabhuraj A; Viraktamath CA; Kumar ARV2000	India
	Holotrichia serrata Spodoptera litura	*Steinernema glaseri Heterorhabditis minutus*	Prabhuraj A; Viraktamath CA; Kumar ARV; Tandon PL; Ballal CR; Jalali SK; Rabindra RJ2003	India
Sweet potato	banded cucumber beetle (*Diabrotica balteata* LeConte)	*Heterorhabditis heliothidis* (Khan, Brooks, and Hirschman 1976) Poinar, Thomas, and Hess 1977, South Carolina strain,	Schalk J.M., Creighton C.S 1989	
	diamond back moth, (*Plutella xylostella* L)	*Steinernema glaseri* (Steiner), *S. carpocapsae* (Welser), *S. feltiae* (Filipjev), *Steinernema* sp. Ecomax strain, *Heterorhabditis bacteriophora* (Pioner), *Heterorhabditis* sp. Ecomax strain, two locally isolated strains called as JFC and TFC	Shilpa Shinde; Singh NP; Shinde S2000	India
Gram	gram pod borer, (*Helicoverpa armigera*)	*Heterorhabditis* sp	Vyas RV; Patel NB; Pharindera Yadav; Ghelani YH; Patel DJ; Yadav P2003	India
	crucifer flea beetle, *Phyllotreta cruciferae* (Goeze)	*Steinernema feltiae* [*Neoaplectana feltiae*]	Morris O.N. 1987	

Contd. Table

Table 27.1. Contd.

Crop	Insect	EPN	Reference	Place
	African bollworm, *Helicoverpa (Heliothis) armigera* and spotted bollworm *Erias vitella*	*Steinernema riobravis, S. carpocapsae* (ALL isolate) and a *Heterorhabditis* sp	Tahir HI; Otto AA; Hague NGM	
	Heliothis armigera (Hubner)	*Steinernema sp.*	Vela JRG; Lara MP; Lobaton CS; Roca AC1998	California
Squash	squash bug (*Anasa tristis*)	*Neoaplectana carpocapsae*	Wu H.J. 1988	
Spruce	spruce budmoth, *(Zeiraphera canadensis)*	*Steinernema carpocapsae*	Eidt D.C., Dunphy G.B. 1991	
Lettuce	*Agrotis segetum*	*Neoaplectana bibionis*	Lossbroek, T.G. & J. Theunissen. 1985.	

REFERENCES

Abdel Razek, A.S. (2003). Pathogenic effects of *Xenorhabdus nematophilus* and *Photorhabdus luminescens* (Enterobacteriaceae) against pupae of the diamondback moth, *Plutella xylostella* (L.). *Anzeiger fur Schadlingskunde.* **76**(4): 108-111.

Akalach, M. and Wright, D.J. (1995). Control of the larvae of Conorhynchus mendicus (Col.: Curculionidae) by *Steinernema carpocapsae* and *Steinernema feltiae* (Nematoda, Steinernematidae) in the Gharb area (Morocco). *Entomophaga.* **40**(3-4): 321-327.

Baur, M.E., Kaya, H.K., Gaugler, R. and Tabashnik B. (1997). Effects of adjuvants on entomopathogenic nematode persistence and efficacy against *Plutella xylostella. Biocontrol Science and Technology.* **7:** 513-525.

Bélair, G., Anderson, R.V. and Boivin G. (1989). Characteristics of a Listronotus strain of *Steinernema* from a carrot weevil host. *Proceedings of the International Congress of Nematology.* **1:** 9.

Belair, G., Fournierm Y. and Dauphinais, N. (2003). Efficacy of steinernematid nematodes against three insect pests of crucifers in Quebec. *J. Nematology.* **35**(3): 259-265.

Bélair, G. and Boivin, G. (1985). Susceptibility of the carrot weevil (Coleoptera: Curculionidae) to *Steinernema feltiae, S. bibionis* and *Heterorhabditis heliothidis. J. Nematology.* **17:** 363-366.

Berry, R.E., Liu, J. and Reed, G. (1997). Comparison of endemic and exotic entomopathogenic nematode species for control of Colorado potato beetle (Coleoptera: Chrysomelidae). *J. Economic Entomology.* **90**(6): 1528-1533.

Boivin, G. and Bélair, G. (1989). Infectivity of two strains of *Steinernema feltiae* (Rhabditida: Steinernematidae) in relation to temperature, age, and sex of carrot weevil (Coleoptera: Curculionidae) adults. *J. Economic Entomology.* **82**(3): 762-765.

Boselli, M., Curto, G.M. and Lazzari, G.L. (1991). Monitoring of sugar-beet weevil (*Temnorrhinus mendicus* Gyll.) (*Coleoptera Curculionidae*) and preliminary biological control trials by entomogenous. *Informatore Fitopatologico.* **41**(7-8): 55-59.

Boselli, M., Curto, GM. and Tacconi, R. (1997). Field efficacy of entomopathogenic nematodes against the sugarbeet weevil Temnorhinus (=Conorrhynchus) mendicus Gyll. (Coleoptera: Curculionidae). *Biocontrol Science and Technology.* **7**(2): 231-238.

Bracken, G.K. (1990). Susceptibility of first-instar cabbage maggot, *Delia radicum* (L.) (Anthomyiidae: Diptera), to strains of the entomogenous nematodes *Steinernema feltiae* Filipjev, *S. bibionis* (Bovien), *Heterorhabditis bacteriophora* Poinar, and *H. heliothidis* (Khan, Brooks, and Hirschmann). *Canadian Entomologist.* **122**(7-8): 633-639.

Caballero, P., Vargas Osuna, E. and Santiago Alvarez C. (1989). Presence of cutworms on various crops in Andalucia and Extremadura and their associated parasitoids and pathogens. *Boletin de Sanidad Vegetal, Plagas.* **15**(1): 3-7.

Capinera, J.L., Epsky, N.D., Menout, G.S. and Pelissier, D. (1988). Control of black cutworm, Agrotis ipsilon (Lepidoptera: Noctuidae), with entomogenous nematodes (Nematoda: Steinernematidae, Heterorhabditidae) *J. Invertebrate Pathology.* **52**(3): 427-435.

Capinera, J.L. and Epsky, N.D. (1992). Potential for biological control of soil insects in the Caribbean basin using entomopathogenic nematodes. *Florida Entomologist.* **75**(4): 525-532.

Choi, D.R., Ishibashi, N. and Tanaka, K. (1988). Possible integrated control of soil insect pests, soil-borne diseases, and plant nematodes by mixed application of fungivorous and entomogenous nematodes Bulletin of the Faculty of Agriculture, Saga University, Saga 840, Japan. **65:** 27-35.

Choo, H.Y., Kaya, H.K. and Reed, D.K. (1988). Biological control of onion maggot and tobacco cutworm with insect-parasitic nematodes, *Steinernema feltiae* and *Heterorhabditis heliothidis. Korean J. Applied Entomology.* **27**(4): 185-189.

Choo, HoYul., Lee SangMyeong., Chung BuKeun., Park YeongDo., Kim HyeongHwan., Choo, H.Y., Lee, S.M., Chung, B.K., Park, Y.D. and Kim, H.H. (1995). Pathogenicity of Korean entomopathogenic nematodes (Steinernematidae and Heterorhabditidae) against local agricultural and forest insect pests. *Korean J. Applied Entomology.* **34**(4): 314-320.

Curto, G.M., Boselli, M., Tacconi, R., Gwynn, R.L., Smits, P.H., Griffin, C., Ehlers, R.U., Boemare, N. and Masson, J.P. (1999). Field effectiveness of entomopathogenic nematodes against the sugarbeet weevil *Temnorhinus* (=Conorrhynchus) *mendicus* Gyll. (Coleoptera: Curculionidae) in Italy. Cost 819-entomopathogenic nematodes: application and persistence of entomopathogenic nematodes. Proceedings held at Todi, Perugia, Italy, 16 to 20 May 1995. 107-115.

Danilov, L.G., Karpova, E.V. and Shilovskaya, T.S. (1995). Entomopathogenic nematodes against cabbage flies *Zashchita Rastenii Moskva.* **7:** 40.

Ellers Kirk, C.D., Fleischer, S.J., Snyder, R.H. and Lynch, J.P. (2000). Potential of entomopathogenic nematodes for biological control of *Acalymma vittatum* (Coleoptera: Chrysomelidae) in cucumbers grown in conventional and organic soil management systems. *J. Economic Entomology.* **93**(3): 605-612.

Fedorko, A. (1971). Nematodes as factors reducing the populations of Colorado potato beetle, Leptinotarsa decemlineata. *Say Acta Phytopathologica Academiae Scientiarum Hungaricae.* **6:** 175-179.

Finney, J.R. (1983). The feasibility of using nematodes (Heterorhabditis sp. and Neoaplectana sp.) for control of Brassica, apple and blueberry pests Proceedings of the Canadian Pest Management Society. **13:** 39-41.

Fodor, A., Sáringer, G. and Georgis, R. (1989). Application of insect pathogenic nematodes (*Neoaplectana carpocapsae* and *Heterorhabditis* spp.) against Colorado potato beetle (*Leptinotarsa decemlineata*) larvae: Small plot experiments and laboratory trials. *Kiserletek Növényvédelem.* **25**(5): 215.

Fuxa, J.R., Agudelo Silva, F. and Richter, A.R. (1988). Effect of host age and nematode strain on susceptibility of *Spodoptera frugiperda* to *Steinernema feltiae. J. Nematol.* **20:** 91-95.

Georgis, R., Poinar, G.O. Jr. and Wilson, A.P. (1983). Practical control of the cabbage root maggot, *Hylemya brassicae,* (Diptera: Anthomyiidae) by entomogenous nematodes International Research Communications System Medical Science: Microbiology, Parasitology and Infectious Diseases. **11:** 322.

Georgis, R., Wojcik, W.F. and Shetlar, D.J. (1989). Use of *Steinernema feltiae* in a bait for the control of black cutworms (*Agrotis ipsilon*) and tawny mole crickets (*Scapteriscus vicinus*). *Florida Entomologist.* **72**(1): 203-204.

Ghode, M.K., Mishra, R.P., Swain, P.K., Nayak, U.K. and Pawar, A.D. (1988). Pathogenicity of *Steinernema feltiae* Weiser to *Heliothis armigera* Hubner Plant Protection Bulletin (India). **40**(2): 7-8.

Glazer, I. and Navon, A. (1989). Enhancement of the persistence and the efficacy of entomoparasitic nematodes against *Heliothis armigera* (Lepidoptera: Noctuidae) on the plant foliage. *J. Cellular Biochemistry.* **13A:** 172.

Hara, A.H., Kaya, H.K., Gaugler, R., Lebeck, L.M. and Mello, C.L. (1994). Entomopathogenic nematodes for biological control of the leafminer, *Liriomyza trifolii* (Dipt.: Agromyzidae). *Entomophaga* **38**(3): 359-369.

Harris, M.A., Begley, J.W. and Warkentin, D.L. (1990) *Liriomyza trifolii* (Diptera: Agromyzidae) suppression with foliar applications of *Steinernema carpocapsae* (Rhabditida: Steinernematidae) and abamectin. *J. Economic Entomology.* **83**(6): 2380-2384.

Hellqvist, S., Johansson, L. and Burman, M. (1988). *Neoaplectana carpocapsae* (Nematoda) against turnip root fly (*Delia floralis*) - *Växtskyddsrapporter, Jordbruk.* **53:** 44.

Hommes, M. (1988). Alternative possibilities of cabbage root fly control after the efficiency of chlorfenvinphos and carbofuran has begun to fail in some growing areas of the Federal Republic of Germany. *Acta Horticulturae.* **219:** 47-51.

Hsiao, W.F. (1997). Pathogenicity of entomopathogenic nematode, *Steinernema carpocapsae* to cabbage webworm and cabbage pyralid. Plant Protection Bulletin (Taipei). **39**(4): 305-312.

Hussaini, S.S., Singh, S.P., Parthasarathy, R. and Shakeela, V. (2000). Virulence of native entomopathogenic nematodes against black cutworms, *Agrotis ipsilon* (Hufnagel) and *A. segetum* (Noctuidae : Lepidoptera). *Indian J. Nematology.* **30**(1): 103-105.

Hussaini, S.S., Singh, S.P., Parthasarathy, R. and Shakeela, V. (2001). Comparison of bioassays to measure the infectivity of some isolates of entomopathogenic nematodes against *Agrotis ipsilon* (Hufnagel) and *A. segetum* (Schiff). *Current Nematology.* **12**(1-2): 63-68.

Inque, T. (1990). Control of lepidopterous insect pests by *Steinernema feltiae* in the fields of cabbage and Japanese radish. Development of biological integrated control of agricultural pests by beneficial nematodes. Ed. Ishibashi N. (Eds). Ministry of Education, Culture and Science, Japan. Saga University, Japan pgs. 76-78.

Ivanova, T.S., Borovaya, V.P. and Danilov, L.G. (1994). A biological method of controlling the potato moth. *Zashchita rastenii Moskva*. No. **2:** 39.

Jansson, R.K., Lecrone, S.H., Gaugler, R. and Smart, G.C. Jr. (1990). Potential of entomopathogenic nematodes as biological control agents of sweet potato weevil (Coleoptera: Curculionidae) *J. Economic Entomology.* **83**(5): 1818-1826.

Jansson, R.K., Lecrone, S.H. and Gaugler, R. (1991). Comparison of single and multiple releases of *Heterorhabditis bacteriophora* Poinar (Nematoda: Heterorhabditidae) for control of *Cylas formicarius* (Fabricius) (Coleoptera: Apionidae) *Biological Control.* **1**(4): 320-328.

Jansson, R.K., Lecrone, S.H. and Gaugler, R. (1993). Field efficacy and persistence of entomopathogenic nematodes (Rhabditida: Steinernematidae, Heterorhabditidae) for control of sweetpotato weevil (Coleoptera: Apionidae). *J. Econ. Entomol.* **86**(4): 1055-1063.

Jansson, R.K. and Lecrone, S.H. (1997). Effect of sweet potato cultivar on efficacy and persistence of entomopathogenic nematodes for control of *Cylas formicarius. Nematropica.* **27**(1): 41-52.

Josephrajkumar, A. and Sivakumar, C.V. (1997). Biological control potential of native entomopathogenic nematodes against *Spodoptera litura* (Fab.). *Annals of Plant Protection Sciences.* **5**(2): 164-167.

Kondo, E. and Ishibashi, N. (1983). Invasion and growth of *Neoaplectana carpocapsae* (DD-136) in larvae of the common cutworm, *Spodoptera litura* . Proceedings of the Association for Plant Protection of Kyushu. **29:** 111-114.

Kondo, E. (1987). Size-related susceptibility of *Spodoptera litura* (Lepidoptera: Noctuidae) larvae to entomogenous nematode, *Steinernema feltiae* (DD-136) *Applied Entomology and Zoology.* **22**(4): 560-569.

Kondo, E. and Ishibashi, N. (1988). Comparison of infectivity and propagation of steinernematid nematodes on the lepidopterous insect larvae, *Galleria mellonella* and *Spodoptera litura.* Proceedings of the Association for Plant Protection of Kyushu. **34:** 154-158.

Kondo, E. (1989) Studies on the infectivity and propagation of entomogenous nematodes, Steinernema spp., (Rhabditida: Steinernematidae), in the common cutworm, *Spodoptera litura* (Lepidoptera: Noctuidae) Bulletin of the Faculty of Agriculture, Saga University, Saga 840, Japan. **67:** 1-88.

Kurashvili, B.E., Kakuliya, G.A. and Gurgenidze, T.V. (1987). Utilization of *Neoaplectana* in insect control Helminths of insects. (Sonin, M.D. Eds.). Amerind Publishing Company. New Delhi, pgs. 113-117.

Lossbroek, T.G. and Theunissen, J. (1985). The entomogenous nematode *Neoplectana bibionis* as a biological control agent of *Agrotis segetum* in lettuce. *Exp. Appl. Nematol..* **39:** 261-264.

Mannion, C.M. and Jansson, R.K. (1992). Movement and postinfection emergence of entomopathogenic nematodes from sweetpotato weevil, *Cylas formicarius* (F.) (Coleoptera: Apionidae). *Biological Control* **2:** 297-305.

Mannion, C.M. and Jansson, R.K. (1993). Within-root mortality of *Cylas formicarius* (Coleoptera: Apionidae) by entomopathogenic nematodes. *J. Economic Entomology.* **86**(3): 722-729.

Morris, M.A. (1989). Sampling methods, biology and management of the mint flea beetle in central Oregon peppermint M.Sc. Oregon State University, Corvallis.

Miklasiewicz, T.J., Grewal, P.S., Hoy, C.W. and Malik, V.S. (2002). Evaluation of entomopathogenic nematodes for suppression of carrot weevil. *Biocontrol.* **47:** 545-561.

Narayanan, K. and Gopalakrishnan, C. (1987). Effect of entomogenous nematode *Steinernema feltiae* (Rhabditida: Steinernematidae) to the pre-pupa, pupa and adult of *Spodoptera litura* (Noctuidae; Lepidoptera) *Indian J. Nematology.* **17**(2): 273-276.

Philipsen, Holger, and Nielsen, Otto. (2003). Host potential of insects from cruciferous crops to entomopathogenic nematodes and augmentation of nematodes through oil seed rape growing. Paper presented at Entomopathogens and parasitic nematodes: Current research and perspectives in pest control. 8th European Meeting in the IOBC/WPRS Working Group, Athens, Greece, 29 May - 2 June; Published in *IOBC Bulletin* **26**(1): page pp. 141-146.

Prabhuraj, A., Viraktamath, C.A. and Kumar, A.R.V. (2000). Field evaluation of entomopathogenic nematodes against brinjal root weevil, Myllocerus discolor Boh. *Pest Management in Horticultural Ecosystems.* **6**(2): 149-151.

Prabhuraj, A., Viraktamath, C.A., Kumar, A.R.V., Tandon, P.L. (Ed.) Ballal, C.R. (Ed.) Jalali, S.K. (Ed.) Rabindra, R.J. (2003). Pathogenicity of two entomopathogenic nematodes against *Holotrichia serrata* (Fabricius) (Coleoptera: Scarabaeidae) and *Spodoptera litura* (Fabricius) (Lepidoptera: Noctuidae). Biological control of lepidopteran pests. Proceedings of the Symposium of Biological Control of Lepidopteran Pests, July 17-18, 2002, Bangalore, India. page pp 205-209.

Ratnasinghe, G. and Hague, N.G.M. (1997). Efficacy of entomopathogenic nematodes against the diamondback moth *Plutella xylostella* (Lepidoptera: Yponomeutidae). *Pakistan J. Nematology.* **15**(1-2): 45-53.

Salmanov, A.A. (1988). Results of attempted infection of the Colorado beetle with the nematode Neoaplectana carpocapsae in laboratory and field conditions. *Izvestiy a Akademii Nauk Azerbaidzhanskoi SSR, Biologicheskie Nauki.* **2:** 45-47.

Schroeder, P.C., Ferguson, C.S., Shelton, A.M., Wilsey, W.T., Hoffman, M.P. and Petzoldt, C. (1996). Greenhouse and field evaluations of entomopathogenic nematodes (Nematoda: Heterorhabditidae and Steinernematidae) for control of cabbage maggot (Diptera: Anthomyiidae) on cabbage. *J. Economic Entomology.* **89**(5): 1109-1115.

Schalk, J.M. and Creighton, C.S. (1989). Influence of sweet potato cultivars in combination with a biological control agent (Nematoda: *Heterorhabditis heliothidis*) on larval development of the banded cucumber beetle (Coleoptera: Chrysomelidae) *Environmental Entomology.* **18**(5): 897-899.

Shilpa Shinde., Singh, N.P. and Shinde, S. (2000). Susceptibility of diamond back moth, *Plutella xylostella* (L) to entomopathogenic nematodes. *Indian J. Experimental Biology.* **38**(9): 956-959.

Stewart, J.G., Boiteau, G. and Kimpinski, J. (1998). Management of late-season adults of the Colorado potato beetle (Coleoptera: Chrysomelidae) with entomopathogenic nematodes. *Canadian Entomologist.* **130**(4): 509-514.

Subramanian, S. (2002). Evaluation of biocontrol agents against *Plutella xylostella* (Linn.). *Insect Environment.* **8**(4): 150-151.

Stepanova, N.E. and Perera, M. (1993). Insect-pathogenic nematodes in the control of the potato moth *Zashchita Rastenii Moskva.* **5:** 20.

Tacconi, R., Boselli, M., Curto, G.M. and Santi, R. (1998). Entomopathogenic nematodes against the sugarbeet weevil (*Temnorhinus mendicus* Gyll., Coleoptera, Curculionidae). *Notiziario sulla Protezione delle Piante.* No. **8:** 213-218.

Tacconi, R., Boselli, M., Curto, G.M. and Santi, R. (1999). Entomopathogenic nematodes in the control of the sugarbeet weevil. *Informatore Fitopatologico.* **49**(6): 20-22.

Tahir, H.I., Otto, A.A. and Hague, N.G.M. (1995). The susceptibility of *Helicoverpa* (*Heliothis*) *armigera* and Erias vitella larvae to entomopathogenic nematodes. *Afro Asian J. Nematology.* **5**(2): 161-165.

Tojo, S. (1993). Suppression of the prophenol oxidase cascade in the larval haemolymph of the turnip moth by an entomopathogenic nematode and its symbiotic bacteria. Detection of beneficial nematodes in Japan and systematized nematode application. (Ishibashi N. Eds). Ministry of Education, Culture and Science, Japan. Saga University, Japan pp. 47-48.

Vyas, R.V. and Yadav, D.N. (1992). Infectivity of entomopathogenic nematode *Steinernema glaseri* to two soil dwelling lepidopterans. *Annals of Biology* (Ludhiana). **8**(1): 59-63.

Vyas, R.V., Patel, N.B., Pharindera Yadav., Ghelani, Y.H., Patel, D.J. and Yadav, P. (2003). Performance of entomopathogenic nematodes for management of gram pod borer, *Helicoverpa armigera. Annals of Plant Protection Sciences.* **11**(1): 107-109.

Vanninen, I., Vainio, A. and Jaakkola, S. (1992). Attempts to control *Delia* spp. with entomopathogenic nematodes Bulletin of the International Organization for Biological and Integrated Control of Noxious Animals and Plants. **15**(4): 143-153.

Vela, J.R.G., Lara, M.P., Lobaton, C.S. and Roca, A.C. (1998). Effectiveness of entomopathogenic nematodes on *Heliothis armigera* (Hubner) (Lep.: Noctuidae) larvae in laboratory. *Boletin de Sanidad Vegetal, Plagas.* **24**(4): 849-852.

Webb, S.E. and Capinera, J.L. (1996). Management of pickleworm with entomopathogenic nematodes. *Proceedings of the Florida State Horticultural Society.* No. **108:** 242-245.

Welch, H.E. (1958). Test of a nematode and its associated bacterium for control of the Colorado potato beetle *Leptinotarsa decemlineata* (Say) Annual Report Entomological Society of Ontario. **88:** 53-54.

Welch, H.E. and Briand, L.J. (1961). Field experiment on the use of a nematode for the control of vegetable crop insects Annual Report Entomological Society of Ontario. **91:** 197-202.

Wu, H.J. (1988). Biocontrol of squash bug with *Neoaplectana carpocapsae* (Weiser *Bulletin of the Institute of Zoology Academia Sinica*. **27**(3): 195-203.

Wu, H.J. and Chow, Y.S. (1989). Susceptibility of *Pieris rapae* crucivora (Lepidoptera: Pieridae) to the imported entomogenous nematode *Steinernema feltiae. Bulletin of the Institute of Zoology Academia Sinica*. **28**(4): 237-244.

Xu, J.L. (1983). Susceptibilities of vegetable insect pests to entomopathogenic nematodes. *Natural Enemies of Insects*. **5**(4): 231-239.

Yokoo, S., Tojo, S. and Ishibashi, N. (1992). Suppression of the prophenoloxidase cascade in the larval haemolymph of the turnip moth, *Agrotis segetum* by an entomopathogenic nematode, *Steinernema carpocapsae* and its symbiotic bacterium. *J. Insect Physiology*. **38**(11): 915-924.

Yoshihara, T. (1991). Effect of entomogenous nematode, *Steinernema feltiae* (DD-136) on several species of vegetable pests and related factors. *Bulletin of the National Research Institute of Vegetables, Ornamental Plants and Tea Series A*. **4:** 15-30.

Zhilyaeva, I.N., Ezhov, G.I. and Kondratov, E.S. (1973) Raising the entomopathogenic nematode *Neoaplectana carpocapsae* agriotos (Veremchuk, 1972) in larvae of the cabbage moth (*Barathra brassicae*). *Materialy Naushnoi Konferentsii Vsesoyuznogo Obshchestva Gel'mintologov*. **25:** 97-102.

28

Pest Management by Tissue Culture

K. Elanchezhyan

Department of Agricultural Entomology, Tamil Nadu Agricultural University, Coimbatore–641 003, Tamil Nadu

Arthropods have become the centre of attraction of scientists due to their importance in agriculture and medicine. They played an important role as intermediate hosts or vectors of pathogens causing diseases in man and pet animals. Invertebrate tissue culture, particularly insect cell lines is one of the most valuable tools in biological, biotechnological, biopharmaceutical and biopesticide research (Khurad *et al.*, 2006). This is true with lepidopteran cells. They are commonly used to study insect viruses (Blissard, 1996) and have been considered for the production of certain virus species, biopesticide or recombinant proteins. The use of insect cell culture systems for cloning and production of virulent insect viruses, production of novel proteins of medical importance using the baculovirus expression systems (BVES) and construction of genetically engineered nuclear polyhedrosis viruses (NPVs) with greater pest control potential has provided a major impetus towards the commercial application of insect cell culture in the fields of insect pest control and biotechnology (Rabindra and Rajasekaran, 1996). With the availability of the modern tools in molecular biology and the advancements made in biotechnology, the indigenous cell lines may prove useful in creating a future without bio-hazardous chemical pesticides as well as producing life-saving pharmaceuticals and vaccines for many diseases (Sudeep *et al.*, 2005). The continued development of new cell culture technology is essential for the future growth and application of insect cell and baculovirus biotechnology (Granados *et al.*, 2007).

HISTORY AND DEVELOPMENT

In the early part of the 20th century, entomologists in several fields of study had a dream of utilizing insect cells grown *in vitro* as a tool. The first attempt to maintain insect tissue *in vitro* was made by Goldschmidt (1915) using explants from the *Cecropia* silk moth *Hyalophora cecropia* early in the century. Then, in the late 1930s, W. Trager made a major breakthrough by developing an insect cell medium based on the knowledge of the physical and chemical properties of the insect haemolymph (Trager, 1935 and 1938). Following this, Wyatt (1956) developed a medium based on a detailed biochemical analysis of silkworm haemolymph and demonstrated that this medium induced the growth of cells from ovarian tissue of *Bombyx mori* and promoted *in vitro* cell growth and survival for extended periods of up to three weeks.

In 1962, Thomas D.C. Grace made a phenomenal breakthrough almost 50 years after Goldschmidt's first studies, by establishing four cell lines from the large Australian emperor gum moth, *Antherea eucalypti* (Grace, 1962). In the next 15 years, cell lines were established from at least five orders of insects representing some 50 different species (Hink, 1976). Thomson and Grace (1963) showed that insect cell lines could be grown on a medium that did not contain homologous haemolymph. Then Yunker *et al.* (1967) demonstrated that a strain of the Grace Antheraea cell line could be grown on medium containing vertebrate sera and the insect haemolymph. Many cell lines have been developed on media free of insect haemolymph. Today, insect cell lines can be grown on media without serum supplements of any kind (Goodwin, 1976; Hink *et al.*, 1977). Since then, over 500 continuous cell lines have been established from over 100 insect species (Lynn, 1999) belonging to the orders Lepidoptera, Diptera, Orthoptera, Hemiptera, Coleoptera and Hymenoptera, many of which are of agricultural and medical importance.

Insect tissue cultures are now commonly used in insect physiology, developmental biology, pathology and molecular biology. Insect pathologists have cells capable of replicating dozens of insect specific viruses (Granados and McKenna, 1995), while plant pathologists and vertebrate pathologists have cells capable of replicating viruses transmitted by insects (Mitsuhashi, 1989). Physiologists have fat body cultures (Lynn *et al.*, 1988) and other cell types, including cultures with neural properties (Sheppard and Lynn, 1996) for studying cell signalling pathways, while developmental biologists have imaginal discs cell lines (Lynn *et al.*, 1982; Ui *et al.*, 1987) and myoblasts (Lynn and Hung, 1991) for studying differentiation. Since the development of the baculovirus expression vector system (Smith *et al.*, 1983), hundreds of proteins of interest in medicine and biology have been produced in insect cells. Virtually, every cell line is a potential source of material for molecular biologists to use in a wide range of studies.

ESTABLISHMENT OF INSECT CELL LINES FROM DIFFERENT EXPLANTS

Cell lines can be established from various tissues viz., embryos, haemocytes, fat bodies, imaginal discs, larval gonads, pupal and adult ovaries (Hink and Ignoffo, 1970; Goodwin, 1975; Lynn, 1989; Pant *et al.*, 1998 and Mitsuhasi, 1998). These tissues are extracted aseptically and introduced into cell culture medium containing foetal bovine serum (FBS). The cells proliferate from the explants, multiply and establish in the medium. The cells grown as monolayer attach to the surface of the cell culture flask. Cells can be cultured in specially made cell culture flasks or roller bottles of bioreactors. Different types of media for culturing cell lines are available commercially either as powders or liquids. Currently, there are many commercial insect cell culture media available and these media may have as many as 30 components in their formulation. The commonly used media are:

- TNM-FH – *Trichoplusia ni* Medium Fred Hing
- IPL 41 – Insect Pathology Laboratory
- TC 100 – Tissue Culture 100
- MM – Mitsuhashi Maramorosch medium
- Grace Antheraea medium etc.

Several types of immature insect tissue have been used to initiate primary cultures including embryos, neonate larvae, imaginal discs and ovaries (Vaughn, 1985). The establishment of cell lines using immature or undifferentiated tissue has been more successful than with mature or differentiated tissues because of the higher proportion of stem cells. Development of cell lines from mature tissue instead of immature tissues has several advantages. First, the identity of the cell types present in the primary cultures is less diverse than immature tissue. Second, it is easy to identify the cell types in mature tissue as the cells have distinct morphologies, while the cells in immature tissues do not. Third, the distinct morphologies allow for assessments of the impact of culture components on the survival of specific cell types in primary cultures. Fourth, changes in cell morphology observed during the transformation process to an immortal cell could provide insight into the identity of the cell types in the cell line. Fifth, marker proteins identified in similar mammalian cells may be useful for establishing the identity of the cells in the insect cell line (Freshney, 1992).

EMBRYOS

The egg masses, within 24 h of hatching, collected from the laboratory reared healthy colonies. They are surface sterilized by immersing in 0.2 per cent sodium hypochlorite for two minutes and rinsed in distilled water. They are then collected in a sterile muslin cloth, transferred to sterile laminar hood and immersed in 70 per cent ethyl alcohol for 2-3 minutes followed by rinsing twice in sterile Milli-Q-water.

Embryo culturing procedure described by Lynn (1989) can be followed. The surface sterilized eggs are crushed along with few drops of TNM-FH medium containing cysteine (60 mg per 100 ml) through a sterile stainless steel gauze using a sterile silicone rubber policeman. Cysteine is added to prevent melanization of the medium. The explants are washed down by adding a few more drops of medium into sterile embryo cup and transferred into tissue culture flasks (25 cm^2) containing five ml of TNM-FH medium. The flasks are incubated at 27.5°C in a cooled incubator. Whenever cell attachment is seen, the medium in such flasks is changed once in 7-10 days.

LARVAL HAEMOCYTES

The mouth and rectum of final instar larvae are ligatured to prevent regurgitation and excretion and surface sterilized with 70 per cent ethyl alcohol for two minutes and rinsed twice with Milli-Q-water. Proleg of the larvae is cut and the haemolymph is collected in an embryo cup containing medium with cysteine. The blood cells are transferred to culture flasks containing five ml of TNM-FH medium containing cysteine. The flasks are kept in the incubator at 27.5°C.

LARVAL AND PUPAL FAT BODIES

Fat bodies are dissected out from larvae and pupae of lepidoptera aseptically and transferred to sterile Hank's balanced salt solution (HBSS). The adhering tracheoles are removed and only the fat body lobes transferred to embryo cups containing sterile TNM-FH medium. The explants are minced repeatedly with the help of micro scissors and transferred to tissue culture flasks containing 5 ml of medium. The flasks are kept in the incubator at 27.5°C.

PUPAL OVARIES AND TESTES

Six to seven-day old pupae are immersed in 70 per cent ethyl alcohol for two minutes and rinsed twice with Milli-Q-water in a sterile hood. The pupae are placed on a sterile dissecting dish containing sterile HBSS and fixed firmly on the dish using sterile stainless steel pins at both ends. The pupae are cut open in the mid dorsal line and the ovaries/testes transferred to embryo cups containing sterile HBSS. The unwanted tissues like fat bodies, malpighian tubules and trachea attached to the gonads are removed and the gonads alone transferred to another embryo cup containing a few drops of TNM-FH medium. The gonads are minced with sterile micro dissecting scissors and the explants transferred to culture flask containing five ml of medium. They are incubated in a cooled incubator at 27.5°C.

MAINTENANCE AND SUBCULTURING

Cells should be maintained at 27°C in a non-humidified environment. Cells can be maintained at room temperature on the bench top or in a drawer, however, a 27°C controlled environment is recommended. Insect cells will grow more slowly below 27°C. They will recover and resume normal doubling times once the temperature is returned to 27°C. Above 30°C, cells may display decreased viabilities. They will not recover if the temperature is returned to 27°C.

The primary and established cultures are maintained in cooled incubator at 27.5°C. Once in 5-7 days, four ml of the spent medium is removed and replaced with fresh medium in primary cultures. For the established cell lines, subculturing is done at the confluence stage. In the case of attached cell line, about one ml of spent medium is retained and the remaining is removed. Then four ml of fresh medium is added and the cells are scraped with a silicone rubber policeman. Then five ml of fresh medium is added, mixed well and split into two culture flasks. The subculturing is repeated at every confluence stage.

CHARACTERIZATION OF CELL LINES

Insect cell lines have been characterized to insect family, but not cell type. This has been accomplished by determining isozyme and/or karyotype profiles (Greene *et al.*, 1972; Harvey and Sohi, 1989). The other information provided for insect cell lines includes descriptions of morphology and growth rate (Vaughn, 1985). There are several approaches that could be taken to identify cell types present in a cell line. First, changes in cell morphology could be documented throughout the development of the cell line using electron microscopy. This approach has been used to follow the transformation of insect cells in primary cultures (Kuroda and Shimada, 1989) and can be expanded to follow the changes in cells during development to a continuous cell line. Second, there may be unique proteins in a cell that could be used as a marker. In mammalian cell culture, cell types are distinguished by the presence of unique membrane and structural proteins (Chew, 1994; Evans *et al.*, 1994). Several marker proteins have been identified in mammalian epithelial cells including intermediate filament proteins, desmosome junctions and cell surface antigens (Freshney, 1992). This information will be used to confirm the presence of epithelial cells in the cell lines developed from midgut epithelial cells (Baines *et al.*, 1994). Third, monoclonal antibodies have been produced from proteins in insect cells (Oddou *et al.*, 1993) and these methods can be applied to insect cell culture to produce antibodies that recognize specific cell types. A similar approach has been taken to confirm the identify of the insect family that contributed the tissue for developing a cell line (Vaughn *et al.*, 1977).

The newly established cell lines were characterized using routine parameters i.e., morphology, karyology, growth curve, species specificity, etc., as done for all other established cell lines in the world. Morphological studies carried out for all the newly established lepidopteran cell lines indicated that epithelial-like cells dominated in most of the cell lines followed by fibroblast-like cells and a few giant or vacuolated cells. Karyological analysis indicated a typical lepidopteran pattern showing a large number of dot-like, darkly stained chromosomes in all the cell lines and their number ranged from 45 to over 300. For morphological and karyological studies, the method described by Schneider (1973) was used. Growth of the cell lines differed individually and ranged from 4- to 10-fold weekly. Species specificity of all the cell lines was determined using isoenzyme profile analysis and heteroduplex analysis. Isoenzyme profile analysis is an important parameter used in the characterization of a new cell line to rule out the chances of cross-contamination from the existing cell lines using the method described by Tabachnik and Knudson (1980). The enzyme profiles (lactate dehydrogenase, malate dehydrogenase, Glucose-6-phosphate dehydrogenase and isocitric dehydrogenase) indicated the cell lines as distinct, new cell lines, and not a contaminant of any of the cell lines maintained in the laboratory. The heteroduplex analysis is a new molecular technique used to authenticate cell lines and used effectively to differentiate cell lines recently (Kshirasagar *et al.*, 1997). The appearance of a homoduplex indicated 100 per cent homology, while heteroduplex indicated cross-contamination. The analysis of all the cell lines gave homoduplex with their parent DNA indicating the cell lines as new, unique cell lines (Shouche *et al.*, 1999).

LEPIDOPTERAN CELL LINES

Lepidopteran cell lines constitute the backbone for studying baculoviral biology *in vitro* and for baculovirus vector based recombinant protein expression systems. The baculovirus expression system combined with insect cell cultures has become more attractive for the expression of many heterologous proteins than other systems viz., bacterial, yeast, vertebrate viruses, etc. due to its unique characteristics. This technology is also being used in the construction of recombinant baculoviruses to use as biopesticides that offered comparatively faster killing of insect pests than by wild type baculoviruses (Wood, 1995; Tracey *et al.*, 1997; Gershberg *et al.*, 1998 and Fuxa *et. al.*, 2002).

In the last two decades, a lot of work has been done in India to develop cell lines from lepidopteran insects due to their potential application in the fields of agriculture and medicine. The cell lines are used as a tool to grow many entomopathogenic viruses, which have the potential as biopesticides. These viruses are highly virulent to susceptible insect hosts and are eco-friendly. Members of the family baculoviridae, in particular, have been projected as biopesticides of the future and are being used in many countries for pest control (Vlak and Hong, 1997).

Baculoviruses have been isolated from the members of Phylum Arthropoda (insects and crustaceans) and no isolations have been reported from vertebrates so far (Heinz *et al.*, 1995). Baculoviruses are extremely virulent to their susceptible hosts and are safe for humans. In nature, baculoviruses play an important role in bringing down the pest population to economic thresholds as and when their population increases (Wood, 1995).

The establishment of the first lepidopteran cell line was reported by Pant *et al.*, 1977. They established a cell line from the embryonic tissue of potato tuber moth (*Phthorimaea operculella*) and characterized the cell line in detail. The cell line was initiated and maintained in MM medium supplemented with 10 per cent FBS, which was later replaced with 0.8 per cent bovine serum albumin supplemented with yeastolate and lactalbumin hydrolysate. This is one of the very few insect cell lines maintained in growth medium without supplementing FBS. The cells could also be grown as suspension culture in conical flasks on a rotary shaker giving higher yield of cells.

In the last decade, though many scientists at different laboratories actively worked on development of new lepidopteran cell lines. In India, lepidopteran cell lines i.e., sf-9, sf-21, etc., are being used for various studies and all these cell lines are procured from abroad. The new, indigenously developed lepidopteran cell lines were characterized in detail using the routine parameters including their susceptibility to several baculoviruses. Some of the cell lines have already been proved useful in propagating baculoviruses due to their high susceptibility and virus yield, while some expressed recombinant proteins with better efficiency than sf-9 cells.

ESTABLISHMENT OF LEPIDOPTERAN CELL LINES

Lepidopteran cell lines were primarily being established to propagate insect viruses as a biopesticide for the control of insect pests. Cell lines were established from different organs/ tissues i.e., embryonic, ovarian, fat body, testis, imaginal discs, dorsal vessel, etc., at different stages of development from lepidopteran insects. This is possible due to the comparatively larger size of these insects. The maximum number of cell lines has been established from the embryonic tissues followed by ovarian tissue (Lynn, 2001). In India, majority of lepidopteran cell lines were established from the ovarian tissue followed by embryonic tissue (Table 28.1). The initiation of cell cultures from different tissues of lepidopteran insect is shown diagrammatically in Fig. 28.1.

Table 28.1. Lepidopteran cell lines established in India and other countries.

Name of species	Designation	Tissue of origin	References
Phthorimaea operculella	Not designated	Embryonic (eggs)	Pant *et al.*, 1977
Spodoptera litura	NIV-SU-992	Larval ovary	Pant *et al.*, 1997
Spodoptera litura	NIV-SU-893	Pupal ovary	Pant *et al.*, 1998
Spodoptera frugiperda	Sf-9 and IPLB-SF-21	Pupal ovary	Vaughn *et al.*, 1977 O'Reiley *et al.*, 1992
Heliothis virescens	IPLB-HvE1 IPLB-HvE6	Embryos	Lynn and Shapiro, 1998
Spodoptera litura	NIV-SU-1095	Larval haemocyte	Pant *et al.*, 2000
Helicoverpa armigera	NIV-HA-1195	Larval haemocyte	Sudeep *et al.*, 2002
Helicoverpa armigera	NIV-HA-197	Embryonic (eggs)	Sudeep *et al.*, 2002
Bombyx mori	NIV-BM-1296	Larval ovary	Sudeep *et al.*, 2002
Bombyx mori	NIV-BM-197	Pupal ovary	Sudeep *et al.*, 2002
Ephestia kuehniella	IPLB-Ek-x4T IPLB-Ek-x4V	Embryo	Lynn and Ferkovich, 2004
Phthorimaea operculella	NIV-PTM-1095	Embryonic (eggs)	Sudeep *et al.*, 2005
Trichoplusia ni	BTI-TN-5B1-4 BTI-TN-MG-1	Embryonic tissue Midgut	Granados *et al.*, 1994 Sin, 2005
Bombyx mori	DZNU-Bm-1	Larval ovary	Khurad *et al.*, 2006

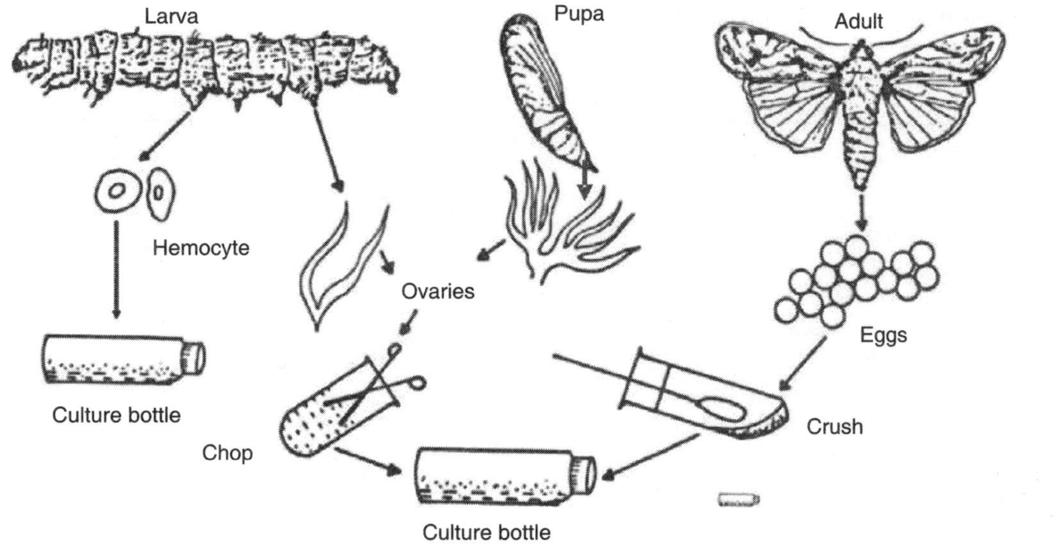

Fig. 28.1. Diagrammatic representation of establishing primary cultures from different tissues of lepidopteran insects (Sudeep *et al.*, 2005).

Susceptibility of the Established Cell Lines to Certain Baculoviruses

(i) *Helicoverpa armigera cell lines*

H. armigera is a polyphagous insect causing damage to over a hundred species of plants especially cash crops i.e., cotton, vegetables and ornamental plants causing damage worth billions of dollars globally (Lua and Reid, 2000). In India, this is a major pest of cotton, vegetables and oilseeds. Therefore, efforts are being made to control this insect using baculoviruses i.e., HaSNPV.

Two cell lines from the *H. armigera* were developed i.e., from the larval haemocyte and embryonic tissue and designated as NIV-HA-1195 and NIV-HA-197, respectively. Both the cell lines were tested for their susceptibility to different baculoviruses (Table 28.2). The NIV-HA-197 cell line was highly susceptible to HaSNPV, yielding a very high titre (2.88×10^7 NPV/ml) on the 10th PID (Sudeep *et al.*, 2002). The cell line was also susceptible to AcMNPV and SpltMNPV yielding high titres (Sudeep, 2002). The HaSNPV grown in the cell culture was bio-assayed in different instar larvae of *H. armigera*. The OBs produced *in vitro* were highly virulent to 2nd and 3rd instar *H. armigera* larvae causing cessation of feeding on the 2nd day and mortality in six days. This cell line is also growing in goat serum (GS) supplemented medium producing a comparable yield of OBs. These OBs were found more virulent than the former and 100 per cent mortality was observed within five days of infection to 2nd instar *H. armigera* larvae (Sudeep, 2005). Goat serum, being cheap and locally available, will help in the large scale production of HaSNPV for use as a biopesticide in future. The successful application of HaSNPV has been reported from China, Australia and other parts of the world.

Table 28.2. Susceptibility of newly developed lepidopteran cell lines to different baculoviruses.

Cell line	Baculoviruses				
	AcMNPV	HaSNPV	SpltMNPV	BmMNPV	PTM-GV
NIV-SU-992	+++	-	++	-	-
NIV-SU-893	++	-	++	-	-
NIV-SU-1095	++	-	+	-	-
NIV-HA-1195	+	++	+	-	-
NIV-HA-197	++	+++	++	-	-
NIV-BM-1296	++	-	-	+++	-
NIV-BM-197	++	-	-	++	-
NIV-PTM-1095	-	-	-	-	?

(Sudeep *et al.*, 2005)

+ Approximately 25 per cent cells showing occulusion bodies (OBs) on 7th PID; ++ approximately 50 per cent cells showing occulusion bodies; +++ approximately 75 per cent cells showing occulusion bodies; - Not susceptible (No OBs); MNPV–Multiple Nucleopolyhedrovirus; SNPV–single nucleopolyhedrovirus; GV–granulosis virus; Ac– *Autographa californica*; Splt–*Spodoptera litura;* Ha–*Helicoverpa armigera*; PTM–*Potato tuber moth* and Bm–*Bombyx mori*

(ii) *Cell lines from Heliothis virescens*

The tobacco budworm, *Heliothis virescens*, is part of a complex of related species in the *Heliothis/ Helicoverpa* genera that are very serious pests of cotton, corn, tomato and tobacco. This insect quickly develops resistance to chemical pesticides and is a prime target for biological controls, including insect viruses (Ignoffo, 1966). While other cell lines have been established from this insect (McIntosh and Ignoffo, 1983; Lynn *et al.,* 1988), the yields of baculovirus occlusion bodies (OBs) in these cells were lower than can be obtained in other lepidopteran lines (McIntosh *et al.,* 1985). All of these lines were established from larval or pupal reproductive tissue.

Seven cell strains have been established from embryos of the tobacco budworm, *H. virescens.* HvE1a, HvE1-1t and HvE6a are each predominantly fibroblast-like, firmly attached cells.

Although the HvE1a and HvE1-1t cells are narrow and form a web-like network, the HvE6a cells are broader and are closely aligned to each other in parallel arrays. Twenty to 30 per cent of HvE6a-1t are similar to HvE6a, but at least 70 per cent of these cells exhibit epithelial like morphology. HvE1s cells are predominantly round and suspended and occur singly or in small clumps, while both (HvE6s and HvE6s-1t) occur in large aggregates of suspended cells with some cells occurring as well-organized vesicles. The low temperature strains were selected because cell lines may develop with a broader host range to various insect viruses at low temperature (Winstanley and Crook, 1988).

Table 28.3. Replication of various nucleopolyhedrovirus in new cell strains from *H. virescens.*

Virus	HvE1a	HvE1s	HvE1-lt	HvE6a	HvE6a-1t	HvE6s	HvE6s1t
HzSNPV	+++	(+)	+++	+++	+++	+++	+++
AcMNPV	++	++	++	+++	++	+++	++
AfMNPV	+++	+++	++	+++	+++	++	++
RoMNPV	++	+++	++	+++	+++	++	++
LdMNPV	-	Nt	-	-	-	-	-
OlMNPV	-	-	-	-	-	-	-
OpMNPV	-	Nt	+	+	+	-	-
AgMNPV	+++	+++	+++	+++	+++	+++	+++

+++, over 90% of cells with OBS; ++, 50-80%; +, 10-20%; nt, not tested; -, no OB's apparent

Six primary cultures were initiated, and from these initial cultures two produced sufficient cell growth to allow subcultivation and eventually led to the establishment of seven cell strains, three of which are maintained at low temperatures (17°C). The strains were compared with a previously established cell line from *H. virescens* by isozyme analysis and shown to be from the same species. All the *H. virescens* strains were inoculated with various baculoviruses, including *Autographa californica* nucleopolyhedrovirus (NPV), *Anagrapha falcifera* NPV, *Anticarcia gemmatalis* NPV, *Rachoplusia ou* NPV, *Lymantria dispar* NPV (LdMNPV), *Orgyia pseudotsugata* NPV (OpSNPV), *O. leucostigma* NPV (OlMNPV), and *Helicoverpa zea* NPV (HzSNPV). All seven strains were highly susceptible to the noctuid NPVs, and large numbers of occlusion bodies (OBs) were produced in most of the inoculated cells. Three of the *H. virescens* strains were also susceptible to OpSNPV, although only 10–20% of the cells produced OBs with this virus (Table 28.3). These new strains appear to have good potential for studying HzMNPV, and with some further optimization of the growth conditions, they may also be useful in production scheme for this virus.

(iii) *Helicoverpa zea cell line*

A *Helicoverpa zea* cell line was adapted to grow in suspension culture using a serum-free medium, SF 90011 and serum supplemented SF 90011. The serum supplemented cells were infected with a wild type HaNPV at different stages of growth. Cultures infected at low level densities produced similar yields of virus (20-40 PIB/cell). Only 60-70 per cent of the viable cells in a culture produced polyhedra as a result of infection (Chakraborty, 1996).

(iv) *Spodoptera litura cell lines*

S. litura is an important agricultural pest in India causing serious damage to tobacco and vegetables. Three new continuous cell lines from larval ovaries (NIV-SU-992), pupal ovaries (NIV-SU-893) and larval haemocyte (NIV-SU-1095) were established and characterized in detail (Pant *et al.*, 1997; Pant *et al.*, 1998 and Pant *et al.*, 2000). Susceptibility of these cell lines to certain baculoviruses was also studied and found susceptible to the homologous virus *Spodoptera litura* multiple nucleopolyhedrovirus (SpltMNPV) as well as *Autographa californica* multiple nucleopolyhedrovirus (AcMNPV). On infection with SpltMNPV, the three cell lines yielded 5.6, 5.3 and 3.0 × 10^6 occlusion bodies (OBs)/ml, respectively on the 7th PID. With AcMNPV, the cell lines yielded 5.5, 3.0 and 1.96 × 10^6 OBs/ml, respectively. However, none of these cell lines supported the growth of *Helicoverpa armigera* single nucleopolyhedrovirus (HaSNPV), *Bombyx mori* multinucleopolyhedrovirus (BmMNPV) and potato tuber moth granulosis virus (PTM-GV).

(v) *Cell lines from Spodoptera frugiperda*

SF-9 and SF-21 cell lines are the traditional cell lines used with baculovirus and originated at the USDA insect Pathology Laboratory. These two cell lines originated from the IPL BSF-21 cell line, derived from the pupal ovarian tissue of the fall armyworm, *Spodoptera frugiperda* (Vaughn *et al.*, 1977 and O'Reilly *et al.*, 1992). Both cell lines are suitable for transfection, generating high titre stocks, plague formation and expression of recombinant proteins (Hink *et al.*, 1991). Sf-9 and Sf-21 share the following characteristics:

- Grow well in monolayer and suspension culture
- Adaptable to serum-free medium

The SF-9 clone and SF-21 cells are similar in their characteristics with the following properties: They are (1) robust cells and easy to culture in monolayer or suspension; (2) highly susceptible to AcMNPV; (3) able to grow rapidly to high cell densities and produce high budded virus (BV) titers; and (4) readily adaptable to serum-free medium and scaleup culture.

Recently, a new cell line was established from the SF-9 cells by selection in serum-free medium supplemented with human insulin. This new line, which is reported to be genetically and morphologically distinct from the parent SF-9 cells, is known commercially as express SF+ (Granados *et al.*, 2007).

An engineered cell line derived from the Sf9 line and engineered to express glycosyltransferases (galactosyltransferase and sialyltransferase) is commercially available from Invitrogen under the trade name "Mimic™ Sf9 Cells" (Granados *et al.*, 2007).

(vi) *Phthorimaea operculella cell line*

Potato tuber moth (PTM) is an important pest of potatoes in India causing extensive damage to the crop. This pest attacks the crop in the field as well as in storage, and efforts to control this insect using conventional methods has not been satisfactory (Dalaya and Patil, 1973). Release of parasitoids and predators has also not helped in the control of this insect and the damage is increasing steeply. PTM-granulovirus (PTM-GV) is a viable option and is being used in the control of this insect pest. PTM-GV is presently being produced in PTM larvae, which is labour intensive and extensive.

The growth of PTM-GV in a newly established PTM cell line was reported by Lery and coworkers in Egypt (Lery *et al.*, 1997). An embryonic cell line (NIV-PTM-1095) was established from the moth and characterized including the susceptibility to PTM-GV. On infection with PTM-GV, a few virus-like particles were observed in the tissue culture fluid in electron microscopic studies (Pant *et al.*, 2002). The cell line did not support the replication of any of the other baculoviruses tested. However, subsequent studies with HaSNPV showed a few occlusion bodies in a few cells. In AcMNPV infected cultures also, the morphology of the cells was found significantly changed in comparison to the control cells, however, occlusion bodies were not observed. Sudeep *et al.* (2005) observed three different cell types at the 40th passage level and comprised of epithelial-like cells (77%), fibroblast-like cells (20%) and giant cells (3%).

(vii) *Cell line from Ephestia kuehniella*

The Mediterranean flour moth, *E. kuehniella* is a serious pest of stored food products, especially whole and milled grains. Two continuous cell strains have been established from embryos of Mediterranean flour moth. Ek-x4T cells are predominantly epithelial-like, firmly attached cells although some cells are more fibroblastoid, especially at lower cell densities. Ek-4xV cells grow in large aggregates of suspended cells with many cells occurring in well organized vesicles (Lynn *et al.*, 1982). Both of the new lines were highly susceptible to five of the viruses originally isolated from noctuid insects (*A. californica, A. falcifera, H. armiger, H. virescens* and *Rachoplusia ou*) as well as viruses from *Galleria mellonella* and *P. xylostella*, while being only slightly susceptible to *A. gemmatalis* NPV, and with no apparent susceptibility to the NPVs for *H. zea* of *Lymantria dispar* (Table 28.4) (Lynn and Ferkovich, 2004). These new cell lines can be useful in biocontrol research.

Table 28.4. Replication of nucleopolyhedroviruses in new cell lines from *Ephestia kuehniella*.

Virus	IPLB-Ek-x4T	IPLB-Ek-x4V
Autographa californica nucleopolyhedrovirus	+++	+++
Anagrapha falcifera nucleopolyhedrovirus	+++	+++
Anticarcia gemmatalis nucleopolyhedrovirus	+	+
Galleria mellonella nucleopolyhedrovirus	+++	+++
Heliothis armigera nucleopolyhedrovirus	+++	+++
Heliothis virescens nucleopolyhedrovirus	+++	+++
Helicoverpa zea nucleopolyhedrovirus	-	-
Lymantria dispar nucleopolyhedrovirus	-	-
Plutella xylostella nucleopolyhedrovirus	+++	+++
Rachoplusia ou nucleopolyhedrovirus	+++	+++

+ Complete replication with the formation of occlusion bodies in < 20% of cells by 1-week post infection; +++ Complete replication with the formation of occlusion bodies in < 80% of cells by 1-week post infection; - No apparent infection by visual inspection with phase contrast or Hoffman modulation contrast microscopy.

(viii) *Trichoplusia ni cell line*

An insect cell line has been established and characterized, derived from embryonic tissue (BT1-TN-5B1-4, ATCC CRL 10859) and midgut (BTI-TN-MG1, ATCC CRL 10860) of *Trichoplusia ni* (cabbage looper). These lines were susceptible to various baculoviruses, including TnSNPV and AcMNPV and may be used to replicate such viruses for use as insecticides (Sin, 2005). The BTI-TN-5B1-4 cells are a clonal isolate of a parental line TN5B1, which was established from *T. ni* eggs for studies of the replication of a *T. ni* granulovirus and NPV. The clonal isolate, Tn5B1-4 normally grows in monolayer culture and has the following characteristics. (1) Highly susceptible to AcMNPV; (2) adaptable to grow in suspension; (3) able to grow to moderate cell densities and produce moderate budded virus titres; (4) readily adaptable to serum-free media and scaleup procedures; and (5) able to produce very high levels of many recombinant proteins. Tn5B1 cell line is available for research purposes under the trade name "High Five" cells. Recently, two novel, clonal cell lines (Tn-H5CL-B and Tn-H5CL-F) were isolated from the High Five cells at passage 90. The clonal lines also exhibited higher resistance to nutrient stress (Li *et al.*, 2003).

(ix) *Bombyx mori cell lines*

B. mori cell lines are effectively being used in the expression of foreign proteins using BEVS. In India, research work has been carried out with *B. mori* cell lines, (BMN, BM5, etc.) brought from abroad for expression studies (Sriram *et al.*, 1997; Saigal *et al.*, 1998). Two cell lines were established from the larval and pupal ovaries of *B. mori*. The newly developed larval and pupal ovary cell lines were designated as NIV-BM-1296 and NIV-BM-197, respectively and characterized. Both the cell lines were susceptible to BmMNPV and AcMNPV. However, the cell lines did not support the multiplication of HaSNPV and SpltMNPV.

Cell lines derived form *B. mori* could be genetically engineered to continuously express high levels of foreign protein *in vitro* (Farrell *et al.*, 1998, 1999). There are some established cell lines of *B. mori* originated from embryonic tissues (Chen *et al.*, 1988; Imanishi *et al.*, 1988; Inoue and Mitsuhashi, 1984; Ninaki *et al.*, 1988; Pandharipande, 1994) and larval and pupal ovarian tissues (Grace, 1967; Quiot, 1982; Sudeep *et al.*, 2002), but only a few can support replication of BmNPV. Such a cell line will be useful in study of BmNPV replication vis-á-vis BmN cell line with respect to effect of strains and effect of the age of the cell lines. Since BmNPV can infect larval ovary of *B. mori* and also transmits to next generation (Khurad *et al.*, 2004).

A new continuous cell line designated as DZNU-Bm-1 from larval ovaries of the silkworm, *Bombyx mori*. DZNU-Bm-1 is a promising candidate for studying the BmNPV replication *in vitro* and can be utilized to express recombinant proteins using BmNPV derived expression vectors (Khurad *et al.*, 2006).

(x) *Cell lines from Leptinotarsa decemlineata*

Four cell lines were established from tissues of Colorado potato beetle. One was initiated from embryonated egg fragments in the M3 medium supplemented with 20 per cent FBS and then transferred to the Ex-cell 400 medium with 20 per cent FBS. Another was initiated from adult haemocytes; one in the Ex-cell 400 with 5 per cent FBS only. These cell lines have been characterized by their morphology with light and electron microscopy, their karyotypes, cell growth and isozyme analysis. Each cell line differs in morphologic, karyologic, growth and isozyme patterns (Charpentier, 2001).

(xi) *Trichogramma cell lines*

Numerous insect cell lines have been developed over the past three decades, but only few of these have been from the order Hymenoptera. The extremely small size of these insects has made physiological and biochemical studies difficult. Now, with the development of the cell lines, a limitless supply of biologically active material is available for a wide variety of basic biological studies. The *Trichogramma confusum* and *T. exigum* cell lines (IPLB-T con 1 and IPLB-Tex 2) were characterized by chromosome and isozyme techniques (Lynn and Hung, 1991). The morphological change in IPLB-Tex 2 is accompanied by an induction of highly contractile cells, which indicates that this cell line may be composed of myoblast cells.

(xii) *Mosquito cell lines*

The establishment of the first mosquito cell line in the world from *Aedes aegypti* mosquitoes was reported by Grace in the year 1966. Subsequent studies using isoenzyme analysis indicated this cell line as a contaminant of *Antherea eucalypti* cell line (Greene, 1972). Singh (1967) established two cell lines at the NIV, Pune from the larval tissues of *Aedes aegypti* and *A. albopictus* in MM medium, which was originally designed for culturing leafhopper cells by Mitsuhashi and Maramorosch, in the year 1964. Singh's cell lines are therefore considered as the first, true mosquito cell lines established in the world. In addition, he used commercially available foetal bovine serum (FBS) as a supplement in place of insect haemolymph, which was considered essential for establishment and maintenance of mosquito cell cultures.

Subsequently, a number of mosquito cell lines from different species were established in India and abroad. Table 28.5 shows the number of mosquito cell lines established at NIV, Pune, India (Sudeep *et al.*, 2005).

The first *Culex* cell line in the country was established from the embryonic tissue of *C. bitaeniorhynchus* mosquitoes (Pant and Dhanda, 1982). This species is considered as one of the important vectors of Japanese Encephalitis in India (Reuben, 1971; Dhanda and Kaul, 1980). Cell lines from other two species of *Culicine* mosquitoes viz., *C. infula*, *C. ambiguous*, which are members of *C. bitaeniorhynchus* complex were also established subsequently (Pant and Dhanda, 1985). The cell line also supported the multiplication of other arboviruses of public health importance in India viz., Chikungunya (CHIK), dengue (DEN) and Sindbis viruses (Pant *et al.*, 1985). This cell line was also used in the epitope analysis of different strains of JE virus using monoclonal antibodies (Ghosh *et al.*, 1989).

A new cell line from the forest mosquito *A. krombeini* was established in the early 1990s, which proved highly sensitive to several arboviruses (Pant *et al.*, 1992). The new cell line supported the replication of 13 mosquito-borne arboviruses (Pant *et al.*, 1992; Sudeep *et al.*, 1997).

Unfortunately, due to the presence of certain cytoplasmic inclusions in the cells (Sreenivasan *et al.*, 1994), the cell line was not used for further research especially for isolation of virus from field collected mosquitoes. Recently, a new cell line from *C. tritaeniorhynchus* mosquitoes was established from the embryonic tissue (Athawale *et al.,* 2002).

Table 28.5. New cell lines established from mosquitoes in India.

Mosquito species	Tissue of origin	Designation	References
Aedes aegypti	Neonate larvae	ATC-10	Singh, 1967
Aedes albopictus	Neonate larvae	ATC-15	Singh, 1967
Aedes vittatus	Neonate larvae	ATC-121	Bhat and Singh, 1970
Aedes walbus	Neonate larvae	ATC-136	Singh and Bhat, 1971
Aedes novalbopictus	Neonate larvae	ATC-173	Bhat and Guru, 1973
Culex bitaeniorhynchus	Embryonic tissue	ATC-415	Pant and Dhanda, 1982
Aedes krombeini	Neonate larvae	NIV-AK-453	Pant *et al.*, 1992
Culex tritaeniorhynchus	Embryonic tissue	NIV-CT-894	Athawale *et al.*, 2002

(xiii) *Tick cell lines*

Ticks are the most economically important haematophagous ectoparasites of domestic animals worldwide, both because of the damage they cause during feeding and as vectors of numerous viral, bacterial and protozoal diseases (Jongejan and Uilenberg, 2004). Ticks are also becoming increasingly important as vectors of zoonotic pathogens affecting humans.

Tick cell lines are valuable tools for the *in vitro* propagation of tick-borne pathogens, and provide model systems for study of tick-pathogen interactions at the cellular and molecular level (Bell-Sakyi *et al.*, 2007). Around 50 cell lines have been established over the past three decades from 13 hard (ixodid) tick species of the genera *Amblyomma*, *Boophilus*, *Dermacentor*, *Hyalomma*, *Ixodes* and *Rhipicephalus*. Moreover, cell lines are now available from two soft (argasid) tick species, *Carios capensis* (Mattila *et al.*, 2007) and *Ornithodoros moubata* (Bell-Sakyi, unpublished).

Most tick cell lines are derived from embryos (eggs), moulting larvae and nymphs have also yielded continuous lines. Establishment can take between 1 and 5 years, and the success rate is generally low. Tick cells are grown in various mammalian culture media supplemented with foetal bovine serum, usually at temperatures between 28° and 34°C. They may grow as monolayers, in suspension, or a mixture of both, and do not exhibit contact inhibition. Some lines comfortably achieve cell densities of 5-10 × 10^6/ml. Providing the medium is changed regularly, tick cells can survive for long periods – months or even years – without subculture; this characteristic is particularly useful for attempted isolation *in vitro* of slow-growing microorganisms.

In some cases, particular cell lines have proved susceptible to infection with a number of bacterial pathogens, which may or may not be transmitted by the parent tick species. For example, *I. scapularis* cell lines support growth of *Anaplasma phagocytophilum*, *A. marginale*, *Ehrlichia canis*, *E. chaffeensis* and *E. ruminantium*, amongst which *I. scapularis* ticks only naturally transmit *A. phagocytophilum*. *E. ruminantium*, only transmitted naturally by ticks of the genus *Amblyomma*, grows in cell lines from *Amblyomma*, *Boophilus*, *Ixodes* and *Rhipicephalus* ticks (Bell-Sakyi, 2004), while *A. marginale*, although naturally transmitted by ticks of the genera *Boophilus* and *Dermacentor*, has only been reported to grow in *I. scapularis* cells (Munderloh *et al.*, 1996). Tick cell culture systems have greatly facilitated research on pathogens such as *A. marginale* (Blouin *et al.*, 2002) for which there is no standardised continuous culture sytem in mammalian cells, and *E. ruminantium*, which produces infection levels in ticks insufficiently high for proteomic analysis (Postigo *et al.*, 2008). Tick cell lines will play an increasing role in functional genomics studies on ticks and tick-borne pathogens *in vitro* (Bell-Sakyi, 2008).

MISCELLANEOUS STUDIES

In addition to lepidoptera, coleoptera, and hymenoptera, Bhat and Singh (1970) developed a cell line from the triatomine bug, *Triatoma rubrofasciatus*. Though three different cell types along with multicellular vesicles were observed in the culture, they degenerated after 30 days. Pant *et al.* (1988) successfully established three cell lines from the bed bug, *Cimex hemipterus* (Hemiptera: Cimicidae) and characterized one cell line (NIVI-CH-440) in detail. The susceptibility of the cell line was studied in important arboviruses *viz.* JE, CHIK, DEN-2, etc., and the rickettsiae, *C. burnettii*.

NOVEL INSECT CELL LINES

Cell lines have been established from species represented in eight orders of insects including Lepidoptera. The order Lepidoptera (moths and butterflies) is the second largest order of insects and has more than 200,000 species in 130 families. Cell lines from Lepidoptera are used primarily for the study of insect pathogenic viruses including baculoviruses and consequently most lines have come from insect species that are serious agricultural pests. Of the 29 baculovirus genomes that have been sequenced, 19 are NPVs (including AcMNPV) which are lepidopteran-specific viruses (Jehle *et al.*, 2006). Of these 19 sequenced genomes, all have been isolated from insects representing only four families. All four families are comprised of moth species, many of which are serious pests of agricultural crops and forests.

The saltmarsh caterpillar, *Estigmene acrea* (Lepidoptera: Arctiidae) is an occasional pest in agriculture and an *E. acrea* hemocyte cell line (BTI EAA) was established for the study of insect viruses (Granados and Naughton, 1975, 1976). The cell line was cloned and the clonal line, Ea-4 was found to produce greater amounts of complex N-linked glycolsylation on recombinant proteins than the SF9 cells (Ogonah *et al.*, 1996; Wagner *et al.*, 1996). Similarly, the Monarch butterfly, *Danaus plexippus*, Lepidoptera: Danaidae) is not a plant pest and an established cell line (DpN1) was reported to be capable of complex N-glycosylation and sialylation of recombinant proteins (Palomares *et al.*, 2003). Cells from unusual, non-pest species may be an important source of new lines for use in basic research and biotechnology.

Insect Cell Cultivation and Bioreactors

Insect cells used in conjunction with the baculovirus expression system (BEVS) are gaining ground as a platform for recombinant protein production. The success of commercially viable production schemes using insect cell cultures depends on several issues including the development and characterization of superior insect cell lines adapted for growth in large-scale vessels, the formulation of inexpensive and convenient insect cell culture media, especially chemically defined serum-free and protein-free media and the design of appropriate large-scale bioreactors together with the optimization of the conditions favouring high productivity in these systems (Agathos, 1993).

Insect cells are cultured in artificial medium containing carbohydrates, vitamins, organic acids, inorganic salts and animal serum. Cells can be cultured in specially made cell culture flasks or roller bottles or bioreactors. Bioreactors are chambers in which microorganisms are cultivated in a liquid or solid medium. Bioreactors can be distinguished on the basis of cell culturing methods (suspension, attachment to a solid surface, immobilization by entrapment or encapsulation) and the mode of bioprocessing (batch, repeated batch and continuous).

(i) *Attached cell culture systems*

Most types of insect cells tend to attach loosely to solid surfaces for growth. The cells can be detached by gentle agitation or with a silicon rubber scraper or similar device ("rubber Policeman"), but can also be removed by enzymatic treatment in a manner analogous to trypsinization of anchorage dependent mammalian cells.

In the interest of producing cost effectively abundant quantities of wild type baculoviruses such as AcNPV for bioinsecticide applications, the roller bottle method was proposed. Roller bottles are attractive because of the simplicity of the set up, the straightforward scale-up and the possibility to contain epidodic contamination in a single individual unit. In combination with semi-automatic systems dispensing medium and inoculum, roller bottles can be important contributors to the overall economy of scale.

The split-flow airlift bioreactor does not require the use of pumps and has good potential for volumetric scale-up of process utilizing cell lines which require a growth surface.

The following table summarises some important factors.

Pros	Cons
Easy and inexpensive to maintain.	Mechanical manipulations required for subculturing may decrease viability.
Allows easy visual inspection under inverted microscope to follow infection course.	Requires multiple flasks (e.g. several 150 cm^2 flasks) for large-scale expressions and high-titer stocks.
All cell lines can be maintained with this method (Sf9, Sf21, High Five.).	Limited cell density/ml due to monolayer. This may limit protein yields/ml of cell culture.

(Version, 2002)

(ii) *Suspension growth systems*

Many groups routinely use either stirred tank or airlift bioreactors are the pilot scale for insect cell cultivation. Cultivating insect cells in suspension using stirred tank or airlift bioreactors with submerged aeration and most insect cell lines can be adapted to grow in suspension (Schmid, 1996). Airlift reactors are reported to be simpler in design and construction than stirred tank reactors, and as a result they lead to reduced capital and maintenance cost and reduced risk of microbial contamination, hence reliable and low cost operation (Rhodes *et al.*, 1991).

Conventional stirred tank systems have been used for large-scale insect cell culture ranging from various volumes of 3 litres of over 150 litres (Bedart *et al.*, 1994). The basic attraction of a stirred tank reactor system is its wide availability and suitability for multipurpose use and the long standing experience in design and understanding of scale-up principles in microbial fermentation.

The following table summarises some important factors.

Pros	Cons
Can generate from 250 ml to greater than 1 litre of high-titer stock or expressed protein expression per spinner flask.	Requires spinners, spin plates, incubator – can be expensive to start up.
Facilitates scale up of protein expression.	
Higher cell densities/ml (up to 3 x 10^6 cells/ml for Sf9 cells). This can increase protein yields per ml of cell culture.	Requires daily cell counts and viability determinations to follow growth patterns and infection courses.
Greater oxygenation and minimal manipulation may increase viability. Cell viabilities are typically 98% or greater.	Can be difficult to maintain sterility.
Sf9 and Sf21 cells require no adaptation and readily switch from adherent to suspension and vice versa.	High Five. cells require adaptation to suspension culture.

(Version, 2002)

Insect Cell Infection with Recombinant Baculoviruses

The baculovirus-insect cell expression system is now considered a highly effective, proven and robust technology for the expression of thousands of recombinant proteins for use in cell and molecular biology research with both insect and mammalian cells (Kost *et al.*, 2005). Commercial production of recombinant proteins is a growing industry and several companies are using this expression system for the commercial development of immuno-therapeutics, animal vaccines and human vaccines.

The baculovirus expression vector system and insect cells have been used extensively for the production of secreted glycoproteins, and this represents one of their most important biotechnological applications. Many or most secreted animal proteins that have been examined are expressed and secreted in abundance in the baculovirus expression system.

The baculovirus expression vector system was developed as powerful tool for eukaryotic recombinant protein production and biopesticide production due to its strong polyhedron promoter and toxin to host insect. There are two ways to infect insect cells in suspension culture, classified into high multiplicity of infection (MOI) and low MOI. In the case of high MOI, a synchronous infection of the insect cells is achieved, the kinetics of which is not complex. However, the design of large-scale implementations involves the scaling up of two parallel processes, one for the cells and one for the infecting virus, as high MOI requires large amounts of viruses. On the other hand, recombinant protein and biopesticide production using low MOI can be carried out on an industrial scale to shorten the production process, reduce the cost of equipment and diminish the possibility of contamination. The main consideration in this approach is that inoculation of the viral stock can be done directly from a well-characterized master bank into a single scaled-up bioreactor (Zhang *et al.*, 2005).

Important Pharmaceutical Proteins Produced in Insect Cells Infected with Recombinant Baculoviruses

- Hepatitis B surface antigen and Hepatitis A virus antigen
- Human interferon
- Insulin-like growth factors
- Human interleukin 2
- Human interleukin 3
- HIV 1, gag, pol, sor, gp 41 and gp120 proteins
- Human b interferon
- Human para influenza virus haemagglutinin
- Human erythroprotein
- Human insulin factor
- Human epidermal growth factor
- Human trosinase hydroxylase etc.

The envelope protein of HIV-1 produced in cells of *S. frugiperda* by the AcMNPV have been approved for testing humans for AIDS vaccination (Maeda, 1989; Treacy *et al.*, 1999).

Quality and Advantages of Proteins Produced in Baculovirus Expression Systems

- Correct functionality of the foreign proteins
- High expression level for many proteins
- Possibility of post-translational modification that is absent in other systems

- Capacity for large-scale DNA insertion
- Simultaneous expression of multiple genes

Commercial Development of Products Produced in Insect Cells

Numerous small to midsize startup biotechnology companies in North America and the Europe are currently using the baculovirus-insect cell technology to produce custom recombinant proteins for research and commercial applications. Currently, there are two animal vaccines that are on the market and several immuno-therapeutic and human vaccines are being developed and produced in SF21, SF9, expressSF+, or High Five cells. In Europe, two commercial subunit vaccines for classical swine fever are produced in *S. frugiperda* cells (Van Oers, 2006). Several human vaccines that are being produced in *S. frugiperda* cells and are in mid to late phase clinical trials are: (1) **Provenge,** an immuno-therapeutic vaccine for prostate cancer, (2) **Flublok,** a vaccine for human influenza virus, and (3) **Chimigen** vaccines for chronic hepatitis B and C.

Commercial products in development using *T. ni* High Five cells include: (1) **FavId** (idiotype/KLH), an immuno-therapeutic vaccine for B-cell non-Hodgkin's lymphoma and (2) **Cervarix**, a vaccine for cervical cancer. Cervarix may be the first insect cell produced vaccine to be commercialized for human use.

Uses of Insect Tissue Cultures

- Insect tissue cultures are widely used in viral diagnosis and biotechnology, for the production of recombinant proteins, viral pesticides and vaccines as well as in basic research in genetics, molecular biology, biochemistry, endocrinology and virology.
- Insect tissue cultures are now commonly used in insect physiology, developmental biology, pathology and molecular biology.
- Lepidopteran cell cultures are increasingly being used in diverse areas of research i.e., in biotechnology, agriculture and medicine. Many lepidopteran cell lines are being used to propagate insect viruses in large scale and are being used as biopesticides.
- Insect cell lines have been widely used in genetic improvement of NPVs. It can also be used to enhance the virulence of insect viruses.
- Baculoviruses have proven useful as both bioinsecticides and as agents facilitating the expression of foreign protein in insect cells (Davis *et al.*, 1993). In India, imported cell lines i.e., *S. frugiperda* cell line (Sf-9 and Sf-21), *T. ni* and *B. mori* cell lines have been used to study the expression of different foreign proteins (Sriram *et al.*, 1997; Saigal, 1998; Palhan *et al.*, 1995).
- The baculovirus expression vector system and insect cells have been used extensively for the production of secreted glycoproteins, and this represents one of their most important biotechnological applications.
- Insect cell lines, such as the ovarian cells of the moth *S. frugiperda*, are utilized by histologist and geneticists to study gene expression. They also seek possible causes and therapies for human genetic disorders such as gouty arthritis using insect cell cultures.
- The Sf-9 cell lines can be used as model for studying insect octapamine receptors.
- Immunofluorescent technique (IFA) is now being used for the detection of virus antigen in infected cell lines and mosquitoes. This new technique has helped to rediscover the mosquito cell lines in the study of arboviruses.

- The IFA technique is proved quicker in detecting the presence of rickettsiae as in other conventional methods like staining with Gimanez staining, etc., the detection was possible only on the 11th post infection day.
- Mosquito cell lines have been used widely in the isolation of arboviruses. These cell lines with the help of modern tools available in molecular biology may find application in solving many of the complex problems associated with mosquito-borne diseases.
- In India, though many mosquito cell lines are available, these are used mainly in the isolation of arboviruses except for a few other studies. Mosquito gained considerable importance in biochemical and genetic research due to their increasing resistance to insecticides.
- Mosquito cell lines are routinely being used to carry out genetic and biochemical aspects. The cell cultures are also being used in the exploration of immune responses in mosquitoes, receptor studies, host parasite interaction, etc. in addition to their various other applications in molecular biology.
- Insect cell lines have been shown to secrete insect hormones (Leob *et al.*, 1989; Lynn *et al.*, 1987) and other novel compounds (Okai, 1985) with potential commercial applications.
- Cell lines are suited to the study of hormonally regulated mechanism of macromolecular biosysnthesis and gene expression.
- Insect cell lines can be used in studies on molecular, physiological, morphological and developmental aspects of ecdysteroid action.
- Insect cells have been used for studying the effects of certain biologically active compounds.
- Insect cells have been used to study the toxicity of crystal proteins from *Bacillus thuringiensis*.
- Cell lines have also been used to study the ion channels.

Drawbacks and their Remedies

- Maintenance of healthy cultures depends on strict avoidance of certain practices. Most insect cell cultures can be maintained at 26-28°C on a weekly subculture interval, but they may require adjusting the split ratios (seeding density).
- Microbial contamination is the major problem of cell culturists. Antibiotics might seem a useful tool for avoiding such problems. Well maintained modern laminar flow hoods or biosafety cabinets and proper aseptic technique should eliminate the need for antibiotics in the maintenance of the stock cultures.
- Cross-contamination of cell lines has been a significant problem since the early days of cell cultures. Use of different bottles of medium for each cell line will help to avoid such a tragedy, but cell cultures should be checked regularly to confirm their identity. Isozyme analysis is a simple, straight forward method and kits are available for performing these assays. Alternatively, DNA fingerprinting is also a viable technique.
- Many insect cell lines have never been characterized sufficiently to identify the specific cell types in the culture.
- Unlike plant cells equipped with environmentally resistant plant walls, insect cells must able to balance internal and external ionic solutions of potassium and sodium. This is particularly problematic for aquatic and wetland species that spend at least a significant portion of their life stages submerged in water.

CONCLUSION

Insect cell cultures are widely used in viral diagnosis and biotechnology, for the production of recombinant proteins, viral pesticides and vaccines as well as in basic research in genetics, molecular biology, biochemistry, endocrinology and virology. A large number of cell lines from dipteran, hemipteran, coleoptern, hymenopteran and lepidopteran insects were established and characterized in India. The US Environment Protection Agency has approved the use of six baculoviruses for application against pests of agriculture and forestry (Sudeep *et al.*, 2005). The new cell lines along with the existing ones will help in future research to develop biological pesticides as well as the large-scale production of important life-saving pharmaceutical products. The continued development of new cell culture technology is essential for the future growth of insect cell and baculovirus biotechnology.

REFERENCES

Athawale, S.S., Sudeep, A.B., Pant, U., Berde, P.V., Jadi, R. and Mishra, A.C. (2002). A new embryonic cell line from the embryonic tissues of *Culex tritaeniorhynchus* and its susceptibility to certain flaviviruses. *Acta Virol.*, **46:** 237-240.

Baines, D., Brownwright, A. and Schwartz, J.L. (1994). Establishment of primary and continuous cultures of midgut epithelial cells of the spruce budworm, gypsy moth and silkworm. *J. Insect Physiol.*, **40:** 347-357.

Bell-Sakyi, L. (2004). *Ehrlichia ruminantium* grows in cell lines from four ixodid tick genera. *J. Comparative Pathology,* **130:** 285-293.

Bell-Sakyi, L., Zweygarth, E., Blouin, E.F., Gouid, E.A. and Jongejan, F. (2007). Tick cell lines: tools for tick and tick-borne disease research. *Trends in Parasitology,* **23:** 450-457.

Bhat, U.K.M. and Singh, K.R.P. (1970). *In Vitro* culture of embryonic cells from and Indian triatomine bug, *Triatoma rubrofasciatus* De Geer (Hemiptera: Reduviidae). *Curr. Sci.*, **39:** 488-489.

Blissard, G.W. (1996). Baculovirus–insect cell interactions. *Cytotechnology*, **20:** 73-93.

Blouin, E.F., Fuente, J., Garcia-Garcia, J., Sauer, J.R., Saliki, J.T. and Kocan, K.M. (2002). Applications of a cell culture system for studying the interaction of *Anaplasma marginale* with tick cells. *Animal Health Research Reviews,* **3:** 57-68.

Buckley, S.M. (1973). Singh's *Aedes albopictus* cell cultures as helper cells for adaptation of Obodhiang and Kotonkhan viruses of the rabies sero-group to some vertebrate cell cultures. *Appl. Microbiol.*, **25:** 695-696.

Carbonell, L.F., Hodge, M.R., Tomalski, M.D. and Miller, M.K. (1988). Synthesis of a gene coding for an insect-specific scorpion neurotoxin and attempts to express it using baculovirus vectors. *Gene,* **73:** 409-418.

Chakraborty, S., Greenfield, P. and Reid, S. (1996). *In Vitro* production studies with a wild-type *Helicoverpa* baculovirus. *Cell Culture Engineering,* **22:** 1-3.

Charpentier, G., Tian, L., Cossette, J., Lery, X. and Belloncik, S. (2003). Characterization of cell lines developed from the Colorado potato beetle, *Leptinotarsa decemlineata* say (Coleoptera: Chrysomelidae). *Applied Microbiology and Biotechnology,* **62(1):** 1-20.

Chen, Q., Li, L., Yu, Z. and Peng, J. (1988). Establishment of cell lines from embryos of the silkworm, Bombyx mori. *In: Invertebrate and Fish Tissue Culture.* (Uroda, Y., Kurstak, E., Maramorosch, K. Eds.). Springer-Verlag, New York, pp. 259-261.

Chew, C.S. (1994). Parietal cell culture: new models and directions. *Annu. Rev. Physiol.*, **56:** 445-461.

Dalaya, V.P. and Patil, S.P. (1973). Laboratory rearing and field release of *Copidosoma kochleri* Blanchard, an exotic parasite for control of *Gnorimoschema operculella* Zeller. *Res. J. Mahatma Phule Agril. Univ.*, **4:** 99-107.

Davis, T.R., Wickham, T.J., McKenna, K.A., Granados, R.R., Shuler, M.L. and Wood, H.A. (1993). Comparative recombinant protein production of eight insect cell lines. *In Vitro Cell. Dev. Biol.*, **29A:** 388-390.

Dhanda, V. and Kaul, H.N. (1980). Mosquito vectors of Japanese encephalitis virus and their bionomics in India. *Proc. Indian Natl. Sci. Acad.*, **46:** 759-768.

Dhanda, V., Mourya, D.T., Mishra, A.C., Ilkal, M.A., Pant, U. and Jacob, P.G. (1989). Japanese encephalitis virus infection in mosquitoes reared from field collected immatures and in wild caught males. *Am. J. Trop. Med. Hyg.*, **41:** 732-736.

Evans, G.S., Flint, N. and Potten, C.S. (1994). Primary cultures for studies of cell regulation and physiology in intestinal epithelium. *Annu. Rev. Physiol.*, **56:** 399-417.

Fallon, A.M. (1988). Ribosome metabolism in mosquito cells. In: *Invertebrate and Fish Tissue Culture* (Kuroda, Y., Kurstak, E. and Marasmosch, K. Eds.). Japan Scientific Societies Press, Springer-Verlag. p. 50-52.

Fallon, A.M. and Sun, D. (2001). Exploration of mosquito immunity using cells in culture. *Insect Biochem. Mol. Biol.*, **31:** 263-278.

Fallon, A.M. and Stollar, V. The biochemistry and genetics of mosquito cells in culture. *Adv. Tissue cult.*, **5:** 97-137.

Farrell, P.J., Lu, M., Prevost, J., Brown, C., Behie, L.A. and Iatrou, K. (1998). High-level expression of secreted glycoproteins in transformed lepidopteran insect cells using a novel expression vector. *Biotechnol. Bioeng.*, **60:** 656-663.

Farrell, P.J., Behie, L.A. and Iatrou, K. (1999). Transformed lepidopteran insect cells: New sources of recombinant human tissue plasminogen activator. *Biotechnol. Bioeng.*, **64:** 426-433.

Freshney, R.H. (1992). Culture of animal cells. Wiley-Liss, New York.

Fuxa, J.R., Richter, A.R., Ameen, A.O. and Hammock, B.D. (2002). Vertical transmission of TnSNPV, TnCPV, AcMNPV and probably recombinant NPV in *Trichoplusia ni. J. Invertebr. Pathol.* **79:** 44-50.

Gershberg, E., Stockholm, D., Froy, O., Rashi, S., Gurevitz, M. and Chejanovsky, N. (1998). Baculovirus mediated expression of a scorpion depressant toxin improves the insecticidal efficacy achieved with excitatory toxin. *FEBS Lett.*, **422:** 132-136.

Ghosh, S.N., Sathe, P.S., Sarathi, S.A., Cecilia, D., Dandawate, C.N. and Athawale, S.S. (1989). Epitype analysis of strains of Japanese encephalitis virus by monoclonal antibodies. *Indian J. Med. Res.*, **89:** 368-375.

Ghosh, S.N., Tongaokar, S.S. and Dandavate, C.N. (1973). Intacellular and extracellular complement fixing antigens of JE and CHP viruses in mosquito tissue cultures. *Curr. Sci.*, **42:** 286-288.

Goldschmidt, R. (1915). Some experiments on spermatogenesis *in vitro. Proc. Nat. Acad. Sci., USA*, **1:** 220-222.

Goodwin, R.H. (1976). Insect cell growth on serum free media. *In Vitro*, **12:** 303.

Grace, T.D.C. (1962). Establishment of four strains of cells from insect tissues grown *in vitro. Nature,* **195:** 788-789.

Grace, T.D.C. (1966). Establishment of a line of mosquito (*Aedes aegypti*, L.) cells grown *in vitro. Nature,* **211:** 366-367.

Grace, T.D.C., (1967). Establishment of a line of cells from the silkworm, *Bombyx mori. Nature,* **216:** 613.

Granados, R.R. and McKenna, K.A. (1995). Insect cell culture methods and their use in virus research. *In: Baculovirus Expression Systems and Biopesticides* (Eds. Schuler, M.L., Wood, H.A., Granados, R.R. and Hammer, N.A.). Wiley-Liss, New York. pp. 13-39.

Granados, R.R. and Naughton, M. (1975). Development of *Amsacta moorei* entomopoxvirus in ovarian and haemocyte cultures from *Estigmene acrea* larvae. *Intervirology*, **5:** 62-68.

Granados, R.R. and Naughton, M. (1976). Replication of *Amsacta moorei* entomopoxvirus and *Autographa californica* nuclear polyhedrosis virus in haemocyte cell lines from *Estigmene acrea*. Academic Press, New York.

Granados, R.R., Li, G. and Blissard, G.W. (2007). Insect Cell Culture and Biotechnology. *Virologica sinica*, **22**(2): 83-93.

Greene, A.E., Charney, J., Nickols, W.W. and Corriel, L. (1972). Species identity of insect cell lines. *In Vitro*, **7:** 313-322.

Gubler, D.J., Kuno, G., Sather, G.E., Valez, M. and Oliver, A. (1984). Mosquito cell cultures and specific monoclonal antibodies in surveillance for DEN viruses. *Am. J. Trop. Med. Hyg.*, **33:** 158-165.

Harvey, G.T. and Sohi, S.S. (1989). Isozyme characterization of 8 hymenopteran and 20 lepidopteran cell lines. In: Invertebrate cell system applications (Mitsuhashi, J. Ed.). CRC Press Inc., Florida. pp. 71-76.

Heinz, K.M., McCutchen, B.F., Herrmann, R., Parella, M.P. and Hammock, B.D. (1995). Direct effects recombinant nuclear polyhedrosis viruses on selected non-target organisms. *J. Econ. Entomol.*, **88:** 259-264.

Hink, W.E. (1976). A compilation of insect cell lines and culture media. In: *Invertebrate Tissue Culture Research Application.* (Ed. Maramorosch, K.), Academic Press, New York, pp. 319-369.

Hink, W.F., and Ellis, B.J. (1971). Establishment and characterization of two new cell lines (CP-1268 and CP-169) from the codling moth, *Carpocapsa pomonella* (with a review of culture of cells and tissues from Lepidoptera). *Curr. Top. Microbiol. Immunol.*, **55:** 19-28.

Hink, W.F., Strauss, E.M. and Lynn, D.E. (1977). Growth of TN-368 insect cells in serum-free media. *In Vitro*, **13:** 177.

Ignoffo, C.M. (1966). Susceptibility to the first instar of the bollworm, *Heliothis zea* and the tobacco budworm, *Heliothis virescens* to *Heliothis* nuclear polyhedrosis virus. *J. Invertebr. Pathol.*, **8:** 531-536.

Ilkal, M.A., Dhanda, V., Rodrigues, J.J., Mohan Rao, C.V.R. and Mourya, D.T. (1984). Xenodiagnosis of laboratory acquired dengue infection by mosquito inoculation and immunofluorescence. *Indian J. Med. Res.*, **79:** 303-306.

Imanishi, S., Sato, S. and Mori, T. (1988). Characterization of cell lines derived from embryos in the silkworm, *Bombyx mori.* In: *Invertebrate and Fish Tissue Culture.* (Kuroda, Y., Kurstak, E. and Maramorosch, K. Eds.). Springer-Verlag, New York, pp. 255-258.

Inoue, H. and Mitsuhashi, J. (1984). A *Bombyx mori* cell line susceptible to a nuclear polyhedrosis virus. *J. Seric. Sci. Jpn.*, **53:** 108-113.

Jehle, J.A., Blissard, G.W. and Bonning, B.C. (2006). On the classification and nomenclature of baculoviruses: A proposal for revision. *Arch. Virol.,* **151:** 1257-1266.

Jongejan, F. and Uilenberg, G. (2004). The global importance of ticks. *Parasitology,* **129:** S3-S14.

Khurad, A.M., Kanginakudru, S., Qureshi, S.O., Rathod, M.K., Raiand, M.M. and Nagaraju, J. (2006). A new *Bombyx mori* larval ovarian cell line highly susceptible to nucleopolyhedrovirus. *J. Invertebrate Pathology,* **92:** 59-65.

Kost, T.A., Condreay, J.P. and Jarvis, D.L. (2005). Baculovirus as versatile vectors for protein expression in insect and mammalian cells. *Nat. Biotechnol.*, **23:** 567-575.

Kshirasagar, S.G., Patole, M.S. and Shouche, Y.S. (1997). Characterization of insect cell lines: Heteroduplex analysis employing a mitochondrial 16S ribosomal RNA gene fragment. *Anal. Biochem.*, **253:** 65-69.

Kuroda, Y. and Shimada, Y. (1989). Electron microscopic studies on *in vitro* differentiated cells from Drosophila embryos. In: *Invertebrate cell system applications* (Mitsuhashi, J. ed.). CRC Press Inc, Florida. pp. 77-89.

Kurti, T.J. and Munderloh, U.G. (1984). Mosquito tissue culture. *Adv. Cell cult.*, **3:** 259-302.

Leob, M.J., Brandt, E.P. and Woods, C.W. (1989). Testes of the tobacco budworm moth: ecdysteroid production by the testis sheath. In: *Invertebrate cell system applications.* (Mitsuhashi, J. Ed.). CRC Press Inc, Florida.

Lery, X., Giannotti, J., Taha, A., Ravale, M. and Abd-Ela, S. (1997). Multiplication of a GV from PTM in a newly established cell line of *Phthorimaea operculella. In Vitro Cell Dev. Biol. (Animal).* **33:** 640-646.

Li, G., Hashimoto, Y. and Granados, R.R. (2003). Growth characteristics and expression of recombinant proteins by new clones derived from *Trichoplusia ni* (BTI Tn5B1-4) High Five cells. Bioprocessing, **5:** 1-4.

Lua, L.H.L. and Reid, S. (2000). Virus morphogenesis of *Helicoverpa armigera* nucleopolyhedrovirus in *Helicoverpa zea* serum free suspension culture. *J. Gen. Virol.,* **81:** 2531-2543.

Luckow, V.A. (1991). Cloning and expression of heterologous genes in insect cells with baculovirus vectors. In: *Recombinant DNA technology and applications.* (Eds. Prokop, A., Bajpai, R.K., Ho, C.S.). McGraw-Hill Inc; New York. p. 97-152.

Lynn, D.E., Miller, S.G. and Oberlander, H. (1982). Establishment of a cell line from lepidopteran wing imaginal discs: Induction of newly synthesized proteins by 20-hydroxyecdysone. *Proc. Nat. Acad. Sci. US Am.* **79:** 2589-2593.

Lynn, D.E. (1999). Development of insect cell lines: Virus susceptibility and applicability to prawn cell culture. *Methods in Cell Science*, **21:** 173-181.

Lynn, D.E. (2001). Novel techniques to establish new insect cell lines. *In Vitro Cell Dev. Biol. (Animal)*, **37:** 330-337.

Lynn, D.E. and Hung, A.C.F. (1991). Development of continuous cell lines from the egg parasitoids, *Trichogramma confusum* and *T. exiguum. Archieves of Insect Biochemistry and Physiology.* **18:** 99-104.

Lynn, D.E. and Ferkovich, F.M. (2004). New cell lines form *Ephestia kuehniella*: Characterization and susceptibility to Baculoviruses. *J. Insect Science*, **4:** 9-13.

Lynn, D.E. and Oberland, H. (1983). The establishment of cell lines from lepidopteran wing imaginal discs: Induction of newly synthesized proteins by 20-hydroxyecdysone. *Proc. Nat. Acad. Sci., USA.* **79:** 2589-2593.

Lynn, D.E. and Shapiro, M. (1998). New cell lines from *Heliothis virescens*: characterization and susceptibility to Baculoviruses. *J. Invertebr. Pathol.*, **72:** 276-280.

Lynn, D.E., Dougherty, E.M., McClintock, J.T. and Loeb, M. (1988). Development of cell lines from various tissues of Lepidoptera. In: *Invertebrate and Fish tissue Culture* (Eds. Kuroda, Y., Kurstak, E. and Maramorosch, K.). pp. 239-242, Japan Scientific Socieites Press, Tokyo.

Maeda S. (1989). Gene transfer vectors of a baculovirus, *Bombyx mori* nuclear polyhedrosis virus, and their use for expression of foreign genes in insect cells. In: *Invertebrate cell system application.* (Ed. Mitsuhashi, J.) CRC Press, Boca Raton. p. 167-182.

Mattila, J.T., Burkhardt, N.Y., Hutcheson, H.J., Munderloh, U.G. and Kurtti, T.J. (2007). Isolation of cell lines and a rickettsial endosymbiont from the soft tick *Carios capensis* (Acari: Argasidae: Ornithodorinae). *J. Medical Entomology,* **44:** 1091.

McIntosh, A.H. and Ignoffo, C.M. (1983). Characterization of five cell lines established from species of *Heliothis. Appl. Entomol. Zool.,* **18:** 262-269.

McIntosh, A.H., Ignoffo, C.M. and Andrews, P.L. (1985). *In Vitro* host range of five baculoviruses in lepidopteran cell lines. *Intervirology,* **23:** 150-156.

Mitsuhashi, J. (2001). Development of highly nutritive culture media. *In Vitro Cell Dev. Biol., (Animal),* **37:** 330-337.

Mitsuhashi, J. and Maramorosch, K. (1964). Leafhopper tissue culture: embryonic, nymphal and imaginal tissues from aseptic insects. *Contrib. Boyce. Thompson Inst.,* **22:** 435-460.

Mourya, D.T., Ilkal, M.A., Mishra, A.C., George Jacob, P., Pant, U. and Ramanujam, S. (1989). Isolation of Japanese encephalitis virus from mosquitoes collected in Karnataka state, India from 1985-1987. *Trans. R. Soc. Trop. Med. Hyg.,* **83:** 550-552.

Munderloh, U.G., Blouin, E.F. and Liu, Y. (1996). Establishment of the tick (Acari: Ixodidae)-borne cattle pathogen *Anaplasma marginale* (Rickettsiales: Anaplasmataceae) in tick cell culture. *J. Medical Entomology,* **33:** 656-664.

Munderloh, U.G., Liu, Y., Wang, M., Chen, C. and Kurtti, T.J. (1994). Establishment, maintenance and description of cell lines from the tick *Ixodes scapularis. J. Parasitology,* **80:** 533-543.

Naik, P.S., Ilkal, M.A., Pant, U., Kulkarni, S.M. and Dhanda, V. (1990). Isolation of Japanese encephalitis virus from *Culex pseudovishnui* Colless, 1957 (Diptera: Culicidae) in Goa. *Indian J. Med. Res.,* **91:** 331-333.

Ninaki, O., Nagayasu, K., Kameoka, H., Fujiwara, H., Miyajima, N., Takada, N. and Maekawa, H. (1988). Establishment of cell lines from *Bombyx mori* and *B. mandrina* and attempt of cell fusion. In: *Invertebrate and Fish Tissue Culture.* (Kuroda, Y., Kurstak, E. and Maramorosch, K. Eds.). Springer-Verlag, New York, pp. 243-246.

Oddou, P., Hartmann, H., Radecke, F. and Geiser, M. (1993). Immunologically unrelated *Heliothis* sp. and *Spodoptera* sp. Midgut membrane proteins bind *Bacillus thuringiensis* Cry1A9b) endotoxin. FEBS, **212:** 145-150.

Ogonah, O.W., Freedman, R.B. and Jenkins, N. (1996). Isolation and characterization of an insect cell line able to perform complex N-linked glycosylation on recombinant proteins. *Biotechnol.,* **14:** 197-202.

Okai, Y. (1985). Insect fibroblast like cells release inhibitory peptides of lymphocyte DNA synthesis in to serum free culture medium. *Insect Biochem.,* **15:** 483-487.

O'Reilly, D.R., Miller, L.K. and Luckow, V.A. (1992). Baculovirus Expression Vectors: A Laboratory Manual. W.H. Freeman and Company, New York. pp. 3-8.

Padbidri, V.S., Pant, U., Kulkarni, S.S. and Dhanda, V. (1984). Multiplication of *Coxiella burnettii* in *Culex bitaeniorhynchus* cell line. III *International Symposium on Rickettisa and Rickettsial diseases,* Smolence, Czechoslovakia, pp. 10-14.

Palhan, V.B., Sumathy, S. and Gopinathan, K.P. (1995). Baculovirus mediated high level expression of luciferase in silkworm cells and larvae. *Biotechniques,* **19:** 97-104.

Palomares, L.A., Joosten, C.E. and Hughes, P.R. (2003). Novel insect cell line capable of complex N-glycosylation and sialylation of recombinant proteins. *Biotechnol. Prog.,* **19:** 185-192.

Pandharipande, T.N., (1994). Two new continuously growing cell lines from *Bombyx mori* (Lepidoptera: Bombycidae). *Appl. Entomol. Zool.,* **29:** 604-607.

Pant, U. and Banerjee, K. (1990). Arthropod tissue culture in India (1967-1988)–A review. *Indian J. Med. Res.,* **91:** 397-407.

Pant, U. and Dhanda, V. (1982). Establishment of a cell line from *Culex bitaeniorhynchus. J. Tissue Culture Methods,* **6:** 61-63.

Pant, U. and Dhanda, V. (1985). Arthropod tissue culture and its uses in arboviral research. *ICMR Bull.,* **16:** 33-38.

Pant, U., Sudeep, A.B., Athawale, S.S. and Vipat, V.C. (2002). Baculovirus studies in new indigenous lepidopteran cell lines. *Indian J. Exptl. Biol.,* **40:** 63-68.

Pant, U., Mascarenhas, A.F. and Jagannathan, V. (1977). *In vitro* cultivation of a cell line from embryonic tissue of potato tuber moth, *Gnorimoschema operculella* (Zeller). *Indian J. Expt. Biol.,* **15:** 244-245.

Pant, U., Banerjee, K., Ahawale, S.S. and Dhanda, V. (1985). Susceptibility of *Culex bitaeniorhynchus* cell line to some arboviruses. *Indian J. Med. Res.,* **76:** 789-794.

Pant, U., Ilkal, M.A., Soman, R.S., Shetty, P.S., Kanojia, P.C. and Kaul, H.N. (1994). First isolation of Japanese encephalitis virus from the mosquito *Culex tritaeniorhynchus* Giles, 1901 (Diptera: Culicidae) in Gorakpur district, Uttar Pradesh. *Indian J. Med. Res.,* **99:** 149-151.

Pant, U., Athawale, S.S. and Sudeep, A.B. (1992). Susceptibility of *Aedes krombeini* cell line to some arboviruses. *Indian J. Med. Res.*, **95:** 239-244.

Pant, U., Dhanda, V., Argade, S.P. and Banerjee, K. (1988). Establishment of three cell lines from the embryonic tissue of *Cimex hemipterus* (F.) (Hemiptera: Cimicidae). *In Vitro Cell Dev. Biol. (Animal).* **24:** 1201-1203.

Paul, S.D., Singh, K.R.P. and Bhat, U.K.M. (1969). A study on the cytopathic effect of arboviruses on cultures from *Aedes albopictus* cell line. *Indian J. Med. Res.*, **57:** 339-348.

Pavri, K.M. and Ghosh, S.N. (1969). Complement fixation tests for simultaneous isolation and identification of dengue virus using tissue culture. *Bull. World Health Organ.*, **40:** 984-986.

Postigo, M., Taoufik, A., Bell-Sakyi, L., Bekker, C.P.J., de Vries, E., Morrison, W.I. and Jongejan, F. (2008). Host cell-specific protein expression in *Ehrlichia ruminantium. Veterinary Microbiology,* **128:** 136-147.

Quiot, J.M. (1982). Establishment of cell line (S.P.C. Bm 36) from ovaries of *Bombyx mori* L. (Lepidoptera). *Sercologia,* **22:** 25-31.

Rabindra, R.J. and Rajasekaran, B. (1996). Insect cell cultures: A Tool in Basic Research, Biotechnology and Pest Control. *In: Biotechnological perspectives in chemical ecology of insects* (Ananthakrishnan, T.N. Ed.). Oxford and IBH Publishing Co. Pvt. Ltd., New Delhi. P. 299.

Reuben, R. (1971). Studies on the mosquitoes of North Arcot district, Madras State, India. Part I: seasonal densities. *J. Med. Entomol.*, **8:** 119-126.

Saigal, D. and Gopinathan, K.P. (1998). Recombinant *Bombyx mori* nucleopolyhedrovirus harboring green fluorescent protein. *Biotechniques,* **25:** 997-1008.

Schneider, I. (1973). Karyology of cell in culture, F. Characteristics of Insect cells. In: *Tissue culture methods and applications* (Eds. Kruse, P.F. and Patterson, M.K.). Academic Press, USA. pp. 788-790.

Sheppard, C.A. and Lynn, D.E. (1996). Immunoreactivities for calcium signaling components and neural-like properties of a Colorado potato beetle cell line. *Archieves of Insect Biochemistry and Physiology*, **33:** 197-209.

Shouche, Y.S., Patole, M.S., Pant, U., Paranjpe, S. and Banerjee, K. (1999). Authentication of two cell lines developed from the larval and pupal ovaries of *Spodoptera litura* by RNA based methods, *In Vitro cell Dev. Biol. (Animal),* **35:** 244-245.

Sin. S. (2005). Cell line isolated from larval midgut tissue of *Trichoplusia ni. Acta Biochim. Biophys.*, **37:** 857-861.

Singh, K.R.P. (1967). Cell cultures derived from larvae of *Aedes albopictus* (Skuse) and *Aedes aegypti* (L.). *Curr. Sci.*, **36:** 506-508.

Singh, K.R.P. (1972). Growth of arboviruses in arthropod tissue culture. *Adv. Virus Res.*, **17:** 187-206.

Smith, G.E., Summers, M.D. and Frazer, M.J. (1983). Production of human beta interferon in insect cells infected with a baculovirus expression vector. *Mol. Cell Biol.*, **3:** 2156-2165.

Smith, G.E., Fraser, M.J. and Summers, M.D. (1983). Molecular engineering of the *Autographa californica* nuclear polyhedrosis genome: Deletion mutations within the polyhedron gene. *J. Virology*, **46:** 584-593.

Sreenivasan, M.A., Pant, U., Athawale, S.S. and Arankalle, V. (1994). Detection by electron microscopy of endogenous viruses in *Aedes krombeini* (H.) line. *In Vitro Cell Dev. Biol. (Animal).* **30:** 142-144.

Sriram, S., Palham, V.B. and Gopinathan, K.P. (1997). Heterologous promoter recognition leading to high level of expression of cloned foreign genes in *Bombyx mori* cell lines and larvae. *Gene.* **161:** 181-189.

Sudeep, A.B. and Pant, U. (1999). *Aedes krombeini* (H.) cell line–A new approach to arboviral research. Paper presented at the IV International Symposlum on Vectors and Vector Borne diseases. Organised by Academy of Vector Borne Diseases, Bhubaneswar and Defence Research and Development Establishment, Gwalior, held at Gwalior, India from 18-20 February. p. 45.

Sudeep, A.B., Mourya, D.T. and Mishra, A.C. (2005). Insect cell culture in research: Indian Scenario. *Indian J. Med. Res.*, **121:** 725-738.

Sudeep, A.B., Pant, U., Dhanda, V. and Banerjee, K. (1997). A new *Aedes krombeini* cell line and its susceptibility to some arboviruses. *In: Invertebrate cell culture: Novel directions and biotechnology applications.* (Maramorosch, K. and Mitsuhashi, J. Eds.). Science Publishers Inc., USA. pp. 11-17.

Sudeep, A.B., Mishra, A.C., Shouche, Y.S., Pant, U. and Mourya, D.T. (2002). Establishment of two new cell lines from *Bombyx mori* (L.) (Lepidoptera: Bombycidae) and their susceptibility to baculoviruses. *Indian J. Med. Res.*, **115:** 189-193.

Tabachnick, W.J. and Knudson, D.L. (1980). Characterization of invertebrate cell lines. II. Isoenzyme analyses employing starch gel electrophoresis. *In Vitro*, **16:** 392-398.

Thomson, J.A. and Grace, T.D.C. (1963). Cytological observations on cell strains established in culture from insect ovarian tissue. *Aust. J. Biol. Sci.*, **16:** 869-876.

Tracey, M.F., All, J.N. and Glidia, G.M. (1997). Effect of ecdysteroid UDP-glycosyl transferase gene deletion on efficacy of a baculovirus against *Heliothis virecens* and *Trichoplusia ni* (Lepidoptera: Noctuidae). *J. Econ. Entomol.*, **90:** 1207-1214.

Trager, W. (1935). Cultivation of the virus of grasserie in silk worm tissue culture. *J. Exp. Med.*, **61:** 501-513.

Trager, W. (1938). Multiplication of virus of equine encephalomyelitis in surviving mosquito tissues. *Am. J. Trop. Med.*, **18:** 387-393.

Ui, K., Ueda, R. and M987, T. Cell lines from imaginal discs of *Drosophila melanogaster. In Vitro Cellular and developmental Biology*, **23:** 707-711.

Van Oers, M.M. (2006). Vaccines for viral and parasitic diseases produced with baculovirus vectors. *Adv. Virus Res.*, **68:** 193-253.

Vaughn, J.L., Goodwin, R.H., Tompkins, J. and McCawley, P. (1977). The establishment of two cell lines form the insect *Spodoptera frugiperda* (Lepidoptera: Noctuidae). *In Vitro*, **13:** 213-217.

Vlak, J.M. and Hong, H.Z. (1997). Biosafety of engineered baculoviruses and cultured insect cells. *In: Invertebrate cell culture: Novel Directions and Biotechnology Applications.* (eds. Maramorosch, K. and Mitsuhashi, J.). Science Publishers Inc., USA. p. 181-191.

Wagner, R., Geyer, H. and Geyer, R. (1996). N-acetylbeta-glucosaminidase accounts for differences in glycosylation of influenza virus hemagglutinin expressed in insect cells from a baculovirus vector. *J. Virol.*, **70:** 4103-4109.

Winstanley, D. and Crook, N.E. (1993). Replication of *Cydia pomonella* granulosis virus in cell culture. *J. Gen. Virol.*, **74:** 1599-1609.

Wood, H.A. (1995). Development and testing of genetically improved baculovirus insecticides. *In: Insect cell cultures: Production of improved bio-pesticides and proteins from recombinant DNA.* (Eds. Shuler, M.L., Wood, H.A., Granados, R.R. and Hammer, D.A.) Wiley-Liss, New York. p. 91-130.

Wyatt, S.S. (1956). Culture *in vitro* of the tissue from the silkworm *Bombyx mori. J. Gen. Physiol.,* **29:** 841-852.

Yunker, C.E., Vaugh, J.L. and Cory, J. (1967). Application of insect cell line (Grace's Antheraea cells) to medium free on insect haemolymph. *Science*, **155:** 1565-1566.

29

Integrated Pest Management in India: Progress, Problems and Prospects

M.C. Diwakar*, Vishwa Nath and D. Prasad

Division of Entomology, Indian Agricultural Research Institute, New Delhi–110 012
Division of Nematology, Indian Agricultural Research Institute, New Delhi–110 012

In agriculture, chemical pesticides have been considered as one of the wonder discoveries that include chemical fertilizers, hybrid vigour, mutation breeding, phytohormones, tissue culture and stem cell technology, etc. The use of chemical pesticides became very popular with the farmers after their introduction in 1942 onwards. In a very short span of time, they became major tools of plant protection world over. In India, the pesticides have been regarded as panacea for all ills, so much so that they became synonymous to the subject of Plant Protection pushing other methods of pest control to the background.

Wide spectrum and instant killing properties of chemical pesticides prompted the farmers for their over and excessive use everywhere, but within only two decades of their usage, cotton crop failures became evident by the mid-sixties in USA, Mexico, Nicaragua, Sudan, Ethiopia, etc. where severe attack of whitefly (Bemicia sps) and American bollworm (Helivoverpa = *Heliothis armigera*) could not be controlled despite repeated pesticidal treatments. The successive crop failures in spite of pesticidal treatments sent an alarm among the farmers and the governments globally. The food and agriculture organization (FAO) of the United Nations, sensing the gravity of situation constituted an IPM panel of experts in 1968 to go into the causes of such crop failures and suggest remedial measures. Prof. S. Pradhan of India was one of the members of the panel. The FAO panel discussed the problems of the cotton crop failures and came to the conclusion that excessive and injudicious use of pesticides has created resistance in the pest species, rendering the pesticides ineffective. The panel also recommended that instead of exclusive pesticidal control, Integrated Pest Control (IPC) strategy will have to be adopted for effective pest management system. The panel also coined the definition of IPM for the first time, which is as follows. "IPM is a pest management system that, in the context of the associated environment and the population dynamics of the pest species, utilizes all

*Ex Director (IPM), Directorate of Plant Protection, Quarantine and Storage, Ministry of Agriculture, Government of India, NH-4, Faridabad–121 001.

suitable techniques and methods in as compatible a manner as possible and maintains the pest population at levels below those causing economic injury."

This definition was later on revised by FAO and other scientists, giving different wordings, the central idea remaining the same.

GENESIS AND PROGRESS

1. Rapid Roving Surveys

High yielding rice varieties such as Taichung Native 1(TN-1) and IR-8 were introduced in India during mid-sixties as a part of adoption of modern agricultural strategy in order to maximize crop production. But these varieties were severely affected by Rice Tungro Virus (RTV) transmitted by rice green leaf hoppers (*Nephotettix viresens*) in the states of Assam, Bihar, Madhya Pradesh (Chhattisgarh region), Orissa, Uttar Pradesh and West Bengal in vast areas inflicting heavy crop losses during 1968 and 1969 Kharif seasons. The Directorate of Plant Protection, Quarantine and Storage, Government of India in collaboration with the Ford Foundation and state departments of agriculture–Assam, Bihar, Madhya Pradesh, Uttar Pradesh, Orissa and West Bengal launched an ad hoc programme of rapid roving rice surveys in order to monitor and detect the population of rice green leafhoppers, the vector of RTV, at an early stage so that timely pesticidal measures could be advised and adopted by the states and farmers. The survey information gathered by the teams during roving surveys proved extremely useful in organizing timely pest control measures based on economic threshold levels (ETLs). The rapid roving pest survey programme was later on extended to all rice growing states and also to other major field crops including cotton, sugar cane, oilseeds, pulses, etc. on the advice of Prof. M.S. Swaminathan, the then secretary to the Government of India.

2. Establishment of Central Surveillance Stations and Central Biological Control Stations

After the commencement of rapid roving rice surveys in 1970, the government of India also sanctioned two national schemes, namely pest and disease surveillance scheme and central biological control schemes in third five year plan. Under these schemes, 19 central surveillance stations and 11 central biological control stations were established during 1972-1977 in various states. The functioning of these stations marked the beginning of IPM implementation in India. The CSSs organized rapid roving surveys and fixed plot surveys on major field crops including rice, wheat, sugar cane, cotton, oilseeds, pulses, etc. and provided information on pest build ups in the form of short-term forewarnings to the states, which helped a great deal in organizing timely and need based control measures instead of prophylactic treatments. Thus, the survey information helped in avoiding blanket spray operations over large areas. Similarly, the CBCSs took up laboratory breeding and field releases of biocontrol agents in the field crops like sugar cane, rice, cotton, vegetable and fruit crops. In addition, these stations also organized intensive crop surveys for assessing biocontrol potential for conservation. It is mentioned that through field to field augmentation and conservation, sugar cane Pyrilla has been successfully managed in India through pest surveillance and biocontrol measures. In nutshell, it could be said that these modest efforts of pest surveillance and biological control programmes laid the foundation of IPM in India.

3. IPM Policy of the Government

The Government of India has declared Integrated Pest Management as a policy on plant protection. The policy declaration was made by the then Union Agriculture Minister Sardar Buta Singh in 1984 while inaugurating national conference on plant protection. At about the same time, the Directorate of Plant Protection, Quarantine and Storage, government of India devised the definition of IPM which is as follows "IPM is a broad ecological approach of managing pests that utilizes available methods/techniques/skills such as genetic, agronomic, mechanical, biological and chemical (as last resort) in a compatible manner so as to suppress pest intensity below economic injury level, based on pest surveillance". This definition has been widely accepted and used in India by all concerned. However, some people have changed the wordings of this definition. Since the declaration of IPM as national policy, it has been the cardinal principle and major plank of the centre and states, though the implementation of IPM has not been to the extent desired due to several problems and constraints.

4. Establishment of Central IPM Centres

The Government of India gave a major thrust to IPM programmes by establishing 28 central IPM centres (CIPMCs) in various states for promotion of IPM activities during VIII plan by amalgamating existing 19 central surveillance stations, 11 central biological control stations and 14 plant protection stations. Four more such centres have been opened during X five year plan. These centres are engaged in promoting IPM approach through IPM demonstrations and conducting farmers field schools on IPM besides organizing season long IPM training programmes for Master Trainers in collaboration with the states.

5. Pesticides Usage in India

In India, the use of chemical pesticides commenced in the locust control programme by importing about 200 MT of formulated DDT. During 1950, around 500 MT of imported formulated DDT was used. With the advent of green revolution since 1970 onwards, the usage of pesticides increased rapidly and it reached around 95000 MT (technical grade) by 1990. Since then, usage has come down to approx. 40000 MT (technical grade) in 2005, mainly due to IPM efforts. The pesticides industry grew very fast in India and now has the installed production capacity of around 150000 MT (technical grade) and has emerged as a major pesticide exporting country in South-East Asia. Around 70000 MT (technical grade) of pesticides worth several hundred crores of rupees are exported to several African, South-East Asian and Latin-American countries. In addition, around 200 biocontrol laboratories in public and private sectors are producing biopesticides (parasitoids, predators and insect pathogens), which is very insignificant keeping in the vast crop area and the requirement.

6. International Cooperation

The FAO launched an inter-country Integrated Pest Control (IPC) programme in rice for south and South-East Asia and Pacific during 1980 in 6 countries including India, Sri Lanka, Bangladesh, Thailand, Malaysia and Indonesia. This inter-country IPC programme continued in its II and III phases up to 1994 and was extended to other countries of the region, namely Myanmar,

Cambodia, Laos, Vietnam and North Korea. Under the programme, intensive IPM field trainings of a fortnight duration each were organized to the subject matter specialists and plant protection officers of centre and states. This programme proved catalytic in promotion of IPM in rice in India. These trainings helped a great deal in the field identification of pests, diseases and biocontrol agents besides exposing trainees to skills and techniques of IPM approach in rice ecosystem. In all about 20 such trainings were organized during 1986-1993, wherein about 500 extension staff were imparted training.

7. IPM Farmers Field Schools

Toward the end of eighties, severe attack of brown plant hopper (BPH) (*Nilaparvata lugens*) on rice in Indonesia has occurred over large areas. The FAO and Indonesian government developed IPM strategy and organized IPM farmers field schools for mass education and training of farmers on community basis. This Indonesian model of IPM in rice gave spectacular results in the management of rice BPH. The FAO advocated and adopted this model of IPM FFSs for other countries and several other crops including cotton, vegetables etc. for IPM training to farmers. In India also this model of IPM FFSs has been tried in rice, cotton, vegetable and other crops and has been found to be useful. In fact, IPM FFSs programmes implies IPM of the farmers, for the farmers and by the farmers, which aims at empowering the farming community in identifying the field problems and evolving their solution through group discussion and making decisions by common consensus. Field observations form major part of this FFSs programmes.

8. FAO/UNDP IPM Project

Taking a cue from the success of FAO inter-country IPC rice programme, the government approved a project entitled "FAO/UNDP IPM project for the development and implementation of IPM in India" during 1993 for a period of 5 years. Under this project, season long IPM field programmes have been organized for creating a pool of IPM trained master trainers (training of trainers-TOTs) at central and state level. About 50 season long training programmes have been conducted in rice, cotton, sugar cane, oilseeds, pulses and vegetable crops during 1994-98. Under this programme, about 2000 extension officials have been exposed to learning IPM skills and methods. Besides, the IPM FFSs have also been organized and about 5000 farmers in 50 villages have been directly benefited. In these trainings, IPM experts drawn from several countries and FAO/UNDP imparted trainings in addition to national experts. Simultaneously, the central IPM centres also organized FFSs on field crops and organized IPM demonstrations under the national IPM programme. It is estimated that in total approximately 20000 extension functionaries both of centre and states and 15000 farmers have been trained in 10000 IPM Farmers Field Schools organized under the FAO/UNDP programme during 1995-2005 in collaboration with centre and states. Looking into vast cropped area, many crops, several pests/diseases and varied agro-ecosystems and over 100 million faming families in India, these efforts have only touched the tip of the iceberg.

9. IPM Research in India

It is indeed a matter of pride that late Prof. S. Pradhan of India was one of the member of FAO panel of experts constituted in 1968. He advocated 'integrated pest control' strategy and advised the scientists to step up research to develop IPM for different crops, pests and regions. But unfortunately his call had very little or no impact on the scientific community in India and pesticidal research continued on various aspects of the chemical pesticides with little or no attention to IPM aspects. This in fact delayed the development and implementation of IPM in India. There is no denial of the fact, however, that a few individuals and institutions did start the research efforts to develop IPM components if not the total packages/models for a crop or pest or region. During IV and V year plan, the Indian Council of Agriculture (ICAR) had initial integrated pest control programmes in rice and cotton crop under 'whole village adoption' and 'lab to land' programmes and recommended tentative IPM packages keeping view the major pest/disease problems. During VIII to X five year plans, the ICAR implemented foreign aided National Agricultural Research Project (NARP) and National Agricultural Technology Project (NATP) wherein there was a lot of scope to develop IPM technology for major crops and important pests, but not much of researches appear to have been taken up in these schemes. Truly speaking, most of the scientific community in India was pesticide minded and continue to hold the view that "the crops cannot be grown without pesticides or pesticides are essential inputs". They did not like to listen to IPM talks during seventies and eighties till the crop failures became evident in rice due to BPH, etc. and in cotton due to whitefly and American bollworm (*Helicoverpa armigera*). IPM researches in India picked up of late during last decade of the previous century and some very useful findings have come out of the same. Now, IPM models for many crops are available including rice, cotton, sugar cane, oilseeds, pulses and vegetables. There is still a need to further intensify research efforts on many crops for different agro-climatic regions. It is hoped that though belated, the research efforts will catch up with time and need of the hour. The ICAR and agricultural universities, however, deserve all commendations and praise for developing several pest and disease-resistant/tolerant varieties of almost all field and commercial crops suited for varied agro-climatic zones with optimum production potential. These resistant/tolerant varieties minimized pest/disease incidence and resultant crop losses. The genetic control being major component of IPM played a significant role in crop production. The development of resistant/tolerant varieties under the all India coordinated crops research projects and by the crop institutions indeed noteworthy and has been applauded worldover as we became self-sufficient and surplus in food production.

PROBLEMS AND CONSTRAINTS

India is a huge and liberal democracy. In our federal system, states enjoy a lot of autonomy and independent powers. Agriculture is a state subject, though plant protection is on concurrent list of our constitution. It is quite unfortunate that agriculture, health and education get very meagre and grossly inadequate budgetary allocation both at centre and state levels. Further, it is a pity that in agriculture, plant protection is considered the last and the least. In addition, ours is a heterogeneous society consisting of multi-religious, multilingual, multi-regional and composite culture. This type of heterogeneity and variability is of course very conducive for development but at a slow pace. It also implies on agriculture and plant protection.

1. Inadequate Infrastructure and Set-up

Both in research and development sectors, the set-up and infrastructure is quite inadequate at centre as well as at states to effectively promote and implement IPM approach among farmers. There is dearth of qualified and trained personnel who can opt for this national endeavour. Transfer of IPM technology at the grass roots level is not anybody's or everybody's job. Only willing and devoted persons with physical and mental stamina can deliver goods. It is an irony that in India, extension and development workers are lowly paid, lowly recognized and lowly rewarded. As such, hardly any qualified and committed people opt for serving in the development and extension departments either in the central or state governments.

2. Lack of Coordination

There is no proper coordination between extension agencies at centre and states to implement the IPM programme effectively. High powered IPM steering committees headed by senior level officials and scientists both for IPM research and extension separately are required to be constituted. Similarly, these steering committees are required in the states. The committees should review the progress periodically and sort out the problems and hurdles coming in the way of IPM research or implementation.

3. Lack of IPM Trained Staff

In a country like ours, an army of properly trained personnel is required to reach each of about 6 lakh village and over 100 million farming families for their education and training. This is a huge task and needs careful consideration and planning by the concerned governments in a time bound manner.

4. Inadequate Pest Surveillance and Monitoring Setup

Pest intelligence is a prerequisite for success of IPM. The role of pest monitoring is like that of a diagnostic centre or pathological clinic in a hospital. The surveillance teams have to act as mobile clinic and diagnostic units to identify the field problems so that proper solution and treatment could be advised and advocated to farmers. On the basis of pest monitoring, a short-term forewarning on likely and impending pest/disease development should be issued at fortnightly intervals. At present, the pest surveillance and monitoring set-up at centre and states is almost non-existent because the staff that was conducting pest surveys and monitoring has, of late, been redeployed for conducting farmers field school and IPM demonstrations. Thus, pest monitoring activity has taken a back seat. Therefore, the centre and state are caught napping in the event pest/disease flare-ups and no timely plant protection measures are possible. The pest surveillance activity should be coordinated at national level and a national pest database and data banks be maintained for quick retrieval and use in the IPM programme. Pest surveillance is highly technical and requires qualified and trained staff for covering major crop belts. Since it is a gigantic task, attention of the governments at higher levels is needed.

5. Lack of Research and Extension in Biological Control

Biological pest management is one of the major components of IPM, perhaps second only to the genetic management. But the research and extension in this respect has been very insufficient, in view of the potential of biocontrol in IPM. Except the management of sugar cane Pyrilla through biological and IPM means, not much efforts appear to have been made in this direction. More than the laboratory breeding and augmentative field releases of parasitoids and predators, conversation of naturally existing biocontrol fauna in the crop ecosystem needs to conserved by minimizing pesticidal treatment and regular field monitoring. Intensive and regular survey in various crop ecosystems by the teams qualified scientist is required to assess and map out biocontrol potential. A suitable and separate set up for such task under ICAR could be taken. Sooner it is done, better it would serve the IPM interests.

6. Research and Technological Gaps

IPM is a knowledge-based and dynamic technology that grows along with the growth of knowledge and information generated through scientific and field researches. There are several crops and pests, which need research efforts to generate information to be included in IPM packages and models. The etiology, life cycle and epidemiology of several pests/diseases is still not well understood and requires further researches. For example, the etiology of mango malformation, coconut rootwilt disease, dieback syndrome in citrus and mango orchards, etc. Fortunately, India has one of the best agriculture research set-ups in the world and there is no reason why we cannot divert our research efforts to the priority areas.

7. Cooperation of Pesticide Industry

In India, pesticides industry is well developed and is quite resourceful. The industry has a well-knit infrastructure and set up throughout the country including rural areas. In case, their cooperation is sought, it will go a long way in promoting IPM technology. The industry has to be convinced and assured that IPM is in no way anti-pesticides as the IPM advocates rational and need based use of pesticides. It is not difficult to get the desired cooperation if efforts are made sincerely.

8. Inadequate Availability of IPM Inputs

Although in IPM, knowledge and information are the best inputs; however, material inputs such as biocontrol agents, biopesticides, insect lures, pheromones, etc. are not readily available. The quantity of the available material, in general, is also not up to the standards. There are genuine difficulties in the production, transportation and storage of bioagents and biopesticides as special handling skills and techniques are required. These are living units. Bioagents and biopesticides cannot be produced without advance planning and firm orders from users. All these factors compound the problem and availability of quality biocontrol material affected. Unless, serious efforts are made in this direction, the situation does not seem to improve.

9. Inadequate IPM Education and Training to Farmers

The present programmes of imparting IPM education and training to farmers through IPM farmers field schools and demonstrations are grossly inadequate, both quantatively and qualitatively. A massive national programme for farmers training in IPM is essential to promote IPM in a phased manner. A lot of funds will have to be allocated for this purposes besides staff and other logistic arrangements.

10. Political Will and Commitment

Having enumerated several problems and constraints, the sum and substance is that political will and commitment is most important for any programme to succeed and IPM is no exception. Strong political will and commitment implies that planning and funds will flow out automatically. Although the Government of India has declared IPM as a national policy on agriculture, no plans or funds commensurate to the magnitude of task have been placed in order to achieve the goal.

11. Inadequate International Cooperation

It has been recognized world over that pesticides are harmful to human health, environment and ecosystem causing soil, water and air pollution. Several international organizations, agencies and NGOs are extending technical and financial assistance to IPM programmes in most of the developing countries. FAO, UNDP, UNEP, ADB, World Bank, etc. are more than willing to extend technical and financial assistance to central and states. In India, only a few small IPM projects have so far been launched with help of FAO UNDP, ADB, etc. Since IPM is an eco-friendly and poverty alleviating programme, it fits well into the criteria of international agencies to extending technical or financial assistance.

PROSPECTS OF IPM

In the developed countries, the driving forces behind the IPM approach have been health and environmental concerns, whereas in India, the driving forces for IPM are maximizing crop production and cost of crop production. Lack of public awareness about the ill-effects of chemical pesticides on human and animal health and environment is also detrimental to the cause IPM. It is hoped that with the economic development and globalization, public education and awareness will increase. Several prevailing circumstances and situations such as crop failures due to pest/disease attacks, increasing cost of inputs, in general, and pesticides, in particular, poisoning and health hazards due to pesticides usage and governments' thrust on sustainable agriculture, etc. are leaning in favour of IPM. The organic farming programmes are getting high priority in the governments and IPM is one of the components in organic farming. Thus, there is no doubt that the future of IPM is very bright and it is bound to be popular with the farmers in the years to come.

CONCLUSION

Though IPM research and extension is still in the embryonic stage, we have certainly made a modest beginning on several fronts of IPM. The IPM journey ahead is, of course, very challenging but rewarding. No doubt, concentrated efforts are needed at higher levels of the governments to remove the impediments in the path of IPM. Many factors such as rapid economic growth, globalization, increased public awareness about ill-effects of pesticides, frequent crop failures due to excessive and improper use of pesticide, exorbitant cost of agricultural inputs specially pesticides, increased health and environmental concerns will dominate the scenario in near future and will act in favour of IPM programme.

NB: The word 'pest' used in this text stands for animal and insect pests, diseases and weeds, etc.

30

Virus Diseases of Orchids: Diagnosis and Management

R.P. Pant, K.B. Pun and R.P. Medhi

[1]*National Research Centre for Orchids, Pakyong–737 106, Sikkim*
[2]*IARI Regional Station, Kalimpong–734 301, West Bengal*
E-mail: rajendrapdpant@yahoo.co.in

Orchidaceae, one of the most highly specialized families of the flowering plants is also one of the largest family comprising more than 800 genera and 30,000 species to which more and more new ones are being added every year. The family is widely distributed from Equator to Arctic Circle and from lowland areas to almost up to snowline regions. Orchids are capable of growing in nearly every environment of the earth's surface and on a variety of substratum ranging from growing on soil; perching on tree trunks; on rock surfaces; growing beneath the surface of the substratum and even on dead organic matter. They may or may not have chlorophyll. Orchids are not parasites, most of the species are epiphytes, which depend on other trees for support only, but they do not draw any nutrition from them. Orchids have adapted themselves to extremes of the environmental conditions, producing thereby great variations in vegetative forms, and one may often find it difficult to identify them as orchids if they are not in flower.

Viruses are widely distributed in vegetatively orchid plantations detrimentally affecting their ornamental value. There are more than 31 different viruses reported infecting different species of orchids in the world (Table 30.1). Among orchid viruses, cymbidium mosaic virus (CymMV) is the oldest virus that is very dangerous and reported from 26 species of orchids. (Francki, 1970). The second important virus is Odontoglossum ring spot virus (ORSV), which infects about 20 genera of orchids all over the worlds (Brunt *et al.*, 1996). CymMV and ORSV are considered the most important viruses due to their prevalence, worldwide occurrence and the severity of symptoms they produce on orchid genera. It generally reduces vigour, lowering floral quality and leading to economic losses. In orchids, mixed infection of viruses is also common as these viruses have similar biological and epidemiological characteristics and are frequently transmitted together (Freitas, 2003).

Table 30.1: Orchid viruses, particle morphology and their mode of transmission.

Sl. No.	Virus	Particle morphology	Transmission
1.	Bean yellow mosaic Potyvirus	Flexuous	Sap and aphids
2.	Brazilian bacilliform Rhabdovirus	Rod-like	-
3.	Calanthe mild mosaic Potyvirus	Flexuous	Sap and aphids
4.	Cerotobium mosaic Potyvirus	Flexuous	-
5.	Clover yellow vein Potyvirus	Flexuous	Sap and aphids
6.	Cucumber mosaic Cucumovirus	Isometric	Sap and aphids
7.	Cymbidium chlorotic mosaic Sobemovirus	Isometric	sap
8.	Cymbidium mild mosaic virus*	Isometric	Sap
9.	Cymbidium mosaic Potexvirus	Flexuous	Sap, tools and water
10.	Cymbidium ringspot Tombusvirus	Isometric	Sap and soil borne
11.	Dendrobium mosaic Potyvirus	Flexuous	Sap and aphids
12.	Dendrobium Rhabdovirus	Large bullet-shaped, bacilliform	-
13.	Dendrobium vein necrosis Closterovirus	Flexuous	-
14.	Dendrobium Rhabdovirus	Short bullet-shaped, bacilliform	-
15.	Cypripedium filamentous *Potyvirus*	Flexuous	-
16.	Grammatophyllum bacilliform *Rhabdovirus*	Short bullet-shaped, bacilliform	-
17.	Habenaria mosaic *Potyvirus*	Flexuous	Sap and aphids
18.	Impatiens necrotic spot *Tospovirus*	Quasispherical, enveloped	Sap and thrips
19.	Masdevallia isometric virus*	Isometric	-
20.	Laelia red leafspot *Rhabdovirus*	Large bullet-shaped, bacilliform	-
21.	Long orchid *Rhabdovirus*	Large bullet-shaped, bacilliform	-
22.	Odontoglssum rinspot *Tobamovirus*	Rigid rod	Sap tools and water
23.	Orchid fleck *Rhabdovirus*	Short bullet-shaped, bacilliform	-
24.	Phalaenopsis *Rhabdovirus*	Bacilliform	-
25.	Short orchid *Rhabdovirus*	Bullet-shaped, bacilliform	-
26.	Tobacco rattle *Rhabdovirus*	Rigid rod	Sap and nematode
27.	Tomato ringspot *Nepovirus*	Isometric	Sap and nematode
28.	Tomato spotted wilt *Tospovirus*	Quasispherical, enveloped	Sap and thrips
29.	Trichopilia isometric virus*	Isometric	-
30.	Turnip mosaic *Potyvirus*	Flexuous	Sap and aphids
31.	Vanilla mosaic *Potyvirus*	Flexuous	Aphids

- = not known; *= group not established

SYMPTOMS

Viruses produce different kind of symptoms depending on the orchid species, host part and environmental conditions. Generally, six types of symptoms have been recorded on orchids upon virus infection (Department of Crop Science, University of Illinois at Urbana Champaign (RPD No. 614, May 1990).

1. Leaf Necrosis, Black Streak

Leaf necrosis, caused by the cymbidium mosaic virus, is probably the most common virus disease of many kinds of orchids. Infected plants have irregular, brown-to-black, elongated spots and streaks of dead tissue on both surfaces of older leaves (Figs. 30.1A and 30.1B). Infected leaves that show symptoms tend to age quickly and dry up. Flowers from such plants the usually symptom less, and open in an unthrifty manner and fewer in number and smaller in size. Leaf necrosis has been reported in *Cattleya* and its hybrids and species of *Angraceum, Cymbidium, Epidendrum, Laelia, Oncidium, Spathoglottis, and Zygopetalum*. Certain

Fig. 30.1. A and B. Symptoms showing mosaic, mottling, necrosis and sunken pits on leaves of Cymbidium sp. due to infection of CymMV and ORSV.

strains of the tobacco mosaic virus also produce leaf necrosis on Cattleya, Cymbidium, and others orchids that closely resemble symptoms produced by the cymbidium mosaic virus.

2. Mild Flower Break

This is caused by a strain of tobacco mosaic virus. It is common in *Cattleya.* Affected plants are malformed and the leaves show mild mosaic symptoms. The means of transmission of this virus is unknown. Mild flower breaking has been reported on about 30 species and hybrids of *Cattleya.* This virus also produces spots and rings of dead tissue on *Odontoglossum* and diamond mottle on *Cattleya, Cymbidium, Odontoglossum*, and *Phalaenopsis*. Diamond mottle is distinguished by elongate chlorotic areas that are often diamond-shaped. Older leaves sometimes develop brown-to-black flecks and streaks.

3. Severe Flower Break

This conspicuous virus disease of Cattleya and its hybrids is characterized by rolling and twisting of the sepals and petals and variegation in the flowers. The normal pigment of the petals, sepals and petioles is replaced by irregular streaks, blotches, ring and line patterns, which are either more or less intense than the normal colour. Leaves that develop after infection are sometimes twisted and mottled with streaks of light and dark green. The dark-green areas are somewhat raised, producing ridges and bumps. The severe flower-breaking virus is transmitted by the green peach aphid (*Myzus persicae*) by grafting, or by juice inoculation. The disease is also reported in *Cattleya* and its hybrids.

4. Symmetrical Flower Break

Symmetrical variegation has been reported in *Cattleya.* The pigment occurs along the sepal margins and over most of the petals except in middle areas that have little or no pigment. The leaves develop an inconspicuous mottle. It is generally sap transmitted.

5. Blossom Necrotic Streak

This kind of symptom is produced by a strain of cymbidium mosaic virus that causes latent streaks or spots on orchid blossoms. The blossoms first appear symptomsless but brown (necrotic) spots and streaks become visible after one or more weeks. Delay in the appearance of flower symptoms after the blossoms open creates a problem for growers who sell their blooms. Infected flowers may develop necrotic symptoms 3 to 5 days after they are cut from plant. Long, yellowish, irregular streaks may develop in the leaves. Transmission of this strain of cymbidium mosaic virus occurs on contaminated pruning tools. It is reported in *Cattleya* and its hybrids and species of *Angraceum, Cymbidium, Epidendrum, Laelia, Oncidium. Spathoglottis,* and *Zygopetalum.*

6. Ringspot

Cymbidium ringspot virus also causes single or concentric ringspot patterns on old and young leaves. The rings may later be brown to purple-black enclose a central light green or yellow "island". Rings may overlap or merge to form large compound patterns. Ringspot may cause death of young shoots and often of the entire. Species of *Cattleya, Cymbidium, Spathoglottis,* and *Trichosma* are affected. Ringspots symptoms may also result from double virus infections, such as a strain of tobacco mosaic virus and cymbidium mosaic virus.

DIAGNOSTICS AVAILABLE FOR ORCHID VIRUSES

Bioassay

Under certain conditions, some diseased orchid plants do not show distinct symptoms and control measures cannot be carried out effectively. In such instances, a simple diagnostic test is required to determine whether a plant is infected with virus, or whether the symptoms are due to other causes. Biological tests can be performed on young indicator plants outside the Orchidaceae. The common indicators plants for orchid viruses are listed in Table 30.2. The greatest advantage of bioassay is that it is cost intensive and requires no special skill. However, it has numerous limitations. It is reliable under certain conditions and is time-consuming, labour intensive and requires lots of space in the glasshouse and also certain environmental conditions.

Table 30.2: List of orchid viruses and their known bioassay hosts.

Sl. No.	Name of the virus	Bioassay host	References
1.	Cymbidium mosaic virus (CymMV)	*Cassia,sp.chenopodium amaranticolor, C.quinoa, Datura stramonium*	Corbett, 1960; Francki, 1970; Hollings,1966; Inouye, 1965 & 1968.
2.	Odontoglossum ringspot virus (ORSV)	*Chenopodium sp. Nicotiana sp. Gomphrena globosa, tetragonium expansa.*	Kado and jenson, 1964; Inouye, 1965; kado *et al.*,1968; Thomson and Smirk, 1967
3.	Bean yellow mosaic virus	*C.quinoa, vicia faba*	Bos, 1970
4.	Dendrobium mosaic virus	*C.quinoa and C.amranticolor*	Inouye, 1976
5.	Clover vein yellow virus	*Viscia faba, P.sativum*	Inouye, 1976
6.	Turnip mosaic virus	*Nicotiana clevelandii and C.quinoa*	Leseman and Vetten, 1985
7.	Tobacco rattle virus	*N.tabaccum cv.xanthi*	Leseman and Vetten, 1985

Table Contd.

Table 30.2: Contd.

Sl. No.	Name of the virus	Bioassay host	References
8.	Cymbidium ringspot virus	*C.amrantocolor, N.clevelendii, Phaseolus vulgaris cv. Prince*	Holling and Stone, 1963; Holling, 1966 & 1977
9.	Cucumber mosaic virus	*C.amranticolor, citrullus vulgaris and D.stramonium*	Inouye, 1969
10.	Cymbidium mild mosaic virus	*C.amranticolor,*	Chang *et al.*, 1978
11.	Tomato ringspot virus	*N.tabaccum cv.kentuckey 35 and Vigna anguiculata*	Goff and Corbett, 1977
12.	Orchis Rhabdovirus	*Chenopodium amaranticolor, C.quinoa, N.clevelandii*	Lesemann and Doraiswamy 1975 a
13.	Orchid flec virus	*Chenopodium amaranticolor, C.quinoa, N.glutinosa, Petunia hybrida, Tetragonium expensa and Vigns sinensis*	Chang *et al.*, 1976, Doi *et al.*, 1977. Kondo *et al.*, 1995
14.	Tomato spotted wilt virus	*Petunia hybrida, N.clevelandii*	Hu *et al.*, 1992
15.	Impatiens necrotic ringspot virus	*P.hybrida and N.clevelandii*	Mayhew *et al.*, 1992

Electron Microscopy

Electron microscopy is an important tool for virus detection. It gives us an idea of shape and size of virus particle in a very short time. Like other viruses, EM has also been used for

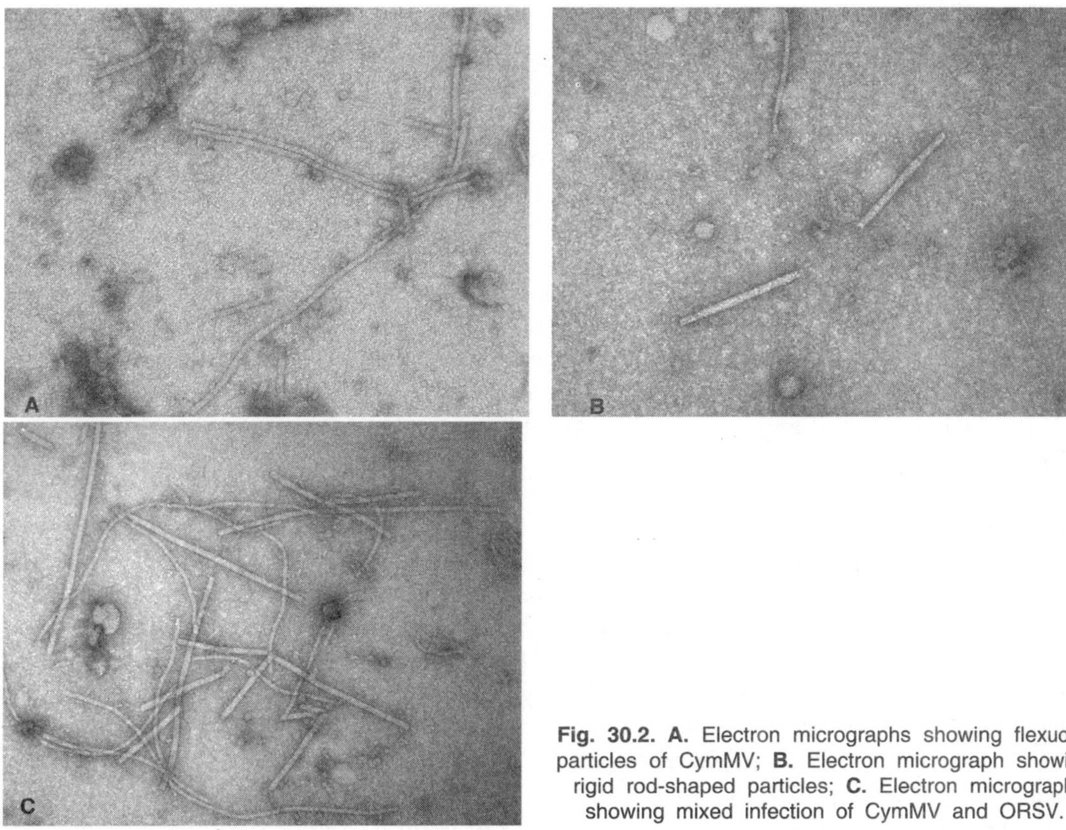

Fig. 30.2. A. Electron micrographs showing flexuous particles of CymMV; **B.** Electron micrograph showing rigid rod-shaped particles; **C.** Electron micrograph showing mixed infection of CymMV and ORSV.

detection of orchid viruses. Leaf dip preparation is a common method used for routine diagnosis or to see the virus particle from crude sap or partially purified/purified preparations. CymMV is a flexuous virus measuring about 450 to 500 nm while ORSV is a rigid rod measuring about 310 nm. In nature, both the viruses occur in mixed infection in many species of orchids. Ultra thin sectioning has also been used for location of viruses in host tissue and virus induced inclusion bodies, which is helpful in virus detection. CymMV and ORSV fail in bioassay and serological tests, but are detected by EM (Lawson and Brannigan, 1986). Immunosorbent electron microscopy (ISEM) was found to be rapid diagnostic method for CymMV and ORSV. The method showed that ORSV was more frequent than CymMV, but neither virus were detected in seedlings from seed or meristem.

Serodiagnostic Techniques

Serological techniques have been extensively used for detection of low concentrations of virus in orchid bulbs. These techniques are especially useful for large-scale testing because many samples can be tested at a time. Immunodiffusion assay in agar gel has been developed and used successfully for detection of CymMV and ORSV using high titer antisera of these viruses. Wisler *et al*., 1982 used this assay successfully at three nurseries in USA and it was found that incidence of CymMV was greater than ORSV. Micro agglutination test has also been used to diagnose CymMV and ORSV in Japan and gave reliable results. Result obtained by biological tests and electron microscopy gave exact agreement and serology tests showed 99.4% agreements with the other tests (Inouye, 1977). The latex agglutination test (L) and a modification of this test, protein A-coated latex (PAL) were also standardized for detection of CymMV. The minimum detectable concentration of virus was similar for both L and PAL reaction at 0.104 mg/ml. The detectable limit in Oncidium sap was similar at a dilution of 1:2560. PAL appeared as more sensitive that L at a higher globulin dilution of 1:512, and in general, PAL demonstrated greater sensitivity than L in the orchid species tested (Samad and Ari, 1993).

With the development of more sensitive serodiagnostic technique of Enzyme Linked Immunosorbent Assay (ELISA) (Clark and Adams, 1977), virus testing became comparatively easier and simpler not only in orchids but also in other plants. Hu *et al*., 1993 used ELISA and indexed 3600 orchid plants representing 44 genera from three orchid collection, 22 commercial farms and six nurseries on the islands of Oahu and Hawaii. CymMV and ORSV were detected in 25% of the plants and among these 20% genera represented double infection of both viruses. Tomato spotted wilt virus (TSWV) was detected in one nursery of *Oncidium.* Most commercial *Dendrobium* orchids grown in Hawaii are seed propagated hybrids produced by the Horticulture Department at the University of Hawaii (UH). Only 4% of Dendrobium hybrids were infected by CymMV and 94% of >3 years old UH Dendrobium was infected with CymMV. ORSV was not detected from any commercial grown UH Dendrobium hybrids. TSWV infection was found in *Oncidium* only in one nursery in Hawaii. *Dendrobium* mosaic *Potyvirus* was detected in *Dendrobium* super bum. Beetham and Moran (1988) used ELISA to index ORSV (using an antiserum of a Victorian Isolate of ORSV) in *Cymbidium* genus from 10 commercial and home garden collections in Metropolitan Melbourne, Victoria, Australia. DAC-ELISA was used in China to test 159 diseased orchid samples representing 20 species collected from 7 sites in Guandong province. CymMV was detected in 40 samples and ORSV was detected in 25 samples and mixed infection of CymMV and ORSV was found to be in 67 samples (HouGen *et al*., 1996).

DAS-ELISA was not found suitable using homo and heterologous antisera of ORSV and TMV for detection of both viruses due to lack of common epitopes on the virions (Wong *et al.*, 1994).

ELISA can be done using monoclonal antibodies (MAbs) or polyclonal antibodies (PAbs). It is recommended that a mixture of polyclonal and monoclonal antibodies in DAS-ELISA is mandatory for reliable detection of both ORSV and TMV-O in orchids (Dore *et al.*, 1987). However, this detection procedure will not find use as a routine method for nursery evaluation owing to cost and technical know-how involved (Ramachandran and Varma, 1991).

Other sensitive and rapid immunological techniques used in the detection of orchid viruses are Dot Immuno Binding Assay (DIBA) on nitrocellulose membranes and Rapid Immuno Filter Paper Assay (RIPA). These two techniques were evaluated to detect CymMV in several orchid genera including *Cymbidium, Cattleya, Dendrobium, Phalaenopsis* and *Vanda*. Minimum detection levels of CymMV using these methods were 100 ng/ml in purified preparations, and CymMV was detected in a 10^{-4} dilution of extracts from infected orchid leaves. Though DIBA took longer time to detect CymMV, it was reliable for diagnosis of a large number of samples. RIPA detected CymMV within a few minutes and sensitivity was similar to that of DIBA. It is suggested that RIPA is suitable as a rapid tool for diagnosing virus diseases in orchids. RIPA could detect both CymMV and ORSV simultaneously in extracts from a plant infected by both viruses (Gara *et al.*, 1997). RIPA was used in large-scale detection of CymMV and ORSV in 43 nurseries in 8 provinces of Thailand (Tanaka *et al.*, 1997).

La *et al.*, (1999) applied Tissue Blot Immuno Assay (TBIA) to detect CymMV and ORSV from infected *Cymbidium* and *Cattleya* plant in Korea. CymMV and ORSV were readily detected in tissue blots made from leaves and roots of infected plants both direct TBIA. CymMV and ORSV were efficiently detected in root than leaf samples by TBIA. Tissue blots made on polyvinylidene difluoride membrane gave much clearer specific colour reactions than those made on nitrocellulose or nylon membrane.

Nucleic acid hybridization based techniques using specific probes have been variously used for detection of orchid viruses. A Tissue Print Hybridization Assay was standardized to detect and locate viruses in orchids. In this method orchid leaves are cut surface are printed on nitrocellulose paper. The nucleic acid on the nitrocellulose membrane is then hybridized to radio labelled cDNAs of CymMV and ORSV and the viruses are detected by autoradiography. The assay is able to trace as little as 20 pg of purified virus that is 2-3 folds more sensitive than ELISA for CymMV. This method is also able to show the distribution of viruses within the plants (Chia *et al.*, 1992). Tissue print hybridization with cDNA probes was applied to detect directly from infected leaves of *Phalaenopsis* species (Wang *et al.*, 2000).

Further improvement took place in the nucleic acid hybridization techniques and a Slot Blot Hybridization Assay was developed and standardized using either crude extracts or total RNA from infected leaves. Prior to that, DIG labelled sense and antisence cRNA probes were synthesized from cDNA clones of CymMV and ORSV for detection in infected plants. This assay detected 50 pg and 20 pg of purified CymMV and ORSV RNA, respectively. As little as 30 mg of *Nicotiana benthanmiana* infected leaves was sufficient to provide positive detection. CymMV and ORSV were detected at 3125 and 625 times dilution of leaf extracts respectively. The advantage of this method is that the DIG labeled cDNA probes is stable for more than one year. This method is sensitive and reliable and is suitable for large-scale routine testing of plant viruses. By using the two DIG labelled cRNA probes *in situ,* CymMV and ORSV were located in systemically infected leaves and stems of *Nicotiana benthamiana* and orchids (Hu *et al.*, 1998).

DOT Blot ELISA was found to be faster and cheaper than all others methods (Albouy *et al.*, 1994). Detection of CymMV by ISEM in crude sap preparation of infected leaves was two times more sensitive than ELISA, while DOT Blot Immuno Assay was 8 times more sensitive than ELISA. Bean yellow mosaic virus isolated from Masdevallia and Calanthe could not be detected by either DAS-ELISA or antigen coated indirect ELISA directly from the infected orchids using polyclonal antisera. But the detection was improved when MAbs were used in a DOT Blot ELISA or nitrocellulose membrane (Hammond and Lawson, 1988).

PCR BASED TECHNIQUES

PCR based techniques were recently introduced for the detection of orchid viruses. This technique is used to amplify the desired region of genome that could be synthesized and used in RT-PCR (Reverse Transcription Polymerase Chain Reaction). As sequences of orchid viruses are known, this technique is routinely used to detect orchid viruses due to its sensitivity and can detect single copy virus genome in the infected plant (Hull, 1993). Polymerase Chain Reaction was successfully used for rapid detection of CymMV using two oligonucleotide primers from regions homologous to potex viruses and CymMV in 30 cycles of PCR, which resulted in amplicons of 313 bp (Lim *et al.*, 1993a). PCR was also tested to detect CymMV and ORSV in plantlets obtained from thin section cultures of infected plants and PLBs with ribavirin treatment (Lim *et al.*, 1993b). A combined assay of Reverse Transcription and the Polymerase from virus infected (RT-PCR) was performed with 20-mer ORSV specific primers and crude nucleic acid extracts from virus (Rye-KiHyun *et al.*, 1995). Similar results were also obtained when cp gene primers from 3′ terminal and 3′ non-coding regions were used for detection of CymMV in cymbidium plants (Loi *et al.*, 1995; Chung *et al.*, 1996). The RT-PCR is 103 times more sensitive, reproducible and time saving method than ELISA. It has been established that RT-PCR leads to quick diagnosis of CymMV and ORSV (Jeong *et al.*, 2001).

Immuno Capture PCR (IC-PCR)

It is an improved version of PCR and emerged as an efficient, reliable and rapid method to detect CymMV and ORSV and compared with binding PCR and ELISA. In this technique, Hawaii strain of each of CymMV and ORSV was purified from Dendrobium and Cattleya orchid, respectively to improve virus detection and to facilitate genetic engineering. The ORSV coat protein (CP) and 54k putative replicase component gene and CymMV and movement protein (MP) genes were cloned and sequenced. The sequences of these genes were compared to those of CymMV and ORSV from Singapore, Japan, Korea and Germany. The high degree of sequence identity (91-99%) at the nucleotide level for all gene sequences analyzed shows that CymMV and ORSV from different countries are closely related (Barry *et al.*, 1996). They observed that IC-PCR is 10-100 folds more sensitive than ELISA and is able to detect CymMV from the equivalent of 100 mg of Dendrobium leaf tissue. Therefore, IC-PCR can be used to confirm samples, which are inconclusive in ELISA.

TaqMan R real-time PCR

Recently, Chreng *et al.* (2000a) developed a highly sensitive novel detection technique called, TaqMan R real-time PCR, which combines RT-PCR with the power of fluorescent detection for simultaneous quantitation of two orchid viruses, CymMV and ORSV. Four TaqMan R probes were synthesized, targeting at the RNA dependent RNA polymerase (RdRp), and coat protein genes of both viruses detect as little as 5 fg each of CymMV and ORSV simultaneously with either the RdRp or CP genes the target. Another identical technique is the development of molecular beacons. It is a new approach to plant virus detection. This approach has been successfully applied to detect as little as 0.5 ng of viral RNA from Oncidium leaves simultaneously infected with CymMV and ORSV.

Matrix Assisted Laser Desorption-ionisation (MALDI)

Based on the molecular weights of coat protein of viruses, another rapid simultaneously detection method of the two most prevalent orchid viruses, CymMV and ORSV, was developed using Liquid Chromatography/Mass Spectrometry (LC/MS) and Mass Spectrometry. Using this method, molecular weights of coat proteins of CymMV and ORSV were accurately estimated and detected from 1g of virus infected flower of Oncidium (Tan *et al.*, 2000). Tan and his associates claimed that since LC/MS and MALDI perform automated analyses of multiple samples with simple preparation steps, both the techniques are ideal for virus detection in large number of samples from infected plant extract.

Molecular Beacons

Molecular beacons are single-stranded nucleic acid molecules with a stem loop conformation (Tyagi and Kramer, 1996). The stem portion consists of complementary sequences at 5' and 3' terminals of the molecule, while the loop portion consists of probe sequences that are complementary to the target sequences of choice. Eun and Wong (2000) used a fluorescent moiety attached to one end and quenching moiety attached to the opposite end. RT-PCR was carried out with primers that amplified specific genome sequences of interest, yielding targets complementary to their respective molecular beacons for subsequent detection. This technology has detected 0.5 ng of viral RNA of both orchid viruses.

Quartz Crystal Microbalance (QCM)

Immunosensors

QCM immunosensors is the adsorption of substances on the surface of a quartz crystal changes its resonance oscillation frequency. A QCM immunosensors has been developed for the detection of CymMV and ORSV by pre-coating the QCM with virus specific antibodies (Eun *et al.*, 2002). Upon binding of virions with immobilized virus antibodies, the increase in mass at the QCM surface resulted in a reduction in the frequency of resonance oscillation in a manner dependent upon the amount of virus bound. It has detected 1 ng of viruses of both viruses.

DNA based Biosensors

Nucleic acid hybridization forms the basis of numerous QCM applications for the detection and analysis of genetic materials (Konig and Gratzel, 1993; Kosslingrt *et al.*, 1992). The nucleic acid based CymMV and ORSV QCM biosensors are more sensitive than the corresponding antibody-based ones (Eun *et al.*, 2002). This could be due to the mechanism of probe and target recognition. QCM immunosensors possess immobilized antibodies, which recognize the epitopes on the viral coat protein, whereas QCM DNA biosensors posses immobilized nucleic acid sequences, which hybridize with their complementary virus CP gene sequences. QCM DNA biosensors offer the capability of detecting specific gene of interest.

Table 30.3: Sensitivity of different diagnostics.

Sl. No.	Method	Sensitivity
1.	Indirect ELISA	50-100 ng
2.	DAS-ELISA	6.25 ng
3.	Rapid Immuno Filter Paper Assay (RIPA)	100 ng
4.	DOT Immuno Binding Assay (DIA)	100 ng
5.	Slot Blot Hybridization Technique 250 ng (ORSV)	50 ng (CymMV)
6.	Tissue print Hybridization Assay	20 ng
7.	PCR	Single copy of virus genome
8.	RT-PCR	10 fg
9.	Immuno Capture PCR	100 ng (CymMV), 10 ng (ORSV)
10.	Taq Man R real-time PCR	5 fg each of CymMV and ORSV
11.	Molecular beacons	0.5 ng of CymMV and ORSV

MANAGEMENT APPROACHES FOR ORCHID VIRUSES

The basic need of any effective programme is quick and reliable detection of the pathogen that needs to be managed. Various methods are now available for virus detection with different levels of sensitivity, which makes the choices more difficult. In large-scale detection particularly from standpoint of international trade and quarantine, what is needed is an easy-to-use, readily available and affordable diagnosis kit, be it based on protein or nucleic acid component for virus diagnosis. For instance, if a grower needs only to evaluate for the mere choose an imprint hybridization using a readily available DIG labelled broad-spectrum probe, which would help screening a large number of samples in a short time (Ramachandran and Varma, 2001).

Cultural practices play a very significant role particularly as all orchid viruses are vegetatively propagated. Strict sanitation practices were recommended for control of CymMV (Hu and Ferreira, 1994). Hot water treatment of infected planting for material was found ineffective for eradicating orchid viruses in Malaysia (Ganapathi *et al.*, 1985). Clean cultivation by avoiding the sources of inoculum is best requiring minimum inputs, but unfortunately this is not practiced by orchid growers as ornamental orchids are by far the most expensive and hardly anyone would agree to discard the virus infected ones. This also hampers any efforts made from the point of view of quarantine. Often, the plants or clones, which need to be multiplied on large scale by tissue culture, are only tested by the grower himself or given to a stats-owned servicing agency.

Propagations of orchids by *in vitro* techniques have been practiced for producing large quantities of planting material. Morel (1960) applied this to obtain virus free clonal material of orchids by standardizing meristem tip culture. It has been observed that majority of plant viruses do not infect the meristematic dome despite severe systemic infection of the plant, and therefore, this region can be excised and regenerated to plants. Since then, the technique has been used routinely for a large number of orchids. But the monopodial orchids such as *Aranda* and *Vanda*, which comprise most or the South-East Asia's cut-flower production, *in vitro* propagation is a difficult option and therefore, large quantities of virus-indexed replacement stock is hard to obtain. Tissue culture raised plants also need to be indexed for freedom from viruses and for the maintenance of germplasm and its exchange.

Chemotherapy has been used either singly or in combination with tissue culture. In most instances, success was not achieved with different substances such as skimmed milk, Ivory liquid detergent, physan, Agribro and ethanol even in laboratory tests for control CymMV. Cholrax (5.25% sodium hypochloride) and 1% NaOH were, however, effective against both CymMV and ORSV (Hu *et al.*, 1994). Albouy *et al.* (1988) used an antiviral agent Virazole (Ribavirin) in solid medium culture of *Cymbidium protocorms*. It has been observed that five subcultures of 18 days interval in the presence of 25 mg/l Virazole yielded 95% virus-free plantlet and increased the efficiency of eradication of both CymMV and ORSV. Virazole was also found effective for obtaining virus-free Dendrobium shoot tips and callus from parent plants infected by CymMV and ORSV (Loi *et al.*, 1991) and in eradication of ORSV from *in vitro* cultures of *Cymbidium* (Toussaint *et al.*, 1993). In another study, meristem culture and thin section culture with chemotherapy were used to generate virus-free orchid plants. It was reported that in contrast to the meristem culture, high percentages of virus-free plantlets were obtained from thin section cultures of infected plantlets and protocorm-like bodies with Ribavirin treatment. Regenerants from thin section cultures without Ribavirin treatment were also free from CymMV and ORSV (Lim *et al.*, 1993a, b). Prakash *et al.* (1996) have also highlighted the importance of thin section culture as an efficient micro propagation method as also an effective procedure for elimination of orchid viruses alone or combination with Ribavirin.

Management through Meristem Culture

Morel (1960) was the first to develop revolutionary techniques of meristem culture to produce virus- free Cymbidiums through meristem culture, which established the techniques as a clonal propagation procedure. Morel standardized meristem tip culture to obtain virus free clonal material of *Cymbidium* (such as *C. talma, C. doris* or *C. candeur*) based on the concept that majority of plant viruses do not infect the meristematic dome despite severe systematic infection of the plants, and therefore, this region can be excised and regenerated to plantlet (Morel, 1964). Since then meristem culture has been routinely done for a large number of orchids to obtain CymMV free plants. Ishii (1974) successfully eliminated CymMV from double infected mericlone tissue pieces grafted onto callus tissue but ORSV remained infections. However, true meristem culture, which uses the actual meristem less than 0.5 mm, takes longer time to regenerate plants. Sometimes, little or no survival may result when tissue pieces are less than 0.5 mm in length. On the other hand, tissue pieces of 0.5-1.0 mm showed an increase in survival but remained virus infected. Consequently, the meristem tip culture, which was once thought to be the best method to obtain CymMV and ORSV free orchids has no longer been accepted as a means of clonal propagation techniques for virus-free plants.

MANAGEMENT THROUGH MERISTEM CULTURE WITH CHEMOTHERAPY

Further attention was paid to develop an alternative of meristem tip culture to get quantity of virus-free orchid stocks. In the early nineties, new concepts of meristem tip cultures with chemotherapy came into the picture. In this improved method, meristematic tip plant tissue of *Cymbidium* hybrid Ayako Tanaka "Honodezuru" is cultured *in vitro* in association with antiviral agent ribavirin (Virazole). Results revealed that all plants regenerated from ribavirin from media were still infected by ORSV, whereas all treatments with ribavirin (with the exception of one) provide at least one virus-free plant. However their efficiency varied with the concentration of the ribavirin, the way it was added to the medium and the type of medium to which it was added, resulting in 8.3-100% virus-free orchids. Thus, association of meristem tips culture and chemotherapy *in vitro* was effective for the production of ORSV free Cymbidium (Freitas and Rezende, 1998).

Kuehnle (1996a) tested virucides at various concentrations and treatment times using protocorms like bodies (PLBs) of *Dendrobium* (Jaquelyn Thomas "Uniwai Mist") infected with CymMV. These chemicals were added during the liquid culture phase of PLBs and also during solid medium phase of plant formation from the PLBs. Among the viricides, ribavirin or dithiouracil eliminated virus infection. In liquid culture chemotherapy, the best results were obtained with 53% virus free tissue in 0.1 mM Ribavirin and 63% virus free tissue in 0.2 mM dithiouracil. A five weeks treatment was sufficient. In addition, rapidly growing cultures in liquid medium also increased the recovery of virus free material with up to 41% of virus control. Most exciting results were obtained with a combination of antiviral compound in liquid and solid medium tissue culture. Mericloning in 0.2 mM dithiouracil (liquid) followed by plant production on 0.1 mM dithiouracil (solid) produced up to 100% virus free platelets. Plants remained free for riding Dendrobium of CymMV at clonal propagation stage and concluded that meristem treatments with ribavirin gave better eradication than that of callus tissues.

MANAGEMENT THROUGH THIN SECTION CULTURE WITH CHEMOTHERAPY

Thin section culture with chemotherapy was established as an efficient method of eradication of orchid virus by various researchers. It was reported that thin section cultures of infected plantlets and protocorms like bodies with chemotherapeutic treatment of Ribavirin appeared to be better than meristem culture with chemotherapy. Lim *et al.* (1993a) demonstrated that regenerates from thin section cultures of infected monopodial orchid hybrid (Mokara Char Kauan Pink) with or without Ribavirin treatment were also free from CymMV and ORSV. However, the percentage of regenerates free from CymMV and ORSV varied. The regenerates from thin section (TS) culture of shoot tips were practically all free from both CymMV and ORSV, whereas regenerates from TS of PLBs were only about 60% free. Prakash *et al.*

(1996) have also tested the suitability of TS cultures with chemotherapy of ribavirin (virazole) for the production of virus-free plantlets in several other orchids including *Aranda, Cattleya, Dendrobium, Oncidium*, and *Vanda* and suggested that TS culture could be used for mass propagation of orchids as well as for the elimination of viruses from the infected plants alone or in combination with virazole.

ERADICATION OF VIRUS THROUGH ANDROGENESIS

Eradication of CymMV from *Dendrobium* through androgenesis coupled with chromosome doubling with colchicines treatment during PLB culture was reported in Hawaii (Kuehle, 1996a). Androgenesis refers to the production of haploid plant with only the parental chromosome set. In *Dendrobium*, ovule androgenesis occurs after fertilization of the egg cell with the elimination of maternal nucleus. It was first observed by Kamemoto and reported in details by Amore and Kamemoto (1991) that amphidiploid plants crossed with the triploid D.x Luois Bleriot as female gave rise to plants identical to the amphidiploid male parent but with half the usual number of chromosomes (i.e. haploid). Since the haploid plants go through seed, they become free virus. Chromosome doubling of the androgenetic diploid with colchine (during PLB culture) then restores the tetraploid chromosome number. This in effect enables recovery of virus-free amphiploid practically, identical to the original parental plants. This natural phenomenon was applied to obtain free breeding plants (Amore *et al.* 1996).

MANAGEMENT THROUGH GENETICALLY ENGINEERED VIRUS RESISTANCE

Genetic engineering coupled with tissue culture techniques offers a useful way to introduce specific genes into plants. The development of a suitable gene transfer system for *Dendrobium* orchids would provide the breeder with greater opportunity to produce commercially desirable hybrids. Several gene transfer methods were tested with *Dendrobium*. Among these micro particles bombardment (Klein *et al.*, 1987) was identified by two separate research groups as highly suitable for transformation of *Dendrobium* based high frequency recovery of transgenic plants. DNA sequences thought effective against CymMV and ORSV were introduced into commercial cut flower varieties of *Dendrobium* (Barry *et al.*, 1995). Selected regenerated, engineered plants were mericloned and are being grown in the greenhouse to permit meaningful replicated disease trials. Dendrobium breeders can now proceed with increased confidence to provide producer with virus-free starting material, and in near future, new varieties possessing genes for virus resistance. With disease oriented molecular breeding programmes already well underway in Singapore or being established in South Korea, New Zealand and elsewhere using a variety of orchid genera (Kuehnle 1996b) and the threat of disease from CymMV promises to be reduced on a global scale in the new millennium.

It is needless to emphasize the economic importance of orchids, which are best known for their most attractive flowers and hence dominate their international cut-flower trade. They are perhaps the most expensive ornamentals known today. The vanilla flavour is derived from seed capsules ('beans') of *Vanilla planifolia* and is second most expensive spice after saffron. Even the wild growing orchids are highly treasured as rate botanical specimen. Each plant or clone thus bears a price tag and is made available to them regarding virus infection and the potential threats arising due to such infection. Few come forward to accept it in terms of eradicating the virus-infected plants from their collection. Therefore, viruses remain widespread in cultivated orchids, even though the requisite to educate the commercial growers about the importance of producing virus-free plants will add value to their produce and prevent further spread of the viruses naturally as well as intentionally (Ramachandran and Varma, 2001).

REFERENCES

Albuoy, J., Flouzat, C., Kusiak, C. and Tronchet, M. (1988). Eradication of orchid viruses by chemotherapy from *in vivo* cultures of *Cymbidium. Acta Hort.* **234:** 413-420.

Albuoy, J., Kusiak, C., Louanchi, M., Wang, W., Mokra, V., Brunt, A., Derk, T. and Zaayen, A.V. (1994). Evaluation of rapid ELISA procedures for detecting viruses inpropagation stock of Pelargonia stock of Pelargonia, Dahlia and Orchids. *Acta Hort.,* **337:** 189-195.

Amore, T. and Kamemoto, H. (1991). Andogenesis in *Dendrobium. Lindleyana*, **6:** 117-120.

Amore, T., Kamemoto, H., Kuehnle, A.R. and Sugii, N. (1996). Rejuvenation of virused amphidiploid Dendrodum breeding plants. *Univ. of Hawaii Hort. Digest,* No. **110:** 3-4.

Barry, K., Hu, J.S., Kuehnle, A.R. and Sugii, N. (1995). Nuclear cloning and sequences analysis of Cymbidium mosaic and Odontoglossum ring spot viruses. *Phytopathology*, **85:** 1184.

Barry, K., Hu, J.S., Kuehnle, A.R. and Sugii, N. (1996). Sequences analysis and detection using Immuno capture PCR of Cymbidium mosaic and Odontoglossum ring spot viruses in Hawaiian orchids, *J. Phyto Path..* **144:** 179-186.

Beetham, P.R. and Morani, J.R. (1988). ELISA indexing of Odontoglossum ring spot virus using urease antibody conjugate and its use in field surveys. *Acta Hort.* No **234:** 243-248.

Bos, L. (1970). Bean Yellow mosaic virus. CMI/AAB. *Description of Plant viruses.* No. 40. Common Wealth Agricultural Bureaux Slough, England.

Brunt, A.A., Crabtree, K., Dallwitz, M.J., Gibbs, A.J., Watson, L. and Zurcher, E.J. (Eds.) (1996). *Plant Virus online*: Descriptions Lists from the Vide Database. Version: 20th August 1996. URL http://biology.anu.edu.au/groupsMES/vide/.

Chang, M.U., Arai, K., Doi, Y. and Yora, K. (1976). Morphology and intracellular appearance of Orchid fleck Virus. *Ann. Phytopath. Soc. Japan.,* **42:** 156-57.

Chang, M.U., Doi, Y. and Yora, K. (1978). Cymbidium mild mosaic virus, isolation Cymbidium mild mosaic virus. *Korean J. Plant Prot.*, **17:** 131-138.

Cherng, Eun-Jin., Seoh, M., SekMan Wong., Eun, J.C.A., Seoh, M.L. and Wong, S.M. (2000a). Simultaneous quantitation of two orchid viruses by TagMan R real time RT-PCR. *J. Virol. Methods*, **87:** 151-160.

Cherng, Eun-Jin., SekMan Wong., Eun, J.C.A., Seoh, M.L. and Wong, S.M. (2000b). Molecular beacons: a new approach to plant virus detection detection. *Phytopath.*, **90:** 269-275.

Chia, T.F., Chan, Y.S. and Chua, N.H. (1992). detection and localization of viruses in orchids bu tissue print hybridization. *Plant Path.* **41:** 355-61.

Chowdhery, H.J. (2001). Orchid Diversity in North East. *J. Orchid Soc. India.* **15:** 1-17.

Chung, S.Y., Yoon, K.E., Ryu, K.H., Park, W.M., Chung, S.Y., Yoon, K.E., Ryu, K.Y. and Park, W.M. (1996). Detection of Cymbidium mosaic virus from orchids by reverse transcriptase and polymerase chain reaction amplification of the viral coat protein of the viral coat protein and 3' non-coding region sequence. *J. Korean Soc. Hort. Sci.* **37:** 158-165.

Clark, M.F. and Adams, A.M. (1977). Characteristics of micro plates method of ELISA for the detection of plant viruses. *J. Gen. Virol.*, **34:** 475-83.

Corbett, M.K. (1960). Purification by density gradient centrifugation, electron microscopy and properties of Cymbidium mosaic virus. *Phytopath.* **50:** 346-251.

Doi, Y., Chang, M.U. and Yora, K. (1977). Orchid fleck virus. CMI/AAB. *Description of plant viruses.* No. 183.

Dore, I., Dekker, E.L., Porta, C. and Regenmortel, M.V.H. (1987). Detection by ELISA of two tobacco viruses in orchids using monoclonal antibodies. *J. Phytopath.*, **120:** 317-32.

Eun, A.J.C. and Wong, S.M. (2000). Molecular beacons: A new approach to plant virus detection. *Phytopathology,* **90:** 269-275.

Eun, A.J.C., Huang, L.Q., Chew, F.T., Li, S.F.Y. and Wong, S.M. (2002). Detection of two orchids viruses using quartz crystal microbalance (QCM) immunosensors. *J. Virol. Meth.* **99:** 71-79.

Francki, R.I.B. (1970). CMI/AAB Description of plant viruses. No. 27. Common wealth Agriculture Bureau Slough, England.

Freitas, A.J. (2003). Orchid viruses: Prevention remains the best defense against these pathogens. *Orchids,* **27:** 578-580.

Freitas, A.J and Rezende, (1998). Odontoglossum ringspot virus free Cymbidium obtained through meristem tip chemotherapy. *Fitopath. Brasi.* **23:** 158-160.

Ganapathi, A., Chinnathambi, C. and Chullakaset. (1985). Some virus free disease of orchids and an attempt at their control. *In: Proc. Reg. Conf. Pl. Quar.Supp.Agric.Devel.* pp.87-97. Kuala Lumpur. Malaysia.

Gara I.W., Kondo, H., Maeda, T., Mitsuhara, K. and Inouye, N. (1996). Further characterization of CymMV from Vanda orchid. *Bull. Res. Inst. Bioreso., Okayama Univ.,* **5:** 39-46.

Goff, L.M. and Corbett, M.K. (1977). Association of Tomato spotted wilt virus with a chlorotic leaf streak of Cymbidium orchids. *Phytopath.,* **67:** 1096-1100.

Hammond, J. and Lawson, R.H. (1988). A strain of bean yellow mosaic virus is phid transmitted. *Acta Hort.,* No **234:** 365-370.

Hollings, M. (1966). Local lesion and other test plants for the identification and culture of viruses.pp 230-241 in (Beemster, A.B.R. and Dijksta, J. Eds.). Viruses of plant. North Holland Pub. Co., Amsterdam.

Hollings, M. (1977). Cymbidium ring spot virus. CMI/AAB. *Description of plant viruses.* No. 178.

Hollings, M. and Stone, O.M. (1963). Cymbidium ring spot (previously undescribed virus). *Glasshouse crops Res. Inst. Ann Report.,* pp. 89.

Hu, J.S., Ferreira, S., Wang, M. and Xu, M.Q. (1993). Detection of Cymbidium mosaic, Odontoglossum ring spot, Tomato spotted wilt virus and Poty viruses infecting orchids in Hawaii. *Plant Dis.,* **77:** 464-68.

Hu, J.S., Wang, M., Ferreira, S. and Ogata, D. (1992). Tomato spotted wilt virus on Oncidium orchids in Hawaii. *Plant Dis.* **76:** 426.

Hu, W.W. and Wong, S.M. (1998). The use of DIG labled cRNA probes for the detection of Cymbidium mosaic Potex virus Odontoglossum ring spot tobamo virus in orchids. *J. Virol. Methods,* **70:** 193-199.

Hull, R. (1993). Nucleic acid hybridization procedures. *In: Diagnosis of Plant viruses* (Ed. Mathews, R.E.F.) pp. 253-71, CRC Press, Boca Raton.

Inouye, N. (1965). Virus diseases of orchids. II. Symptoms and properties of viruses in *Cymbidium. Japan Orchid Soc. Bull.,* **11:** 1-6.

Inouye, N. (1968). Some experiment on transmission of Cymbidium mosaic virus and Odontoglossum ring spot virus. *Nogaku Kenkyu,* **52:** 89-97.

Inouye, N. (1969). Cucumber mosaic virus isolated from *Dendrobium. Agric. Res. (kurashiki).,* **53:** 49-60.

Inouye, N. (1976). Dendrobium mosaic virus, *Ber. Ohara Inst. Landw. Biol. Okayama Univ.* **16:** 165-174.

Inouye, N. (1977). Serological diagnosis methods for Cymbidium mosaic and Odontoglossum ring spot virus in orchids. *Noga Kenk.,* **56:** 1-13.

Ishii, M. (1974). Partial elimination of virus from doubly infected orchids by meristem explant culture. *Acta Hort.,* No. **36:** 229-33.

Joeng J.H., Yang, D.C. and Paek, K.Y. (2001). Comparison of ELISA and RT-PCR for efficient diagnosis if Cymbidium mosaic and Odontoglossum ring spot viruses in *Cymbidium* hoosai Makino. *J. Korean Soc. Hort. Sci.* **42:** 336-340.

Kado, C.I. and Jensen, D.D. (1964). Cymbidium mosaic virus in *Phalaenopsis. Phytopath.,* **54:** 974-977.

Kado, C.I., Van Regenmortel, M.H.V. and Knight, C.A. (1968). Studies on some strain of tobacco mosaic virus in orchids. I. Biological, chemical and serological studies. *Virology,* **34:** 17-24.

Klein, T.M., Wolf, E.D., Wu, R. and Sandford, J.C. (1987). High velocity microprojectiles for delivering nucleic acids into living cells. *Nature,* **327:** 70-73.

Kondo, H.J., Matsumoto, J., Maeda, T. and Inouye, N. (1995). Host range some properties of orchid fleck virus isolated from oriental Cymbidium in Japan. *Bull. Res. Inst. Biores., Okayama Univ.* **3:** 151-161.

Kondo, H., Maeda, T., Mitsuhara, K. and Iinouye, N. (1996). Detection of the viruses occurring in Oriental Cymbidium in Japan. *Bull. Res. Inst. Bioreso., Okayama Univ.* **4:** 149-162.

Kuehnle, A.R. (1996a). Toward control and eradication of virus in Dendrobium by hybridization, tissue culture and Genetic engineering. *J. Orchid Soc. India,* **10:** 43-51.

Kuehnle, A.R. (1996b). Molecular Biology of Orchid: *In: Biology: Review and Perspectives,* VII (Ed. J. Additti) Kluwer Acadamic.

La, Ya. J., Han, J.H., Sim, G.B., Kim, B.D. and K.K. (1999). Detection of Cymbidium mosaic virus and Odontoglossum ring spot virus in orchid plants by Tissue Blot Immuno assay. *J. Korean Soc. Hort, Sci.,* **40:** 481-484.

Lawson, R.H. and Brannigan, M. (1986). Virus diseases of orchids. pp. 2-49 in *Hand Book on Orchid Pests and Diseases* (revised edition) Amer. Orchid Soc. Bill. West Plam Beach, Florida.

Lesemann, D.E. and Doraiswamy, S. (1975a). Bullet-shaped virus-like particles in chloritic and necrotic leaf lesions of orchids. *Phytopathol. Z.* **83:** 27-39.

Lesemann, D.E. and Doraiswamy, S. (1975b). Detection and electron microscopy characterization of long bullet shaped virus-like particles in cultivated orchids. *Phytopathol. Z.* **84:** 201-214.

Lesemann, D.E. and Vetten, H.J. (1985). The occurrence of tobacco rattle and turnip mosaic viruses in Orchids sp., and of an unidentified Potyvirus in *Cipripedium calculus. Acta Hort.,* **164:** 45-54.

Lim, S.T., Wong, S.M., and Goh, C.J. (1993a). Elimination of Cymbidium mosaic virus and Odontoglossum ring spot virus from orchid by meristem culture and thin section culture with chemotherapy, *Ann. Appl. Biol.,* **122:** 189-197.

Lim, S.T., Wong, S.M., Yoeng, C.Y., Lee, S.C. and Goh, C.J. (1993). Rapid detection of Cymbidium mosaic virus by Polymerase chain reaction (PCR). *J. Virol. Methods,* **41:** 37-46.

Loi, J.S., Lee, S.M., Lam Can, L.T., Fan, S. and Yiang W.H. (1991). Eradication of orchid viruses in *Dendrobium* sonia using virazole. *Singapore J. Primary Industries,* **19:** 16-22.

Mayhew, D.E., Cooke, A.L. and Raabe, R.D. (1992). Special report. A new virus is reported from *Phalaenopsis. Amer. Orchid Soc. Bull.,* **61:** 574-577.

Morel, G.M. (1960). Producing virus free Cymbidium. *Amer. Orchid Soc. Bull.,* **29:** 495-97.

Morel, G.M. (1964). Anew means of clonal propagation of orchids. *Amer. Orchid Soc. Bull.,* **31:** 437-77.

Prakash, L., Lee, C.L. and Goh, C.J. (1996). *In vitro* propagation of commercial orchids: an assessment of current methologies and development of a novel approach–thin section culture. *J. orchid Soc. India.* **10:** 31-41.

Ramachandran, P. and Varma, A. (2001). Viral diseases of orchids and their management. In: *Orchid: Science and Commerce.* (Eds. Pathak, P., Sehgal, R.N., Skhar, N., Sharma, M. and Sood, A.) Bishen Singh Mahendra Pal Singh, Dehradun, India, pp. 520.

Ryu, KiHyum, WonMok Park., Ryu, K.H. and W.M. (1995b). Detection by PCR of Cymbidium mosaic virus in orchids. *J. Phytopath.,* **143:** 643-646.

Samad, N.A. and Ari, Z. (1993). The use of antibody sensitized latex to detect Cymbidium mosaic virus in orchids. *Pertanika. J. Trop. Agric. Sci.,* **16:** 157-160.

Tan, WeiLing, SekMan Wong, Kini, R.M., Tan, W.L.S. and Wong, S.M. (2000). Rapid simultaneous detection of two orchid viruses using LC-and/or MALDI-Mass Spectrometry. *J. Virol. Methods,* **85:** 93-99.

Tanaka, S., Nishi, H., Ito, S., Iwaka, M. Kameya and Sommartya, P. (1997). Detection of Cymbidium mosaic Potex virus and Odontoglossum ring spot Tobamovirus from Thai orchids by Rapid Immuno filter paper assay. *Plant Dis.,* **81:** 167-202.

Thompson, A.D. and Smirk, A.A. (1967). An unusual strain of tobacco mosaic virus from orchids. *New Zealand J. Bot.,* **5:** 197-70.

Tuossaint, A., Kummert, J., Maroquin, C., Lebrun, A. and Roggemans, J. (1993). Use of Virazole to eradicate ORSV from *in vitro* cultures of *Cymbidium Sw. Plant Cell Tiss. Org. Cult.,* **32:** 303-309.

Tyagi, S. and Kramer, F.R. (1996). Molecular beacons: Probes that fluoresce upon hybridization. *Nature Biotechnol.* **14:** 303-308.

Wang, H.I., Wang Y.T. and Liao, L.J. (2000). Detection of Cymbidium mosaic Potex viruses by hybridization with cDNA probes or RT-PCR. *Plant Protection Bull. Taipei,* **42:** 25-42.

Wisler, G.C., Zettler, F.W. and Mu, L. (1987). Virus infection of *Vanilla* and other orchids in French Polnesia. *Plant Dis.,* **71:** 1129.

Wisler, G.C., Zettler, F.W. and Purclfull, D.E. (1982). A serodiagnostic technique for detecting Cymbidium mosaic virus and Odontoglossum ring spot viruses. *Phytopathology,* **72:** 835-37.

Yap, M.L., Lam, L.T.C., Yik, C.P., Kueh, J., Lee, S.M. and Loh, H.M. (1999). Eradication of virus from commercial orchids. *Singapore J. Primary Industries,* **27:** 11-19.

Zettler, F.W., Wisler, G.C., Elliott, M.S. and Ko, N.J. (1987). Some new, potentially significant viruses of orchids and their probable means of transmission. *Amer. Orchid Soc. Bull.,* **56:** 1044-1051.

31

Trichoderma as Biocontrol Agent for Disease Management

O.P. Singh and Anil K. Jain

Division of Plant Pathology, IARI, New Delhi–110 012
Janta Vedic College, Baraut, Baghpat–250 611, U.P.

It was as early as 1932 that Weindling (1932) showed that *T. lignorum* could be used to control certain plant pathogenic fungi. Since then *Trichoderma* spp., which are predominantly found in the soil throughout the world, have been tested for their biocontrol potential. Their strong, antagonistic behaviour towards other fungi and easy multiplication in the laboratory has resulted as the most favoured fungi for application as biocontrol agents.

TAXONOMY AND MORPHOLOGICAL CHARACTERS

The genus *Trichoderma* was introduced into mycological literature by Person in 1974. The taxonomic confusion surrounding this hypomycete genus existed for several years, until Rifai (1969) claimed that it is possible to recognize species of genus *Trichoderma* on the basis of branching systems of conidiophore, the manner of phialide disposition and the characters of the phialospores. He constructed a key to establish nine species of *Trichoderma* viz. *T. piluliferum, T. koningii, T. polysporum, T. hamatum, T. aureoviride, T. harzianum, T. longibrachiatum, T. pseudokoningii* and *T. viride.*

In culture the colonies of *Trichoderma* spp. are either loosely fluccose or completely tufted and usually grow rapidly. The colour of the colony of *T. viride* is dark green, but older species show yellowish or light green colouration. The colour of the colony is largely due to pigments secreted by the phialospores. The fungus produces hyaline, septate, much branched and smooth walled mycelial hyphae. Chlamydospores are intercalary or rarely terminal on a short side branch of the hyphae. They are globose or ellipsoidal, colourless and smooth walled. In majority of species, the conidial areas are produced as distinct and characteristic ring like zones. When the colonies are a little older, the zonation often becomes less obvious. The main branches of the conidiophores produce numerous shorter side branches, which are terminated by phialide. These phialides bear slimy phialospores at the apex. *T. viride* has distinct rough walls whereas in other species no spore marking is visible.

BIOCONTROL POTENTIAL

A number of *Trichoderma* species have a promising potential for biological control of plant pathogenic fungi. *T. viride* showed antagonistic activity against soil-borne *Pythium debaryanum, Sclerotium rolfsii, Fusarium lini, F. culmorum* and *Fomus anosus* (Wright, 1956). *T. viride* has been favoured in commercial orchard oils for the control of *Armillaria mellea* (Bliss, 1951). In *in vitro* and *in vivo* tests, *T. viride* was found to be an antagonist against *Venturia inequalis,* the casual agent of apple scab (Andrews, 1983; Lindow, 1985).

T. viride was found to be an antagonist to *Drechslera sorokiniana*, the casual agent of root rot, seedling and foliar blights of wheat (Prasad *et al.*, 1978; Kirvoshchakova and Mishchank, 1990). Mostafa (1993) applied this biocontrol agent on barley seeds and suppressed infection of *Helminthosporium* on coleoptiles by 87 per cent. Singh *et al.* (1991) reported significant reduction in germination of teliospores and sporidia of *N. indica* with culture filtrate of *T. viride.* Mondal *et al.* (1995) obtained inhibition of germination of chlamydospores and suppression of mycelial growth of *Ustilago segtum tritici* with culture filtrate of *T. viride*. Field tests also showed the biocontrol potential of this agent against loose smut of wheat (Aggarwal *et al.*, 1993a).

T. harzianum has been shown to suppress the growth of *S. rolfsii,* the casual agent of root rot of various crops. Wells *et al.* (1972) applied inoculum of this species along the rows of tomato seedlings and protected the crop from *S. rolfsii.* It has also shown potential for biocontrol of *Macrophomina phaseolina* (Elad *et al*., 1986), *F. oxysporum* f.sp. *vasinfectum, F. culmorum* and *F. oxysporum* f.sp. *radius lycopersici*, the casual agents of seedling blight of cotton, melon, wheat and crown rot of tomato respectively (Siven and Chet, 1986; Sivan *et al*., 1987). Biocontrol potential of *T. harzianum* against *F. solani* (Calvet *et al*., 1990), *F. oxysporum* f.sp. *ciceri* (Kaour and Mukhopadhyay, 1992), *Pythium aphanidermatum* (Bhardwaj and Gupta, 1987), *Rhizoctonia solani* (Wilson *et al*., 1988; Cole and Zrinyka, 1988) and *S. rolfsii* (Deb and Dutta, 1988; Kaur and Mukhopadhyay, 1992) has also been demonstrated.

The study of the effect of *T. harzianum* on fungal pathogens infecting wheat and black oat straw showed that the antagonist resulted in reduction in the incidence of *Cochliobolus sativus* (Fernandaz, 1992). Culture filtrate of this antagonist reduced germination of teliospores and spordia of *N. indica* (Singh *et al*., 1991) and chlamydospores of *U. segetum tritici* (Mondal *et al*., 1955).

The biocontrol potential among other *Trichoderma* species has been shown by *T. koningii* against *U. segetum tritici* and *Drechslera sorokiniana* (Mondal *et al*., 1995), *T. reesi* against *D. sorokiniana* (Mandal, 1995) and *T. pseudokoningii* against karnal bunt (Amer, 1995).

MECHANISM OF ACTION OF *TRICHODERMA* SPP.

Antagonists act against pathogens through antibiosis, competition or parasitism. It has been shown that antagonists produce toxic metabolites near the infection court of the host pathogen and thus they inhibit them by antibiosis (Dandurand and Knudsen, 1993). Antagonistic properties of *Trichoderma* spp. due to the production of effective antifungal compounds make them useful as biocontrol agents as can be seen from the under mentioned tabulated account.

Species	Compound(s)	References
Trichoderma viride	6-pentyle pyrone	Collins & Halim, 1972; Kikuchi *et al.*, 1974; Moss *et al.*, 1975
T. viride (UC 4785)	Dermadine Heptelidic acid Trichodermin	Pyke & Kietz, 1966; Itoh *et al.*, 1980; Godfredsen & Vangedal, 1965; Brewer *et al.*, 1987
T. koningii	Isonitrin Homothallin II Heptelidic acid	Fujiwara *et al.*, 1982; Anderson *et al.*, 1978; Endo & Kanbe, 1981; Endo *et al.*, 1985
T. hamatum	Trichoviridin Isocyanin Isoniterin A	Godfredsen & Vangedal, 1965; Ollis *et al.*, 1980; Brewer *et al.*, 1982; Edenborough and Herbert, 1988
T. longibranchiatum	Trichodermin Trichobrachin	Watts *et al.*, 1988; Bruckner *et al.*, 1990
T. harziamum	Trichorzianines 6-pentyl-alpha pyrone Homothallin II	Horvath *et al.*, (1995) Scarselletti and Faull (1994); Faull *et al.*, (1994)

PURIFICATION OF ANTIFUNGAL COMPOUNDS

Homma and Suzui (1989) eluted the active fractions of antibiotic compounds from *Pseudomonas cepacia* strain RB 425 by TLC and used them against several soil-borne plant pathogens including *R. solani* and *Verticillium dahliae*. Infection of *Septoria tritici* and *Puccinia recondita* of wheat was suppressed by an antibiotic producing fluorescent pseudomonad (Levy *et al.*, 1989). The fractionation and analysis of two inhibitory compounds from cultures of this strain in TLC plates yielded two antibiotic compounds viz., the phenazine-antibiotics, 1-hydroxyphenazine (PhOH) and chlororaphin.

Harman *et al.* (1993) purified a variety of cell wall degrading chitinolytic enzymes from *T. harzianum* including N-acetyl-beta-D-glucosaminidase, chitin 1,4-beta-chitobasidases and endochitinase. Lorito *et al.*, (1993) also reported that *T. harzianum* produce cell wall degrading chitinolytic enzymes, endochitinase and chitobasidase. Subsequently Lorito *et al.* (1994) purified a glucan 1, 3-beta-glucosidase and an N-acetyl-beta-glucosaminidase from the culture filtrate of *T. harzianum* strain P1. The enzymes showed a synergistic inhibitory effect on both spore germination and germ tube elongation of *Botrytis cinerea*, Cruz *et al.* (1995) purified and characterized an endo-beta- 1, 6-glucanase from *T. harzianum*. The enzyme was specific for beta 1, 6 linkages.

Rhizosphere Competence

Baker, 1988, mutated *T. harzianum* for the tolerance to 100 μg a.i. benomyl and found that such strains were more rhizosphere competent and exhibit higher cellulase activity than non competent strain. Baker, 1989 suggested that higher cellulase activity enables the fungus *T. harzianum* to colonize the mucigel covering the plant roots.

Genetic Improvement of Biocontrol Agents

Papavizas (1987) suggested some genetic manipulations to improve the effectiveness of biocontrol agents for plant disease control, such as exposure to fungicides, mutagenesis and selection and sexual, and parasexual hybridization using biocontrol agents and fungicides.

Papavizas *et al.* (1982) were the first to report tolerance induction in *T. harzianum* to benomyl by UV irradiation. Selection for benomyl resistant biotypes of *T. viride* (T-1-R9) proved effective against Fusarium wilt of chrysanthemum (Locke *et al.*, 1985) and scurf of potato. Graeme-cook *et al.* (1991) induced the altered antibiotic production in mutants of *T. harzianum* using UV light and they found the production of two new antifungal compounds in addition to antibiotic produced by the parental strain. Cuevas *et al.* (1994) developed the benomyl tolerant mutants of *T. harzianum* using UV light. They found that the production of carboxy methyl cellulase and Beta glucanase activities of mutant strains were higher than those of parental strain. Frank *et al.* (1994) reported the development of morphologically different mutants from that of parental strain using cadmium as mutagenic agent.

Antagonistic Property

Trichoderma spp. are found in soils worldwide (Domsch *et al.*, 1980; Christensen, 1981) and have a promising potential for biological control of plant pathogenic fungi (Papavizas, 1985). Antagonistic effect of *T. lignorum* against *Rhizoctonia solani, Phytophthora parasitica, Pythium* spp. and *Sclerotium rolfsii* the causal agents of damping off of citrus, was shown by Weindling as early as in 1932. Subsequently, Weindling and his co-workers (Windling, 1934; Weindling and Fawcett, 1936; Weindling and Emerson, 1936) explored the possibility of controlling certain pathogenic fungi by augmenting soil with an abundance of *T. lignorum* and suggested that this mycoparasite might serve as a biocontrol agent against soil-borne pathogens. Thus, the idea of biological control of plant pathogens by mycoparasitism was borne.

Among the mycoparasites, *Trichoderma* species have been studied most frequently. In 1956, Wright (1956) found antagonistic activity of *T. viride* against a number of soil-borne pathogens such as *Pythium debaryanum, Sclerotium rolfsii, Fusarium lini, F. culmorum, Armillaria mellea* and *Fomes anosus. T. viride* has been favoured in commercial orchard soils for the control of *Armillaria mellea* (Bliss, 1951; Baker and Cook, 1974). Kudryavtseva (1980) and Sychev and Shaposhnik (1982) reported *T. viride* as highly antagonistic to *Fusarium solani* and *F. oxysporum* in field and pot tests. *T. viride* was found to be an antagonist *in vitro* and *in vivo* tests against *Venturia inequalis,* the causal agent of apple scab (Andrews *et al.*, 1983; Lindow, 1985).

Under experimental conditions *T. harzianum* suppressed *S. rolfsii* a causal agent for root and stem diseases of several agricultural crops (Backman and Rodriguez-Kabana, 1975; Chet *et al.*, 1979; Elad *et al.*, 1980; Wells *et al.*, 1972) but only when applied at relatively high inoculum rate. Mathur and Sarbhoy (1978) introduced *T. viride* and *T. harzianum* in soils and succeeded in controlling root-rot of sugarbeet caused by *S. rolfsii. T. harziamum* also has been shown to be effective agent in suppressing *Rhizoctonia solani* (Chet and Baker, 1980; Chet *et al.*, 1979; Elad *et al.*, 1980; Hader *at al.*, Liu and Baker, 1980). Antagonistic effect of *T. harzianum* against *F. solani* (Bhardwaj and Gupta, 1987; Cole and Zrinyika, 1988) *F. oxysporum* (Calvet *et al.*, 1990), *Pythium aphanidermatum* (Bhardwaj and Gupta, 1987), *R. solani* (Wilson *et al.*, 1988; Cole and Zrinyika, 1988), *S. folfsii* (Dev and Dutta, 1988; Singh and Dwivedi, 1987) has been demonstrated. Kaur and Mukhopadhyay (1992) effectively controlled 'Chickpea wilt complex' caused mainly by *F. oxysporum* f. sp. *cicer, R. solani* and *S. rolfsii* by *T. harzianum.*

T. hamatum, a more recently recognized species with biological control potential (Chet and Baker, 1981), is similarly effective against *R. solanii* and also against *Pythium* spp. (Chet and Baker, 1981; Chet *et al.*, 1981; Harman *et al.*, 1980; 1981). Elad *et al.*, 1982 noted that damping off of cotton was decreased when seeds were coated with *T. hamatum* as well as *T. harzianum.*

Antagonistic effect of *Trichoderma* has also been explored against certain fungal pathogens causing diseases in wheat. Butler, 1953, reported that antibiotics produced by *Trichoderma* spp. inhibited the growth of *Cohliobolus sativus. T. harzianum* was found to be an effective biocontrol agent of *Fusarium roseum* culmorum when applied in soil or as seed treatment (Sivan and Chet, 1986). In a preliminary test of *Trichoderma* isolates from field soil against *F. culmorum* on spring wheat, it was found that the higher number of effective antagonistic isolates belonged to *Trichoderma* sp. followed by *Gliocladium* sp. and *Fusarium* sp. Earlier, Krisvoshchekova and Mishchenk (1990) observed that a treatment with *Trichoderma* sp. considerably suppressed the growth of *F. culmorum* on wheat seed.

In *in vitro* test, *T. viride* was found to be antagonist to *S. rolfsii* on wheat (Manjappa, 1979). Similary, antagonistic action of *Trichoderma* sp. on *Drechslera sorokiniana,* the causal agent of root rot, seedling blight and follar blight of wheat was observed by Prasad *et al.* (1978). Krisvoschchekova and Mishchenk (1990) reported that *Trichoderma* sp. produced trichodermin and gave complete control of *Helminthosporium* and *Fusarium* rots of wheat. The study on the effect of *T. harzianum* on fungal pathogens infesting wheat and black oat straw showed that the antagonist resulted in reduction in the incidence of *Cochliobolus sativus, F. graminearum* and other *Fusarium* spp. (Fernadez, 1992). Aggarwal *et al.* (1992) made use of certain biocontrol agents as an alternate strategy to combat loose smut of wheat. They reported antagonistic effect of *T. viride, Gliocladium deliquescens* and *Racillus subtilis* on loose smut infection.

Singh *et al.* (1991) tested the antagonistic effect of *Trichoderma* spp., *Gleocladium* spp. and *Bacillus subtilis* on tellospores of Karnal bunt (*Neovossia indica*) of wheat. It was observed that after 15 days of treatment, *T. viride*, *T. harzianum* and *G. deliquescens* significantly reduced teliospore germination.

In addition to *Trichoderma,* bacterial isolates have also been introduced as a coating of cells or spores on seeds or other planting materials to inhibit pathogens (Kommedahl and Windels, 1981). *Pseudomonas fluorescens* Z-79 (NRRLB-15132) isolated originally from the rhizosphere of wheat had been shown to suppress take-all (*Gaeumannomyces graminis* var. *tritici*), a major root and crown disease of wheat. de Freitas and Germid (1991) showed the antagonistic effect of *Pseudomonas cepacia* (R 55 and R 85), *P. putida* (R 104) against *R. solani* in wheat when used as seed inoculants in winter wheat.

Antifungal Compounds

Biological control of fungal diseases of plants including seed borne diseases using non-pathogenic fungi and bacteria has received increasing attention of plant pathologists and soil microbiologists all over the world. Antagonistic microorganisms applied to seeds prior to planting, colonize rhizosphere of seedling and thus are present at or near the pathogen's infection court, where they act by producing antifungal compounds through hyperparasitism, or by competitively colonizing spermosphere and rhizosphere substrate (Dandurand and Knudsen, 1993).

During the 1920s and 1930s, it was generally assumed that microorganisms that interfered with root disease development in soil did so by means of antibiosis. This possibility was studied extensively during the 1950s and 1960s following the discovery of numerous antibiotic substances that were produced by microorganisms isolated from soil (Baker, 1987). The discovery of penicillin by Fleming in 1928 and its purification and use in medicine in 1939 greatly stimulated studies of antagonists of plant pathogens. Grossbard in 1948-1952, Wright in 152-57, and other demonstrated that antibiotics were produced in soil by *Penicillin, Aspergillus, Trichoderma* and *Streptomyces* spp. Antibiotics produced in direct contact between microorganisms may function at continuous low concentration and be less subject to biotic decomposition (Baker, 1987).

Basically two mechanisms have been postulated in case of *Trichoderma* spp. One envisages the involvement of inhibition through antibiosis and the other involves penetration, colonization and subsequent lysis of the pathogen propagules through hyperparasitism/mycoparasitism. Hyphal invasion *Trichoderma* sp. *in vitro* was noticed in *R. solani* (Lewis and Papavizas, 1980). Similarly, hyphal contact with *Pythium ultimatum* and penetration were responsible for antagonistic effect of *T. viride* (Dennis and Webster, 1971b). Tight coiling around the pathogens, i.e., *R. solani* and *Lentinus edodes* by *T. polysporum* and *T. viridei* respectively were reported (Dennis and Webster, 1971c; Komatsu, 1968). In *in vitro* tests, hyphal interference and lysis of mycelium of *S. rolfsii* and *R. solani* by *T. harzianum* was recorded (Wells *et al.*, 1972; Elad *et al.*, 1980; Liu and Baker, 1980; Hadar *et al.* 1979; Henis *et al.*, 1978; 1979).

Mechanism(s) by which *Trichoderma* spp. control or suppress the activities of plant pathogens are not fully known. Earlier, it was established that antibiotics were produced by various species that were lytic to a number of fungi (Weindling 1934; Weindling and Emerson, 1936; Weindling and Fawcett, 1936). 'Viridin' produced by a pigment forming strain of *T. viride* inhibited the germination of conidia of *Botrytis alli* at 200 times dilution (Brain *et al.*, 1946). A sequiterpene antibiotic, trichodermin was extracted from *T. virdie* by Godtfredsen and Vangendal (1965). Suzukacillin (Oaka *et al.*, 1966), alamethicine (Meyer and Reusser, 1967) and U-21963, i.e., Dermadin (Pyke and Dietz, 1966; Meyer, 1966) are some of the antibiotics extracted from culture filtrates of *T. viride.* Suzukacillin and alamethicine are peptide antibiotics with antifungal and antibacterial properties. U-21963 (Dermadin) is an unsaturated monobasic acid active against Gram negative and gram bacteria and a wide range of fungi.

Dennis and Webster (1971 a, b, c) discovered non-volatile antibiotics, mostly chloroform soluble antibiotics, including trichodermin and peptide antibiotics, as well as volatile compounds like acetaldehyde, alamethicine and polypeptide antibiotics. More recently *Trichoderma* spp. have been shown to produce various cell wall degrading enzymes (β-(1-3) glucanase, chitinase) (Chet and Baker, 1981; Jones and Watson, 1969; Elad *et al.*, 1982; Ridout *et al.*, 1986). Claydon *et al.* (1987) extracted an antibiotic (6-n-pentyl pyrone) from culture filtrate of *T. harzianum*, which protected lettuce seedlings from damping off caused by *R. solani.* Production of chitinolytic enzymes, endochitinase and chitobasidase by *T. harzinum* strain P1 was discovered by Lorito *et al.*, (1993).

ANTAGONISTIC INTERACTION OF BIOCONTROL AGENT

The antagonists parasitize hyphae of host fungi and bring about morphological changes like coiling, haustoria, lysis of host cells, penetration of host cells, etc. Use of light microscopy to observe parasitism, or more sophisticated approaches through scanning electron microscopy, transmission electron microscopy, fluorescence microscopy have been applied to study the process of mycoparasitism (Ayers and Adams, 1981; Elad *et al.*, 1983; Elad *et al.*, 1987). Durrell (1968) using phase contrast and electron microscopy produced informative photographs showing haustoria and hyphae of *T. viride* within hyphae of Phycomycetes. The phenomenon of physical contact followed by the disorganization of the host cells have been established for *Trichoderma* species (Elad *et al.*, 1983; Elad *et al.*, 1982).

Quantitative Enumeration and Isolation of *Trichoderma*

Members of genus *Trichoderma* are known to be common inhabitants of soils. Martin's medium (Elad *et al.*, 1981) failed to eliminate other fungi and *Trichoderma* itself are too fast growing for suitable enumeration. Dextrose-yeast extract agar was used (Danielson and Davey, 1973c) for isolation and enumeration of *Trichoderma*, but the medium failed to restrict fast growing isolates.

An improved technique was developed by Davet (1979), where soil agar discs were placed on specific medium containing among others, streptomycin sulphate, allyl alcohol and vinclozolin, which provided the specificity to the medium. Devet's (1979) technique had low efficacy of recovery (Maiti, 1982) and was liable to other contaminants. Liu and Baker (1980) used a modification of Martin's Rose Bengal streptomycin agar by adding 100 mg PCNB.

Elad *et al.* (1982) developed an improved medium fortified with Dexon (p-dimethyl amino benzene diazo-sodium sulphonate) and Pentacholor (PCNB) along with other additives to improve specificity to the medium. This medium resulted in good recovery of *Trichoderma* from soil. Papvizas (1981) and Papvizas and Lumsden (1982) developed a selected medium on V-8 juice agar. Maiti (1986) used *Trichoderma* selective medium originally developed by Elad *et al.* (1981), replacing Dexon with methyl orange (Dimethyl-amino-azobenzene sodium sulphonate) and obtained more precise results.

Effects of Environmental Conditions on Antagonistic Properties of *Trichoderma* spp.

The living creatures interact continuously under fluctuating environmental conditions, maintaining or increasing stability within the ecosystem. Some of the interactions may be beneficial to a particular microbial population, others may be detrimental to it. It is the detrimental or negative interactions, which form the basis for the biological control of pathogens. Successful biological control of plant pathogens can be achieved only through a fundamental understanding of ecological relationship of the diverse microbial populations including plant pathogens and biological control agents. In a given environment, all microbial populations including pathogens, saprophytes, antagonists and hyperparasites must compete for substrates for growth and survival. The success of a microbial population in a particular ecosystem requires maximum growth during active growing conditions as well as the ability to tolerate and survive adverse conditions. Thus, biological control of plant pathogens can be successful only if the organisms introduced into environment are able to survive, grow, and carry out active metabolism and, more important, suppress the pathogen population by competition, antagonism, or hyperparasitism, in the right place at the right time.

(i) Effect of temperature

Many studies on the interaction between fungi have been carried out at temperature between 20° and 25°C (Ikediugwu and Webster, 1970 a, b, Dennis and Webster, 1971 a, b, c). However, isolates of *Trichoderma* species exhibit a wide range of temperature relation. Danielson and Davey (1973a) studied the growth of 16 *Trichoderma* isolates using poly-temp Thermostat. They reported that there was a continuum of temperature maxima ranging from 28 to 42°C and optima ranging from 22 to 34°C. *T. pseudokoningii, T. saturnisporum* and three isolates, of *T. harzianum* had high temperature maxima. A majority of isolates did not grow at 35°C and a small number had maxima above 38°C. Gutter (1963) and Kamatsu and Hashioka (1966) observed that white *Pachybasium* (*Trichoderma*) strains had maximum temperature between 30 and 31°C. In another study, 54 isolates of *Trichoderma* were tested at different temperatures ranging from 2° to 20°C. It was observed that strains of *T. viride* and *T. polysporum* can grow at temperatures as low as 2°C, whereas *T. harzianum* and *T. longi-brachiatum* do not grow below 5°C.

Kohl and Scholsser (1989) determined the germination rates of 24 isolates of *Trichoderma* at 5°, 10° and 24°C. The average germination rate was more than 94 per cent at 24°C after 24 h. At 10°C, conidia of all isolates were able to germinate within 48 h. High cold tolerant isolates started to germinate at 5°C after 4 days. From these results it was emphasized that biological control of *Botrytis cinerea* could be possible using cold tolerant strains of *Trichoderma* at higher altitudes. *T. harzianum* has been effective in diverse field conditions both at high and low temperatures or in early planting.

Metabolites that inhibit target organisms may be elaborated only under certain temperature and nutrient regimes. Tronsmo and Dennis (1978) mentioned that the isolates of *Trichoderma* spp. could produce volatile and non-volatile antibiotic metabolites at 10°C and 5°C and showed hyphal interaction with mycelium of test fungi.

(ii) Effect of pH concentration

The optimum H-ion concentration for the growth of *Trichoderma* ranged from pH 3.7 to 4.7 (Danielson and Davey, 1973d). With the exception of *T. harzianum* and *T. hamatum*, growth was less than 10 per cent of the maximum in media at pH 6.2. The growth of *T. pseudokoningii* and *T. saturnisporum* was adversely affected by either extreme acidity or condition approaching neutrality. In buffered liquid culture, *Trichoderma* was best adopted to a substrate between pH 4 and 5.

Fungal propagule's density depends on pH of the soil. According to Martin *et al.* (1985), the level of propagule of the antagonist should be high in acidic soil. It was suggested that the level of *Trichoderma conidia* of 106-108/g soil be effective in acidic soil. Chet and Baker (1981) reported that highly organic soil from Columbia contained 108 propagules/g soil. Clay loam soil having pH 6 and amended with 10 *Trichoderma* propagules/g soil was conducive for diseases.

(iii) Effect of carbon dioxide and bicarbonate

Trichoderma has been observed to be slightly inhibited by 3-5 per cent CO_2 (Waid, 1962) and relatively unaffected by CO_2 levels upto 10 per cent or 20 per cent (Macauley and Griffin,

1969a). The relative growth response of *Trichoderma* to high level of CO_2 as compared to growth in air was much greater on a slightly basic medium than on an acidic medium (Danielson and Davey, 1973b, d). Macauley and Griffin (1969b) previously showed that the effect on growth, either positive or negative, was more extreme at high pH values than at low values. They felt that HCO_3^- was more influential in altering growth than CO_2. HCO_3^- may be responsible for determining the composition of fungal populations in the root zones. The addition of HCO_3^- to an alkaline medium strongly inhibited *Trichoderma* while having little affect on fungi isolated from roots (Danielson and Davey, 1973b).

(iv) Effect of light on antagonist

Most species of *Trichoderma* were photosensitive, sporulating readily on many natural and artifical substrates in a concentric pattern of alternating rings in response to diurnal alternation of light and darkness with conidia being produced during the light period (Gressel and Hartmann, 1968). They also observed maximum photoinduction activity at around 380 nm and 440 nm, with sporulation not occurring below 254 nm or above 1,100 nm. Betina and Spisiakova (1976) observed that the best photoinduction of phialoconidiogenesis occurred with exposure to daylight for three minutes or to near ultraviolet radiation (366 nm) for 10-30 seconds.

(v) Compatibility of Trichoderma spp. with biocides/fungicides

The usefulness of biocontrol agent depends on its integration possibility with other control measures. Research in biological control of soil-borne pathogens has indicated that biological control is often not highly effective when deployed as a single control strategy (Martin *et al.*, 1985). Under IPM system, an antagonist insensitive to biocide or fungicide may be combined with it, extending the duration and effectiveness of control. *Trichoderma* spp. are known to have greater tolerance for broad-spectrum biocides than many other soil microorganisms, and colonize the treated soil more rapidly than other soil competitors (Munnecke, 1972). The dominance of *Trichoderma* in soil following the application of certain biocides like methyl bromide (Elad *et al.*, 1982), metham sodium (Henis and Papvizas, 1983), etc. has been reported.

Positive compatibility of *Trichoderma* spp. with different fungicide treatment like Thiram (Davet, 1981), Metalaxyl (Papavizas, 1981) has been shown. It was also reported that some fungicides like Captan, chlorothalonil, chloroneb, iprodione, procymidone, vinclozolin were not inhibitory to *Trichoderma* (Abd-El Moity *et al.*, 1982; Ordentlich and Chet, 1989).

Combined effect of PCNB and *T. harzianum* decreased the inoculum potential of *R. solani* and increased disease control of damping off of radish (Henis *et al.*, 1978). Low concentrations ineffective alone improved control of *R. solani* damping off when applied with *T. harzianum* (Hadar *et al.*, 1979). Recently, Kaur and Mukhopadhyay (1992) reported that chickpea wilt complex was effectively controlled by *T. harzianum* alone or in combination with fungicides namely Vitavax-200 (carboxin+thiram and Ziram.)

BIOACTIVITY OF NEEM PRODUCTS AGAINST PLANT PATHOGEN

The scientific renaissance regarding fungicidal nature of neem products came into existence quite recently. Singh and Pandey (1966) reported that the chopped leaves and leaf composts of neem and other plant product added to the infested soil at the rate or 5 per cent (w/w) inhibited the population of *P. aphanidermatum.* Dath (1982) also reported the inhibitory effect of green manure (neem) as soil amendments on the survival of sclerotia of *Corticium sasakii.*

Singh *et al.* (1980) reported the fungicidal effect of neem on some soil-borne pathogens of chickpea, namely *Rhizoctonia solani, Sclerotium rolfsii* and *Sclerotinia sclerotium.* The growth of these fungi was inhibited in liquid medium containing extracts of leaf, trunk bark, fruit pulp and oil of neem.

Antifungal activity of neem leaf extracts was evaluated against *Curvularia lunata* (Bhowmick and Vardhan, 1981), *Fusarium equisetii* and *F. semitectum* on 'parwal' and bottle gourd (Krishna *et al.*, 1986) and *Aspergillus niger* (Sinha and Saxena, 1987).

Ghewande (1980) studied the effect of leaf extract of neem along with some fungicide against *Phaeoisariopsis personata (Mycosphaerella berkeleyi)* and *Puccinia arachidis* on groundnut and got 15.4-39.8 per cent increased yield. The effect of neem leaf extract along with other plant extract was studied against *Aspergillus niger* and *A. flavus* and groundnut (Bansal and Sobti, 1990), *Pyricularia oryzae, R. solani, C. lunata, A. niger* and *Fusarium moniliforme* (Mishra and Tewari, 1990).

An exhaustive study on the utility of neem oil against fungi like keratinophilic fungi (Jain and Agarwal, 1978), *Erysiphe polygoni F. moniliformi, A. niger, Drechslera rostratum* and *Macrophomina phaseolina* (Vir and Sharma, 1985a) has been reported.

The effects of volatile and non-volatile fractions of neem were also studied. Dwivedi and Dubey (1986) studied the effect of volatile and non-volatile fractions of neem on the germination of sclerotia of *Macrophomina phaseolina*, and reported that deleterious effects of volatile fractions of hydrodistillates of neem on the germination of these sclerotia were more pronounced as compared to non-volatile fractions.

Muthusamy *et al.* (1988) studied the neem products against rust disease of groundnut and reported that the seed extract and oil were highly effective in inhibiting spore germination by 96.2 and 96.0 per cent respectively, whereas the neem cake extract showed 87.6 per cent inhibition.

Ramarao and Usharaja (1983) reported that the neem and groundnut cake extracts increased the mycelial growth of *Sclerotium rolfsii,* causal organism of wheat root-rot, over control at lower concentration.

Chakrabarti and Sen (1991) reported that suppression of *Fusarium* wilt of muskmelon (*F. solani)* was affected in the presence of mustard cake by (2%), groundnut cake (1%), neem cake (2%), and saw dust (1%) and also noted that neem cake reduced the wilt by 80 per cent.

Khan *et al.* (1974) studied the effect of water-soluble fraction (WSF) of different oil cakes and bitter principles from neem viz., nimbidin and thionemone and some fungi, e.g. two isolates of *Rhizoctonia solani, Fusarium oxysporum* f.s.p *lycopersici. Helminthosporium nodulosum, Alternaria tenis* and *Curvularia tuberculata.* It was noted that all the concentration of WSF of oil cakes of neem and bitter principles were found to be highly inhibitory for the growth of all the rest fungi.

Singh (1993) reported the significant antifungal properties of oil from the seeds of neem against *Sclerotium rolfsii,* the cause of collar rot of pigeonpea. He noted that the growth of mycelium was checked completely at 5000 ppm concentration of oil but could not destroy the viability of sclerotia. Oil was also found to have stimulatory effect on seed-germination and seedling-growth of pigeonpea. He also mentioned that when pigeonpea seeds were treated with

T. harzianum spores and neem oil, mortality of seedling was 26 per cent as against 98 per cent in the control.

Khan and Kumar (1990) reported the antifungal activity of leaf extract of neem on wheat seed mycoflora. The treatment caused a marked reduction in seed mycoflora and enhanced seed germination.

REFERENCES

Abd-El Moity, T.H., Papvizas, G.C. and Shatla, M.N. (1982). Introduction of new isolates of *Trichoderma harzianum* tolerant to fungicides and their experimental use for control of white rot of onion. *Phytopathology,* **72:** 396-400.

Aggarwal, R., Srivastava, K.D., Singh, D.V., Bahadur, P. and Nagarajan, S. (1992). Possible biocontrol of loose smut of wheat. *Indian J. Bio. Control.* **6:** 111-112.

Andrews, J.H. (1983). Future strategies for integrated control. In *challenging problems in Plant Health* (Eds. Kommedahl, T. and Williams, P.H.) *Amer Phytopathol. Soc.* St. Paul, MN.

Ayers, W.A. and Adams, P.B. (1981). Mycoparasitism and its application to biological control of plant diseases. In *Biological Control in Crop Production* (Ed. G.C. Papavizas) *Beltsville Symp. Agric. Res.* **5:** 91-103. Totowa, NJ: Allanheld, Osmun.

Backman, P.A. and Rodriguez-Kabana (1975). A system for the growth and delivery of biological control agents in soil. *Phytopathology* **65:** 819-821.

Baker, K.F. (1987). Evolving concepts of biological control of plant pathogens. *Annu. Rev. Phytopathol.* **25:** 67-85.

Baker, K.F. and Cook, R.J. (1974). *Biological Control of Plant Pathogens.* W.H. Freeman, San Francisco. pp. 433.

Bansal, R.K. and Sobti, A.K. (1990). An economic remedy for the control of two species of *Aspergillus* on groundnut. *Indian Phytopath.* **43:** 451-452.

Betina, V. and Spisiakova, J. (1976). Suppression of photo-induced sporulation in *Trichoderma viride* by inhibitors. *Folia Microbio.* **21:** 362-370.

Bhardwaj, S.S. and Gupta, P.K. (1987). *In vitro* antagonism of *Trichoderma* spp. against fungal pathogens associated with rhizome rot of ginger. *Indian J. Pl. Path.* **5:** 51-42.

Bhowmic, B.N. and Vardhan, V. (1981). Antifungal activity of some leaf extracts of medicinal plants on *Curvularia lunata. Indian Phytopath.* **34:** 385-386.

Bliss, D.E. (1951). The destruction of *Armillaria mellea* in citrus soils: *Phytopathology* **41:** 655-683.

Brain, P.W., Curtis, P.J., Hemming, H.G. and McGowan, J.G. (1946). The production of viridin by pigment forming straining of *Trichoderma viride. Ann. Appl. Biol.* **33:** 190-200.

Butler, F.C. (1953). Saprophytic behaviour of some cereal root-rot fungi II. Factors influencing saprophytic colonization of wheat straw. *Ann. Appl. Biol.* **40:** 298-304.

Calvet, C., Pera, J. and Brea, J.M. (1990). Interaction of *Trichoderma* spp. with *Glomus mosseae* and two wilt pathogenic fungi. *Agri. Ecosystem Environ.* **29:** 59-65.

Chakrabarti, S.K. and Sen, B. (1991). Suppression of Fusarium wilt of muskmelon by organic soil amendment. *Indian Phytopath.* **44:** 476-479.

Chet, I. and Baker, R. (1980). Induction of suppressiveness of *Rhizoctonia solanii* in soil. *Phytopathology* **70:** 994-998.

Chet, I., Hadar, Y., Elad, Y., Katan, J. and Henis, Y. (1979). Biological control of soil-borne plant pathogens by *Trichoderma harzianum.* pp. 585-592. *In: Soil Borne Plant pathogens* (Eds. Schippers, B. and Gams, W.), Academic Press, N.Y.

Chet, I., Harman, G.E. and Baker, R. (1981). *Trichoderma hamatum:* Its hyphal interactions with *Rhizoctonia solanii* and *Pythium* spp. *Microb. Ecol.* **7:** 29-38.

Christensen, M. (1981). Species diversity and dominance in fungal communities. In *The Fungal Community: its Organization and Role in the Ecosystem* (Eds. Wicklow, D.T. and Carroll), Marcel Dekker, Inc. New York, pp. 201-232.

Cole, J. and Zrinyika, S. (1988). Integrated control of *Rhizoctonia solani* and *Fusarium solani* in tobacco transplant with *Trichoderma harzianum* and trlademenol. *Pl. Pathol.* (London), **37:** 271-277.

Dandurand, L.M. and Knudsen, G.R. (1993). Influence of *Pseudomonas fluorescens* on hyphal growth and biocontrol activity of *Trichoderma harzianum* in the spermosphere and rhizosphere of pea. *Phytopathology* **83:** 265-270.

Danielson, R.M. and Davey, C.B. (1973b). Nutritional factors affecting the growth of *Trichoderma* in culture. *Soil Biol. Biochem.* **5:** 495-504.

Danielson, R.M. and Davey, C.B. (1973c). Carbon and nitrogen nutrition of *Trichoderma. Soil Biol. Biochem.* **5:** 505-515.

Dath, A.P. (1982). Effect of soil amendment with some green manures on the survival of sclerotia of *Corticium sasaki. Indian Phytopath.* **35:** 523-525.

Davet, P. (1981). Effect, de quelques pesticides sun la colonisation d'un substract par le *Trichoderma harzianum* Rifai en presence des autres champignons du sol. *Soil Bio. Biochem.* **13:** 513-517.

de Freitas, J.R. and Germida, J.J. (1991). *Pseudomonas cepacia* and *Pseudomonas putida* as winter wheat inoculants for biocontrol of *Rhizoctonia solani. Canad. J. Microbiol.* **37:** 780-784.

Dennis, C. and Webster, J. (1917a). Antagonistic properties of species groups of *Trichoderma. Trans. Brit. Mycol. Soc.* **57:** 25-39.

Dennis, C. and Webster, J. (1971b). Antagonistic properties of species groups of *Trichoderma. Trans. Brit. Mycol. Soc.* **57:** 41-48.

Dennis, C. and Webster, J. (1971c). Antagonistic properties of species groups of *Trichoderma. Trans. Brit. Mycol. Soc.* **57:** 363-369.

Deve, P.R. and Dutta, B.K. (1988). *Trichoderma* as a biocontrol agent against *Scelerotium rolfssii. Microbios. Lett.* **37:** 107-114.

Domsch, K.H., Gams, W. and Anderson, T.H. (1980). *In: Compendium of Soil Fungi.* Vol. 1 Academic Press, New York.

Durell, (1968). Hyphal Invasion by *Trichoderma viride. Mycopath. Mycol. Appl.* **35:** 138-144.

Dwivedi, R.S. and Dubey, R.C. (1986). Effects of volatile and non-volatile fractions of two medicinal plants on germinations of *Macrophomina phaseolina* sclerotia. *Trans. Brit. Mycol. Soc.* **87:** 326-328.

Elad, Y., Chet, I. and Henis, Y. (1982). Degradation of plant pathogenic fungal *Trichoderma harzianum. Canad. J. Microbiol.* **28:** 719-725.

Elad, Y., Chet, I. and Katan, J. (1980). *Trichoderma harzianum*—a biological agent effective against *Sclerotium rolfsii* and *Rhizoctonia.*

Elad, Y., Harder, Y., Chet, I. and Henis, Y. (1982). Prevention with *Trichoderma harzianum* Rifai aggr., of reinfection by *Sclerotium rolfsii* sacc and *Rhizoctonia solanii* kuhn of soil fumigated with methyl bromide, and improvement of disease control in tomatoes and peanuts. *Crop Prot.* **1:** 199-211.

Ferna ndez, M.R. (1992). The effect of *Trichoderma harzianum* on fungal pathogens infesting wheat and black oat straw. *Soil Biol. Biochem.* **24:** 1031-1034.

Ghewande, M.P. (1989). Management of foliar diseases of groundnut *(Arachis hypogaea)* using plant extracts. *Indian J. Agr. Sci.* **59:** 133-134.

Godffredsen, W.O. and Vangedal, S. (1965). Trichodermin, a new sesquiterpene antibiotic. *Acta Chemica Scandinavica* **19:** 1088-1102.

Gressel, J.B. and Harmann, K.M. (1968). Morphogenesis in *Trichoderma* Action spectrum of photoinduced sporulation. *Planta* **79:** 271-274.

Gutter, Y. (1963). Contributions to the biology of *Trichoderma viride* Pers. ex Fries, Pathogenic to citrus fruits, *Israel J. Bot.* **12:** 27-40.

Hadar, Yl., Chet, I. and Henis, Y. (1979). Biological control of *Rhizoctonia solani* damping off with wheat bran culture of *Trichoderma harzianum. Phytopathology,* **69:** 64-68.

Harman, G.E., Chet, I. and Baker, R. (1981). Factors affecting *Trichoderma hamantum* applied to seeds as a biocontrol agent. *Phytopathology,* **71:** 569-572.

Harman, G.E., Chet, I. and Baker, R. (1980). *Trichoderma hamatum* effects on seed and seedling disease induced in radish and pea by *Pythium* spp. or *Rhizoctonia solani. Phytopathology* **70:** 1167-1172.

Henis, Y. and Papavizas, G.C. (1983). Factors affecting germinability and susceptibility to attack of scelrotial of *Sclerotium rolfsii* by *Trichoderma harzianum* in field soil. *Phytopathology,* **73:** 1469-1474.

Henis, Y., Ghaffar, A. and Baker, R. (1978). Integrated control of *Rhizoctonia solani* of radish: Effect of successive plantings, PCNB, and *Trichoderma harzianumi* on pathogen and disease, *Phytopathology,* **68:** 900-907.

Ikediugwu, F.E.O. and Webster, J. (1970b). Hyphal interference in a range of comprophilous fungi. *Trans. Brit. Mycol. Soc.* **54:** 205-210.

Jain, P.C. and Agarwal, S.C. (1978). Note on the activity of some odoriferous orgamc compounds against some keratinophilic fungi. *Proc. Nat. Acad. Sci.* **44**(B): 273-276.

Jones, D. and Watson, D. (1969). Parasitism and lysis by soil fungi of *Sclerotinia sclerotiorum* (Lib.) de Bary, a phytopathogenic fungus. *Nature* (London) **224:** 287-288.

Joshi, L.M., Singh, D.V. and Srivastava, K.D. (1985). Status of rusts and smuts during dwarf wheat era in India. *Rachis* **4:** 10-16.

Khan, M.I. and Kumar Rishi (1990). Antifungal activity of leaf extract of *neem* on seed microflora of wheat. *Indian J. Appl. Pure Biol.* **5:** 13-14.

Kohl, J. and Schlosser, E. (1989). Decay of sclerotia of *Botrytis cinerea* by *Trichoderma* spp. at low temperature. *Phytopathology,* **125:** 320-326.

Komatsu, M. (1968). *Trichoderma viride* as an antagonist of wood-inhibiting hymenomycetes. VII on the distribution of *Trichoderma, Pachybaridum* and *Gliocladium* in the fields of shiitake mushroom of three genera, *Trichoderma, Pachybasium, glioladium* and other sterile forms. Rep. *Tottori Mycol. Inst.* **6:** 18-28; 29-42.

Krishna, A. Prasad, and Ojha, N.L. (1986). Antifungal evaluation of leaf extracts for the control of some cucurbitaceous fruit rot diseases. *Indian Phytopath.* **39:** 153.

Krivoshchekova, T.G. and Mishchenk, V.S. (1990). The effectiveness of trichodermin. *Zashchita Rastenii* (Moskova) No. 11, 22 (Ru).

Kudryavtseva, K.I. (1980). The possibility of using trichodennin in the control of root rot of cucumbers in glass house. In *Sbornik Nauchnykh Trudov Institute, USSR, Lenin Acad. Agric. Sc.* p. 111-119.

Lewis, J.A. and Papavizas, G.C. (1980). Integrated control of *Rhizoctonic* fruit rot of cucumber. *Phytopathology,* **70:** 85-89.

Lindow, S.E. (1985). Foliar Antagonists: Status and Prospects. *In: Biological Control in Agricultural IPM Systems* (Eds. Hoy, M.A. and Herzog, D.C.), Academic Press, Inc. (London), pp. 395-413.

Lorito, M., Harman, G.E., Hays, C.K., Broadway, RM., Tronsmo, A., Woo, S.L. and Di Pietro, A. (1993). Chitinolytic enzyme produced by *Trichoderma harizianum: Antifungal* activity of purified endo-chitinase and chitobasidase. *Phytopathology,* **83:** 302-307.

Maiti, D. (1982). *Role of Nutrient Amendments on Soil Antagonism towards Sclerotium rolfsii* Sacc M.Sc. thesis, B.C.K.V. Kalyanal, W.B.

Manjappa, B.H. (1979). *Studies on Survival and Variation in Sclerotium rolfsii* Sacc M. Sc. (Agril.) Thesis, VAS, Bangalore, pp. 86.

Martin, S.B., Abawi, G.S. and Hoch, H.C. (1985). Biological control of soil-borne pathogens with antagonists. In *Biological Control in Agricultural IPM Systems* (Eds. Hoy, M.A and Herzog, D.C.). Academic Press, Inc. (London), pp. 433-454.

Mathur, S.B. and Sarbhoy, A.K. (1978). Biological control of sclerotium root rot of sugarbeet. *Indian Phytopath.* **31:** 365-367.

Meyer, C.E. (1966). U21963, a new antibiotic. II Isolation and characterization *Appl. Microbiol.* **14:** 511-512.

Meyer, C.E. and Reusser, F. (1967). A polypeptide antibacterial agent isolated from *Trichoderma viride, Experimentia,* **23:** 85-86.

Mishra, M. and Tewari, S.N. (1990). Ethanolic extract toxicity of three botanicals against five fungal pathogens of rice. *Nat. Acad. Letters.* **13:** 409-412.

Munecke, D.E. (1972). Factors affecting the efficacy of fungicides in soil. *Annu. Rev. Phytopathol.* **10:** 375-398.

Muthusamy, M., Eswaramurthy, S., Muthusamy, S. and Mariappan, V. (1988). Evaluation of neem products against rust disease of groundnut. *Neem Newsl.* **5b:** 48.

Ooka, T., Shimojima, Y., Akimoto, T., Takeda, I., Senoh, S. and Abe, J. (1966). A new antibacterial peptide 'Suzukacillin'. *Agril. Biol. Chem.* **3:** 700-702.

Ordentlich, A. and Chet, I. (1989). Biological control of soil-borne plant pathogenic fungi by antagonistic *Trichoderma. Israel Agresearch* **3:** 1-2.

Papavizas, G.C. (1981). Survival of *Trichoderma hazianum* in soil and in the pea and the bean rhizosphere. *Phytopathology,* **72:** 121-125.

Papavizas, G.C. and Lumsden, R.D. (1982). Improved medium for isolate of *Trichoderma* spp. from soil. *Plant Dis.* **66:** 1019-1020.

Prasad, K.S.K., Kulkarni, S. and Siddaramaiah, A.L. (1978). Antagonistic action of *Trichoderma* spp. and Streptomyces sp. on *Drechslera sativum (Helminthosporium sativum)* Pam. King and Bakke) Subram. and Jain. *Curr. Res.* **12:** 202-203.

Pyke, T.R. and Dietz, A. (1966). U219632, a new antibiotic I. Discovery and biological activity. *Appl. Microbiol.* **14:** 506-510.

Ramarao, P. and Usharaja, (1983). Effect of organic amendments in soil on foot and root rot of wheat caused by *Sclerotium rolfsii Indian Phytopath.* **36:** 196-197.

Ridout, C.J., Coley-Smith, I.R. and Lynch, J.M. (1986). Enzyme activity and electrophoretic profile of extracellular protein induced in *Trichoderma* spp. by cell walls of *Rhizoctonia solani J. Gen. Microbiol.* **132:** 2345-2352.

Singh, D.V., Srivastava, K.D. and Aggarwal, R. (1991). Significant findings under PL-480 Karnal bunt Project at Delhi centre. 30th *All India Wheat Res. Workers' Workshop,* Kanpur (India), pp. 1-17.

Singh, H.B. (1993). Antifungal activity of *neem* oil and its efficacy in biocontrol measures in combination with *Trichoderma harzianum. World Neem Confer.* (Abstr.) 24th to 28th Feb. 1993, Bangalore.

Singh, R.K. and Dwevedi, R.S. (1987). Studies on biological control of *Scelorotium rolfsii* Sacc. causing foot rot of barley. *Acta Botanica Indica.* **15:** 160-164.

Singh, U.P. and Pandey, K.R. (1966). Effect of green and mature plant residues and compost on population of *Pythium aphanidermatum in Soil. Indian Phytopath.* **19:** 367-371.

Sinha, P. and Saxena, S.K. (1987). Effect of *neem* leaf powder and extract on the development of fruit root caused by *Aspergillus niger. Neem Newsl.* **4:** 45-47.

Sivan, A. and Chet, I. (1986). Biological control of *Fusarium* spp. in cotton, wheat and muskmelon by *Trichoderma harzianum. J. Phytopath.* 116.

Sychev, P.A. and Shaposhnik, Yu, A. (1982). Antagonistic effects of *Trichoderma viride* Fr. in relation to some pathogens of *Cucumis sativum* L. *Mikolog-Fitopatol.* **16:** 154-159.

Thonsmo, A. and Dennis, C. (1978). Effect of temperature on antagonistic properties of *Trichoderma* species. *Trans. Brit. mycol. Soc.* **71:** 469-474.

Vir, D. and Sharma, R.K. (1985a). Efficacy of fungicides. XXXII. Evaluation of *neem* oil for control of plant pathogens. *Asian Farm Chem.* **1**(7/8): 23-24.

Waid, J.S. (1962). Influence of oxygen upon growth and respiratory behaviour of fungi from decomposing ryegras roots. *Trans. Brit. Mycol. Soc.* **45:** 479-487.

Weindling, R. (1934). Studies on a lethal principle effective in the parasite action of *Trichoderma lignorom, Rhizoctonia solani* and other soil fungi. *Phytopathology,* **24:** 1153-1179.

Weindling, R. and Emerson, O.H. (1936). The isolation of a toxic substance from the culture of a *Trichoderma Phytopathology,* **26:** 1068-1070.

*Weindling, R. and Fawcett, H.S. (1936). Experiment in the control of *Rhizoctonia* damping off of citrus seedling. *Hilgardia* **10:** 1-16.

Wells, H.D., Bell, D.K. and Casimir, A.J. (1972). Efficacy of *Trichoderma harzianum* as a biocontrol for *Sclerotium ro/fsii. Phytopathology,* **62:** 442-447.

Wilson, M., Crawford, E.K. and Campbell, R. (1988). Biocontrol by *Trichoderma harzianum of* damping off lettuce caused by *Rhizoctonia solani. Bull. Oepp.* **18:** 83-90.

Wright, J.M. (1956). The production of antibiotics in soil III. Production of gliotoxin in wheat straw burided in soil. *Annu. Appl. Biol.* **44:** 461-466.

32

Attraction Behaviour of *Discolaimus major* to Prey Nematodes and Factors Influencing Attraction

Rashid Pervez*

Department of Zoology, Aligarh Muslim University, Aligarh–202 002, U.P.

The chemical cues of prey nematode and behavioural responses of predatory nematodes to these cues, therefore, appear to play a major role in all aspects of their survival and should be helpful in their inter-specific interactions and food finding activities. In case of predaceous nematodes, the prey searching and chemo-attraction are important attributes that may influence their predatory potentials. Several studies have been made to explain food searching abilities of bacteriophagous (Young *et al.*, 1996), plant-parasitic (Bilgrami, *et al.*, 1986b, Pervez and Bilgrami, 2000, 2001a, 2001b) and insect nematodes (Grewal *et al.*, 1994; Anderson *et al.*, 1997a; 1997b). Moreover, information on attraction responses of predatory nematodes towards prey kairomones is also insufficient to be used for any application. Speculations about attraction of predators towards their prey (Esser, 1963; Yeates, 1969; Bilgrami *et al.,* 1985) were, however, substantiated later by Bilgrami and Jairajpuri (1988) when they studied attraction and aggregation behaviour of diplogasterid predators, *Mononchoides longicaudatus* and *M. fortidence.*

Keeping in view the inadequate information available on the prey searching and attraction behaviour of predatory nematodes, following experiments were made using a dorylaim nematode, *Discolaimus major* as predator and *Hirschmanniella oryzae* used as prey.

D. major culture were maintained in Petri dishes containing 1 per cent water-agar using *Rhabditis* sp. as food of *D. major*. Five mg infant milk powder (Lactogen) was spread over the surface of agar to permit growth of a mixed bacterial lawn that served as food for the *Rhabditis* sp.

***Present address:** Indian Institute of Pulses Research, Kanpur-284 024, U.P.

PREY SEARCHING BEHAVIOUR OF PREDATORY NEMATODES

The prey searching behaviour of *D. major* was studied in observation chamber using *Hirschmanniella oryzae* as prey. Observation chamber as shown in Fig. 32.1 consisted of a glass slide glued with 0.5 cm high glass pieces. A glass chamber, 4 × 2 × 0.5 cm thus, formed was filled with 2 mm thick layer of 1 per cent autoclaved agar. A 4 mm high plastic straw pipe was placed vertically so as one of its end, glued with 400 mesh nylon sieve, remained within the agar. This is referred as incubation chamber, and the point where the predators were released as inoculation point. The prey searching behaviour of the predators was observed after incubating twenty-five individuals of *H. oryzae* in the incubation chamber for 16 h at 28±2°C. The adult predators were tested, one at a time to study and analyse their prey searching behaviour. Each observation was made fifteen times.

ATTRACTION RESPONSE OF *D. MAJOR* TOWARDS EXCISED AND NON-EXCISED *H. ORYZAE*

Attraction responses of *D. major* were tested in 5.5 cm diameter Petri dishes, divided into inner, middle and outer zones by two concentric circles, 0.5 and 2.5 cm in diameter at the bottom. A plastic straw pipe 5 mm high and 5 mm in diameter, one end sealed with a piece of 400 mesh sieve (to restrict prey nematodes within the tube), was placed vertically in the inner circle so as the sealed end remained in contact with the bottom of the dish. Autoclaved 1 per cent water agar was poured into the Petri dish and straw pipe to make a 2 mm thick agar layer. Twenty-five excised (cut into two pieces) and non-excised (live) adult individuals of *H. oryzae*, extracted fresh from the soil and surface sterilized with 0.01% mercuric chloride and 0.01% streptomycin solutions, were inoculated separately in the straw pipe and incubated for 16 h at 28 ± 2°C. Five adult *D. major* were released at the periphery of the middle circle. The distribution of predators was recorded after 2, 4, 6, 8, 10 and 12 h and scores were obtained by summing up the product of the number of nematodes of each zone with their

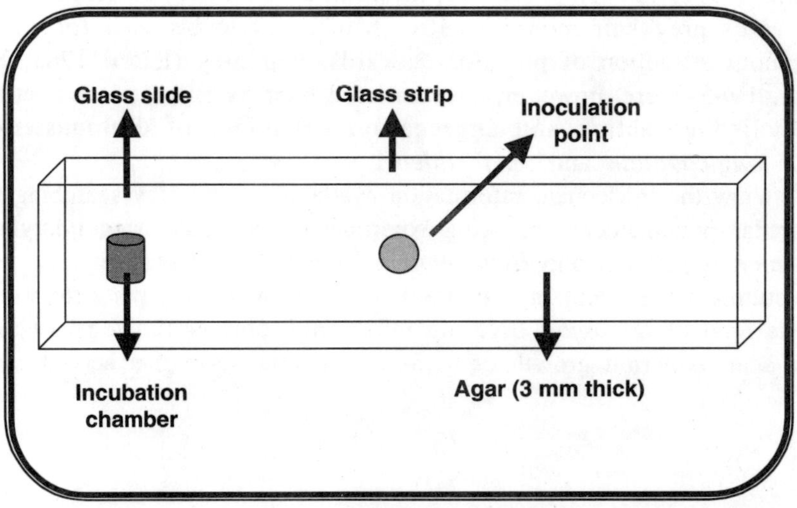

Fig. 32.1. Observation chamber to study prey searching and attraction behaviour of predatory nematodes.

corresponding weighting factors. The weighting factors were obtained by dividing the area of the outer zone by that of each of the three zones (Bilgrami and Jairajpuri, 1988). The scores were then converted into log scores. The log score of the 40 replicates was summed up and mean was calculated, which is referred here as mean log score. Mean log score, therefore, is the mean attraction response of the predators and the scale at which the attraction response of the predators was measured and compared.

Mean attraction response of predators towards kairomones emitted by different prey trophic groups: To test the attraction responses of *D. major* towards kairomones/attractant dispersed by the prey nematodes belonging to different trophic groups, Petri dish experiments, as described above, were performed. On the basis of Yeates *et al.* (1993) and Bilgrami (1993 and 1995), seven trophic groups of prey nematodes viz., saprophagous, epidermal feeders, endodermal feeders, predatory nematodes, migratory semi-endodermal feeders, virus vectors and cortical feeders were defined and described (Table 32.1). Using the above Petri dish experiments, mean log score for each species of prey belonging to different trophic groups was determined. The Group Mean Log Score (GMLS) of a trophic group was then calculated as follows and presented.

$$\text{GMLS} = \frac{\text{Sum of mean log scores of all prey species belonging to a trophic group}}{\text{Total number of prey species belonging to respective trophic group}}$$

Table 32.1. List of nematodes used as prey.

Prey Trophic Group	Genera
Saprophagous	Mesorhabditis, sp., Acrobeles sp., Rhabditis sp.
Epidermal feeders	*Tylenchorhynchus mashhoodi, Basiria graminicola*, Aphlenchoides sp.,
Migratory semi-endodermal feeders	*Hirschmanniella oryzae, Hoplolaimus indicus, Helicotylencus indicus.*
Predators	*Mesodorylaimus bastiani, Aquatides thornei,* Laimydorus baldus
Cortical feeders	*Hemicriconemoides mangiferae, Hemicycliophora dherendi*
Endodermal feeders	Second stage juveniles of *Meloidogyne incognita, Heterodera mothi, Anguina tritici*
Virus vectors	Longidorus sp., *Paralongidorus sali, Xiphinema americanum*

Factors influencing the attraction of *D. major* towards *H. oryzae*: Different biotic and abiotic factors viz., temperatures, incubation periods of prey, prey density, starvation periods of predators, agar concentrations and agar thicknesses affected the attraction of *D. major* towards *H. oryzae*.

Effect of temperatures: Effect of different temperature regimes viz., 5, 10, 15, 20, 25, 30, 35 and 40°C on the attraction of *D. major* towards kairomones emitted by *H. oryzae*.

Effect of periods of incubation of prey nematodes: Various incubation period of *H. oryzae* viz., 4, 8, 12, 16, 20 and 24 h were tested to determine their influence on attraction of *D. major*.

Effect of prey density: Effect of prey numbers (25, 50, 75, 100, 125, 150, 175, 200, 225 and 250) on the attraction of *D. major* towards individuals of *H. oryzae*.

Effect of periods of starvation of predators: *D. major* starvation period viz., 0, 1, 2, 3, 4 and 5 day was tested separately towards *H. oryzae.* Day 0 refer to fresh predators. Predators were transferred to fresh water each day.

Effect of agar concentrations: Effect of different concentration of agar viz., 1, 2, 3, 4, 5 and 6 per cent on the degree of attraction of *D. major* towards *H. oryzae* was tested.

Effect of agar thickness: Predatory nematodes, *D. major* were released in Petri dishes containing 2, 4, 6, 8, 10 and 12 mm thick agar layer to determine their influence on the attraction responses of predators.

Observations on the prey searching and attraction behaviour of *D. major* revealed their foraging and chemotactic behaviour towards prey nematodes. The phenomenon that explains the movement of the predator towards prey is related to its sensory mechanisms.

Minimum perceptible attraction gradient may be defined as the attractant diffusion gradient perceptible to predators at a minimum concentration from a particular distance. Minimum perceptible attraction gradient depends upon the difference between the rates of attractants emitted and lost in the medium. It should, therefore, be conformed when the rate of attractants emitted is higher than the rate of attractant lost in medium.

Minimum response threshold may be defined as the ability of predators to respond to attractants at a minimum perceptible attraction gradient. Minimum response threshold of an individual may, therefore, be a parameter that demarcates between stimulated (directional) and non-stimulated (non-directional) locomotion of predators.

Chemotactic response of predators may be defined as the change in the activity and behaviour of the nematodes that occur in response to prey kairomones, allomones or other chemicals present in the surroundings. Chemotactic response of predators may be either homogeneous (if the response is towards the same species) or heterogeneous (if response is towards different species).

PREY SEARCHING BEHAVIOUR OF PREDATORY NEMATODES

Depending upon minimum perceptible attraction gradient, minimum response threshold and chemotactic response threshold, prey searching and attraction behaviour of predator may be classified into six phases viz., (1) random movement, (2) exploration and probing, (3) perception of attractant, (4) switched response, (5) co-ordinated, and (6) chemotactic response of predators.

Random movement: Predators moved randomly in presence of prey nematodes till minimum perceptible attraction gradient is formed in agar and perceived by them at their minimum response threshold (Fig. 32.3A). The random movement of predators may be distinguished by shortening of the wavelength, elevation of the body amplitude, less side to side head movements (5-18/m), and the forward and backward crawling (6-9/m).

Exploration and probing: Following the attraction diffusion gradient (Fig. 32.3B) the predators initially foraged and exhibited exploration and probing behaviour (Fig. 32.3C) by showing more lateral head movements (15-18/m), lip rubbing (10-12/m) and head lifting (11-14/m).

Perception of attractants: Increase in the concentration of attractants resulted in formation of minimum perceptible attraction gradient (Fig. 32.3D), which was perceived by the predators at their minimum response threshold to respond. The predators then exhibited forward switched response by showing increased lip probing (14-15/m) and oesophageal pumping (13-14/m).

Switched response: The switched response may be defined as the change of uncoordinated locomotory behaviour to co-ordinated locomotion upon stimulation by attractants or vice versa. Switched response may be divided as forward switched response (when uncoordinated locomotion changes into co-ordinated either due to decrease in the minimum perceptible attraction gradient that was formed by the predators) (Fig. 32.3E) and reverse switched response (when predators exhibited change in their locomotory behaviour from co-ordinated to uncoordinated due to increase in the minimum perceptible attraction gradient or minimum response threshold) (Fig. 32.3L). Forward switched response lead predators to stimulated or co-ordinated locomotion towards prey kairomones (Fig. 32.3F).

Co-ordinated locomotion (Stimulated locomotion): While the approach of predators became directional towards prey attractants (Fig. 32.3G), wavelength formed by them increased, body amplitude decreased, lateral and sideways head movements and forward and backward crawling increased. Predators thus, indicated their movement to up gradients (Fig. 32.3H).

Chemotactic response: While moving to up gradients, the predators reached to the source of attraction (straw pipe containing prey individuals) (Fig. 32.3I). *D. major* exhibited heterogeneous attraction response as they migrated towards other species of nematodes. As minimum perceptible attraction gradient decreased and became unperceivable, predators started moving to down gradients (Fig. 32.3J), exhibiting non-directional movements (Fig. 32.3K) by showing entered into reverse switched response (Fig. 32.3L), by showing short wavelength, high amplitude, less lateral head and sideways movements.

During attraction, if minimum perceptible attraction gradient (Fig. 32.3N), the predators reversed back to show uncoordinated locomotion (Fig. 32.3M) below the minimum response threshold level of predators (Fig. 32.3O). The predators were then reverted back either to move randomly again in the medium on a fresh search mission or abundant its search for a prey.

Attraction response of *D. major* towards excised and non-excised *H. oryzae* (Fig. 32.2A): *D. major* exhibited positive and significant response towards kairomones emitted by the excised and non-excised individuals of *H. oryzae.* The excised prey individuals elicited maximum response of the predators. Attraction of *D. major* increased significantly from minimum to maximum when period of observation was increased from 2 to 8. Observation periods more than 8 h yielded lesser degree of attraction of *D. major*.

Mean attraction response of predators towards kairomones emitted by different prey trophic groups (Fig. 32.2B): Prey members belonging to different trophic groups elicited differential attraction response of *D. major*. All prey trophic groups were more attractive to predators when their members were tested as excised individuals.

D. major was most responsive to kairomones emitted by the excised and non-excised members of epidermal feeders (GMLS = 2.45 for excised and 2.13 for non-excised). The virus vector was least attractive to *D. major* both when its members were tested either as excised (GMLS = 1.99) and non-excised (GMLS = 1.71). Other prey trophic groups were moderately attractive to predators.

Factors influencing the attraction of predators towards prey: Temperatures, incubation periods of prey, prey density, starvation periods of predators, agar concentrations and agar thickness governed attraction of *D. major* towards kairomones emitted by *H. oryzae.*

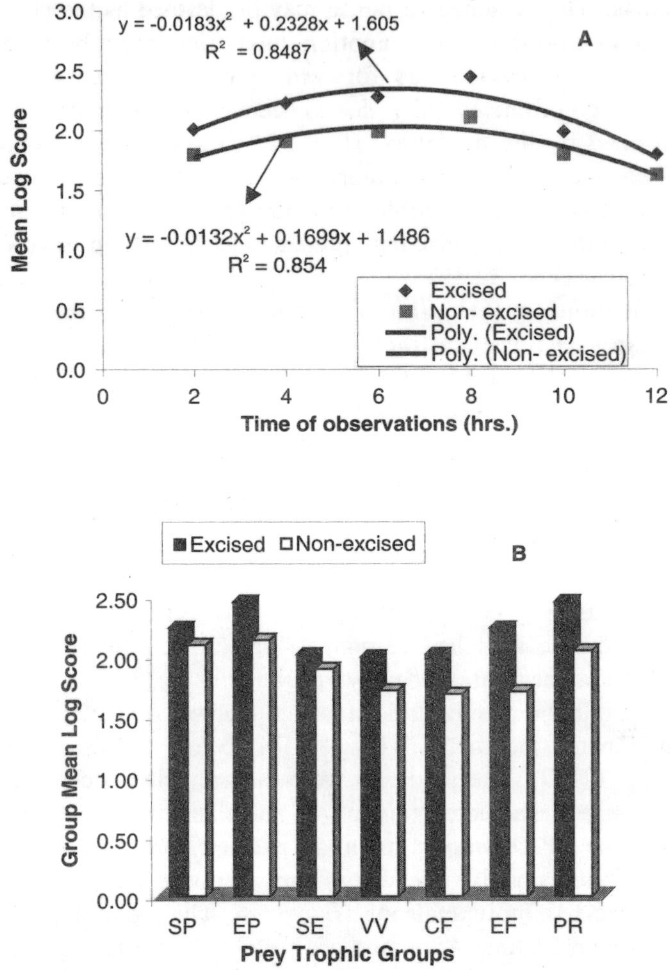

Fig. 32.2. A. Attraction of D. major in response to kairomones emitted by *H. oryzae*. **B.** Mean attraction response of predators towards kairomones emitted by different prey trophic groups of nematodes. **SP:** Saprophagous; **EP:** epidermal feeders; **SE:** migratory semi-endodermal feeder; **VV:** virus vectors; **CF:** cortical feeders; **EF:** endodermal feeders; **PR:** predatory nematodes.

Effect of temperatures (Fig. 32.4A): The positive and significant correlation was found between the degree of attraction of predator and different temperature regimes ($r = 0.63$, PE(r) $= 0.168$, $p < 0.05$). *D. major* was attracted maximally towards attractants emitted by *H. oryzae* at 30°C (MLS = 2.36 ± 0.196; CV = 7%). Attractiveness of prey kairomones to predators was either lost or suppressed at lower temperatures (5-15°C) and above 30°C.

Effect of periods of incubation of prey (Fig. 32.4B): Short (4h) period of incubation of prey attractants yielded minimum attraction response of *D. major* (MLS = 1.56 ± 0.431, CV = 22%). Incubation period (16h) of prey nematodes yielded maximum response of *D. major*

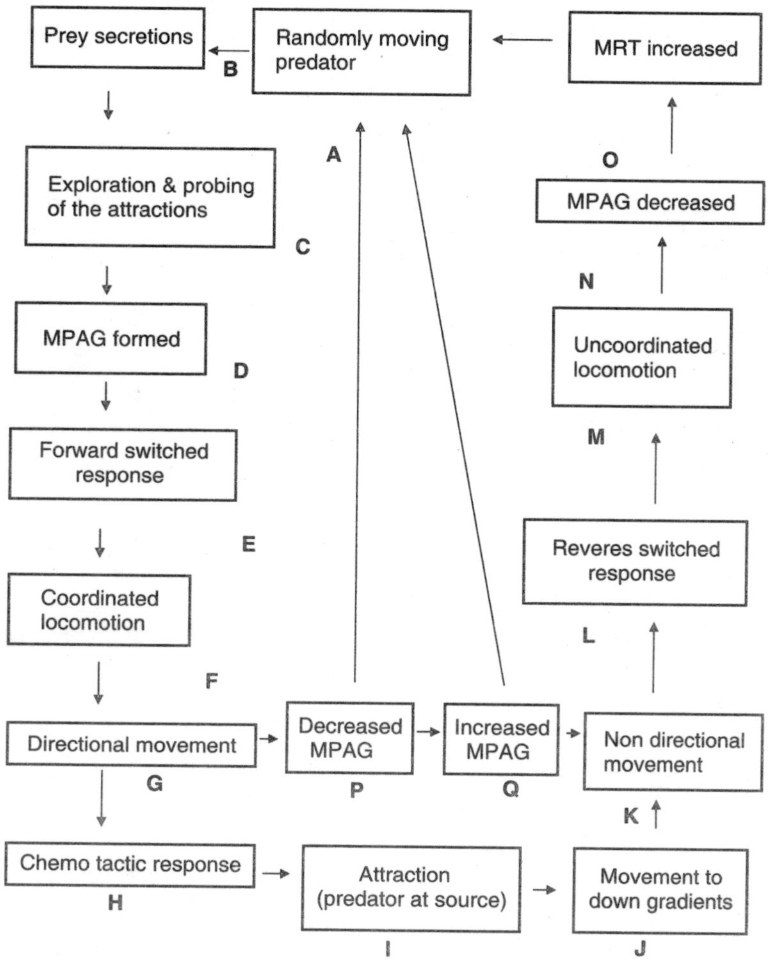

Fig. 32.3. Schematic representation of prey searching and attraction behaviour of *Discolaimus major* against the *Hirschmanniella oryzae*.

(MLS = 2.30 ± 0.187, CV = 8%). The significant correlation was found between attraction of predator and period of prey incubation (r = 0.58, PE(r) = 0.185, $p < 0.05$).

Effect of prey density (Fig. 32.4C): *D. major* exhibited significant correlation between attraction of predators and density of prey (r = 0.78, PE(r) = 0.085, $p < 0.05$). Maximum attraction of predators was recorded when 200 individuals of *H. oryzae* (MLS = 2.45 ± 0.111, CV = 5%) were incubated as prey.

Effect of periods of starvation of predators (Fig. 32.4D): The period of starvation of predators stimulated their urge to search their prey more actively. Attraction of predators increased significantly from 0 day (fresh) to 5th day of their starvation (r = 0.67, PE(r) = 0.154, $p < 0.05$). Kairomones emitted by *H. oryzae* were most attractive to 5 day starved individuals of *D. major* (MLS = 2.45 ± 0.112, CV = 5%).

Effect of agar concentrations (Fig. 32.4E): Attraction of *D. major* was inversely correlated with agar concentrations (r = –0.88, PE(r) = 0.063, p < 0.05). Predators were attracted maximum towards prey kairomones at 1 per cent concentration of agar (MLS = 2.38 ± 0.270, CV = 11%). Higher concentrations of agar inhibited attraction responses of predators towards prey.

Effect of agar thickness (Fig. 32.4F): Attraction of *D. major* decreased as the thickness of agar layer increased. The negative but significant correlation was found between attraction of predator and thickness of agar layer (r = –0.97, PE(r) = 0.016, p < 0.05). A 2 mm thick agar layer was optimum where maximum attraction of predators was recorded (MLS = 2.32 ± 0.193,

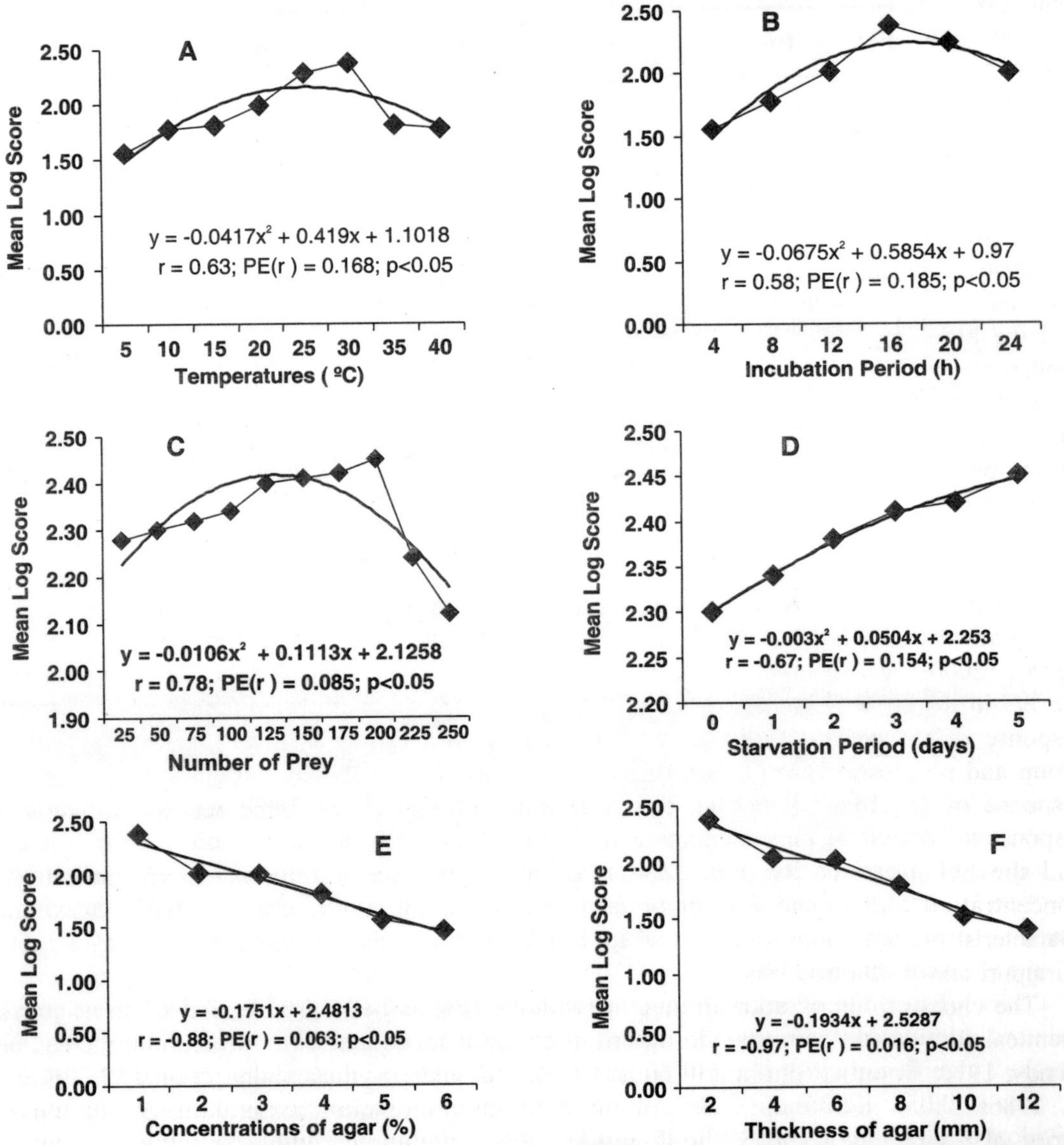

Fig. 32.4. Factors influencing attraction of *D. major* towards the *H. oryzae*. **A**—effect of temperature; **B**—effect of time of prey incubation; **C**—effect of prey density; **D**—effect of period of starvation of predator; **E**—effect of agar concentration; **F**—effect of agar thickness.

CV = 10%). Prey kairomones were least attractive to predators at 12 mm thick agar layer (MLS = 1.39 ± 0.336, CV = 21%).

Attraction behaviour of *D. major* explains relatively early movements of the predator and their migration toward prey kairomones due to their sensing mechanisms (chemotactic response). Following attraction diffusion gradient, the predator foraged initially but exhibited stimulated movements (directional) along the path of the attraction diffusion gradient, the minimum perceptible attraction gradient is developed and perceived by the predator at their minimum response threshold. The orientation of the predator was determined by the probing movement of the head (Ward, 1973; Seymour *et al.,* 1983) and location of the prey upon papillae and the amphids (McLaren, 1976; Wright, 1983). The prey searching behaviour appeared to have depended upon the attraction diffusion gradient, minimum perceptible attraction gradient and the minimum response threshold beside the other physical, chemical and behavioural prey cues. The point where the minimum perceptible attraction gradient was developed and perceived by the predators demarcated random and chemotactic behaviour of the predators. Search parameters like the head probing, head rubbing, stylet thrusting, long wavelengths, more lateral head movements, forward and backward locomotion and the low amplitudes of predators may be designated as the signal of the formation of minimum perceptible attraction gradient which the predator could perceive at its minimum response threshold. In contrast, the sort wave length and the high amplitudes of predator signified their random movements. The reason for more predators remaining in the outer circle could be attributed to the absence of prey kairomones or to the predators inability to perceive attraction diffusion gradient possible due to their high minimum response threshold or non-formation of the minimum perceptible attraction gradient by the prey nematodes. The foraging behaviour of *D. major* and their chemotactic response to localized food (prey) may result in population fluctuation of predator even in the soil environment. Such fluctuation may be triggered separately or in combination by the food depletion, overcrowding, minimum perceptible attraction gradient and minimum response threshold.

Present studies on chemosensory response of *D. major* towards kairomones emitted by *H. oryzae* and member belonging to prey trophic group are in conformity with Bilgrami and Pervez (2000) who reported attraction of two species of dorylaim predator, *M. bastiani* and *A. thornei* in response to kairomones/attractants dispersed by the prey individuals. The individual response of the predators and those reported here are different as they belong to different group and possess different requirement for their survival strategies. The high mean attraction response of *D. major* during an observation period of 8 h and their differential attraction response to various prey trophic group may be attributed to the differential individual requirement and the behaviour of the predators, their preference for a prey, chemical composition, concentration and quantity of prey kairomones/attractants (Bilgrami, 1993; 1995). Such characteristics show inter-specific and intra-specific variations (Doncaster and Seymour, 1973; Jairajpuri and Bilgrami. 1990).

The chemical composition of prey individuals may be determined by a number of physio-chemical factor more specially the ageing and starvation of prey individuals (Cooper and Van Gundy, 1970; Reversat, 1981a; 1981b) and food and feeding habits of the predator (Bilgrami *et al.*, 1986a; Yeates and Wardle, 1996), so the attraction of *D. major* decreased after an observation period of 8 h may be related to the decrease in the minimum perceptible attraction gradient due to the increase in the rate of loss of attractant in the medium. The attractants in this case survived for a total period of 24 h (prey incubation (16 h) + period of observation (8 h) to elicit maximum attraction response of *D. major*. Differential response of *D. major* when used as predator or when put as prey nematode, may have been due to specific difference in their minimum response threshold in their former role and in minimum perceptible attraction gradient in their later role. It is, therefore, possible to make considerable deductions here about the properties of the attractants and the behaviour of nematodes in a particular environment if basis assumptions are accepted on both the diffusion of the prey attractants, i.e. minimum perceptible attraction gradient and the sensory sensitivity of the predatory nematodes, i.e. minimum response threshold.

ACKNOWLEDGEMENTS

The author is grateful to the Chairman, Department of Zoology, Aligarh Muslim University, Aligarh for providing necessary laboratory facilities to carry out the present research work. Thanks are also extended to Director, IIPR and Dr. S.S. Ali, Principal Scientist, Division of Crop Protection, IIPR, Kanpur.

CONCLUSION

Observations were made on a dorylaim predator, *Discolaimus major* to study and analyse its prey searching and attraction behaviour using nematodes as prey belonging to different trophic groups viz., saprophagous, epidermal feeders, endodermal feeder, migratory semi-endodermal feeders, predators, cortical feeders and virus vectors. The prey searching and attraction mechanisms of *D. major* were classified into seven phases depending upon attraction diffusion gradient (ADG) and minimum perceptible attraction gradient (MPAG) of prey kairomones and minimum response threshold (MRT) and chemotactic response threshold (CRT) of predators. The six phases viz., (1) random locomotion; (2) exploration and probing; (3) perception of attractant; (4) switched response; (5) co-ordinated locomotion, and (6) chemotactic response of predators appeared to have depending upon ADG, MPAG, MRT, CRT, besides other physical, chemical and behavioural prey cues. The high group mean attraction response (GMAR) of *D. major* towards kairomones emitted by members belonging to epidermal feeders might be attributed to the inert behaviour of predators, their preference for prey nematodes, chemical concentration, composition, quantity of prey kairomones/attractants.

REFERENCES

Anderson, A.R.A., Saleeman, B.D., Young, I.M. and Griffiths, B.S. (1997a). Nematode movement along chemical gradient in a structurally heterogenous environment–II Theory. *Fundamental and Applied Nematology*, **20:** 165-172.

Anderson, A.R.A., Young, I.M., Saleeman, B.D., Griffiths, B.S. and Robertson, W.M. (1997b). Nematode movement along a chemical gradient in a structurally heterogenous environment–I Experiment. *Fundamental and Applied Nematology*, **20:** 157-163.

Bilgrami, A.L. (1993). Analysis of relationships between predation by *Aprocelaimellus nivalis* and different prey trophic categories. *Nematologica*, **39:** 356-365.

Bilgrami, A.L. (1995). Numerical analysis of the relationship between the predation by *Mesodorylaimus bastiani* (Nematoda : Dorylaimida) and different prey trophic categories. *Nematologie Mediterranea*, **23:** 81-88.

Bilgrami, A.L., Ahmad, I. and Jairajpuri, M.S. (1985). Predatory behaviour of *Aquatides thornei* (Nygolaimina : Nematoda). *Nematologica*, **30:** 457-462.

Bilgrami, A.L., Ahmad, I. and Jairajpuri, M.S. (1986a). A study on the intestinal content of some mononchs. *Revue de Nematologie,* **9:** 191-194.

Bilgrami, A.L., Ahmad, I. and Jairajpuri, M.S. (1986b). Attraction and feeding behaviour of adult *Hirschmanniella oryzae. Indian J. Nematology,* **15:** 209-216.

Bilgrami, A.L. and Pervez, R. (2000). Prey searching and attraction behaviour of *Mesodorylaimus bastiani* and *Aquatides thornei* (Nematoda : Dorylaimida). *International J. Nematology,* **10:** 199-206.

Copper, A.F. and Vangundy, (1970). Metabolism of glycogen and neutral lipids by *Aphelenchus avenae* and *Caenorhabditis* sp. in aerobic, micro-aerobic and anaerobic environments. *J. Nematology,* **2:** 305-315.

Doncaster, C.C. and Seymour, M.K. (1973). Exploration and selection of penetration site by tylenchids. *Nematologica,* **19:** 137-145.

Esser, R.P. (1963). Nematode interactions in plates of non-sterile water-agar. *Proceeding of Soil Crop Science & Society. Florida,* **23:** 121-138.

Grewal, P.S., Gaugler, R. and Selven, S. (1994). Host recognition by entomopathogenic nematode: behavioural response to contact with host feces. *J. Chem. Eco.,* **19:** 1219-1230.

Jairajpuri, M.S. and Bilgrami, A.L. (1990). Predatory nematodes. In: *Nematode bio-control: Aspect and prospects,* pp. 95-125 (Eds Jairajpuri, M.S., Alam, M.M. and Ahmad, I.) New Dehli, India: CBS Publishers and Distributors Pvt. Ltd.

Mclaren, D.L. (1976). Nematode sense organs, 195-265. *In: Advanced parasitology* (Ed. Dames, B.). London: Academic press.

Pervez, R. and Bilgrami, A.L. (2000a). Some factors influencing chemoattraction behaviour of dorylaim predator, *Laimydorus baldus* and *Discolaimus major* towards prey kairomones. *International J. Nematology,* **10**(1): 41-48.

Pervez, R. and Bilgrami, A.L. (2000b). Chemosensory responsdes of predatory nematodes towards kairomones emitted by prey nematodes belonging to different trophic groups. *Nematologie mediterenea,* **28:** 267-277.

Pervez, R. and Bilgrami, A.L. (2001). Role of prey attractant in food finding behaviour of dorylaim predator *Laimydorus baldus* and *Discolaimus major* (Nematoda: Dorylaimida). *Indian J. Nematology,* **31**(1): 1-7.

Reversat, G. (1981a). Effect of ageing and starvation on respiration and food reserve contents in adult *Hirschmanniella spinicaudata. Revue de Nematologie,* **4:** 125-130.

Reversat, G., (1981b). Consumption of food reveres by starved second stage juveniles of *Meloidogyne javanica* under condition including osmobiosis. *Nematologica,* **27:** 509-539.

Seymour, M.K., Wright, K.A. and Doncaster, C.C. (1983). The action of anterior feeding apparatus of *Caenorhabditis elegans* (Nematoda : Rhabditida). *J. Zoology,* **201:** 509-538.

Ward, S. (1973). Chemo taxis by the nematode *Caenorhabditis elegans.* Identification of attractants and analysis of the responses by use of mutants. *Proc. Nat. Acad. Sci., USA,* **70:** 817-821.

Wright, K.A. (1983). Nematode chemosensilla form and functions. *J. Nematology* **15:** 151-158.

Yeates, G.W. (1969). Predation by *Mononchoides potohikus* (Nematoda: Diplogasteridae) in laboratory culture. *Nematologica,* **15:** 1-9.

Yeates, G.W., Bongers, T., De Goede, R.G.M., Freckman, D.W. and Georgieva, S.S. (1993). Feeding habit in soil nematode families and genera–an outline for soil ecologists. *J. Nematology,* **25:** 315-331.

Yeates, G.W. and Wadley, D.A. (1996). Nematodes as predators and prey: relationships to biological control and soil processes. *Pedobiologia,* **40:** 43-50.

Young, I.M., Griffiths, B.G. and Robertson, W.M. (1996). Continuous foraging by bacterial feeding nematodes. *Nematologica,* **42:** 378-382.

33

Standardization of Pathotype Numbers of *Colletotrichum falcatum* on Sugar Cane

Vijai Singh, S.N. Srivastava, Sunita Lal, S.K. Awasthi and B.B. Joshi

Crop Protection Division, Indian Institute of Sugarcane Research, Lucknow–226 002, U.P.

Colletotrichum falcatum Went causes a devastating red rot disease in sugar cane. Identification of highly virulent pathotype is necessary for developing resistant varieties. Ten pathotypes have been identified, so far, from India including 6 from subtropics and 4 from tropics. Their reactions were different on 13 varieties/genotypes (Co 1148, Co 7717, Co J 64, Co 419, Co 997, CoC 671, Khakai, Co 975, Co 62399, Baragua, *CoS* 767, Bo 91 and SES 594), but clear cut characterization of each pathotype has not been achieved, so far. Naming of pathotype merely by putting Cf 01-10 does not leave any impact of potency of the pathotype. It would be more useful if their virulence is known on serially arranged differentials in nomenclature of the pathotype. The description of pathotypes would attain figures as (I) 1, 3. 5, 6, 7, 8 and 9, (II) 2. 5, 6 and 7, (III) 3, 5 and 7, (IV) 1, 2, 3, 4, 6, 7 and 8, (V) 5 and 6. (VI) 3. 5, 6, and 10. (VII) I, 3, and 7, (VIII) 1, 2, 3 ,5 ,6, 7, 8 and 9 (IX) 1, 3, 5, 7 and 11, (X) 2, 4, 5, 6, 8 and 10. Pathotype achieving more numbers of differentials viz., 1, 3, 5, 6, 7, 8 and 9 (I); 1, 2, 3, 4, 6, 7 and 8 (IV); 1, 2, 3, 5, 6, 7, 8 and 9 (VIII) and 1, 3, 5, 7 and 11 (X) appear to be more virulent pathotypes.

Evaluation of varieties with identified pathotypes is a very important aspect for selection of resistant genotypes. Ten pathotypes of *C. falcatum* based on mainly breakdown of resistance and reaction on 13 differentials have been identified through AICRP programmes (Anon. 1991, Satyavir *et al.,* 1984, 2001, Satyanarayana and Rao, 1984, Viswanathan *et al.,* 2003). Very limited work on serological and molecular differences has been undertaken (Alexander and Jothi, 1995, Mohanraj *et al.,* 2002, Singh *et al.,* 1993, Singh, 1995, Suman *et al.* 2005). The nomenclature adopted as Cf 01-10, however, does not reflect any details of the pathotype. It is, therefore, a refined method by indicating the capability of infection to differential is proposed in the communication. This will lead to better understanding of highly potent pathotypes to be utilised in breeding programme.

EARLIER RECOGNITION OF PATHOTYPES OF *C. FALCATUM*

Earlier, on the basis of colour of the mycelium and capability of spore production, light and dark strains of red rot pathogen were recognized. Dark coloured isolates were produced on resistant varieties and light ones on susceptible varieties (Srinivasan, 1962). Sugar cane plant extract has affected the stability of the isolate, however, deviation in reaction could not be clearly explained. Later on the basis of morphology, A, B, C, D, E, F, G, H and I variants of *C. falcatum* were reported (Rafay and Padmanabhan, 1941). Presence of distinct isolates (R-103, R-115, R-117, R-121, R-123, R-135, R-141, R-146, R-158, R-183, R-191, R-200) was also reported by several workers (Kirtikar, 1959; Singh *et al.,* 1984; Pandey and Sakal, 1974 and Gupta *et al.,* 1980). Recently, binary and decanary values of infection were also proposed (Singh *et al.*, 1993).

CURRENT RECOGNITION OF PATHOTYPES OF *C. FALCATUM*

On the basis of reactions on common set of 13 differentials viz., Co 1148, C0 7717, CoJ 64, Co 419, Co 997, CoC 671, S. *sinense* (Khakai), Co 975, Co 62399, *S. officinarum* (Baragua), CoS 767, Bo 91 and *S. spontaneum* (SES 594), uniform method of recognition of pathotypes was evolved by AICRP workshop during 1991. Variety CoS 8436 has also been included in the list of differentials under AICRP programme from 2006 onwards. Three pathotypes viz., Cf 01, 02 and 03 affecting the varieties Co 1148, Co 7717 and Co J 64, repectively, were identified. Later, three new pathotypes viz., Cf 07, 08 and 09 were recognized on Co J 64 and CoS 767 from Western UP, Punjab and Haryana (Beniwal *et al.,* 1989; Anon, 1991, Satyavir *et al.,* 2001, Satyavir, 2003). Cf 04, 05, 06 and 10 were recognized on Co 419, Co 997, CoC 671 and 85 A 261, repectively from coastal AP and tropical India (Satyanarayana and Satyanarayana, 1976; Satyanarayana and Rao, 1984 Satyanarayana and Rao, 1995, Viswanathan *et al.,* 2003). Reactions (Table 33.1) indicated that Cf 02 and 03 failed to infect Co 1148. Cf 02 and 05 produced moderately resistant reaction on CoJ 64 but rest could easily infect it. Cf 01, 03, 05 and 06 failed to infect Co 419. Comparatively, Co 997, CoC 671 and *S. sinense* (Khakai) were susceptible to most of the pathotypes, however, Cf 04 produced MR reaction on Co 997 and Cf 03, 07, 09 produced MR reaction on CoC 671 and Cf 05 and 06 produced MR reaction on Khakai. Cf 02, 03, 05, 07 and 09 failed to infect Co 975. Cf 03, 04, 05, 06, 07 and 09 failed to infect Co 62399 but the variety was MR by Cf 02. Cf 01, 02, 03, 04,05, 07, 08 and 09 failed to infect *S. officinareum* (Baragua). Cf 01, 02, 03, 04, 05, 06, 07, 08 failed to infect CoS 767 but Cf 09 produced MR reaction. Bo 91 and *S.spontaneum* (SES-594) were established to be resistant to all Cf 01-10, so far. The specific selective infectivity under identical conditions has been accepted widely as major criteria for identification of *C. falcatum* pathotypes (Kumar *et al.,* 2001). Malathi *et al.,* 2006 have reported infection of Co 7717 by Cf 01 and Co 1148 by Cf 02 by trying several subculturing on the extract of resistant host.

Table 33.1: Modified nomenclature of pathotypes of *C. falcatum.*

Pathotype	Infectivity potential
Cf 01 (Co 1148)	**S**: 1,3,5-9; **R**: 2,4,10-13 and **MR**: Nil
Cf 02 (Co 7717)	**S**: 2,5,6,7; **R**: 1,8,10-13 and **MR**: 3,4,9
Cf 03 (CoJ 64)	**S**: 1,3,5,7; **R**: 1,2,4,8-13 and **MR**: 6
Cf 04 (Co 419)	**S**: 1-4,6-8, **R**: 9-13 and **MR**: 5
Cf 05 (Co 997)	**S**: 5-6; **R**: 2,4,8-13 and **MR**: 1,3,7
Cf 06 (CoC 671)	**S**: 3,5,6,10; **R**: 2,4,9,11-13 and **MR**: 1,7,8
Cf 07 (CoJ 64)	**S**: 1,3,5,7; **R**: 2,8-13 and **MR**: 4
Cf 08 (CoJ 64)	**S**: 1-3,5-9; **R**: 10,12,13 and **MR**: 4,11
Cf 09 (CoS 767)	**S**: 1,3,5,7,11; **R**: 2,8-10,12,13 and **MR**: 4,6
Cf 10 (85A 261)	**S**: 2,4-6,8,10; **R**: 1,7,11,13 and **MR**: 3,9,12 (not known)

After: Satyavir *et al.,* 2001; Satyanarayana and Rao, 1995
Arrangement of host differentials: 1- Co 1148, 2- Co 7717, 3- CoJ 64, 4=Co 419, 5- Co 997, 6- CoC 671, 7-Khakai, 8- Co 975, 9- Co 62399,10- Baragua, 11- CoS 767,12- Bo 91 and 13- SES 594.

SEROLOGICAL AND MOLECULAR DIAGNOSIS OF PATHOTYPES

Besides host reaction, some work has also been carried out on serology and biotechnology. Serological relationship of pathotypes capable of causing infection in common hosts was established. Molecular characterization of some pathotypes was also made, but precise identification was not possible by serology or RAPD based DNA band pattern (Mohanraj *et al.*, 2002, Suman *et al.*, 2005, Alexandar and Jothi, 1995, Singh, 1995). Genetic relationship indicated that pathotypes prevalent in north-western zone fall in 3 groups (Suman, *et al.*, 2005). However, Cf 01 was found to be molecularly quite distinct from Cf 02 (Mohanraj, *et al.*, 2002). On the basis of similar RAPD-DNA band pattern, *C. falcatum* isolate from CoC 90063 was similar to Cf 09 for causing infection in CoS 767 (Mohanraj *et al.*, 2002). More precise molecular work would be required to differentiate all the pathotypes identified on host reaction basis.

IMPROVEMENT IN NOMENCLATURE OF PATHOTYPES

Nomenclature as Cf number does not leave any impact of virulence of the pathotype. In several crops where genes for resistance are not known, new pathotypes/races are identified by describing infection potential/binary/decanary values (Habgood, 1970). Similarly, a system for identification of races of *Pyricularia oryzae* was proposed on the basis of dichotomous arrangement of susceptible and resistant reactions on differentials. The race number was sufficient to understand the isolate without going into details of chart. In sugar cane-red rot system genes for resistance are not known. A new methodology for designation was proposed. Differentials achieving reactions of 4.1-9.0 grade were put under susceptible and 0.0-4.0 as resistant (Singh and Singh, 1993, Singh *et al.*, 1993, Srinivasan and Bhat, 1961). More potent ones achieved higher decanary values. Putting serial number of susceptible host differential would be still easier to understand the virulent pathotypes. Modified names of 10 identified pathotypes would be as follows.

1- 1,3,5-9 (Cf 01); 2- 2,5-7 (Cf 02); 3- 3,5,7 (Cf 03); 4- 1,2-4,6-8 (Cf 04); 5- 5,6 (Cf 05); 6- 3,5,6,10 (Cf 06); 7- 1,3,5,7 (Cf 07); 8- 1-3,5-9 (Cf 08); 9- 1,3,5,7,11 (Cf 09) and 10- 2,4-7,8,10 (Cf 10). The infectivity expressed in numerical number suggests that Cf 01, 04, 08, 09 and 10 were more potent. Comparison of values of any isolate can be made easily. Less potent can be marked quickly.

REFERENCES

Alexander, K.C. and Jothi, R. (1995). Current status of *Colletotrichum falcatum* Went. In: Proc. Natl. seminar on sugar cane production constraints and strategies for research and management of red rot, (Eds. Agnihotri, V.P. *et al.*) Army printing Press, Lucknow, pp. 273-284.

Anonymous, (1991). *Proc. AICRP*, 1990-91. IISR, Lucknow.

Beniwal, M.S., Satyavir, and Virk, K.S. (1989). Pathogenic variability in *Colletotrichum falcatum* incitant of red rot of sugar cane. *Indian Phytopath.* **42:** 95-98.

Duttamajumder, S.K., Singh, N. and Agnihotri, V.P. (1990). Behaviour of *Colletotrichum falcatum* under waterlogged condition. *Indian Phytopath.* **43:** 227-229.

Gupta, S.C., Singh M.P. and Upadhyay, U.C. (1980). A new biotype of *Colletotrichum falcatum* Went. *Curr. Sci,* **49:** 600-601.

Habgood, R.M. (1970). Designation of physiological races of plant pathogens. *Nature.* **227:** 1268-1269.

Kirtikar, (1959). Physiological specialization in red rot pathogen. *Indian Sugar.* **8:** 797-799.

Kumar, A., Satyavir, and Niwas, R. (2001). Weather factors in relation to incidence of red rot of sugar cane and development of multiple linear models for its prediction. *Indian J. Mycol. and Pl. Pathol.* **31:** 238-239.

Ling, K.C. and Qu, S.H. (1969). Standardization of the international race numbers of *Pyricularia oryzae. Phytopathology,* **59:** 339-342.

Malathi, P., Viswanathan, R. and Jothi, R. (2006). Specific adaptation of *Colletotrichum falcatum* pathotypes to sugar cane cultivars. *Sugar Tech.* **8:** 5458.

Manocha, M.S. and Vasudeva, R.S. (1964). Factors affecting the pathogenecity of *Colletotrichum falcatum,* Went causing red rot of sugar cane. *Indian* J. *Agric. Sci.* **34:** 264-272.

Mohanraj, D., Kumaresan, S. and Sreenivasan, T.V. (2002). Molecular characterization of isolates of sugar cane red rot pathogen, *Colletotrichum falcatum* Went. *Indian Phytopath.* **55:** 147-151.

Pandey, L.N. and Sakal, R. (1974). New pathogenic strain of *Glomerella tucumanensis* (Speg.) Arx and Muller in Uttar Pradesh. *Indian Sugar,* **24:** 707-709.

Rafay, S.A. and Padmanabhan, S.Y. (1941). Strains of *Colletotrichum falcatum. Curr. Sci.* **10:** 25-26.

Satyavir, (2003). Red rot of sugar cane: Current scenario. *Indian Phytopath.* **56:** 245-254.

Satyavir, Beniwal, M.S. Virk, K.S. and Maheswari, S.K. (1984). Breakdown of red rot resistance in sugar cane cultivars Co. 1148. *Indian Phytopath.* **37:** 407.

Satyavir, Singh, N., Virk, K.S., Nageshwararao, G.V., Singh, H. and Misra, S.R. (2001). Pathogenic variability in sugar cane red rot system In: Proc. National Symposium on role of resistance in intensive agriculture, (Eds. Nagararajan. S. and Singh, O.P. Kalyani; Publisher, Ludhiana. pp. 109-114.

Satyanarayana, Y. and Satyanarayana, B. (1976). Variation in the light race isolate of red rot fungus in Andhra Pradesh. *Cooperative Sugar,* **7:** 389-392.

Satyanarayana, Y. and Rao, A.M. (1984). Evidence on the existence of two physiological different strains in the light race isolates of *Physalospora tucumanensis* Speg. In Andhra Pradesh. *Sugar cane* **3:** 6.

Satyanarayana, Y. and Rao, A.M. (1995). Investigations on the race flora of *Colletotrichum falcatum, incitant* of red rot disease of sugar cane. In: Proc. National seminar on sugar cane production constraints and strategies for research and management of red rot (Eds. Agnihotri, *et al.,* IISR, Lucknow, pp. 285-293.

Singh, M.P., Upadhyay, U.C., Verma, K.P. and Singh, H.N. (1984). Pathogenic variability in red rot isolates of sugar cane in U.P. *Indian Sugar* **33:** 669-671.

Singh, N. and Singh, V. (1993). *Colletotrichum falcatum* race designation: A methodology. *Curr. Sci.* **66:** 777-779.

Singh, N., Singh, V., Tewari, S.D. and Verma, K.P. (1993). Occurrence and identification of different physiological races of red rot pathogen of sugar cane in Uttar Pradesh. *Indian J. Sugar cane Technol.* **12:** 54-59.

Singh, V. (1995). Serological affinity between different strains/races of *Colletotrichum falcatum* in sugar cane. *Indian J. Sugar cane Technol.* **15:** 77-78.

Singh, V., Duttamajumder S.K. and Singh, N. (1993). Possible threat for the cultivation of CoS 767. *Indian J. Sugar cane Technol.* **8:** 81-82.

Srinivasan, K.V. (1962). Some observations on variation in the red rot pathogen *Glomerella tucumanesis* (Speg.) Arx and Mullar. Proc. ISSCT, **11:** 795-802.

Srinivasan, K.V. and Bhat, N.R. (1961). Red rot of sugar cane. Criteria for grading resistance. *J. Indian Bot. Soc.* **11:** 566-577.

Suman, A., Lal, Snita, Shasany, A.K., Gaur, A. and Singh, P. (2005). Molecular assessment of diversity among pathotypes of *Colletotrichum falcatum* prevalent in subtropical Indian sugar cane. *World J. Microbiol. and Biotechnology.* **21:** 1135-1140.

Viswanathan, R., Malathi, P. and Padmanabhan, P. (2003). Variation in sugar cane red rot pathogen (*Colletotrichum falcatum* Went). *In:* Frontiers of fungal diversity in India (Eds. Rao, G.P. *et al.,*) International Book Distributing Co., Lucknow. pp. 639-667.

34

Strategies to Combat Challenges for Management of Red Rot in Sugar Cane

Vijai Singh, S.N. Srivastava, B.B. Joshi and S.K. Awasthi

Crop Protection Division, Indian Institute of Sugarcane Research, Lucknow, U.P.

Red rot caused by *Colletotrichum falcatum* Went (Perfect stage *Glomerella tucumanensis* (Speg), Arx & Muller) is a devastating disease of sugar cane in India. Several epidemics of red rot have caused tremendous loss in yield sugar. Unlike several other crops, fungicides were not absorbed to the desired concentration and were degraded quickly in cane plant. Heat treatment failed to protect deep-seated infection and secondary spread of red rot. Host resistance has frequent breakdown. Biocontrol through *Trichoderma harzianum* and *T. viride* has given encouraging results. The biopesticide applied at the time of planting @ 20 kg FYM based TMC/ha was helpful for preventing the secondary spread of the disease. Metabolites of *T. harzianum* absorbed through roots were effective for longer period (150-200 d). *Trichoderma* biopesticide not only controls secondary spread of red rot but also works as biomanure and improves the growth and yield of sugar cane. The biopesticide may be useful to prolong the life of variety and maintain the fertility of soil.

Red rot can be managed by integrated approach. Diseased cane may not be allowed to move for preventing introduction of the pathogen in new areas. Seed cane should be planted after moist hot air treatment (MHAT) in first year followed by the three-tier seed programme. Setts may be treated with bavistin + acetic acid (0.2% each). *Trichoderma* biopesticide may be applied at the time of planting. Rouging of diseased plants followed by spot application of bleaching powder (5-10 g/clump) may be helpful in reducing *C. falcatum* load.

Sugar cane, a long duration clonally propagated crop, is grown in about 3.66 million hectares with an average yield of 64.8 t/ha for the production of crystal sugar, khandsari and gur. The crop is affected by red rot, a most devastating disease (Agnihori, 1996). Several varieties of sugar cane have been phased out from cultivation due to red rot disease. Since, red rot produces complete drying of affected cane, even minor incidence may cause considerable loss in yield. It is often called cancer of sugar cane. Red rot was first recorded by Went (1893) from Indonesia and Barber (1901) from India. Several epidemics of red rot from U.P., Bihar, Haryana, Punjab, Andhra Pradesh, Kerala, Tamil Nadu, Northern Karnataka and Southern Maharashtra have been reported. Popular varieties viz., Co-213 and Co-312 were badly hit by the red rot in UP and Bihar. Co 1148 in the subtropical region has virtually gone out of cultivation due to red rot. CoJ 64, a high sugar and early maturing has also been phased out in subtropical India. Several other cultivars viz., Co 205, 213, 281, 290, 312, 313, 419, 421, 453, 527, 658, 975, 997, 1148, 7717, CoC 671, CoH 82, 84, CoJ 64, 82, 84, Colk 8001, 8102, CoS 510, 767, 770, 802, 8436, Bo 3, 10, 11,14, 17, 24, 34, 54 have also gone out of cultivation due to red rot.

CHARACTERISTIC SYMPTOMS FOR DIAGNOSIS OF THE DISEASE

The disease is recognized by withering of 3rd or 4th leaf of the crown either at the tip or along the margin or both. In severe cases the whole crown dries up. Affected stalks initially exhibit purple discoloration of rind. Pith portion of internodes turns reddish brown and displays white patches. An affected tissue emits sour smell. Additional symptoms such as mortality of buds resulting in the failure of germination; mortality of settling of partially diseased cane; midrib infection on upper leaves in pre-monsoon season and laminar symptoms as small, pin-head, reddish brown lesions have also been noticed.

Spindle infection appears on lower midrib of upper leaves even in cases the plants are affected from the bottom. Infected spindle produces lot of spores, which can readily infect the cane stalk.

SURVIVAL OF THE PATHOGEN AND DISEASE EPIDEMIOLOGY

The pathogen penetrates the host surface directly or via wounds. The primary infection of the crop comes through infected setts but rarely through soil. Survival of *C. falcatum* in soil is reported to be 60-90 days. The inoculum in the form of conidia, thick-walled mycelia, chlamydospores, etc. is carried in setts both internally and externally. The pathogen penetrates through root primordial leaf scars and bud scales and advanced hyphae induce gum formation in moderately resistant genotypes. The pathogen may attack the cane plant at any stage of crop growth starting from germination. Full expression of the disease depends on the prevailing weather conditions, i.e., high humidity, 90-95%, and moderate temperature, 28-31°C. Diagnostic symptoms are observed during monsoon or post-monsoon period. The secondary spread of the fungus during monsoon is mediated through conidial dispersal by irrigation, rainwater, rain splash at temperature 25-30°C, relative humidity 95% and results in the infection of midrib, lamina, leaf sheath and stalk, while in winter air currents help in the spread of the pathogen (Satyavir, 2003).

DISEASE MANAGEMENT

Red rot disease may be managed by quarantine, cultural practices, heat therapy, chemical control, biocontrol and growing resistant varieties. Biocontrol agents viz., *Trichoderma, Pseudomonas, Bacillus* and *Acetobacter* have been recently found effective for protecting the disease. The bioagents have been noticed to induce systemic resistance in susceptible varieties and may prolong the breakdown of resistance.

QUARANTINE

Quarantine measures are duly adopted during exchange of sugar cane varieties between countries; however, proper care is not been taken while movement of cane seed material between the states. The indiscriminate movement of cane facilitates introduction of dreaded pathogen into a new area. The recent introduction of the disease in Karnataka and Maharashtra was observed because seed material of highly susceptible varieties CoC 92061 and CoC 671 was brought from red rot endemic areas. Thus, strict regulatory measures are required for inter-state and zonal movement of seed material. It is required that seed should be moved with proper phytosanitory certificates (Viswanathan and Samiyappan, 2000).

HEALTHY SEED AND FIELD SANITATION

Healthy seed, crop rotation, field sanitation and maintenance of efficient drainage have been recommended to reduce inoculums in the field and minimize losses due to the red rot. The seed must be disease free and preferably taken from heat treated crop raised under three-tier seed programme (Singh, 1983). One year crop rotation is effective in eliminating soil-borne inoculums of *C. falcatum* in India.

HEAT THERAPY

Heat therapy is suitable for control of sett-borne infection. Elimination of sett-borne infection of red rot was achieved by hot air treatment at 54°C for 2:30 h (Singh,1973). Hot water at 50°C for 2 h was also equally effective (Singh and Agnihotri, 1987), but moist hot air treatment (MHAT) was more efficient in deep penetration. It was thoroughly tested in U.P. and Bihar and ascertained as a better practice (Singh, 1983).

CHEMICAL CONTROL

The fungicides have not given satisfactory control because of impervious nature of rind and presence of fibrous nodes of sugar cane. This prevents the low level of absorption of fungicide. Reduction in red rot incidence has been reported when diseased setts were treated with Carbendazim and benomyl for 1-2 h (Satyavir, 2003). Bleaching powder (0.5%) was effective in reducing red rot incidence (Agnihotri *et al.,* 1995). Application of bleaching powder in soil after uprooting the diseased plant reduced the load of inoculums of *C. falcatum.* Secondary spread of red rot by spraying dithiocarbamate and copper fungicides may be reduced (Kirtikar and Verma, 1963). Bavistin (0.2%) reduces sett borne infection of the red rot (Agnihotri *et al.,* 1995). The efficacy of bavistin was improved by adding acetic acid @ 0.2% in solution.

RED ROT MANAGEMENT THROUGH RESISTANT VARIETIES SOURCES OF RESISTANCE

Red rot resistance has been identified in basic *Saccharum* spp. Seven clones of *S. officinarum* were identified as resistant among which 2 were of typical natural hybrid type and 3 were of non-flowering type, which may not be used in breeding. The *S. robustum,* a wild species is close to *S. officinarum* and has 2 resistant and eight moderately resistant clones. Among

S. *sinense* and *S. barberi*, only 3 clones were resistant to red rot. *S. barberi* originated in the area of maximum variability of *Colletotrichum falcatum* and had adequate level of resistance but succumbed to new virulent races of the pathogen. As a result, there was no chance of further genetic recombination and evolution of genes for resistance. The low levels of red rot resistance in cultivars may be due to *S. barberi* initially used in breeding programme. *S. spontaneum* originated from sub-Himalayan region may be utilized for biotic and abiotic stresses. Eastern U.P. and Bihar are among the diversity centres of red rot pathogen. *S. spontaneum* from this region will serve as the natural source of resistance to virulent pathotypes of *C. falcatum* (Sreenivasan, 1995, Viswanathan and Samiyappan, 2000). *S. spontaneum* has comparatively larger no. of sources of resistance (109) than 13 in other species out of 1099 screened (Alexander, 1995).

RESISTANT VARIETIES USED FOR RED ROT BREEDING PROGRAMME

Bo, 12, Bo 91, Bo 102, Bo 104, Co 7314, Co 89003, Co 89015, Co 89016, CoS 91269, CoS 96268, Co Pant 94211 have been utilized for evolving varieties for red rot resistance (Satyavir, 2003).

VERTICAL AND HORIZONTAL RESISTANCE

Red rot resistance is vertical or race specific in sugar cane. Beniwal (1985), studied slow red rot resistance using 22 genotypes. Out of these 4 genotypes viz., Co 1158, CoS 802, S 77/467 and CoJ 64 possessed useful levels of slow red rot. Slow drying of tops, mild spore over the rind and less secondary spread were observed in above varieties. These parameters could be utilized in assessing horizontal resistance of sugar cane. CoJ 64 exhibited useful level of horizontal resistance but it is succumbed to disease in Punjab and Haryana and Western U.P., due to emergence of new virulent pathotypes (Satyavir, 2003). The cultivation of Co J 64 has been partly phased out.

PATHOGENIC VARIABILITY

C. falcatum, a facultative saprophyte, produces new pathotypes in nature due to heterokaryosis, parasexuality and mutation. On the basis of reaction on 13 differentials, 10 pathotypes have been identified so far from India. But few studies were made to identify pathotype with specific host pathogenicity. Work conducted at HAU, Hissar indicated the presence of Cf 01, 02 and 03 (Satyavir *et al.,* 2001). Cf 04, 05 and 06 three pathotypes were reported from Anakapalle, A.P (Satyanarayana and Satyanarayana 1976, Satyanarayana and Rao, 1995). Differentials viz., Baragua (*S. officinarum*), Khakai (*S.sinense*), SES 594 (*S. spontaneum*), Co S 767, Co 975, BO 91, CoC 671, Co 7717, Co 997, CoJ 64, Co1148, Co 419, and Co 62399 have been utilized for identification of pathotypes (Anon, 1991, Singh and Singh 1993). Recently, Cf 07, Cf 08 and Cf 09 were identified from north Western zone and Cf 10 from tropical and east, coastal zone (Singh *et al.,* 1993, Satyavir *et al.,* 2001). Search for better differentials is still required to explain specific host reaction. Identification of filter plants to retain specific pathotypes may be more useful. Precise molecular identification would be of greater importance.

INHERITANCE OF RESISTANCE

Sugar cane plant is a highly heterozygous and complex polyploid crop with meiotic irregularities. Conclusive studies on inheritance of red rot resistance are lacking. Chaudhary *et al.,* 1995 reported that resistance to red rot in sugar cane is governed by the dominant inhibitor genes from *S. officinarum.* Segregation of resistance in progenies was not found to be a Mendelian trait. Indiscriminate distribution of resistance or susceptibility among the progenies was recorded. It is essential to know the combining ability of resistant parents for developing resistant lines. More precise information would be needed to go for directed breeding.

MORPHOLOGICAL FEATURES OF HOST FOR RED ROT RESISTANCE

Studies on the thickness of epidermis, cuticle, bud scale, rind, relative abundance of vascular bundles under the rind and the presence of septa in vascular bundles have been reported as morphological barriers for red rot entry and movement of *C. falcatum* in sugar cane. But the morphological characters after the origin of new pathotype have not been precisely correlated with the resistance (Duttamajumder, 2002).

BIOCHEMICAL CHARACTERS OF HOST FOR RED ROT RESISTANCE

In resistant varieties, a gummy substance is formed that seal off further spread of the pathogen, but it is not produced in susceptible varieties. Total phenolic contents increased in resistant but decreased in susceptible genotypes. Though a slight increase at initial stage after infection in susceptible varieties was also noticed. Phenol alone may not be responsible for resistance (Singh *et al.,* 1976). Activities of polyphenol oxidase (PPO) and flavone-glycosides were linked with the resistance (Srinivasan, 1995). Activity of phenylanine ammonia lyase (PAL) and tyrosine ammonia lyase (TAL) was higher in resistance but lower in susceptible varieties (Madan *et al.,* 1991). But once the resistance of the variety was broken, the activities of PAL and TAL had no correlation. Viswanathan *et al.* (1996) indicated the role of the red rot phytoalexin (3- de oxyanthocyanin) in disease resistance. Once the pathogen starts to behave in saprophytic mode, it indicates the higher susceptibility of the genotype.

BIOCONTROL THROUGH *T. HARZIANUM*

Different biocontrol agents viz., *Trichoderma, Pseudomonas, Bacillus* and *Acetobacter* have been found highly antagonistic to *C. falcatum* under *in vitro* conditions. *T. harzianum* applied to the crop through soil and sett treatment was found effective in protecting the plants from infection of red rot. *T. harzianum* strains viz., T 27, T 37 and T 38 were more efficient in protecting the infection (Table 34.1). *Trichoderma* may be applied through soil @ 2 kg/ha mixed with 20 kg sterilized FYM at the time of planting and tiller stage. The sett treatment with spore suspension (10^6 conidia/ml) was equally effective whereby red rot was reduced by 40-50% in the susceptible variety. Although *T. viride* was less effective for red rot control, it was at par for promoting the growth of sugar cane (Tables 34.1 and 34.2). *T. viride* metabolites (culture filtrates) 2.5% solutions applied at tiller initiation of ratoon produced significantly higher yield (Srivastava *et al.,* 2006). The formulations of *Trichoderma* biopesticide are non-hazardous, eco-friendly and highly economical (Singh *et al.,* 2006, Srivastava *et al.,* 2006).

Table 34.1: Biocontrol of red rot in Co Lk 7701 by *Trichoderma* species.

Strain	Sett treatment (spore suspension)			Soil treatment (TMC)		
	Complete control	Partial control	Canes inoculated	Complete control	Partial control	Canes inoculated
T 6	28	72	100	23	77	100
T 24	24	76	100	24	76	100
T 27	45	55	100	27	73	100
T 37	49	51	100	46	54	100
T 38	46	54	100	31	69	100
Check	00	3	100	00	13	100

T6 (*T. viride*) and T 24, T 27, T 37 and T 38 (*T. harzianum*)

Table 34.2: Biomanurial effects of *Trichoderma* species in CoS 94257.

Parameter	Sett treatment (spore suspension)			Soil treatment (TMC)			CD at 5%
	Check	Treated	Increase (%)	Check	Treated	Increase (%)	
Germination (%)							
T 6	21.3	25.4	4.1	21.3	28.8	7.2	
T 24	20.8	27.4	6.5	20.8	28.4	7.5	2.20
Tillers (000)							
T6	83.6	103.7	24.2	83.6	109.2	30.3	
T24	85.8	104.2	21.1	85.8	11.3	29.0	3.36
NMC (000)							
T6	65.5	68.9	5.1	65.5	73.8	13.5	
T24	67.0	71.5	7.6	66.9	77.2	15.3	2.18
Yield (t/ha)							
T6	59.3	62.9	6.0	59.3	68.3	15.1	
T24	59.3	63.7	7.4	59.3	68.3	15.1	2.24

T6 (*T. viride*) and T 24 (*T. harzianum*)

RESEARCH AREAS FOR FUTURE

The following areas for future research are important as also envisaged by Yadav, 2006 in his vision document.

- Prevailing *C. falcatum* pathotypes should be catalogued after molecular characterization of selective pathogenic behaviour.
- Molecular and biochemical studies on host-parasite interactions are required to understand the operational mechanism of pathogenesis and its utilization for control of red rot.
- The work on broadening of genetic base/gene pyramiding for red rot resistance from *S. spontaneum* is also important.
- Disease prediction models with reference to forecast the epidemic's need to be worked out for judicious adoption of control measures.
- Integrated disease management strategy with the utilization of bioagents and/or chemicals for induced systemic resistance would be more useful.

ACKNOWLEDGEMENTS

Sincere thanks are due to Dr. R.L. Yadav, Director IISR, Lucknow for encouragement and providing facilities for research experiments.

REFERENCES

Agnihotri, V.P. (1996). Current sugar cane disease scenario and management strategies. *Indian Phytopath.* **49:** 109-126.

Agnihotri, V.P., Singh, N., Lal, R.J. and Singh, V. (1995). Management of red rot disease of sugar cane. In: Proc. *National seminar, Sugar cane production constraints and strategies for research and management* of *red rot.* IISR, Lucknow, (Eds. Agnihotri, V.P. *et al.,*) 1995, pp. 294-312.

Agnihotri, V.P., Lal, R.J. and Singh, V. (1997). Integrated management of red rot disease of sugar cane. *Indian Phytopath.,* MEZ., Ann. Meet. Kanpur, Dec. 18-19, pp. 55.

Agnihotri, V.P., Lal, R.J., Singh, N. and Singh, V. (1997). Effect of bavistin and bleaching powder on conidial germinationof *Colletotrichum falcatum* Went. and development of red rot in sugar cane. *Indian J. Sugar cane Technol.* **14:** 108-111.

Alexander, K.C. (1995). An overview of constraints in sugar cane disease management. *In: Proc. National Seminar, Sugar cane production constraints and strategies for research and management of red rot.* IISR, Lucknow, pp. 121-129.

Anonymous. (1991). Proc. AICRP on sugar cane, IISR, Lucknow.

Barber, C.A. (1901). Sugar cane diseases in Godaveri and Ganjam districts. *Madras, Deptt. Land Records and Agic. Bull.* **43:** 181-194.

Beniwal, M.S., Satyavir, and Virk, K.S. (1989). Pathogenic variability of *Colletotrichum falcatum,* incitant of red rot of sugar cane. *Indian Phytopath.* **42:** 95-98.

Chaudhary, B.S., Mahila, A.S., Virk, K.S. and Rishipal. (1995). Breeding for red rot resistance in sugar cane *(Saccharium* hybrid complex): Review, concept and theory. Proc. National seminar, *Sugar cane productions constraints and strategies for research and management of red rot.* IISR, Lucknow, (Eds. Agnihotri, V.P. *et al.*) pp. 221-232.

Duttamajumder, S.K. (2002). A century of red rot disease of sugar cane in India. *In: Sugar cane crop management,* (Eds. Singh, S.B. *et al.,*) Sci. Tech. Publishing LLC, Houston, Texas (USA), pp. 52-108.

Kirtikar, and Verma, H.S. (1963). Role of chemicals in sugar cane disease control. *Indian Sugar* **13:** 139-141.

Madan, V.K., Soni, N., Nigam, M., Solomon, S. and Agnihotri, V.P. (1991). Enzyme activity and cane genotype resistant to red rot. *Sugar cane* **2:** 6-8.

Padmanabhan, P., Mohanraj, D., Viswanathan, R., Madhusudanarao, M., Prakasam, N., Joshi, R. and Alexander, K.C. (1996). Differential interaction of sugar cane clones to pathotypes of *Colletotrichum falcatum. Sugar cane* **4:** 16-20.

Satyavir, (2003). Red rot of sugar cane—Current scenario. *Indian Phytopath.* **56:** 245-254.

Satyavir, Singh, N., Virk, K.S., Nageswara Rao, G.V., Singh, H. and Mishra, S.R. (2001). Pathogenic variability in sugar cane red rot system. *In: Proc. National symposium on role of resistance in Intensive Agriculture* (Eds. Nagarajan, S. and Singh, O.P.) Kalyani Publishers, Ludhiana, pp. 109-114.

Satyanarayana, Y. and Satyanarayana, B. (1976). Variation in the light race isolates of red rot fungus in Andhra Pradesh. *Coop. Sugar.* **7:** 389-392.

Satyanarayana, Y. and Rao, A.M. (1995). Investigations on the race flora of *Colletotrichum falcatum,* incitant of red rot disease of sugar cane. In: Proc. National symposium *Sugar cane production constraints and strategies for research and management of red rot.* IISR, Lucknow, (Eds. Agnihotri, V.P. *et al.,*) pp. 285-293.

Singh, K. (1973). Hot air therapy against red rot of sugar cane. *Pl. Dis. Reptr.* **57:** 220-222.

Singh, K. (1983). Seed-piece transmissible diseases of sugar cane and three-tier seed programme. *In: Recent advances in Plant pathology* (Eds. Husain, A. *et al.,*) Print House (India), Lucknow, pp. 331-343.

Singh, K., Singh, R.P. and Agnihotri, V.P. (1976). Phenolics in relation to sugar cane resistance against red rot disease. *Sugar cane Pathol. News L.* **15/16:** 37-41.

Singh, N. and Singh, V. (1993). *Colletotrichum falcatum* race designation: A methodology. *Curr. Sci.* **66:** 777-779.

Singh, N., Singh, V., Tewari, S.D. and Verma, K.P. (1993). Occurrence and identification of different physiological races of red rot pathogen of sugar cane in Uttar Pradesh. *Indian J. Sugar cane Technol.* **8:** 26-30.

Singh, R.P. and Agnihotri, V.P. (1987). Thermotherapy of sugar cane for disease control. *Tropical Plant Pathology.* **4:** 305-329.

Singh, V., Dattamajumder, S.K. and Singh, N. (1993). Possible red rot threat to the cultivation of CoS 767. *Indian J. Sugar cane Technol.* **8:** 81-82.

Singh, V., Duttamajumder, S.K. and Lal, R.J. (2006). Biocontrol of red rot of sugar cane through induced resistance by isolates of *Trichoderma harzianum. Indian J. Agric. Sci.* (under revision).

Sreenivasan, T.V. (1995). Research strategies for breeding for red rot resistance in sugar cane. In: Proc. National Seminar, *Sugar cane production constraints and strategies for research and management of red rot,* (Eds. Agnihotri, V.P. *et. al.,*) pp. 233-239. IISR, Lucknow.

Srivastava, S.N., Singh, V. and Awasthi, S.K. (2006). *Trichoderma* induced improvement in growth, yield and quality of sugar cane plant and ratoon. *Sugar Tech.* **8:** 166-169.

Virk, K.S., Satyavir, and Chaudhary, B.S. (1985). A source of red rot resistance in sugar cane. *Indian Phytopath.* **38:** 551-552.

Viswanathan, R. and Samiyappan, R. (2000). Red rot disease in sugar cane: Challenges and prospects. *Madras J. Agric. Sci.* **87:** 549-559.

Went, F.A.F.C. (1983). Het rood Snot. Arch. *Java Suikeridustrie* **1:** 265-282.

Yadav, R.L. (2006). Research vision to manage red rot disease of sugar cane in India. *Sugar Tech.* **8:** 99-100.

35

Use of Biocontrol Agents and their Mechanisms of Action

D. Prasad and O.P. Singh[1]

Division of Nematology, I.A.R.I., New Delhi–110 012
[1]Division of Plant Pathology, I.A.R.I., New Delhi–110 012

The biological control of plant disease was defined by Garrett (1965) as 'any condition under which or practice whereby survival or activity of a pathogen is reduced through the agency of any other living organisms (excluding the role of the host and man) with the result that there is a reduction in the incidence of the disease caused by that pathogen'. Keeping in view the functional relationship between the pathogen, the host and the environment in the development of the disease, Baker and Cook (1974) have given a broader definition of biological control. According to them 'biological control is the reduction of inoculum density or disease producing activities of a pathogen or parasite in its active or dormant state, by one or more organisms, accomplished naturally or through manipulation of the environment, host or antagonist, or by mass introduction of one or more antagonists.

In view of hazardous impact of pesticides and other agrochemicals on the ecosystem, the biocontrol of plant diseases as an alternate strategy has received increasing attention in recent years (Papavizas, 1985; Baker, 1987; Lockwood, 1988). In this context about 35 genera of fungal and bacterial species have been used as biocontrol agents against various plant pathogens (Cook and Baker, 1983). *Trichoderma* species, *Gliocladium* spp., *Bacillus* spp., *Pseudomonas* spp., and *Actinomycetes* represent the most thoroughly studied antagonistic microorganisms. The following pages deal with some potential biocontrol agents.

Normally, the biocontrol agents have been selected for *in vitro* antagonism, but were ecologically unsuited to the environments where pathogens grow. Deacon (1991) felt that it is difficult to predict whether inoculant biocontrol agent selected *in vitro* test will contribute significantly to control soil-borne pathogen in future. Therefore, it is suggested that there is urgent need to adopt an ecological approach, with emphasis on interactions in microsites and the design of appropriate screening strategies. Similarly, Deacon and Berry (1993) also expressed the view that only a few soil-borne plant pathogens have been controlled successfully by commercial formulations of biocontrol agents, but many attempts to develop biocontrol inoculants, although promising under experimental conditions, have met with difficulties in practice. They thought that the future of biocontrol of soil-borne plant pathogens probably lies in integrated control system that combines the development and deployment of crop cultivars, fungicides and cropping practices in a way that is intended to exploit both natural and inoculant biocontrol agents not alone but in concert.

ANTAGONISTIC PROPERTIES OF BIOCONTROL AGENTS

Trichoderma Species

Trichoderma are found in soils worldwide (Domsch *et al.,* 1980; Christensen, 1981). A number of *Trichoderma* spp. have a promising potential for biological control of plant pathogenic fungi (Papavizas, 1985). Weindling (1932) used *T. lignorum* against *Rhizoctonia solani, Phytophthora parasitica, Pythium* spp. and *Sclerotium rolfsii,* the causal agents of damping off of citrus and proved the antagonistic property. Subsequently, Weindling and Emerson (1936) augmented soil with abundance of *T. lignorum* and opined that this mycoparasite might serve as a biocontrol agent against soil-borne pathogens.

T. viride showed antagonistic activity against soil-borne *Pythium debaryanum, Sclerotium rolfsii, Fusarium lini, Fusarium culmorum, Armillaria mellea* and *Fomes anosus* (Wright, 1956). This species has been favoured in commercial orchards soils for the control of *A. mellea* (Bliss, 1951; Baker and Cook, 1974). Kudryavtseva (1980) and Sychev and Shaposhnik (1982) reported *T. viride* as highly antagonistic to *Venturia inaequalis,* the causal agent of apple scab (Andrew *et al*., 1983; Lindow, 1985).

T. harzianum has been shown to suppress the growth of *S. rolfsii,* the causal agent of root and stem diseases of various crops. Wells *et al*. (1972) applied inoculum of *T. harzianum* along rows of tomato seedlings and protected the crop from *S. rolfsii.* It has also shown potential for biocontrol of *Macrophomina phaseolina* (Elad *et al*, 1986), *F. oxysporum,* f.sp. *vasiinfectum, F. culmorum* and *F. oxysporum* f.sp. *radius-lycopersici* the causal agents of seedling blight of cotton, melon, wheat and crown rot of tomatoes, respectively (Sivan and Chet, 1986 and Sivan *et al*., 1987). Antagonistic effect of *T. harzianum* against *F. solani* (Bhardwaj and Gupta, 1987; Cole and Zrinyika, 1988), *F. oxysporum* (Calvet *et al*., 1990), *Pythium aphanidermatum* (Bhardwaj and Gupta, 1987), *Rhizoctonia solani* (Wilson *et al*., 1988; Cole and Zrinyika, 1988) and *S. rolfsii* (Deb and Dutta, 1988; Singh and Dwivedi, 1987) has also been demonstrated. Recently, Kaur and Mukhopadhyay (1992) effectively controlled chickpea wilt complex caused by *F. oxysporum* f. sp. *ciceri, R. solani* and *S. rolfsii* using *T. harzianum. T. viride, T. harzianum* and a *Trichoderma* sp. were capable of reducing the incidence of onion white rot caused by soil-borne sclerotial fungus, *Sclerotium cepivorum* (Kay and Stewart, 1994).

Antagonistic effect of *Trichoderma* has also been explored against certain fungal pathogens causing diseases in wheat. In *in vitro* test, *T. viride* was found to be an antagonist to *S. rolfsii* on wheat (Manjappa, 1979). Similarly, antagonistic effect of *Trichoderma* sp. on *Drechslera sorokiniana,* the causal agent of root rot, seedling blitght and foliar of wheat was observed by Prasad *et al*. (1978). Krisvoshchekova and Mishchenk (1990) reported that *Trichoderma* sp. effectively controlled *Helminthosporium* and *Fusarium* rots of wheat. The study on the effect of *T. harzianum* on fungal pathogens infecting wheat and black oat straw showed that the antagonist resulted in reduction in the incidence of *Cochliobolus sativus, F. graminearum* and other *Fusarium* spp. (Fernandez, 1992). Singh *et al.* (1991) tested the antagonistic effect of *T. viride* and *T. harzianum* on teliospores of *Neovossia indica*, causal agent of Karnal bunt of wheat and observed significant reduction in spore germination. *T. viride* when applied to soil or earheads, reduced Karnal bunt infected grains from 1.63% in control to 0.8% and 0.65% respectively under natural infection. Aggarwal *et al.* (1992) reported antagonistic effect of *T. viride* on the incidence of loose smut (*Ustilago segetum tritict*) infection under field conditions. Recently, it was shown that the culture filtrates of *Trichoderma* species viz. *T. viride* (TV-5 strain), *T. koningii, T. harzianum* and *T. lignorum* caused significant inhibition in germination of chlamydospores of *Ustilago segetum tritici.* It was also noticed that these species also suppressed the mycelial growth of the test pathogen (Mondal, 1994).

Gliocladium Species

Gliocladium is known to have a broad range of hosts (Barnett and Lilly, 1962). *G. catenulatum, G. virens, G. roseum* and *G. deliquescens* have been exploited for their antagonistic potential against a variety of pathogenic fungi. In fact, Papavizas (1985) considered that *Gliocladium* species have similar antogonistic effect as *Trichoderma* species.

G. catenulatum caused distortion of host cells of *Sclerotinia sclerotiorum, F. equiseti, F. oxysporum, F. poae* and *F. sporotichoides* (Huang, 1978). A decrease from 1514 sclerotia to 401 in 60 m of row of S. *sclerotiorum* by application of *G. catenulatum* was recorded by Huang in 1980. Similarly, Zazzerini and Tosi (1985) mentioned that *G. catenulatum* reduced the germination ability of sclerotia of *S. rolfsii* as high as 91-100 per cent.

Mycoparasitism of *G. roseum* on *Helminthosporium sativum* was shown by Banett and Lilly (1962). Moody and Gindral (1977) showed that sclerotia of *Phomopsis sclerotoides,* the cause of black rot of cucumber, were rotted by G. *roseum,* hence it has been used successfully in biological control of the disease.

Tu and Vaartaja (1981) mentioned that *G. virens* parasitized *R. solani* and inhibited the growth of *Pythium ultimum* and *Phytophthora megasperma* var *sojae.* Howell (1982, 1991) obtained reduction in damping off of cotton seedlings when *G. virens* was applied to soil along with seed. Damping off of zinnia and cabbage caused by *P. ultimum* and *R. solani* was also controlled by application of this antagonist (Lumsden and Locke, 1989). *G. virens* parasitized and decayed sclerotia of some fungi, i.e. *S. sclerotiorum,* S. *minor, Botrytis cinerea, S. rolfsii* and *Macrophomina phaseolina* (Phillips, 1986; Hussain *et al.*, 1990). The antagonist reduced the Phytophthora root rot of apple seedlings (Roiger and Jeffers, 1991) and chestnut (Chambers and Scott, 1993). Mathur *et al.* (1993) proved the effectiveness of *G. virens* against rhizome rot of ginger caused by *F. solani* and *P. myriotylum. G. virens* strain 41 grown in chitin containing medium was strongly inhibitory to mycelial growth of *P. ultimum. P. megasperma, R. solani* and *B. cinerea* (Di Pietro *et al.*, 1993).

Gliocladium spp. showing potential for control of *F. culmorum* and *F. nivale,* causal agent of seedling blight of wheat was reported by Etheridge and Parry (1993). Aggarwal *et al.* (1992) made use of *G. deliquescens* and considerably reduced the infection of loose smut of wheat by seed treatment and soil application with this antagonist. Antagonistic effect of *G. deliquescens* and *G. penicilloids* (as biocontrol agent) on teliospores of Karnal bunt of wheat has been studied by Singh *et al.* (1991). *G. deliquescens* reduced teliosporic germination significantly, whereas *G. penicilloids* increased the teliospore germination.

Aspergillus Species

In vitro activity of *Aspergilllus niger* against cotton root rot of fungi, *Macrophomina phaseolina* and *R. solani* was noticed in 1940 by Vasudeva and Sikka (1941). Since then *A. niger* has been extensively used as biocontrol agent against soil-borne plant pathogens. It was reported to be a strong antagonist against cotton wilt pathogen, *Verticillum dahliae* (Bora, 1975). Bora (977) further recorded strong antagonism exhibited by *A. niger* both in culture and on host against damping off of eggplant caused by *F. solani, Alternaria alternata* and *R. solani. A. niger* was also used for biocontrol of *R. solani* in coffee seedlings by Venkatasubbiash and Safeeulla (1984). Sharma (1989) used a strain of *A. niger* as effective biocontrol agent against *F. solani* incitant of wilt of muskmelon. *A. niger* strain A6 was the best biocontrol agent against *F. oxysporum,* the causal agent of wilt of watermelon. *In vitro,* seedlings raised from antagonist (strain A6) coated seed could sustain the disease and the reduction in disease intensity was 64 per cent (Naik, 1989).

Among the other species, *A. terreus* was reported to block the capacity of production of toxin by *F. oxysporum* f.sp. *vasinfectum* due to its antibiotic effect upon the pathogen (Chadova *et al.*, 1980). Tashlieva (1980) also observed antagonistic properties of *A. terreus* and *A. flavus* against cotton wilt pathogen, *F. oxysporum* f.sp. *vasinfectum.* Akhtar *et al.* (1982) could notice the reduction in wilt incidence of linseed by the use of *A. flavus. A. flavus* and *A. niger* antagonized *F. udum* causing wilt of pigeon pea (Upadhyay and Rai, 1983).

BACTERIAL ANTAGONISTS

Antagonistic bacteria have been extensively studied as biocontrol agents effective against soil-borne pathogens. Among 20 genera of bacteria, *Bacillus* spp., *Pseudomonas* spp. and *Actinomycetes* (*Streptomyces* spp.) are widely used for their abilities as biocontrol agents.

Bacillus spp. are appealing candidates for biological control because they produce endospores that are tolerant to heat and desiccation. *B. subtilis* inhibited *Streptomyces scabies* on potato (Weinhold *et al.*, 1964) and reduced seedling damping off caused by *P. ultimum* and *R. solani* (Olsen, 1964; Olsen and Baker, 1968; Broadbent *et al.*, 1971). It has also been used to protect carnation seedling against *F. roseum* (Aldrich and Baker, 1970). *B. subtilis* was found to be antagonistic to *Phytophthora cinnamomi* (Broadbent and Baker, 1974).

Infection of leaf scars and wounds caused by *Nectria galligena* in apple trees was checked by spray with *B. subtilis* (Swinburne, 1973; Croke and Hunter, 1979). This antagonist has been identified for its ability to grow with the advancing root when applied to seeds or seed pieces. *B. subtilis* inhibited germination of sclerotia of *Sclerotium cepivorum* in soil and also gave fairly good seasonal protection when introduced on the onion seeds (Utkhede and Rahe, 1980).

Recent investigations have shown the inhibitory effect of *B. subtilis* against a number of fungi, viz., *Fomes durissimus, Phytophthora cactorum, Althernaria alternata, Armillaira mellea, Botrytis allii, B. cinerea, Colletotrichum lindemuthianum, Monilinia fructicola, Penicilium expansum, Rhizoctonia solani, Sclerotinia sclerotiorum, S. ceptivorum, Verticullium dahliae, Venturia inaequalis, Eutypa lata, Fusariumculmorum, F. nivale* (De, 1982; Utkhede, 1984; Utkhede and Sholberg, 1986; Mckeen *et al*, 1986; Utkhede and Li, 1989; Ferreira *et al.*, 1991; Fiddaman and Rossal, 1993; Silveira and Mariano, 1993; da Luz, 1993; Etheridge and Parry, 1993).

Other effective antagonistic *Bacillus* spp. i.e., *B. cereus* and *B. magaterium* have been identified as biocontrol agents against *Puccinia allii* (Utkhede, 1984) and *Colletotrichum corchori* (Purakayastha and Bhattacharya, 1982) respectively. Cultures and culture filtrates of *B. cereus* suppressed damping off of alfalfa caused by *Phytophthora medicaginis* (Silo-suh *et al*., 1994).

Research on *Pseudomonas* spp. has been vigorous in the past few years. The worldwide interest in this group of bacteria was sparked by studies initiated at University of California. Berkeley during 1970s. Several crops have been investigated and include flax (Kloepper *et al*., 1980), wheat (Wong and Baker, 1984; Weller and Cook, 1983, 1986; Cook and Rovira 1976; Sivasithamparam *et al*, 1979). cotton (Howell and Stipanovic, 1980; Loper, 1988; Kempf *et al*., 1993), banana (Sivamani and Gnanamanickam, 1988), chickpea (Kaiser and Hannan, 1989; Myatt *et al*, 1993), tulips (Weststeijn, 1990), mushroom (Nair and Fahy, 1976), potato (Burr *et al*, 1978), tobacco (Ked *et al*., 1989), rice (Sakthivel and Gnanamanickam, 1986; Ganesan and Gnanamanickam, 1987), pea (Parke *et al*., 1991; Stephens *et al*., 1993), etc. In all these crops *Pseudomonas* spp. suppressed the disease caused by various pathogens such as *Gneumannomyces graminis* var. *tritici, G. graminis* var. *avenae, Phoma betae, Pythium ultimum, F. oxysporium* f. sp. *cubense, R. solani* and *Drechslera dictyoides.* Bacterium *Pseudomonas putida* 06909, grew in the rhizosphere of citrus roots and inhibited the growth of *Phytophthora parasitica* (Yang *et al*., 1994).

Pseudomonas fluorescensi was suppressive to *Sclerotinia sclerotiorum, Phymatotrichum omnivorum, Phytopathora megasperma, P. aphanidermantum* (Misaghi *et al*., 1982), *Thielaviopsis basicola* (Ahl *et al*., 1986), *F. oxysporum* f.sp. *cauglutinans, F. oxysporum* f.sp. *dianthi* (Leeman *et al*., 1993; Duijff *et al*., 1993), *F. culmorum* and *H. sativum* (Kropp and Adetuyi, 1993). Zhou and Paulitz (1993) observed that treatment of roots of 4-day old cucumber seedlings with *P. fluorescens* reduced the attraction, encystment and germination of zoospores of *Pythium aphanidermatum. P. ceptacia* strain B37w showed inhibitory action against *F. sambucinum,* the causal agent of dry root in potato due to the production of pyrrolnitrin (Burkhead *et al*., 1994).

Actinomycetes are second only to bacteria in maintaining a satisfactory biological balance in soil, as they are good antibiotic producers. Several species are known antagonists to several plant pathogens. *Streptomyces albus* was antagonistic to *Corticum centrifugum* in unsertile soil (Knouye *et al*., 1959). Antagonistic effect of certain cultures of Actinomycetes to fungal pathogens of cucumber including *F. oxysporum* and *R. solani* was reported by Sieiketov and Kuldybaev (1974).

The antagonistic activity of several *Streptomyces* spp. against species of *Pythium* and *Phytophthora* was shown by Knauss (1976) and Valais *et al*. (1993). Effective inhibition of *F. oxysporum* f.sp. *cucumerinum* (Ogura and Yahata, 1979); *F. culmorum* and *F. sulphureum* (Tahvenen, 1982); *F. oxysporum* f.sp. *albedinis* (Sabaou *et al*., 1983); *F. oxysporum* f.sp. *vasinfectum* (Chung and Hong, 1993).

Streptomyces longisporus was antagonistic to *Helminthosporium oryzae* and *Alternaria solani* (Chattopadhyay and Nandi, 1982). Rothrock and Gottlieb (1984) mentioned that *Streptomyces hygroscopicus* var. *galdanus* was antagonistic to *R. solani* in soil.

Turham and Grossmann (1986) investigated 300 isolates of actinomycetes against six different test fungi and reported that *S. sclerotiorum* was completely inhibited by more than 90% of tested isolates, *Rhizoctonia solani* and *A. alternata* were suppressed only by 17 and 14% isolates respectively, while *Pythium debaryanum, Cochliobolus sativus* and *Macrophomina phaseolina* held an intermediate position. Tu (1988) found that *Streptomyces griseus* was antagonistic to *Colletotrichum lindemuthianum.*

ANTIBIOSIS

It was generally assumed during 1920s and 1930s that microorganisms that interferred with root disease development in soil did so by means of antibiotics. This possibility was studied extensively during 1950s and 1960s (Gottlieb, 1976), following the discovery of numerous antibiotic substances that were produced by microorganisms isolated from soil (Baker, 1981). The discovery of penicillin by Fleming in 1928 and its purification and use in medicine in 1939 greatly stimulated studies in 1948-1952. Grossbard (1952), Wright (1956) and others demonstrated that antibiotics were produced in soil by *Penicillium, Aspergillus, Trichoderma* and *Streptomyces* spp.

Baker (1987) emphasized that antibiotics produced in direct contact between microorganisms may function at low concentration and be less subject to biotic decomposition. According to Dandurand and Kundsan (1993), antagonistic microorganisms applied to seeds prior to planting colonize in rhizosphere of seedling and thus are present at or near the pathogen's infection court, where they act by producing antifungal compounds through hyperparasitism or by competitively colonizing spermosphere and rhizosphere substrate.

The mechanism by which *Trichoderma* spp. control or suppress the activities of plant pathogens is fully known. Earlier, it was established that antibiotics were produced by various *Trichoderma* spp. that were lytic to a number of fungi (Weindling, 1934; Weindling and Emerson, 1936; Weindling and Fowcett, 1936). In 1946 Brain *et al.* (1946) identified antifungal compound, viridin, produced by a pigment forming strain of *T. viride* that inhibited the germination of conidia of *Botrytis allii* at 200 time dilution. A sesquiterpene antibiotic, trichodermin was extracted from *T. viride* by Godtfredsen and Vangedal (1965). Suzukacillin (Ooka *et al.*, 1966), alamethicine (Meyer and Reusser, 1967) and U-21963 i.e. dermadin (Pyke and Dietz, 1966; Meyer, 1966) are some of the antibiotics extracted from culture filtrates of *T. viride.* Suzukacillin and alamethicine are peptide antibiotics with antifungal and antibacterial properties. Dermadin is an unsaturated monobasic acid active against Gram-negative and Gram-positive bacteria and a wide range of fungi.

Dennis and Webster (1971a, b) discovered non-volatile antibiotics, mostly chloroform soluble antibiotics including trichodermin and peptide antibiotics as well as volatile compounds like acetaldehyde, alamethicine and polypeptide antibiotics. *Trichoderma* spp. have also been shown (Chet and Baker, 1981; Elad *et al.*, 1982; Ridout *et al.*, 1986) to produce various cell wall degrading enzymes (β-(1-3) glucanase, chitinase). Claydon *et al.* (1987) extracted an antibiotic (6-n-pentylpyrone) from culture filtrate of *T. harzianum* which protected lettuce seedlings from damping off caused by *R. solani.* Production of lytic extracellular enzymes (β-1-3) glucanase and chitinase, active against *S. rolfsii* by *T. harzianum* was also reported by Elad *et al.* (1982, 1983). On the other hand, Papavizas *et al.* (1982) extracted an unidentified heat stable metabolite from this antagonist. Members of cyclopentyl isocyanide class of metabolites, isonitrin I, E and F were isolated from *T. harzianum* (Fujiwara *et al.*, 1982; Hanson *et al.*, 1985). Dermadin, trichovirdin, isonitrins 10a, 11 and 12 and heptelidic acid were isolated from strains of *T. koningii* (Tamura *et al.*, 1975; Fujiwara *et al.*, 1982; Itoh *et al.*, 1980a, b). *T. kningii* also produced a pyrene compound with antibiotic properties (Simon *et al.*, 1988). Recently, koninginin-A (Culter *et al.*, 1989), koninginin-B and cyclonerodial (Culter *et al.*, 1991) produced by *T. koningii* have been discovered. Other reports describe the production of isonitrites in *T. hamatum* (Fujiwara *et al.*, 1982; Okuda *et al.*, 1982) trichodermin and trichobrachin in *T. longibrachiatum* (Watts *et al.*, 1988; Bruckner *et al.*, 1990) trichodermaol in *T. polysporum* and T-2 toxin in *T. lignorum* (Godtfredsen and Vangedal, 1965; Banburg and Strong, 1969). Ghisalberti and Sivasithamparam (1991) have documented the ability of *Trichoderma* spp. to produce antibiotics and have also discussed the nature and possible ecological relevance of the antibiotics produced by these fungi. Lorito *et al.* (1993a) purified endochitinase and chitobiosidase enzymes from *T. harzianum* strain P1. Combining the two enzymes resulted in a synergistic increase of antifungal activity towards *Botrytis cinerea, Ustilago avenae, Uncinula necator* and *Fusarium solani.* Inhibitory effects on spore germination and germtube elongation of *B. cinerea, F. solani* and *U. necator* was reported by Lorito *et al.* (1993b) by combining strains of *Enterobacter* and *Pseudomonas* and above mentioned chitinolytic enzyme of *T. harzianum.*

Pachenari and Dix (1980) mentioned that *Gliocladium roseum* produces toxins and wall degrading enzymes. Gliovirin, an antibiotic with high activity against *P. ultimum* was identified by Howell and Stipanovic (1983) from culture filtrate of *G. virens.* This antagonist is reported to produce antifungal metabolites like gliotoxin, glioviri, gliocladic, heptalidic (avocetin), viridin, viridiol, valinotricin and a mixture of fatty acids and phenolics including ferutic acid (Taylor, 1986; Lumsden *et al.*, 1992). Parasitism of *G. catenulatum* and *G. deliquescens* appears to be through antibiosis as penetrations of host fungi by these antagonists were not observed (Hashioka and Fukita, 1960; Huang, 1978 and Martin *et al.*, 1985).

Di Pietro *et al.* (1993) reported that *G. virens* strain 41 contained the antibiotic gliotoxin and different types of chitinolytic enzymes, which were inhibitory to *Phytophthora megnasperma. P. ultimum, Botrytis cinerea* and *Rhizoctonia solani.*

Fluorescent pseudomonads have received much attention in recent years as potential biological control agents for suppression of root pathogens. Bacterial colonization of rhizosphere, substrate competition, niche exclusion and the production of secondary metabolites including antibiotics and siderophores may contribute to control pathogens by these bacteria (Handelsman and Parke, 1989, Weller *et al.*, 1988). Silo-suh *et al.* (1994) purified the two antibiotics from *B. cereus* strain UW85 and designated them as antibiotic B and zwittermicin A. Both antibiotics prevented disease of alfalfa seedlings caused by *P. medicaginis.*

Neilands (1973) pointed out that pseudobactin produced by isolates of pseudomonads is involved in competition for iron, and it is in fact a siderophore and has been mislabeled as an antibiotic. According to Kloepper *et al.* (1980), mechanism of action presumably involves antibiosis and/or competition in case of pseudomonads. The fluorescent pseudomonads associated with plant growth promotion are to compete with other rhizoplane and rhizophere microorganisms by production of siderophores which have greater iron-binding affinity than siderophores produced by displaced competitors. Another line of evidence supplied by Kloepper and Schroth (1981) indicates that ferric-pseudobactin producing isolates of *Pseudomonas* spp. increased plant growth whereas mutants not producing the compound did not. Scher and Baker (1982) opined that biological control realised by the activity of these fluorescent pseudomonads operates in the rhizosphere and this emphasises the principal that inhibitory compounds operate in localised areas. An antibiotic strongly inhibitory to *R. solani* was isolated from *P. fluorescens* cultures and identified as pyrrolnitrin (3-chlorophenyl) pyyrole. The antibiotic also inhibited growth of other fungi associated with the cotton seedling disease complex (Howell and Stipanovic, 1979). Woolford (1972) reported that *Pseudomonas fluorescens* strain 175 produced antibiotic effective against Gram-negative and Gram-positive bacteria.

Misaghi *et al.* (1982) studied the fungistatic activity of water-soluble fluorescent pigment of fluorescent pseudomonads and mentioned that it inhibited the growth of *Geotrichum candidum.* Other fungal species that showed sensitivity to the pigments produced by four out of 156 Pseudomonads isolates were *R. solani, Sclerotinia sclerotiorum, Phymatotrichum omnivorum, Phytophthora megnsperma* and *Pythierum aphanidermatum.* A direct correlation was established between the quantity of a partially purified pigment obtained from isolate of *P. fluorescens* and its fungistatic activity against *P. aphanidermatum.*

Ahl *et al.* (1986) reported that a *Pseudomonas fluorescens* strain (CHAO) is involved in suppression of black root rot of tobacco caused by *Thielaviopsis basicola* in the field. It inhibited the growth of the host pathogen by producing siderophores, cyanic acid and several antibiotics.

Thomashow and Weller (1988) reported that *Pseudomonas fluorescens* 2-79 (NRRL B-15132) and its rifampicin resistant derivative 2-79 RN are suppressive to take-all, a major root disease of wheat caused by *Caeumannomyces graminis* var. *tritici.* Strain 2-79 produces the antibiotic phenazine 1-carboxylate, which is active *in vitro* against *G. graminis* var *tritici* and other fungal root pathogens.

Homma and Suzui (1989) studied the antibiotic production by *Pseudomonas cepacia* strain RB425 obtained from a lettuce root and active as a biological control agent against several soil-borne plant pathogens, and reported that the active fractions were detected by bioautography on TLC plates using *R. solani* and *Verticillium daliliae* and sensitive indicators of antibiotic activity following each fractionations. On the basis of UV ir., NMR and mass spectra, the antibiotics were shown to be similiar or identical to pyrrolinitrin and two kinds of pseudane, 2-(2-heptenyl)3-methyl-4- quinolinol and 2-(2-noneyl) 3-methyl-4-luinolinol. The antibiotics showed high activity against several species of fungal pathogens including *Pyricularia oryzae* as well as *R. solani* and *V. dahliae,* but relatively less antibacterial activity.

Levy *et al.* (1989) studied the suppression of *Septoria tritici* and *Puccinia recondita* of wheat by an antibiotic producing fluorescent pseudomonad and reported that fractionation and analysis of two inhibitory compounds from cultures of strain LEC 1 on TLC plates yielded the phenazine-antibiotics, 1-hydroxyphenazine (PhOH) and chlororaphin. 1-hydroxypehnazine reduced the growth of *Pythium aphanidermatum, F. oxysporum* f. sp. *vasinfectum, Sclerotinia sclerotiorum* and *Sclerotium rolfsii.*

Recently, Keel *et al.* (1992) mentioned that *Pseudomonas fluorescens* strain CHAO suppress *Thielaviopsis basicola* induced black root rot of tobacco and *G. graminis* var *tritici* induced take-all of wheat. Strain CHAO produces 2,4 diacetyl phloroglucinol, a metabolite with antifungal, antibacterial and phytotoxic activity. McLoughlin *et al.* (1992) noted that in a field experiment, *Pseudomonas capacia* J82 rif and J51 rif increased sunflower emergence in the presence of the fungus *Sclerotinia sclerotiorum.* Pyrrolnitrin, aminopyrrolnitrin and monochloroamino-pyrrolnitrin were isolated from J82 and identified by TLC, HPLC, electron impact mass, UV and i.r. spectroscopy. Fridlender *et al.* (1993) mentioned that a β-l,3 glucanase producing bacterium identified as *Pseudomonas cepacia* decreased the incidence of disease caused by

R. solani, S. *rolfsii* and *P. ultimum* by 85, 48 and 71%, respectively. Determination of lytic enzymes revealed no chitinolytic activity. However, an active and stable β-1,3 glucanase was detected.

Inouye *et al.* (1959) studied the antagonistic activity of *Streptomyces albus* to *Corticium centrifugum* with a new antibiotic imoticidin. The antibiotic SF2A strongly inhibitory to *P. capcisi* was isolated from the culture filtrate of *Streptomyces violaceoniger* (Hwang and Ahn, 1993).

MYCOPARASITISM—MODE OF ACTION

Mycoparasitism is an act where one fungus parasitizes on another. This term has been generally used with hyperparasitism, direct parasitism or interfungus parasitism (Boosalis and Mankau, 1965). The mycoparasitism includes different kinds of interactions, viz., coiling of hyphae, penetration, production of haustoria and lysis of the hypae.

Hyphal invasion by *Trichoderma* sp. *in vitro* was noticed in *R. solanii* (Lewis and Papavizas, 1980). Similarly, hyphal contact with *P. ultimum* and penetration were responsible for antagonistic effect of *T. viride* (Dennis and Webster, 1971c). Tight coiling around the host pathogen i.e. *R. solani* and *Lentinus edodes* by *T. polysporum* and *T. viride* respectively were reported by Dennis and Webster (1971c) and Komatsu (1968). In *in vitro* tests, hyphal interference and lysis of mycelium of *S. rolfsii* and *R. solani* by *T. harzianum* was recorded (Wells *et al.*, 1972; Elad *et al.*, 1980; Liu and Baker, 1980; Hadar *et al.*, 1979; Henis *et al.*, 1978, 1979). Elad *et al.* (1983) mentioned that *T. hamatum* was attached to the host (*S. rolfsii* or *R. solanii)* by hyphal coils, hooks or appressoria and caused lysis of host cells. Recent observations under SEM demonstrated coiling of antagonist *T. harzianum* around its host *R. solani.* Ultrastructure observations of hyphal interactions showed that contact between the two fungi was mediated by a fine, extracellular matrix originating from cells of *R. solani.* It was followed by a series of degradation events in the host such as retraction of the plasma membrane and cytoplasm aggregatin (Benhamou and Chet, 1993).

Mode of parasitism of plant pathogens by *Gliocladium* species has been shown to be through nectrophic mycoparasitism or antibiosis (Martin *et al.*, 1985). Huang (1978) showed coiling, distortion of host cell, killing of cells and vegetative hyphae, macroconidia and conidiophores of *Fusarium equiseti, F. oxysporum, F. poe* and *F. sporotrichoides* by parasitizing *G. catenuatum.* Tu and Vaartaja (1981) found that hyphal tips of *G. virens* grew towards and alongside the hyphae and sclerotia of *R. solani. G. virens* hyphae produced specialized structures that directly penetrated and produced intracellular hyphae in sclerotia and hyphae of *R. solani.* Parasitized hyphae and sclerotia of *R. solani* eventually collapsed and became non-viable.

Mycoparasitism through contact of host cells and formation of intracellular hyphae to cause death of host cells was observed with *G. roseum* on *Botrytis allii* (Walker and Mande, 1975) and *Eutypa armeniacea* (Ricard *et al.*, 1974).

Tu (1988) showed that *Streptomyces griseus* produces an antibiotic substance against *C. lindemuthianum* in dual culture. SEM showed that hyphae of *S. griseus* in contact with *C. lindemuthianum* produced appressorium like swellings or simply grew over the hyphal surface. Internal parasitism was evidenced by the presence of hyphae and conidia of the mycoparsite inside the host hypha, resulted in formation of several blebs. Cell wall of the parasitized hyphae degenerated having a spong-like texture.

Backhouse and Stewart (1989) studied the ultrastructural changes in hyphae of *Sclerotium ceptivorum* antagonized by *Bacillus subtilis* in culture. At early stages, mitochondria became swollen, but retained their longitudinal orientation. Antagonized hyphae became more vacuolated than controls. Cell death followed rupture of hyphal walls and leakage of cytoplasm. The walls of many burst hyphae had abnormal ultrastructure. It is suggested that antibiotics act on the fungal metabolism rather than on membranes or cytoskeleton.

Nelson *et al.* (1986) mentioned that attachment of *Enterobacter cloacae* to hyphae of *P. ultimum* is a possible role in the biological control of *Pythium* preemergence damping-off, and they reported that growth inhibiton of *P. ultimum* was associated with the binding of *E. cloacae* to fungal hyphae as determined by SEM and agglutination of cell wall fragments of *P. ultimum* by *E. cloacae.*

Lysis and degradation of the mycelia of *P. aphanidermatum* by Actinomycetes was shown by Dish and Yagi (1973) and Kusakari and Ueyama (1975). Sabaou *et al.* (1983) showed antibiotic, mycolytic and parasitic action of two Actinomycetes towards *F. ozysporum* f.sp. *alpedinins* and other formae specials. *Actinoplanes missouriensis* and *Humicola fuscoatra* and other actinomycetes also showed hyperparasitism of some Peronosparales (Sutherland and Lockwood, 1984).

COMPATIBILITY OF ANTAGONISTS WITH FUNGICIDES

The usefulness of biocontrol agents depends on their integration possibility with other control measures. It has been indicated that biological control is often not highly effective when deployed as a single control strategy (Martin *et al.*, 1985). Hence, under IPM system, an effective biocontrol agent should be insensitive to common seed treatment and soil fungicides for long survival in the ecosystem. However, very few attempts have made to improve the efficiency of biological control by inducing fungicide resistant strains.

Positive compatibility of *Trichoderma* spp. with Thiram was shown by Richardson (1954) and Davet (1981). Curl *et al.* (1977) found that lower doses (2 μg/g of soil) of PCNB together with *Trichoderma* sp. were more efficient against *R. solani* than the antagonist alone. Similarly, Henis *et al.* (1978) observed that combined effect of PCNB at lower doses and *T. harziamum* decreased the inoculum potential of *R. solani* and increased disease control of damping off of radish. Hadar *et al.* (1979) showed that a combination of sublethal dose of PCNB and *T. harzianum* reduced the incidence of root rot of eggplant caused by *R. solani* from 40% to 13%, which *T. harzianum* alone reduced it to 26 per cent. Strashnov *et al.* (1985) used *T. harzianum* with methyl bromide to control *R. solani* in carrot and bean.

Troutman and Matejka (1978) induced tolerance of *T. viride* to benomyl by gamma radiation. Similarly, *Trichoderma* strain resistant to benomyl for control of *R. solani* in cotton was produced by Papavizas *et al.* (1982). Ahmad and Baker (1987) developed a benomyl tolerant mutant of *T. harzianum* which was rhizosphere competent when benomyl was added @ 10 μg/g of soil. Zhou and Reeleder (1990) obtained an *Epicoccum purpurascens* strain tolerant to iprodione and vinclozolin. Abd-E1 Moity *et al.* (1982) also developed a *Trichoderma* strain resistant to iprodione. Some fungicides like metalaxyl, captan, chlorothalonil, chloroneb, procymidone and vinclozolin were not inhibitory to *Trichoderma* spp. (Papavizas, 1981; Ordentilich and Chet, 1989).

Trichoderma isolates tolerant to metataxyl for controlling of *P. aphanidermatum* in sugarbeet, tobacco, tomato and brinjal have been developed by Mukhopadhyay *et al.* (1986). Recently,

Kaur and Mukhopadhyay (1992) reported that chickpea complex was effectively controlled by *T. harzianum* alone or in combination with fungicides, namely vitavax 200 and ziram. Mukherjee and Mulkhopadhyay (1993) developed a mutant of *G. virens* by gamma radiation insensitive to carboxin but sensitive to carbendazim. Mondal (1994) evaluated six isolates of *Trichoderma* for their compatiblity with bavistin, vitavax and raxil. *T. viride* and *T. hamatum* showed incompatibility with these seed-dressing fungicides used for loose smut of wheat. However, *T. koningii* showed compatibility with vitavax @ 200 and 500 ppm. But it showed incompatibility with rest of the fungicides. *T. lignorum* and *T. harzianum* were found to be compatible with vitavax.

Plant growth promoting rhizobacteria *(Serratia, Pseudomonas* and corineform-like bacteria) were found to be tolerant to vitavax and capton. Fiddamana and Rossal (1993) used thiram and carboxin as seed dressing fungicides in combination with *B. subtilis* and found that the combination improved plant growth.

EFFECT OF BIOCONTROL AGENTS ON GROWTH AND YIELD OF THE CROP

Soil microorganisms when applied as soil/seed inoculants stimulate the growth of plants. Broadbent *et al.* (1971) indicated that *Azobacter chroococcum, Bacillus megaterium* var. *phosphaticum* stimulate the growth of several agricultural crops. Merriman *et al.* (1974) have also demonstrated that isolates of *Bacillus* spp. and *Streptomyces* spp. selected for biological control of root pathogens enhanced the growth of plants. In addition to adverse effect of certain isolates of bacteria and actinomycetes on *R. solani,* their application increased plant growth and dry matter of chickpea (Singh and Malhotra, 1980).

In an experiment, Kloepper and Schroth (1981) observed that plant growth promoting rhizobacteria (PGPR) caused significant increase ranging from 300-500 per cent in total weight of potato plants grown in field soil in the greenhouse. This increase in plant growth was attributed to antibiosis that occurs in root zone and subsequent displacement of certain root colonizing microorganisms.

Addition of *Trichodermai* spp. to autoclaved soil resulted in increased seedling emergence of tomato and tobacco, and also enhanced root (213-275%) and shoot (295-318%) weight over control (Windham *et al.*, 1986). Earlier, Marios *et al.* (1982) while working on biological control of *Verticullium* wilt of brinjal using *Talaromyces flavus,* noticed increase by 18-54% (by weight) and 22-71% by number of fruits. Recently, Hyakumachi *et al.* (1993) has recorded increase in dry weight as a result of soil treatment with plant growth promoting fungi in bent grass (908%), ryegrass (439%), wheat (151%), tomato (505%), cucumber (715%) and radish (189%).

Rhizosphere colonization and growth promotion by selected strains of *Pseudomonas* spp. was observed by Suslow and Schroth (1982) in case of sugarbeets. Increase in the fresh weight of sugarbeets with PGPR treatment ranged from 20-85 per cent. Inoculation with *Bacillus polymyxa* resulted in statistically significant seedling biomass increase after 8 weeks of pine growth (Holl and Chanway, 1992).

Weller and Cook (1986) tested fluorescent pseudomonads for their ability to increase the growth of wheat. Out of 64 strains 17 strains increased height of wheat seedlings, whereas four strains also increased number of ear-heads, grain yield when tested as seed treatment on winter wheat grown in naturally infested soil with *Pythirum* sp. De Freitas and Germida (1991) studied the effect of 36 known bacteria and 75 isolates obtained from the rhizosphere of winter wheat and revealed that two known bacteria and 12 isolates stimulated growth of the crop. Of these, the most effective were rune isolates that significantly ($P<0.01$) increased plant height, root and shoot biomass and number of tillers. Of these strains, three were tentatively classified as *P. aeruginosa* and one each as *P. cepacia* and *P. fluorescens.*

ORGANIC AMENDMENTS IN BIOLOGICAL CONTROL

Amendments of soil with decomposable organic matter is recognised as an effective method of changing the soil and rhizosphere environment thereby adversely affecting the pathogens. It is suggested that organic amendments rich in carbon help the microorganism in abundant release of CO_2 due to respiratory activity which is deterimental to pathogens sensitive to CO_2. Decomposition of organic matter in the soil also produces a number of fungitoxic substances, especially fatty acids and phenolics. Such substances have a sanitory effect in the soil by causing death or inhibition of parasitic organisms. Liberation of ammonia by organic substances and its accumulation can shift soil pH towards alkalinity which results in disturbance in soil microbial equilibrium and may harm the pathogens. In addition to the changes in soil texture and fertility, organic soil amendments enable the plants to resist their attack through better vigour and/or altered root physiology (Singh and Singh, 1984).

Organic amendements rich in carbon and deficient in nitrogen controlled the 'Take all' disease of wheat caused by *Gaeumannomyces graminis* var. *tritici.* In this case it was emphasised that amendments resulted in considerable liberation of CO_2 by soil saprophytes which suppressed the pathogenic activity of this fungus.

Cook and Schroth (1965) emphasised that available nitrogen in the amended soil was rapidly utilized by fast growing microorganisms. This created acute nitrogen scarcity which adversely affected the growth of *Fusarium solani.*

Singh *et al.* (1988) evaluated the effect of some soil amendments, viz., crop straws (gram, soyabean, guar, pigeon pea), green manures (maize, dhaincha, sanai, bajra, eucalptus leaves), oil cakes (mustard, cotton seed, groundnut) and other organic chemicals (cellulose, gelatin) for their suppressive effect on *Fusarium* wilt of potato. Among crop straws, gram and pigeon pea showed maximum suppressive effect on *F. oxysporum* population, maize and eucalyptus leaves amongst green manures and groundnut cake amongst the cakes. Crop straws were better over other amendments. Byadgi and Hegde (1988) studied the factors affecting survival of *R. bataticola* in amended soil and found that the wheat straw and paddy hulls as soil amendments not only reduced the infection of the bean seedlings, but also supported good multiplication of fungi, bacteria and actinomycetes, antagonistic to the fungus. One *Trichoderma* sp., three *Bacillus* spp. and a *Streptomyces* sp. showed antagonistic effect and suppressed the growth of the fungus in culture.

Composed organic household waste showed a suppression of soil-borne plant pathogens, thereby reducing the incidence of root rot on peas, beans and beet roots caused by *Pythium ultimum* and *R. solani* (Schuler *et al.*, 1989). Chakrabarti and Sen (1991) investigated the suppression of Fusarium wilt of muskmelon by organic soil amendments such as mustard cake (2%), groundnut cake (1%), margosa cake (2%) and sawdust (1%). Results indicated inhibitory effect on mycelial growth of *F. solani in vitro* where mustard cake and sawdust retarded spore germination of the pathogen and margosa cake reduced wilt by 80%. Recently, Singh *et al.* (1993) used neem cake (1%) as seed treatment and compared its effectiveness with 7 fungicides to control black-scurf of potato caused by *R. solani.* The treatment of neem cake inhibited 83.67 per cent disease as compared with 62.8, 70.0, 47.5, 67.3, 89.6, 69.2 and 59.2 per cent with agrozim, bavistin, captaf, derosal, emisan, jksterin and rovral respectively. Recently, Asirifi *et al.* (1994) studied the incidence of sclerotina soft rot of lettuce and recorded that survival rate of sclerotia of *Sclerotinia sclerotiorum* were reduced by the addition of stable manure, fowl manure and lucerne hay. Incorporating the amendments into soil also increased the number and fresh weight of marketable lettuce.

REFERENCES

Adam, D.B. and Pugsley, A.T. (1935). A yellow bacterium associated with 'halo blight' of beans. *Aust. J. Exp. Biol. Med. Sci.* **13:** 157-164.

Anonymous. (1966). Progress of Regional Pulse Improvement Project (1966). U.S. Department of Agriculture, Agricultural Research Service, U.S. Agency for International Development Cooperating. ICAR, New Delhi.

Anuratha, C.S. and Gnanamanickam, S.S. (1987). *Pseudomonas fluorescens* suppress development of bacterial blight (BB) symptoms. *International Rice Research Newsletter* **12:** 17.

Beer, S.V. and Rundle, J.R. (1980). Inhibition of *Erwinia amylovora* by bacteriocin like substances. *Phytopathology* **70:** 459 (Abstr.).

Beniwal, M.S. and Parasnar, R.D. (1979). Studies on bacterial leaf spot of mung (*Vigna radiata* L.). *Geobios* **6:** 26-27.

Beniwal, M.S. and Parashar, R.D. (1981). Evaluation of antibiotics and fungicides for the control of bacterial leaf spot of mung. *Haryana Agric. Uni V. J. Res.* **11:** 36-39.

Brown, M.E. (1972). Plant growth substances produced by microorganisms of soil and rhizosphere. *J. Apple Bact.* **35:** 443-451.

Burkholder, W.H. (1926). A new bacterial disease of bean. *Phytopathology,* **16:** 915-927.

Burr, T.J., Schroth, M.N. and Suslow, T.V. (1978). Increased potato yields by treatments of seed pieces with specific strains of *Pseudomonas fluorescens* and *P. putida. Phytopathology* **68:** 1377-1883.

Chakravarti, B.P., Leben, C. and Daft, G.C. (1972). Numbers and antagonistic properties of bacteria from buds of field grown soybean plants. *Can. J. Microbiol.,* **18:** 696-698.

Chand, J.N., Yadav, O.P. and Yadav, H.C. (1977). Reaction of germplasm lines of moong bean to bacterial leaf spot. *Indian Phytopath.* **30:** 565.

Chatterjee, A.K., Gibbins, L.N. and Carpenter, J.A. (1969). Some observations on'the physiology of *Erwinia herbicola* and its possible implications as a factor antagonistic to *Erwinia amylovora* in the fire blight syndrome. *Can. J. Microbiology,* **15:** 640-642.

Chowdhury, H.D. (1978). Role of phylloplane bacteria and heat killed cells of *Xanthomonas malvacearum* on the development of bacterial blight of cotton. Ph.D. Thesis, IARI, New Delhi-110 012, pp. 168.

Clark, F.E. (1940). Notes on types of bacterial associated with plant roots. *Trans. Kansas Acad. Science,* **43:** 75-84.

*Clark, F.E. (1949). Soil microorganisms and plant growth. *Advanc. Agron.* **1:** 241.

Colin, J., Pussemier, L. and Diouf, A.M. (1984). Demonstration and applications of the antagonism of fluorescent *Pseudomonas* spp. towards phytopathogenic bacteria. *Mendedeligen Van de Faculteit Landbouwnetens chappen Rijksumversiteit Gent.* **49:** 587-596.

Colyer, P.D. and Mount, M.S. (1984). Bacterization of potatoes with *Pseudomonas putida* and its influence on post-harvest soft rot diseases. *Plant Disease* **68:** 703-706.

Crosse, J.E. (1976). Plant Pathology. *In*: East mailing Res. St. Reprt. for 1977, Maidstone, Kent, U.K. pp. 107-176.

*Curl, E.A. and Truelove, B. (1986). The Rhizospbere (Advanced series in Agric. Sciences 15). Springer Verlag, Berlin, Heidelberg.

Darbyshire, J.F. and Greaves, M.P. (1967). Protozoa and bacteria in the rhizosphere of *Sinapsis alba* L., *Trifolium repens* L., and *Lolium perenne L. Can. J. Microbiol.* **13:** 1057-1068.

Devadath, S. (1970). Effect of mixture of inoculations of virulent and less virulent isolates of *Xanthomonas oryzae. Curr. Sci.* **39:** 420-421.

Dhanavantari, B.N. (1976). Biological control of crown gall of peach in South Western Ontario. *Plant Dis. Reptr.* **60:** 549-551.

Dickinson, C.H. (1973). Interactions of fungicides and leaf saprophytes. *Pesticide Science*, **4:** 563-574.

Dye, D.W. 1962. The inadequacy of the usual determinative tests for identification of *Xanthomonas* spp. *N.Z.J. Sci.* **5:** 393-416.

Dye, D.W. (1978). Genus IX *Xanthomonas* Dowson 1939. In Young, J.M., Dye, D.W., Bradbury, J.F., Panagopoulos: C,G., Robbos, C.F. 1978. A proposed nomenclature and classification for plant pathogenic bacteria. *N.Z.J. Agric. Res.* **21:** 153-177.

Fang, C.T., Chen, H.Y. and Chu, C.L. (1964). A comparative study of the species of the genus *Xanthomonas* from leguminous plants. *Acta Phytopath. Sinica* **7:** 21-31, Rev. Appl. Mycol. **44:** 187.

Farabee, G.J. and Lockwood, J.L. (1957). Inhibition of *Erwinia amylovora* by bacteria isolated from fire blight canker. *Phytopathology*, (Abstract). **47:** 10.

Farabee, G.J. and Lockwood, J.L. (1958). Inhibition of *Erwinia amylovora* by *Bacterium* sp. isolated from fire blight cankers. *Phytopathology*, **48:** 209-211.

Fujii, H. (1976). Variation in the virulence of *Xanthomonas oryzae*. Interaction between avirulent mutants and virulent wild type isolates. *Ann. Phytopath. Soc., Japan* **42:** 526-532.

Gantotti, B.V., Kindle, K.L. and Beer, S.V. (1981). Transfer of the drug resistance transposon Tn5 to *Erwinia herbicola* and the induction of insertion mutation. *Curr. Microbiol.* **6:** 377-381.

Gaur, R.B. and Ahmad, S.R. (1983). Studies on the chemical control of bacterial blight of moong incited by *Xanthomonas phaseoli.* (Smith) Dowson. *Pesticides* **17:** 22-23.

Gaur, R.B., Ahmad, S.R. and Kataria, P.K. (1984). Controlling *Xanthomonas phaseoi* (Smith) Dawson of moong seed through chemotherapy and heat therapy. *Hindustan Antibiotics Bull.* **26:** 23-26.

Gee1s, F.P. and Schippers, B. (1983). Reduction in yield depressions in high frequency potato cropping soil after seed tuber treatments with antagonistic fluorescent *Pseudomonas* spp. *Phytopathol Z.* **108:** 207-214.

Goodman, R.N. (1964). The influence of avirulent forms of *Erwinia amylovora* and saprophytic species on the virulence of *Erwinia amylovora in vivo* (Abstract). *Phytopathology*, **54:** 894.

Goodman, R.N. (1965). *In vitro* and *in vivo* interaction between components of mixed bacterial cultures isolated from apple buds. *Phytopathology*, **55:** 217-221.

Goodman, R.N. (1967). Protection of apple stem tissues. against *Erwinia amylovora* infection by avirulent strains and three other bacterial species. *Phytopathology*, **57:** 22-24.

Griffin, D.M. (1972). Ecology of Soil Fungi, Chapman and Hall, London.

Haage, W.L. and Vidaver, A.K. (1974). Purification and characterization syringacin 4-A a bacteriocin from *Pseudomonas syringae* 4-A. *Antimicrob. Agents Chemother.* **6:** 76-83.

Hale, M.G. and Moore, L.D. (1979). Factors affecting root exudation II. 1970-1978. *Advances in Agronomy* **31:** 93-124.

Howie, W.J. and Echaudi, E. (1983). Rhizobacteria: Influence of cultivar and soil type on plant growth and yield of potato. *Soil Biol. Biochem.* **15:** 127-132.

Hsieh, S.P.Y. and Buddenhagen, I.W. (1974). Suppressing effect of *Erwinia herbicola* on infection by *Xanthomonas oryzae* and on symptom development in rice. *Phytopathology*, **64:** 1182-1185.

Isenbeck, M. and Schulz, A. (1985). Biological control of Fire-blight (*Erwinia amylovora* [Burr.] Winslow *et al.*) on ornamentals. I. Control of the Pathogen by Antagonistic Bacteria. *Phytopath. Z.* **113:** 324-333.

Isenbeck, M. and Schulz, F.A. (1986). Biological control of Fire-blight (*Erwinia amylovora* [Burr.] Winslow *et al.*) on ornamentals. *J. Phytopathology* **116:** 308-314.

Ishimaru, C.A., Klos, E.J. and Brubaker, R.R. (1988). Multiple antibiotic production by *Erwinia herbicola. Phytopathology*, **78:** 746-750.

Jindal, J.K. (1976). Variability in Xanthomonads of pulse crops. Ph.D. Thesis submitted to Botany Department, Aligarh Muslim University. pp. 181.

Jindal, J.K. and Patel, P.N. (1980). Variability in Xanthomonads of grain legumes. I. Pathogenic behaviour and taxonomy. *Phytopath. Z.* **99:** 332-356.

Jindal, J.K., Patel, P.N. and Khan, A.M. (1981). Variability in Xanthomonads of grain legumes. II. Pathogenic variability in *Xanthomonas phaseoli* mungbean strain, *X. vignicola* and *X. phaseoli* var. sojense. *Phytopath. Z.* **100:** 1-9.

Jindal, K.K. and Thind, B.S. (1990). Microflora of cowpea seeds and its significance in the biological control of seed-borne infection of *Xanthomonas campestris* pv. *vignicola. Seed Sci. Technol.* **18:** 393-403.

Jones, D.A. and Kerr, A. (1989). *Agrobacterium radiobacter* strain K 1026, a genptically engineered derivative of strain K 84 for biological control of crown gall. *Plant Dis.* **73:** 15-18.

Keen, N.T. and Kennedy, B.W. (1974). Hydroxyphaseollin and related isoflavanoids in the hypersensitive resistance reaction of soybean to *Pseudomonas glycinea*. *Physiological Plant Pathology,* **4:** 37-44.

Kerr, A. (1980). Biological control of crown gall through production of agrocin 84. *Plant Dis.* **64:** 25-30.

Kerr, A. and Htay, K. (1974). Biological control of crown gall through bacteriocin production. *Physiological Plant Pathology,* **4:** 37-44.

Klement, Z. (1963). Rapid detection of the pathogenicity of Phytopathogenic pseudomonads. *Nature,* **199:** 299-300.

Klement, Z. (1968). Pathogenicity factor in regard to relationships of phytopathogenic bacteria. *Phytopathology,* **58:** 1218-1221.

Kloepper, J.W. (1983). Effect of seed piece inoculation with plant growth-promoting rhizobacteria on populations of *Erwinia carotovora* on potato roots and daughter roots. *Phytopathology,* **73:** 217-219.

Kloepper, J.W., Leong, J., Teintze, M. and Schroth, M.N. (1980). *Pseudomonas* siderophores: a mechanism explaining disease suppressive soils. *Curr. Microbiol.* **4:** 317-20.

Kloepper, J.W., Hume, D.J., Scher, F.M., Singleton, C., Tipping, B., Laliberte, M., Frauley, K., Kutchaw, T., SImonson, C., Lifshitz, R., Zaleska, I. and Lee, L. (1988). Plant Growth Promoting Rhizobacteria on Canola. *Plant Dis.* **72:** 42-46.

Kroulik, J.T., Burkey, L.A. and Wiseman, H.G. (1955). The microbial population of the green plants and of the cut florage prior to ensiling. *J. Dairy Sci.* **38:** 256-262 and 263-272.

Laba, G.S. (1990). Role of fluorescent pseudomonads in the management of cotton diseases. M.Sc. Thesis, IARI, New Delhi. pp. 64.

Lambert, B., Layens, F., Rooyen, L. Van., Gossele, F., Papon, Y. and Swings, J. (1987). Rhizobacteria of Maize and their antifungal activities. *Appl. Environ. Microbiol.* **53:** 1866-1871.

Lelliott, R.A., Billing, Ev. and Hayward, A.C. (1966). A determinative scheme for the fluorescent plant pathogenic Pseudomonads. *J. Appl. Bacterial.* **29:** 470-489.

Liao, C.H. (1989). Antagonism of *Pseudomonas putida* strains PP. 22 to phytopathogenic bacteria and its potential use as biocontrol agent. *Plant Dis.* **73:** 223-226.

Maino, A.L., Schroth, M.N. and Vitanza, V.B. (1974). Synergy between *Achronobacter* sp. and *Pseudomonas phaseolicola* resulting in increased disease. *Phytopathology,* **64:** 277-283.

Marimuthu, T. (1983). Effect of sulphur containing amino acids on growth, toxin production of *Xanthomonas campestris* pv. *phaseoli* and leaf blight developnent in green gram. *Indian Phytopath.* **36:** 47-51.

Marimuthu, T. and Kandaswamy, T.K. (1980). Bacterial leaf blight of green gram (*Vigna radiata* var. *aureus*) caused by *Xanthomonas phaseoli* and reaction of certain varieties to tbe disease. *Madras Agri. J.* **67:** 413.

New, P.B. and Kerr, A. (1972). Biological control of crown gall : field measurement and glasshouse experiments. *J. Appl. Bact.* **35:** 279-287.

Nikanishi, K. and Watanabe, M. (1977). Studies on the mechanisms of resistance of rice plants against *Xanthomonas oryzae.* III. Relationship between the rate of production of antibacterial substances and of multiplication of pathogenic bacteria in infected leaves of resistant and susceptible varieties. *Ann. Phytopath. Soc., Japan.* **43:** 265-269.

Nwigwe, C. (1973). Influence of epiphytic yellow bacteria on the pathogenicity of *Xanthomonas oryzae* on rice leaves. *Riso* **22:** 119-122.

Omer, M.E.H. and Wood, R.K.S. (1969). Growth of *Pseudomonas phaseolicola* in susceptible and resistant plants. *Ann. Apple Biol.* **63:** 103-116.

Patel, P.N. and Jindal, J.K. (1970). Bacterial diseases in seed legumes in 1968 and 1969. Paper presented at Fourth Annual Workshop Conference on Pulse Crops held at Ludhiana in April, 1970.

Patel, P.N. and Jindal, J.K. (1972). Bacterial leaf spot and halo blight diseases of mungbean (*Phaseolus aureus*) and other legumes in India. *Indian Phytopath.* **25:** 517-525.

Patel, P.N., Jindal, J.K. and Singh, D. (1972). Studies on resistance in crops to bacterial diseases in India IV. Resistance in mungbean (*Phaseolus aureus*) to *Xanthomonas phaseoli. Indian Phytopath.* **25:** 526-529.

Rangaswami, G. and Gowda, S. (1963). On some bacterial diseases of Ornamentals and Vegetables in Madras State. *Indian Phytopath.* **16:** 74-85.

Riggle, J.H. and Klos, E.J. (1970). Inhibition of *Erwinia amylovora* by *Erwinia herbicola. Phytopathololgy* (Abstract). **60:** 1310.

Riggle, J.H. and Klos, E.J. (1972). Relationship of *Erwinia herbicola* to *Erwinia amylovora. Can. J. Bot.* **50:** 1077-1083.

Rouatt, J.W. and Katznelsm. (1961). A study of the bacteria on the root surface and in the rhizosphere soil of crop plants. *J. Apple Bact.* **24**(2): 164-171.

Sabet, K.A. (1969). Studies on the bacterial diseases of Sudan crops III. On the occurrence, host range and taxonomy of the bacteria causing leaf blight diseases of certain leguminous plants. *Ann. Apple Biol.* **47:** 318-331.

Saini, L.C. and Parashar, R.D. (1982). Suitability of stable bleaching powder and other chemical formulations for the control of bacterial leaf spot of green gram. *Indian J. Agric. Sci.* **52:** 330-334.

Santhi, D.P., Unnamalai, N. and Gnanamanickam, S.S. (1987). Epiphytic association of *E. herbicola* with rice leaves infected by *Xanthomonas campestris* pv. *oryzae* and its interaction with pathogen. *Indian Phytopath.* **40:** 327-333.

Scherff, R.H. (1972). Interaction of *Pseudomonas glycinea* and yellow bacteria in the development of bacterial blight of soybean. *Phytopathology,* **62:** 787.

Scherff, R.H. (1973). Control of bacterial blight of soybean by *Bdellovibrio bacteriovorus. Phytopathology,* **63:** 400-402.

Schmidt, D. (1988). *Pseudomonas fluorescens* and *Erwinia herbicola* reduce wilt of grasses caused by *Xanthomonas campestris* pv. *graminis. J. Phytopathology* **122:** 245-252.

Schroth, M.N. and Hancock, J.C. (1981). Selected topics in biological control. *Ann. Rev. Microbiol.* **35:** 453-476.

Schroth, M.N. and Hancock, J.G. (1982). Disease suppressive soil and root colonizing bacteria. *Science,* **216:** 1376-81.

Shekhawat, G.S. and Patel, P.N. (1977). Seed transmission and spread of bacterial blight of cowpea and leaf spot of green gram in summer and monsoon seasons. *Plant Dis. Reptr.* **61:** 390-392.

Shiv Prasad., Singh, V.P.N. and Singh, D. (1987). Strategy for increasing pulse production. *Agric. Sti. India* **41:** 875-888.

Sierra, G. (1957). A simple method for detection of lipolytic activity of microorganisms. *Antonie van Leewenhoek* **23:** 15.

Sinha, P.P. (1981). Protective effect of epiphytic bacteria on the infection caused by *Xanthornonas malvacearum* (E.F. Smith) Dowson, Ph.D. Thesis, IARI, New Delhi-12, pp. 165.

Singh, D. (1972). Screening and inheritance of resistance to bacterial blight in cowpea and bacterial leaf spot in mung. Ph.D. Thesis submitted to P.G. School, IARI, New Delhi-12.

Singh, D. and Patel, P.N. (1977). Studies on resistance in crops to bacterial disease in India VIII. Investigations on inheritance of reactions to bacterial leaf spot and yellow mosaic diseases, and linkage, if any, with other characters in mungbean. *Indian Phytopath.* **30:** 202-206.

Sivamani, E., Anurath, C.S. and Gnanamanickam, S.S. (1987). Toxicity of *Pseudomonas fluorescences* towards bacterial plant pathogens of banana (*Pseudomonas solanacearum*) and rice (*Xanthomonas campestris* pv. *oryzae*). *Curr. Sci., India.* **56:** 547-548.

Society of American Bacteriologists. (1957). "Manual of Microbiological Methods". McGraw Hill Book Co. New York, pp. 315.

Sood, A.K., Thind, B.S. and Singh, N. (1976). Occurrence and control of bacterial leaf spot of mung (*Phaseolus aureus* Roob.) in Punjab State. *J. Res. PAU.* **13:** 182-185.

Stein, J.I. and Beer, S.V. (1980). Partial purification of bacteriocin from *Erwinia herbicola* (Abstr.) *Phytopathology,* **70:** 469.

Takur, R.P. (1976). Studies on resistance to bacterial leaf spot of mungbean and biochemical mutants of the pathogen *Xanthomonas phaseoli* mungbean strain. Ph.D. Thesis, submitted to P.G. School, IARI, New Delhi, p. 92.

Thomson, S.V., Schroth, M.N., Moller, W.J. and Reil, W.O. (1976). Efficacy of bacterioides and saprophytic bacteria in reducing colonization and infection of pear flowers by *Erwinia amylovora. Phytopathology,* **66:** 1457.

*Teliz-ortiz, M. (1967). *In vitro* and *in vivo* relationships between an antagonistic bacterium and *Pseudomonas phaseolicola* (Burk) Dowson. *Diss. Abstr.* **28:** 42.

Teliz-ortiz, M. and Burkholder, W.H. (1960). A strain of *Pseudomonas fluorescens* antagonistic to *Pseudomonas* and other bacterial plant pathogens. *Phytopathology,* **50:** 119-123.

*Turner, J.T. Jr. (1987). Relationships among plant growth, yield and rhizosphere ecology of peanuts as affected by seed treatment with *Bacillus subtilis.* Ph.D. Dissertion. Auburn Univ. pp. 108.

Unnamalai, N. and Gnanamanickam, S.S. (1984). *Pseudomonas fluorescens* is antagonistic to *Xanthomonas citri* (Hasse) Dye, the incitant of citrus canker. *Curr. Sci., India.* **53:** 703-704.

Verma, J.P. and Singh, R.P. (1974). Recent Studies on the bacterial diseases of fibre and oilseed crops in India. In: Current Trends in Plant Pathology (Raychaudhuri, S.P. and Verma, J.P. Eds.) pp. 134-154.

Verma, J.P. and Singh, R.P. (1976). Races of *Xanthomonas malvacearum,* loss in virulence and the protective effect of virulent strains, heat killed cells and phylloplane bacteria. *Z. Pflkrank Pflschutz* **83:** 748-757.

Verma, J.P., Chowdhury, H.D. and Singh, R.P. (1978). Interaction between phylloplane bacteria and *Xanthomonas malvacearum*. In: Proc. IVth Int. Conf. Plant Pathogenic Bacteria II. Angers, pp. 979.

Verma, J.P., Singh, R.P. and Chowdhury, H.D. (1980). Relationship of epiphytic bacteria to bacterial diseases of plants. *In*: Recent Advances in the Biology of Microorganisms. Ed. K.S. Bilgrami and K.M. Vyas. Publ. Bishen Singh Mahendra Pal Singh, Dehradun, India, p. 344.

Vidaver, A.K. (1976). Prospects for control of phytopathogenic bacteria by bacteriophages and bacteriocins. *Ann. Rev. Phytopath.* **14:** 451-465.

Vidaver, A.K., Mathys, M.L., Thomas, M.E. and Schuster, N.L. (1972). Bacteriocins of the phytopathogens *Pseudomonas syringae, Pseudomonas glycinea* and *Pseudomonas phaseolicola. Can. J. Microbiol.* **18:** 705-713.

Wang, J.S., He, C.Y. and Fang, Z.D. (1988). Bacteriocinogenic strain detection and bacteriocin types of soft rot *Erwinia. J. of Biol. Control* **4:** 55-58.

Watanabe, M., Nakanishi, K., Kumakura, K., Shimoyama, N. and Furukawa, M. (1976). Studies on the mechanism of resistance in the rice plants infected with *Xanthomonas oryzae*. I. Inhibition of lesion length enlargement by pre-inoculation with the non-pathogenic bacteria. *Ann. Phytopath. Soc., Japan.* **42:** 295-303.

Weller, D.M. (1988). Biological control of soil borne plant pathogens in the rhizosphere with bacteria. *Ann. Rev. Phytopath.* **26:** 379-407.

Wrather, J.A., Kuc, J. and Williams, E.B. (1973). Protection of apple and pear fruit tissues against fire blight with non-pathogenic bacteria. Phytopathology **63:** 1075-1076.

Xu, G.W. and Gross, D.C. (1986). Field evaluation of the internations among fluorescent pseudomonads, *Erwinia carotovora*, and potato yields. *Phytopathology*, **76:** 423-430.

Yadav, H.C. and Chand, J.N. (1981). Inheritance of resistance in mungbean (*Vigna radiata* L.) Wilczek to bacterial leaf spot (*X. campestris* pv. *phaseoli*), 5th Int. Conf. on Plant Pathogenic Bacteria, 1981.

Young, J.M., Dye, D.W., Bradbury, J.F., Panagopoulos, C.G. and Robbs, C.F. (1978). A proposed nomenclature and classification for plant pathogenic bacteria. N.Z.J. *Agric., Res.* **21:** 153-177.

---

*Original not seen.

36

Bacterial Antagonists in Biocontrol of Plant Parasitic Nematodes

Satyandra Singh and S.C. Dhawan

Division of Nematology, Indian Agricultural Research Institute, New Delhi–110 012

Biocontrol of economically important plant parasitic nematodes has become the most admired and intensive part of research in the present scenario. According to Kerry (2000), continuous environmental problems associated with the use of nematicides have introduced a sense of urgency into the search for alternative methods of nematode management. According to Sasser and Freckman (1987), crop losses by nematodes range from 8 to 20% on major crops around the world. Chemical control is a widely accepted practice for management of plant parasitic nematodes; however, these are now less attractive due to environmental concerns, high cost, non availability and diminished effectiveness by repetitive applications. Exploitation of natural enemies of nematodes for realistic use in the management of plant parasitic nematodes is gaining weight over demerits of hazardous and unfriendly poisons. Also, public concerns for the quality of environment and food safety have increased the need for alternative control measures for plant-parasitic nematodes. In this line of work, use of biocontrol agents could reduce nematode damage to crops and raise the yield prospective. Of the biocontrol agents, bacteria have attracted the attention of researchers globally in the past decade or so. As a group of important natural enemies of nematode pests, these exhibit diverse modes of action, which these include parasitizing; producing toxins, antibiotics, or enzymes; competing for nutrients; inducing systemic resistance of plants; and promoting plant health. They act synergistically on nematodes through reduction in density of nematodes, promoting plant growth, and facilitating the rhizosphere colonization and activity of microbial antagonists. Studies made on the bacterial bio agents have made significant progress and quite a good number of bacteria have been identified as parasite or toxins producing that are antagonistic to phytonematodes. Root-knot and cyst forming nematodes are two important groups of plant parasitic nematodes that are considered as important constraints in the output of a number of agricultural crops. Since both nematode and bacteria dwell in the same niche in the soil and roots, therefore, interaction among them is a common attribute. A number of bacterial species belonging to *Agrobacterium, Bacillus, Pasteuria* and *Pseudomonas* have been recognized to suppress the nematode populations by one or another mechanism.

HISTORY

Compared to the noteworthy volume of work reported with the use of bacteria as biological control agents of soil-borne pathogenic fungi, considerably less work has been done examining the potential of plant-associated bacteria to control phytonematodes. A few reports indicate that both rhizobacteria and endophytic bacteria have the potential to reduce plant damage to nematodes. Suggested mechanisms include induction of systemic resistance and production of antagonistic compounds; however, definitive mode-of-action studies have not been done with these systems. It was for the first time described by Thorne (1940) as the protozoan, *Duboscqia penetrans*. Later, the same organism was found in the population of root-knot nematode. Prasad and Mankau (1969) observed minute parasite in a greenhouse population of *Meloidogyne javanica,* which resembled *D. penetrans*. Imbriani and Mankau (1977) established the prokaryotic nature of the organism. Bacteria of the genus *Pasteuria* have long been recognized as promising biological control agents for plant-parasitic nematodes (Chen and Dickson, 1998; Bishop and Ellar, 1991; Sayre, 1980). They are known for their ability to disrupt reproduction in their nematode hosts (Ahmad and Gowen, 1991), and have been shown to reduce root-knot nematode damage below economic threshold levels and to improve crop yields (Brown, *et al.*, 1985). Sayre and Starr (1985) renamed *Bacillus* as *Pasteuria.* Becker *et al.* (1988) have shown that plant roots can be protected from fungal and, possibly, nematode attack by the presence of certain bacteria in the rhizosphere. The most probable mechanism is antibiosis, since the bacteria produce compounds that are lethal to the plant pests. Considerable work has been conducted on attempting to isolate the effective rhizobacteria, culture them, and inoculate seedling plants, or to introduce the bacteria at appropriate intervals in irrigation water.

HOST RANGE AND GEOGRAPHICAL DISTRIBUTION

Pasteuria species are cosmopolitan in distribution and have been recorded from at least 12 states of the United States, in 51 countries, 5 continents as well as islands in Atlantic, Pacific and Indian Oceans. The bacterium has been reported from about 204 different nematode species belonging to some 95 nematode genera in 10 orders. The host includes free-living, predacious, plant parasitic, and entomopathogenic nematodes. *Pasteuria* spp. have been reported on a variety of nematode hosts and in many different climates and environments throughout the world (Hewlett *et al.*, 1994; Sayre & Starr, 1988; Stirling, 1988).

BIOLOGY

The biology and ecology of these parasites have been extensively reviewed (Sayre and Starr, 1988; Chen and Dickson, 1998). Life cycle of *Pasteuria* species starts with endospore, the infective stage of the bacteria, which perpetuates year after year in the soil and is released on decomposition of nematode body of adult females and males or in plant root tissues. The life cycle consists of three distinct stages: spore germination, vegetative growth, and sporulation. Spores being non-motile remain in the soil as such and the nematode while moving through the soil matrix picks up the spores as and when it comes in contact with the spores present in the soil. Passive movement of spores, however, occurs by irrigation water, tillage practices, and human beings. Number of abiotic factors has been reported to affect attachment. Once the spores are attached on the cuticle of the host nematode, spore germination occurs. In case of

P. penetrans and *P. nishizawae*, germination of spores occurs only after the encumbered second-stage juveniles invade the plant root and begin to feed. The germ tube emerges through a central opening in the basal attachment layer and penetrates the cuticle of the nematode and finally reaches pseudocoelom. The process of penetration seems to be enzymatic, but what triggers the spore to germinate is unknown. The germ tube, in the pseudocoelom, develops into vegetative microcolony consisting of dichotomously branched septate mycelium. The intercalary cells in the microcolony lyse and give rise to daughter colonies. The process continues, resulting in a large number of daughter colonies containing fewer but larger vegetative cells. Eventually, quartets of developing sporangia predominate in the nematode's body cavity. These quartets separate into doublets of sporangia and finally separate into a single sporangium that will eventually contain an endospore. Endospores of *Pasteuria penetrans* are released into the soil when the plant root with its complement of parasitized root-knot nematode females decompose, and in case of *P. nishizawae*, endospores are released by breakdown of infected cysts wall. Sporogenesis is divided into seven stages. In stage 1, mycelial terminal cells elongate and become fully septate. Stage II is characterized by the formation of a transverse septum that separates the forespore from the endospore mother cell. In stage III, the forespore is engulfed by the endospore mother cell and condensation of the forespore protoplasm occurs. Parasporal fibres are initiated in stage III. Formation of the cortex, coat, and exosporium occurs in stages IV to VI. Mature endospores are formed in stage VII. The dichotomously branched vegetative mycelium is assigned as stage 0.

The endospore-to-endospore cycle was found to depend on soil temperature and species of *Pasteuria*. Temperature determines not only the duration of the cycle but also the number of endospores produced per female root-knot nematode/cyst forming nematode. A temperature of 25°C was found favourable for development of 2.2×10^6 endospores per root-knot female in 38 days. At 20°C, the duration of the nematode life cycle increased from 85 to 100 days with 2.4×10^6 endospores per female. At 30°C, the bacterium rapidly colonised the female, resulting in a lower number of endospores, 1.4×10^6 endospores per female. In *H. cajani*, it takes 23-28 days at 27-38°C.

OCCURRENCE OF STRAINS

A number of strains are reported to exist in *P. penetrans*. Dutky (1978) observed that an isolate of the parasite infecting *Meloidogyne incognita* did not encumber *Pratylenchus brachyurus*. At the same time, two isolates that adhered to *P. brachyurus* from Florida and Maryland (USA) could not adhere to *M. incognita*. Singh and Dhawan (1990) demonstrated the occurrence of two strains through cross inoculation tests, i.e., one infecting *M. incognita* but not *H. cajani* and the second infecting *H. cajani* but not *M. incognita* and the strains are host specific.

PATHOGENESIS AND SURVIVAL

No reports in the literature involve the application of Koch's postulates to any bacterial disease of plant-parasitic nematodes. Consequently, pathogenicity in the strict sense has not been firmly established for a species of *Pasteuria* that causes a disease in a nematode species because of obligate nature of the bacterium. Cultivation of *P. penetrans* in *M. incognita* females on oligoxenic excised tomato root culture has been reported in at least two instances (Verdejo *et al.*, 1988;

Verdejo and Jaffee, 1988). This three-member cultivation may bring the bacterium a bit closer to axenic culture but not enough for use in establishing Koch's postulates. Several attempts have been made to cultivate species of *Pasteuria*, but success has always eluded the researchers.

Little is known about the long-term survival of endospores of *Pasteuria* species. In Florida, *P. penetrans* maintained suppressive levels in a peanut field for *Meloidogyne arenaria* over 10 years (Dickson *et al.*, 1991). In the laboratory, endospores of *P. penetrans* remained viable for more than 1 year at 10°C to 36°C (Mani, 1988). Kamra and Dhawan (1995) have also reported greater viability of spores of *P. penetrans* after three months of storage at 15, 30°C, than at 50°C. It has been demonstrated that endospores can survive high temperature and desiccation. Endospores have been found to remain viable up to 3 years after storing them in dry soil kept in the laboratory.

BIOCONTROL POTENTIAL

Bacterial antagonists consist of rhizobacteria, endophytic bacteia, rhizobia and bacterial symbionts EPN. Rhizosphere contains many microorganisms than the bulk soil. Rhizospheric bacteria are the most plentiful microorganisms in the root zone and can modify this and affect the nematode or the host-parasite interrelationship (Neipp and Becker, 1999). Except the antagonistic fungi and actinomyces, rhizobacteria are the only group in which biological nematicides have been reported. Reduction in root galls, nematode multiplication, egg hatching or juvenile survival, killing of nematodes, toxin production are some of the mechanisms to suppress the nematode. *Pasteuria* spp., have great potential as biological control agents. Field observations (Bird and Brisbane, 1988; Dickson *et al.*, 1991; Dickson *et al.*, 1994; Minton and Sayre, 1989), and glasshouse and microplot experiments (Brown and Smart, 1985; Channer and Gowen, 1988; Chen *et al.*, 1996; Davies *et al.*, 1988; Oostendorp *et al.*, 1991) demonstrate that these organisms can provide effective and significant control of plant-parasitic nematodes.

Some rhizobacteria isolated from a field site suppressive to peach tree short life were suppressive to population increases of the ring nematode, *Criconemella xenoplax* (Kluepfel *et al.*1993). In a separate study, some endophytic bacterial strains from cotton significantly reduced galling of cotton roots by the root-knot nematode, *Meloidogyne incognita* (Hallmann *et al.* 1997). They further reported that nematode populations were correlated with establishment of endophytic bacteria within roots, suggesting that root-penetrating nematodes provide increased entry sites for endophytic bacteria. Endophytes are an interesting group of plant-associated bacteria that live inside plants and show neutral or beneficial interaction with the host plants. The structure of the bacterial community in the endosphere and endorrhiza of field-grown potato was analysed in comparison to the rhizosphere and phyllosphere by a multiphasic approach over a period of two years in Germany. Results of the cultivation-independent approach revealed that T-RFLP profiles of the two endophytic microenvironments were more closely related than those from ectophytic habitats. Using cultivation, composition, diversity and richness of bacterial antagonists confirmed specificity for each microenvironment and an outstanding and promising role of the endorrhiza. In an approach to measure the biological control potential of isolates, a total of 2648 bacteria were evaluated for biological control and plant growth promotion by a hierarchical combination of assays using the soil-borne plant pathogens *Verticillium dahliae* and *Rhizoctonia solani* as target. An average of 14.4% of examined bacteria expressed antifungal properties. The strains were characterized by their antagonistic mechanisms *in vitro* as well as their production of the plant growth hormone indole-3-acetic acid. Complementary, the plant growth promoting effect by antagonistic bacteria was determined using a newly developed microplate assay on the basis of lettuce seedlings. Seven endophytic isolates selected according to *in vitro* criteria were evaluated in greenhouse and field trials regarding their efficiency to control *R. solani* in lettuce, sugarbeet, and potato. In addition, they were screened for their biological control activity against the pathogenic nematode *Meloidogyne incognita*. The most promising candidates *Pseudomonas fluorescens* 2Re2-6 and *Serratia plymuthica* 3Re4-18 will be commercialized as biological control agents Berg *et al.* (2004). There is only one genus *Pasteuria* reported parasitic on plant parasitic nematodes. Two Gram-positive bacterial species, *Pasteuria peneterans* and *P. nishizawae* have been reported parasitic on root-knot and cyst forming nematodes, respectively. Both the species are obligate parasites that produce spores that attach to the cuticle of nematodes. Of the two species, *P. penetrans* has been the subject of intensive studies because of its wide occurrence.

Although the term endophyte is most commonly associated with fungi, there is sufficient literature pertaining to bacteria as endophytes. *Kocuria varians* and *Microbacterium esteraromaticum* (Sturz and Kimpinski 2004) have been found antagonistic towards root-lesion nematode (*Pratylenchus penetrans*) through the attenuation of nematode proliferation. Endophytic bacteria exist in a variety of tissue types within numerous plant species, suggesting a ubiquitous existence in most if not all higher plants (Kobayashi and Palumbo 2000). Rhizobia are soil bacteria that display symbiotic interactions with specific legume hosts and nodulate plant roots. Apart from their nitrogen-fixing capabilities, some rhizobia also benefit economically important crops by protecting them from various soil-borne plant pathogens including root-knot nematodes by suppression of juvenile survival (Siddiqui and Shaukat, 2002). Sharma and Sharma (1989) reported bacterial infection in cysts of *Heterodera cajani, H.graminis* and *H.zeae*. They further observed 100% reduction in egg contents in infected cysts and reduction ranging from 50 to 72 % in penetration of different *Heterodera* species. Bhattacharya (1987) observed maximum reduction in number of cysts of *H.avenae* and improvement in plant growth characters when 1 ml of bacterial suspension having 4.7×10^7 spores per ml was added. Davies *et al.* (1988) reported that the invasion of tomato roots by infected juveniles of *Meloidogyne* sp. was reduced by 86% as compared to healthy ones. Two genera of symbiotic bacteria, *Photorhabdus* (Hu *et al.* 1999) and *Xenorhabdus* (Grewal *et al.* 1999), have been reported antagonistic towards plant-parasitic nematodes. Among all the reported bacteria, *Pasteuria penetrans* (formerly *Bacillus penetrans*) is the most studied one. *P. penetrans* exerts various degrees of nematode biocontrol under greenhouse and field conditions.

Pasteuria endospores are the infective agent applied to crops for nematode control. They are very desirable as nematode control products due to their inherent resistance to heat, drying and mechanical shearing. The major impediment to their use as a commercial nematode control product has always been the inability of researchers to grow them in the absence of a nematode host, thus precluding the use of large-scale, low-cost production methods (Chen and Dickson, 1998; Bishop and Ellar, 1991). Interest in biological control of nematodes using *P. penetrans* gained momentum only after 1980, when Stirling and Wachtel discovered a technique for mass production of the bacterium. They found that dried root powder containing spores of *P. penetrans*, when incorporated in soil @ 100 mg/1 kg soil could result in 99 per cent infection of second-stage juveniles of *Meloidogyne javanica* within 24 hours. Subsequently, several researchers established the potentiality of *P. penetrans* as a biological control of nematodes (Table 36.1). Although *P. penetrans* has been observed in the nematode suppressive soils (Bird & Brisbane, 1988; Minton & Sayre, 1989), the population dynamics of this antagonist has not been clearly understood. It has been associated with high as well as low nematode densities. Oostendorp *et al*., (1991) found that the population density of *P. penetrans* increased over time from the relatively low levels of spores, added initially in a microplot experiment where summer and winter cover crops were grown in infested fields of *Meloidogyne arenaria*. Similar results of increase in spore concentration was obtained by Singh & Dhawan (1996) when low levels of spores were added initially in pots where cowpea was grown successively thrice infested with *Heterodera cajani*.

Table 36.1. *Pasteuria* species as a potential biocontrol agent of root-knot and cyst forming nematodes.

***Pasteuria* species**	**Target nematode**	**Effect**	**Reference**
P.penetrans	*Meloidogyne* species	Larval mobility affected	Mankau and Prasad, 1977 Sayre, 1984; Sell and Hansen, 1987; Stirling, 1984; Giannakou *et al*. 2007; Sankaranarayan *et al*. 2005
		Reduction in nematode Penetration	Brown and Smart, 1985; Sekhar and Gill, 1990;
		Egg inhibition	Sayre and Starr, 1985; Shekar and Gill, 1990; Carneiro *et al*. 2007; Das *et al*, 2007;
		Increased crop yield	Brown *et al*. 1985; Javed *et al*, 2008;
		Decreased galls and Egg masses	Mukhtar *et al*. 2005
		Healthy nursery	Kumar *et al*. 2005
	Heterodera cajani	Reduced penetration	Singh and Dhawan,1992, 1994; Srivastava and Dhawan, 2001
		Reduction in J2 and cyst population in soil and improve plant growth	Singh and Dhawan, 1994, 1996, 1998;
P.nishizawae	*H.glycines*, *H.Elachista*, *H.trifolia* *Globodera Rostochiensis*		Sayre *et al*. 1991

A significant less root-galling in the *P. penetrans* combined with neem cake treatment at the end of the third crop and this treatment also had the greatest effect on the growth of the tomato plants. At the end of the third crop, 30% of the females were infected with *P. penetrans* in those treatments where spores had been applied at the start of the experiment. The effects of neem leaves and neem cake on the nematode population did not persist through the crop sequences, but the potential for combining the amendments with a biological control agent such as *P. penetrans* is worthy of further evaluation (Javed *et al*., 2007). Mukhtar *et al*. (2005) also reported significant reductions in the number of galls (24.68%), egg masses (39.92%) and nematode density (29.41%) in the micro plots where *P. penetrans* was applied. As a result, 22.57% increase in yield was recorded in *P. penetrans*-treated plots compared to the untreated plots. After the first crop, 18.5% nematodes extracted from soil were encumbered with the endospores of *P. penetrans* which reached to 50.5% at the harvest of the final crop. *P. penetrans* infection was recorded on 7.5% females after the first crop which increased to an infection level of 23.75% females in the *P. penetrans*-amended micro plots after the final crop. Kumar *et al.* (2005) observed that the application of *P. penetrans* at 1×10^4 spores/g nursery soil was as effective as the application of carbofuran at 2 kg a.i./ha. At 1×10^5 spores, the yields were comparable to healthy nursery plants.

Encumbered juveniles (5-10 spores/J2) of *M. incognita, M. graminicola* and *M. javanica* caused 41.76% reduction in invasion of tomato roots, 24.58 and 33.13% reductions of barnyard grass and sorghum roots and 11.20 and 19.55% reductions of tomato and aubergine roots, respectively. The reductions in invasions caused by the juveniles encumbered with >11 spores/ J2 were: 46.47% in tomato by *M. incognita*; 25.54 and 38.40% in barnyard grass and sorghum by *M. graminicola* and 23.87 and 43.73% in tomato and aubergine by *M. javanica*, respectively Ahmad and Mukhtar (2007). Giannakou *et al.* (2007) reported that the combination of soil solarization with the bionematicide improved nematode control gave results similar to chemical nematicide. The effectiveness of biological control (Pf 10 isolate of *P. penetrans*) was evaluated by reduction of egg numbers/root system ranging from 60% (0.5 g in clay sandy soil), 70% (5.0 g in clay soil) and 80% (0.5 g and 5.0 g in sandy soil) by Carneiro *et al.* (2007). Das *et al* (2007) reported that spore load of *P.penetrans* as well duration played important role in nematode development.

Although several types of fermentation processes are commonly used for commercial production processes, submerged (liquid culture) fermentation is commonly used for the production of biopesticides such as *Bacillus thuringiensis*, and is well-understood as an economical production method for a huge variety of industrial products (Bailey and Ollis, 1986; Hsu and Wu, 2002). Bionem is the commercial preparation of *Bacillus firmus* from Israel used against Root knot and other nematodes including *Heterodera avenae.*

BACTERIAL METABOLITES TOXIC TO ROOT-KNOT AND CYST FORMING NEMATODES

This is another interesting area of research that has stimulated researchers worldwide to exploit the saprophytic rhizospheric bacteria for the control of nematodes. A great deal of knowledge has been generated on the compounds produced by the rhizobacteria that are toxic to nematodes in general and root-knot and cyst forming nematodes in particular.

Several workers have demonstrated that a subset of rhizobacteria, termed as plant growth promoting rhizobacteria (PGPR), exert beneficial effects on plant development through growth promotion and/or biological control, and may help reducing plant damage caused by phytoparasitic nematodes. The rhizobacteria which have been so far investigated for nematode biocontrol include *Agrobacterium, Acetobacter, Alcaligenes, Azotobacter, Clostridium, Desulfovibrio, Bacillus, Psedomonas, Serratia and Streptomyces.* Of the various rhizosphere bacteria reported, mostly species of *Bacillus* and *Pseudomonas* have been investigated. Various formulations of *Bacillus thuringiensis*, normally used against insect pests, have also been found toxic to eggs and larvae of *Meloidogyne javanica* in *in vitro* conditions (Prasad *et al.*, 1972). Ignoffo & Dropkin (1977) reported toxic effect of exotoxin produced by *B. thuringiensis* to the populations of *Meloidogyne*, *Panagrellus* and *Aphelenchus*. Two strains of *B. thuringiensis* sub sp. *kurstaki*

(Dipel and SAN 415) were found to supress populations of *Meloidogyne javanica* and *Tylenchulus semipenetrans*.

Among the various rhizobacteria that have been so far, 7-10% have shown antagonistic potential against root-knot and cyst forming nematodes (Sikora, 1992). These have been divided into two groups: i) plant growth promoting rhizobacteria (PGPR) and ii) plant health promoting rhizobacteria (PHPR). Successful attempts have been made for the control of root-knot and cyst forming nematodes using PHPR with seed and planting material. A decrease in root galling and an increase in shoot weight of at least 10% was assessed for 331 and 23% of bacterial isolates from tomato and cucumber, respectively (Zavaleta and Van Gundy, 1987). Similarly, a reduction in penetration by *Heterodera schachtii* and a significant increase in sugarbeet's yield as seed treatment with selected strains of *Pseudomonas fluorescens* (Oostendorp and Sikora, 1992). Racke and Sikora (1992) reported reduction in penetration in roots of potato by *Globodera pallida* when *Agrobacterium radiobacter* and *Bacillus sphaericus* were sprayed independently, while combined application had no effect on penetration in the greenhouse tests. In contrast, in field, the combined application of the two bacteria caused a significant reduction in root penetration. A list of bacteria having detrimental effect on root-knot and cyst forming nematodes is presented in Table 36.2.

Table 36.2: Bacteria producing metabolites toxic to root-knot and cyst forming nematodes.

Bacterium	Nematode affected	Effect	Reference
Agrobacterium radiobacter	*Globodera pallida*	Reduced nematode infection with increase in yield	Racke and Sikora, 1992
Azotobacter chrococcum	*Meloidogyne incognita*	Inhibited hatching Decreased nematode multiplication	Chahal and Chahal, 1986
Bacillus liciniformis	*M. incognita*	Reduced nematode penetration & improved crop growth	Siddiqui and Hussain, 1991
B. subtilis	*M. incognita*	Toxic to eggs and juveniles	Siddiqui and Mahmood, 1993
B. firmus	*Mi*	Suppressed nematode population, J2 motality	Giannakou *et al.* 2007
B. thuringiensis (Dipel)	*M. javanica*	Suppressed nematode population	Prasad *et al.*, 1972
B. thuringiensis sub sp. *kurstaki* (SAN 415)	*Meloidogyne* sp.	Reduction in galling and nematode multiplication	Osman *et al.*, 1988
B. thuringiensis (CR-371)	*Meloidogyne* sp.	Suppressed nematode growth and reproduction	Zukerman, 1993
Bradyrhizobium japonicum	*M. incognita*	Reduction in nematode penetration and increase in yield	Carappo and Pelagatti, 1988
Pseudomonas fluorescens	*Heterodera schachtii*	Reduction in nematode population	Oostendorp and Sikora, 1989
Pseudomonas fluorescens *P. mindocina*	*M. incognita*		Siddiqui and Hussain, 1991

BACTERIAL METABOLITES ANTAGONISTIC TO NEMATODES

Some soil bacteria produce nematicidal compounds (Sayre and Starr, 1988), and anthelmintics based on the toxin produced by the actinomycete *Streptomyces avermitilis* have been commercialized. Several researchers have screened large number of rhizosphere bacteria and reported that some bacteria inhibit egg hatching and mortality of *Rotylenchulus reniformis* (Niknam and Dhawan, 2002a, b and c), while others reduce root-knot galling and populations of semi-endoparasitic and endoparasitic nematodes (Becker *et.al.*, 1988; Zuckerman *et al.*, 1993; Oostendrop and Sikora, 1992; Spiegel *et al.*, 1991; Niknam and Dhawan, 2003a, b and c etc).

Fluorescent pseudomonads have been reported to produce iron chelating siderophores (Kloepper *et.al.*, 1980; Leong, 1986), antibiotics and hydrogen cyanide (Ahi *et al.*, 1986), and these compounds were implicated in reducing pathogenic rhizosphere microorganisms, and promoting environment for root growth (Kloepper *et.al.*, 1980; Leong, 1986). The increase in plant growth might be associated with secretion of auxins, gibberellins and cytokinins (Ramamoorthy *et al.*, 2001).

An alternative mechanism for biological control by PGPR is induced systemic resistance. This occurs through the action of an inducing agent, which systemically activates the host plant's chemical or physical defense mechanisms, leading to reduced symptoms to pathogens which infect tissues distal to the inducer, a process referred to as 'Induced Systemic Resistance' (ISR) or 'Systemic acquired resistance' (Kloepper *et al.*, 1999). For the first time, Hasky-Gunther *et al.* (1998) found that *Agrobacterium radiobacter* and *Bacillus sphaericus* caused induced resistance mechanism of action against a potato cyst nematode, *Globodera pallida.* Subsequently several researchers have demonstrated induced systemic resistance resulting in reduced penetration and multiplication of nematodes (Hallmann *et al.*, 1998; Hasky-Gunther *et al.*, 1998; Niknam and Dhawan, 2002c, 2003).

Numerous biotic factors such as plant genotypes, microbial competition and rhizobacterial communities, and abiotic factors such as soil temperature, moisture, pH and clay content may influence the efficacy, survival and establishment of the bacteria and their interaction with the pathogen. Inoculum delivery methods also greatly affect the distribution and population sizes of PGPR on roots (Schroth and Becker, 1990). Population sizes of PGPR are greater on roots when inoculum is incorporated into soil as granules than when applied as a seed treatment (Bahme *et al.*, 1982).

Besides, various formulations of *Bacillus thuringiensis*, normally used against insects, have been reported toxic to eggs and larvae of *Meloidogyne* species (Prasad *et al.*, 1972; Osman *et al.*, 1988; Zuckerman *et al.*, 1993; Chen *et al.*, 2000; Dhawan *et al.*, 2004), *Heterodera glycines* (Sharma and Gomes, 1996 a, b). Marroquin *et al.* (2000) reported that the action of Bt toxins on *Caenorhabditis elegans* is similar to toxin effects in insects.

Isolation of Rhizobacteria (*Bacillus* and *Pseudomonas*)

Roots of host plants collected are first shaken gently to remove loose soil particles and then these are cut into small segments. One g of the root segments is agitated in 10 ml of 0.1M $MgSO_4$ for 15 min to separate bacteria from the roots. The suspension obtained is processed for isolation of *P. fluorescens* and *B. subtilis*. Three different media, namely, King's medium B,

Nutrient Agar, and Fluorescent Pseudomonad Agar are used for isolation of *P. fluorescens*. Nutrient Agar, Tryptic Soy Agar and YGCA media are used for isolation of *B. subtilis*.

The bacterial suspension obtained is aseptically streaked on the media contained in Petri plates. These are then incubated for 48 h at 28°C. For isolation of spore forming bacterium, a part of the bacterial suspension is heated at 90°C for 10 min. The heated suspension is streaked on Tryptic Soy Agar medium adjusted to a pH of 7.2, incubated at 28°C for 48 h. All morphologically distinct colonies from each Petri plate are picked up and aseptically streaked onto a new medium. This process is repeated at least thrice to get a pure culture. Pure cultures are subsequently identified for *Bacillus* and *Pseudomonas* characters. These cultures are kept in test tube slants at 4-5°C in refrigerator and subculturing is done at monthly interval.

Pasteuria spp.

Unexpected decline of root-knot and cyst forming nematode field populations which were previously above damaging level have been found to have niche of *P. penetrans* and by knowing the previous history of nematode damage (Bird and Brisbane, 1988; Dickson *et al.* 1994; Minton and sayre, 1989; Nishizava, 1987 and Weibelzahl-Fulton *et al.*, 1996). According to Mankau (1980) it can be detected in fields that had ideal conditions for the development of severe disease from root-knot nematodes but persistent to generate quality yields even though planted with susceptible crops. The first indication of presence of *Pasteuria* spp. on nematodes extracted from soil in samples is the clumping of nematodes (Mankau and Prasad, 1977). The vegetative stages of Pasteuria spp and females can be stained dark blue by using lactophenol with 1% cotton blue (v/w) as processed by Serracin *et al.*, 1997. mature endospores can be stained by applying Brilliant Blue satin @ 1 mg/ml solution in distilled water (Bird, 1988).

Preparation of Bacterial Cells Inocula

Bacterial colonies raised on nutrient agar are streaked on a series of Petri plates containing the same medium and incubated at 28°C. After 48 h, the bacterial growth is scrapped from the agar surface with a sterile needle and suspended in sterile 0.1 M $MgSO_4$. The number of cells per ml is counted using dilution plating method and desired dilutions (10^{-4}, 10^{-5}, 10^{-6}, 10^{-7}, 10^{-8}, 10^{-9} and 10^{-10}) of inocula are prepared for experimental use.

The counting of cells is done by taking 0.1 ml aliquot from each of 10^{-4}, 10^{-5}, 10^{-6}, 10^{-7}, 10^{-8}, 10^{-9} and 10^{-10} dilutions. It is spread over the media contained in Petri plates with the help of a sterilized glass spreader and incubated at 28°C for 48 h. Three plates are maintained for each dilution. The bacterial colonies grown on the media are counted and the number of colony forming units (cfu) per ml of the suspension is calculated to divide average number of counted colony forming units by the product of amount plated and dilution factor.

PREPARATION OF CELL-FREE FILTRATE

A single colony of the bacteria is cultured in flask containing nutrient broth incubated at 28°C on a shaker at 100 rpm for 2-3 days. The number of cells per ml of the liquid culture is determined by dilution plating method as described above. The culture is subsequently passed through sterilized Whatman filter paper Nos.1 and 42, respectively. This is then concentrated by centrifugation at 6000 g for 10 min to precipitate the bacterial cells. The supernatant is collected aseptically and finally passed through Millipore filter of 0.22 μm (Sartorius, Minisart) to get cell free filtrate of the desired bacteria. This is designated as standard filtrate (S).

In vitro Study

For *in vitro* study, a replicated trial is conducted in which two ml each of 10^4 to 10^{10} cells/ml of the bacteria is placed into separate wells of a 24 multiwell plate (Iwaki Glass, Japan). Wells containing only 2 ml of sterile 0.1 M $MgSO_4$ served as control. Five egg masses of root-knot nematodes or 5 cysts in case of cyst forming nematodes are handpicked and placed in each well. Keep at least 4 replications for each dilution and control (water/media). Three such sets of multiwell plates for each treatment are kept at 28°C. After 24,. 48 and 72 h exposure, one set of multiwell plates is removed and observations on number of juveniles hatched are counted. The number of unhatched eggs is also counted after dissolving the gelatinous matrix of egg masses in 0.5% NaOCl solution for 10-15 min or breaking the cyst wall. The hatching percentage is calculated to dividing number of hatched juveniles by (Number of hatched juveniles + Number of unhatched eggs) multiplied by 100.

Prepare 10^4 to 10^{10} dilutions of the bacterial cells. Pour 1 ml of each (10^4 to 10^{10}) dilution into a separate well of the 24-well plate containing 50 second-stage juveniles of the root-knot and cyst forming nematodes in 1 ml sterile distilled water. Wells containing equal number of juveniles in 2 ml sterile distilled water served as control. Keep at least 4 replications for each dilution and control. Three such sets are kept at 28°C. The number of inactive (dead) and active (alive) juveniles was recorded after 24, 48 and 72 h of interval and calculate the per cent mortality to divide number of inactive juveniles by (number of inactive juveniles + Number of active juveniles) and multiplied by 100.

The inactive juveniles are transferred separately from each dilution into sterile distilled water and kept overnight to check whether mortality caused is permanent or temporary. Repeat the experiment twice.

Two ml each of undiluted filtrate (S) and their dilutions to S/4, S/8, S/16 and S/20 are placed into each well of the 24-well plate. Besides, liquid medium (in which bacterial culture is raised) and sterile distilled water is used as two controls. Five handpicked egg masses of root-knot/5 cysts of cysts forming nematodes, wash them three times in sterile distilled water, place them in each well. Keep at least four replications for each dilution and controls (broth medium and water). Three similar sets for each treatment are kept at 28°C. After 24, 48 and 72 h exposure, each set is removed and observations are recorded as mentioned above.

Take one ml of each of double strength of S/4, S/8, S/16 and S/20 as prepared above and pour it separately into each well of the 24-well plates containing 50 second-stage juveniles in 1 ml of root-knot/cyst forming nematode suspension. For S dilution 2 ml is taken into each well and 50 handpicked second-stage juveniles are added to it. Besides, equal number of juveniles in the liquid medium (in which bacterial culture was raised) and sterile distilled water are kept as two controls. Keep at least four replications for each dilution and controls. Three such sets are kept for each dilution. After 24, 48 and 72 h exposure, each set is removed and observations on the number of inactive (dead), active (alive) juveniles and their percentage of mortality is calculated as described above. The inactive juveniles are then simultaneously transferred from each dilution into sterile distilled water for overnight to check whether mortality caused is permanent or temporary.

In vivo

Seeds of test plant are surface-sterilized with 2.4% sodium hypochlorite (NaOCl) solution for 2-3 min, followed by rinsing in sterile distilled water and dried for 2 h under a sterile air stream (laminar flow). Bacterial inocula to be used for seed bacterization is prepared by inoculating plates of TSA (Tryptic Soy Agar) and King's B media with a bacterial suspension of *B.subtilis* and *P. fluorescens*, respectively. These are incubated at 28°C for 2-3 h to allow bacterial growth to begin. Plates are then injected at the centre with 0.2 ml of sterilized (with disposable millipore filters of 0.22 μm) streptomycin sulphate (CDH ®, India) solution containing 100 μg of active ingredient per ml to get antibiotic resistant mutant. These are again incubated at 28°C for 48 h. The stability of the antibiotic resistance trait is confirmed for both the bacteria by streaking thrice on the two media (without antibiotic) mentioned above. Single colony of the bacteria is taken and streaked on the media containing 100 ppm streptomycin sulphate and incubated at 28°C for 48 h to compare the growth of mutants with that of the wild types (original cultures).

Seed Treatment

The bacterial growth of the mutants is scrapped off from the agar surface with a sterile needle and mixed with 10 ml of a 0.1% (w/v) sterilized aqueous methyl cellulose solution in sterile distilled water as an adhesive. The surface sterilized seeds are exposed to the suspension of respective bacterium (10^{10} cells/ml) for 30 min and then dried under a laminar flow for at least 3 h. Surface sterilized seeds treated with sterile methyl cellulose served as non-bacterized controls.

Bacterial colony forming units (cfu) per seed are determined by washing 10 seeds after dipping them in sterile distilled water using dilution plating method.

Bacterized and non-bacterized seeds are sown separately in sterilized soil-sand mixture contained in earthen pots. Data on the percentage of germination is recorded daily up to one week. If nursery is raised, then data can be collected on total fresh weights of seedlings, lengths of shoots and roots of both bacterized and non-bacterized seedlings at the time of transplanting.

Root Colonization

Seedlings raised from bacterized seeds with streptomycin-resistant mutants are carefully removed after 7, 14 and 21 days of sowing. The bacterial populations on the roots are ascertained by cutting the roots into 1cm segments. One g of root segments is shaken vigorously in 10 ml sterile solution of 0.1 M $MgSO_4$ for 10 min to disperse the bacteria into the solution. The required dilutions of *B. subtilis* and *P. fluorescens* suspension are, respectively, streaked on Tryptic Soy Agar and King's B media containing 100 μg ml^{-1} streptomycin sulphate. These are incubated at 28°C, and after 48 h bacterial populations of *B. subtilis* and *P. fluorescens* are determined by counting cfu using dilution plating technique.

Besides, the population of the mutants per cm length of the roots of 10 plants/seedlings raised from bacterized seeds with streptomycin-resistant mutants selected randomly is also estimated after 10 days of direct sown crop and 21 days in transplanting crop. Root segments of 1.0 cm length were cut from four regions (immediately near the seed, the main root, the primary root tip and a side root) of each plant separately and weighed. Root lengths from each region were mixed separately and examined by shaking them vigorously in 10 ml of sterile solution of 0.1 M $MgSO_4$ for 10 min to disperse the bacteria into the $MgSO_4$ solution for estimation of cfu per cm using dilution plating method as mentioned above.

Bare Root Dip

At transplanting time, roots of three-weeks-old seedlings, raised from bacterized seeds with streptomycin-resistant mutants are immersed separately for 2 h each in suspension of *P. fluorescens* and *B. subtilis* (10^{10} cells/ml). Treated seedlings are immediately transplanted carefully into 15 cm earthen pots containing 1 kg steam-sterilized soil-sand mixture. Roots of seedlings dipped only in sterile distilled water served as control.

Soils Drench (Soil application)

Twenty-five ml of bacterial suspension (10^{10} cells/ml) is poured into the soil around each plant at transplanting time as soil drench followed by covering with removed soil. Bacterial suspension of *P. fluorescens* and *B. subtilis,* before using them, is prepared by scrapping the bacterial colonies from the media and thoroughly mixed in sterile distilled water. The cfu are determined using dilution plating technique as per method described above. Plants receiving 25 ml of sterile distilled water served as control.

Compatibility of *P. penetrans* with Nematicides and other Biological Control Agents

Most studies have shown compatibility of the bacterium with large number of standard nematicides (Maheshwari *et al*., 1987; Singh and Dhawan, 1994; Mahapatra and Mohanty (2002); Jaykumar *et al.* (2003); Haseeb *et al.* 2006) with exception to DBCP (Mankau and Prasad, 1977); Methyl Bromide (MBr) + Chloropicrin. The bacterial parasite is also found compatible with other biological control agents and this was demonstrated by Dube and Smart (1987); Maheshwari *et al.* (1987); Rao *et al.* (2004); Haseeb and Kumar (2006); Rao (2007); Chen *et al.* (2000) and Haseeb *et al.* (2007) who reported effective control of root-knot nematode, *Meloidogyne.* Subramanium *et al.* (2002) reported management of pigeonpea cyst nematode, *Heterodera cajani* in pulses by biocontrol agents.

The requirement of endospore counts needed to control plant parasitic nematode is reliant on several factors e.g., susceptibility of host nematode to *Pasteuria* spp./isolate, distance and time to reach the target, nematode population and soil texture and temperature. The endospore densities between 10^4 to 10^5 endospores/g of soil are needed to reduce damage caused by root-knot significantly (Chen *et al*., 1995; Stirling, 1988; Stirling *et al*., 1990).

Current Status and Prospects of Rhizobacteria (*Pasteuria penetrans*, pseudomonads, etc.) in Nematode Management

The term rhizobacteria, PGPR (plant growth-promoting rhizobacteria) comprise those bacteria that induce beneficial effects on plants during colonization. Benefits of PGPR include promotion of plant growth and biological control of plant diseases caused by fungi, bacteria and plant parasitic nematodes. Besides increasing yield, specific PGPR often promote early season growth promotion that can be reflected in various forms such as enhanced seedling emergence, increased biomass of root and/or foliage, etc. Some PGPR are responsible for biocontrol by producing metabolites or by triggering induced systemic resistance in the plant. There are numerous bacteria found in the rhizosphere that function as PGPR, but the most widely studied are species of *Pseudomonas*, *Bacillus* and a parasitic bacterium, *Pasteuria*.

CONCLUSION

The soil houses numerous bacteria in the rhizosphere of plants, which on interaction produce metabolites that are toxic to nematodes, in general, and root-knot and cyst forming nematodes, in particular. Scanty information is available on the identification of toxic metabolites, which need immediate attention so that these can be used against the nematodes. Also, more and more investigations are required to explore the most potential strains of rhizobacteria that can be exploited in the management of nematodes. Besides the role of parasitic bacterium, *Pasteuria* species as a potential biocontrol agent of root-knot and cyst forming nematodes has been well documented, but the only difficulty involved in the culturing of the bacterium is that it is an obligate parasite. This can be overcome by understanding the genomic constituent of the bacterium and finally transferring it into culturable bacterium or producing transgenic plants.

The available literature on the use of *Pasteuria penetrans*, a bacterial parasite, is undoubtedly the most potential biological control agent in suppressing the major plant parasitic nematodes. Since the bacterium is very resistant to various adverse environmental conditions in soil, such as high and low temperature, chemicals, and humidity changes, *P. penetrans* has a great scope to substitute chemical nematicides. Unfortunately, there is no report on the cultivation of *P. penetrans* on synthetic medium, although attempts are going on in different laboratories for its mass multiplication. However, before exploiting this bacterium, some of the key issues like methods and strategies for introducing the bacterium and monitoring its population in the soil and in an integration crop approach need to be investigated under different ecological conditions. Attempts should also be made to explore strains of the bacterium that are efficient and has wide host range to make its availability throughout the year on different phyto-parasitic nematodes to keep the nematode populations below threshold levels.

REFERENCES

Ahi, P., Voisarad, C. and Defago, G. (1986). Iron bound siderophores, cyanic acid and antibiotics involvement in suppression of *Thielaviopsis basicola* by *Pseudomonas fluorescens* strain. *J. Phytopathology*, **116:** 121-134.

Ahmad, R. and Gowen, S.R. (1991). Studies on the Infection of *Meloidogyne* spp. with isolates of *Pasteuria penetrans*. Nematologia mediterranea, **19:** 233.

Ahmad, R. and Mukhtar, T. (2007). Root invasion of different plant hosts by juveniles of *Meloidogyne* species encumbered with *Pasteuria penetrans* spores. *Pakistan J. Nematology*. 2007; **25**(1): 199-206.

Bailey, J. E. and D. F. Ollis. (1986). Biochemical Engineering Fundamentals. McGraw-Hills Publishing Co. New York.

Becker, J.O., Zavaleta-Mejia, E., Colbert, S.F., Scroth, M.N., Weinhold. A.R., Hencock, J.G. and Gundy, S.D. Van. (1988). Effects of rhizobacteria on root-knot nematodes and gall formation. *Phytopathology*, **78**(11): 1466-1469.

Berg, G., Krechel, A., Lottmann, J., Faltin, F., Ulrich, A., Hallmann, J. and Grosch, R. (2004). Endophytes: a new source for multi-target biological control agents? Bulletin OILB/SROP. 2004; **27**(8): 401-405.

Bhattacharya, D. (1987). Pathogens of cyst nematodes of the genus *Heterodera* and their effect on nematode biology and control. Ph.D Thesis, PG School, IARI, New Delhi-12, pp. 92.

Bird, A.F. (1988). A technique for staining the endospores of *Pasteuria penetrans.* Revue de Nematologie, **11:** 3.

Bird, A.F. and Brisbane, P.G. (1988). The influence of *Pasteuria penetrans* in field soils on the reproduction of root-knot nematodes. *Revue de Nematologie*, **11:** 75-81.

Bishop, A.H. and Ellar, D.J. (1991). Attempts to culture *Pasteuria penetrans in vitro. Biocon. Sci. Tech.*, **1:** 101-114.

Brown, S.M. and Smart, G.C. Jr. (1985). Root penetration by *Meloidogyne incognita* J2 infected with *Bacillus penetrans. J. Nematology*, **17:** 123-126.

Brown, S.M., Kepner, J.L. and Smart, G.C. Jr. (1985). Increased Crop Yields Following Application of *Bacillus Penetrans* to Field Plots Infested With *Meloidogyne Incognita. Soil Biol. Biochem.* **17:** 483-486.

Carneiro, R.M.D.G., Mesquita, L.F.G., Cirotto, P.A.S., Mota, F.C. and Almeida, M.R.A. (2007). Biological control of *Melodogyne incognita* on coffee plants using the isolates P10 of *Pasteuria penetrans.* 21st International Conference on Coffee Science, Montpellier, France, 11-15, September, 2006. pp. 1409-1414.

Channer, A.G. and Gowen, S.R. (1988). Preliminary studies on the potential of *Pasteuria penetrans* to control *Meloidogyne* species. Proceedings of Brighton Crop Protection Conference, Pests and Diseases. Surrey, England: The British Crop Protection Council.

Chen, J., Abawi, G.S. and Zuckerman, B.M. (2000). Efficacy of *Bacillus thuringiensis, Paecilomyces marquandii,* and *Strepromyces costaricanus* with and without organic amendments against *Meloidogyne hapla* infecting lettuce. *J. Nematology,* **32:** 70-77.

Chen, Z.X. and Dickson, D.W. (1998). Review of *Pasteuria penetrans*: Biology, ecology, and biological control potential. *J. Nematology,* **30:** 313-340.

Chen, Z.X., Dickson, D.W., Mitchell, D.J. and McSorley, R. (1995). Biological control of *Meloidogyne arenaria* using *Pasteuria penetrans* in Florida. *J. Nematology*, 27, 496 (Abstr.).

Chen, Z.X., Dickson, D.W., McSorley, R., Mitchell, D.J. and Hewlett, T.E. (1996). Suppression of *Meloidogyne arenaria* race 1 by soil application of endospores of *Pasteuria penetrans. J. Nematology*, **28:** 159-168.

Das, M., Gogoi, B.B. and Neog, P.P. (2007). Effect of *Pasteuria penetrans* on the penetration and development of *Meloidogyne incognita. Crop Research Hisar.* 2007; **34**(1/3): 238-242.

Davies, K.G., Danks, C. and Lander, F.B. (1992). Proteins involved in the attachment of a hyperparasite, *Pasteuria penetrans*, to its plant-parasitic nematode host, *Meloidogyne incognita. J. Invertebrate Pathology*, **59:** 18-23.

Davies, K.G., Kerry, B.R. and Flynn, C.A. (1988). Observations on the pathogenecity of *Pasteuria penetrans*, a parasite of root-knot nematode. *Annals of Applied Biology*, **112:** 491-501.

Dhawan, S.C., Kaur, Sarvjeet, and Singh, Iqbal, (2004). Effect of *Bacillus thuriengiensis* on the mortality of root-knot nematode, *Meloidogyne incognita. Indian J. Nematol.* **34:** 98-99.

Dickson, D.W., Mitchell, D.J., Hewlett., T.E., Oostendorp, M. and Kannwischer-Mitchell, M.E. (1991). Nematode-suppressive soil from a peanut field. *J. Nematology*, **23:** 526 (Abstr.).

Dickson, D.W., Oostendorp, M., Giblin-Davis, R.M. and Mitchell, D.J. (1994). pp. 575-601 in (Rosen, D., Bennett, F.D. and Capinera, J.L. Eds). Pest management in the subtropics, biological control. A Florida perspective. Andover, Hampshire, UK: Intercept.

Dube, B. and Smart, G.C. Jr. (1989). Biological control of *Meloidogyne* by *Paecilomyces lilacinus* and *Pasteuria penetrans. J. Nematology*, **19**(2): 222-227.

Dutky, EM. (1978). Some factors affecting infection of plant parasitic nematodes by a bacterial spore parasite of *Bacillus penetrans.* Thesis, University of Maryland, College Park, Maryland. pp. 46.

Giannakou, I.O., Anastasiadis, I.A., Gowen, S.R. and Prophetou Athanasiadou, D.A. (2007). Effects of a non-chemical nematicide combined with soil solarization for the control of root-knot nematodes. *Crop Protection.* 2007; **26**(11): 1644-1654.

Grewal, P.S., Lewis, E.D. and Venkatachari, S. (1999) Allelopathy: a possible mechanism of suppression of plant-parasitic nematodes by entomopathogenic nematodes. *Nematology,* **1:** 735-743.

Hallmann, J., Rodríguez-Kábana, R. and Kloepper, J.W. (1997). Nematode interactions with endophytic bacteria. In *Plant Growth-Promoting Rhizobacteria—Present Status* and *Future Prospects* (Eds. Ogoshi, A., Kobayashi, K., Homma, Y., Kodama, F., Kondo, N. and Akino, S.), pp. 243-245. Nakanishi Printing, Sapporo.

Haseeb, A., Amin, A. and Sharma, A. (2007). Management of *Meloidogyne incognita-Fusarium oxysporum* disease complex of *Pisum sativum. Indian J. Nematology.* 2007; **37**(1): 19-22.

Haseeb, A. and Kumar, V. (2006). Management of *Meloidogyne incognita—Fusarium solani* disease complex in brinjal by biological control agents and organic additives. *Annals of Plant Protection Sciences*. 2006; **14**(2): 519-521.

Haseeb, A., Kumar, V., Shukla, P.K. and Ahmad, A. (2006). Effect of different bioinoculants, organic amendments and pesticides on the management of *Meloidogyne incognita-Fusarium solani* disease complex on tomato cv. K-25. *Indian J. Nematology*. 2006; **36**(1): 65-69.

Hasky-Günther K., Hoffmann-Hergarten, S. and Sikora, R.A. (1998). Resistance against the potato cyst nematode *Globodera pallida* systemically induced by the rhizobacteria *Agrobacterium radiobacter* (G12) and *Bacillus sphaericus* (B43). *Fundamental and Applied Nematology*, **21:** 511-517.

Hewlett, T.E., Cox, R., Dickson, D.W. and Dunn, R.A. (1994). Occurrence of *Pasteuria* spp. in Florida. *J. Nematology,* **26:** 616-619.

Hsu, Y.L. and Wu, W.T. (2002). A novel approach for scaling-up a fermentation system. *Biochem. Eng. J.* **11:** 123-130.

Hu, K.J., Li, J.X. and Webster, J.M. (1999) Nematicidal metabolites produced by *Photorhabdus luminescens* (Enterobacteriaceae), bacterial symbiont of entomopathogenic nematodes. *Nematology*, **1:** 457-469.

Ignoffo, C.M. and Dropkin, V.H. (1977). *J. Kans. Entomol. Soc.*, **50:** 394.

Imbriani, J.L. and Mankau, R. (1977). Ultra structure of the pathogen *Bacillus penetrans*. *J. Invertibrate Pathology,* **30:** 337-347.

Javed, N., Hassan, E.I., Gowen, S., Pemproke, S.B. and Inam ul Haq, M. (2007). The potential of combining *Pasteuria penetrans* and neem (*Azadirachta indica*) formulations as a management system for root-knot nematodes on tomato. *European J. Plant Pathology*. 2008; **120**(1): 53-60.

Jayakumar, J., Rajendran, G., Ramakrishnan, S. and Ramakrishnan, S. (2003). Compatibility of avermectin with nematicides and bio-agents for reniform nematode, *Rotylenchulus reniformis* management in okra. Proceedings of National Symposium on Biodiversity and Management of Nematodes in Cropping Systems for Sustainable Agriculture, Jaipur, India,11-13 November, 2002, pp. 208-212.

Kamra, A.D. and Dhawan, S.C. (1995) Effect of storage temperature on viability of *Pasteuria penetrans* spores infecting *Heterodera cajani. Indian J. Nematol.* **25:** 116-119.

Kerry, B.R. (2000). Rhizosphere interactions and the exploitation of microbial agents for the microbial agents for the biological control of the olant parasotoc nematodes. *Annual Revue of Phytopathology*, **38:** 423-441.

Kloepper, J.W., Rodriguez-Kabana, R., Zehuder, G.W., Murphy, J.F., Sikora, E. and Fernandez, C. (1999). *Australian Plant Pathology*, **28:** 21-26.

Kloepper, J.W., Scroth, M.N. and Miller, T.D (1980). Effects of rhizosphere colonization by plant growth-promoting rhizobacteria on potato plant development and yield. *Phytopathol.* **70:** 1078-1082.

Kluepfel, D.A., McInnis, T.M. and Zehr, E.I. (1993). Involvement of root-colonizing bacteria in peach orchard soils suppressive of the nematode *Criconemella xenoplax. Phytopathology,* **83:** 1240-1245.

Kobayashi, D.Y. and Palumbo, J.D. (2000). Bacterial endophytes and their effects on plants and uses in agriculture. *In:* White, J.F. and Bacon, C.W. (Eds.). Microbial Endophytes. Marcel Dekker, Inc., New York, NY, pp. 199-235.

Kumar, A., Walia, R.K. and Ashima Kapoor, (2005). Field evaluation of *Pasteuria penetrans* as nursery bed application against *Meloidogyne javanica* infecting brinjal. *International J. Nematology*. 2005; **15**(2): 183-186.

Leong. J. (1986). Soderophores: their biochemistry and possible role in the biocontrol of plant pathogens. *Annual Review of Phytopathology*, **4**(11): 285-288.

Mahapatra, S.N. and Mohanty, K.C. (2002). Management of root-knot nematode in brinjal using *Pseudomonas fluorescens* and its compatibility with carbofuran. Proceedings of National Symposium on Biodiversity and Management of Nematodes in Cropping Systems for Sustainable Agriculture, Jaipur, India, 11-13 November, 2002; pp. 135-137.

Maheshwari, T.U., Mani, A. and Rao, P.K. (1987). Combined efficacy of the bacterial parasite, *Pasteuria penetrans* and nematicides in the control of *Meloidogyne javanica* on tomato. *J. Biological Control*, **1:** 53-57.

Mani, A. (1988). Studies on the bacterial parasite *Pasteuria penetrans*: I. spore viability after storage II. Culture on citrus nematode *Tylenchulus semipenetrans. International Nematological Network Newsletter.* **5**(2): 24-25.

Mankau, R. and Prasad, N. (1977). Infectivity of *Bacillus penetrans* in plant-parasitic nematodes. *J. Nematology,* **9:** 40-45.

Mankau, R. (1980). Biological control of *Meloidogyne* populations by *Bacillus penetrans* in West Africa. *J. Nematology,* **12:** 230.

Minton, N.A. and Sayre, R.M. (1989). Suppressive influence of *Pasteuria penetrans* in Georgia soils on reproduction of *Meloidogyne arenaria. J. Nematology,* **21:** 574-575 (Abstr.).

Mukhtar, T., Pervaz, I., Ahmad, R. and Khan, H.U. (2005). Management of *Meloidogyne javanica* by cropping sequence and effects on *Pasteuria penetrans. Pakistan J. Nematology.* 2005; **23**(1): 173-180.

Neipp, P.W. and Becker, J.O. (1999). Evaluation of biocontrol activity of rhizobacteria from *Beta vulgaris* against *Heterodera schachtii. J. Nematology,* **31**(1): 54-61.

Niknam, G.R. and Dhawan, S.C. (2002). Effect of cell concentrations and cell-free filtrates of *Pseudomonas fluorescens* (isolate Pf1) on hatching and mobility of *Rotylenchulus reniformis Indian J. Nematology,* **32:** 37-43.

Niknam, G.R. and Dhawan, S.C. (2002). *In vitro* study on the efficacy of *Bacillus subtilis* strain Bst cell concentrations and cell-free filtrates on hatching and mobility of *Rotylenchulus reniformis. Indian J. Nematology,* **32:** 9-15.

Niknam, G.R. and Dhawan, S.C. (2002). Systemic resistance induced by *Pseudomonas fluorescens* isolate Pf1 in tomato against *Rotylenchulus reniformis. International J. Nematology,* **12:** 203-208.

Niknam, G.R. and Dhawan, S.C. (2003). Effect of seed bacterization and methods of application application of *Pseudomonas fluorescens* on the control of *Rotylenchulus reniformis* infecting tomato. *Nematologia Mediterranea,* **31:** 231-237.

Niknam, G.R. and Dhawan, S.C. (2003). Induction of systemic resistance by *Bacillus subtilis* isolate Bst against *Rotylenchulus reniformis* in tomato. *Nematologia Mediterranea,* **31:** 239-243.

Niknam, G.R. and Dhawan, S.C. (2003). Effect of three application methods of *Bacillus subtilis* isolate Bs_t on penetration and multiplication of *Rotylenchulus reniformis* infecting tomato. *International J. Nematology,* **13:** 93-99.

Nishizawa, T. (1987). A decline phenomenon in a population of upland rice cyst nematode, *Heterodera elachista,* caused by bacterial parasite, *Pasteuria penetrans. J. Nematology,* **19:** 546 (Abstr.).

Oostendorp, M., Dickson, D.W. and Mitchell, D.J. (1991). Population development of *Pasteuria penetrans* on *Meloidogyne arenaria. J. Nematology,* **23:** 58-64.

Oostendorp, M. and Sikora, R.A. (1992). Seed treatment with antagonistic rhizobacteria for the suppression of *Heterodera schachtii*—early root infection of sugar beet. Revue de Nématologie, 1989, Vol. 12, Num. 1, p. 77-83.

Prasad, N. and Mankau, R. (1969). Studies on sporozoan endoparsite of nematodes. *J. Nematology,* **1:** 301-302.

Prasad, S.S.V., Tilak, K.V.B.R. and Gollakota, K.G. (1972). *J. Invertebrate Pathology,* **20:** 377.

Racke, J. and Sikora, R.A. (1992). Wirkung der pflanzengesundheitsfördernden Rhizobakterien *Agrobacterium radiobacter* und *Bacillus sphaericus* auf den *Globoderapallida*-Befall der Kartoffel und das Pflanzenwachstum. Zeitschrift für Pflanzenkrankheiten **134:** 198-208.

Rao, -M-S. (2007). Production of bio-agent colonized vegetable seedlings for sustainable nematode management. *Acta. Horticulturae.* 2007; **752:** 545-548.

Rao, -M-S., Dhananjay-Naik, and Shylaja, -M. (2004). Bio-intensive management of root-knot nematodes on bell pepper using *Pochonia chlamydosporia* and *Pseudomonas fluorescens. Nematologia Mediterranea.* 2004; **32**(2): 159-163.

Sasser, J.N. and freckman, D.W. (1987). A world prospective on Nematology: the role of society. *In:* Veech, J.A. and Dickson, D.W. (eds.). Vistas on Nematology: a commemoration of the twenty fifth anniversary of the society of nematologists. Society of nematologists, Lakeland, FL, pp. 7-14.

Sayre, R.M. (1980). Biocontrol: *Bacillus penetrans* and Related Parasites of Nematodes. *J. Nematol.* **12:** 260-269.

Sayre, R.M. and Starr, M.P. (1985). *Pasteuria penetrans,* a mycelial and endospore forming bacterium parasite in plant parasitic nematodes. *Proc. Helminthological Soc. Washington,* **52:** 149-165.

Sayre, R.M. and Starr, M.P. (1988). Bacterial diseases and antagonisms of nematodes. In Diseases of nematodes (eds. Poiner, G.O. Jr. Jansson, H.B. Boca Ratan Fla USA): CRC Press, 1988, V-1, p. 69-101.

Serracin, M., Schuerger, A.C., Dickson, D.W. and Weingartner, D.P. (1997). Temperature-dependent development of *Pasteuria penetrans* in *Meloidogyne arenaria. J. Nematology,* 29.

Sharma, R. and Sharma, S.B. (1989). Sticky swarm disease of *Heterodera cajani* and *Meloidogyne javanica* caused by *Pasteuria penetrans. International Pigeon Pea Newsletter.* **10:** 26-27.

Siddiqui, I.A. and Shaukat, S.S. (2002). Mixtures of plant disease suppressive bacteria enhance biological control of multiple tomato pathogens. *Biol. Fertil. Soils.* **36:** 260-268.

Sikora, R.A. (1992). Management of the antagonistic potential in agricultural ecosystems for the biological control of plant parasitic nematodes. *Annual Review of Phytopathology* **30:** 245-270.

Singh, B. and Dhawan, S.C. (1996). Suppression of *Heterodera cajani* by *Pasteuria penetrans* during three successive plantings of cowpea. *Indian J. Nematology* **26**(2): 216-221.

Singh, B. and Dhawan, S.C. (1990). A new bacterial strain of *Pasteuria penetrans,* its host range and effect of temperature on spore attachment to second-stage juveniles of pigeon-pea cyst nematode, *Heterodera cajani. Indian J. Nematology,* **20**(2): 161-166.

Singh, B. and Dhawan, S.C. (1994). Effect of *Pasteuria penetrans* on the penetration and multiplication of *Heterodera cajani* in *Vigna unguiculata.* Nematol. Medit. **22:** 159-162.

Singh, B. and Dhawan, S.C. (1998). Compatibility of *Pasteuria penetrans* with carbofuran in controlloing *Heterodera cajani* on cowpea. *Indian J. Nematol.* **28**(2): 110-113.
Singh, B. and Dhawan, S.C. (1999). Comparison of different methods of application of *Pasturia penetrans* for the control of *Heterodera cajani* on cowpea. *Indian J. Nematol.* **29**(2): 118-120.
Stirling, G.R. (1988). New host and geographical records of nematode-parasitic bacteria of the *Pasteuria penetrans* group. *Nematologica*, **34:** 350-356.
Stirling, G.R. and Wachtel, M.F. (1980). Mass production of *Bacillus penetrans* for the biological control of root-knot nematodes. *Nematologica*, **26:** 308-312.
Stirling, G.R.K., Sharma, R.F. and Perry, J. (1990). Attachment of *Pasteuria penetrans* spores to the root-knot nematode *Meloidogyne javanica* in soil and its effects on infectivity. *Nematologica*, **36:** 246-252.
Sturz, A.V. and Kimpinski, J. (2004). Endoroot bacteria derived from marigolds (Tagetes spp.) can decrease soil population densities of root-lesion nematodes in the potato root zone. *Plant Soil* **262:** 241-249.
Subramanian, S., Sivakumar, M. and Rajendran, G. (2002). Management of pigeonpea cyst nematode, *Heterodera cajani* in pulses by biocontrol agents. Proceedings of National Symposium on Biodiversity and Management of Nematodes in Cropping Systems for Sustainable Agriculture, Jaipur, India,11-13 November, 2002; 148-150.
Thorne, G. (1940). *Duboscqia penetrans* n. sp. (Sporozoa: Microsporidia, Nosematidae), a parasite of the nematode *Pratylenchus pratensis* (de Man) Filipjev. *Proc. Helminthol. Soc. Wash* **7:** 51-53.
Verdejo, S. and Jaffee, B.A. (1988). Reproduction of *Pasteuria penetrans* in atissue culture system containing *Meloidogyne javanica* and *Agrobactrium rhizogenes* transformed roots. *Phytopathology*, **78**(10): 1284-1286.
Verdejo, S., Jaffee, B.A. and Mankau, R. (1988). Reproduction of *Meloidogyne javanica* on plant roots genetically transformed by *Agrobactrium rhizogenes. J. Nematology*, **20**(4): 599-604.
Weibelzahl-Fulton, E., Dickson, D.W. and Whitty, E.B. (1996). Suppression of *Meloidogyne incognita* and *M. javanica* by *Pasteuria penetrans* in field soil. *J. Nematology*, **28:** 43-49.
Zavaleta-Meija, M.F. and Van Gundy, S.D. (1987). Revista Mexicana de Fitipatologia **5:** 38-44.
Zukerman, B.M., Dicklow, M.B. and Acosta, N. (1993). A strain of *Bacillus thuringiensis* for the control of plant parasitic nematodes. *Biocon. Sci. Tech.*, **3:** 41-46.

37

Weed—An Important Pest

Rajvir Sharma[1], R.A. Marathe[2], Suresh Pal[3] and U.C. Sharma[4]

[1]Division of Agronomy, Indian Agricultural Research Institute, New Delhi–110012
[2]NRC on Pomegranate, Solapur (Maharashtra)
[3]ASRB, KAB-I, ICAR, Pusa, New Delhi–110012
[4]NAIP, KAB-II, ICAR, Pusa, New Delhi–110012

There are four natural hazards, e.g., weather, weeds, insect pests and plant diseases including nematodes, which stand between humans and continuing food supply since the man started cultivating crops about 10,000 BC. Sometimes these hazards work independently and many times they work hand in hand. Weeds have been existing on the earth ever since the man started domesticating/cultivating plants and animals. This was the beginning of farming and the farmers began to replace hunters about that time, and the process continues till today. Weeds have been recognized as a problem since then and the battle against them is a never ending one and often the costliest agronomic input for the success in crop production. They cause higher reduction/loss in the yield of crops than other pests and diseases. Yet they are the most underestimated crop pests in tropical agriculture. The concept of a weed is very subjective. A plant may be a weed to one person and a useful plant to another or a weed in one place but not in another, or a weed at one time of year and a valuable plant at another season. Weeds are, therefore, defined in different ways by different authors. Jethro Tull, a British farmer was the first person who defined the weed as a plant growing where it is not desired in his first book entitled *The New Horse Hoeing Husbandry* published in 1731. Since then several people have defined weeds in different ways. But a comprehensive, scientific and most relevant one is "Weeds are the plants growing where and when they are not desired". For example, Bermudagrass (*Cynodon dactylon*) is a valuable plant in pasture, but it is a troublesome weed in crop field. This definition clearly indicates that no portion of plant kingdom can be identified as weedy. Weedy nature of a plant is decided by the situation in which it occurrs and the individual interest in that situation. Finally, it is said that all weeds are unwanted plants but all unwanted plants may not be weeds.

Basically, plants exist in three classes in nature : i) wild plants (which grow naturally outside the human disturbed habitat), ii) weeds (which thrive in habitats that are continuously disturbed by humans), and iii) domesticates/crops (which are artificially propagated and often require cultivation and care by humans in order to grow and make optimum use of environmental resources). Some estimates suggest that out of 3,00,000 species (some people believe 2,50,000 species) of angiosperm (closed-seeded plants) under 450 different botanical families present on the globe, nearly 30,000 species (some people believe 25,000 species) of plants behave as weeds (Maheshwari, 1963).

Weed distribution usually falls in 60 different botanical families, but nearly 70% of these weeds occur in 12 families and over 40% of the total weeds are found in only 2 families, *Poaceae* and *Asteraceae*. There are about 200 or 250 most serious weeds (< 0.1 or nearly 0.1 % of the total angiosperms) in the world. The *Poaceae* has the largest number of serious weed species, and the weeds that often resemble crop plants both morphologically and physiologically (crop mimicry), and thus contribute to the difficulty of weed control.

Usual distribution of some important weeds in crop fields compiled over regions/locations is presented in Table 37.1. Weeds listed under a category do not necessarily imply that all of these weeds are present in all locations, districts or states of the country (Table 37.1). Rather it is found that certain weed species could pose dominance in certain agroclimatic or cropping situation. The same is true for a specific crop or for a specific season, rainy/wet (*kharif*), winter/dry (*rabi*) or summer (*zaid*). Yet some loose generalization based on record and history of weed infestation in crops cannot be ruled out.

At present distribution and biology of 76 worst weeds of the world has been described These 76 weed species are distributed in 30 botanical families. Thirty-nine of them are serious weeds of the tropics. The eighteen worst weeds of the world are listed in Table 37.2. About 70% of the world's worst weeds occur in 9 families and over 50% of these species are found in only 2 families, *Poaceae* and *Asteraceae*. Of the 30 families that contain the world's worst weeds, 5 of them, *Poaceae*, *Solanaceae*, *Convolvulaceae*, *Euphorbiaceae* and *Fabaceae*, also supply 75% of the world's food supply. From the known weeds, 44% are perennial and 56% annual.

Table 37.1. Distribution of weeds in the crop fields in India.

Crops	Grass weeds	Broad-leaved weeds	Sedge weeds
1. Rice (i) Upland direct-seeded rice culture	*Echinochloa colona* *Echinochloa crusgalli* *Commelina benghalensis*	*Galinsoga parviflora* *Trianthema portulacastrum* *Trianthema monogyna* *Scoparia dulcis*	*Cyperus iria* *Cyperus difformis* *Cyperus rotundus* *Cyperus esculentus*
	Commelina communis *Leptochloa chinensis* *Panicum sp* *Paspalum conjugatum*	*Celosia argentea* *Amaranthus spinosus* *Bidens pilosa* *Portulaca oleracea* *Physalis minima* *Amaranthus viridis* *Croton sp* *Phyllanthus niruri*	
(ii) Rainfed direct-seeded low-land rice culture	*Monochoria vaginalis* *Echinochloa colona*	*Ludwigia octovalvis* *Ludwigia parviflora*	*Fimbristylis miliacea* *Scirpus supinus*
	Echinochloa crusgalli *Nymphaea sp*	*Eclipta alba* *Ammania baccifera* *Ipomoea reptans* *Marsilea quadrifoliata*	*Cyperus iria* *Cyperus difformis* *Protomegoton sp* *Eleocharis sp*

Table Contd.

Table 1. Contd.

Crops	Grass weeds	Broad-leaved weeds	Sedge weeds
		Hydrilla verticilliata	*Typha sp* *Chara zeylanica* *Fimbristylis littoralis*
(iiii) Puddled direct-seeded and transplanted rice culture	*Monochoria vaginalis* *Echinochioa colona* *Echinochloa crusgalli* *Cyanotis axillaris* *Caesulia axillaries*	*Ludwigia octovalvis* *Ludwigia parviflora* *Sphenoclea zeylanica* *Oxalis corniculata* *Ammania baccifera* *Ipomoea aquatica* *Eclipta alba* *Rungia repens* *Lippia nudiflora* *Marsilea quadrifoliata*	*Cyperus iria* *Cyperus difformis* *Eleocharis sp* *Fimbristylis miliacea* *Scirpus maritimus* *Fimbristylis littoralis*
2. Maize, sorghum, pearl millet, green gram, black gram, pigeon pea, groundnut and soybean [all wet-season (*kharif*) crops]	*Commelina benghalensis* *Commelina communis* *Rottboellia cochinchinensis* *Eleusine indica* *Acrachne racemosa* *Digitaria sanguinalis* *Dactyloctenium aegyptium* *Echinochloa colona* *Echinochloa crusgalli* *Setaria verticillata/plumeri* *Cynodon dactylon* *Digitaria abyssinica*	*Striga asiatica (=lutea)* *Trianthema portulacastrum* *Trianthema monogyna* *Amaranthus retroflexus* *Digera arvensis* *Tribulus terestris* *Celosia argentea* *Phyllanthus niruri* *Corchorus sp* *Solanum nigrum* *Ageratum conyzoides* *Portulaca oleracea* *Xanthium sp* *Acalypha indica/ciliate* *Physalis minima*	*Cyperus rotundus* *Cyperus esculentus* *Cyperus iria*
3. Wheat, barley, oat, chickpea, rapeseed, mustard, pea and lentil [all winter-season (*rabi*) crops]	*Avena sterilis ssp ludoviciana* *Phalaris minor* *Phalaris paradoxa* *Polypogon monspeliensis* *Asphodelus tenuifolius* *Lolium temulentum* *Poa annua* *Cynodon dactylon*	*Chenopodium album* *Chenopodium murale* *Melilotus alba/indica/ parviflora* *Convolvulus arvensis* *Coronopus didymus* *Cirsium arvense* *Spergula arvensis* *Carthamus oxyacantha* *Anagallis arvensis* *Medicago denticulate* *Fumaria parviflora* *Argemone mexicana* *Vicia sativa/hirsuta* *Lathyrus aphaca* *Orobanche sp* (mustard) *Polygonum aviculare* *Sisymbrium irio* *Galium aparine* *Stellaria media* *Papavar rhoeas*	*Cyperus rotundus* *Cyperus esculentus*

Table 37.2: Details of the eighteen worst weeds of the world.

SN	Species	Family	Origin	Distribution
1.	*Cyperus rotundus*	Cyperaceae	India	Tropics and Subtropics
2.	*Cynodon dactylon*	Poaceae	Tropical Africa, Indo-Malaysia	Tropics and Subtropics
3.	*Echinochloa crus-galli*	Poaceae	Europe & India	Worldwide except Africa
4.	*Echinochloa colona*	Poaceae	India	Tropics
5.	*Eleusine indica*	Poaceae	China, Japan, India, Malaysia	Tropics
6.	*Sorghum halepense*	Poaceae	Mediterranean region	Sub-temperate
7.	*Imperata cylindrical*	Poaceae	Old World tropics	Tropics and Subtropics
8.	*Eichhornia crassipes*	Pontederiaceae	Brazil	Tropics and Subtropics
9.	*Portulaca oleracea*	Portulacaceae	Europe, North Africa	Tropics and Subtropics
10.	*Chenopodium album*	Chenopodiaceae	———	Temperate
11.	*Digitaria sanguinalis*	Poaceae	Europe	Tropics and Subtropics
12.	*Convolvulus arvensis*	Convolvulaceae	Europe	Sub-temperate
13.	*Avena fatua* and other wild oats	Poaceae	———	Sub-temperate
14.	*Amaranthus hybridus*	Amaranthaceae	North America	Tropics and Subtropics
15.	*Amaranthus spinosus*	Amaranthaceae	Tropical America	Tropics and Subtropics
16.	*Cyperus esculentus*	Cyperaceae	———	Temperate
17.	*Paspalum conjugatum*	Poaceae	Tropical America	Tropics
18.	*Rottboellia exaltata/ cochinchinensis*	Poaceae	India	Tropics

FEW IMPORTANT CHARACTERISTICS OF WEEDS

1. Weeds are prolific breeder and seeder. They have sexual (seed propagation), asexual and vegetative propagation. Annual weeds have more sexual propagation (Table 37.3), whereas perennial weeds have more vegetative propagation.

Table 37.3: Prolificacy of major weed species.

Weed species	Seeds/plant	Weed species	Seeds/plant
Striga asiatica (= lutea)	90,000	*Chenopodium album*	72,000
Striga hermonthica	42,000	*Trianthema portulacastrum*	52,000
Orobanche cernua	12,00,000	*Cyperus esculentus*	820
Cuscuta campestris	16,000	*Leucas aspera*	5,000
Parthenium hysterophorus	15,000	*Echinochloa colona*	42,000
Amaranthus spinosus	2,35,000	*Digitaria horinzontalis*	12,000
Argemone mexicana	20,000-30,000	*Eleusine indica*	50,000-135,000
Bidens pilosa	3,000-6,000	*Amaranthus viridis*	10,000

2. Of the total annual loss of agricultural produce from various pests in India, weeds roughly account for 37%, insects for 29%, diseases for 22% and other pests for 12%. As shown in Fig. 37.1, they decrease quantity and quality of produce/food, fibre, oil, forage/fodder, animal products (meat and milk) and cause health hazards for humans and animals. Estimates also indicate that weeds generally reduce yield by 15-30% in wheat, 30-35% in rice and 18-85% in maize, sorghum, pulses and oilseeds.

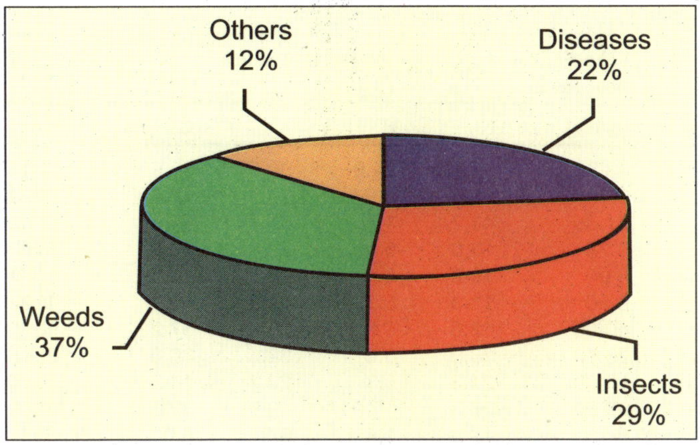

Fig. 37.1. Losses caused by various agricultural pests.

Table 37.4: Crop yield reductions due to unchecked weed infestation in India.

Crops	Yield (%)	Crops	Yield reduction (%)
Rice (upland)	50-60	Chickpea/lentil	10-15
Rice (transplanted)	15-20	Cotton (Rainfed)	30-50
Dwarf wheat/barley/pea	10-25	Green gram/black gram/cowpea	15-20 (some cases more than 80%)
Tall wheat	10-25	Groundnut	50 (erect varieties); 25 (spreading varieties)
Maize	20-30	Sugarcane	20-70
Sorghum	20-30	Jute	30-70
Pearl millet	20-30		

3. Weeds generally remove 30-40 kg N, 10-15 kg P_2O_5 and 20-40 kg K_2O per hectare in irrigated agro-ecosystem, although the magnitude of nutrient removal varies based on crops (Tables 37.5 and 37.6), weed species composition, density and growth, crop growing season, and varying agroclimatic situations. Unchecked weed growth in wheat fields removed about 20 kg N/ha and reduced the grain yield to the tune of 12 q/ha (Gill, 1974). Based on recent estimate, rice grain production in India is reported to suffer a yearly loss of 15 million tonnes due to weed competition (Parthiban and Kathiresan, 2002).

Table 37.5: Nutrient depletion (kg/ha) by major crops.

Crops	N (kg/ha)	P (kg/ha)	K (kg/ha)
Wheat	15-60	4-15	15-60
Rice	30-85	5-20	20-60
Maize	20-60	4-15	15-50
Sugar cane	30-85	10-40	50-200
Soybean	15-65	5-15	20-102
Onion	16-60	4-20	10-77
Potato	30-80	3-15	30-95

Table 37.6: Nutrient depletion by weeds in major field crops.

SN	Crops	Weed biomass	Major nutrients (kg/ha)				Micronutrients (g/ha)		
		(q/ha)	N	P	K	Cu	Fe	Mn	Zn
1.	Upland rice	43.70	93.1	24.9	90.5	36.9	6415	422.6	154.8
2.	Transplanted rice	5.17	11.0	2.9	10.7	12.3	2524	166.3	60.8
3.	Maize	31.35	41.7	19.0	67.0	44.8	4765	189.7	117.6
4.	Sorghum	9.75	13.0	5.9	20.8	13.9	1482	59.0	36.6
5.	Pearl millet	9.00	19.2	5.1	18.7	12.9	1371	54.6	33.8
6.	Finger millet	8.50	11.3	5.1	18.1	12.1	1292	51.4	31.8
7.	Black gram	40.00	62.0	6.0	85.0	57.2	6080	242.0	150.0
8.	Soybean	25.10	35.0	5.0	32.3	36.0	3815	151.8	94.1
9.	Groundnut	10.00	21.3	5.7	20.7	14.3	1520	60.5	37.5
10.	Cotton	22.96	28.3	13.9	48.9	32.8	3489	138.9	86.1
11.	*Wheat	8.13	12.0	1.4	6.3	6.0	813	37.5	9.4
12.	**Wheat	30.60	30.6	4.5	50.3	22.6	3060	141.0	35.5
13.	Gram	20.00	29.0	3.4	15.5	14.8	2000	92.0	23.2
14.	Potato	24.00	63.0	11.0	80.0	18.0	2400	110.6	28.0
15.	Lentil	16.50	24.0	3.0	13.0	12.0	1650	76.0	19.0
16.	Mustard	8.00	11.6	1.4	6.2	6.0	800	37.0	9.3

(Gupta, 2002)

4. Weeds serve as hosts of several diseases, nematodes and insect pests as mentioned in Table 37.7.

Table 37.7. Weeds serving as hosts of diseases, nematodes and insect pests of crops.

SN	Weed hosts	Diseases, nematodes and insect pests (with crop hosts)
		DISEASES
1.	*Leersia oryzoides*	Bacterial leaf blight (rice)
2.	*Cyperus rotundus*	Blast disease (rice)
3.	*Echinochloa colona*	Tungro and Hoja blanca virus (rice)
4.	*Eleusine indica*	Blast, stem rot and tungro virus (rice)
5.	*Dactyloctenium aegyptium*	Smut (wheat)
6.	*Agropyron repens*	Black rust and bacterial foot rot (wheat)
7.	*Saccharum spontaneum*	Downey mildew (maize)
8.	*Chenopodium album*	Mildew (spinach)
9.	*Cenchrus ciliaris*	Ergot (pearl millet)
10.	*Plantago* sp, *Vernonica persica*	Tobacco mosaic virus (TMV) (tobacco, tomato)
		NEMATODES
1.	*Brachiaria deflexa*	Root-knot nematodes
2.	*Digitaria sanguinalis, Cynodon dactylon*	Sting nematode (vegetables)
		INSECT PESTS
1.	*Cyperus distans, Echinochloa* sp, *Panicum* sp, *Pennisetum purpureum*	Stem borer (rice)
2.	*Paspalum orbiculare*	Gall midge (rice)
3.	*Chenopodium album*	Gram caterpillar (tomato, brinjal, red gram, okra, peas); stalk borer (tomato, maize, lilly, aster); greasy cut-worm (potato, gram, pea)
4.	*Brassica kaber*	Aphids (cole crops: radish, spinach, brinjal); melon aphis (melons, cucumber, squash)

Table Contd.

Table 37.7. Contd.

SN	Weed hosts	Diseases, nematodes and insect pests (with crop hosts)
5.	*Solanum* sp	Greasy cut-worm (potato, gram, pea); stalk borer, flea beetle and potato beetle (potato, tomato); weevils (bell pepper)
6.	*Amaranthus* sp	Gram caterpillar (tomato, brinjal, red gram, okra, peas); caterpillar (tobacco, crucifers)
7.	*Datura* sp	Caterpillar (tobacco, crucifers)
8.	*Trianthema portulacastrum*	Greasy cut-worm (potato, gram, pea)
9.	*Achyranthes* sp	Leaf-eating caterpillar (maize, alfalfa)
10.	*Digera arvensis*	Caterpillar (tobacco, crucifers)

5. Weeds growing in canals and ponds are also a big menace as they choke irrigation systems resulting in poor delivery of water to field crops. Thick mat of aquatic weed like water hyacinth impede light penetration, depletes oxygen supply to fishes, thereby affecting pisciculture. The spread invasive weeds like *Parthenium hysterophorus*, *Lantana camera* and *Phalaris* minor seriously affect crop production, livestock farming and have become the biggest threat to biodiversity.

METHODS OF WEED MANAGEMENT

Traditional methods of weed control being practised since inception of agriculture like hand weeding, hoeing and other intermethod operations are expensive, labour intensive and ineffective in some situations due to frequent flushes especially during rainy season. Under these conditions, use of herbicides can be the only viable alternative.

Chemical weed control has largely been a post-World War II development and it has made a rapid growth. Apart from the introduction of Dinoseb in 1935, MCPA and 2,4-D in 1945, it is now over 50 years since the first effective herbicide was marketed. Curiously these first herbicide compounds MCPA, 2,4-D and Dinoseb are still widely used today.

In early sixties, the search for new herbicides grew more intense and met with considerable success. More than 200 herbicides are now available to cater to the various needs.

The immense benefits offered by these herbicides have been most fully harnessed by the developed countries, while their impact on agricultural scenario in the developing countries has been abysmally marginal.

HERBICIDES SCENARIO IN INDIA

In India, although 2,4-D was first tested as early in 1946, the chemical concept for weed control struck deeper roots only when tea growers began using herbicides on a commercial scale in late 1970s. Total pesticide use particularly of insecticides/fungicides continuously increased up to 1990-91, but after that there was a sharp decline due to adoption of Bt cotton, as infestation of pest in cotton decreased drastically (Fig. 37.2). However, it took almost a decade for herbicide technology to make progress in the field crops, particularly in wheat and rice crops, and in the recent past, the use of herbicides is increasing at a faster rate. Even then adoption and growth of herbicides has been slow and concentrated in few crops and that also in isolated areas (Punjab and Haryana) in our country.

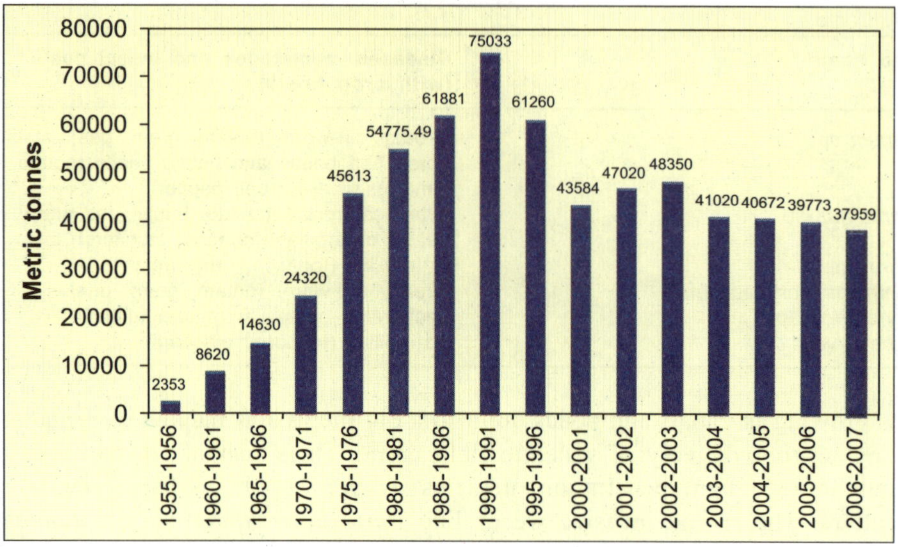

Fig. 37.2. Pesticide consumption in India (1955-2007).

The use of herbicide in crop production account for only 17% of the total pesticide usage compared with 55% at global level. Global agrochemicals market vis-à-vis the total amount of herbicide (17%) used in India is not properly used in all crops. Crop-wise use of herbicides in India is shown in Fig. 37.3.

INTEGRATED WEED MANAGEMENT—A NEW APPROACH

Weed management strategy could be effective only if it is planned considering the occurrence of nature of weed flora. Weeds may be classified on the basis of morphological characteristics

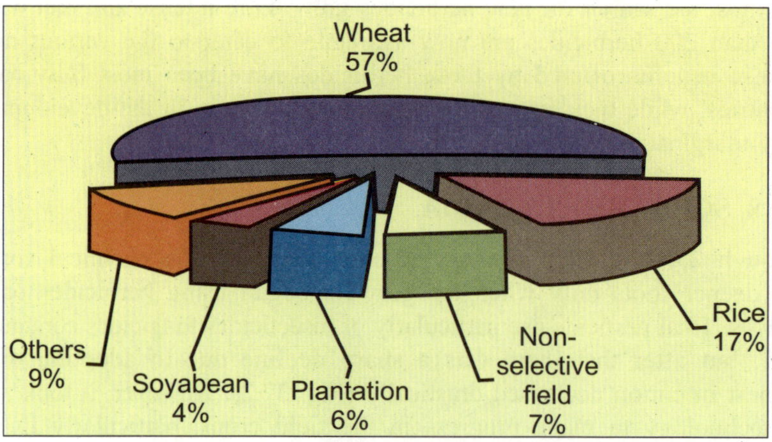

Fig. 37.3. Crop-wise herbicide usage in India.

and season. Knowledge of their classification according to season and their identification is an important prerequisite for their successful management. Dissemination of weed seeds, being mode of reproduction plays an important role in the spread of an individual weed species. Sometime admixture of farm produce with weed seed (objectional weeds) may become the cause of rejection of some important consignment for export purpose. Knowledge of weed seed and their proper identification by seeing structure, colour is essential and could be effective component of integrated approach.

Extent of loss caused by weeds varies from crop to crop, season to season and nature and intensity/density of weeds. Use of herbicides offers selective and economic weed control only if they are applied in right dose and at right time with recommended method. Generally, a herbicide is needed to be persisted in the soil for about 2 months for having desirable level of weed control. But repeated herbicide use alone may result in shift in weed flora toward more persistent perennials, build-up of herbicide residues in the soil and consumable product. There are reports that some herbicides like triazine and sulfonyl urea are known to persist in the soil for more than 2 years and limit the choice of farmer growing certain crops due to residue left over in the soil. Atrazine residues are also known to cause depression effect on bacterial plants.

Therefore, sole reliance on use of single method of weed management would never fulfill all the requirements of farmers satisfactorily. Now the new concept of integrated weed management is only the viable option for effective, economical and eco-friendly weed management. Major objectives and components of this new concept are given below.

- To increase agricultural production
- To reduce economical losses
- To reduce risk to human health
- To reduce potential damage to flora and fauna
- To improve the safety and quality of the environment
- To develop sound strategy to encourage a judicious use of herbicides in integration with other safe, effective and eco-friendly and sociologically acceptable control tactics.

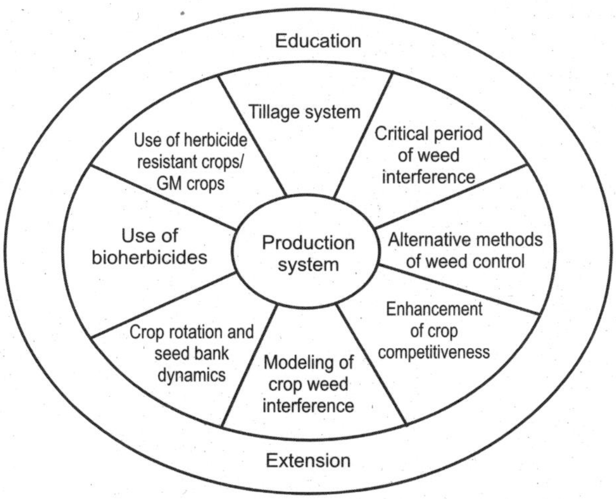

Fig. 37.4: Major components of IWM.

To achieve the above-mentioned objectives and implement effective and economic weed management without any ill effects and use of herbicides in the main and succeeding crop, an attempt has to be made for adopting/integrating several approaches for weed management in order to realize potential yield of major field crops.

REFERENCES

Gill, K.S. (1974). Wheat agronomy of dwarf wheat. Research on dwarf wheat, ICAR, New Delhi.

Gupta, O.P. (2002). Modern weed management–A textbook and manual. Published by Agro Botanica, pp. 19.

Maheshwari. J.K. (1963). The Flora of Delhi (Reprinted in 1990). Council of Scientific and Industrial Research, New Delhi, pp. 447.

Parthiban, C. and Kathiresan, R.M. (2002). Use of certain plant materials for weed management in transplanted rice. *Indian J. Weed Sci.*, **34**(3&4): 187-191.

38

Integrated Weed Management in Major Field Crops

Rajvir Sharma[1], R.A. Marathe[2], Suresh Pal[3] and U.C. Sharma[4]

[1]*Division of Agronomy, Indian Agricultural Research Institute, New Delhi–110012*
[2]*NRC on Pomegranate, Solapur (Maharashtra)*
[3]*ASRB, KAB-I, ICAR, Pusa, New Delhi–110012*
[4]*NAIP, KAB-II, ICAR, Pusa, New Delhi–110012*

India is a densely populated country that exploited the reserves of nutrients in soil since long. At present, the estimated removal of major nutrients is about 28 mt, resulting in a net negative balance of about 10 mt. This is a serious soil health hazard that needs an urgent attention of all concerned. The present level of food production would not be sufficient for next 10 years for the ever increasing population. Recently, the Government of India has emphasized the need to increase the productivity of field crops with due care to rainfed/dry land crops. It implies that for increasing the productivity, the nutrient removal would be more than double of the present level of about 56 mt, since nutrient requirement for the incremental production would be more. So this large gap between nutrient supply through all sources and removal would further increase from the present level of food production. This gap can be filled either by supplying additional nutrients in the form of fertilizers (organics and inorganics) or by reducing the removal/depletion of NPK reserves by crops/weeds so that fertilizer use efficiency, i.e., the proportion of applied nutrient recovered by the crop, could be increased (Table 38.1) (Tandon and Pratap Narayan, 2000; Tiwari, 2002). It is commonly expressed as a percentage of fertilizer used by the crop or alternatively in terms of crop yield per unit of fertilizer applied (e.g., kg grain per kg of applied nutrient). Since nutrients are costly inputs and all Indian farmers are not able to apply these nutrients at the recommended doses at proper time and in adequate amount, therefore, the scope of balancing the removal and addition of nutrients through supplying additional nutrients in all the crops in all the regions is not much viable. Other way of reducing the gap is to reduce the depletion of nutrients by plants other than crops.

Table 38.1: Gross nutrient removal by crops and nutrient addition through fertilizers in India.

Year	Food grain prod. (mt)	Nutrient removal (mt)				Nutrient consumption (mt)			
		N	P_2O_5	K_2O	Total	N	P_2O_5	K_2O	Total
1961	82	3.1	1.3	1.3	9.5	0.21	0.05	0.03	0.29
1971	108	4.1	1.7	6.5	12.3	1.48	0.54	0.24	2.26
1986	150	5.9	2.6	10.3	18.8	5.66	2.00	0.81	8.47
1989	170	6.6	3.0	11.6	21.2	7.25	2.72	1.07	11.04
1997	198	7.7	3.5	17.5	24.4	10.30	2.98	1.03	14.31
2004	212	12.4	4.9	20.2	36.2	11.08	4.12	1.59	16.79

WEEDS—MAJOR SOIL NUTRIENTS DEPLETER

Weeds have been present ever since the man started cultivating crops about 10,000 B.C. and the same were undoubtedly recognized as a problem from the beginning. The battle against weeds is a never ending one and often the costliest agronomic input for successful crop production. The concept of a weed is very subjective (Swarbrick and Mercado, 1987). A plant may be a weed to one person and a useful plant to another or a weed in one place but not in another, or a weed at one time of year and a valuable plant at another season. Weeds are, therefore, defined in different ways by different authors. But the most scientific and most relevant one is "Weeds are the plants growing where and when they are not desired". Sometime it is said that all weeds are unwanted plants, but all unwanted plants are not weeds.

Out of 3,00,000 species of angiosperm present on the globe, about 30,000 species of plants may behave as weed directly or indirectly (Maheshwari, 1963). There are about 200 species which are designated as most serious weed species, as they often resemble crop plants both morphologically and physiologically (crop mimicry), and this contributes to the difficulty in weed control.

Therefore, weeds are considered the major biotic constraint in achieving the potential yield of our field crops and come under the category of potential pest causing 33% yield loss of field crops when compared to 26% due to diseases, 20% due to insects, 15% loss due to storage and miscellaneous pest and 6% due to rodents, based on the estimates made during 1992-1995. Crop losses due to weeds are estimated to be about Rs. 20,000 crore annually in India. Recent estimates show that average crop yield reduction due to weeds vary from 12 to 72% depending on the crop. At the global level, the world food loss is reported to be more than 287 million tonnes due to weeds accounting to 11% of the total food production (Mukhopadhay, 1993). However, yield losses due to weeds in major *rabi* crops account for about 16% in wheat, 32.9% in pea, and 68% in onion, while in *kharif* crops, weeds account for about 36.5% losses due to availability of adequate moisture. Apart from the direct losses, weeds also deplete substantial quantities of nutrients from soil. Since the weeds alone cause nearly 33% of the total losses, every attempt has to be made to control the weed menace/weed management to uphold the challenging task of increasing the agricultural production to an amazing requirement of 260 million tonnes by 2030. So control of weeds in all crops should be given due care.

Extent of Nutrient Depletion by Weeds

Weeds absorb more nutrients than the crop plants. Since the weeds have competitive advantage in water uptake, they also have competitive advantage in nutrient uptake. Weeds accumulate more nutrients than the associated crop plants. Among the major nutrients, N is the one for which competition is most severe. Crops show the deficiency symptoms for N in heavily weed infested fields. Unchecked weed growth in wheat fields removed about 20 kg N/ha and reduced the grain yield to the tune of 12 q/ha (Gill, 1974). When supplemented N is applied to fields of upland rice infested with *Cyperus rotundus* L. weeds benefited more than rice. This is because the weeds compete more effectively for soil moisture under high N condition. However, addition of higher N level to cotton in competition with annual weeds resulted in a substantial increase in the height of the plants. This gave cotton an advantage in competition for light not realized at low N level. In some areas of Karnal and Kurukshetra, farmers discarded their wheat crop and used as fodder because of intensive population and excessive growth of grassy weeds. A population of 200 plants per square metre of *Phalaris minor* is reported to result in complete failure of wheat crop at farmers field in Haryana State (Malik and Singh, 1995). Besides these weeds remove 30-40 kg N, 10-20 kg P and 20-40 kg K/ha from the soil and transpire 250-300 tonnes of water per hectare (Malik and Singh, 1995). One of the reasons for low yield of rice is weed problem. Uncontrolled weeds compete with transplanted rice and reduce yield up to 76% (Mishra and Gautam, 1995).

Major Weeds Infesting *rabi* Crops

Adoption of large-scale cultivation of input-responsive dwarf high-yielding varieties and hybrids resulted in the heavy infestation of grasses, sedges and broad leaf weeds. Among them the infestation and competitive stress of grassy weeds particularly *Phalaris minor* and *Avena ludoviciana* in wheat is very serious. Knowledge of nature and type of weed flora is the essential prerequisite for achieving effective, safe, and economical weed management. Major weed flora infesting *rabi* crops are enlisted.

Grassy weed: *(1) Phalaris minor, (2) Avena ludoviciana, (3) Poa annua, (4) Polypogon monspeliensis*

Broadleaved weeds: *(1) Chenopodium album, (2) Chenopodium murale, (3) Rumex dentatus, (4) Anagallis arvensis, (5) Coronopus didymus, (6) Convolvulus arvensis, (7) Cirssium ravense, (8) Cannabis sativa, (9) Argemone mexicana, (10) Melilotus indica, (11) Spergula arvensis, (12) Fumaria perviflora*

Sedge: *Cyperus rotundus, Cyperus esculentus*

Major Weeds Infesting *kharif* Crops

Grasses: *(1) Echinochloa colonum, (2) E. crusgalli, (3) Dactyloctenium aegyptium, (4) Leptochloa chinensis, (5) Digitaria sanguinalis, (6) Paspalum distichum, (7) Cynodon dactylon*

Broadleaf: *(1) Commelina benghalensis, (2) Amaratnhus, spinosus, (3) Triantherma portulacastrum, (4) Eclipta alba, (5) Digera arvensis, (6) Parthernium hysterophorus, (7) Portulaca oleracea, (8) Ammannia baccifera.*

Sedges: *(1) Cyperus rotundus, (2) C. Iria, (3) Cyperus difformis* L *(4) Cyperus esculentus* L., *(5) Scripus tuberosus* (Desf).

Aquatic weeds: *Chara* spp, *Spirogyra* spp., *Najas* spp., *Typha* spp, *Salavania* sp:

(a) (b) (c) (d)

Fig. 38.1. Grassy weeds.

(a)

(b)

(c)

(d)

(e)

(f)

(g)

(h)

(i)

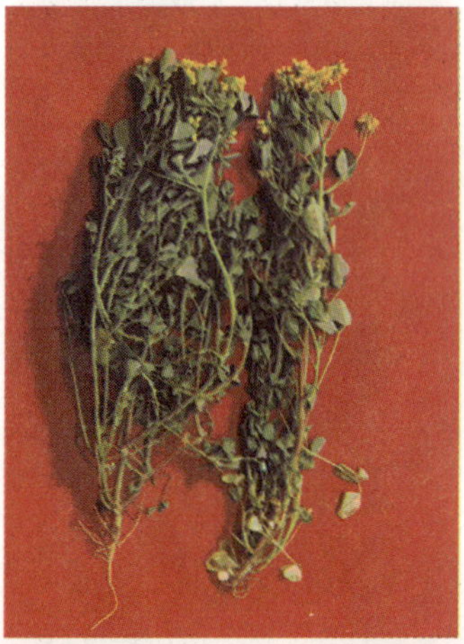

(j)

(k)

(l)

Fig. 38.2. Broad leaved weeds.

Fig. 38.3. Grasses.

Fig. 38.4. Broad leaved weeds.

Fig. 38.5. Sedges.

WHEN TO CONTROL WEEDS?

The magnitude of loss caused by weeds would depend on the type of weeds and duration of competition with the crop. Removing weeds at any time during the growing season may not be beneficial and economical. The period during which crop-weed competition is causing the highest economic losses has been identified for each crop.

Generally, in *kharif* season, weed problem is more due to frequent weed flushes. Manual weeding often leads to delayed weeding due to high demand of manual labour resulting in poor yield. Therefore, timely control of weeds during critical crop-weed competition period is important for minimizing losses caused by them. Critical period refers to the duration of crop-weed competition and can be defined as the shortest span of time in the ontogeny of the crop growth when weeding will results in higher economic returns. The crop yield level obtained by weeding during this short span should provide crop yield sufficiently close to that obtained by full crop season free from weeds. But growers or farmers often assume that removing the weeds any time during the growing season solves the problem of crop-weed competition. So time of removal of weed is as important as removal itself. Extent of reduction in the yield due to unchecked weed growth based of the analysis of 10 years data is given in Table 38.2 (Sahoo and Saraswat, 1988).

Table 38.2: Critical period of crop-weed competition in major field crops.

Crop	Critical period	Reduction in yield (%) (Days after sowing)
Wheat	30-45	20-40
Rapeseed-Mustard	15-40	15-30
Safflower	15-45	35-60
Gram	30-60	15-25
Pea	30-45	20-30
Lentil	20-45	30-40
Sugarcane	30-120	20-30
Potato	20-40	30-60
Cole crops	30-45	50-60
Onion	30-75	60-70
Rice (Direct seeded)	20-45	20-90
Rice (Transplanted)	30-45	15-40
Maize	15-45	40-60
Bajra	30-45	15-60
Sorghum	15-45	15-40
Mung	15-30	25-50
Pigeonpea	15-60	20-40
Black gram	15-30	30-50
Cowpea	15-30	15-30
Soybean	20-45	20-60
Groundnut	40-60	40-50
Sunflower	30-40	30-50
Cotton	15-60	40-50
Bhindi	15-30	40-50
Tomato	30-45	40-70

HOW TO CONTROL WEEDS?

Prevention (inhibition of weeds establishment), eradication (seeds cleaning and protective measures) and control are the major methods of weed management.

Preventive methods are employed for preventing the introduction and spread of weeds in new areas, which are used in advance of the occurrence of weed problem like use of weed seed free crop seed for sowing and removal of weed plant before producing seeds from the field, use of weed seed free FYM so that weeds seeds bank may be reduced and magnitude of weed infestation may be reduced to some level. These practices restrict the introduction, propagation and spread of weeds on a local or regional level. Wherever possible measures can be taken to minimize the risk of new weed introduction, although it is almost impossible to guard against such entry. But this is very much true that some weeds, which were introduced in past either deliberately or accidentally, have become one of the most deadly plant species of the country today. The curative methods include control and eradication (complete destruction of weeds). In case of field crops eradication/total elimination of a weed species is neither practical, economical or nor required as presence of weeds causes economic loss only during a specific period. Eradication involves the methods like manual weeding/physical, mechanical method, herbicides use and biological method which are generally used for controlling weeds.

Amongst these methods, although use of herbicide assumes to be an important strategy but it should not be encouraged much alone because it invites several problems like herbicide resistant weed. Soil solarization, stale-seed bed technique, sowing on furrow irrigated raised beds (FIRB) and adoption of suitable crop establishment technique, suitable crop rotation, closer spacing and higher seed rate are some of the techniques, which directly or indirectly influence the intensity of weeds in the crop field. Research workers from time to time have suggested various cultural, mechanical, chemical and biological control measures, but even then many weeds like Purple nutsedge (*Cyperus rotundus* L.) continue to infect vast productive land and still remains as the tropical scourge in cultivated crops. Over reliance on use of herbicides is also associated with many environmental related issues like groundwater contamination. So knowledge of successful cultural practices needs to be expanded and used to supplement mechanical and chemical control practices. Apart from above methods, their proper utilization as compost/vermicompost, biogas production, biopesticide and green manuring owing to have high mineral composition or in other way before seed setting particularly for those weeds which reproduce through seeds may be effective and viable method for their control on large scale. Considering the magnitude of menace of weeds and wide diversity in distribution of weeds flora and critical–period for crop-weed competition in field crops, problem of weeds and their management is discussed with individual crops.

RICE

In India, rice is the most important and extensively grown food crop, occupying about 43 million ha area with the production of 91.05 million tonnes and having average productivity of 2084 kg/ha (Anonymous, 2007). As discussed earlier, the infestation of heterogeneous weed flora under different crop establishment techniques (rice cultures) in rice field is one of the serious limitation in rice production. Since most of the associated weeds in rice field are C4 plants, they are more vigorous and aggressive to compete for nutrients, moisture, space and sunlight with crop plant and create an extremely adverse environment, which results in poor growth owing to low yield (Sharma and Upadhyay, 2002). Based on recent estimate, rice grain production in India is reported to suffer a yearly loss of 15 million tonnes due to weed competition. (Parthiban and Kathiresan, 2002). However, the extent of loss depends upon system of cultivation of rice, nature and intensity of weeds and their competition duration. Generally, weeds offer severe competition from the beginning of seeding of rice in upland direct seeded rice as compared to other system of rice cultures. The extent of yield reduction due to weeds is 50% in direct seeded upland rainfed rice, 51-74% in rainfed lowland rice, 30-35% in direct seeded puddled rice and 15-20% in puddled transplanted rice. In rice crop, weed free period of 30 days is required to avoid the significant loss in rice yield due to weeds. Weed infestation depleted the soil of 24.7 kg/ha nitrogen, 5.8 kg/ha phosphorus and 63.4 kg/ha potassium in one season. Therefore, their timely control is essential to realize the yield potential of rice crop. Weeds can be controlled by several methods, but integration of two or more methods, i.e., integrated weed management is effective in order to avoid ill effects of the use of signal method.

Agrotechniques

In areas of Chhattisgarh and some parts of Orissa, rainfed rice is grown in upland condition where farmers have to be very cautious right from the beginning in adopting integrated approach of weed management. Drilling method of sowing of sprouted seeds is reported to help in reducing overall weed infestation. Summer ploughing and deep ploughing before rice cultivation during the hottest month of the year is also known to reduce weed density. Localized placement of fertilizer through split application is also encouraged for providing competitive advantage to the rice crop. Two manual weedings, one at 15 DAS and other at 40-45 DAS is found adequate season long weed free environment. In areas where rainfed lowland rice is being practiced, a mechanical operation called beushening or biassi or bidahan (practice of cross ploughing the field with desi plough one month after sowing in 5-10 cm standing water) is widely practiced to reduce weed population in states of Orissa and Bihar. Weed problem is lesser in puddle direct seeded rice and puddle transplanted rice in comparison to other rice cultures. Weed species in both puddled direct sown and transplanted rice are almost common. Here 1-3 puddling decrease weeds drastically. Moisture regimes also affect the occurrence of weeds nature and their intensity. Continuous submergence with 5-10 cm water is found better than wetting and drying for reducing growth of grasses and sedge. Manual weeding at 20 and 40 DAS is good enough in absence of herbicide use. Weeds can also be controlled with the integration of herbicide along with other cultural practices. However, alone use of these methods is cumbersome, laborious, costly and less efficient.

Weed Control through the use of Herbicides: Being effective and economical, the use of herbicides assumes a significant role in controlling weeds in rice due to their selective action. Herbicides like butachlor (1-2 kg a.i./ha), oxadiazon (0.5-1.0 kg/ha), thiobencarb (1.5-2.0 kg/ha), oxyfluorfen (0.3-0.4 kg/ha), pendimethalin (1-1.25 kg/ha), pretilachlor (0.75 kg/ha) and anilofos (0.4 kg/ha) are effective in controlling weeds in rice through their pre-emergence application. Use of Almix (readymade mixture of chlorimuron and metsulfuron (50% each) @ 20 g/ha (20% EC) was also found suitable for effective control of weeds in rice (Vandana and Reddy, 1999). Besides these herbicides, some new herbicides of low dose nature and possessing

Table 38.3: Details of herbicides recommended for weed control in different cultures of rice.

Rice culture	Type of weeds to be controlled	Name of herbicides	Dose (g a.i./ha)	Time of application	Method of application
Areas where rainfed upland/ lowland rice is being grown	Grassy weeds as well as broad-leaved weeds	Butachlor	1500	1-2 days after sowing	Dissolve the required quantity of herbicides in 500 litre of water and made uniform spray over the soil surface for broad spectrum weed flora
		Anilofos	500	—do—	—do—
		Thiobencarb	1500	—do—	—do—
		Oxadiazon	750	—do—	—do—
		Oxyfluorfen	400	—do—	—do—
		Pendimethalin	1000	—do—	—do—
	For grassy and sedges	Propanil	4000	15-20 DAS	Dissolve the required quantity of herbicides in 500 litre of water and ensure uniform application at 2-3 leaf stage
	For broad leaved	2-4-D	750	—do—	—do—
Area where rice is grown through direct seeding/transpl-anting under puddle condition	Grassy weeds	Butachlor	1000	1-2 days after sowing	Dissolve the required quantity of herbicides in 500 litre of water and made uniform spray over the soil surface for broad spectrum weed flora
		Anilofos	400	—do—	—do—

Table 38.3: Contd.

Rice culture	Type of weeds to be controlled	Name of herbicides	Dose (g a.i./ha)	Time of application	Method of application
		Thiobencarb	1000	—do—	—do—
		Oxadiazon	500	—do—	—do—
		Oxyfluorfen	300	—do—	—do—
		Pendimethalin	750	—do—	—do—
		Propanil	3000	15-20 DAS	Dissolved the required quantity of herbicides in 500 litre of water and ensure uniform applicaiton at 2-3 leaf stage
		2-4-D	500	—do—	—do—
For rice nurseries	All weeds	Pretilachlor + safner	500		
	SOME	**NEW**			**HERBICIDES**
Transplanted rice	Complex weed flora	Ethoxysulfuron	20	—do—	Dissolve the required quantity of herbicides in 250 litre of water and made uniform application.
		Oxadiargyl	70-90	—do—	—do—
		Pyrazosulfuron	25-30	—do—	—do—

high soil activity like ethoxysulfuron (20 g/ha), oxadiargyl (70-90 g/ha) and pyrazosulfuron (25 g/ha) were also found suitable for effective control in rice (Raju and Reddy, 1995).

Some herbicides like propanil (3-4 kg/ha) are applied as post-emergence for controlling grassy weeds, particularly when weeds are of 2-3 leaf stage (15-20 DAS/DATP), while 2-4-D is applied as post-emergence about 20-25 DAS/DATP for controlling broadleaved weeds.

In rice nurseries sown under puddle condition, application of pretilachlor @ 0.5 kg/ha + safner 2-3 days after sowing of sprouted seeds give effective control of weeds.

WHEAT

Wheat is the most important cereal crop of India after rice and considered the most important part of the most remunerative cropping system, i.e., rice-wheat occupying about 12.10 million ha area in Indo-Gangetic plains of India. Adoption of dwarf genotypes of wheat coupled with adequate availability of nutrients and moisture led to widespread invasion of *Phalaris minor* in rice-wheat cropping system in north-west India. Even a moderate infestation of *P. minor* alone causes 15-20% reduction in grain yield of wheat (Brar and Walia, 1989). The conventional methods of its control are less effective, inefficient, labour and time-consuming because of its morphological similarity with wheat. Use of only isoproturon herbicide available in Indian market provided an acceptable level of control of *P. minor* for over one and a half decades since 1980s. But in recent years, the efficacy of isoproturon against *P. minor* has reduced in many parts of Haryana and Punjab due to the development of resistance to this herbicide. Adoption of zero tillage for sowing of wheat after rice has reported 50-60% higher wheat grain yield over conventional tillage at village Amarsipur, District Chandauli, U.P. Moreover, adoption of zero tillage for sowing of wheat after rice crop has been proved very effective strategy for the management of resistant biotypes of *P. minor* in different districts of Haryana. Zero tillage besides being economical, time saving and energy efficient, manages weeds, conserves soil along with its moisture and advances sowing time also. As a result of it, zero tillage technology of wheat sowing is becoming increasingly popular in rice-wheat cropping system of Indo-Gangetic Plains as adoption of zero tillage significantly reduced the infestation of *Phalaris minor*. Some other cultural/mechanical practices like stale-seed bed, higher seed rate, criss-cross sowing, selection of competitive dwarf varieties and crop establishment technique help in reducing weed problem in the wheat crop. Hand weeding once at 25-30 days after sowing or after first irrigation and second at 45-50 DAS is a very common practice adopted by farmers for controlling weeds in wheat.

Soil Solarization

Adoption of soil solarization in hot summer month before sowing soybean crop has shown excellent results against (Table 38.4) a wide variety of weeds. In this technique, soil temperature is enhanced by 8-12°C to be lethal to weed seeds by spreading polyethylene transparent film of 0.05 millimicron (μm) during the month of May (hottest month).

Adoption of FIRB System

Sowing of wheat crop under FIRB (furrow irrigated raised bed) system has now been proved an important component of integrated weed management in wheat crop. Under this system,

Table 38.4: Details of herbicides recommended for weed control in wheat.

Area	Type of weeds to be controlled	Name of herbicides	Dose (g a.i./ha)	Time of application	Method of application
Areas where wheat is being grown after rice (Herbicide resistance prone area)	Grassy weeds	Sulfosulfuron	25	25-30 days after sowing	Dissolve the required quantity of herbicides in 200-250 litre of water and made uniform spray over the crop.
		Clodinofop	60	—do—	—do—
		Fenoxaprop-p-ethyl	100-120	—do—	—do—
		Tralkoxydim	350		
		Metribuzin	200	—do—	Dissolve the required quantity of herbicides in 500 litre of water and ensure uniform application.
Areas where wheat is grown after other than rice crops	Grassy weeds	Isoproturon	1000	—do—	—do—
All wheat growing areas of the country	Broadleaf weeds	2,4-D (Sodium salt)	750	—do—	Dissolve the required quantity of herbicides in 500 litre of water and made uniform application. But avoids its use in some sensitive cultivars like HD 2009, WH 283, WH 416.
		Metribuzin	200	—do—	Dissolve in 500 litres of water and spray uniformly.
		Metsulfuron-methyl	4-6	—do—	Dissolve the required quantity of herbicides in 200-250 litres of water and spray uniformly.

2-3 rows of wheat are sown on the top of bed, 70 cm wide and irrigation is done through the furrows. In this system, overall infestation of *P. minor* is less and input use efficiency is reported higher.

Stale Seed Bed Technique/Use of Total Weed Killers

Considerable degree of weed control could be obtained by irrigating the field to encourage germination of the weeds and subsequently killing them with a shallow tillage. This technique is called stale seed bed technique. It has proved to be a very practical approach for weed management in all *rabi* crops as well as *kharif* crops. In case of shortage of time in view to avoid further delay in sowing the next crop, one can use the non-selective herbicides like Paraquat or Glyphosate @1.0 kg a.i./ha by mixing in 500 litres of water per hectares before planting/sowing any *rabi/kharif* crop as chemical ploughing.

Chemical Weed Control

Physical/manual removal of weeds is time and labour intensive and sometime not possible due to unavailability of labour. Under these situations, use of herbicides can be an alternative for effective, quick, economical and selective control of weeds. But optimum herbicide rates are dependent on cultural practices, soil type and environmental condition. Herbicide programme for all the crops depends upon the herbicide itself, weed flora, application techniques and time, crop tolerance and cost-effectiveness. The inconsistency in the rate of herbicide application is due to varying weed species, weed population and nature of weeds. Details of well promising herbicides including dose, time and method of application recommended for weed control in wheat are given in Table 38.4. The continuous use of a single herbicide year after year in the monocropping system may lead the evolution of herbicide resistance in a particular weed species, which may further result in weed shift from grasses to non-grasses, i.e. sedges and broadleaf weeds. In view to avoid any more instances of herbicide resistance, herbicide use of different mode of action may be rotated depending upon the selectivity of the concerned crops.

MAIZE

Maize is an important *kharif* cereal crop of India. This crop has a very high yield potential, particularly the hybrids, which may produce as high as 6 t/ha, but the average yield (2.5 t/ha) is considered to be very low due to various constraints. Wide row spacing in maize yield coupled with favourable environment allows luxuriant weed growth, which may reduce the yield by 30-90%. In addition, maize is generally raised under marginal conditions with meager inputs, which make it a poor competitor. Timely weed control in maize has become essential for realizing its potential yield (58.0 q/ha).

Major Weed Flora: Knowledge of the type and nature of weed species commonly occurring in maize crop is the prerequisite for their effective and economical management, as single method of weed control may not be able to work for the control of all weed species. Generally, three types of weeds species are found in maize crop.

Grassy Weeds: *Echinochloa colonum, Echinochloa crusgalli, Acrachne racemosa, Digitaria sanguinalis, Dactyloctenium aegyptium, Paspalum dialatum* and *Cynodon dactylon.*

Broadleaved Weeds: *Trianthema portulacastrum, Trianthema monogyna, Digera arvensis, Commelina benghalensis, Phylllanthus niruri, Xanthium strumarium, Boerhaavia diffusa, Oxalis corniculata* and *Parthenium hysterophorus.*

Sedges: *Cyperus rotundus, Cyperus esculentus* and *Cyperus iria.*

Suitable Time for Weed Control

Occurrence of weeds in cropped field is always harmful as they are competing with crop for all the natural resources. But their presence/unchecked growth during certain period of crop growth causes the highest losses in crop yield. That period of crop growth is called critical period for crop-weed competition, which is different for different crops. It is well established that first 30 days after sowing are critical period of weed competition in maize. Therefore, maize crop should be kept weedfree for at least first 30–45 days after sowing to obtain full potential of newly developed high-yielding cultivars.

How to Control Weeds?

Weeds can be controlled through the use of several methods, viz. mechanical method, cultural method, biological method and herbicidal (chemical) method. But application of no single method has generally proved effective and economical. But several flushes of weeds due to incessant heavy rains, non-availability and high wages of labour make manual weeding difficult, tedious, back breaking, unpractical and finally uneconomical crop husbandry techniques like stale seed bed technique (a practice of allowing to germinate the weed seeds and killing them subsequently before sowing crop), tillage practices (conservational/conventional), soil solarization (mulching) planting techniques (raised furrows) could be combined in an integrated way to make weed control strategy effective and also economical. Practice of growing maize with cowpea (1:1) as intercrop also helps in reducing the weed growth. Cowpea is known as cover crop, which does not leave the space in between two rows of maize for weeds emergence. In maize crop, several flushes of weeds appear due to adequate moisture, which becomes the major constraint for its productivity. Therefore, weeds should be removed through intercultural operation at 20 and 40 days after sowing in order to obtain the potential yield of *kharif* maize (60 q/ha).

Chemical method of weed control: Application of chemical for the control of weeds is quick and cheaper than mechanical method of weed control. Choice of herbicides, their dose and time of application is very important for obtaining higher weed control efficiency. Several herbicides along with their dose and time of application recommended for controlling weeds in maize are given in Table 38.5. Results of the field experiments conducted at Research Farm of I.A.R.I. for evaluating the efficacy of various herbicides alone and tank mix application revealed that tank mix application of atrazine + pendimethalin (500 + 750 g/ha) before the emergence of both weeds and crops recorded the highest increase in productivity of maize over unweeded situation (Table 38.6). However, alone application of atrazine (1000 g/ha pre-em.) and pendimethalin (1000 g/ha pre-em) was found equally effective in controlling weeds but inferior to their tank mix application.

Table 38.5: Name, dose and time of application of recommended herbicides for weed control of maize.

Name	Dose (g a.i./ha)	Time of application	Remarks
Atrazine	1000	Apply with 500 litre/ha of water before emergence of crop as well as weeds	Control all weeds except *Dactyloctenium aegyptium* and *Cyperus rotundus*
Atrazine	1000-1500	Can be safely applied upto 15 days after sowing of maize crop	-do-
Pendimethalin	750-1000	Apply with 500 litre/ha of water before the emergence of crop as well as weeds	Control all weeds except *Commelina benghalensis* and *C. rotundus*
Metribuzin	200-300	It can be safely applied both pre-emergence and post-emergence (upto 15 days after sowing)	It is effective against both broadleaved and grassy weeds.

Table 38.6. Effect of pre-emergence herbicides on weed growth and maize yield.

Treatments	Dose (g a.i./ha)	Weed population (No/m^2)	Weed dry wt.(g/m^2)	Grain yield (q/ha)
Weedy check	-	124.3	84.5	28.1
Hand weeding	-	16.6	5.8	42.7
Atrazine	1000	42.6	32.2	39.1
Pendimethalin	1000	53.3	45.2	37.6
Atrazine+ Pendimethalin	500+750	19.6	8.7	42.9
FOE(5043)	150	86.0	62.7	31.6
FOE(5043)	300	75.3	54.9	31.7
FOE(5043)+ Metribuzin	150+200	47.3	39.7	40.1

FOE- Flufenacet

In other field experiments, it has also been proved that atrazine (1000 g/ha) can also be applied as post-emergence (15DAS) without any toxic effect on maize plant, if it could not have been done as pre-emergence (Table 38.7).

Table 38.7: Effect of post-emergence herbicides on weed growth and maize yield.

Treatments	Dose (g a.i./ha)	Grain yield(q/ha)
Weedy check	-	22.3
Atrazine 15 DAS	1000	33.5
Atrazine 30 DAS	1000	30.9
Atrazine 15 DAS	1500	35.1
Atrazine 30 DAS	1500	34.9
Metribuzin 15 DAS	100	29.7
Metribuzin 30 DAS	100	28.1
Metribuzin 15 DAS	200	35.3
Metribuzin 30 DAS	200	31.5

SORGHUM

Sorghum is an important crop of the semi-arid tropics and a major staple food grain in several parts of India. Weeds are a serious problem in *Kharif* sorghum as the sowing of crop coincides with the commencement of monsoon and the congenial temperature condition. The traditional surface irrigation methods encourage severe weed growth as the inter row spaces and unwanted areas are also irrigated (Patel, 1999). Weeds in general cause 45% of annual loss in agricultural production (Singh, 1999), and their competitive ability and aggressiveness increase with increasing resource availability like nutrients and water.

Nature of Weed Problem

The dominant weeds found associated with this crop are *Cynodon dactylon* **(*L.*)** Pears., *Dinebra retraflexa* Panz, *Eleusine aegyptiacum* (L.) Desf., *Sorghum halepense* L., *Commelina benghalensis* L., *Acalypha indiaca* L., *Digera arvensis* Forsk., *Euphorbia hiruta, Marremia emarginata* Hallier.f., *Corchorus* spp., *Striga* spp. *and Cyperus rotundus* L in *Kharif* sorghum and ***S**onchus arvensis* L., *Chrozophora rottleri Klotzrch* and *Phyllanthus maderaspatensis* L. in *rabi* sown sorghum.

Weed Control

Manual weeding and other cultural practices may be done within six weeks after sowing of crop, as early weed competition in the crop is more serious. However, in sorghum sown in *rabi*, one hand hoeing within 25-30 DAS is found sufficient. But no weeding is required for the crop sown of fodder purposes.

Among herbicides pre-emergence application of atrazine (0.75-1.0 kg/ha), isoproturon (0.75 kg/ha), alachlor (1.0 kg/ha) and pendimethalin (1.0 kg/ha) found safer in sorghum sown for grain and fodder purpose. 2-4-D and atrazine can also be applied as early post-emergence for controlling established weeds. Integration of herbicide use (metolachlor @ 1.5 kg/ha as pre-em.) along with intercropping of black gram resulted the highest increase the grain yield of sorghum (Sundari and Kathiresan, 2002).

Striga, a parasitic weed, can be controlled effectively by stimulating germination as suicidal germination of its seeds by growing some catch/trap crops like cotton, sunflower, groundnut and castor. However, directed application of 2-4-D Amine salt at the rate of 0.5-0.75 kg/ha is a very practical alternative for breaking future population of striga.

Pearl Millet

Pearl millet [*Pennisetum glaucum* (L.) R.Br. emend Stuutz] is widely cultivated for grain and fodder in arid and semi-arid parts of our country as rainfed crop. India is the largest producer (7.7 mt) of pearl millet with an area of 9.7 m ha. It is sown at the onset of rainy season and becomes infested with grassy and broadleaved weeds, which cause severe competition with the crop resulting in yield reduction upto 60% (Malik *et al.*, 1980). Amongst various constraints weeds emergence and depletion of huge amount of moisture further increase the scarcity of water in dry land areas. Therefore, poor weed management is one of the factors limiting the productivity of the crop.

The predominant weeds generally associated with this crop are *Cynodon dactylon* (L.) Pears., *Echinochloa colonum* L., *Digitaria* spp, *Eleusine aegyptiacum* (L.) Desf., *Cenchrus* spp., *Tribulus terristris* L., *Tephrosia purpurea* (L.) Pears. *Celosia argentea L., Commelina benghalensis* L., *Digera arvensis* Forsk., *Euphorbia hirta, and Phullanthus niruri* L. *Trianthema monogyana* Linn., *Cyperus compresses* L., *Bulbostylis barbata* (Rottb) Cl. and *Cyperus rotundus* L.

Weed Control

In this crop two hand weeding, one at 15-20 days after sowing and another at 35-40 days are found sufficient to check weed growth. New flushes of weeds are taken care of by the smothering potential of this crop. But the safety range for herbicide use in this crop is comparatively less. Of the various pre-emergence herbicides, atrazine is very effective at 0.50-0.75 kg/ha. However, it can also be applied as early post-emergence. 2,4-D Na salt can also be applied as post–emergence against broadleaf weeds and sedges (Sandhu *et al.*, 1988).

Weed Management in Oilseed Crops

India is one of the largest oilseed producing countries that covers nearly one-fifth of the entire area under this group of crops and yields nearly one-tenth of the total oilseeds production of the nine annual oilseed crops in the world. In India, oilseeds are the second largest agricultural commodity after cereals. These occupy about 13.5% of the gross cropped area in the country and account for 5% of the gross national product and 10% of the value of all agricultural products. But for last few years, the area and production has become stagnated despite set up of a Technology Mission on Oilseeds by the Government of India in May 1986. Amongst several constraints recognized in increasing productivity of oilseeds, severe competition offered by weeds for nutrients, water and energy is one the most important factors responsible for deplorably low yield.

In India, nine annual oilseeds crops, namely, groundnut, sesame, niger, castor, linseed, mustard and rapeseed, sunflower and safflower are grown in various states. In rainy season, weed problem is comparatively higher and more flushes of weed emerged in comparison to winter season. Details of major weed control practices adopted in major oilseed crops are discussed here.

GROUNDNUT

Groundnut (*Arachis hypogaea* L.) is generally sown at the onset of monsoon or just before the onset of monsoon. Initial slow growth combined with prostate nature of its growth and hot humid climate prevailing during the *Kharif* season permit early and severe crop-weed competition resulting in loss of yield to the tune of 35 to 80% (Ghosh, 2000). Rathi and Sharma (1987) noted strong negative correlation of weed biomass/ha with pod yield and reduction in pod yield was predicted to the extent of 74.4 kg/ha with increase of 100 kg weed dry weight/ha.

The major weed flora found in groundnut crop is:

Echinochloa crusgalli L., *Echinochloa colonum* L., *Cynodon dactylon* (L.) Pears *Digitaria* spp, *Eleusine aegyptiacum* (L.) Desf., *Dactyloctenium aegyptium* L., *Brachiaria* sp, *Cenchrus* spp., *Tribulus terristris* L., *Tephrosia purpurea* (L.) Pears. *Celosia argentea* L., *Commelina benghalensis* L., *Digera arvensis* Forsk., *Euphorbia hirta,* and *Phyllanthus niruri* L. *Trianthema monogyana* Linn., *Cyperus compresses* L., *Bulbostylis barbata* (Rottb) Cl., *Cyperus rotundus* L. and *Cyperus iria* L.

Weed Control

Groundnut heavily suffers from weed stress during 4-8 weeks after seeding. Raghvani et al. (1984) reported first 15 to 60 days as most critical period of weed competition. Use of semi-spreading (110 kg/ha seed rate) and bunch type (125 kg/ha), reduced spacing to 20 cm and selection of fast growing varieties like GAUG1 and AK 12-24 are some of the techniques which help in suppressing the weed growth. Infestation of some perennial weeds like *Cynodon dactylon* (L.) and *Saccharum spontaneum* L. can also be reduced through deep summer ploughing. Generally, two hand weeding, one at 20 days and other at 40 days after sowing before pegging are sufficient for weed control.

Presently alachlor (1.5-2.0 kg/ha), pendimethalin (0.75-1.5 kg/ha), metalachlor (1-2.0 kg/ha) and oxyflurofen (0.20-0.30 kg/ha) as pre-emergence and fluchloralin and trifluralin (0.75-1.0 kg/ha) as pre-planting incorporation are reported suitable for broad spectrum weed control in this crop. Some herbicides like fluzifop-p-butyl and haloxyfop-methyl (0.25-0.50 kg/ha) can also be applied as post-emergence herbicides at 20-25 days after sowing.

RAPESEED-MUSTARD

Mustard (*Brassica juncea* (L) (Czern. & Coss) and Rapeseed (*Brassica campestris* var. toria duth) are commonly known as rai and laha and toria respectively. Besides these two, there are yellow sarson, brown sarson, gobhi sarson and taramira. These are grown during winter season. Invasion of weeds in mustard reduced yield ranging from 38 to 49% (Singh and Singh, 1993). Weeds compete with the crop plant and remove nutrients faster in relatively larger amount than crop, resulting in significant yield loss. About 30-40% removal of applied nutrients by weeds has been reported by Dryden and Krishnamoorthy (1977).

The major weeds that cause considerable losses in mustard-rapeseed are *Phalaris minor, Avena ludoviciana, Cynodon dactylon,Chenopodium album, Chenopodium murale, Vicia sativa, Lathyrus aphaca, Cichorium intybus, Anagallis arvensis, Coronopus didymus, Convolvulus arvensis, Cannabis sativa, Argemone mexicana, Melilotus indica, Spergula arvensis, Fumaria perviflora, Cyperus rotundus, Cyperus esculentus.* However, toria, which is being sown early in the month of September, is also infested with some of the *Kharif* season annuals, viz. *Trianthema portulacastrum, T. monogyna, Digera arvensis, Amaranthus viridis* L. *Phyllanthus niruri* L. and *Dactyloctenium aegypticum.*

Apart from the above weed flora, mustard is also heavily infested with broomrape (*Orobanche aegyptiaca* Pers.), a holoparasitic weed particularly in the light and unirrigated and in low fertile soils.

Weed Control

The period between 20 and 40 days after sowing is identified as critical period of weed competition for mustard. Therefore, weeds should be removed in third week after sowing. To control weeds one hand weeding at 20 DAS and other at 35-40 DAS is considered necessary. Amongst cultural practices higher seed rate of 20 kg/ha under unirrigated and 10 kg/ha under irrigated condition is more effective for weed control. Drilling of seeds in narrow row spaced at 30 cm for closer short statured varieties will also help in smoothering the weeds. Sometimes time of sowing of mustard in the first fortnight of October and of toria during early September to mid of October also help in the suppression of weeds.

In areas where farm labour is not available or is expensive, herbicides can be used for effective control of weeds in rapeseed-mustard fields. A pre-plant incorporation of fluchloralin or trifluralin @1.0-1.25 kg/ha is recommended for broad spectrum weed control (Pandey, 1980). Anand and Vaishya (2002) at Faizabad recorded the highest yield of mustard (1357 kg/ha) with pre-plant incorporation of fluchloralin at 1.0 kg/ha owing better weed control. At Ludhiana, Gill *et al.* (1982) found greater effectiveness of pre-emergence application of pendimethalin and oxadiazon at 0.5 and 0.75-1.0 kg/ha. Post-emergence application of isoproturon (1.0 kg/ha) was also reported effective in controlling weeds with tolerable injury to crop (Ahuja and Yaduraju, 1993). Isoproturon appears also to be good for wheat-mustard mixed/intercropping (Rai *et al.* 1993). Two hand weedings at 25 and 50 days after sowing and integration of fluchloralin at 0.75 kg/ha supplemented by one hand weeding at 25 days after sowing found significantly superior in terms of reducing weed density and dry weight of weeds over fluchloralin at 1.0 kg/ha and weedy check. However, maximum net profit and benefit: cost ratio were recorded with fluchloralin at 0.75 kg/ha + one hand weeding at 25 days after sowing (Singh, 2006). For the control of *Orobanche* hand weeding at weekly intervals is effective in preventing its further spread, but it is too costly an operation. Herbicidal control of Orobanche has not been found possible in mustard as herbicides found effective against *Orobanche* are not selective to mustard crop. However, spray or tip contact with certain vegetable oils and terbutol were reported effective against the growth of *Orobanche* (Krishnamurthy, 1992). Repeated applications of these oils were needed to destroy the new flushes of the weed and it is labour intensive. Kumar (2002) reported dessication and prevention of seed setting of the treated part of *Orobanche* with mustard, sunflower, castor, groundnut and neem oil but subsequently provoked the emergence of other shoots of the broomrape on the mustard crop and resulted in no reduction in the dry weight of the broomrape when compared with untreated crop.

Other Oilseed Crops: In other remaining crops nature of occurrence of weed flora is almost similar depending upon the season. Description of all practices for each crop would be quite lengthy. However, the detail of herbicides recommended for different oilseed crops is given in Table 8. A range of dose is suggested where lower doses are employed in light soils and higher doses in heavy soils. Instead of attempting to control weeds with a high dose, it is desirable to use lower dose followed by a manual weeding for better control and higher productivity.

Table 38.8: Details of herbicides recommended for weed control in oilseed crops.

Crop	Name of herbicides	Dose (kg a.i./ha)	Time of application
All crops	Fluchloralin (Basalin 45% EC)	0.75-1.5	Pre-planting Incorporation
All crops	Trifluralin (Treflan 48% EC)	0.75-1.0	—do—
Soybean	Imazathepyr (Pursuit 10% EC)	0.10-0.150	—do—
Groundnut, Niger, Soybean, Sesame Sunflower, Safflower	Alachlor (Lasso 50% EC)	1.5-2.0	Pre-emergence
Soybean	Chlorimuron (Kloben 25% WP)	0.009-0.012	—do—
	Imazathepyr (Pursuit 10% EC)	0.10-0.150	—do—
	Metribuzin (Sencor 70% WP)	0.250-0.500	—do—
All crops	Pendimethalin (Stomp 30% EC)	0.75-1.50	—do—
Soybean, Groundnut, Sunflower, Mustard, Safflower, Niger, Linseed	Oxadiazon (Ronster 25% EC)	0.5-1.0	—do—
Soybean, Groundnut, Sunflower, Mustard, Linseed	Oxyfluorfen (Goal 23.5% EC)	0.075-0.150	—do—
Mustard, Toria	Isoproturon (Arelon 75% WP)	0.5-0.75	—do—
Linseed	Isoproturon (Arelon 75% WP)	0.75-1.0	—do—
Soybean	Chlorimuron (Kloben 25% WP)	0.009-0.012	Post-emergence
	Imazathepyr (Pursuit 10% EC)	0.10—0.15	—do—
	Fenoxaprop (Puma super 10% EC)	0.15—0.20	—do—
Mustard	Diclofopmethyl (Illoxan 28% EC)	0.50—0.75	—do—
Linseed	Isoproturon (Arelon 75% WP)	0.75-1.25	—do—
Mustard, Toria,	Isoproturon (Arelon 75% WP)	0.50-0.75	—do—
All crops	Fluazifop-butyl (Fusilade 28% EC)	0.25-0.50	—do—

WEED MANAGEMENT IN PULSE CROPS

Pulse are an important component of food grain crops because of their high nutritive value (protein content ranging from 17-27%) and adaptability to a wide range of agro-ecological, cropping system and management variables. There are several pulse crops considered important at various locations throughout the region. Mungbean, urdbean, pigeonpea, cowpea, chickpea, lentil and peas contribute more than 80% to the total acreage and production of pulse in the region.

Lack of adequate weed management has been found to cause 30-80% reduction in the yield of pulses cultivated during the monsoon season (9Ali, 1984), 21-30% during the winter season (8Anonymous, 1978-84) and 70-80% during the summer and spring season (6Anonymous, 1981-92). The degree of reduction in yield of these pulse crops due to weeds depends upon the density and duration of weed species and fertility status of soil.

The slow initial growth characteristics of grain legumes necessitate adoption of sound weed management practices at the early growth stages of these crops.

The Major Weeds

The magnitude of weed problem in pulses varies with the agro-ecological condition and the level of management. Major weeds associated with pulses are listed in following table:

Season	Predominant weeds
Kharif	
Annual grasses and sedges	*Echinochloa crusgalli* L., *Echinochloa colonum* L *Cynodon dactylon* (L.) Pears *Digitaria* spp, *Eleusine aegyptiacum* (L.) Desf., *Dactyloctenium aegyptium* L., *Cyperus compresses* L., *Bulbostylis barbata* (Rottb) Cl., *Cyperus rotundus* L. and *Cyperus iria* L.
Broadleaved weeds	*Tribulus terristris* L., *Tephrosia purpurea* (L.) Pears. *Celosia argentea* L., *Commelina benghalensis* L., *Digera arvensis* Forsk., *Euphorbia hirta, and Phyllanthus niruri* L. *Trianthema monogyana* Linn., *Trianthema portulacastrum* L.., *Amaranthus viridis* L., *Cleome viscosa*
Perennials	*Cyperus rotundus* L. and *Cyperus iria* L. *Cynodon dactylon* (L.) Pears, *Sorghum halepense* and *Saccharum spontaneum*
Rabi	
Annual grasses	*Phalaris minor* Retz., *Avena ludoviciana,* Dur, *Cynodon dactylon, and Poa annua* L.
Broadleaved weeds	*Chenopodium album* L. *Chenopodium murale, Vicia sativa, Rumex dentatus Lathyrus aphaca, Medicago denticulate, Cuscuta* spp. *Cichorium intybus, Anagallis arvensis, Coronopus didymus, Convolvulus arvensis, Cannabis sativa, Argemone mexicana, Melilotus indica, Spergula arvensis, Fumaria perviflora,*
Perennials	*Cirsium arvense, Cyperus rotundus* L. and *Cynodon dactylon (L.)* Pears

Weed Control : The weed management in pulses requires an integrated approach that utilizes effective cultural, mechanical, biological, ecological and chemical methods of weeds control in a mutually supported manner into the crop production system. Emergence of weeds in pulses begins simultaneously with the crops leading to severe competition between weeds and the crops right from the very early stages. In pulses grown during monsoon season, weeds emerge in succession almost throughout the crop season because of favourable environment condition and frequent rain. Therefore, identification of critical duration of weed-crop competition is of utmost importance to make weed management practices more effective and economical.

The initial 30-40 days period was found to be most critical for weed-crop competition in summer and rainy season pulse crops. While in winter pulses, the critical–period was found to be 30-60 days after sowing. So pulse crops are very sensitive to weed competition during this period and effective weed control measures are required to be taken on priority basis.

Inter-culture operations are performed chiefly to destroy the weeds present in the crops. Hand hoeing and manual weeding are the most common practices performed for weed control in almost all pulse crops. Numerous studies are conducted to find out the optimum frequency and stage of manual weedings or intercultivation in different pulse crops at several locations. Invariably, 1-2 weedings done during critical period of weed-crop competition provide satisfactory weed control in all pulses. The number of weedings may vary with the nature of weeds. Generally, extra weedings may be required when fields are infested with perennial weeds.

Weed management through manual weedings, hoeings or inter-cultivation thought effective in reducing weed competition are time and labour intensive. During monsoon season, unfavorable soil and weather condition do not permit these operations at proper stage resulting into delayed weed control. Under such conditions use of herbicides may become an essential component for integrated weed management. Pre-plant incorporation of fluchloralin at 0.75-1.0 kh/ha has been found to be very effective against most of the annual grasses and non-grassy weeds in mungbean, urdbean, lentil, chickpea, pea and pigeonpea. Alachlor at 2.0-2.5 kg/ha applied pre-emergence provides effective weed control in many pulses during monsoon season. Oxadiazon has been tested at several locations in pigeonpea, chickpea, and lentil. Pre-emergence application at 0.5 kg/ha provided effective control of most of the annual grasses and broadleaf weeds associated with these crops.

Weed management in other field and vegetable crops: By using following herbicides, we can control the weeds in the following crops.

Table 38.9. Details of herbicides recommended for weed control in other field crops.

Crop	Name of herbicides	Dose (kg a.i./ha)	Time of application	Method of application
Potato	Metribuzin	0.5	1-3 DAS	Dissolve the herbicide in 500 litre of water and apply unifomily
	Pendimethalin	1.0	—do—	
Sugar cane	Atrazine	2.0	7-10	—do—
	2,4-D	0.75	30-40 DAS	—do—
Cole crops Cabbage/ Cauliflower	Fluchloralin	1.0	Pre-planting incorporation	Apply the herbicide before sowing the crop and mixed well in soil.
	Pendimethalin	1.0	1-3 DAS	Dissolve the herbicide in 500 litre of water and apply unifomily
Onion/Garlic/ Brinjal	—do—	1.0	1-3 DAS	—do—
Cotton	Pendimethalin	1.0	1-3 DAS	—do—

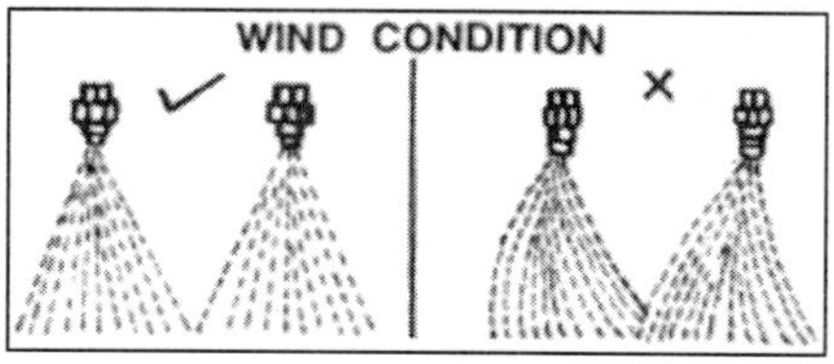

Fig. 38.6.

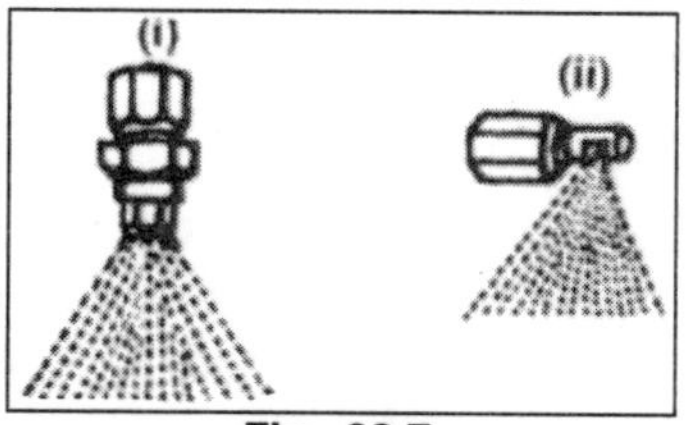

Fig. 38.7.

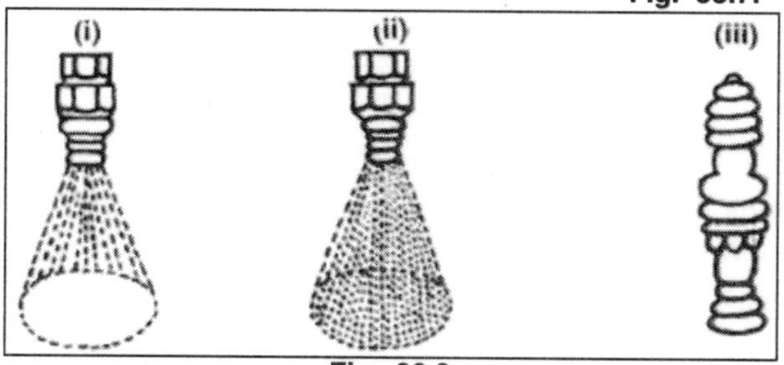

Fig. 38.8.

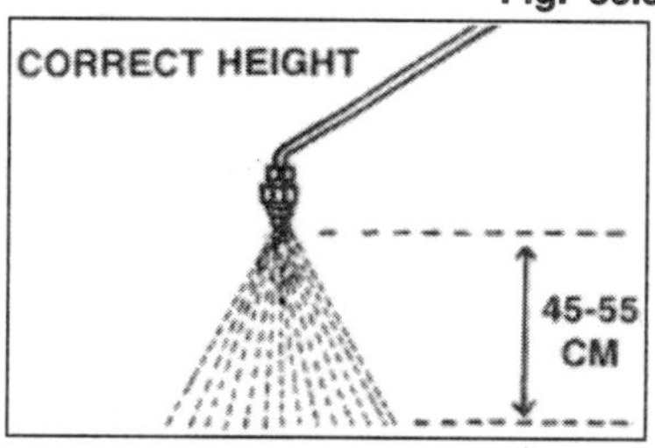

Fig. 38.9.

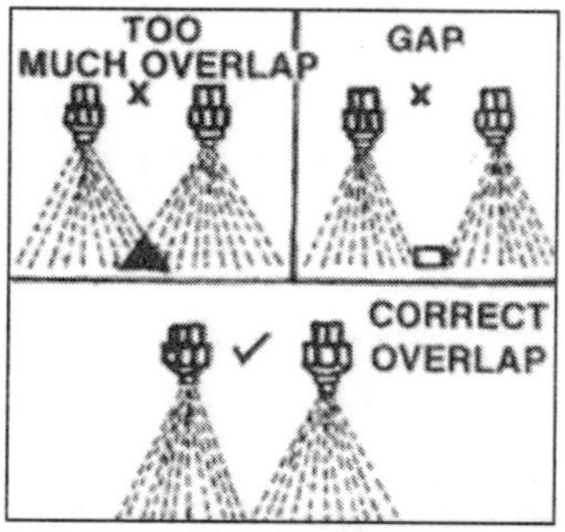

Fig. 38.10.

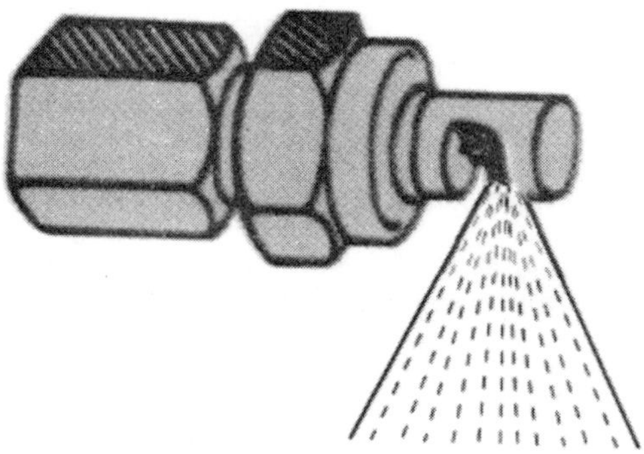

Fig. 38.11. Floodjet nozzle.

Fig. 38.12. Flatfen nozzle.

PRECAUTIONS TO BE TAKEN WHILE PERFORMING SPRAY OF HERBICIDES

Selection of suitable herbicides, calculation of their optimum dose and procurement of right type of spraying machine would not make your spray successful till the right nozzle and required spray pattern is not obtained. Besides this some important points to be followed during spraying of herbicides are listed below along with direction of spray with different types of nozzles (Fig. 38.7 and Figs. 38.11 & 38.12).

- Follows all the instruction given on the packing of herbicide properly.
- Hand weeding may be given wherever necessary in addition to herbicide treatment.
- Use lower rates in light soils and higher rates in heavy soils.
- These recommendations are based on the sole cropping. For herbicide use in intercropping system, consult the weed scientist.
- Apply herbicide using a sprayer with recommended amount of water.
- Apply spray solutions immediately after preparation. If delay is inevitable, stir the solution or shake the sprayer thoroughly before application to prevent settling down of herbicide in the bottom.
- Flat fan nozzle should be used for spraying the herbicide.
- Separate sprayer need to be used for herbicide and for insecticide spray.
- Avoid spraying on windy days as drifting may damage the neighbouring susceptible crops (Fig. 38.6).
- Do not spray against the wind. Avoid smoking or eating during spray application.
- Wash sprayer thoroughly after each spray.
- Handle chemicals carefully. Always keep them away from children and food articles.
- Dispose of the empty container by burrying in the soil or by burning.
- Use protective clothing (gum boots, gloves, etc.), goggles and mask wherever possible.

REFERENCES

Anonymous (2007). Data Book-2007 Published by IASRI and ICAR, DIPA, KAB-I, Pusa, New Delhi, pp. 211.

Ahuja, K.N. and Yaduraju, N.T. (1993). Response of rapeseed-mustard to different herbicides. *Proceedings of International Symposium of Indian Society of Weed Science,* Hisar, 126-127.

Anand, Kul Bhooshan and Vaishya, R.D. (2002). Effect of weed control measures on weeds, nitrogen uptake by weeds and yield of mustard. *Indian J. Weed Sci.,* **34**(3&4): 305-306.

Brar, L.S. and Walia, U.S. (1989). Economic evaluation of weed control system in different field crops. *Pesticides,* **23:** 27-30.

Dryden, R.D. and Krishnamoorthy, C.H. (1977). Year round tillage. *Indian J. Weed Science.* **9:** 14-18.

Ghosh, D.C. (2000). Weed management in rainfed groudnut (*Arachis hypogaea*). *Indian J. Weed Sci..* **32**(1&2): 92-93.

Gill, K.S. (1974). Wheat agronomy of dwarf wheat. Research on dwarf wheat, ICAR, New Delhi.

Gupta, O.P. (1998). Modern weed management–A textbook and manual. Published by Agro Botanica, pp. 19.

Keeling, W. (1995). Weed and cover crop management in conservation tillage for the south-west. In: *Proc. Baltwide Cotton Conf., San Antonio, Texas, USA.* 4-7 *January.* **1:** 73-74.

Krishnamurthy, G.V.G. (1992). Two new methods to control *Orobanche* on tobacco. *Indian Farming.* **426**(1): 7-8.

Kumar, Subhash. (2002). Preliminary studies on the control of broomrape (*Orobanche aegyptiaca*) in mustard. *Indian J. Weed Sci.* **34**(3&4): 303-304.

Maheshwari. J.K. (1963). The Flora of Delhi (Reprinted in 1990). Council of Scientific and Industrial Research, New Delhi, pp. 447.

Malik, R.K. and Singh, S. (1995). *Weed Tech.,* **9:** 419-425.

Malik, D.S., Balyan, R.S. and Dhankar, R.S. (1980). Cultural and chemical weed control in bajra. *Ann. Arid Zone* **19**: 103-109.

Mishra, J.S. and Gautam, K.C. (1995). *Indian Fmg.*, **33:** 36-38.

Pandey, J. (1980). Higher yield of rai (*Brassica juncea*) with herbicides and nitrogen in calcareous soil. *Indian J. Weed Sci.* **17:** 178-180.

Parthiban, C. and Kathiresan, R.M. (2002). Use of certain plant materials for weed management in transplanted rice. *Indian J. Weed Sci.*, **34**(3&4): 187-191.

Patel, V.B. (1999). Using waste sensibly. In: *The Hindu Survey of Indian Agriculture.* pp.165-168.

Rai, B., Singh, R.P., Singh, Y., Singh, R. and Singh, R.K. (1993). Technology for increasing rapeseed and mustard production in Eastern Uttar Pradesh. *Indian Farming,* **43:** 3-6.

Raju, R.A. and Nageswara Reddy, M. (1995). Performance of herbicide mixtures of weed control in transplanted rice (*Oryza sativa* L.). *Indian J. Weed Sci.,* **27**(1&2): 106-107.

Rao, A.S. and Rao, R.S.N. (2006). *Indian J. Weed Sci.* **38**(1&2): 33-36.

Rathi, G.S. and Sharma, R.S. (1987). An analysis of correlation between weed parameters and yield in rainfed groundnut. *Indian J. Weed Sci.*, **19**(3&4): 243-246.

Tandon, H.L.S. and Pratap Narayan. (1990). Fertilizer in Indian Agriculture – Past, present and future (1950-2000). FDCO, New Delhi, India pp. 160+ viii.

Tiwari, K.N. (2002). *Fert. News* **47**(8): 23-52.

39

Role of Composting in Organic Farming and Plant Protection

Livleen Shukla, Archana Udai Singh[1] and D. Prasad[1]

Division of Microbiology and [1]Division of Nematology,
Indian Agricultural Research Institute, New Delhi–110012
E-mail: arch_212@yahoo.com

Composting is the microbiological production of humus and increases availability of plant nutrients from the conversion of organic wastes. Organic wastes include all materials of plant and animal origin developing in or applied to soil. The organic components in soil undergo dynamic changes and is in various stages of decomposition. Organic manures and compost contribute a great deal to the humus content of soil and are essential inputs for soil management. (Conghos *et al.*, 2003). Recycling of organic wastes improves the soil structure, waterholding capacity, drainage, base exchange capacity and checks soil erosion. (Abdel-Azeem, 2001). Humus substances have been found to increase the efficiency of bio and chemical fertilizers. (Riegel and Noe, 2000). Organic wastes such as straw, organic manure and neem seed cake are known to check proliferation of plant nematodes. Also humified organics have been found to reduce the toxic effects of pesticides to crop.

WHAT CAN BE COMPOSTED?

India has a vast potential of manurial resources like animal wastes, which include shed wastes of cattle, buffalo dung, urine, spoiled fodder, sheep and goat droppings, by-products of slaughter houses, blood meat wastes, bones, leather and hair wastes. Another category is crop residues, tree wastes and aquatic weeds like wheat, paddy, bajra, jowar, grams, mung, urad, arhar, etc. water hyacinth, forest litter, urban waste include city garbage, sewage, sullage and rural waste FYM, urine, litters, green manure like sunhemp, dhancha, senji, clusterbeans, cowpea, horsegram, berseem, etc. (Beary *et al.*, 2002; Eldin *et al.*, 2001; Anonymous, 2005). Marine wastes is another area that includes fishmeal and sea weeds. Organic recycling can be recycled by composting, direct incorporation in soil or through organic mulch.

HISTORICAL BACKGROUND

The oldest reference of origin of composting relates to the Akkadin empire of Mesopotamian Valley, i.e. before Moses was born. Later, Verro (40 BC) highlighted stress on the complete decay of manure before applying to crops. Fermentation was brought about by stering the manure in moist condition. Columella (AD 90) highlighted the use of pits for keeping up the optimum moisture condition. Renaissance literature also has the reference of use of stable manure in composting (18th century).

The scientific study of composting was initiated by Hutchinson and Richards of Rothamsted (1921) and elaborated by Sir Albert Howard, who is also known as father of the 'organic method'. Howard developed the Indore method of composting, which is named after place Indore where he was working. This process has been widely used in India by Acharya and co-workers.

Some countries like India, Korea, Nepal, Thailand, Vietnam, Indonesia have taken a lead in recycling, and research work has been going in IARI from many years.

ORGANISMS IN COMPOSTING

Compostable waste materials normally contain a large number of many different types of bacteria, fungi moulds and other living organisms. During decomposition, marked changes takes place in the nature with abundance of biological population. The microbial population fluctuates during aerobic composting. Chang and Hudson (1967) have described a typical pattern. The fungi and acid producing bacteria appear during the mesophilic stage. As the temperature increases above 40°C, these organisms are replaced by thermophilic bacteria, actinomycetes and fungi. Spore forming bacteria develop at temperatures above 70°C. Finally, mesophilic bacteria and fungi reappear as the temperature falls again.

Many aerobic bacteria are initially present and multiply, but after their increased activity, as the temperature rises, their number decreases. A minimum is reached at 55–65°C, but their numbers increase again as the temperature drops below 50°C.

The role of mesophilic bacteria may be involved in raising of temperature of compost for development of the thermophilic bacteria, which colonise the mass. Mesophilic bacteria that flourish during a limited time consume most of the readily degradable carbohydrates and proteins. *Bacillus* spp. are especially involved in degradation of proteins, amino acids, peptones and blood meal.

Actinomycetes degrade starch very actively and also bring about large losses of water soluble fractions. Thermophilic bacteria do not appear to be important in degradation of cellulose and lignin but attack protein, lipids and hemicelluloses. Lipids are degraded to great extent, whereas degradation of celluloses and hemicelluloses is comparatively slow and intermediate. Lignin is the compound most resistant to decomposition. Actinomycetes (*Thermomonospora curvata*) may be important in cellulose decomposition. Mesophilic fungi are present as compost warms up, they are quickly replaced by thermophilic fungi, but reappear in large numbers as heap cools below 40°C. It appears that they persist in the outer layers of heap during the thermophilic stage and become active again as the temperature drops. The mesophilic fungi can utilize cellulose and hemicellulose but not as efficiently as do the thermophilic fungi.

Thermophilic fungi occur when the temperature ranges between 40°C and 60°C. Like mesophilic fungi, they survive in the periphery of the compost heap when the temperature exceeds this range.

IMPORTANT FACTORS IN COMPOSTING

1 (a). **Raw Material:** Composts are prepared from straw, paddy husk, sugar cane trash, leaves, baggase, city garbage, sewage, and vegetable wastes. All the material undergoes decomposition under thermophilic and mesophilic conditions in heaps or pits and finally results in a brown to dark colour humified material.

(b). **Shredding or Grinding:** Raw materials give better results when they are shredded for composting. Shredding exposes a greater surface area, which makes it more susceptible to microbial attack. Shredding material makes it more homogenous and produces beneficial initial aeration and also makes the material more responsive to moisture control and aeration as well as moving and handling. Shredded waste heats more uniformly and also withstands excessive drying at the surface of the pile and resists moisture penetration from rain better than unshredded waste. Fly control is always better when waste is shredded or pulverized. Compost users find that shredded or ground material can be applied more readily and uniformly to land.

Normally, particle size less than 2 inches is the best but larger particles can be composted satisfactorily. Vegetative and herbaceous matter should not be grounded because it becomes soggy and high moisture content of these materials make them unsatisfactory for aerobic composting. Shredding and grinding materials shorten the decomposition.

2. **C : N ratio:** Organic matter decompostion is affected by the presence of carbon and nitrogen, as the formation of manure is mainly microbiological and influenced by carbonaceous and nitrogenous materials present in organic waste. Microbes need carbon for growth and nitrogen for protein synthesis. For efficient composting, C/N ratio of 30 is most desirable and 26–35 is for rapid composting. If the carbon is excess, decompostion decreases as nitrogen is used up and many organisms die. Another factor is available nitrogen in soil, which is used by microorganisms for degradation and term used is "robbing" the soil of nitrogen and delaying the nitrogen availability to growing plants. When the carbon is less, the required organisms use maximum available carbon and get rid of excess nitrogen as ammonia. The release of ammonia to the atmosphere produces a loss of nitrogen from compost pile and should be kept to minimum.

Plant residues are made up largely of:

(a) Sugar, starch, simple proteins (decompose rapidly)
(b) Crude protein (decompose slowly)
(c) Hemicellulose (decompose slowly)
(d) Cellulose (decompose slowly)
(e) Lignin, fat, wax (decompose slowly)

C/N ratio of some compostable materials is highlighted and easily available in the rural areas of India.

Compostable material	% N	C/N
Farm residues (wastes)		
Rice straw	0·3–0·5	80–130
Wheat straw	0·3–0·5	80–130
Barley straw	0·3–0·4	100–120
Maize stalk & leaves	0·8	50–60
Cotton stalks	0·6	70
Sugar cane trash	0·3–0·4	110–120
Lucerne residue	2·55	19

Materials high in carbon relative to nitrogen (C/N > 30) such as straw and sawdust will decompose very slowly unless a source of N is added. Materials with a low C/N ratio are good sources of N and include manure, inorganic fertilizer, vegetable table scraps and grass dippings.

1. **Moisture and aeration:** Moisture content and degree of aeration are important factors in composting. If the amount of moisture in the compost pile is below 40%, decomposition will be aerobic but slow. Aerobic decomposition will occur at any moisture content between 30–100% if adequate turning is provided, but higher moisture contents should be avoided. The optimum moisture level for aerobic composting is 50–60%, however, a range of 40–80% may be quite satisfactory depending upon the nature of the material to be composted.
2. **Temperature:** Temperature is a function of pile size, oxygen and moisture content. When the temperature rises above 40°C, the mesophilic microorganisms are replaced by thermophilic organisms. The temperature in the middle of the pile goes up to 55–70°C, which it gradually cools to ambient values. Decomposition is fastest in the thermophilic range. The optimum temperature for organic matter decomposition into carbon oxide and water has been found to be 60°C by Wiley and Pierce (1955), whereas Wiley (1957) showed that maximum temperature of 71°C was best but should not be allowed to exceed 71°C for long as decomposition will be slowed due to thermal kill of microorganisms.
3. **Reaction:** The initial pH value in compost heaps is generally slightly acidic to neutral, around 6–7. After 2–4 days, the pH usually begins to rise and will level of at between 8·0 and 9·0 towards the end of the process. The control of pH in composting is seldom a problem requiring attention if the material is kept aerobic, but large amounts of organic acids are often produced during anaerobic decomposition on a batch basis. Lime may be applied for counteracting acidity, but nitrogen can be conserved without using limestone. Studies from the University of California (1953) have shown that when calcium carbonate is added to compost, significant losses of nitrogen occur and the alkaline conditions favour volatilization of ammonia.
4. **Destruction of Pathogenic Organisms:** Destruction of pathogenic organisms is a most important aspect and problem of compost containing sewage or other highly contaminated materials. Experiments have shown that aerobic composting at high temperatures is effective in destroying pathogenic organisms.

The absence of health hazards is the characteristic of well managed composting operations in many parts of the world. This is significant evidence of the effectiveness of thermophilic composting.

Acharya (1948) found that pathogenic organisms in rural solid wastes may lead to serious outbreaks of disease and can be destroyed by composting. He suggested that wastes should be composted before being applied to the land. Gotaas (1956) recommended a temperature of 60°C for thermal kill of common pathogens. Nandakishore (1980) observed that pile height and the degree of aeration influence significantly the temperature profiles of compost heaps.

The extent and duration of high temperatures in compost heaps as well as antibiosis as a result of a mixed population of microorganisms provide reasons to believe that pathogens and parasites are not able to survive the composting process. However, to achieve this, at least two or three turnings of the heap are necessary to create conditions for high temperature.

Anaerobic composting in the mesophilic temperature range does not affect good destruction of parasites in an anaerobic environment. The biological antagonisms will eventually eliminate them, but this will generally take at least six months. Anaerobic composting involves sewage or night soil should be preceded by aerobic composting for at least a week.

METHODS OF COMPOSTING

1. **The Indore Method:** An important advance in the practice of composting was made at Indore, India, by Howard in collaboration with Jackson and Wad during the period 1924 to 1926. The traditional procedure was systematized into a method of composting now known as the 'Indore Method'.

Raw materials: The materials needed are mixed plant residues, animal dung, urine, earth, wood, ash and water. All vegetable wastes available on a farm such as weeds, stalks, stems, fallen leaves, prunings, chaff, fodder remnants, green matter, and so on are collected and stacked in a pile. Hard woody materials are first spread on the farm road and crushed under vehicles such as tractors or bullock carts before being piled. Green materials, which are soft and succulent, are allowed to wilt for 2–3 days to remove excess moisture before stacking and stacking material is spread in layers of about 15 cm thickness until the heap is about one and a half meters high.

The heap is then cut into vertical slices and about 20–25 kilogram is put under the feet of cattle in the shed as bedding for the night. The next morning, the bedding along with the dung and urine and urine-earth is taken to the pits where the composting is to be done.

2. **Pit Method:** The site selected for the compost pit should be at high level so that no rain-water gets in during the monsoon season, and it should be near to cattle shed and water source. The pit should be about 1m deep, 1.5–1.2 m wide and of any suitable length.

Filling the pit: The material brought from the cattle shed is spread evenly in the pit in layers of 10 cm to 15 cm. On each layer is spread a slurry of dung made with 4.5 kg dung, 3.5 kg urine-earth and 4.5 kg of inoculum taken from 15 days old composting pit. A sufficient quantity of water is (nearly 90%) sprinkled over the material in pit to wet it. The pit is filled in this way layer by layer and it should not take longer than one week to fill. Care should be taken while compacting the material in any way.

Turning: The material is turned three times during the whole period of composting—first after 15 days from filling the pit, second after another 15 days and third after another month. At each turning, the material is mixed thoroughly, moistened with water and replaced in the pit.

3. **Heap method:** During rainy seasons or in regions with heavy rainfall, the compost may be prepared in heaps above ground and protected by a shed. When sufficient nitrogenous material is not available, a green manure or leguminous crop like sunhemp is grown on the fermenting heap by sowing seeds after the first turning. The green matter is then turned in at the time of the second mixing.

Dimensions: The basic Indore pile is about 2 m wide at the base, 1.5 m high and 2 m long. The sides are tapered so that the top is about 0.5 m narrower in width than the base. A small bund is sometimes built around the pile to protect it from wind, which tends to dry the heap.

Forming the heap: The heap is usually commenced with a 20 cm layer of carbonaceous material such as leaves, hay, straw, sawdust, woodchips and chopped corn stalks. This is thin covered with 10 cm. of nitrogenous material such as fresh grass, weeds or garden plant residues, garbage, fresh or dry manure or digested sewage sludge. This pattern is followed until the pile is 1.5 m high and they are normally wetted so that they feel damp but not soggy. The pile is sometimes covered with soil or hay to retain heat and is turned at 6 and 12 week intervals.

If materials are limited, the alternate layers can be added as they become available. Also all materials may be mixed together in the pile if one is careful to maintain the proper proportions. While shredding, the material speeds up decomposition considerably.

Advantages and limitations: This method of composting is recommended where land and labour are easily available. The process can also be partially mechanized to make it economical when large quantities of materials are to be handled.

There is lack of protection from rain and wind. A considerable amount of water is needed and so the method is not suitable in areas of scanty rainfall. The intense aerobic decomposition to which the material is subjected no doubt shortens the period of composting, but leads to heavy losses of organic matter and nitrogen. Therefore, the C/N ratio should be maintained between about 30–40 to reduce such losses.

4. **The Bangalore Method:** This method of composting was developed at Bangalore in India by Acharya (1939). The method is basically recommended when night soil and refuse are used for preparing the compost. The method overcomes many disadvantages of the Indore method, but the time involved in production of a finished compost is much longer. The method is suitable for areas with a scanty rainfall.

Preparation of the pit: Trenches or pits about 1m deep are dugged, the breadth and length of the trenches can be made depending on the availability of land and the type of material to be composted. The selection of site for the pits is made as in the Indore method. The trenches should preferably have sloping walls and a floor of 90 cm slope to prevent water-logging.

Filling the pit: Organic residues and night soil are put in alternate layer and after filling, the pit is covered with 15 cm–20 cm thick layer of refuse. The materials are allowed to remain in the pit without turning and watering for three months. During this period the material settles down due to reduction in volume of the biomass, and additional night soil and refuse in alternate layers are placed on top and plastered or covered with mud or earth to prevent loss of moisture and breeding of flies. After the initial aerobic composting, which is for about 8–10 days, the material undergoes anaerobic decomposition at a very slow rate and it takes about 6–8 months to obtain the finished product.

Advantages and Limitations: The recovery of the finished product is much greater as compared to aerobic composting and loss of nitrogen is negligible. Labour requirement is less than the Indore method as turning of materials is not done. Labour is needed only for digging and filling the pits.

The method requires a long time to produce a finished compost and so takes up more land use. A uniform temperature is not assured in the biomass. Problems of odour and flybreeding need to be attended to.

5. **Rural Composting in India:** The pits can be partitioned into two with partitions made from hardy straw materials. The sources of material are generally cattle dung mixed with straw and stubble and used as bedding for the animals and different kinds of farm weeds in addition to crop residues.

The straw materials, rural garbage, and so on are spread in the pit layer and rock phosphate or superphosphate is added. Moisture is maintained by sprinkling water on each layer while filling the pit, or sometimes farmers use a slurry of cattle dung for accelerating the process of decomposition. When the pit is completely filled, it is covered with a layer of earth or is plastered with mud or covered with wood ash. When the compost is ready for use, it is removed and transported to the fields by bullock cart.

6. **High Temperature Compost:** This form of compost is prepared mainly from night soil, urine, sewage, animal dung and chopped plant residues at a ratio of 1 : 4. The materials are heaped in alternate layers starting with chopped plant stalks and followed by human and animal wastes and water is added to optimum amount.

At the time of making the heap, a number of bamboo poles are inserted for aeration purposes. After the heap formation is complete, it is sealed with 3 cm of mud plaster. The bamboo poles are withdrawn after one day of composting, leaving the holes for aeration of the heap. Within four or five days, the temperature rises to 60–70°C and the holes are then sealed. The first turning is usually done after 2 weeks and the moisture is made up with water or animal or human excreta, the turned heap is again sealed with mud. The compost is ready for use within two months and is considered free from pathogens.

7. **The Berkley Method:** This method of composting was developed at the University of California at Berkley. Compost can be prepared within two weeks by this method.

Raw materials: For rapid decomposition, a mixture of organic residues in the proportion of two parts of cellulose rich materials to one part of nitrogen rich materials should be used. However, pure cellulose materials such as paper waste should not form more than 10% of the raw material to be composted. It is recommended that leaves of all kinds, grass clippings and dry manure should be mixed and shredded with a mechanical device before the heaps are made. The C/N ratio of the mixed Berkley organic wastes averaged 30 : 1.

Dimensions of the Berkley Composting Heap and Procedure: The material is composted in heaps of 2.4 m by 2.2 m with a height of about 1.5 m and is composted for two weeks.

The turning and mixing is repeated on the seventh and tenth days. At this stage the temperature normally starts to decrease and within two weeks the compost is ready for use.

Advantages and limitations: The main advantage of the Berkley method is speed, since large quantities of compost can be prepared in a relatively short time. The turning schedule is tight and labour intensive. There is a limitation of using materials with a narrow C/N ratio.

EFFECTS OF COMPOSTING

1. **Role of Compost in Soil Fertility:** In view of the universal and pervasive energy shortage, energy must be conserved through efficient management of rural wastes resources. Composting is one of the best recycling methods for converting organic residues into inputs for growing food crops.

Composts are known to improve crop production directly and indirectly by improving soil properties. There are several reasons for using compost, but its main contribution is a supplier of humus substances to soil thus supplementing major plant nutrients (NPK), and being a good source for providing micronutrients whose deficiencies are seen up with intensive cropping and high productivity targets.

Physical Effects: One of the important contributions of compost is the high organic matter fraction, which improves the physical condition of poor soils. The addition of compost improves soil structure, texture and tilth (Hesse and Misra, 1982). Gaur and Subba Rao (1975) isolated bacterial germs and studied their effect on soil aggregation. Polysaccharides of *Rhizobium trifolii*, *Bacillus pulvifaciens*, *Beijerinckia india* and *Agrobacterium* had a positive effect on aggregate stability and content of carbohydrate, uronic acid and protein. Soils with more aggregates or granules are less sticky, have better permeability and a greater waterholding capacity than soil with poor structure.

Chemical Effects: Composting conserves organic resources by returning the nutrients in soil and indirectly saving the energy which would otherwise have been consumed for the production of mineral fertilizers through industrial processes to provide the same inputs. As compost is prepared from plant residues and their by products, it contains all elements of which the plant is made up. Its addition, therefore, increases the total stock of these elements in the soil. It provides both macro and micronutrients.

Humus forms compounds with certain elements, which are then rendered more available for plant growth. A positive correlation is found between content of organic matter in soil and extractability of trace elements. The positive effect of adding iron chelates on iron chlorosis is well known (Steward and Leonhard, 1957 and Bould, 1955). The addition of water soluble, chelate forming organic compounds to soil can enhance the availability of micronutrients to plants. (Perkins and Purvis, 1954; Holmes and Brown, 1955; Butler and Bray, 1956). Both organic acids and humic fractions are efficient in releasing phosphorus from rock phosphates and in reducing fixation of phosphorous by soils (Gaur, 1969, 1972; Basu, 1973; Pareek and Gaur, 1973). Humus also neutralizes the toxic effect of certain mineral elements on plants by reducing their uptake.

Biological Effects: Compost carries with it a very large population of actinomycetes, fungi and bacteria, and their incorporation into the soil not only adds millions of microorganisms but those already present in the soil are stimulated by the fresh supply of humic material (Gaur *et al.*, 1971). The application of compost helps the microorganisms to produce gummy substance which build good soil structure. Ammonification and nitrogen fixation are increased due to improved microbiological activity. Compost also stimulates the mycorrhizae which live in symbiotic association with the roots of plants and trees and play an important role in transferring certain nutrients from soil to plant. There is also the probability that compost carries small quantities of growth promoting substances similar in nature to hormones. Recent technology for inoculating compost with *Azotobacter* and phosphate solubilizer (Gaur, 1979) not only improves the quality of the compost, but also supplies agriculturally useful and efficient microorganisms, which will further multiply and function in the soil and so improves soil fertility on a permanent basis.

Effect of humic substances on plant growth and nutrition: Many investigators have reported the favourable effect of humic substances on growth yield and nutrient uptake by plants in nutrient solutions and sand or soil cultures (e.g., Chaminade, 1966, 1968; Charreau, 1975). The findings showed that strictly humic substances besides affecting the growth and development of plants, also improved the efficiency of biofertilizers such as *Azotobacter*, *Rhizobia*, blue-green algae and phosphobacteria.

Role of compost in control of plant nematodes: It has been reported that high rates of nitrogenous fertilizer increases the nematode population. Since soil fumigation methods to control plant-parasitic nematodes are expensive, the application of organic manures such as compost or farmyard manure to soil will be useful for reducing the incidence of plant diseases. Addition of organic manure, even in the presence of applied NPK, suppressed the numbers of plant parasitic nematodes. Neem cake (*Melia azadarichta*) was a most effective organic material for lowering the population of nematodes (Gaur and Prasad, 1970). Turner (1988) monitored the nematode numbers during compost production at three major production plants in Northern Ireland. They were present at all stages of production up to the end of phase-I, the most probable source of contamination being from damp straw. The majority of nematodes were from the order Rhabditida. Peak heating (phase-II) did not allow nematode survival if done under recommended conditions.

Farmyard manure, oilcakes of mustard, cottonseed and groundnut were incorporated in powdered form into the compost and casting soil separately for the control of *Aphelenchoides composticola* infesting mushroom crop. The mustard oilcake at 5 to 10 g/kg of wet compost was the most effective in increasing the yield of mushroom buttons and controlling the nematode. Incorporation of organic amendments into the compost was better than that of their incorporation into the casing soil. The farmyard manure proved the least effective. Higher doses of the amendments (10 g/kg) caused maximum mortality of the nematode and resulted in maximum yield of mushrooms. (Kumar *et al.*, 1993)

Role of compost in mitigating the toxic effects of pesticides: It is a general observation that pesticides applied to one crop may damage the growth of the next crop. There has been concern about the toxic effect of pesticides on the soil microorganisms of importance in soil productivity (Fletcher and Bollen, 1954; Eno and Everette, 1958; Bardiya and Gaur, 1968, 1970, Gaur and Prasad, 1970; Gaur, 1980).

Gaur and Pareek (1969) found that one per cent farmyard manure applied to alluvial soil alleviated the toxic effect of high doses of DDT on nodulation and leg haemoglobin content of green gram. They obtained better nodulation and yield of Bengal gram by application of farmyard manure at three levels to an alluvial sandy loam. The investigation also confirmed the role of applied organic manure in detoxification of Lindane (benzene hexachloride) used as a soil insecticide.

REFERENCES

Acharya, C.N. (1939). Comparison of different methods of composting waste materials. *Ind. J. Agril. Sci.,* **9:** 565.

Acharya, C. N. (1939). Composting versus direct ploughing in of waste materials. *Indian Farming,* **9:** 125.

Abdel-Azeem, H.H.M. (2001). Some factors affecting the composting process of rice straw. *Anal. Agrl. Sci.* Cairo. **46**(2): 525-542.

Anonymous (2005). In: Economic Survey of India, Directorate of Economics and Statistics, Dept. Agril. Coop., Ministry Agril, Govt of India.

Bardiya M.C. and Gaur, A.C. (1968) Influence of insecticides on CO_2 evolution from soil. *Ind. J. Microbiol.,* **8:** 233.

Basu, S.N. (1973). Solubilization of rock phosphate. *Fert. News,* **18:** 30.

Beary, T.P., Boopathy, R. and Templet, P. (2002). Accelerated decomposition of sugarcane crop residue using a fungal-bacterial consortium. *Int. Biodeterior. Biodegra.* **50:** 41-46

Bould, C. (1955). Chelated iron compounds for the correction of lime induced chlorosis in fruit. *Nature,* Lond., **175:** 90.

Butler, P.C. and Bray, R.H. (1956). Effect of zonc chelate of EDTA on plant uptake of Zn and other heavy metals. *Soil Sci. Soc. Amer. Proc.,* **20:** 348.

Chaminade, R. (1966). Nouvelles recherches sur les composes phospho-humiques. *Ann. Agron.,* **16:** 191.

Chaminade, R. (1968). Role specifique de la matiere organique sur la nutrition et le rendement des vegetause. *Pontificiae Acad. Scient. Scripta Varia,* **32:** 777.

Change, Y. and Hudson, H.J. (1967). The fungi of wheat straw compost. *J. Ecol. Studies. Trans. Br. Mycol. Soc.,* **50:** 649.

Charreau, C. (1975). Organic matter and biochemical properties of soil in dry tropical zone of W. Africa. *In Soils Bull.* 27 FAO Rome. 313.

Conghos, M.M, Aguirre, M.E. and Santamaria, R.M. (2003). Biodegradation of sunflower hulls with different nitrogen sources under mesophilic and thermophilic incubations. *Bio. Fertil. Soils.* **38:** 282-287.

Eldin, S.M.S.B. and Abosedera, S.A. (2001). Acceleration of composting of sugar beet haulms using two highly effective cellulose decomposing microorganisms. *Egyptian J. Microbiol.* **36:** 161-174.

Eno, C.F. and Everette, P.H. (1958). Effect of soil application of ten chlorinated hydrocarbon insecticides on soil microorganisms and the growth of stringless black valentine beans. *Soil Sci. Soc. Am. Proc.,* **22:** 235.

Fletcher, D.W. and Bollen, W.B. (1954). The effect of aldrin on soil microorganisms and some of their activities related to soil fertility. *App. Microbiol.* **2:** 349.

Gaur, A.C. (1979). Organic recycling prospects in Indian agriculture. *Fert. News* **24:** 59.

Gaur, A.C. (1980). Effect of pesticides on symbiotic nitrogen fixation. *Microbial.* **20:** 362.

Gaur, A.C. and Pareek, R.P. (1969). Effect of DDT on leghemaglobin content of *Phaseolus aureus* (green gram). *Experimentia,* **25:** 777.

Gaur, A.C. and Prasad, S.K. (1970). Effect of organic matter and inorganic fertilizers on plant parasitic nematodes. *Ind. J. Ent.* **32:** 186.

Gaur, A.C., Sadasivam, K.V., Vimal, O.P. and Mathur, R.S. (1971). A study on the decomposition of organic material in alluvial soils : CO_2 evolution, microbiological and chem. transformations. *P1. Soil,* **35:** 17.

Gaur, A.C. and Subba Rao, R.V. (1975). Note on isolation characterization and aggregate stabilizing effects of extracellular polysaccharides from some bacteria of agric. importance. *Ind. J. Agric. Sci.,* **45:** 86.

Gotaas, H.B. (1956). Composting-sanitary disposal and reclamation of organic wastes. W.H.O. Geneva.

Hesse, P.R. and Misra, R.V. (1982). 'Mineral or Organic' Project Field Doct. No. RAS/75/004 FAO/UNDP project on organic recycling.

Holmes, R.S. and Brown, J.G. (1955). Chelates as correctives for chlorosis. *Soil Sci.,* **80:** 167.

Kumar, K., Chhabra, H.K. and Kaul, V.K. (1993). Efficacy of organic amendments on the yield and control of *Aphelenchoides composticola* infesting mushroom, *Agaricus bisporus* (Lange) Sing. *Indian. J. Nematol.* **23**(1): 82-86.

Nandakishore, T.G. (1980). Studies on urban solid waste composting. M.Sc. thesis. Univ. Agril. Sciences, Bangalore, India p. 166.

Pareek, R.P. and Gaur, A.C. (1973). Release of phosphate from tri-calcium phosphate and rock phosphate by organic acids. *Curr. Sci.*, **42:** 278.

Perkins, H.F. and Purvis, E.R. (1954). Soil and plant studies with chelates of EDTA. *Soil Sci.*, **78:** 325.

Riegel, C. and Noe, J.P. (2000). Chicken litter soil amendment effects on soil-borne microbes and *Meloidogyne incognita* on cotton. *Plant Disease* **84:** 1275-1281.

Stewart, J. and Leonhard, C.D. (1957). Use of chelates in citrus production in Florida. *Soil Sci.* **84:** 87.

Turner, S.J. (1988). Effects of compost production on nematode survival. *Mushroom Journal.* No. **183:** 505-507.

Univ. of California (Sanitary Eng. Lab.) (1953). Reclamation of municipal refuse by composting. *Tech. Bull.* No. 9 Berkley. California.

Walker, S.A. and Crawford, D.V. (1968). The role of organic matter in adsorption of the triazine herbicides by soils. In isotopes and radiation in soil organic matter studies. IAEA, Vienna. 91.

Weed, S.B. and Weber, J.B. (1974). Pesticide-organic matter. In Pesticides in Soil and Water. Ed. W.D. Guenzi. *Soil Sci. Soc.* Amer. Madison Wis. 39.

Wiley, J.S. (1957). Report on high rate composting studies. *Proc. 12th Ind. Wastes Conf. Purdue Univ. Engr. Bull.* Vol. 17 p. 596.

Wiley, J. and Pierce, G.A. (1955). Preliminary study of high rate composting. *Proc. Am. Soc. Civil Eng. J. Sanit. Engr.* Div., **81:** 846.

40

Applications of Nanoscience and Nanotechnology: A Revolution in Agriculture

Nishant Iyengar and Dr. (Mrs.) S. Anju Iyengar

M.Tech. Nanoscience and Nanotechnology, Department of Physics and Astrophysics & Department of Chemistry, University of Delhi, Delhi-110007, India and Institut de Chimie Moléculaire de Grenoble, Université Joseph Fourier, Grenoble, France. E-mail: nishant.iyengar@yahoo.com

Indian Council of Medical Research, Headquarters, New Delhi-110029, India. E-mail:iyengars.anju@yahoo.in

INTRODUCTION

Nanoscience and nanotechnology is about the precise and purposeful manipulation of matter at the atomic level. Nanoscience is the study of atomic and molecular assemblies with dimension ranging from 1 nm to about 100 nm and nanotechnology is concerned with incorporation of such assemblies into devices.

Nano is a Greek word which means dwarf or very small. The dictionary meaning of nanotechnology is the techniques, machines, tools, and processes needed to manipulate matter size of atoms and molecules. Molecular nanotechnology will allow very complete control over the placement of individual atoms and replace the conventional manufacturing techniques manipulating billions of atoms at a time using large-scale deformation methods like pounding and chipping.

The main object of nanotechnology research programme is to reduce complex and sophisticated machinery into very small operational units. Nanotechnology describes technologies where the component parts can measure just a few atomic diameters (generally a millionth of a millimetre). The name comes from nanometer, which is a length equal to one-billionth of a metre. The technique is best suited for fabricating devices on the sub-micron scale originated in semiconductor processing technology for the production of integrated circuits, nanoscale devices and are all based on semiconductors.

The hypothesis of nanotechnology was the discovery of an American engineer Kim Eric Drexler (1955-), who was a student of genetic engineering at the Massachusetts Institute of Technology in 1971, envisioned the potential for building machines and materials with atoms as basic blocks just as living matter is constructed with deoxyribonucleic acid. Atom became visible in 1981 when the scanning tunnelling microscope was built. The microscope has the ability to scan through the clouds created by electrons around atoms so the individual atom can be seen. Richard Errett Smalley, an American chemist and physicist (1943-2005), a professor with Rice University, USA opened the doorway to nanotechnology in 1985 and he found a way to produce carbon in a third form that is crystalline and possesses incredible strength properties. The other two natural forms of carbons are graphite and the diamond. The molecule containing 60 carbon atoms with a shape like soccer ball or a geodesic dome built by an American architect Buckminster Fuller (1895-1983), thus the molecule was named the buckyball in Fuller's honour.

Nanotechnology is generally referred to as "bottom up" manufacturing as it aims to start with smallest possible building materials, atoms using them to create a desired product that is working with individual atoms which allow atom by atom design of structures. This process will allow us to have every atom in the right place and make almost any structure consistent with most law of physics and chemistry that we can specify in atomic details and get manufacturing costs not much exceeding the cost of the required raw materials and energy, eliminating unwanted byproducts.

SIZE OF NANOSCALE OBJECTS

Nanometre (abbreviated nm), derived from the Greek word for midget. "Nano" is metric prefix and indicates a billionth part (10^{-9}). A micron is a millionth of a metre, which is the scale that is relevant to building logic devices. A nanometre is one thousandth of a micron and a thousandth of a millionth of a metre (a billionth of a metre). A nanometre is about the width of six bonded carbon atoms and approximately 40,000 are needed to equal the width of an average human air. Table 40.1 shows size of few nanoscale objects.

Table 40.1: Size of nanoscale objects – nature vs. fabrication.

Size of Nanoscale Objects – Nature vs. Fabrication	
Objects	***Diameter***
Hydrogen atom	0.1 nm
Buckminsterfullerene (C_{60}) (carbon third form)	1.0 nm
Six carbon atoms aligned	1.0 nm
DNA (width)	2.0 nm
Nanotube	3-30 nm
Proteins	5-50 nm
Quantum Dots (CdSe)	8.0 nm
Dip pen nanolithography features	10-15 nm
Dendrimers	10 nm
Microtubules	25 nm
Ribosome	25 nm
Virus	75-100 nm
Nanoparticles range from	1-100 nm
Semiconductor chip features	90 nm

SPECIAL ABOUT NANOTECHNOLOGY

The special about nanotechnology are as under:

1. Small size (High surface to volume ratio), therefore requires self-assemblers.
2. Significantly higher hardness, breaking strength and toughness at low temperatures and super plasticity at high temperatures, the emergence of additional electronic states, high chemical selectivity of surface sites and significantly increased surface energy.
3. New entry ways (high mobility in human body, plants and environment)[1].

NANOTECHNOLOGY – A MULTIDISCIPLINARY RESEARCH

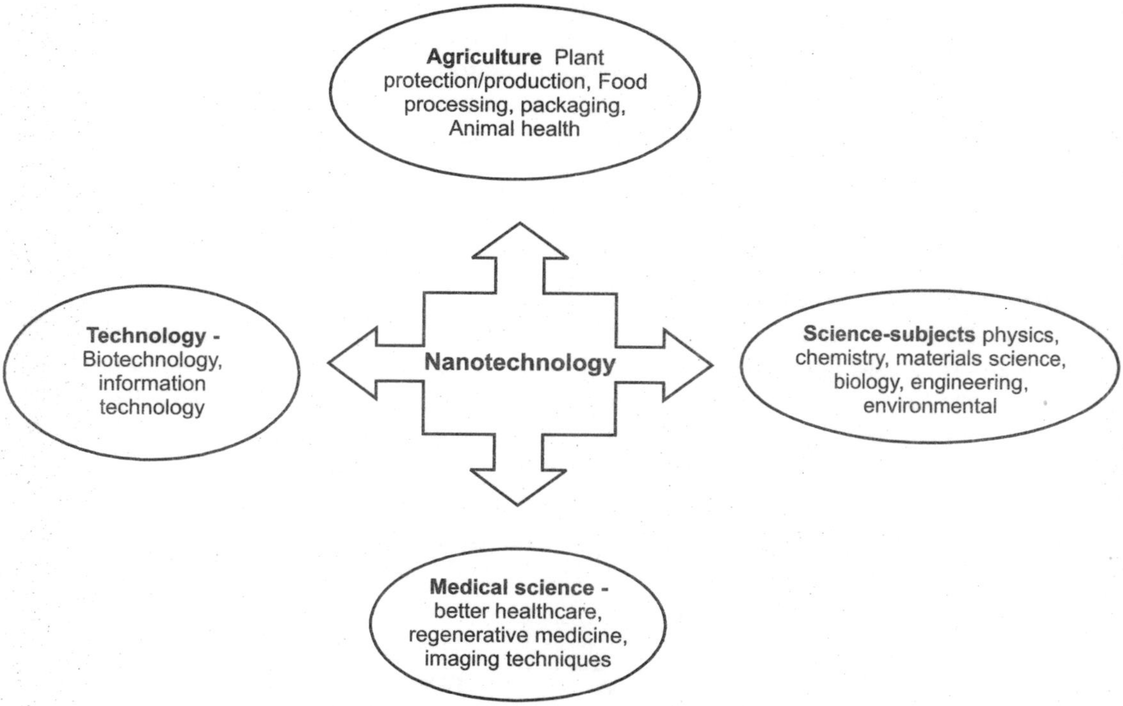

Fig. 40.1. Nanotechnology – a multidisciplinary research.

Growth of Scientific Research Publications in Nanotechnology

Technology and innovation management research related to nanotechnology has been rather limited since the technology is still in its fluid stage of development. The past few years have seen an explosion of interest in the area of science and technology labelled as nanotechnology. Scientific disciplines, which were not receiving much funding for their research, have now taken advantage of the increased funding, are relabelled their research topics as nanotechnology, rather than the general disciplines, e.g., chemistry, physics, materials science and biology. The meaningful evolution of nanotechnology research across the world, approximately 15 countries where researches in nanotechnology are being conducted have been investigated and analysis, the conflated data are shown in Table 40.2 to 40.5 country-wise in the discipline of chemistry, physics, materials science and biology. In terms of disciplines which have conflated into nanotechnology, the following three approaches have been considered:

1) A search for the nanoscientific output of general disciplines' journals categorized by Essential Science Indicators (ESI) and the Journal Citation Reports (JCR) classification schemes (finding the extent to which nanotechnology papers were coming from scientific disciplines of the countries across the world).
2) An investigation of reference citations of nanoscience and nanotechnology papers by each disciplines journals (classify the references into disciplines' journals using ESI and JCR journal-based classification) and
3) Identified the key factors that drive multidisciplines to conflate into nanotechnology.

The analysis is based on the selection of top-ranked English language journals of each discipline classified by the Institute for Scientific Information (ISI) using ESI and JCR in terms of publications. The ISI Web of Science [an expanded version of Science Citation Index (SCI)] database, one of the most representative collections of peer reviewed science and technical articles from the leading journals as well as proceedings of the International Society Symposiums were used as a sample search. The data collected by using specialist keywords, were derived from previous technical publications of the Nanoscience and Technical Institute.

Nanotechnology Research Conflation

Many countries regard nanotechnology as an interesting and important area for future exploitation and have set up national initiatives in order to prepare for the technological challenge in the decade 2000-2010. Bibliometric indicators are used as a measure of output of the research system in order to analyze the performances. Volumes of scientific publications are a commonly accepted indicator of scientific performance: they help illustrate the existing status and forecast future developments of technology. Table 40.2 demonstrates the overall shares of nanotechnology articles in the field of chemistry authored in the respective countries and subsequently compares them with the share of the most active countries (top 15 countries) indexed by SCI for the period (2000-2004).

Table 40.2: Percentage share of nanotechnology research output in chemistry.

Country	2000	2001	2002	2003	2004
Singapore	6.09	5.74	10.83	17.16	33.52
PR China	9.48	13.42	11.43	20.14	26.94
South Korea	6.25	9.31	14.94	17.33	22.12
Taiwan	5.37	3.47	5.98	11.21	16.20
Israel	11.53	12.40	13.15	14.23	14.92
Australia	5.03	6.64	7.71	9.70	14.42
USA	7.05	7.86	8.76	11.46	13.58
Japan	5.27	6.44	8.61	10.75	13.28
Germany	6.84	6.12	9.43	10.93	12.66
Switzerland	10.75	10.07	9.62	9.00	12.32
Canada	4.87	5.35	5.46	8.94	10.83
Italy	6.06	4.84	7.71	9.11	10.60
India	6.44	7.59	6.32	7.05	10.28
England	4.82	5.69	7.93	7.75	9.84
France	5.90	6.25	9.35	9.68	9.08

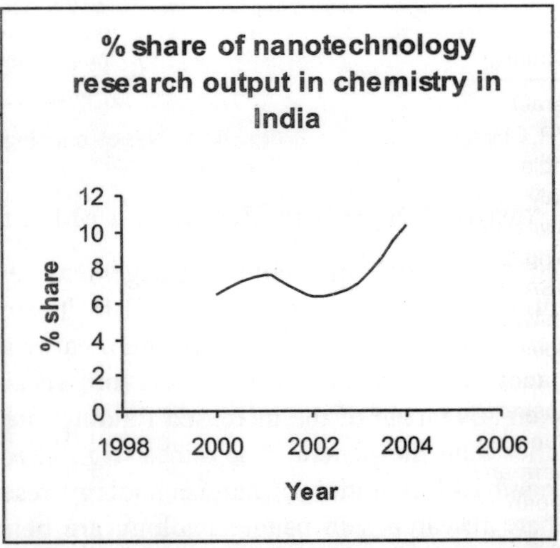

Fig. 40.2. Graph shows the publication growth in chemistry.

Compared to Western countries, the volumes of nanotechnology related articles in Asian countries (e.g., Singapore, China, South Korea, Taiwan and Israel) appear to be substantial. Similarly, Table 40.3 and Table 40.4 show the overall shares of nanotechnology in the physics and materials science respectively in the same way as appeared in Table 40.2.

Table 40.3: Percentage share of nanotechnology research output in physics.

Country	2000	2001	2002	2003	2004
Singapore	14.38	16.02	21.58	17.34	23.18
PR China	11.19	13.24	15.53	17.48	17.59
South Koea	6.04	7.87	13.92	13.78	14.85
Taiwan	4.46	4.87	9.11	10.22	14.36
Japan	7.82	8.87	10.94	9.58	12.39
USA	6.44	7.51	9.64	10.77	12.07
India	5.19	6.00	6.37	8.92	10.06
Germany	5.97	6.34	7.21	8.38	10.02
France	5.66	6.35	7.63	9.21	9.16
Australia	5.16	5.47	5.75	8.44	8.91
Italy	3.74	4.22	5.95	6.76	8.30
Canada	4.27	3.29	6.48	6.24	8.09
England	4.77	5.99	6.95	6.87	7.70
Israel	4.77	5.75	6.95	5.72	7.60
Switzerland	4.24	5.78	6.59	5.87	6.62

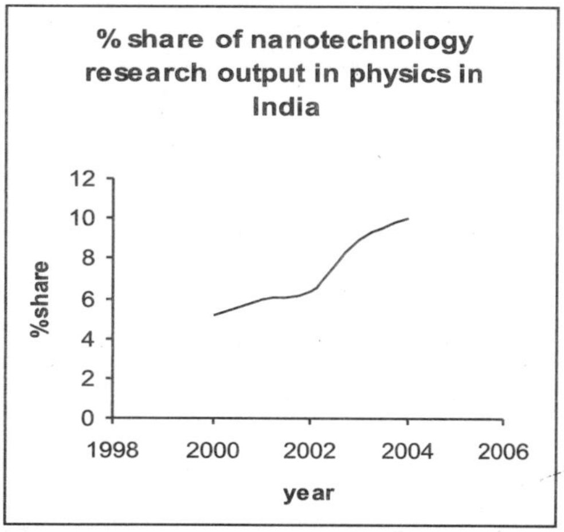

Fig. 40.3. Graph shows the publication growth in physics.

Table 40.4 depicts the overall shares of nanotechnology articles in the materials science field authored in the respective countries and subsequently compares them with the most active countries share indexed by SCI for the same period. Compared to other countries, the volumes of nanotechnology-related articles in Asian countries (e.g., Israel, China and India) including the European countries appear to be substantial. This growth corresponds to the higher level of nanotechnology research conflation within materials science compared to the chemistry and physics.

Table 40.4: Percentage share of nanotechnology research output in materials science.

Country	2000	2001	2002	2003	2004
Israel	35.10	21.51	50.70	39.18	37.28
PR China	21.01	24.05	28.56	32.04	35.05
India	15.12	20.83	16.54	18.58	34.14
Italy	12.88	14.21	20.35	20.97	29.97
Switzerland	18.09	21.18	32.72	23.36	28.47
Singapore	16.91	23.02	17.85	19.57	26.81
USA	14.76	17.98	20.60	24.54	26.80
Taiwan	12.02	13.04	16.99	10.22	25.62
South Korea	7.11	12.42	13.56	22.75	24.55
France	19.42	13.57	20.82	20.71	24.42
Australia	13.49	12.88	17.02	18.83	23.24
Canada	15.52	17.48	16.74	18.36	22.33
Germany	15.88	16.92	16.93	20.11	21.92
Japan	11.17	12.69	14.58	15.18	21.10
England	8.64	9.81	10.15	14.03	19.49

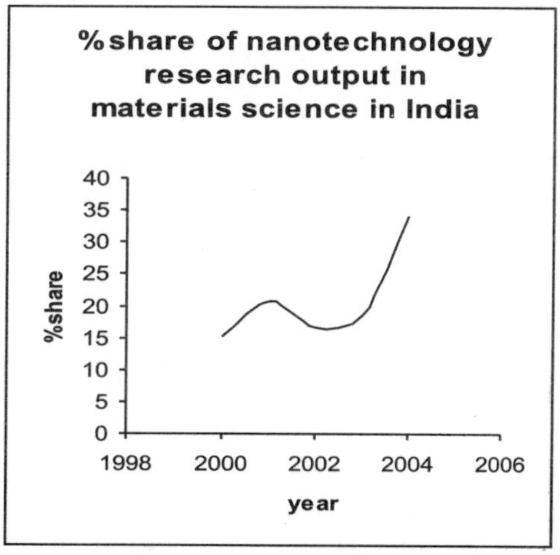

Fig. 40.4. Graph shows the publication growth in materials science.

Table 40.5 shows the overall shares of nanotechnolgy articles in the biological science field authored in the respective countries and subsequently compares them with the most active countries' share indexed by SCI for the same period.

Table 40.5: Percentage share of nanotechnology research output in biological sciences.

Country	2000	2001	2002	2003	2004
Israel	0	2.46	1.66	0	6.32
PR China	1.47	0	3.10	4.22	5.96
Taiwan	0	0	0	0	5.40
Germany	1.82	2.64	2.64	1.63	3.77
Switzerland	0	4.51	2.07	2.00	3.62
India	0	0	1.31	1.77	2.70
USA	1.11	1.42	1.36	1.62	2.41
France	0.83	2.31	1.95	2.69	2.13
Canada	0.52	1.16	1.21	2.27	2.09
Italy	1.50	0.75	0.67	1.00	1.99
Japan	0.52	1.37	1.09	1.42	1.85
South Korea	0	0.85	0	0.67	1.58
England	1.51	2.52	1.59	1.52	1.43
Australia	0.97	0.92	0.99	0	1.22
Singapore	0	0	0	6.89	0

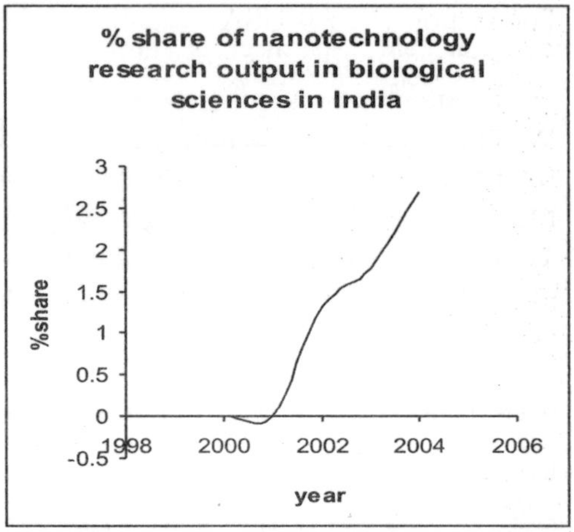

Fig. 40.5. Graph shows the publication growth in biological sciences.

The government of Singapore has been backing nanotechnology research in the areas of high density of data storage, highly integrated chips and biomedical applications. These specific initiatives may have had an impact on it becoming the top player in physics and chemistry disciplines related to nanoscientific output. Taiwan and South Korea are in turn pushing R&D of nanorelated to nanomaterials and electronics, particularly computer memories and logic devices. In China (including Hong Kong), the nanotechnology policy strongly emphasizes the establishment of national R&D laboratories and industrial parks: these top-down initiatives have helped China to lead in all the areas of nanotechnology research in Asia.

NANOSCIENCE AND NANOTECHNOLOGY: ITS INITIATIVES AND POTENTIAL APPLICATIONS

In India, the government through the Department of Science and Technology (DST), New Delhi had established the Nanoscience and Technology Initiative in the later part of 2001 and invested Rs. 350 crores (2002-06). Indian nanoscience and nanotechnology efforts cover a wide spectrum of areas including micro-electromechanical systems (MEMS), DNA chips, quantum computing electronics, carbon nanotubes and biomedical applications. In May 2007, the Cabinet of the Government of India granted approval for the Nanomission and sanction Rs. 1000 crores for five years starting from 2007.

Several academic institutions in both public and private sectors in India have initiated nanoscience and nanotechnology research and development (R&D). The Council of Scientific and Industrial Research, New Delhi, India's premier public R&D institution, holds numerous nanotechnology related patents, including novel drug delivery systems, production of nano-sized chemicals and high temperature synthesis of nano-sized titanium carbide. Impressive research initiatives have been taken in India, but still the research in the area of nanoscience and nanotechnology is in its infancy compared to the degree of sophistication in R&D already achieved by the developed countries.

Roco and his group promoted nanoscience and nanotechnology research as a national priority in the United States of America (USA) during mid 1990s. The National Science Foundation (NSF), USA, established the National Nanotechnology Initiative (NNI) in 2000 to support research in the area of nanotechnology. Developing countries, which found themselves on the sidelines watching the exitement of technological innovation developed in the advanced Western countries are now poised to rewrite the script in nanoscience and nanotechnology. Countries like China, South Korea, India, Thailand, Philippines, South Africa, Brazil, Chile and Mexico have already started investing resources in the area of nanoscience and nanotechnology. These countries have set up laboratories in numerous research institutions, developed national government funding programmes and introduced a few products of nanoscience and nanotechnology in the market.

Nanoscience and Nanotechnology Products

Commercial applications of nanoscience and nanotechnology have begun to appear in various countries. For example, some US automotive companies like GM Motors and Chevrolet are already using clay nanoparticles, which make the materials stronger, lighter, more durable and often transparent. Lee, a textile company that produces jeans, has developed stain-resistant khakis. In the area of food, nanoscience and nanotechnology is being applied to study improved flavour delivery and maintenance of freshness.

In the areas of electronics and information technology, it is claimed that nanoscience and nanotechnology has the potential for smaller and faster computers with large memories. In the areas of healthcare, it would be possible to create artificial organs and implant them in human bodies. Targeted drug delivery would be possible. In military, space and security, researchers are investigating smart materials that would be responsive to the conditions of their environment, sensors to detect chemical or biological warfare agents and lightweight bulletproof materials.

NANOTECHNOLOGY APPLICATIONS IN AGRICULTURE

This technology has emerged with the new challenges in sector like growing demand for healthy, safe food, an increasing risk of disease in plants, threats to fishery production due to changing weather patterns. This technology has the potential to revolutionize the agriculture and food industry with new tools for the molecular treatment of diseases, rapid disease detection and enhancing the ability of plants to absorb nutrients. Smart sensors and smart delivery systems will provide help to this industry to fight against viruses and other crop pathogens. The nanotools and devices will be available to increase the efficiency of pesticides and herbicides allowing lower doses to be used. This technology will protect the environment indirectly through the use of alternative (renewable) energy supplies and filters or catalyst to reduce pollution and clean up existing pollutants. An agricultural methodology widely used in the USA, Europe and Japan by efficiently utilizing the modern technology for crop management, which is called Controlled Environment Agriculture (CEA). CEA is an advanced and intensive way of hydroponically-based agriculture. Plants are grown within a controlled environment so that horticultural practices can be optimized. The computerized system that regulates and monitors localized environments such as fields of crops. CEA technology provides an excellent platform for the introduction of nanotechnology to agriculture.

Precision Farming

This farming has been long-desired goal to maximize output (i.e., crop yields) while minimizing input (i.e., fertilizers, pesticides, herbicides, etc.) through monitoring environmental variables and applying targeted action. This farming makes use of computers, global satellite positioning systems and remote sensing devices to measure highly localized environmental conditions to determine whether the crops are growing at maximum efficiency or precisely identifying the nature and location of problems. The data so collected from the said sources, the scientists can utilize the same to determine the soil conditions and plant development, seeding, fertilizer, chemical and water used to reduce the production cost and increase the crop yield to benefit the farmers. Precision farming can help to reduce the agricultural waste and keep the environmental pollution to a minimum. Although, this farming is in the process of implementation, the nanotechnlogy devices like tiny sensors and monitoring systems will have a large impact on methodologies. Such developed nanosensors are already being used in parts of the USA and Australia. The initial cost of setting up such a system is justified by the favourable result that best grapes were grown which in turn produce finer wine that command a premium price. The scientists are in the process of commercially marketing wireless nanosensors equipment of increased sensitivity to allow enhanced productivity in agriculture by providing accurate information to the farmers to take better decisions.

Plant Protection

Plant protection is a vital and important part in the field of agriculture to make it healthy and get maximum yield from the crop. The scientists in agricultural universities and research stations have been doing researches globally to make the plants healthy from harmful insects, make efficient pesticides and fertilizers, disease diagnostics, management of nematodes and weeds. The utility of nanotechnology in health, electronics and several other biological and industrial sectors is already well appreciated but its potential for transforming agriculture, especially in crop protection is just starting.

Nanocides: pesticides via encapsulation

The advantage of the nano-formulation is that the pesticide dissolves more easily in water that is to simplify application to crops and is more stable and the killing capacity of the chemical such as herbicide, insecticide or fungicide is optimized. Pesticides containing nano-scale active ingredients are on the market and many of the world's leading agrochemicals firms are conducting research on the development of new nano-scale formulations of pesticides.

BASF of Germany, an agrochemical manufacturing company is conducting basic research and has applied for a patent on a pesticide formulation, "Nanoparticles Comprising a Crop Protection Agent", that involves an active ingredient whose ideal particle size is between 10 and 150 nm.

Bayer Crop Science of Germany has applied for a patent on agrochemicals in the form of an emulsion in which the active ingredient is made up of nano-scale droplets in the range of 10-400 nm.

Syngenta, headquarters in Switzerland, is the world's largest agrochemical corporation already sells pesticide products formulated as emulsions containing nano-scale particles. Nanocapsules would help to avoid phytotoxicity on the crop by using systemic herbicides against parasitic weeds. Syngenta has two newly developed products Primo MAXX Plant Growth Regulator, which is designed to keep golf course turf grass from growing too fast and Banner MAXX fungicide that is meant for treating golf course turf grass are oil-based pesticides mixed with water and then heated to create emulsion. Syngenta claims that both products' extremely small particle size of about 100 nm (or 0.1 micron) prevents spray tank filters from clogging and the chemicals mix so completely in water that they won't settle out in the spray tank. Banner MAXX fungicide will not separate from water for up to one year, whereas fungicides that contain larger particle size ingredients typically require agitation every two hours to prevent misapplications and clogging in the tank. According to Syngenta, the particle size of this formulation is around 250 times smaller than typical pesticides particles and is absorbed into the plant's system and cannot be washed off by rain or irrigation.

Nanoprecipitation technique for the encapsulation of agrochemical active ingredients

French scientists have published a research article "Nanoprecipitation technique for the encapsulation of agrochemical active ingredients" in Journal of Microencapsulation to explain the study performed in 1997 to assess the ability of nanospheres (NS) to improve the biodelivery of new insecticide to plants. This biological study was performed on cotton plants infested with aphid to estimate the direct contact efficacy of NS formulations on the insects and the systemicity of the encapsulated active ingredient and its level of penetration through the plant, compared to a classical suspension used as a reference. It was concluded that NS formulation performed better than the reference by enhancing the penetration due to their small size.

Nanotechnology for parasitic plant control

The development of nanodevices as smart delivery systems to target specific sites for controlled chemical release for improvement of existing crop management is achieved on the basis of medical applications. Nanocapsules would help to avoid phytotoxicity on the crop by using systemic herbicides against parasitic weeds. Nanoencapsulation can also improve herbicide application, providing better penetration through cuticles and tissues and allowing slow and constant release of the active substance. Nanoparticles have a great potential as magic bullets loaded with herbicides, chemicals or nucleic acids targeting specific plant tissues or areas to release their charge. Viral capsids can be altered by mutagenesis to achieve different configurations and deliver specific nucleic acids, enzymes or antimicrobial peptides acting against parasites. Many unresolved problems such as increasing the scale of production processes and lowering the costs as well as toxicological issues remain, but the foundations of a new plant treatment concept have been laid and applications in the field of parasitic plant control can be started.

Nanosilica-from medicine to pest control

Scientists from Zoology Department, University of Delhi have attempted to study the impact of nanosilica on basic biology, medicine, agro-nanoproducts and use of amorphous nanosilica as biopesticide in their article "Nanosilica-from medicine to pest control" published in Parasitology Research. The basis of their study was the extensive applications of silica in various industries. Such as, application of gold-coated silica has been used in the treatment for benign and malignant tumor. Surface-modified hydrophobic as well as lipophilic nanosilica could be effectively used as novel drugs for treatment of chicken malaria and nuclear polyhedrosis virus (BmNPV), a scourge in silkworm industry.

Role of pumpkin plant in environment

Pumpkin plants grown in an aqueous medium containing magnetite (Fe_3O_4) nanoparticles can absorb, translocate and accumulate the particles in the plant tissues. The results of the research conducted at Department of Physics and Astronomy, University of Delaware, U.S.A. suggest that plants as an important component of the environmental and ecological systems, need to be included while evaluating the overall fate, transport, and exposure pathways of nanoparticles in the environment.

Nanoparticle penetration and transport in living pumpkin plants: in situ subcellular identification

A research article appears as "Nanoparticle penetration and transport in living pumpkin plants: in situ subcellular identification" published in BMC Plant Biology issue of 2009 indicates that nanoparticles applied in plants using two application methods, injection and spraying and magnets were used to retain the particles in movement in specific areas. This research was conducted using nanoparticles, which contained iron-carbon and adopted the said methods for the specific areas of the plant for the treatment of disease in Cucurbita pepo. The main experimental approach, using correlative light and electron microscopy provided evidence of intracellular localization of nanoparticles and their displacement from the application point. Long range movement of the particles through the plant body was detected, particles having been found near the magnets used to immobilize and concentrate them and the cell response to the nanoparticles presence was detected. It was concluded that the nanoparticles were capable of penetrating living plant tissues and migrating to different regions of the plant, although movements over short distances seemed to be favoured.

NANOBIOTECHNOLOGY

DNA in Nano World

The DNA molecule has appealing features for use in nanotechnology: its minuscule size with a diameter of about 2 nm, its short structural repeat (helical pitch) of about 3.4 - 3.6 nm and its stiffness with a persistence length (a measure of stiffness) of round 50 nm. Nanotechnological constructions are of two types like top-down and bottom-up. As a chemically based assembly system, DNA will be a key player in bottom-up nanotechnology. This approach was originated in early 1970s, when *in vitro* genetic manipulation was performed by taking together molecules with 'sticky ends'. A sticky end is a short single-stranded overhang protruding from the end of a double-stranded helical DNA molecule. Like flaps of Velcro, two molecules with complementary sticky ends – that is, their sticky ends have complementary arrangements of the nucleotide bases adenine, cytosine, guanine and thymine – will cohere to form a molecular complex. Sticky-ended cohesion is arguably the best example of programmable molecular recognition: there is significant diversity to possible sticky ends (4N for N-base sticky ends), and the product formed at the site of this cohesion is the classic DNA double helix. Likewise, the convenience of solid support-based DNA synthesis makes it easy to programme diverse sequences of sticky ends. Thus, sticky ends offer both predictable control of intermolecular associations and predictable geometry at the point of cohesion. Perhaps one could get similar affinity properties from antibodies and antigens, but in contrast to DNA sticky ends, the relative three-dimensional orientation of the antibody and the antigen would need to be determined for every new pair. The nucleic acids seem to be unique in this regard, providing a tractable, diverse and programmable system with remarkable control over intermolecular interactions, coupled with known structures for their complexes.

Nanobiotechnology: molecular biology complementing nanotechnology

The credit for the term nanobiotechnology goes to Lynn W. Jelinski, a biophysicist at Cornell University. Nanobiotechnology joins the breakthroughs in nanotechnology to those in molecular biology. Molecular biologists help nanotechnologists understand and access the nanostructures and nanomachines designed by 4 billion years of natural engineering and evolution – cell machinery and biological molecules. Exploiting the extraordinary properties of biological molecules and cell processes, nanotechnologists can accomplish many goals those are difficult or impossible to achieve by other means. For example, rather than build silicon scaffolding for nanostructures, DNA's ladder structure provides nanotechnologists with a natural framework for assembling nanostructures and its highly specific bonding properties bring atoms together in a predictable pattern to create a nanostructure.

DNA has been used not only to build nanostructures but also as an essential component of nanomachines. Most appropriately, DNA, the information storage molecule, may serve as the basis of the next generation of computers. As microprocessors and microcircuits shrink to nanoprocessors and nanocircuits, DNA molecule mounted onto silicon chips may replace microchips and with electron flow-channels etched in silicon. Such biochips are DNA-based processors that use DNA's extraordinary information storage capacity. Conceptually, they are very different from the DNA chips. Nanobiotechnology is an emerging area of opportunity that seeks to fuse nano/microfabrication and bio-systems to the benefit of both.

Potential applications of nanobiotechnology in agriculture

- High throughput DNA sequencing and nanofabricated gel-free systems.
- Microarrays and expression profiling.
- Increasing the speed and power of disease diagnostics.
- Creating bio-nanostructures for getting functional molecules into cells.
- Miniaturizing biosensors.

NANODEVICES: REVOLUTION IN AGRICULTURE

A change in agriculture has just begun with the application/introduction of nanotechnology such as nanosensors, nanoemulsions, nanoparticles, particle farming etc. These devices can be used to detect the plant diseases, for treatment and to detect the condition to do sowing. The device like nanosensors, which could be distributed throughout the field where they can monitor soil conditions and crop growth. The nanoscale devices are capable of responding to different situations, taking appropriate remedial action and also identify plant health issues before they become visible to the farmers.

NANOTECHNOLOGY: AS REMEDY FOR ENVIRONMENTAL CONTAMINANTS

The exploitation of natural resources for fulfilling the human needs has given rise to new pollutants at abundant levels in air, water and soil, which are above the cleaning capacity of environment. This is the result of industrialization, urbanization, modernization, agricultural development and energy generation globally. The scientists are working to improve the quality of treatment technologies to make it more effective and efficient to deal with the present pollution problems. At the same time, our scientists are giving more emphasis on the repercussions of nanotechnology, both good and bad since it is about the precise and purposeful manipulation of matter at the atomic level. Hence, the applications and advancement of nanotechnologies is interdisciplinary to all sciences like natural science – chemistry, physics, biology and astronomy; social science – economics and sociology and applied science – agriculture, engineering. The Government of India has set up a regulatory board in advance to avoid any devastations arising out of the application of nanotechnology.

Recent advances in the field of nanoscience and nanotechnology has made it possible to employ environmental cleanup technologies by developing nanodevices and nanosensors for prevention and control of pollution in the environment. The strategy of pollution prevention refers to source reduction and other practices that efficiently use raw materials, energy, water and other resources to reduce or eliminate creation of wastes.

Nanoparticles have high surface area compared with the bulk equivalent. Thus, nanoparticles are being explored as a means to remove or destroy toxic substances from the environment. Nanoparticles are more reactive and react to a great extent because of small size. They have more surfaces available for chemical interactions and can also exhibit higher chemical reactivity due to unusual crystal shape and lattice order. Nanoscale iron particles represent a new generation of environmental remediation technologies that could provide cost-effective solutions to some of the most challenging environmental cleanup problems.

Nanotechnology could be used for green chemistry to eliminate or reduce wastes by synthesizing improved and new catalysts at atomic level for industrial processes, building information into molecules, self-assembling molecules with the foundation for new chemicals and materials and building molecules in microscale reactors. The research at the University of Oregon used biometric methods of organizing metal particles 1-5 nm in diameter. This approach, 'biomolecular nanolithography' involves assembling these small metal particles on a biopolymer template or scaffold stretched out on a surface. The goal is to organize the nanostructures into well-defined chip architectures, such as lines and grades and replace current process chemicals and processes that are harmful to the environment. Ideally, nanotechnology greatly enhances environmental protection by providing a cost-effective way to reduce toxic chemicals such as persistent bioaccumulative toxins, hazardous air pollutants and volatile organic compounds.

Nanotechnology can be applied to remediation that is removing a toxic spill or hazardous material, such as benzene, chlorinated solvents, toluene, dichloro-diphenyl-trichloroethane (DDT) and heavy metals like mercury and lead. Researchers at Lehigh University have used nanoscale bimetallic particles for *in situ* groundwater remediation. These particles are made of iron and palladium and have been shown to be very effective reductants for wide array of chlorinated hydrocarbons. Metallic iron technology has been used in groundwater treatment for many years. Iron acts as an electron donor, reducing environmental contaminants, for example, trichloroethylene can be reduced to ethane:

$$C_3HCl_3 + 4Fe^0 + 5H^- = C_2H_6 + 4Fe^{2+} + 3Cl^-$$

This reaction can be enhanced by coupling ion to a noble metal. Most current research on this topic concerns granular or microscale iron (50 μm). Moving to the nanoscale makes the particle more reactive and together with the smaller size makes more flexible as a remediation technique. Fe and Pb nanoparticles could be added to slurry reactors for treatment of contaminated soil and sediment since they can penetrate intraparticle pores of soil and sediment aggregates those are inaccessible even to microorganisms. These nanoparticles could also be anchored onto granular activated carbon and other media in waste treatment applications.

Pseudomonas from bacterial family plays role as microbial nanosensors for degradation of pollutants. The typical Pseudomonas bacterium in nature might be found in a biofilm, attached to some surface or substrate, or in a planktonic form, as a unicellular organism, actively swimming by means of its flagellum. Many evidences pointed to a dominant, but poorly understood, role of attached microbial activity within the context of biofilm growth in aquifers. Using microbial nanosensors, fundamental understanding can be developed that how biofilms and related biodegradation actively respond to environmental conditions and investigating the genetic control of cell attachment to the solid surfaces. This will assist in achieving the overall aim of 'the cell mineral – interface programme' in understanding the interaction between the components of the complex underground system; mineral, water, microbes and pollutants, in order to predict and potentially manage biodegradation activity within contaminated aquifers and soils.

Nanoscience and nanotechnology: ethical, legal, social and environmental issues

The world continues to witness the controversies concerning various types and degrees of risk associated with many technologies, highlight the possibility of analyzing the question of development of technology (including nanoscience and nanotechnology) in terms of stakeholders opinion and highlight the risks associated with a technology.

A new scientific or technological development has the usual concomitant associated with ethical issues, specifically regarding containment and regulation. The arrival of nanotechnology raises the question whether and how ethical thinking could be integrated with the scientific research into the opportunities presented by the technology. There is no doubt that ethics will play a major role in the forming of nanotechnology. The ethical issues are more pronounced with nanotechnology due to the sharp divide between those who see its great potential and opponents who express fears. Nanotechnology supporters believe that it has potential to transform our lives dramatically, while opponents fear that self-replicating "nanobots" could escape from laboratories and reduce all life on earth.

NANOTECHNOLOGY IN THE FOOD INDUSTRY

The meaning of nanofood can be enunciated as application of nanotechnology or use of nanomachines, nanotools, nanodevices or nanoparticles during cultivation, production, processing or packing of the food. The nanotechnologists are optimistic about potential change in the existing food processing and ensure the safety of food products, creating a healthy food culture, enhancing the nutritional quality food through additives and improvement to the way the body digests and absorbs food and food packaging.

The impact of nanotechnology in the food industry has become apparent over the last few years. The companies hesitant to reveal their research programmes in nanofood have now gone public announcing plans to improve existing products and develop new ones to maintain market dominance. The types of application include: smart packaging, on demand preservatives and interactive food. The concept of interactive food is to allow consumers to modify food depending on their own nutritional needs and tastes, which is built on the idea of "on-demand" food. The second idea is based on the taste of the consumers in such a way that thousands of nanocapsules containing flavour or colour enhancers, or added nutritional elements (such as vitamins) would remain dormant in the food till the consumer triggers it to release. The major food giants like Nestle, Kraft, Heinz and Unilever support specific research programmes to capture a share of the nanofood market in the coming years.

Food Processing

Nanotechnology has already made an impact on the development of functional or interactive foods, which respond to the body's requirements and can deliver nutrients more efficiently.

Several applications of nanotechnology in food industries have already been developed in number of products in different countries. Example:

- One of the leading bakeries in Western Australia has successfully incorporated nanocapsules containing tuna fish oil (a source of omega 3 fatty acids) in their top selling products 'Tip-Top' Up bread. The nanocapsules are designed in such a way that they break open only when they have reached the stomach, thus avoiding the unpleasant taste/smell of the fish oil.
- The technology has been adopted and marketed by Shemen Industries to deliver the product 'Canola Activa oil, which it claims reduces cholesterol intake into the body by 14% by competing for bile solubilisation. This is a potential application in the pharmaceutical industry.

- The USA based Oilfresh Corporation marketed a new nanoceramic product which reduces oil use in restaurants and fast food shops by half. As a result of its large surface area, the product prevents the oxidation and agglomeration of fats in deep fat fryers, thus extending useful life span of the oil. Another benefit is that the oil heats up more quickly, reducing the energy required for cooking.

Packaging and food safety

The application of nanotechnology in packaging of food would optimize the shelf-life, repair small holes/tears, respond to environmental conditions (e.g., temperature and moisture changes) and alert the customers regarding the contamination. Kraft Foods, a business organization along with researchers at Rutgers University, USA are developing an 'electronic tongue', to be included in the packaging. This is made up of an array of nanosensors, which are extremely sensitive to gases released by food as it spoils, causing the sensor strip to change colour as a result giving a clear visible signal of whether the food is fresh or not.

Bayer Polymers has developed the Durethan KU2-2601 packing film, which is stronger, lighter and more heat resistant than those available in the market. This new film is known as hybrid system that is enriched with an enormous number of silicate nanoparticles. These particles reduce the entrance of oxygen and other gases, exit of moisture, thus preventing the food from spoiling. Kodak, a commercial organization, is developing antimicrobial film that have the ability to absorb oxygen from the contents of the package, impeding food deterioration.

Breweries would ideally use plastic bottles to ship beer, as these are lighter than glass and cheaper than metal cans but alcohol in beer reacts with the plastic used for the bottles severely shortening shelf-life. To overcome this problem, two institutions, Voridan and Nanocor have developed a nanocomposite containing clay nanoparticles called 'imperm', which minimizes loss of carbon dioxide from the beer and the ingress of oxygen to the bottle. The bottle is both lighter and stronger, less likely to shatter and keeping the beer fresher and giving it up a six-month shelf-life. This technology has been adopted by several companies including Miller Brewing Co. Honeywell Speciality Polymers has ssuccessfully designed plastic beer bottles that incorporates nanocomposites giving an extended shelf-life upto 26 weeks.

A research group at the University of Leeds in UK has come to a decision that nanoparticles of magnesium oxide and zinc oxide are highly effective at destroying microorganisms. These are much cheaper to manufacture than silver nanoparticles. This could have tremendous applications in food packaging.

Product	Developer/manufacturer	Nano content in the product	Purpose of the product
Nano agrochemicals under development			
Herbicide (Raj 2006)	Tamil Nadu Agricultural University, India and Technologico de Monterry, Mexico	Nano formulated	Designed to attack the seed coating of weeds, destroy soil seed banks and prevent weed germination
Pesticide, including herbicide (Invest Australia 2007)	Australian Commonwealth Scientific and Industrial Research Organization	Nano-encapsulated active ingredients	Very small size of nano-capsules increases their potency and may enable targeted release of active ingredients
Current use of nanomaterials in agriculture, food and food packaging			
Primo MAXX **(Plant growth regulator)**	Syngenta	100 nm particle size emulsion (micro-emulsion concentrate).	Extremely small particle size allows Primo MAXX to mix completely with water and not settle out in a spray tank.
Geohumus **(Soil wetting agent)**	Geohumus	Biocompatible high performance polymer	Soil enhancer with water storage capacity based on nanotechnology.
Nanoceuticals 'mycrohydrin' powder **(Nutritional Supplement)**	RBC Lifesciences	Molecular cages 1-5 nm diameter made from silicamineral hydride complex	Nano-sized mycrohydrin has increased potency and bioavailability. Exposure to moisture releases H- ions and acts as a powerful antioxidant.
Oat Chocolate Nutritional Drink Mix **(Nutritional Drink)**	Toddler Health	300 nm particles of iron (SunActive Fe)	Nano-sized iron particles have increased reactivity and bioavailability.
Nano silver cutting board **(Food contact material - cooking equipment)**	A-Do Global	Nanoparticles of silver	Nano-sized silver particles have increased antibacteial properties.
Nano silver baby mug **(Food contact material - crockery**	Baby Dream	Nanoparticles of silver	Nano-sized silver particles have increased antibacteial properties.
Antibacterial Kitchenware **(Food contact** material - kitchenware	Nanocaretech/NCT	Nanoparticles of silver	Nano-sized silver particles have increased antibacteial properties.

Product	Developer/manufacturer	Nano content in the product	Purpose of the product
Adhesive for McDonald's burger containers **(Food packaging)**	Ecosynthetix	50-150 nm starch nanospheres	These nanoparticles have 400 times the surface area of natural starch particles. When used as an adhesive they require less water and thus less time and energy to dry.
Durethan® KU 2-2601 plastic wrapping **(Food packaging)**	Bayer	Nanoparticles of silica in a polymer-based nanocomposite	Nanoparticles of silica in the plastic prevent the penetration of oxygen and gas of the wrapping, extending the product's shelf life.
Aquasol Preservative **(Food additive)**	AquaNova	Nanoscale micelle (capsule) of lipophilic or water insoluble substances.	Surrounding active ingredients within soluble nanocapsules increases absorption within the body (including individual cells).

Post-harvest Management

Nano bar codes and identity preservation

Usage of bar code is the essential feature for selling almost all products in the international and national market. The bar code is a sticker having a number of black and white bars with certain digits written at the bottom. The bar code is an electronic data depicting several parameters such as date of production, place of packing, prices, etc. and reading this code requires an electronic data reader. With the advent of nanotechnology, nano based bar codes are also available which can do the same function as that of conventional bar codes, thereby helping in tracking and controlling the quality of food products and give all relevant details in a minute.

A huge amount of shipments of livestock and other agricultural products are moved all over the world every day and it is becoming difficult to keep a track on critical control points of the production, shipment and storage processes. Lack of funds is a limiting factor to appoint number of inspectors at these critical control points. An identity preservation system can be installed, which will provide the consumers with information about the practices and activities used to produce an agricultural product. It is possible to provide stakeholders and consumers with access to information records and supplier protocol regarding the farm of origin, environmental practices used in production, food safety and security and information regarding animal welfare issues.

NANOTECHNOLOGY FOR ANIMAL HEALTH

Nanovaccines

Vaccination is a method of prevention of disease in advance by developing antibody against the particular pathogen. Most of the vaccines are applied as a fluid form and generally injected into bloodstream. These vaccines require cool temperature to be stored and they also have limited life span within which they are to be utilized. These two limitations have prevented the utility of vaccine particularly in the rural area where availability of electricity, fridge and in many cases the veterinarian are uncertain. Therefore, strong, healthy and durable vaccines are the only solution for the successful eradication of particular disease.

Veterinary Vaccine

The requirements for veterinary vaccines vary for different animals since the diseases differ among these species. Animals can tolerate the more side effects than by humans. Because the public pressure on animal welfare increases, less reactogenic vaccine preparations are favoured. In wildlife species, knowledge of targeting molecules (cell surface markers and immune modulators) is minimal, which is a limiting factor. Environmental issues dominate in the use of vaccines for wildlife. The route of delivery differs from oral administration in pets to long distance ballistic intramuscular delivery in wild animals. The size of the particles can range from a few nanometers to 1 cm: biobullets (1-2 cm) shot with an air rifle at wild animals, microparticles (1-100 ìm) for DNA vaccination and nano-beads (10-500 nm) as adjuvants in injected formulations.

Nano-bead

The antigen conjugated to the bead and the size of the bead play a major role in eliciting a combined response of humoral and cell-mediated immunity. Compared to alum as an adjuvant, the investigation shows nano-beads adjuvant induced substantial cell mediated responses along with moderate humoral responses. No adverse reaction were seen at the site of immunization in the sheep. Therefore, nano-bead adjuvants in veterinary species may be useful for the induction of immunity to viral pathogens where a cell mediated response is required.

Nanoapoptosis

Like human beings, cattle are also prone to tumour and in some cases of cancer. This particular disorder in many times becomes the reason for death of animals. Researchers at Rice University, USA are using nanoshells injected into the animals' bloodstream with targeted agents applied to the nanoshells to seek out and attach to the surface receptors of cancer cell. Illumination of the body with infra-red light raises the cell temperature to about 55 degree centigrade which burns and kills the cancer cells.

Paramagnetic nanoparticles

These nanoparticles are being tried for both diagnostic and therapeutic purposes. Some research groups have been experimenting smart super paramagnetic nanoparticles. The nanoparticles are made from iron oxide that when subjected to a magnetic field enhances the ability of nanoparticles to locate tumour cells. The nanoparticles when injected in the bloodstream target tumour receptor cells. At the site of the tumour, the nanoparticles emit an attached drug to kill the cancer cells.

Quantum Dots (QD)

These are tiny crystals measuring around 2-10 nm, which can be made to fluorescence when stimulated by ultraviolet light. QD may also be injected into the blood stream of animals and they may detect cells those are malfunctioning. As the QD respond to light, it may be possible to illuminate the body with light and stimulate the QD to heat up enough to kill the cancerous cells.

Animal Breeding

Efficient breeding of cattle is one of the most important factors for successful development of breeds, which mostly involve distant parents and their oestrus is some times very difficult to detect and it leads to repeat breeding (failure of fertilization after repetitive insemination). In oestrous cycle, a period called oestrus, or heat, occurs in female animals. The female is sexually excited throughout oestrus and, among most species of mammals, will mate only during this period. The problem of repeat breeding is very common in rural areas where even artificial insemination fails to fertilize the egg as eggs are still under developed. Therefore, the management of breeding is an expensive and time-consuming process for dairy and swine farmers. Scientists are making effort to solve this problem by implanting nanotube under the skin to provide real time measurement of changes in the level of estradiol (female sex hormone) in the blood. The nanotubes are used as a means of tracking oestrus in animals because these tubes have the capacity to bind and detect the estradiol antibody at the time of oestrus by near infrared fluorescence. The signal from this sensor will be incorporated as a part of a central monitoring and control system to actuate breeding.

NEED FOR REGULATORY BOARD

There is a need for regulation of probable and possible risk associated with nanoscience and nanotechnology in research. The Government of India has set up a regulatory board to enforce safety regulations, since there are no standards for judging what makes a product dangerous. The regulatory body will have its own set of challenges. The board will examine all new nanotechnology devices before they are commercially marketed. The board will have experts from all different fields including medicine and agriculture. This news item appeared in a leading daily English newspaper 'Hindustan Times'.

FUTURE TRENDS

A strong appeal for developing a R&D base in nanotechnology has been made by subgroup set up by Planning Commission to bring forward the suggestion for the adoption and generation of relevant technologies and to spread the same to the farmers. Dr. Mangala Rai, former Director General of Indian Council of Agricultural Research, India and head of this group has put together a detailed model for nanotechnology initiate under the national agricultural research system with a National Institute of Nanotechnology in Agriculture (NINA) as its hub. This institute is proposed to have linkages/cross-linkages with other R&D bodies in relevant disciplines as well as the private sector.

"Considering the rapid progress in nanotechnology R&D being made in advanced countries, it is imperative that India makes strategic investment in this frontier area of science and technology in the 11th Five Year Plan so that benefits can be reaped in the foreseeable future. Nanotechnologies have huge scope in the basic and strategic research in genomics, besides in areas like the efficient use of water, plant nutrients and plant protection chemicals and post-harvest operations", the subgroup set up by the Planning Commission has stated in its report. Several Indian institutions are actively engaged in nanotechnology R&D, prominent among them include the Indian Institutes of Technology, the Indian Institute of Science, the National Physical Laboratory, universities like University of Delhi, Banaras Hindu University, Gurunanak Dev University, Jamia Hamdard, Guru Jambheshwar University of Science and Technology and various agricultural and other universities and apex scientific and research bodies.

The unique aspect of nanotechnology is that it seeks to gainfully tap some of the extraordinary properties that the minutest of the minute parts of the atom or cell tend to possess, which normally happen to be relatively far more potent than the properties of the atom or cell as a whole. It can, thus, be used to develop diagnostic kits than can detect the exact causes of the disease and their cure or prevention through drugs and vaccines. In vaccine administration too nano-particles can help dispense the precise amount of vaccine's active ingredient at the specific location for speedy action. It can combat dreaded plant diseases including the foot-and-mouth disease, which the country is striving to eradicate.

REFERENCES

Barik, T.K., Sahu, B. and Swain, V. (2008). Nanosilica-from medicine to pest control. *Parasitol Res.* **103(2):** 253-8.

Boehm, A.L., Martinon, I., Zerrouk, R., Rump, E. and Fessi, H. (2003). Nanoprecipitation technique for the encapsulation of agrochemical active ingredients. *J. Microencapsul.* **20:** 433-41.

Corredor, E., Testillano, P.S., Coronado, M.J., Gonzalez-Melendi, P., Fernandez-Pacheco, R. and Marquina, C., *et al.* (2009). Nanoparticle penetration and transport in living pumpkin plants: *in situ* subcellular identification. *BMC Plant Biol.* **9:** 45.

Florczyk, S.J. and Saha, S. (2007). Ethical issues in nanotechnology. *J. Long Term Eff Med Implants.* **17(3):** 271-80.

Fulekar, M.H., Singh, A., Geetha, M. (2009). Nanotechnology – an approach for remediation of environmental contaminants. *Indian J. Environ. Prot.* **29(1):** 26-29.

http://www.nanoforum.Org/dateien/temp/nanotechnology%20in%20agriculture%20 and %20food_pdf

http://www.ptonline.com/articles/200602fa2.html

http://www.iasri.res.in/ebook/EBADAT/6-Other%20Useful%20Techniques/10-nanotech_in_Agriculture_BM_Prasanna_1.2.2007.pdf

Islam, N. and Miyazaki, K. (2009). NanoSI: exploring nanotechnology research conflation and nano-innovation dynamism in the case of Japan. Sci Public Policy. **36(3):**170-82.

Iyengar, N. and Iyengar, S.A. (2010). Nanotechnology in crop protection. In: 8th National Symposium on Problems and perspectives in eco-friendly innovatives to plant protection, held on 24-25 January, 2010 at C.S. Azad University of Agriculture and Technology, Kanpur, India. Abstract no. 12, p. 19.

Nandedkar, T.D. (2009). Nanovaccines: recent developments in vaccination. *J. Biosci.* **34(6):** 995-1003.

Nanotechnology and its applications in crop improvement.http://www.scribd.com/doc/7640458/Nanotechnology-and-Its-Applications-in-Crop-Improvement

Nanotechnology watchdog soon. Hindustan Times 2010; February **23:** 1-15.

Patra, D., Ejnavarzala, H. and Basu, P.K. (2009). Nanoscience and nanotechnology: ethical, legal, social and environmental issues. *Curr. Sci.* **96(5):** 651-7.

Perez-de-Luque, A. and Rubiales, D. (2009). Nanotechnology for parasitic plant control. *Pest Manag Sci.* **65:** 540-5.

Raj, K.S. and Asha, K.K. (2009). Nanotechnology in agriculture. *Trendz BioTech.* 27-8.

Scheerlinck, J.P.Y., Gloster, S., Gamvrellis, A., Mottram, P.L., Plebanski, M. (2006). Systemic immune responses in sheep, induced by a novel nano-bead adjuvant. Vaccine. **24:** 1124-31.

Sud, S.A. 'Nano' step: farm view. http://www.business-standard.com/India/news/- %60nano%60-step/320749/

The Gale Encyclopedia of Science. 4th ed. New York: Thomson - Gale, 2008. pp. 2917-21.

Zhu, H., Han, J., Xiao, J.Q. and Jin, Y. (2008). Uptake, translocation and accumulation of manufactured iron oxide nanoparticles by pumpkin plants. *J. Environ Monit.* **10(6):** 713-7.

Index